cold regio [barcode: M000275921]

utilities monograph

Third Edition

Technical Editor:	D.W. Smith
Steering Committee and Contributing Authors:	D.W. Smith, Co-Chair W.L. Ryan, Co-Chair V. Christensen J. Crum G.W. Heinke

Special Contributions:

L. Barber	A.J. Hanna	D.H. Schubert
E. Bjornstad	T. Heintzman	A. Shevkenek
R.H. Boon	R. Kent	M. Stafford
J.J. Cameron	W.O. Mace	S.J. Stanley
S. Cheema	C. Marianayagam	W. Tobiasson
R. Dalton	M. Mauser	J. Vogel
G. Eddy	J. O'Neill	J. Warren
K. Egelhofer	D. Prince	
R. Feilden	R.L. Scher	

Editor:	N. Low
Principal Proponents:	Technical Council on Cold Regions Engineering American Society of Civil Engineers Cold Regions Engineering Division Canadian Society for Civil Engineering

Published by

 American Society of Civil Engineers

1801 Alexander Bell Drive
Reston, Virginia 20191-4400

ABSTRACT:

The concepts related to the design, construction and operation of infrastructure components for the delivery of water, the removal of liquid and solid wastes, and the provision of power for services are relatively new in the extremely cold environment found in the far north or south. The objective of this monograph is to introduce the basic principles of cold region environmental engineering. While it does not cover the basics of the environmental engineering field, it does present a great deal of introductory information related to cold regions engineering and the special geotechnical considerations that influence the design of utilities systems. The book is divided into 17 different sections that cover such topics as: 1)Planning, geotechnical, and thermal considerations; 2) water source development; 3) water storage; 4) wastewater collection; 5) utilidors; 6) central facilities; 7) remote camps; 8) fire protection; and 9) energy management. There are also appendices that provide additional information on piping options, vehicle-haul systems, snowdrifting and snow loads, and freeze protection.

Library of Congress Cataloging-in-Publication Data

Cold regions utilities monograph / technical editor, D.W. Smith; editor, N. Low ; principal proponents, Technical Council on Cold Regions Engineering, American Society of Civil Engineers [and] Cold Regions Engineering Division, Canadian Society for Civil Engineering. — 3rd ed.
p.   cm.
First ed. published 1979 under title: Cold climate utilities delivery design manual.
ISBN 0-7844-0192-6
1. Underground utility lines—Arctic regions. 2. Sanitary engineering—Cold weather conditions. I. Smith, D. W. (Daniel W.) II. Low, N. (Nola) III. Technical Council on Cold Regions Engineering. IV. Canadian Society for Civil Engineering. Cold Regions Engineering Division. V. Title: Cold climate utilities delivery design manual.
TD168.C65 1996   96-28438
628.1'0911—dc20   CIP

Cover: Ends of various types of insulated pipes showing different inside pipe materials, different insulation materials, different thaw tubes and conduits, and different external protective jackets. Location: Norman Wells, N.W.T. Photographs by Daniel W. Smith. Graphics by Dennis Weber

PREFACE

The concepts related to the design, construction and operation of infrastructure components for the delivery of water, the removal of liquid and solid wastes, and the provision of power for services are relatively new in the extremely cold environment found in the far north or south. The development of these systems was originally limited to the communities founded in the early 1900s during the initial period of gold exploration. A second period of significant interest started during the Second World War with the need to protect the northern part of North America from invasion and to supply allies with equipment and support. The Cold War lead to a sustained engineering effort in the creation of community industrial and military infrastructure.

Early efforts at the provision of community utilities in Alaska and the Canadian Arctic regions were plagued with many challenges and system failures. The Arctic Health Research Center in Alaska, an arm of the U.S. Public Health Service along with the U.S. Army Cold Regions Research and Engineering Laboratory, the Department of Public Health, and later the Northern Technology Centre in Canada became the center of each country's research, design and development efforts.

In an attempt to share the knowledge being gained through the various civilian and military-based studies made by both governments, a variety of professional conferences and meetings dealing with permafrost were initiated. These gatherings also included many geotechnical and engineering issues. In the 1960s and early 1970s a few professional conferences were held that focused on engineering issues alone. In 1978, the first in a series of conferences based on utilities systems was held in Edmonton, Alberta. Over the following ten years four professional conferences were held on related issues (three in Edmonton, one in Fairbanks). In a related effort, a series of two cold regions engineering conferences was held at the University of Alaska in Fairbanks. In 1979 after the American Society of Civil Engineers (ASCE) formed the Technical Council on Cold Regions Engineering under the leadership of Amos "Joe" Alter, the first of a continuing series of cold regions engineering conferences was held in Anchorage, Alaska. That series of professional conferences was expanded to address issues common to Alaska, Canada, and the northern tier of the United States. With global networking and communications the exchange of technical information has expanded to the point that new developments anywhere in cold regions can now be shared throughout the world.

This document developed out of an initial effort at the first utilities delivery conference in 1976 in Edmonton, Alberta. The need to document the results of the many research, design, construction and operation investments was obvious. Following that conference, the Northern Technology Centre was funded to initiate the planning of a reference manual for utilities systems in cold regions. In 1976 a group of Canadian and United States engineers agreed to cooperate in the development of such a document. The manual was to be a reference document for professional engineers trained in more temperate climates. It was to introduce the issues of planning design, construction and operation of utilities systems in cold, freezing conditions. That objective continues in this third edition of the document.

The Cold Climate Utilities Delivery Manual (1979) was originally published by Environment Canada (in both English and French) and republished by the U.S. Environmental Protection Agency. The second edition, entitled the Cold Climate Utilities Manual (1986), was produced by Environment Canada, the Government of the Northwest Territories and the Canadian Society for Civil Engineering (CSCE), and was marketed by the CSCE. Through a cooperative agreement between the American Society of Civil Engineers and the Canadian Society for Civil Engineering, the Cold Regions Utilities Monograph has been produced and marketed by ASCE. The Technical Council on Cold Regions Engineering sponsored the monograph as a part of their monograph series. The series includes the following volumes:

- Design for Ice Forces

- Cold Regions Construction

- Frost Action and Its Control

- Freezing and Thawing of Soil-Water Systems

- Thermal Design Considerations in Frozen Ground Engineering

- Embankment Design and Construction in Cold Regions

- Arctic Coastal Processes and Slope Protection Design

- Cold Regions Hydrology and Hydraulics

This third edition of the cold regions utilities document has been made possible through the guidance, direction and contributions of the Steering Committee members and the individual contributions of the many persons that revised sections or subsections of the monograph. A sincere thank you is extended to all of the participants in this monograph.

It is appropriate to acknowledge the assistance of Ms. K.M. Emde, Mr. R.M. Facey and Dr. G. Putz in the preparation of some parts of the document.

The production costs and time spent by each author who prepared a section of the manuscript was supported by their respective companies and is acknowledged. The ASCE Technical Council on Cold Regions Engineering provided financial support for the meetings of the Steering Committee. Financial support for the production of the manuscript was provided by the CSCE, the Department of Municipal and Community Affairs of the Government of the Northwest Territories and the research funds of Daniel W. Smith.

Daniel W. Smith

Table of Contents

SECTION 1

INTRODUCTION

3rd Edition Steering Committee Coordinators

D.W. Smith

3rd Edition Principal Author

D.W. Smith

Section 1   Table of Contents

1 INTRODUCTION

1.1 Objective

The opportunity to use water, wastewater, solid waste and energy components of community infrastructure is key to the improvement of the quality of life. In the colder regions of the world the provision of these components is more difficult due to the added complexity of long periods of extremely low temperatures, frozen ground conditions, small populations and limited access. Each of these adds to the difficulty of providing the desired community infrastructure.

The objective of this monograph is to introduce the basic principles of cold region environmental engineering. It has been prepared for persons with a good understanding of environmental engineering; it does not present the basics of that field. The monograph does present a great deal of introductory information related to cold regions engineering and the special geotechnical considerations that influence the design of utilities systems. No design standards are presented.

Many components of conventional utilities planning, design, construction and operation can be used in cold regions. However, no conventional components should be used without careful analysis of the effects of cold stress on them. Furthermore, the analysis should assume that the system will freeze. As a result, all systems should incorporate a plan for the recovery of a frozen component or system.

In many cases the engineering requirements will exceed the depth of technical description provided in this document. Many references have been provided which may assist in resolving technical problems. In some cases, specialists will have to be consulted.

The following subsections introduce the type of information to be found in the sections and appendices of this monograph.

Cold Regions

The phrase "cold regions" is applied to a large portion of the world. Assuming it is the region where the mean monthly temperature of one month per year is below 1°C yields a portion of the earth where over one billion people live. Using definitions involving longer periods of temperatures below the freezing of water can dramatically reduce the region involved and the population affected.

In North America the term is often applied to arctic and subarctic regions of the United States and Canada. They include Alaska, the Yukon, the Northwest Territories, northern parts of some of the Canadian provinces, and some areas of the northern contiguous United States. The fundamentals presented in this monograph may also be helpful for other cold regions, such as the Antarctic, Greenland, Scandinavia and the northern Asian continent.

In North America, the cold region stretches from the Bering Sea to the Davis Strait (some 5,000 km) and from the northern parts of the Canadian provinces to near the North Pole (about 2,500 km). This region is inhabited by relatively few people. The region also contains a many resources such as fisheries, minerals, oil and gas, gemstones and forests, the development of which often involves the development of community and industrial infrastructure.

In these regions minimum and maximum temperatures can range from as low as -50°C to as high as 35°C. Mean annual total precipitation (rain and snow) varies from about 150 to over 450 mm.

Environmental Engineering

The objectives of environmental engineering include:

- protection of public health;
- protection of environmental components;
- protection of aesthetic qualities; and
- efficient design, construction, operation and maintenance of systems.

The provision of utility services is influenced by each of these objectives and has an impact related to each of them. However, utilities services focus these objectives to:

- improving the quality of life (including standards of health, comfort, convenience and aesthetics);
- providing fire protection;
- facilitating socioeconomic development of the community; and
- protecting the environment.

Although the fundamental objectives, and the environmental and regulatory constraints in cold regions are similar to other areas, the appropriate and economical technical solutions vary in general and from site to site.

1.2 Planning and Management Considerations

Many communities were located and developed without anticipating utility services and their present population. This reality of development has contributed to the high cost and delays in providing a standard of utilities in these communities. Since the provision of services tends to anchor the location and fix the layout of the community, the long-term physical and socioeconomic implication of all decisions must be considered.

Planning the construction of utilities in cold regions is particularly challenging for a number of reasons:

- utility systems provide a critical life support function and there are severe physical and social consequences of failure;

- the cost of construction and operation is high due to such factors as high energy and transport costs; and

- environmental characteristics such as low temperature, permafrost, and isolation impose technical constraints.

The remoteness of many communities results in high transportation costs. Most of the materials used for services must be imported. The spread-out, low-density layout of existing settlements results in further high costs. Replanning of a settlement, including the relocation of roads and houses should be a prerequisite to construction of piped water and sewer systems. In some cases, complete relocation of a settlement may be the most economical solution for provision of utility services in the community. Upgrading existing housing and construction of low- and high-density housing for permanent and transient populations may be required. Central commercial, educational and recreational facilities must be incorporated in the community plan. Any useful and practical plan, whether for housing, schooling or servicing, must be both technically and economically sound, and most importantly, socially acceptable to all groups of the community, both natives and newcomers. The special problems of planning and preliminary engineering are addressed in Section 2.

1.3 Geotechnical Considerations

Permafrost occurs where the mean annual ground temperatures are below 0˚C for several years. Its thickness varies with location and can be up to 600 m. The existence of permafrost causes special difficulties for the construction of buildings, services and facilities, and are discussed in detail in Section 3.

This section provides some of the basic geotechnical information necessary for the planning, design, and construction of utilities in cold regions.

Utility systems in cold regions must function under severe climatic conditions. Climatic and geotechnical site information is necessary to design for freeze-protection, foundation stability, thermal stress, and economy.

The limited data available must usually be modified either to estimate the extreme climatic conditions or to allow for highly variable surface changes within a small area. Site-specific surveys consistent with the thermal and structural design considerations are imperative. Of primary concern is the movement and possible structural damage due to the freezing and thawing of the soil. Therefore, the maximum thickness of this active layer, and the thermal properties and frost susceptibility of the soil must be determined. In permafrost areas, particularly in soil with a high ice content, the soil survey must extend to the maximum range of anticipated major thermal effects; surface conditions and drainage patterns must also be noted.

1.4 Thermal Considerations

In cold regions the thermal considerations are as important as the hydraulic and structural features of utility systems and must be included in the selection of materials, components, and processes and in the design of systems. Thermal analyses are necessary to design for freeze protection, foundation stability, thermal stress, and economy. In Section 4, the methods for calculation of frost and thaw depths for steady state conditions are presented. Major emphasis is placed on the thermal design of piped water and sewer systems. Thermal aspects are discussed, and simple equations with illustrative examples are presented for solving some of the thermal problems encountered in utility system design.

1.5 Water Sources

All traditional sources of water are present in most parts of the cold regions. However, the conditions peculiar to these regions require that special considerations be made of these sources prior to the selection of a community water supply.

Groundwater is an excellent source of water in remote northern locations. However, in continuous and discontinuous permafrost regions a reliable groundwater source normally must be obtained from beneath the frozen zone or from under lakes or rivers. This often means the construction of expensive wells is necessary due to the thermal protection requirements. In addition, the well water from under permafrost may be highly mineralized. In a few locations, ice and snow are melted for water, but the high cost of fuel or electricity makes this source of water uneconomical, if adequate quantities are to be provided.

Although there are many lakes, they are generally quite shallow and many freeze to the bottom, or their effective storage capacity is severely reduced by the thick ice that forms each winter. This freezing also concentrates minerals in the unfrozen water, which may render it unsuitable for consumption. Surface water is often highly colored from the organic material washed into lakes by runoff. Because of low precipitation, the water contained in lakes may be the result of many years' accumulation. Using large amounts of this water may drain the lake and result in the loss of the supply. During the winter, clean water can be obtained from below the ice of large rivers, but during break-up, floating ice and other debris carried by flood waters are a hazard to any permanent installation, such as a water intake. Following break-up, the silt content of the river could be extremely high, making the control of water treatment more difficult. Small rivers and creeks often freeze to the bottom, and so their use as a water source is limited to certain times of the year. Natural or anthropologic contamination of water must also be dealt with in some locations. Reservoirs are difficult to construct where the underlying soil has a high ice content; any attempt to pond water will cause melting of the underlying permafrost and possible sinking of the dikes. Many issues related to water sources are addressed in Section 5.

1.6 Water Treatment

Water treatment requirements in North America are dictated by specified minimum treatment of surface water and shallow groundwaters as well as by specified maximum allowable contaminant concentrations. The challenges for design revolve around low termperature, widely varying raw water quality, limited supply and remote, small systems. Section 6 addresses these issues in a way that identifies the unique factors that require special attention.

1.7 Water Storage

Water storage must be considered an important component of a community water supply system where the following conditions exist:

- an adequate water supply cannot be assured;
- flow equalization is necessary; and
- fire protection requirements must be met.

Storage requirements can vary from large-scale facilities such as open earth impoundments or insulated steel tanks meeting annual community needs, to small tanks for in-house use meeting individual family needs. These issues are presented in Section 7.

Water consumption rates show large variation depending on the method of distribution and the plumbing facilities available to the users. A water supply objective of 60 litres per person per day (L/(p•d)) is generally considered minimum for adequate drinking, cooking, bathing and laundering. Only piped systems or a well equipped trucked water system can meet this objective. In many communities currently served by a vehicle-haul water system, about 20 L/(p•d) is supplied and used.

1.8 Water Distribution and Sewage Collection

The dominant concern of cold-region utility systems is the need to prevent both the water and sewage lines from freezing. Heat may be added to the water or to the mains, and continuous circulation maintained to prevent freezing. The degree of freeze protection required depends on whether the pipes are buried or built above ground. Buried water and sewer lines are preferred for community planning, aesthetic and engineering reasons. In areas with subarctic climate (arbitrarily defined as where for one to three calendar months the mean monthly temperatures rises above 10°C), such as Anchorage, Fairbanks, Whitehorse and Yellowknife, underground systems are used. They differ from southern systems in that various methods of freeze protection are provided, such as insulation, heating, recirculation, and water wasting. Insulation around pipes also prevents thawing of ice-rich permafrost and consequent settling of pipes. In the past, underground services were considered to be technically and economically unfeasible in ice-rich permafrost areas. Therefore, above-ground utility systems were constructed in such areas. They are generally more expensive, cause difficulties with roads and drainage, are subject to vandalism, and are not desirable from community planning and aesthetic points of view. Engineering developments and materials now allow underground construction in areas where this was previously thought impossible. This has reduced the use of above-ground utilidors. However, above-ground utility systems may still be necessary in thermally sensitive, ice-rich permafrost areas, or where excavation equipment is unavailable for installation and maintenance. Above-ground utilities are also used for temporary facilities.

Trucked delivery of water and collection of sewage is an alternate means of providing service. Water storage tanks used for homes vary from open used oil drums (180 L) to proper pressure tanks of 1,200 L capacity. In some settlements, water is delivered only to some homes. Other individuals must pick up water in pails from water storage tanks within the settlement. When the house is equipped with complete indoor plumbing, all wastes are generally discharged to a holding tank, which must be pumped out regularly. When indoor plumbing is not available, toilet facilities may consist of pit privies or chemical toilets of various design; however, they usually consist of plastic bags in a container under the toilet, termed "honey bags". The bags are picked up daily or several times per week on a community-wide basis. Washwater wastes, kitchen sink wastes, and laundry water has often been disposed of to the ground surface in the immediate vicinity of the home, contributing to localized drainage and health problems. Water distribution and wastewater collection are covered in Sections 8 and 9.

1.9 Waste Disposal

Very few communities have sewage treatment plants, although there are now package-type plants at industrial facilities in the Northwest Territories, Yukon, and Alaska which were built as a consequence of mineral and oil exploration and pipeline construction. Sewage lagoons are the most common method for the treatment of wastewater. In some regions these meet recommended design and operating criteria. In the past, they were often constructed by utilizing existing lakes or low areas with the suitable addition of dikes. For most of the year they provide long-term storage of wastewater, with some anaerobic decomposition taking place. It is during the summer that extensive biological activity occurs, with the resulting reduction in organic matter. Over summer retention insures highly effective treatment by lagoons without the need of skilled and costly operation and maintenance. In certain subarctic locations, septic tanks and absorption fields are used. In communities near the ocean, disposal

of wastewater is often to the sea. This practice may not be permitted without a special study of local conditions or some degree of treatment. Wastewater treatment systems used in cold regions are presented in Section 10.

1.10 Utilidors

In established communities, pipes for water supply and sewage may be housed in a utilidor for effective frost protection. A utilidor is a structure whose function is to contain the utility piping and other services of a community or camp. Utilidors may enclose water and sewer pipes as well as central heating, fuel oil, natural gas, electrical and telephone conduits, and others.

Utilidors have been used in stable and unstable seasonal frost and permafrost soils as well as in snow. They may be large enough to provide access for maintenance purposes or for use as an enclosed walkway, or they may be compact with no air spaces.

Most utilidors have some mutually beneficial heat transfer between the enclosed pipes. If central heating pipes are included, heat loss from these can be sufficient to replace utilidor heat losses and prevent freezing. But with this arrangement, temperature control within the utilidor is difficult and can lead to inefficiency and undesirably high temperatures in the water pipes. The special problems of utilidors are discussed in Section 11.

1.11 Central Facilities and Individual Systems

Some cold region communities are located where it is difficult and very expensive to construct and operate conventional, piped water supply and wastewater systems. Ice-rich permafrost, rock and low-density housing are among the factors limiting the use of conventional piped water supply and wastewater systems.

An alternative for these communities is to provide a single, central facility where people can obtain sanitation services: drinking water, a sanitary means of waste disposal, and facilities for laundering clothes and bathing. Such systems are described in Section 12.

Community services are not always possible or feasible in cold regions, therefore many dwellings and other isolated buildings must depend on individual haul systems for water supply and waste disposal. These systems can range from completely independent on-site water supply and waste disposal to buildings that have independent internal water distribution and waste collection but depend on vehicle haul for water delivery and waste removal. Various alternatives are presented in Section 14.

1.12 Camps

There are many permanent and temporary camps in northern areas, all with their special servicing problems. With the search for oil and gas came a large number of industrial camps.

At first camps tended to allow conditions unsupportive of an adequate lifestyle; the living quarters were crowded and wastes were dumped without concern for environmental effects.

Concern by government, labor unions, environmental groups and companies led to greatly improved conditions in remote camps and minimized detrimental effects on the environment. In new camps, housing is comfortable, dining facilities are excellent, recreational facilities are provided, water is safe to drink, package waste disposal systems are available for solid and liquid wastes, and incineration of combustible waste is practiced before disposal to controlled land sites.

In general the difficulties in servicing these camps are fewer than those experienced in communities, because they are planned, designed and constructed as a total facility, and because trained personnel are available to maintain and operate the facilities. Section 13 addresses the special needs of camp services.

1.13 Fire Protection

One of the most serious hazards to life and property in remote northern locations is fire. The seriousness of fire is aggravated by the problems of providing adequate quantities of water to fight fires during the extremely cold periods of the year, the extreme dryness of wood and organic materials, and the dependence on heating systems for survival. Fire protection techniques must be approached at all levels: the use of fire retardant materials, development and use of early warning devices, the provision of fire control equipment and personnel, and educational efforts towards fire prevention. Detailed information in fire protection and prevention is provided in Section 15.

The problems concerning fire protection in cold regions arise primarily from:

- the small size and isolation of communities;
- the lack of finances (tax base) in most places;

- some social conditions resulting from the changing culture;
- the harsh climate; and
- the inadequate water supply for firefighting.

It takes a minimum amount of equipment and trained personnel to provide fire protection for a community. The small size of communities therefore makes per capita costs of fire protection very high. The isolation makes sharing of equipment through mutual aid agreements impossible. It also makes sharing of personnel difficult, if not impossible. The provision of fire protection systems must therefore primarily involve preventive measures. Since very few, if any, serious fires occur in small communities during a year, keeping the need for fire protection in the minds of the public is difficult. This lack of interest contributes to a high turnover among volunteer firefighters, a lack of maintenance of equipment, and a general deterioration of fire prevention efforts.

Most northern communities have no tax base and therefore must depend on government assistance. In communities with a large industry, such as a mine, help can often be obtained for the purchase of firefighting equipment. Usually such communities are sufficiently well organized to collect taxes. With even a very small mil rate, the presence of such an industry adds substantially to the community's ability to raise money.

The effectiveness of a fire protection system depends to a large extent on the general public and to a lesser extent on the fire department. Most fires can be prevented through reasonable care and by avoiding dangerous practices. This, in many cases, cannot be legislated and therefore depends on public awareness. In the small, traditional, native villages the number of fires reported each year per capita is quite small. In larger northern towns with a larger transient population and a less traditional lifestyle, the number of fires per capita is large.

1.14 Solid Waste Management

Solid waste disposal has, until recently, been the most neglected area of sanitation. Where a community collection system exists, disposal is generally at a dump. Because of the lack of proper cover material, sanitary landfill cannot be practiced in most places. Few landfills are fenced, contributing to widespread and uncontrolled dumping. Human feces in honey bags are often disposed of in a landfill without any treatment. In some cases, pits have been constructed; in other cases, the contents are emptied into a lagoon which also receives wastewater. In some cases, ill-chosen sites contribute to water pollution. The objective of solid waste management is to collect and dispose of wastes in the most economical manner for a given situation without creating hazards, nuisances, or aesthetic blights for people or in the environment. To achieve this objective, each component of the system, namely storage in or at the source, collection and treatment/ disposal, must be properly carried out.

So far, forms of refuse disposal such as incineration, shredding, baling, etc. have found limited application due to their high maintenance and operational costs. See Section 16 for detailed information.

1.15 Energy Management

In cold regions a large amount of energy is consumed to attain an adequate standard of living. The energy is used for heating, electricity production, and transportation.

Even though considerable coal, oil, and natural gas deposits have been confirmed in these regions, energy costs are high. This is due to the high cost of production and distribution. In some cases raw materials are transported to southern processing plants then returned to northern locations. Good energy management fosters the use of less energy and savings in the cost of heating and electricity. When designing utilities in cold regions, local weather conditions, energy sources and distribution, and methods of conservation require careful consideration.

Concomitant with an improved standard of living is the improvement in the quality and safety of the housing environment. The provision of heat and electricity is often considered of equal, if not greater, importance than the provision of safe water and sanitation services. In reality, all four utilities along with a properly designed and constructed building are essential constituents of a good, healthy living environment.

Conservation of energy becomes especially important in cold regions. Heating expenses can be reduced by increased insulation of buildings and increased use of waste heat from generating stations. Oil, gas, coal, wood and hydrogenerated electricity are now most commonly used as energy sources. These may be supplemented by geothermal, solar, and wind energy where and when feasible. Section 17 provides information on energy management considerations.

1.16 Summary and Future Needs

An evaluation of the present level of sanitation in most of the smaller northern communities shows that the overall situation is in some cases more primitive than in comparably sized southern communities, but that the situation is steadily improving. Goals have been set out broadly, and implementation of these goals will, to a large extent, depend on available funds.

Development of the northern areas of Canada and of Alaska will undoubtedly increase in the next few decades, but the scale and pace of development is uncertain. In the course of resource development, hydroelectric works and transportation facilities such as pipelines, roads, railroads, bridges, harbors and airports will be built. Several different types of communities will have to be developed, such as construction and permanent camps, the expansion and improvement of existing communities, and the building of new towns. The engineering community must be ready to plan, design, construct and operate facilities appropriate for the conditions encountered.

SECTION 2

PLANNING CONSIDERATIONS

3rd Edition Steering Committee Coordinators

Vern Christensen

James A. Crum

3rd Edition Principal Author

James J. Cameron

Section 2   Table of Contents

Section 2   List Of Figures

Section 2   List of Tables

2 PLANNING AND MANAGEMENT CONSIDERATIONS

This section outlines planning and management considerations in providing water, wastewater and solid waste services (i.e., utilities) in cold regions, primarily within a community context. The objectives of utility services and the characteristics of various utility system options are outlined. Considerations in evaluating utility options are presented. As well, this section outlines general considerations that are common to various utility systems and facilities including site conditions, project management and operations and maintenance. Other sections of this monograph contain more detailed information on specific utility system components and facilities.

Planning utilities in cold regions is important and challenging for the following reasons:

- Utility systems provide a critical life support function with severe physical and social consequences of failure.

- The cost to construct and operate utilities is typically two to five times higher than in warmer climates, due to such factors as high energy costs, high transport costs, low temperature, permafrost and isolation.

- The constraints imposed by environmental characteristics such as low temperature, permafrost and isolation require focused attention to detail, contingencies for failure, design innovation and simplification, and cost effectiveness.

- Adequate and reliable utilities are fundamental to the physical, social, economic and political development of communities.

- The mixture of cultures, values and lifestyles results in an uncommon diversity of aspirations that affect utility objectives and priorities.

- Costs for local utility systems are often funded by senior government (i.e., national, state, provincial or regional government) resulting in a need to develop joint decision making and local management.

- Utilities can dominate the physical development of communities and the design of housing and buildings.

- The somewhat short history and limited experience with developing modern communities and utility services in cold regions compared to more temperate areas requires planners and designers to be innovative and less reliant on "tried and true" solutions.

Inadequate planning has lead to social and physical failures. Inadequate planning has also contributed to the high cost and delay in providing a high standard of utilities to residents in cold regions.

The utilities required to serve community and industrial activities, whatever their duration, function or scale, must be considered within the context of the following:

- political and socioeconomic conditions and policies at the local, regional and national levels;

- users' needs and lifestyle;

- financial, technical and managerial capabilities of the users, private sector and government;

- existing conditions and future development plans; and

- local environment.

2.1 Objectives

The objectives in providing utility services are to:

- improve the quality of life (i.e., safeguard public health and improve comfort, convenience and aesthetics);

- protect the environment;

- facilitate socioeconomic development of the community;

- provide fire suppression capability; and

- provide equitable services among residents.

The pursuit of these objectives is the justification and rational for providing public assistance to individuals and communities for utility services and systems. These objectives provide the fundamental basis for the evaluation and design of utilities. Although the objectives are the same in cold regions as in more temperate regions, the appropriate methods to achieve these objectives in cold regions may be unique (Alter, 1972). Furthermore, the socially appropriate, politically supported and economical solutions to achieve these objectives vary across cold regions, even though the environmental constraints may be similar.

2.1.1 Health. The protection and improvement of public health is usually the primary objective of utility services and is the strongest argument for public assistance for utility systems.

Water-related health problems brought on by self-contamination within the household and the community are a major concern. Since pathogens survive longer in cold environments and may remain viable indefinitely when frozen, the potential health hazard from wastewater discharges is greater in cold regions (Heinke and Prasad, 1976; Lapinleimu et al., 1984). Contamination of rivers and lakes is a concern, especially for casual water users.

Residents in small, remote, subsistence economy communities, particularly the aboriginal people, have historically experienced poorer health than residents in larger, developed communities in cold regions and residents in the "south" (Weller and Manga, 1987). However, environmentally-related diseases have been reduced where water, wastewater and solid-waste services and housing are improved (World Health Organization, 1982; Robinson and Heinke, 1990; Brocklehurst and Heinke, 1985; Michael, 1984; Robinson and Heinke, 1991).

Providing high-quality water affords protection against adverse chemical constituents and waterborne diseases such as gastroenteritis, hepatitis, typhoid and giardiasis. To reduce the occurrence of gastrointestinal disorders caused by direct fecal contact as well as impetigo and other skin diseases, a sufficient quantity of water must be conveniently available for personal hygiene, laundry and cleaning. Sanitary collection and disposal of excreta and washwater are necessary to prevent contamination in homes, around the community and in water supplies. Several hepatitis outbreaks have been traced to indiscriminate or accidental disposal of excreta and washwater onto the ground.

The relationship between health, and water and sanitation service is complex and difficult to quantify, especially with respect to predicting the impact of specific infrastructure improvements on health of individuals (Saunders and Warford, 1976; Feachem et al., 1983). Water-related diseases have different transmission mechanisms and therefore different appropriate prevention strategies (Saunders and Warford, 1976; Feachem et al., 1983; McGary et al., 1980). Utility services oriented to health improvements should respond to specific problem diseases,

their likely transmission routes, existing hygiene practices and socioeconomic norms of the community (McGary et al., 1980). Disease preventive strategies include:

- improving the quality of water supplied and, if stored within the house, maintain the quality;

- discouraging the direct use of unimproved water sources, particularly water sources that are close to the community and potentially contaminated;

- increasing the quantity of water available;

- improving the availability and accessibility of water and the convenience in using water;

- improving personal hygiene;

- implementing sanitary collection, treatment and disposal practices for excreta, greywater (washwater) and solid waste; and

- controlling insect vectors (e.g., surface water management).

Utility services and systems that the people do not effectively use will not protect health. Factors that influence use are:

- perceptions and knowledge of the causes of illness and the benefits of water and sanitation services (Tester, 1976);

- attitudes and perceptions of waste management and community living (Tester, 1976);

- water use practices, i.e., personal hygiene (Martin, 1982);

- acceptance of water quality, particularly chlorinated waters;

- confidence in the service and system;

- type of plumbing and appliances; and

- ability and willingness to pay for services.

Health protection and improvement depend on the type of utility system and the level of service. Providing a safe source of water within a community, but not delivery water to the homes, can improve health where the alternate water sources are distant or where local water sources are contaminated. However, health improvement is often limited for the following reasons:

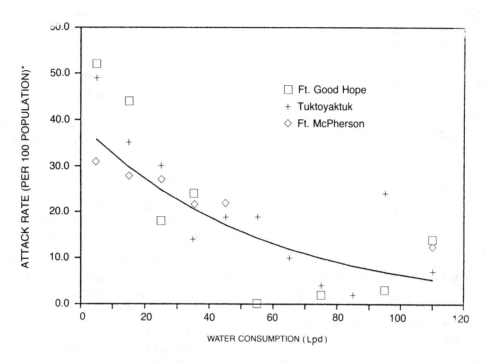

*Gastro-Intestinal and Skin Diseases.

Figure 2-1   EFFECT OF WATER CONSUMPTION ON THE ATTACK RATE OF GASTROINTESTINAL AND SKIN DISEASE (Michael, 1984)

- Some people continue to use an unsafe source of water, particularly when it is closer and unchlorinated.

- Where people must haul their water they use water primarily for drinking and cooking and sparingly for personal hygiene, cleaning and laundry.

- Although safe water is supplied at a water point, water can become contaminated during transport, storage, or use in the home.

Levels of health are higher where water delivery and waste collection service or sanitary on-site disposal is provided to each house (Schliessmann et al., 1958), and where houses have pressurized hot and cold running water and plumbing fixtures and appliances to allow convenient personal hygiene and cleaning (Brocklehurst and Heinke, 1985). Disease rates for gastrointestinal and skin diseases are lower in houses with plumbing fixtures and where water use is higher (Figure 2-1) (Robinson and Heinke, 1990; Michael, 1984; Heinke, 1984). Statistical analysis of three communities in the Northwest Ter-

ritories showed a clear and significant increase in the risk of diarrhea (an important indicator of the relationship between health and municipal services) associated with low water use (Robinson and Heinke, 1990). Water use less than 30 litres per capita per day (L/(p·d)) was shown to be coincident with serious health conditions but there was no statistical decrease in health conditions where water use was over 65 L/(p·d). The utility systems in the study communities included vehicle-haul water and honey-bag collection, vehicle-haul water and pumpout wastewater collection, and piped service. Independent of water use, the type and combination of utility system did not have a statistical impact on the risk of diarrhea. However, it should be noted that in these communities honey bags were collected at least twice per week.

Studies show that improvements in municipal services in cold regions communities have brought improvements in health (World Health Organization, 1982; Robinson and Heinke, 1990; Brocklehurst and Heinke, 1985; Michael, 1984). However, it is not possible to rank specific service improvements with respect to potential health improvement or cost effectiveness (Robinson and Heinke, 1991). Nor is it practical to attempt a benefit-cost analysis of utility and health improvements, particularly on a community or project basis.

Utility services and health should not be considered in isolation. The provision of utility services is necessary but not a sufficient condition for public health improvement. Programs that must be integrated into a comprehensive public health improvement program include:

- water and sanitation services;

- housing and household plumbing (Martin, 1982);

- nutrition;

- public health education (Tester, 1976);

- medical services; and

- sociopolitical development (Reinhard, 1976; Young, 1982).

2.1.2 Environmental Protection. The arctic and subarctic ecosystems of the cold regions are generally characterized by relatively low biological activity rates and low species diversity. Typical environmental conditions include:

- large seasonal variations in energy input and climate;

- low nutrient levels;

- low temperatures (average and seasonal); and

- short growing season.

Bodies of fresh water in the cold regions typically have:

- low nutrient levels;

- low temperature;

- long period of ice cover;

- low dissolved oxygen level during winter; and

- large seasonal variations in water quality and quantity.

The assimilative capacity of the waters and lands in cold regions is generally somewhat lower than in temperate areas. However, due to the limited agricultural and industrial activity and the low population density, water pollution problems and issues are generally confined to specific areas and cases

(James F. MacLaren Ltd., 1980). Pollution originating in other areas of the earth affects the cold regions. Pollutants such as pesticides, radioactive particles and heavy metals are carried over long distances by winds and sea currents. Significant environmental impacts caused by waste discharges from the typical small communities in cold regions are generally limited to where the receiving water is a small river, lake or confined bay and to local effects within the mixing zone. In cold regions, the timing, method and point of discharge are often more environmentally important than the level of wastewater treatment. Waters in cold regions are most sensitive to wastewater discharges during the ice-covered period when reoxygenation is impeded and flows are low.

The uncertainty in predicting environmental impacts in cold regions makes it prudent to monitor the effects of waste discharges and to plan flexible responses into any waste management systems.

Aesthetic and psychological reasons for maintaining the pristine environment are important and justifies wastewater treatment in cold regions.

Wastewater generated from activities and communities in cold regions is generally more difficult and expensive to treat (see Section 10).

2.1.3 Socioeconomic Development of a Community. The World Health Organization has stressed the importance of adequate water, sewer and solid-waste services in the social and economic development of a community. There is likely to be poor health in a community without basic utility services and there is little prospect of social or economic development without healthy residents (World Health Organization, 1982). Infrastructures such as water and sanitation systems, energy, communications and transportation are essential elements in the economic development of communities. Utility services should match the development potential and objectives of the community.

The level of service and the types of utility systems in a community are influenced by the size, nature, function and economic base of the community and by specific utility needs of current or prospective commercial, institutional and industrial operations. In urban growth centres (though usually small by southern standards), resource communities and industrial camps, a standard of service and type of system equal to temperate areas is provided since industry, government and residents demand this level of convenience and they can sustain the high cost. In these communities and camps piped water and wastewater services are common.

In small, remote, predominantly aboriginal communities with a small formal economy the residents often cannot afford southern-type services. And southern-type services may not be appropriate for sustainable development. Utility services should strengthen the local culture, lifestyle and economy (including the informal economy). The economic benefits of utility expenditures should be captured within the community. Emphasis should be placed on creating and filling stable, permanent jobs by qualified local residents and programs should be in place to allow the residents to become qualified (Rees, 1986).

In cold regions, the need for water and sanitation services is one of the first issues that must be faced in a permanent, organized community. Participating in the planning, construction, operation and maintenance (O&M), and management of utilities provides an important opportunity for individuals and communities to become involved with their own development and to build technical skills and organizational capacity within the community.

Direct social and economic benefits through employment and training are important considerations in the evaluation of utility options and in the design, construction, operation and maintenance, and management of utility systems.

2.1.4 Fire Protection. Fire protection aspects include fire prevention and fire suppression. Compared to southern regions on a per capita basis, fires are generally only slightly more frequent. However, the loss of life, injuries and property damage are all significantly higher in communities in cold regions.

Water systems, including self-haul, vehicle-haul and piped, have different inherent or design capabilities with respect to fire suppression. The selection and design of utility systems and the physical planning of communities must incorporate fire protection measures. In cold regions, particularly in small communities, passive fire resistance in buildings (e.g., fire walls, sprinklers, noncombustible materials) and an active fire prevention program are often more important than fire suppression equipment or the water supply capacity of the utility system in reducing the loss of life and the destructiveness of fires. Fire protection considerations are detailed in Section 15.

2.1.5 Convenience and Aesthetics. Convenience and aesthetics are often the users' most important objectives for comparing and implementing

utility services, particularly once basic survival and health requirements are met.

User convenience factors related to the utility system include:

- effort (i.e., time, energy, etc.) required by users to get water to the house and to remove wastewater;

- effort required by users to use water for personal hygiene and cleaning (i.e., the type of plumbing and fixtures); and

- rationing, interruptions in service, failures, or other conditions that require users to adjust their water use activities.

A very inconvenient utility service can impede achieving other objectives such as health and socioeconomic development.

User aesthetic factors may include:

- maintaining a sanitary, unpolluted and uncluttered environment within and around the community;

- offensive activities, such as self-haul of excreta;

- toilets with exposed excreta or odor;

- smells when pumpout tanks are emptied; and

- taste and color of the supplied water.

Acceptability of water quality for drinking and making tea is a common issue with aboriginal residents (Sims, 1982). Chlorine (and by-products of disinfection), iron and other chemicals at concentrations within drinking water standards' levels may still impart a taste to water and tea. If the water provided through the utility system is not aesthetically pleasing to the users, their needs will not be satisfied and some people may use uncontrolled and potentially polluted water sources.

Perceptions and priorities for convenience and aesthetics are largely based on culture and experience, and they are not static. Therefore standard objectives cannot be specified. Convenience and aesthetic objectives should be made explicit when they are used to evaluate utility services and systems, so they can be verified. These objectives should be established by the users (Campbell, 1980).

2.1.6 Equity. Equity is a sociopolitical objective which varies with social and political structure and philosophy of the region or nation. Equity has been interpreted various ways, including ensuring that:

- basic utility services for survival and to safeguard public health are available to all residents;

- utility services are equitable to the services enjoyed by people living in the more temperate parts of the same country or in other communities within the same cold region; and

- utility services do not discriminate or limit the service available to residents within the same community.

The equity objective is based on correcting the biases of geography and society and is addressed through programs to make water and sanitation services available and affordable. Equity is an important, if not always explicit, social objective of government assistance programs for utility services and systems. Equity is an inherent component of other utility service objectives; for example, the equitable potential for good health and the equitable opportunity for socioeconomic development.

Providing equitable utility services to residents should not be interpreted as providing equal utility systems or technology. The utility systems that are appropriate and cost effective vary within and between communities.

2.2 Government Assistance Programs

The combination of high cost, low income and lack of local expertise are formidable obstacles to providing even basic levels of utility services in many communities in cold regions, particularly in the small, remote, subsistence-economy communities. The need for assistance and the types and levels of assistance that are available to individuals and communities in cold regions are generally greater than the assistance available in the south (i.e., more temperate areas of the same nation).

National, regional and local governments and some nongovernment organizations in cold regions provide a variety of assistance programs to:

- help communities construct utility infrastructure;

- help communities provide utility services (e.g., operation and maintenance assistance); and

- help users receive and afford services.

Assistance programs are commonly justified to promote the objectives outlined in Section 2.1, particularly where the benefits of utilities accrue beyond the immediate users or community (for example, public health and environmental protection). Assis-

tance programs are political actions and as such vary across cold regions depending on the socio-political philosophy and structure in the region or nation. Assistance programs of the different national and regional governments in cold regions vary substantially and include the following:

Financial Programs

- Capital grants for utility systems or specific infrastructure (e.g., water intake, wastewater treatment or specific equipment). Grants are often based on a percentage of the cost (often 100%) or a fixed contribution. Eligibility or amount is often varied based on the community's resources or political status.

- Operations and maintenance grants for general operations or specific items such as labor or equipment. Grants are usually based on a percentage of the cost.

- Rate subsidies for all or certain customers (e.g., private homeowners) and for all or a basic level of service. Subsidized rates include fixed-unit rates or service charges or a percent of the actual cost.

- Grants for utility planning.

Technical Programs

- Direct responsibility or participation of senior government (i.e., national or regional governments) to provide utility services, such as when there is no local government with the authority or capability to manage utilities.

- Direct responsibility or participation of senior government in the construction of all or certain infrastructure.

- Direct responsibility or participation of senior government in the operation and maintenance of all or certain services or infrastructure.

- Training programs for operation and maintenance, management, accounting, etc.

Indirect Programs

- Financial and technical assistance for house improvements including plumbing, service connections, and building tanks.

- Financial assistance for transportation, energy, labor and other costs.

- Public housing, with subsidized utility costs.

The type and level of assistance programs that are available to a community are often the most significant factors that influence the level of service and the type of systems that are implemented in a community. For example, if financial assistance is limited to capital grants, then service options with high operating costs (e.g., vehicle-haul services) will not be as attractive to the users. If technical operating assistance is available, then systems that may be beyond the community's current ability to operate can be constructed.

Although assistance programs are essential, they can artificially distort the selection of systems and they detract, to varying degrees, from local control and self-reliance. The types and levels of assistance to be provided are political decisions. The evaluation of assistance programs is beyond the scope of this monograph.

2.3 Legislation, Regulations, Policies, Standards and Criteria

National and regional governments regulate many aspects of cold-regions utilities including water withdrawals, waste discharges, land use, environmental impacts and public water supplies. Many permits or clearances must be obtained before a project can proceed. Requirements differ in individual Canadian provinces and territories and in the United States.

Agencies providing financial assistance for community utilities usually have program standards and criteria that may specify eligibility criteria, planning and evaluation requirements, performance criteria, design standards, operating requirements or other requirements that must be considered in utility planning and design.

Senior governments give local governments the authority and responsibility to provide local water, wastewater and solid-waste services. The local government usually owns and operates the local water and sanitation system. Exceptions to this include communities in the Northwest Territories without a property-tax base where the Government of the Northwest Territories owns and operates the utility infrastructure and in Inuvik, NWT where a utility agency operates the utility system. In the United States both local governments and private utility companies are present. Local governments control physical development, including utilities, through an official plan, zoning regulations and development standards. Local governments can control and regulate the use of their utility systems through utility regulations (i.e., municipal bylaws or ordinances)

(Municipal and Community Affairs, 1991b). Many small communities in cold regions do not have proper regulations or methods to enforce the regulations.

Some national, industrial and professional regulations and standards that apply to utilities are universal in their coverage but they do not adequately consider the physical, environmental, social or economic conditions in cold regions. Thus, they must not be used in cold regions without adequate assessment.

2.4 Planning Process

The planning process should inspire, guide and help the residents become aware of problems and conditions of their lives and to plan and act to change adverse conditions for the better (Cameron, 1985). Planning provides a forum and information to facilitate informed debate and rational choice. Adequate time and resources must be allocated to the planning stage to permit proper planning.

2.4.1 Community Considerations. The planning process in large urban communities in cold regions is essentially the same as in the south, although the solutions and technology must be appropriate to the local environment. However, utility planning in the small, remote, subsistence-economy, aboriginal communities in cold regions have special characteristics and considerations, including the following:

- There is a high level of senior government involvement in the financing and planning of community utilities. Agencies that provide significant financial and technical assistance share in the decision making, specify various requirements for planning process and evaluation criteria, and often initiate community utility improvements and planning.

- The current level of service may be low or even nonexistent. Systems that provide only basic levels of service may be considered. Utility planning in cold regions may entail evaluating levels of service and system options with respect to health benefits, management and technical skill requirements, and affordability to an extent that is more common in utility planning in less developed countries (e.g., Saunders and Warford, 1976; Cairncross et al., 1980; Kalbermatten et al., 1980).

- Aboriginal peoples' relationships with the environment (hunting, fishing and trapping) are typically important cultural and economic activities. There are various aboriginal cultures in cold regions and the level of influence or strength of the aboriginal culture varies by individual and community. Therefore it is inappropriate to stereotype planning in aboriginal communities. Nevertheless, in all communities, the utility planning process and utility planners must adapt to the local culture, lifestyle, decision-making process (formal and informal), and authorities and leaders (e.g., the municipal government, aboriginal organization, elders, etc.). Utility planners must adapt the planning process and their activities to the local language, knowledge, values, communication style and methods used in interpersonal relations (Simon et al., 1984).

- In small communities, the local government seldom has professional staff to guide, interpret or manage the utility planning. Utility planners and designers must deal directly with the local government and residents.

- In many situations the participation in utility planning or the utility technologies being considered may be new for the residents and community. Planning therefore assumes a teaching role. Planners must present the issues, data, analysis and technologies in ways and language the residents can understand. A major challenge is to integrate the scientific, engineering knowledge of the utility planner with the traditional knowledge and values of the aboriginal residents.

2.4.2 Level of Planning. The decisions to be made, the information requirements, the evaluation criteria and the planning process all vary with the level of planning. Levels of planning include:

Policy Planning. This is concerned with establishing regulations, policies, standards or assistance programs to meet specific objectives (e.g., health, environmental protection). Policy planning is a political function, usually at the national or regional level, and is not specifically addressed in this monograph.

Planning for Utility Services. This is planning at the community level to select the level of service and type of utility system (e.g., central facility, vehicle- haul, piped) and to develop a utility plan (e.g., utility master plan) that addresses the total utility requirements for the community over a planning horizon of at least 20 years. Planning for utility services is concerned with integrating utility services with the physical, social, economic and political development objectives of the community and meeting the objectives in providing utilities. Planning for utility services is best undertaken during the development of, or within the context of, a community development plan.

Planning of Utility Systems. This is planning to implement or construct systems (e.g., water supply system) or components (e.g., water storage tank). Planning of utility systems is primarily concerned with engineering considerations (e.g., performance, reliability, efficiency) which are detailed in various sections in this monograph. Planning of systems or components is best undertaken within the context of an overall utility plan.

2.4.3 Community and Public Participation. In this context "community participation" means participation of the formal local government, other community organizations or interest groups, or the community as a whole. "Public participation" means participation of individual residents.

Inadequate community participation in utility planning will almost certainly result in failure, either social or physical failure, at the project level.

Utility planning should always provide adequate opportunities for public participation. Public participation should be actively sought when utility planning decisions involve the desires, preferences, knowledge, commitment and values of individual residents.

Public participation should be more extensive when planning for utilities since decisions will be made on the level of service and long-term utility system. Public participation is necessary in the planning of utilities to establish design and implementation requirements and facility siting preferences.

Community and public participation should always be under the auspices of the community government. Soliciting public participation is appropriate when the local government may not wish to, or may not represent the range of individual concerns or the extent of local knowledge. Public participation should not be used to bypass the decision-making authority and responsibilities of elected representatives.

Participation Methods. The approaches for public participation are different for large, urban communities compared to small, remote, aboriginal communities. Nevertheless, personal contacts and dialogue are important in both situations.

Community participation is necessary to solicit local preferences and concerns:

- to determine locally-perceived objectives and needs;

- to determine the level of the local government's commitment, to manage and operate various options; and

- to facilitate integration of utilities with community development objectives.

Community participation can be achieved through formal and informal meetings with individuals and groups that have responsibilities and authority related to utilities (such as municipal politicians and committees, cultural leaders, public health staff, municipal and utility staff, housing organizations, etc.) as well as with local interest groups and individuals that express an interest in utilities.

Public participation is necessary to obtain information and support from individuals within the community. Public participation can be solicited in many ways including formal and informal meetings and through surveys of all or representative residents. A survey of individual residents may be appropriate to determine views on service and system options related to:

- potential for and impediments to achieving objectives (e.g., health benefits);

- potential for and impediments to using specific levels of services and types of systems; and

- service and system design information, particularly related to individual building and on-site facilities.

A survey of individual residents is necessary to obtain information on:

- desire for utility improvements and perceived needs and objectives;

- ability and willingness to participate and pay for utility improvement options;

- perceived relationships between health, sickness, nuisance, utility services and personal hygiene practices;

- attitudes to level of service including convenience, aesthetics and reliability of service;

- preferences for water source and quality, waste disposal locations, etc.;

- present utility services (individual and community) and use patterns of existing utilities;

- housing conditions and improvement plans, particularly plumbing; and

- preference and attitude toward communal facilities (i.e., central facilities).

2.5 Community and Site Considerations

The planning, design and management of utility systems in cold regions are greatly influenced by the community and site conditions. Community and site conditions influence the social and political appropriateness, the technical feasibility and practicality, the ultimate design, management requirements and the cost of utility services and systems.

Information requirements depend on the types of decisions to be made. For example, planning for utility services requires information adequate to assess current conditions and to define and evaluate feasible options. General community information is required. Planning of utility systems and design of utility infrastructure requires more focused and detailed information; site information is required. In cold regions, it is often necessary to gather relatively detailed community and site information in the planning stages to ensure that options are feasible, practical, effective, efficient and to estimate costs. Of particular importance in cold regions are:

- the demographic, economic, health and housing conditions;

- the potential for growth and future needs;

- the local management and technical capabilities; and

- the access, soil conditions, topography, hydrology and natural resources.

The following outlines general considerations and information required to plan, design and manage utilities. The relative importance of these factors will vary. Not all factors need to be considered for a specific community or for all utility options. Specific considerations and data required to design utility components are provided in the various sections of this monograph.

2.5.1 General Community Considerations. If there is a community development plan it should address the general community considerations and contain information required to plan for utilities including:

- community development: history, trends, projections and any formal community development plan;

- demographic data (current and projected): population, household size and composition, ethnic composition;

- housing (current and projected): types (single family, multifamily, apartment), occupancy, density, tenure (rent, own, public) and condition (particularly the current plumbing and potential for improved utilities);

- nonresidential buildings (current and projected): types, density, utility requirements; and

- land use (current and projected): land use, ownership and zoning restrictions.

2.5.2 Physical Infrastructure.

Access to Community and Site.   Access influences, and can dictate, the choice of equipment and materials that can be practically or economically transported to a community or project site. Detailed information on transportation means, constraints and cost are provided in the discussion on project management (Section 2.8).

By Air.   Virtually all communities in cold regions have air access but there are severe restrictions on the size, weight and type of materials that can be transported, particularly on scheduled airlines. Larger aircraft can be chartered for special equipment. Local runway conditions may limit the aircraft. Temporary runways can be constructed on ice to accommodate larger than normal aircraft. Helicopters can be used to access virtually any remote site (e.g., during investigations).

By Water.   Ships or barges are a major transportation system to many of the cold regions, especially for heavy and bulky items. However, shipping is limited to a short period. Missing shipping deadlines can cause a full year's delay in construction. Ships and barges often arrive in isolated communities late in the summer construction season. Shipping size and weight limitations are usually based on the dock and unloading facilities at the community or site.

By Land.   Where available, road and rail access provide the easiest and least restrictive access to a community or site. In some remote locations, winter roads provide seasonal access. Information concerning construction and maintenance of winter (snow and ice) roads is available in Section 3 and other reports (Adam et al, 1984; Clark et al., 1973; McFadden and Bennett, 1991; Grey and Male, 1981). Off-road vehicles can be used to access remote sites or facilities.

Other Utilities.   Other utilities within the community should be considered with respect to their availability, capacity, reliability and cost to supply the needs of water, wastewater and solid-waste systems. Such utilities include:

- communications to, from and within the community; and

- energy sources including electricity, oil, gasoline, natural gas and wood (see Section 17.3).

In small communities the energy required to pump and heat water can be a significant portion of the total community energy needs and proposed utility facilities may overtax the existing capacity. All community utility needs must be coordinated to ensure adequate capacity and backup. Sometimes the community utilities may not be adequate and upgrades or independent utilities must be provided to support the water, wastewater and solid-waste systems (e.g., oil storage, electricity, etc.).

Community Layout.   Community layout considerations include:

- existing land uses, land status (ownership), location of buildings, and utility service needs of sites and areas;

- formal land use plans and zoning regulations, anticipated redevelopment and expansion areas;

- existing road or trail conditions, patterns and capacity;

- local preferences or restrictions with respect to locating utility infrastructure including buildings, pipelines, water supply and waste disposal sites; and

- utility easements.

2.5.3 Socioeconomic Conditions

Health:

- health problems in the community and relative importance of water and sanitation related diseases, likely disease transmission routes;

- attitudes, perceptions and knowledge of the causes of illness, health and sanitation relationships, and the benefits of improvements;

- personal hygiene and water-use practices;

- health facilities and health education activities.

Level of Interest:

- residents' perceptions of present conditions and interest in improvements;

- residents' perception of objectives and need for utilities;

- community leadership commitment to improvements;

- community leadership commitment to accept responsibilities (e.g., for ownership, operation, management).

Willingness and Ability to Pay:

- individual customers' (households) willingness and ability to pay for utility service options;

- income levels and disposable income available for utility services;

- customers' expenditure patterns (particularly for other utilities);

- ownership of land and house.

Economy:

- current economic base and economic outlook, as a measure of needs and ability to pay;

- utility needs of specific current and future economic activities (e.g., a fish-processing plant);

- potential economic activities arising from improvements in utilities;

- employment created through utility construction, operation and management.

Labor Force:

- work force availability with respect to constructing, operating and managing utility services and systems.

2.5.4 Political and Institutional:

- local leadership, authorities and decision-making process;

- local government's administration and management experience and capabilities;

- allocation of responsibility and authority for providing utility services, construction, operation, administration and management;

- financial, technical and management assistance programs available.

2.5.5 Utility Management:

- utility management and administration requirements;

- utility organization authority and responsibilities;

- utility master plan;

- financial status (debt, surplus);

- accounting practices and records, budgeting process, cost recovery (rate-setting) procedures;

- utility regulations (i.e., bylaws or ordinances);

- utility operation and maintenance requirements and capabilities;

- skill and trades requirements (job descriptions, qualifications);

- works management system (e.g., O&M schedules, procedures, manuals, operational data collection and reporting practices).

2.5.6 Utility Service and Systems:

- existing levels of service, systems and infrastructure;

- service and system capacity;

- current utility use patterns, service demands, trends, projections including:

  - water source(s), formal and informal;

  - water consumption, volume and uses;

  - wastewater generation, volume and strength;

  - solid-waste generation, volume and characteristics; and

  - customers, types (residential, commercial, industrial) and demands.

- deficiencies with respect to:

  - meeting current and projected needs (e.g., water quantity, quality);

  - meeting regulations (e.g., water quality, environmental); and

- Efficiency and reliability.

2.5.7 Physical Environment

Climate:

- temperature (profile, extremes, average, freezing index heating index, frost-free period);

- precipitation (snow, rain);

- wind (direction, velocity, wind rose);

- sunlight.

Soil Conditions. Soil conditions are a very important consideration in locating, developing, or expanding a site, camp or community and in the planning and design of utility systems. Often the soil conditions will make certain development and utility options or designs impractical or unfeasible. Section 3 provides detailed information on geotechnical considerations.

At the planning stage it may be possible to obtain adequate, reliable information through aerial photo interpretation, site inspection, surface vegetation (Johnston, 1981) and interpolating from previous soil surveys in the area. At the design stage more detailed site information is required. In cold regions it is often necessary to initiate a soils investigation early in the planning stage because soils information is necessary to determine the feasibility and cost estimates of options and because of the lag time in obtaining detailed information on the soils.

A soil investigation is necessary to identify:

- foundation conditions for buildings, pipelines, roads, etc.;

- soil conditions for water supply (e.g., wells) and wastewater disposal (e.g., septic fields);

- sources of borrow material (sand, gravel, rock); and

- other factors depending on the project (e.g., wastewater lagoons, water reservoirs).

A soil investigation should identify:

- soil cover (vegetation, organic) and types;

- permafrost;

- moisture/ice content;

- susceptibility to thaw settlement;

- temperature;

- rock outcrops and bedrock;

- depth of frost or thaw penetration (active layer);

- other factors depending on the project.

Soil surveys must consider that soil conditions can vary considerably over a short distance and that more structures have failed because of inappropriate foundation design than for any other reason.

Soil borings should be made throughout the project site. The size and complexity of the utilities or structures to be built and the variation in ground conditions determines the number of necessary soil borings.

Ground-temperature information indicates the presence or absence of permafrost and the depth of the active layer. Temperature profiles are extremely important for foundation design but few locations have instrumentation and temperature records. Temperature instrumentation should be installed as soon as possible.

Changes in ground temperatures and the thickness of the active layer due to site development and disturbance of the natural ground and vegetation cover must be considered in planning, design and operations.

Permafrost. Permafrost is the state of any material that stays below freezing (0°C) for two or more years. Thawing permafrost that is predominantly ice will result in catastrophic soil failure. However, thawing frozen, solid rock would rarely create an unusual construction problem since it contains little or no water. Definitions of permafrost characteristics are provided in the Glossary.

Thawing of permafrost can result from a disturbance to the surface layer or soil conditions. Such disturbance can be caused by excavation, removal of the surface organic layer, or heat from buildings or warm pipes placed on or in the soil.

The construction method used in permafrost areas depends on the structure or facility being designed and the ground stability upon thawing. Active and passive construction methods have been developed for use in permafrost areas. Active construction consists of prethawing the permafrost and/or removing the permafrost and replacing it with a non-frost-susceptible soil. Passive construction is based on maintaining the permafrost by construction above the ground on piles, installing refrigeration (such as thermopiles), or insulating the ground surface.

In discontinuous permafrost areas, it may be possible to avoid permafrost areas for roads, land development, buildings and pipelines. For example, south-facing slopes are often free of permafrost. Sites with soils that have little or no ice content are preferable since any thawing of the permafrost will result in less settlement and instability.

Geotechnical considerations in permafrost are detailed in Section 3.

Topography.   Topographic factors include:

- land contours and slopes;
- surface drainage and water courses;
- land stability and areas of active erosion; and
- floodplains and shorelines.

Sites with a gentle slope (e.g., about one percent) are preferred, since they provide good drainage. The slope of the site should be used to prevent ponding of water. Ponding causes thermal degradation of the permafrost. South-facing slopes receive more incident sunlight and solar energy and provide better protection against cold north winds. Natural obstacles that would promote snowdrifting should be avoided.

Most communities in the North are located along rivers, lakes or ocean. Therefore flooding is often a problem. Serious consideration must be given to:

- the frequency and extent of flooding;
- the cause of flooding (e.g., ice jam, snow melt, storms, or rain);
- water and ice forces that would cause damage to surface structures (Figure 2-2);
- erosion by rivers or wave action; and
- thermal erosion.

Natural Resources.   The local availability of natural resources, particularly construction materials, will influence the selection, design and cost of utility systems.

For most construction, the availability of an adequate quantity and quality of sand and gravel is important. Sand and gravel are required for foundation pads, concrete, roads and backfilling around pipes. In many communities along Alaska's west and north coast, gravel is nonexistent. In Bethel, for example, barges must travel nearly 200 kilometres upriver to obtain construction gravel. Where bedrock or rock outcrops are present, they can be quarried and crushed to supply fill material.

FIGURE 2-2   ICE DEPOSITED BY RIVER – NAPASKIAK, ALASKA

Trees suitable for building logs, lumber and piles can be an important construction material and reduces dependence on importing.

Local energy sources, such as coal, wood, oil, gas, hydroelectricity, wind and solar radiation can increase self-sufficiency and reduce energy costs.

Environment. The need to protect the local environment can place restrictions and design criteria on utility systems including:

- timing of activities;

- location of infrastructure and restoration requirements; and

- location, timing, quality and quantity of discharges to the environment.

Potential environmental impacts of utilities include:

- disturbance of seasonally sensitive conditions (e.g., migration patterns of birds or mammals) or destruction of the environment through the construction of roads, pipelines, reservoirs, lagoons, etc.;

- pollution of the environment through waste discharges including wastewater, solid waste and sometimes air pollution; and

- water withdrawal, where natural flows are significantly altered.

Information on the local ecology is necessary to:

- assess environmental impact of options;

- assess compliance with regulations (land use, effluent quality limits, etc.); and

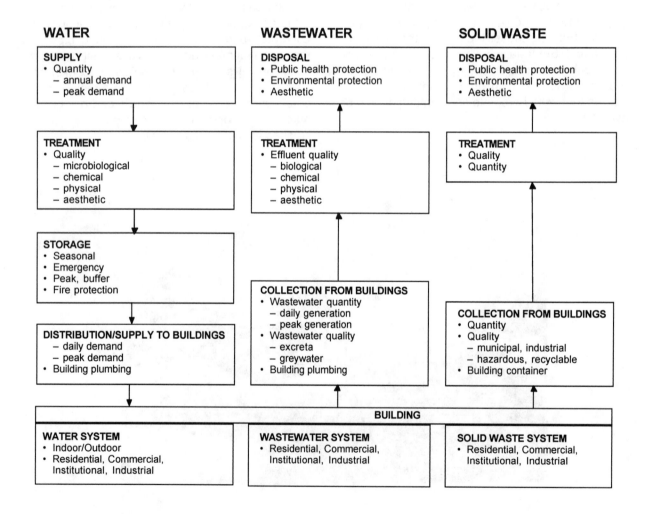

FIGURE 2-3   ASSESSMENT OF NEEDS

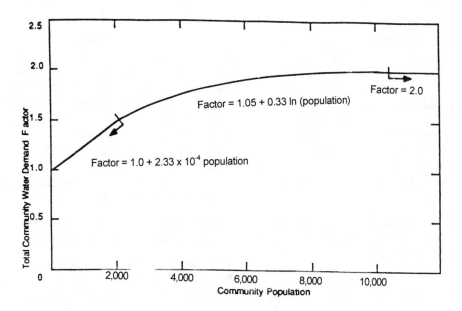

FIGURE 2-4   TOTAL COMMUNITY WATER DEMAND FACTOR
(Government of the Northwest Territories, 1986)

- identify the need for mitigation measures and monitoring.

2.6   Utility Considerations

2.6.1   Assessment of Needs.   The need for utility services and systems is established through the planning process and the assessment of community and site conditions. An assessment of current and future needs is necessary to plan utility services and components of utility systems. Needs for utility components are illustrated in Figure 2-3.

In many cold-regions communities, future needs should not be simply extrapolated from current demand and population growth for the following reasons:

- Building plumbing may be improved (e.g., from nonpressure to pressure water; from honey bag to flush toilet).

- Water use habits may change (e.g., more frequent laundry, bathing, etc.).

- Level of services may improve (e.g., eliminate rationing; increase reliability).

- Types of utility systems may change (e.g., from self-haul to vehicle haul; from vehicle haul to piped).

- New activities (e.g., tourism, government decentralization) can have a significant impact on the growth, population and utility needs of small communities.

- New buildings (e.g., school, hospital, fish-processing plant) can have a significant impact on the utility needs of small communities.

The quantity of water demand and the resulting quantity of wastewater generated are critical to assessing needs for most utility system components. Water demand for various buildings, activities and utility systems are provided in Sections 5, 8 and 14. The total community water demand is the sum of the residential and nonresidential (i.e., commercial, institutional and industrial) demand. The nonresidential activities can vary significantly but they tend to increase in proportion to the community population. Where existing water demand is not appropriate to use for long-term planning, the Government of the Northwest Territories estimates the total community water demand by multiplying the residential water demand by the total community water demand factor illustrated in Figure 2-4 (Government of the Northwest Territories, 1986). For planning purposes, the residential water demand will be the projected future demand. This formula assumes that the type of system and the level of plumbing is the same for the residential and nonresidential buildings, which may not be the case today but is often appropriate for long-term planning.

Water conservation measures can have a significant impact on water use and consequently the need for additional infrastructure, the design capacity required for future infrastructure, the cost of service (particularly for vehicle-haul systems), and the evaluation of utility system options (particularly the relative economics of vehicle-haul versus piped systems). Water-demand forecasts should be based on the efficient and economical use of water. Water-demand management policies and measures include:

- meter water use and establish customer charges based on the volume of water used, rather than a fixed charge;

- install low-water-use plumbing fixtures and appliances (see Cameron and Armstrong, 1980 and Section 14.3);

- eliminate bleeding (i.e., wasting water to prevent freezing) on building service connections and mains; and

- consumer education.

2.6.2 System Options. Utility system options are described in this section. Individual components (e.g., water source development, wastewater treatment, etc.) and facilities which make up a utility system are dealt with in the appropriate sections of this monograph.

The primary options for the various components of water, wastewater and solid-waste systems in cold regions are illustrated in Figure 2-5. The primary water and wastewater system options can be categorized as individual, central facility/self-haul, vehicle haul and piped. These system concepts are not unique to cold regions but the system designs are unique. The utility options presented in this monograph reflect current practice in cold regions but should not stifle creativity or innovation.

Many communities have mixed or multiple utility systems. An example of a mixed system is where there is a water supply pipeline to the community and vehicle-haul water distribution to the buildings. An example of multiple systems is when some buildings in a community have piped services while other buildings are serviced by vehicle haul or a central facility. In some communities mixed or multiple systems are economical while in others they are necessary to meet different user needs or building requirements. However, different levels of service within a community can foster resentment and discrimination.

In some situations, consideration should be given to phasing systems and facilities. For example, the first phase could be a central watering point, with a piped or vehicle-haul system following a few years later. This phasing would allow time to improve house plumbing, construct roads, increase community interest in further improvements, or develop community operating or management capabilities. Phasing may also be necessary if sufficient funding is not immediately available, particularly funding assistance from senior government.

The primary system options are individual and community systems. These are described below.

Individual Systems. Individual water, wastewater and solid-waste systems are necessary for isolated buildings or camps and may be appropriate for rural community development. However, in cold regions the typical southern rural systems of wells, pit privies and septic fields are often not feasible (e.g., in areas with permafrost since the ground is frozen) or practical. Self-contained building systems such as compost toilets or wastewater treatment/reuse are attractive concepts for buildings in cold regions. However, these systems tend to be complex, costly and require considerable user attention and commitment. Individual on-site systems are discussed in Section 14.

Community Systems. The primary community utility systems are self-haul, central facility, vehicle haul and piped. These are described below.

Self-Haul. Self-haul water systems involve the development of a safe water point from which individuals can haul their own water. More than one water point may be needed in a large or spread-out community. Water can be supplied to a water point directly from the source (e.g., a well), from a water pipeline, or from a tank which is filled by a vehicle. Fetching water is time consuming and arduous and the resulting low volume of water available in the house inhibits adequate personal hygiene and household sanitation.

Self-haul waste systems consist of users hauling their own wastewater (usually only excreta) to a central point (e.g., waste bunker) or directly to a community treatment/disposal site. Although self-haul of excreta to a community facility may be safer than indiscriminate dumping on the ground, there are serious health concerns with these systems (Bond, 1985). Treatment and disposal of undiluted

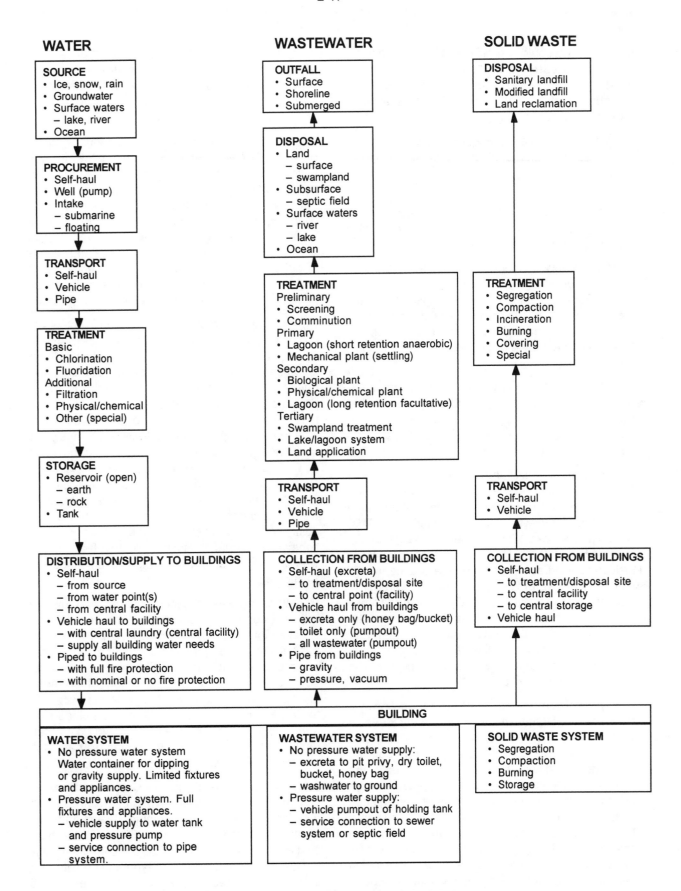

FIGURE 2-5   WATER, WASTEWATER AND SOLID WASTE OPTIONS

human waste could consist of a sludge pit lagoon, trench, or treatment plant (see Section 16.5). Washwater (greywater) is generally disposed of on the ground surface, a practice that should be avoided for health reasons.

Solid waste that is not managed on-site (e.g., burning) is hauled by individuals to a community disposal site.

Self-haul systems are relatively low in initial cost but provide the least health improvement and level of convenience compared to other community utility system options.

Central Facility. A central facility has a water point as well as other public services and facilities which may include a means to deposit wastewater (excreta) and solid waste as well as facilities for laundry, bathing, sauna and restrooms. The fixtures and facilities in the central facility are usually coin operated.

The initial cost for a central facility depends primarily on the services provided, the degree of water treatment necessary, the type of wastewater treatment and refuse disposal selected. Central facilities are discussed in Section 12.

Vehicle Haul. Vehicle-haul water delivery systems (or trucked systems) use vehicles to deliver water to buildings and to collect wastewater from buildings. Vehicles vary from all terrain vehicles (ATVs) which pull a small 250 litre tank to large trucks mounted with 10,000 litre tanks. Water delivery vehicles are filled at the water source or a central fill point and the water is then delivered to water storage tanks in the buildings, usually one to three times per week. Vehicle-haul water systems can complement a central facility with laundry and bathing facilities or it can provide full water needs to the individual buildings. Vehicle-haul water delivery systems are discussed in Section 8.

Two types of vehicle-haul wastewater collection systems are used. One type is to collect only the toilet waste (excreta) from buildings. Excreta is collected using buckets, bags (honey bags) or from building tanks designed only for toilet waste. Usually the buildings being served do not have pressure water systems (running water), they have very limited plumbing and fixtures, and washwater (greywater) from the buildings is discharged onto the ground around the building. The other type of vehicle-haul system is where all of the wastewater generated in the building flows into a building wastewater holding tank from which it is pumped out into a vehicle-

haul tank and removed, usually once or twice a week. The buildings usually have full plumbing and pressure water systems. Vehicle-haul wastewater collection systems are discussed in Sections 9 and 16.

The cost of vehicle-haul systems depends primarily on volume of water delivered and wastewater collected. Water use depends on the level of household plumbing (i.e., whether there is running water, hot or cold water, and the types of fixtures and appliances). Costs for vehicle-haul systems include labor, vehicle and garage capital, and vehicle fuel and maintenance costs. Adequate roads and access to buildings are necessary. The community layout and building density do not have a very significant impact on the cost to operate vehicle-haul systems.

Piped. Piped water distribution and wastewater collection systems in cold regions are the same in concept as piped systems in other locations. Water is transported through pipes from the source to the buildings and wastewater is collected and transported through pipes from the buildings to a treatment/disposal facility. Pipes are located above or below ground. Piped water systems can be designed to provide water requirements for fire suppression as well as consumption. The wastewater collection system could be gravity, vacuum or pressure operated. Building service pipe connections to the main water and wastewater pipe network are critical and costly parts of piped systems. Piped water distribution systems are discussed in Section 8 and piped wastewater collection systems are discussed in Section 9.

Construction costs for piped systems vary greatly with soil conditions, topography and cold-regions construction cost factors. The cost of piped service per building or per capita depends primarily on the community layout and the building density and occupancy. Initial capital costs are higher than those of other service options, but the greatest health protection and convenience is offered with piped systems. Piped systems have lower O&M costs than vehicle haul systems where buildings have full plumbing and all wastewater is collected.

2.6.3 Community Planning Considerations.
Community layout and the types of buildings have a major influence on the planning, design and cost of utilities. At the same time, because of the high cost and importance of utilities, the community layout and housing design (e.g., type and density) are strongly influenced by the physical and economic

conditions imposed by the utilities. Community planning considerations are discussed below.

Location of Utility Facilities. Facilities such as water tanks, wastewater treatment and solid waste disposal sites should be located to allow for future expansion of the facility and of the community. Tank farms for oil or other chemicals should be located where they can easily be filled and used, yet away from the existing community and future growth areas. Secondary containment is generally mandatory around all tanks containing hazardous products (e.g., petroleum).

Water sources must be protected from possible contamination. Waste disposal areas should be located where they will not create aesthetic, odor or vector problems.

Minimizing Problems Caused by Snowdrifts. Problems arising from drifting snow can often be avoided or reduced by the proper location and design of structures.

Runways and roads built above the surrounding terrain usually blow free of snow, thus reducing snow removal efforts and costs. Roadway cuts perpendicular to the wind should be avoided where possible as these will fill with snow.

Several factors should be considered in locating buildings, storage tanks or similar structures. Taller structures should be located so they do not shade smaller ones from either sun or wind. Where snowdrifting is a potential problem, placing small structures on the windward or leeward sides of larger ones must be avoided. One storm can completely bury a house located leeward of a large building or storage tank. The size of snowdrifts can be reduced by orienting buildings so the long axis is parallel to the prevailing wind. In some cases it is necessary to consider moving existing structures.

Snowdrifting considerations are detailed in Appendix C.

2.6.4   Utility Layout Considerations.

A Compact Community Layout is Cost Effective. Pipe services in cold regions are much more expensive than comparable services in the south. Construction and operating costs for piped services can be reduced if the community layout is compact. Compactness is not critical to the operating cost for vehicle-haul services; however, a compact layout

FIGURE 2-6   MEDIUM DENSITY APARTMENT BUILDINGS IN GODTHAAB, GREENLAND

will reduce other infrastructure costs such as road construction and maintenance.

A compact layout can be achieved through high-density buildings, efficient land use and lot design, and efficient layout of buildings, roads and utilities. For example:

- Increase the density such as multifamily row housing and apartments (multistory) and multiuse buildings (Figure 2-6).

- Unserviced areas and large open spaces such as parks, school playgrounds and industrial yards should be located to maximize the use of expensive utility pipes.

- Land uses and buildings that do not require piped services should be located on the outskirts of a camp or community.

- "Double loadi" water/sewage pipes (i.e., locate buildings on both sides of the pipes).

- Orient lots and buildings with their long axis perpendicular to utility pipes to be more efficient.

- Reduce the lot width and separation of buildings, while considering fire protection requirements.

- Avoid cul-de-acs or dead-end roads since dead ends are difficult and expensive to pipe service (i.e., to provide circulation).

High-density buildings and housing (typically apartment buildings or high-density row housing) and compact layout are necessary to make central heating systems practical and economical. Central heating can reduce energy consumption by utilizing waste heat or cogeneration. The risk of building fires caused by heating systems, which is a major cause of accidental deaths in cold regions, is also reduced. Disadvantages to central heating include the high capital cost, the difficulty of expanding these systems if the community or camp unexpectedly expands, and the severe consequences of failure of centralized systems.

Length of Water and Sewer Pipes Required.
The total length of water and sewer mains required depends on the site conditions, housing density and community layout. The length of water pipes required is often higher in cold regions where water must be circulated. Small communities in cold regions tend to be relatively spread-out and predominantly single-family housing. The total length of water and sewer mains required per person tends to be higher in small communities.

Figure 2-7 illustrates a community layout that considers the high cost of piped services and the need

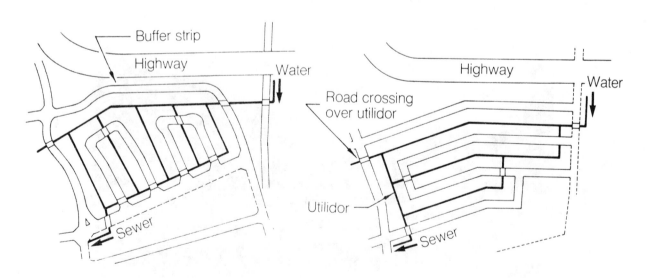

Original development proposal   Improved development proposal

FIGURE 2-7   EXAMPLE OF EFFICIENT LAYOUT

to circulate water resulting in a 12% reduction in the capital cost for piped services. Ongoing O&M costs would also be reduced. This example shows the importance of addressing utility layout considerations early in the physical planning of communities and subdivisions.

Location of Water and Sewer Pipes. Water and sewer mains are usually located in the road right-of-way or along rear lot-lines. In cold regions, service lines are the major source of freezing problems. To reduce freezing problems, service lines should be as short as practical. Therefore, buildings should be located as close to the water and sewer mains as possible, and preferably within 18 metres.

Advantages of placing water and sewer mains along rear lot-lines are as follows:

- The snow is usually not removed, as it would be in the street, thereby providing moderated and warmer ground temperatures. The frost penetration is significantly greater under cleared or packed snow areas such as roadways.

- Utility pipes are not as likely to be damaged by vehicles.

- Service pipes to both rows of houses are shorter, because the right-of-ways for alleys or utilities do not need to be as wide as the streets in front of the lots.

- Houses can be placed close to service pipes since back yards are not used very much in northern communities, especially in the winter.

- Access vaults, lift stations and valve boxes can be terminated above grade.

Major disadvantages of rear of lot servicing are:

- Access may be restricted during part of the year because of snow or poor soil conditions.

- Residents may store supplies and unneeded materials in the right-of-way.

Above- vs Below-Ground Pipes. In cold regions, as elsewhere, below ground piped services are preferable. However, poor soil conditions (e.g., ice-rich thaw-unstable permafrost) and lack of non-frost-susceptible backfill material may necessitate above-ground construction.

Advantages of below-ground pipes (and disadvantages of above-ground pipes) are:

- minimizes exposure to vandalism and traffic damage;

- heat loss, the cost of energy (heat) and the risk of freezing (i.e., time to freeze) are relatively less for below ground pipes;

- grades for gravity sewer pipes from the buildings to the sewer mains are easier to achieve whereas with above-ground pipes buildings may have to be elevated to achieve adequate sewer drainage;

- eliminates the need for piles;

- eliminates the need for elevated road and sidewalk crossings; and

- allows for more normal community layout whereas above-ground pipes can artificially dissect the community.

Disadvantages of below-ground pipes (and advantages of above-ground pipes) are:

- in areas with high ice content permafrost, below-ground pipes are difficult to design, expensive to construct, and control of adjacent activities and development is essential to ensure long-term stability;

- more expensive and extensive equipment (e.g., excavation equipment) is necessary to install, repair and maintain below-ground pipes;

- above-ground pipes are often easier to install, remove (salvage) and maintain; and

- pipe breaks and leaks in below-ground pipes are harder to detect and locate, particularly in frozen ground.

Heat loss is directly proportional to the difference between the inside and outside temperatures. An above-ground utilidor or pipeline has a much higher maximum rate of heat loss (and therefore shorter time to freeze) compared to a similar pipeline buried just below the ground surface. Thermal considerations are covered in detail in Section 4.

Excavating frozen ground to find and repair a leak can be difficult. A pipeline leak can travel under the frozen surface layer and surface through a ground tension crack several metres away.

2.7 Evaluation of Utility Options

Evaluating options is a critical function in planning utility services and systems. Options are evaluated with respect to technical, social, political and economic criteria. Evaluation criteria vary and may include;

- objectives (e.g., as outlined in Section 2.1);

- specific legislation, regulations, policies, standards and criteria that are applicable to the location (region or site) or activity (service, system or project);

- community and site consideration (e.g., as outlined in Section 2.5);

- utility considerations (e.g., as outlined in Section 2.6);

- project management considerations (e.g., as outlined in Section 2.8);

- operations and management considerations (e.g., as outlined in Section 2.9); and

- specific design considerations detailed invarious sections of this monograph.

A comparison of water and wastewater system options with objectives is summarized in Table 2-1. A comparison of water and wastewater system options with various characteristics that are important in cold regions is summarized in Table 2-2. These tables provide only descriptive comparisons of system options. No ranking or recommendation is attempted.

In Tables 2-1 and 2-2, water and wastewater systems are categorized as individual, self-haul/central facility, vehicle haul (with excreta-only collection and with all-wastewater collection) and piped. More than one type of system can be provided in a community. For example, piped services may be provided to high-density, high-water-use buildings while vehicle-haul service is provided to low-density housing and houses without pressure water systems. Another common example is where a water-supply pipeline carries water from a distant water source to a vehicle-fill station or central facility in the community.

It is often necessary to consider staging and upgrading of services and systems over time (i.e., within the planning horizon of the planning study). Upgrading, for example from central facility to vehicle haul, might depend on house plumbing improvements and users' ability to pay. Staging, for example from vehicle-haul to piped system, might become economical in the future if demand (per capita demand and population) increases and/or housing density increases.

Quantitative evaluation of options can be achieved using an assessment matrix approach (also called a weighted factor analysis). Each criterion is assigned a relative numerical weight and the degree to which each option satisfies each criterion is assigned a numerical score. The total weighted performance figures for the options indicate their rank. Only options that meet any specific minimum criterion (i.e., the "must needs") should be evaluated. Ranking options in this way requires agreement on the criteria and their relative weights and is appropriate to evaluate options with respect to meeting "want needs".

2.7.1 Economic Evaluation. It is desirable to rank the options that pass the technical, social, political or other essential screening criteria with respect to their cost. Economic analysis of options in cold regions, as elsewhere, involves calculating the present value of the annual expenditures, both capital and O&M (i.e., life-cycle costing), and the benefits associated with an option over a planning period (usually a minimum of 20 years) (Cameron, 1982; Cameron, 1979).

Ideally a cost-benefit analysis would be used to rank alternatives. Unfortunately it is not practical to quantify benefits such as public health or user convenience, particularly at the community or project level of evaluation (Robinson and Heinke, 1991). Attempts at cost-benefit analysis should be restricted to policy-level evaluation.

Where options have different levels of benefits or service, the option with the lowest direct cost will not necessarily be the overall economical optimum. For example, it is misleading to compare the cost of a central facility with the cost of a piped system since the level of service and the health benefits are significantly higher with piped service. A simple least-cost analysis will not provide sufficient information to select the most economical or appropriate option. Nevertheless, it will indicate the cost tradeoffs corresponding to different levels of service. The community and users can use this information to determine their willingness to pay for various levels of service. Government can use this information to determine the financial assistance required to meet their utility service objectives.

Cost effectiveness (least-cost) analysis is appropriate when options have the same benefits. For example, comparing vehicle haul to a water-supply pipeline to deliver the same volume of water to a community. To allow comparison of vehicle-haul and piped system options, the Government of the North-

TABLE 2-1   COMPARISON OF WATER AND WASTEWATER SYSTEMS WITH OBJECTIVES (Cameron, 1982)

Type of System	Public Health			Environmental Protection	Socioeconomic Development	Fire Protection	Convenience and Aesthetics	Equity (Level of Service)
	Potential Water Contamination	Water Availability and Consumption	Potential Sewage Contamination					
INDIVIDUAL	Varies. High for self-haul to low for wells.	Varies. Generally low, especially in winter when surface water sources freeze.	Varies. Generally high, but risk is confined for isolated buildings.	Varies. Potential pollution but impact is reduced for isolated buildings.	Low density, site and service restrictions can impede development.	Individual responsibility.	Generally provides minimum or basic service but can provide convenient service.	Varies.
CENTRAL FACILITY (Self-Haul)	High potential for contamination due to user handling (filling, transport, emptying) and from containers for transport and storage.	Very low availability and consumption due to self-haul, no pressure water in buildings and limited building plumbing.	Very high. Potential for contamination from excreta during storage and user transport and for contamination of ground around buildings by washwater.	Potential problems disposing of concentrated excreta and plastic bags (where used). Pollution from washwater difficult to control.	Impedes activities requiring water/sewage service to buildings.	Water delivery to fires not integral part of system. Buildings require independent fire suppression systems. Central facility may contain fire water storage.	Provides only minimum basic service. Requires significant user effort to obtain water/sewage services. Self-haul of excreta offensive.	Level of service much below North American norms.
VEHICLE HAUL a) Water Delivery: Excreta Collection	Potential contamination through careless handling of fill hose and with storage in building. Periodic vehicle and building tank cleaning required.	Low consumption due to limited building plumbing and no pressure water.	High. Potential for contamination from excreta during storage and collection and for contamination of ground around buildings by washwater.	As above.	Impedes activities requiring more than basic water/sewage service to buildings.	Haul vehicles can support fire fighting. Water delivery to fire limited by vehicle availability and size and water supply. Large buildings require independent fire suppression systems.	Provides only minimum basic service to buildings. Severe water conservation required results in user inconvenience. Excreta toilets and washwater disposal to ground offensive.	Level of service much below North American norms.
b) Water Delivery: Wastewater Pumpout Collection	As above but lower potential contamination with sealed building water storage tanks.	Delivery system supports full building plumbing and indoor water uses. Efficient water use necessary.	Low. Potential for leakage during collection and transport.	All wastewater collected facilitates management. Relatively dilute wastewater suitable for lagoon or other conventional treatment system.	Can impede high water use activities (e.g., outdoor water uses) and high-density development.	As above but larger vehicles can provide more water to fires.	Varies with service reliability, building plumbing and tank size. Limited water available and need to conserve can result in user inconvenience. Odors during pumpout offensive.	Level of service potentially equitable with North American norms for indoors water uses.
PIPED WATER AND SEWER	Very low since no user handling or storage required.	No practical limits on water demands. Efficient water use desirable.	Very low. Potential contamination during failures.	All wastewater collected. Very dilute wastewater suitable for conventional treatment. Large quantity of wastewater for disposal.	Facilitates activities with high water demand and wastewater generation.	Protection varies with system design – pipe size, hydrants, storage, fire pumps. Large buildings should have independent fire suppression systems.	No practical limits on user water use activities.	Level of service and system equal to North American norms.

TABLE 2-2   CHARACTERISTICS OF WATER AND WASTEWATER SYSTEMS (Cameron, 1985)

Type of System	System and Building Characteristics			Feasibility/ Practically	Simplicity/ Reliability	Vulnerability	Flexibility – Level of Service	Flexibility – Physical	Economics (Relative)	Employment Requirements	Self Reliance
	Water System	Sewage System	Building Plumbing								
INDIVIDUAL	Surface water, well, rainwater, ice. Self-haul or individual building supply, treatment systems.	Pit toilet, compost toilet, septic tank, land disposal. Self-haul or individual building treatment, disposal systems.	Valves. Very low to full.	Wells and sub-surface disposal not feasible in permafrost. Viable only for isolated and rural buildings.	Varies.	Buildings are independent.	Low to high level of service possible.	Not viable for permanent, urban communities.	Varies.	None. Individual user responsibility.	Very high. Buildings and users are independent.
CENTRAL FACILITY	Self-haul from central facility.	Self-haul of excreta to central facility. Washwater disposal to ground.	No pressure water system. No laundry or bathing facilities.	Viable in small communities and camps.	Central facility can vary from simple to complex. Central location facilitates O&M. Building facilities very simple.	Central facility vulnerable to failure. Alternative user service possible.	Only low level of service available (unless combined with service to buildings).	Central facility location fixed. Service only to nearby, low-density houses.	Low capital cost. Low operating cost.	Low. Facility operator required. Skill level varies with facility complexity.	Potential for local technical, financial and administrative responsibility and control and user finance.
VEHICLE HAUL											
(a) Water Delivery: Excreta Collection	Delivery to building water container. Vehicles typically 500 to 5,000 litres.	Collection of excreta in plastic bag, bucket, or from building tank for toilet only. Washwater disposal to ground.	No pressure water system. Bucket toilet or very low flush toilet. Washwater disposal to ground.	Requires vehicle access to buildings and trails, boardwalk or roads adequate for vehicles used.	Relatively simple technology, easily repaired or replaced. Service vulnerable to weather and road conditions.	Buildings and building systems are independent and designed for periodic service. Alternative building service possible (e.g., self-haul).	Limited response to demand for higher level of service. Can upgrade buildings.	Responsive to changes in community layout, location of water source, etc. Service only to low density houses.	Low capital costs. High operating costs.	Continuous requirement for drivers and vehicle mechanics.	As above.
(b) Water Delivery: Wastewater Collection	Delivery to building water container. Vehicles typically 2,500 to 10,000 litres.	Collection of all building wastewater from pumpout tank. Vehicles typically 2,500 to 10,000 litres.	Require building pressure water system and tanks. Low to full plumbing. All wastewater to pumpout tank.	Requires vehicle access to buildings and all-weather roads adequate to support large vehicles.	As above.	As above but buildings are vulnerable to prolonged service interruption.	Service can match individual building level of service needs and quantity of service demand. Limited response to unexpected high demand.	Responsive to physical changes. Can service low to moderate density houses and buildings.	Low capital costs. High operating costs. Key economic factor is water use.	Continuous requirement for semi-skilled drivers and skilled vehicle mechanics.	As above. High levels of service (water use) may require financial assistance.
PIPED	Piped water to building. System may incorporate fire suppression (hydrants, large mains, storage and pumping capacity).	Piped wastewater collection by gravity, pressure or vacuum.	Pressure water from system. Full plumbing. Flush toilet. Require service connection to mains.	Requires freeze protection. Special foundation requirements in permafrost (especially high ice content, thermally sensitive soils).	Complex, sophisticated technology. Usually reliable but difficult to repair or replace in winter.	Buildings highly vulnerable to any service interruptions. Consequence of failure potentially highly disruptive.	Provides only high level of service. High response to changes in quantity of service.	Fixed facilities are unresponsive to physical changes. Can service high density housing and buildings.	High capital costs. Low operating costs. Key economic factor is building density.	Continuous requirement for skilled workers to construct, operate and maintain system.	Often requires technical and financial assistance.

west Territories specifies that in utility planning studies, a cost-effectiveness analysis of vehicle-haul services is to be based on a residential water consumption of 90 L/(p•d) (Government of the Northwest Territories, 1986). This quantity of water is assumed to provide a level of service and benefits (particularly health benefits) equivalent to that of a piped system. Under these assumptions, the social desirability and benefits can be left out of the evaluation and a cost-effectiveness analysis can be conducted to optimize and rank piped and vehicle-haul services. Service options such as self-haul and bagged-excreta collection are not included in the cost-effectiveness analysis since they are assumed not to be able to provide a comparable level of service and benefit.

Two types of economic analysis can be conducted; economic and financial.

Economic Costing. The economic cost is the total cost of service. It is based on including all the costs associated with an option, no matter who incurs them, and ignoring any financial assistance and taxes since these represent transfer of money within the economy rather than a cost to it. For example, economic costing would include the costs for a piped service connection or a building storage tank even though these costs may be borne directly by the building owner.

Because governments have sociopolitical objectives that may be only indirectly related to economic objectives, some market prices may not represent the real economic cost. For example, a "shadow price" for labor may be considered where there is a large pool of unemployed labor. The discount rate used in economic analysis is the government's opportunity cost of capital. Governments usually specify the discount rate to be used in the economic analysis of publicly- funded projects. The discount rate is net of inflation and is usually about eight percent.

Economic costing is usually of primary interest to agencies that provide financial assistance for utility services. It is used to identify the least-cost options with respect to the national or regional economy.

Financial Costing. The financial cost is the cost that will be recovered in the user charges for utility service. It is based on including only the costs associated with an option that are incurred by the community (or utility company) and is net of any financial assistance. Financial costs are subject to government financial assistance programs and loan (debenture) interest rates.

Financial costing results are of primary interest to the community (or utility company) and the customers that pay for services. It identifies the least-cost options with respect to the customer charges.

Vehicle-Haul vs Piped Systems. The evaluation of vehicle-haul vs piped systems is a common planning issue in cold regions. The characteristics of each system are summarized in Tables 2-1 and 2-2. Only vehicle-haul service to buildings with full plumbing and all wastewater collection should be compared to piped service in terms of level of service and cost.

The economic comparison between piped and vehicle-haul systems is a classic economic choice between a high capital and low annual cost alternative (piped) and a low capital and high annual cost alternative (vehicle haul). The typical breakdown of the capital and O&M costs for vehicle-haul and piped systems is presented in Table 2-3. Costs for roads are not included in the cost breakdown in Table 2-3. Roads might be considered a community requirement and common benefit, independent of the utility system. The cost for roads required only for the utility service might be considered as a cost to the utility option, for example a road to a water source. Also, administration costs are not included in Table 2-3 as they are approximately the same for piped and vehicle-haul systems.

Variable costs account for approximately 60% of the total economic cost of vehicle-haul systems. Conversely, the fixed costs for piped systems account for approximately 70% of the total economic cost. The labor portion of the total economic cost is approximately 45% for vehicle-haul systems (drivers, mechanics) and 15% for piped systems (construction workers, operator).

The key economic factors are opposite for these systems. For piped systems, the key cost factor is the housing (building) type, density and layout (i.e., the length of water and sewer pipe required per capita), whereas water use is not very significant. Table 2-4 shows how the cost for piped services declines significantly when housing density increases. For vehicle-haul service the key cost factor is the total quantity of service (i.e., water use per capita), whereas the density is not very significant. The effect of water use on the total annual cost of vehicle-haul service is shown in Table 2-4. The economic evaluation often reduces to a choice of toilet (water use) and house (density) (Cameron, 1985).

TABLE 2-3   *COST BREAKDOWN FOR VEHICLE-HAUL AND PIPED SYSTEMS*
(Cameron, 1982)

System and Cost	Capital (%)	O&M (%)	Total (%)
Piped System			
Building service connections	15.5	10.1	25.6
Water and Sewer mains	34.3	8.9	43.2
Fixed facilities	19.8	11.4	31.2
Total	69.6	30.4	100.0
Vehcile-Haul System			
Building containers	14.5	2.5	17.0
Garages	3.5	2.1	5.6
Vehicles	8.9	17.0	25.9
Labor (drivers)	–	32.9	32.9
Fixed facilities	19.8	11.4	18.6
Total	46.7	65.9	100.0

NOTES:
1. Adapted from Cameron, 1984.
2. Capital costs are the equivalent uniform annual cost of capital expenditures amortized at 8% over a twenty-year period.
3. O&M is the average annual operations and maintenance costs.
4. Does not include road costs or administration costs.
5. Fixed facilities include water supply, treatment, storage, distribution and waste disposal (lagoon). Based on a community of approximately 1,000 persons.

In communities in the Northwest Territories the economic cost is generally less to provide vehicle-haul services to single-family houses (Kalbermatten et al., 1980; Feilden et al., 1988). Larger communities tend to have more high-density housing, are more compact and have more nonresidential customers, all of which tends to lower the average cost for piped services, thereby making piped services more economical.

2.8   Project Management

The function of project management is to provide the physical facilities required to provide utility services. Project management activities include planning, design, construction as well as budgeting and scheduling of the activities. Project management responsibilities and processes are generally established by the client and agencies providing funding. Project management must consider the local access, climate, material, equipment and labor to develop the appropriate design, schedule and construction technique.

2.8.1   Scheduling. Scheduling is an important part of all activities in cold regions but it is particularly important and challenging in project management activities. Scheduling constraints include:

- **Transportation.** Barges normally arrive in Barrow, Alaska and the eastern Arctic of the Northwest Territories in early September, which is near the end of the summer construction season. Therefore materials and equipment are often shipped into the site one year in advance in order to get an early start at construction the next summer.

- **Weather.** It is usually critical to get buildings and facilities closed-in before severe winter condition set in so that work can proceed in a protected enclosure.

TABLE 2-4   EFFECT OF HOUSING DENSITY AND WATER DEMAND ON
HOUSEHOLD SERVICE COSTS (Cameron, 1982)

Service / Housing Type and Water Use		Total Service Cost Per Household	
Piped Service			
Detached housing			
Existing layout		$4,343	100%
Redeveloped layout		$3,970	91%
Row housing			
Duplex		$3,200	74%
Quadplex		$2,610	60%
Apartment building			
Varies, maximum =		$1,300	30%
Vehicle-Haul Service			
Household water use (litres per person per day)			
130	144%	$4,971	130%
110	122%	$4,360	114%
90	100%	$3,830	100%
67.5	75%	$3,140	82%
60	67%	$2,890	75%
50	56%	$2,500	67%
45	50%	$2,400	63%
30	33%	$1,920	50%

- **Seasonal.** Soil testing, excavation and foundation work schedules are often determined by seasonal soil characteristics.

- **Limited equipment and labor available.** In small communities there is often limited equipment and labor available for projects, particularly during critical times (e.g., summer, when barges arrive). Project schedules should be coordinated with other community construction activities so as not to exceed the capacity, to spread out the benefits and to maximize local involvement.

These constraints combine to produce a narrow window in which to complete activities. If schedules are not met, a project can be delayed to the next season and/or result in very high costs to get back

on schedule (e.g., overtime or flying-in materials, equipment, workers). Scheduling problems are often in proportion to the amount of materials, equipment and workers brought to the site. Therefore, scheduling problems can often be alleviated by using local resources.

2.8.2 Transportation Means and Cost. Moving people, materials and equipment to a community or site in cold regions can be a major challenge in project management. Transportation restrictions must be considered in the design, construction scheduling and budgeting of projects.

Transportation costs are higher for projects in remote communities and sites. However, the actual cost of transportation is often not as important as the other restrictions imposed by the access and transportation system including:

- scheduling construction activities to meet transportation schedules;

- risk of transportation failures (not arriving) and the loss or damage of supplies in transit;

- need to store materials and equipment at the site until they can be used or removed;

- loss of control over materials (damage, loss) in transit and if stored on the site over winter.

Factors to consider when selecting the transportation means are:

- means available (and possibility of initiating);

- restriction (e.g., time of year, types of materials);

- size, volume and weight of material or equipment to be shipped;

- susceptibility to damage during transit;

- reliability of delivery; and

- costs.

Transport by Water Routes. Materials and equipment are delivered to many construction sites in cold regions by barge or ship. It is very important to investigate the following to avoid delays in project construction:

- Check the scheduled departure and arrival dates.

- Investigate the chances of nondelivery due to shore ice not moving out, rivers being too low, or other natural factors.

- Check the rates charged and the criteria on which they are based (volume or weight). There are often demurrage charges on shipping company containers. Also a charge for unscheduled stops may be levied.

- Check reservation requirements. On many scheduled barges, space must be reserved in advance;

- Check the requirements for delivery of materials to the embarkation point. Usually materials are accepted only during a specific and limited period.

- Check the capacity of barges and the capacity of off-loading docks and equipment, before ordering materials and equipment.

- Determine the facilities available for off-loading at the site. If there are no docks, are the beaches suitable for landing the barge? What equipment (such as cranes, trucks, tractors, etc.) is available? Can the barge or ship be brought into shore or must landing barges be used to shuttle materials between the ship and shore? At ocean coastal locations, landings or docking must usually be coordinated with the tide.

- Investigate chartering. Chartering can be economical depending on the amount of materials to be shipped. Chartering might be necessary if the site is not on an established route.

Transport by Air Routes. Air transportation is the most common means to transport people but it is often competitive for material transport. It is the only transportation means to some sites in the Arctic and Antarctic. The following items should be investigated:

- Determine the flight schedules, the available aircraft and their capabilities. Determine the limitations on types (e.g., paint, chemicals) and size of material that can be transported (maximum material size is usually defined by the size of the loading door), cargo space available, weight limitations and performance limitations for the different sizes and types of planes.

- Check the rates charged (both flying time and standby time) and the criteria on which they are based (volume or weight).

- Compare the available facilities for loading and unloading available at the site with the requirements of the aircraft.

- Determine the landing strip capabilities. Is the strip useful year-round or just in the winter? Even year-round strips usually have seasonal limitations, such as soft spots in the spring, and depending on the aircraft, crosswind limitations year-round. Temporary strips can be constructed on lakes, river or sea ice or on compacted snow during the winter (Clark et al., 1973; McFadden and Bennett, 1991; Grey and Male, 1981). Lake ice is usually smoother than river ice and is thus suited for a wider range of aircraft and heavier loads.

- Investigate chartering aircraft for heavy, bulky or special materials. The economics of chartering depends on the amount of material to be shipped and the type of aircraft used for scheduled flights.

Transport by Land Routes. Similar considerations are necessary in areas served by roads or railroads, or both. Are the roads useful year-round or are they seasonal, and do they have load restrictions during the spring? Rates by the weight or volume and the size and weight restrictions must be considered. Check the demurrage charges on railcars or containers.

Transportation Costs. When comparing the costs and benefits of the different transportation means available, consider:

- the gain in time and flexibility in scheduling construction that can be achieved by using air transportation;

- the number of times the materials and equipment must be handled. The amount of damage during shipping is directly related to the number of loadings and off-loadings needed;

- the true shipping costs (i.e., including demurrage cost, lighterage costs and long-shoring); and

- the difference in costs for chartering and using scheduled carriers.

Protection Against Damage During Shipping. Investigate the shipper's or carrier's liability for loss and damage of the shipment. Loss or damage of one important component could delay the completion of a project by a year or more. Materials and equipment must be packed to prevent damage during shipment. Items must usually be handled several times. Crating and protection needed during transit must be specified; for example, protection of materials and equipment against salt water while on barges and protection against freezing for items such as foam insulation, paints, polyelectrolytes and adhesives.

2.8.3 Construction Season. The summer construction season in the cold regions is relatively short. Once it has ended, inclement weather sets in, the daylight hours shorten markedly and some soils freeze solid making excavation nearly impossible without special equipment. Some construction activities are uneconomical, inadvisable or impractical (i.e., virtually impossible) to carry out during the extreme winter conditions experienced in cold regions. However, some activities are much easier during the winter, for example installing a water intake from the ice surface or excavating thaw-unstable soils.

Knowledge of local conditions, construction practices and insightful project management is necessary to optimize construction activities where there are seasonal constraints. Some means to extend the construction season are discussed below.

Length of Construction Season. The construction season depends on the type of project and activity and varies with factors such as soil conditions, length of daylight and climate. The length of a "normal" construction season varies from two to three months along the arctic coast to six or eight months in the more southern areas within cold regions.

Compensating For Lack of Daylight. A chart showing the hours of daylight at various latitudes is included in Section 17. Daylight values may be shorter in a location with mountains or other local terrain variations. The lack of daylight can be partially corrected by using artificial light.

Excavation of Soils. Soils with low moisture content (i.e., coarse-grained soils) are relatively easy to excavate when they are frozen and shoring may not be necessary to prevent trenches from caving in. Ice-rich silt is often easier to excavate when frozen since when it thaws the soupy material is difficult to handle. Some muskeg and highly organic soils are easier to excavate when they are frozen. When thawed, such soils often will not support equipment or even people. On the other hand, the water in saturated frozen soils cements the soil particles together and to excavate this soil it must be fractured into small pieces. Saturated frozen gravel is much like concrete. Special excavation equipment (proper design and usually larger) is required. Alternatively, the frozen soil can be thawed or drilled and blasted.

Effects of Inclement Weather on Productivity. The weather (temperature, wind, snow, whiteouts etc.) greatly affects the amount of work accomplished. Temperature affects workers and equipment even when other environmental factors are considered ideal (Figure 2-8). Darkness, wind and whiteouts reduce productivity still further. The effects of cold temperatures on oils, fuel, antifreeze solutions, steel, wood and plastics are shown in Section 17. It

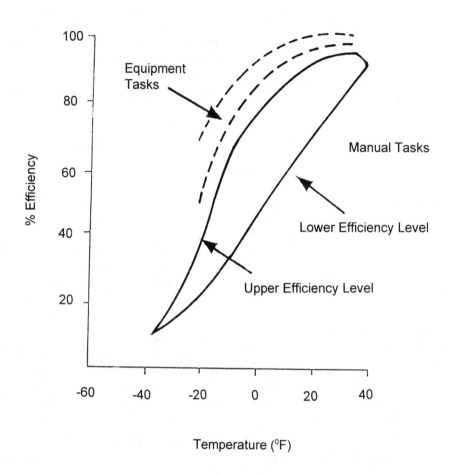

FIGURE 2-8   TEMPERATURE EFFECTS ON CONSTRUCTION
PRODUCTIVITY (after U.S. Army Cold Regions Research and Engineering Laboratory)

is important that these effects be understood before cold temperature construction is attempted.

Construction Activity During Winter.   Winter
construction, particularly outdoor work, generally is
more difficult, less efficient, harder on equipment,
inherently more dangerous, more difficult to obtain
a high quality of work, takes longer and costs more.
It may be economical to avoid winter construction
by accelerating construction activities during the
summer; for example, by increasing the work force
and equipment or by working longer hours. When
possible, the most temperature-sensitive activities
(e.g., roofing, welding, concrete work, painting)
should be scheduled for the best weather.

Nevertheless, construction is often done in winter
to meet completion schedules or to minimize the
construction time and total cost (i.e., rather than shut
down during the winter and start up again the next

year). Measures can be taken to extend the construction season. These include enclosing and heating the construction site, building or work area (e.g.,
an air-supported 100 m by 150 m structure was used
to enclose the construction site for a water treatment plant for Anchorage, Alaska), winterizing equipment, and artificial lighting (McFadden and Bennett,
1991). Usually the exterior shell of buildings or facilities is scheduled for completion before severe
winter conditions set in so that the interior work can
be completed in a heated and lighted space.

Winter construction inside an enclosure can create
severe moisture problems. Portable heaters used
to keep the interior warm produce considerable
amounts of water in the combustion process and,
unless they are vented to the outside, this water
accumulates in the building interior where it can
cause problems for years to come.

Concrete work in the winter should be avoided since quality control is particularly difficult to achieve. Where concrete work must be done in winter, aggregate should be excavated and stockpiled before the ground freezes. Aggregates must be kept from freezing or thawed before use. Concrete work must be protected from freezing.

2.8.4 Equipment. The use of existing local equipment will benefit the community and eliminate the cost and time to bring in equipment. Where equipment must be brought in, the type of equipment selected and the preparation of equipment for work in cold weather are very important with respect to time and cost. Equipment that is to be used during the winter must be properly winterized with low temperature lubricants and coolants, engine heaters, heated personnel compartments, etc. (McFadden and Bennett, 1991).

Reliability of Equipment. Equipment must be in good condition before it is sent to or used in a remote location. Equipment and tools for repair work are usually limited in remote areas and importing mechanics and parts is expensive as is the downtime. A large inventory of critical spare parts is recommended. Standardizing the equipment to reduce the required inventory of parts can prove economical. Standby units should be provided for critical equipment so that a job is not stopped completely. Preventive maintenance is extremely important. Schedules should be established and followed. Good maintenance records are important as they will:

- provide valuable data when the time comes to decide which brand of equipment to select for future projects;

- help pinpoint weaknesses and reduce the spare parts inventory; and

- provide checks to ensure that preventive maintenance is done on schedule.

The initial cost of a piece of equipment is usually not as important as its reliability. Equipment should be selected for which there are reliable dealers with adequate parts inventory, service facilities and staff nearby.

Flexible Use of Equipment. Equipment requirements should be planned carefully, especially large pieces that cannot be flown in later. For mobility, select equipment that can be flown from one location to another. Equipment should be adaptable to various uses to reduce the number of pieces needed.

2.8.5 Use of Local Materials. Local materials should be used wherever practical. For example, the use of local logs for building facings can reduce the import of materials and also provide an aesthetic design and match the local building style. Unfortunately there is often little natural construction material available in cold regions, particularly in the Arctic.

Local sources of fill material, sand and aggregate are critical for concrete, roads, backfilling and other earthworks. These materials will be very expensive to import. If they are not available locally, the utility system selection and design should minimize the need for them.

2.8.6 Labor. Labor costs typically make up about 40 percent of the total cost of a facility construction project. Client policies and preferences regarding using local labor must be considered early in the project planning so that appropriate design, materials, construction methods and project management strategies can be developed to maximize the use of local labor and to develop appropriate training programs. It is often better to use local labor whenever practical.

Advantages of Using Local Labor:

- Local people are more familiar with, and adapted to, local conditions, including weather.

- If they live near the project site, camp facilities will not be required.

- If they live in the community receiving the facilities they have more at stake because they have to live with the final facilities. Knowledge gained during the installation will also facilitate repairs.

- Local hiring provides an economic boost to the area in addition to the benefits provided by the product (i.e., the facilities).

- Local people can be trained as equipment operators, carpenters, plumbers etc., which will help them obtain jobs in the future.

- Construction by the local residents may increase local interest and pride in the facility.

Potential Limitations to Using Local Labor:

- In some locations not enough skilled and experienced workers are available.

- Local customs and politics must be recognized and honored.

- Labor force may be sporadically unavailable due to traditional subsistence hunting and fishing activities.

- Supervisors and specialists will probably have to be imported. They must be selected carefully and they must be able to work and communicate with local people. They must also be willing to tolerate the remote conditions, a hardship usually offset by high wages.

- Union contracts may preclude using nonunion local labor.

2.8.7 Inspection and Safety. Proper inspection during construction is extremely important, particularly when the site is remote. Once imported construction equipment and specialized workers have been removed, it is very expensive to return to repair mistakes. A small mistake, or a case of poor-quality work can cause an entire facility to fail. Quality control can be difficult to ensure when there are severe weather conditions or pressure to complete work to meet shipping or seasonal deadlines.

Working in the extreme cold is inherently dangerous. Reaction times are slower when people are cold and bulky clothing interferes with movement, vision and hearing. Darkness, wind, whiteout and slippery footings all contribute to a hazardous working environment. Equipment is more difficult to operate and visibility is often impaired. Safety precautions must be developed and followed. Minimizing the chances of accidents is very important in remote locations because of the limited local medical facilities and the delays to evacuate the injured for emergency treatment.

2.8.8 Utility Design Considerations. Utility design must consider the community and site conditions (Section 2.5), utility considerations (Section 2.6), project management considerations (Section 2.8) and operations and maintenance considerations (Section 2.9). The design considerations for specific components are provided in the other sections of this monograph. The following are general design considerations that apply to all utility designs in cold regions.

Design Concept. The needs, objectives and realities of the local conditions and users should be the basis for design. System concepts and designs from the south, or even other cold regions, should not be mechanistically adopted.

Design concepts that should be strived for in cold regions include:

- maximize self-reliance of the users and community (and minimize the reliance on outside financial or technical support);

- maximize the use of local resources (and minimize the need to import resources, physical and human);

- minimize the disruption in service and damage due to freezing;

- minimize energy and water requirements; and

- simplicity and reliability.

In cold regions it is prudent to anticipate that systems and components will fail and to design for "safe failure". It is usually not economical or realistic to design for "fail safe" system.

These design concepts are difficult to achieve and are sometimes incompatible while meeting the users' objectives for utility service and the regulatory requirements.

Reliability of Service. Reliability is a critical planning and design objective for utility systems, facilities and works in cold regions. Full utility services should be strived for and essential service must be available at all times. Essential utility service may be defined as follows:

- water supply to provide:

    - uninterrupted water supply to pipe-serviced buildings; and

    - normal water delivery to vehicle-haul serviced buildings;

- availability of water requirements for fire suppression;

- public health is not endangered, for example through wastewater overflows at wastewater lift stations; and

- the integrity of water and sanitation works and facilities is not endangered.

To ensure reliable utility services, it is essential that the equipment, facilities and works used to provide utility services are themselves reliable. Standby equipment or designs that facilitate quick repair are essential. Components and materials must be rugged enough for the harsh conditions to which they are exposed without the need for frequent or specialized repairs. Equipment or systems that need servicing by specialized tradespeople (not available locally), equipment or parts that are not available locally or are difficult to obtain, are undesirable, though at times they may be necessary.

Backup measures should be incorporated where necessary to provide reliable essential service.

Backup Measures. The need for backup measures should be evaluated based on the following:

- the consequence (and cost) of interruption in service, considering the duration of the interruption in service with respect to essential service and full service;

- the risk of interruption in service for various durations (i.e., the likelihood of various events that could lead to interruption in service); and

- the cost of measures to prevent interruption in service, including backup measures.

Backup measures include standby power supply, duplicate component (e.g., duplicate pumps), alternate component (e.g., alternate water source) and alternate system (to provide service, e.g., vehicle haul). Backup measures for a facility should be coordinated with other backup measures within the water and wastewater system and with backup facilities within the community (usually power) in order to maximize the effectiveness of backup measures and to minimize duplication and costs.

If vehicle-haul serviced buildings have adequate water and wastewater storage capacity in each building, short-term (i.e., one to two days) interruptions in service might not compromise essential water delivery and wastewater collection service.

Freeze Protection Measures. All pipelines and other works and buildings should be designed and operated to:

- prevent freezing under design conditions;

- provide adequate time before freezing in the event of cessation of flow, heat or other condition to allow measures to be taken to minimize damage;

- minimize the potential damage when freezing occurs, including providing for the timely and convenient draining of pipes including pipes within buildings, service connections and mains (i.e., design for safe failure); and

- provide measures to minimize the resources and time required to thaw and reinstate a frozen pipeline or facility.

Monitoring and Control Measures. Instrumentation and monitoring should be incorporated into systems and facilities to allow for safe and efficient operation. All critical facilities should have instrumentation, monitoring, alarm and notification systems to indicate critical conditions that would result in:

- cessation of essential service (e.g., monitor flow, temperature, pressure); or

- damage to facilities (e.g., monitor for fire, low temperature).

Critical conditions (e.g., flow, pressure, pump operation, etc.) should be monitored to prevent failure and to notify staff in the event of failure. Communication from remote instrumentation may be via telephone line or radio transmission, whichever is cost effective.

Siting. Siting of utility facilities (including buildings, pipelines, works, water source, waste-disposal sites) should consider the following:

- national and regional regulations, particularly where the facilities are located outside the community boundaries and jurisdiction;

- community development plan and zoning regulations;

- suitability from a community planning and development perspective;

- land development costs; and

- possibility of complexing with other utility or community facilities.

Utility facilities should be located on a site of sufficient area to:

- contain the facility and allow for future replacements and expansion; and

- provide adequate access and parking space for vehicles and equipment required for operation and maintenance.

Utility facilities must be designed and oriented to:

- minimize snowdrifting effects on the facility, particularly around entrances;

- minimize snowdrifting effects on nearby buildings and roads; and

- provide efficient access for service vehicles (e.g., water, wastewater and solid waste).

Utility facilities should be located within a public road allowance, within a utility easement or on land controlled by the community. Land ownership and easements should be verified in the planning stage to identify conflicts and requirements. Land ownership and jurisdiction can be in conflict due to aboriginal land claims, particularly lands outside a municipal boundary.

Aesthetics. Utility facilities should be visually consistent with the functional requirements of the facility and to the local topography and adjacent buildings. Facilities should be designed and located to minimize the impacts of offensive emissions including odors, smoke and noise.

Complexing. Wherever practical, utility facilities should be located in a common structure to reduce construction and heating costs and to centralize operations and controls. Sometimes it is practical and economical to combine utility facilities with other community facilities, particularly in small communities. For example, by constructing a water storage tank in a school building to provide both building and community water storage requirements or designing a community swimming pool to provide water for fire reserve.

Where facilities are complexed, the increased vulnerability to fire or other disasters must be considered.

Security and fire protection measures must be included in the design of multipurpose buildings.

Standardization. Communities in cold regions with similar physical conditions and needs will require facilities with the same functional capabilities. In some cases standard design concepts or components can be established; for example, a vehicle-fill station, a building-service connection, or a type of pipe material. Using standard design concepts or components can simplify and reduce the time and cost required to plan, design and construct facilities. Standard design concepts and components are also used to ensure a required performance (e.g., experience may indicate a design concept, component or material with proven success) or to facilitate O&M (i.e., to allow common training and parts). Most government agencies that provide assistance for utilities in cold regions have developed standard design concepts and components. It is important to ensure that standard designs and components are appropriate to the local physical conditions (e.g., access, climate and site conditions) and preferences (e.g., aesthetics).

Energy Efficiency. The efficient use of energy is a planning and design objective for utility design in cold regions. Reducing energy requirements is important where energy costs are high, particularly where electricity is generated from fuel oil that must be transported to remote locations. Energy efficiency evaluations for utility systems include pumping, heating systems, heat loss from pipes, tanks, buildings and facilities, lighting (see Sections 4 and 17). Energy efficiency measures should be evaluated using life-cycle costing procedures assuming the full unsubsidized cost of energy.

Water Efficiency. The efficient use of water is a planning and design objective for utility design in cold regions. Plumbing and fixtures in utility buildings should be evaluated using life-cycle costing procedures assuming the full unsubsidized cost of water and wastewater services (see Section 14). Water and wastewater facilities and works, including building service connections, should be designed to operate without the need to waste (or bleed) water to prevent freezing.

2.8.9 Utility Building Design Considerations. Utility buildings include small enclosures, such as a pump station, truck fill station or wastewater lift station, to relatively large buildings for a water treatment plant or water distribution pumphouse. The primary function of utility buildings is to contain and protect equipment. They are often only occupied periodically.

Building Design. Building design should strive to:

- minimize life-cycle costs;

- minimize the amount and need for specialized equipment and trades required to construct and maintain the facility;

- minimize the gross floor area and volume necessary to accommodate the needs (e.g., by complexing or efficient layout). Cubical buildings are more efficient to heat and insulate;

- provide easy access to equipment for maintenance, upgrading, replacement, etc. (e.g., overhead doors, knockout panels); and

- enable expansion as simply as possible.

Buildings must be designed for the local climate and site conditions as well as building codes and regulations. In cold regions the design temperatures are extreme (e.g., -50°C to +30°C) and the heating index is high. Some areas, especially above the tree line, have high winds and snowdrifting. Careful detailing is necessary to ensure airtight, energy efficient buildings and reliable building systems. The design of buildings in cold regions requires the services of a person especially qualified to work in that field. Information on building design is provided in other sections of this manual (e.g., Sections 12, 13, 17 and Appendices C and D) and in other references (McFadden and Bennett, 1991; Department of Public Works and Services, 1993; Hutcheon and Handegord, 1983).

Snowdrifting. Snowdrifting can impede access to and from buildings. Snowdrifting should be managed through careful siting and design so that problems can be avoided or reduced without the need to introduce wind-control devices, such as scoops or accelerators, or snow fences. Although these devices can be effective, they are expensive alternatives to proper siting to take advantage of natural wind scouring. Entrances should be located so that predominant winds scour the area. Snowdrifting information is provided in Appendix C.

Building Structure and Foundation. Consideration of permafrost, transportation and available local equipment limit the choices that are appropriate for the structural design and materials for a site. Prefabricated metal buildings and steel framing are common for utility buildings; however, wood frame construction may be appropriate to provide opportunities for local labor to use or develop construction skills. Also, wood is less susceptible to damage in transit than prefabricated assemblies. The use of concrete should be limited and generally discouraged where granular materials, mixing equipment and testing facilities for quality assurance are not available. The design of foundations for permafrost soils requires the services of a person especially qualified to work in that field. Building foundation information is available in Section 3 and other references (McFadden and Bennett, 1991; Johnston, 1981; Sanger, 1969).

Building Envelope. The building envelope (walls, roof, floor) in the cold regions fulfils the same basic functions as in the south but greater temperature differentials, higher humidity differentials and greater exposure to high winds necessitate modifications to southern design practices.

In cold regions additional thermal upgrading of the building envelope is necessary to reduce heat loss and energy costs. A hydrophobic or low-water-absorption insulation should be used in damp or wet locations. Thermal bridging must be recognized and thermal breaks should be provided where there is a continuous connection to the cold exterior (e.g., structural framing, doors, windows, foundations).

Even greater attention must be paid to the integrity and continuity of the air/vapor barrier in cold regions due to the high difference in vapor pressure and air pressure. Penetrations of the air/vapor barrier should be minimized. Joints, penetrations (e.g., plumbing and electrical) and any holes in the air/vapor barrier must be thoroughly sealed.

Roofs probably cause more problems than all other building components combined (Tobiassion, 1971). Flat roofs should be avoided as they cannot shed or drain water effectively and are susceptible to ice damming. Rigid closed-cell polystyrene foam insulation is preferable since its effectiveness is only minimally affected by moisture. By locating the air/vapor barriers and insulation outside the roof structural framing, ventilation problems can be avoided.

Windows have high heat loss and maintenance, therefore, they should be avoided in utility buildings unless they perform a critical function. Windows should be designed to provide the most benefit with the smallest area (Flanders et al., 1981). Window orientation toward the south, multiple glazing, shutters and insulated frames with thermal break reduce heat loss.

Heating Systems. Qualified or experienced tradespeople are not always available in small communities. Therefore, heating systems must be reliable, simple and tamper proof. Hydronic heating systems are commonly used to heat large areas with multiple heating zones.

Ventilation and Humidity. Buildings in cold regions should be relatively air-tight to reduce heating costs. Ventilation requirements depend on the function of the utility building and the activities (i.e., occupancy, chemicals). Open-water surfaces should be avoided. In humid areas, such as wastewater lift stations, moisture control or dehumidifiers are necessary.

2.8.10 Site Rehabilitation and Landscaping.
Existing drainage and vegetation should be protected and maintained wherever practical. After construction, site grading, landscaping and revegetation may be required to control erosion, control dust, insulate permafrost, protect public safety and provide a pleasing appearance. Landscaping using lawns, flower beds, trees and shrubs is not practical in much of the cold regions, especially above the tree line. Vegetation added to the site must be suitable to the site, hardy and require little or no maintenance. Transplanting of local vegetation is encouraged. Grass seeding and fertilizer procedures

suitable for tundra areas have been developed (Communications, 1974). Synthetic mats have been successfully used to control erosion and allow vegetation to start.

2.8.11 Design Life for Facilities and Equipment. The design life for facilities and equipment in cold regions is usually shorter than that for the same units operated in more temperate climates. This is especially true of equipment which operates outside in the winter. Actual life depends in large part on the appropriate design (i.e., design is appropriate for the actual conditions and does not fail under design conditions) and the quality of operation and maintenance. The design life used in economic analysis or capital planning should be based on local experience. Table 2-5 provides some commonly used values for design life in cold regions, assuming appropriate design and O&M.

2.8.12 Construction Considerations.

Responsibility. The responsibility for the construction of community utility facilities and works is usually the community government, an agency that provides funds, or a combination. The actual construction is carried out by the owner, an agency providing assistance, a contractor or a combination of these.

Local Involvement. Involving local workers, contractors and suppliers in construction projects can be facilitated through construction financial and management policies and practices. These include:

- tender bid adjustment to evaluate bids by qualifying regional and/or local contractors or suppliers;

TABLE 2-5   *DESIGN LIFE FOR EQUIPMENT AND FACILITIES*

Component	Design Life (years)
Wells	30
Storage tanks	20 to 40
Water distribution pipes	20 to 40
Wastewater collection pipes	20 to 40
Service connections	10 to 30
Building tanks	10 to 20
Lift station (not pumps)	20 to 30
Pumps and controls	5 to 15
Boilers and furnaces	5 to 15
Meters	5 to 15
Valves	10 to 20
Septic tanks	5 to 10
Drainfields	5
Haul vehicles (wheeled)	7 to 10
Haul vehicles (tracked)	3
Buildings	10 to 30
Utility truck	5
Heavy equipment (backhoe, loader, grader, bulldozer)	12 (or 16,000 hours)

NOTE: Design life is based on appropriate design and proper operation and main.

- invitational tenders restricted to qualifying regional and/or local contractors or suppliers;

- split projects into phases or segments so small local contractors will have an opportunity to bid;

- split materials into grouping so small local suppliers will have an opportunity to bid;

- schedule utility construction and coordinate with other community construction activities to allow regional and/or local involvement; and

- provide long-term contracts to allow regional and/or local manufacturers, contractors or suppliers to recover start-up costs.

Type of Construction Management. The two most frequent types of construction management are force account and contract. Some projects combine force account and contract management. For instance, the owner may do most of the work with their own staff but contract out specialized tasks; for example welding of steel storage tanks, erection of prefabricated metal buildings, installation of water treatment plants and installation of piling.

Force-Account Construction. Force account construction means the community or an agency (e.g., senior government agency) is directly responsible for construction management. This may entail the community or agency using its own forces, hiring crews, purchasing materials and obtaining equipment to accomplish the construction. The community or agency assumes all financial and legal responsibility. It should be noted that some communities and agencies do not have the capacity or mandate to allow force-account construction.

Some advantages of force-account construction are:

- tends to use local labor and contractors to a greater extent;

- usually lower in cost for small and routine projects;

- training of the operators and users can more easily take place during construction;

- equipment used during construction can more easily be left at the site for operation and maintenance of the completed facility;

- is easier and less expensive to phase a project or make major changes during construction, should unforeseen difficulties arise;

- shorter lead times are usually possible for materials and equipment ordering (i.e., do not have to develop and tender contracts); and

- there is more flexibility to adapt to the local environment and social conditions.

Some disadvantages of force-account construction are:

- changes in design, intent, quality control and scope made as the project progresses may not be consistent with the overall functions of the facility;

- construction may require more time and therefore cost more (and exceed budgets);

- staff must be trained and made familiar with project management, material purchasing and transport, cost control, quality control and personnel management, etc.; and

- the owner assumes liability for accidents, damage, cost overruns, defects etc., during and after construction.

Contract Construction. Contract construction means the funding agency contracts with a private firm to do the construction. With design build, both the design and the construction are contracted together.

Some advantages of contracting construction are:

- costs are more clearly defined at the start of construction;

- there is usually more control over the project costs, design intent, project risk;

- completion times are more likely to be met since they are specified in the contract; and

- the owner or funding agency does not take as high a risk; and

- the owner or funding agency can concentrate on quality control.

Some disadvantages of contract construction are:

- tender documents and tendering procedures must follow strict legal procedures, often causing delays;

- mobilization and demobilization costs may be high;

- contractors may not be available at the desired time;

- contractors may not be willing to take the risks to complete the project within available budgets;

- tends to import labor to a greater extent; and

- less flexibility and greater cost to make design changes during construction.

Construction Techniques. There are three main construction techniques: on-site fabrication, prefabrication and modular. With on-site fabrication construction the materials are shipped to the site and all cutting and fabricating is done on-site. With prefab construction, parts of a structure and equipment are prefabricated at the point of manufacture and are assembled at the site. Modular construction means large components are constructed at the point of manufacture and shipped to the site already constructed, e.g., a truck-fill station, access vaults.

Each technique can be useful under different circumstances. Modular construction has a definite advantage where the construction season is short or where labor is in short supply or is expensive at the site. Shipping limitations may limit modular construction. Modular construction may allow high quality control with most of the defects worked out at the point of manufacture, but field changes to fit varying site conditions are more difficult and expensive to make.

Construction Cost Factors. Cost adjustment indices have been developed for the cold regions of North America. Table 2-6 lists a Northwest Territories cost index for capital construction and for operations and maintenance with a value of 1.0 assigned to Yellowknife, NWT. This index applies to facilities with similar design. Table 2-7 lists construction cost factors for Alaska for the construction of repetitive type facilities, such as buildings. A value of 1.0 is assigned to Washington, D.C. This index applies to facilities to perform the same function. Cost indices should only be used for preliminary estimates.

2.9 Operation and Maintenance

2.9.1 Operation and Maintenance Responsibility.
The responsibility for operations and maintenance may be local (e.g., the community government or private utility) or central (e.g., a regional utility organization, a senior government agency).

Various O&M responsibility approaches are used in cold regions, generally based on the regional governments' assistance policies. In Alaska, local utilities are owned and operated by the communities.

In communities in the Northwest Territories without a property tax base, the Government of the Northwest Territories often operates facilities (e.g., water intakes, pumphouses,) and piped systems on behalf of the community under an operating agreement. However, the community government operates and maintains vehicle-haul equipment (i.e., trucks) and manages the service.

Where the community operates the local utility system, successful operation depends on the training and dedication of the operator and the support of the community residents. The operator's dedication and acceptance of responsibility are usually direct indicators of successful operation. High quality operation and maintenance are facilitated when the responsibility rests with an agency which has a large pool of staff and equipment available within the community or region to provide assistance and specialized technical support.

For small, remote communities in cold regions, technical and management help must always be available to the operator and community, no matter where the responsibility for operation lies. This help is usually provided by the funding agencies. Government agencies often provide central training and management support, especially for utility facilities in small communities.

Responsibilities, duties, minimum qualifications, performance requirements etc. for utility staff positions should be clearly defined (e.g., through a position description). This will facilitate hiring, managing and evaluating staff.

2.9.2 Operation and Maintenance Considerations.
Because of the severe consequence and high cost of failure, it is critical that utility systems are properly operated and maintained. Operation and maintenance (O&M) considerations in cold regions include the following.

Access. Access to the community dictates the ease with which parts, materials, equipment and technical help can be secured. This dictates the degree of self-sufficiency required; for example, the equipment, inventory of spare parts and skills required locally. This in turn must be considered in the planning and design of facilities.

Local Labor and Resources. The availability of local skilled labor and resources (equipment) will dictate the level and types of operation and maintenance that can be expected. This in turn dictates the type and design of equipment and facilities and the type of utility system that is appropriate.

TABLE 2-6   *NORTHERN COST INDEX FOR THE NORTHWEST TERRITORIES*

Community	O&M	Capital
FORT SMITH REGION		
Dettah	0.99	1.00
Enterprise	0.96	0.98
Fort Liard	1.35	1.01
Fort Providence	1.20	0.98
Fort Resolution	1.11	1.02
Fort Simpson	1.23	1.02
Fort Smith	1.02	0.99
Hay River	0.96	0.95
Hay River Indian Reserve		
Jean Marie River	1.54	1.05
Kakisa Lake	1.21	1.02
Whati	1.59	1.10
Lutsel K'e	1.45	1.09
Nahanni Butte	1.50	1.03
Rae Edzo	1.18	1.08
Rae Lakes	1.51	1.03
Snare Lake	1.39	1.06
Trout Lake	1.31	1.07
Wrigley	1.37	1.04
Yellowknife	1.00	1.00
INUVIK REGION		
Aklavik	1.39	1.10
Tsiigehtchie	1.57	1.09
Colville Lake	2.16	1.13
Déline	1.42	1.13
Fort Good Hope	1.40	1.09
Fort McPherson	1.40	1.10
Tulita	1.38	1.09
Inuvik	1.25	1.10
Norman Wells	1.22	1.07
Paulatuk	1.67	1.19
Sachs Harbour	1.54	1.21
Tuktoyaktuk	1.37	1.13

Community	O&M	Capital
BAFFIN REGION		
Arctic Bay	1.50	1.35
Broughton Island	1.61	1.33
Cape Dorset	1.51	1.30
Clyde River	1.48	1.32
Grise Fiord	1.54	1.33
Hall Beach	1.53	1.30
Igloolik	1.39	1.33
Iqaluit	1.33	1.29
Kimmirut	1.58	1.30
Nanisivik	1.47	1.35
Pangnirtung	1.46	1.31
Pond Inlet	1.48	1.38
Resolute Bay	1.39	1.40
Sanikiluaq	1.45	1.31
KEEWATIN REGION		
Arviat	1.45	1.18
Baker Lake	1.49	1.20
Chesterfield Inlet	1.64	1.21
Coral Harbour	1.51	1.20
Rankin Inlet	1.44	1.22
Repulse Bay	1.49	1.34
Whale Cove	1.68	1.24
KITIKMEOT REGION		
Cambridge Bay	1.52	1.24
Kugluktuk	1.49	1.24
Gjoa Haven	1.64	1.30
Holman Island	1.68	1.20
Pelly Bay	2.07	1.75
Taloyoak	1.64	1.28

NOTE:
1. Index for communities is established relative to Yellowknife set at 1.00.
2. Capital index is to be used for projects with similar design.

TABLE 2-7   CONSTRUCTION COST FACTORS FOR ALASKA

Location	Factor	Location	Factor
Aleutian Islands	3.0	Inland Area, North Aleutians	4.0
Adak	3.0	Juneau	1.8
Attu	3.0	Kanakanak	2.1
Cold Bay	3.0	Kenai Peninsula	2.1
Dutch Harbor	2.5	Kodiak	2.5
Shemya	3.1	Kotzebue	3.0
Anchorage	1.7	Naknek	2.1
Barter Is., North Coastal Area	5.2	Nome	2.3
Bethel	3.8	Northway, Highway Area	2.3
Clear	1.7	Point Barrow	4.6
Coastal Area, North of Aleutians	3.5	Tanana	3.2
Eielson AFB	1.9	Whittier	1.9
Elmendorf AFB	1.7	Galena	3.0
Fairbanks	1.9	Seattle	1.0
Fort Greely (Big Delta)	2.2	Washington, DC	1.0
Fort Yukon	2.6		

NOTE:

The location factors are applicable to the construction of repetitive type facilities, such as dormitories, dining halls, administrative buildings, fire stations, warehouses, etc., which make maximum use of local skills, materials, methods and equipment. They are not applicable to more complex facilities.

The location factors are multiples of the basic value of 1.0 assigned to Washington, D.C.

Reliability.   Reliability of equipment and facilities is important because of the expense to replace or repair items in remote areas and because of the severe consequences of failures. Catastrophic failure of an entire facility has been caused by the failure of an inexpensive control. Reliability must be a major consideration in selecting components and equipment. This includes the availability of parts and service.

Where electric power is unreliable, particularly in small communities, standby electric generating capacity must be provided for all critical components of a facility. The integrity of the facilities and utility system, in case of power failure, must be ensured.

Automation.   An automated system can be effective and efficient, particularly where the operator is not competent or available at all times. However, automated systems can be highly vulnerable when failure occurs, in part because it has not been necessary for the operator to get to know the system.

Simplicity.   Simplicity and reliability are somewhat synonymous. Systems and equipment that are relatively simple to understand and to service are more likely to be operated correctly and more likely to be repaired with minimum disruption and damage. The types of equipment and variety of materials should be minimized to facilitate training and to minimize the inventory of spare parts and repair material.

Systems or equipment that require servicing by specialized tradespeople or require parts that are difficult to obtain, are inherently less reliable and are not desirable, though at times may be necessary.

Contingency Plan.   A contingency plan must be in place to deal with natural disasters or failures within the utility system. These include measures to minimize the damage (to the utility system and building systems), reinstate service and provide emergency utility services. These measures should be coordinated with the community emergency measures.

Community Support. Community support for the operations and the operator are essential. If the utility service does not satisfy the users' needs, failure is likely since there will be limited priority and attention placed on proper operations and maintenance to ensure service. If the community does not support the requests of the operator (e.g., budget requests or staff support), the physical reliability of the facilities may be jeopardized and the morale of the operator will be undermined.

2.9.3 Works Management. A systematic, comprehensive and efficient method to plan operations and maintenance activities for utilities is particularly desirable in cold regions because of:

- the importance of reliability and severe physical and social consequences of failure or interruption in service;

- the high cost of replacement, due to failure or reduced life; and

- the high cost of operations and maintenance.

The activities in managing operations and maintenance are illustrated in Figure 2-9. A works management system is a generic tool that can be used for any or all community works (e.g., roads, buildings) including water, wastewater and solid-waste services and infrastructure. A works management system provides a planned process to assess needs, allocate resources and determine human resource requirements and training needs. The phases in developing and implementing a works management system are as follows:

Phase 1

- Inventory works (infrastructure) and materials.

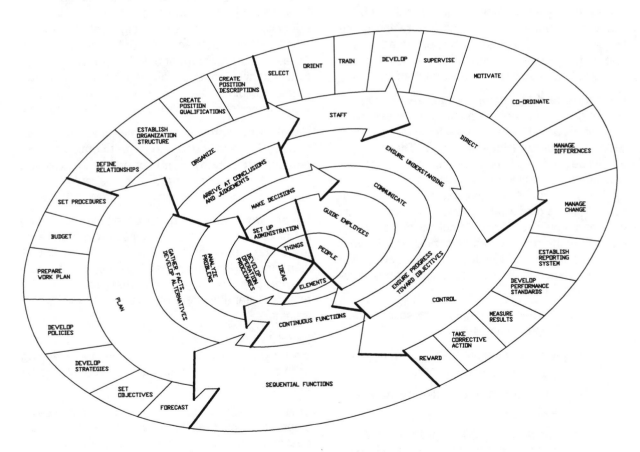

FIGURE 2-9   MANAGING OPERATIONS AND MAINTENANCE (developed from Harvard College, 1969)

- Determine O&M tasks (activities) and define task descriptions.

Phase 2

- Set standards.
- Establish work schedules.
- Prepare budget.

Phase 3

- Do the work.
- Report work activities.
- Review schedules and budget.

The works management system must be adapted to the operations and maintenance requirements of the utility systems in the community and to the characteristics of cold regions (Municipal and Community Affairs, 1991a). A works management system for small communities must be simple.

Preventive Maintenance. The reliability of utility systems depends mainly on the quality and quantity of preventive maintenance performed by the operator. Preventive maintenance is best applied as part of an overall works management system.

Operation and Maintenance Manuals. Operation and maintenance manuals are necessary to allow the operators to care for their facilities correctly and quickly (Squires, 1977). The manuals should be prepared during the design and construction phase, so that they are available for system start up. They should include pictures of all the components that are discussed. Arrows can show exactly what part is being discussed. Sections of the manual describing the operator's duties must also be written according to the individual operator's education and experience level. The operations and maintenance manual should include:

- a comprehensive, simple index so the operator can quickly find items;
- accurate and complete as-built drawings (layout of system, piping, diagram, etc.);
- technical information provided by the equipment suppliers;
- complete parts list and suppliers' names and addresses for all equipment, chemicals, etc., that must be maintained;
- step-by-step trouble-shooting lists of all possible problems;

- step-by-step repair and maintenance lists for all equipment and parts;
- the names and phone numbers of people the operator can call day or night in case of an emergency for which help is required;
- a definition of each part of the facility, what its function is, and why it is important;
- photographs of components and their construction or installation;
- criteria for sizing and designing components and their capacity and expansion provisions; and
- inspection and maintenance requirements.

Community and Operator Training. Training must be geared to the operator's education level. In remote areas, experience has shown that training carried out with the individual operator at his own plant is more successful (also, more expensive) than bringing several operators to a central location or educational institution for training. Probably a combination of individual and group training is desirable. At least two operators should be trained for each facility to allow for backup when the main operator is not available. Operator training should be provided in progressive levels, such as the following:

Level 1 would provide basic emergency measures to minimize facility damage (e.g., to drain the system or start standby pumps or boilers in case of a failure of an important part). It is desirable to have several people in the community trained to this level.

Level 2 would provide, beyond the Level 1, training for minor repairs to boilers, pumps, chlorinators, etc., to get them back on line. This level would not include much preventive maintenance, but it would provide primarily day-to-day operation. Both the main and standby operator should have this level of training.

Level 3 would provide, beyond Level 1 and 2, training for preventive maintenance, such as keeping boilers in adjustment. The main operator should be trained to this level at least.

Level 4 would provide capable and interested Level 3 operators with training to a level where they could qualify for formal certification as water and wastewater treatment operators.

Operators in communities with relatively simple utility systems may not need Level 4 training. At construction camps it is probably desirable to train operators to standards higher than those above, because facilities are usually more sophisticated and salaries are higher to attract trained operators.

Ideally, the main operator and standby operator for a facility or system should be selected at the beginning of construction, so they can receive on the job training throughout construction. The selection should be made by the community, with technical help from the construction, training or funding agency, because the community must accept and support the operators after they assume O&M responsibilities for the facilities. The operator should be selected for dedication and abilities. Operator training is not a one-time event. It is ongoing. Operators move on to other jobs, community administrations change and new operators are appointed, etc. These circumstances require a continuous training program. The operators must be paid enough to keep them on the job and to provide a comfortable living.

The community administration must be trained in bookkeeping, billing, employment forms (tax withholding, worker's compensation, etc.) and other functions needed to operate a utility.

Community education is also an important part of the training process. The community must realize the importance of the operators and support them both morally and financially. Community appreciation of the health and convenience benefits of health the utility services bring should promote this support.

Homeowners must often be trained in the correct use and care of new facilities installed in their home (e.g., building service connections, water tanks, special toilets).

2.9.4 Technical and Management Support and Backup.
Technical and management assistance is essential to help operators in small communities with complex utility facilities or equipment. This assistance is usually provided by a central agency. This assistance may involve support and backup including:

- training, both technical and management;

- helping the operator get parts, materials, equipment, etc.;

- being readily available to help the operators with technical advice and work out complex problems;

- responding with qualified people, specialized materials and equipment during emergencies; and

- making regularly scheduled visits several times a year to discuss any problems, help the operators maintain a sound preventive maintenance program and help in specialized activities.

In cold regions, one of these visits must be made in the fall, so that any problems can be solved or repairs made before winter, especially those that could jeopardize the integrity of the system. During these visits, all critical components, including backup or standby facilities, should be inspected to ensure that they are operating properly. Another visit should be undertaken during the summer when all pipelines can be flushed and fire hydrants operated and cleaned.

2.10 Utility Management

The management activities required for the satisfactory use, operation and maintenance and financing of utilities in communities in cold regions are essentially the same as elsewhere (American Water Works Association, 1980), for example:

- works management;

- operation and maintenance work;

- training;

- records and manuals management;

- budgeting;

- personnel;

- training;

- safety program;

- emergency response planning;

- accounting, financing and budgeting;

- regulating utilities;

- utility regulations (bylaws or ordinances);

- approval and inspection process;

- establishing rates; and

- customer and public relations.

In a small community there may be only a few administrative staff and utility staff, but for good management all the above management activities must still be programmed and carried out. In a small community the utility operator may help in administrative tasks such as collecting user fees.

2.10.1 Service Rates. Service rates in cold regions should generally be based on the cost of service and customer charges should be based on the volume of water used. Although flat rates are simple they provide no economic incentive to use water efficiently. Volume rates are more complex with respect to billing and require meter installation and reading (i.e., building meters or vehicle meters). However, they provide the customers with the information and incentive to use water efficiently.

2.10.2 Utility Regulations. The basic requirements for municipal water, wastewater and solid-waste utility regulations (i.e., utility bylaws or ordinances) are the same as elsewhere (American Water Works Association, 1980). In addition, in cold regions the regulations for piped services should include (Municipal and Community Affairs, 1991b):

- Define responsibilities for installing service connection pipes and service connections to the mains. Where there are no local contractors, the community may have to install service connections.

- Define responsibilities for investigating and repairing service connection leaks or failures. In frozen ground it is often difficult to identify the location of failures and the cost to investigate and repair is high.

- Define the responsibilities for installing building water meters. Where there are no local contractors the utility organization may have to supply and install meters.

- Define bleeder (wasting water) restrictions (e.g., when allowed) and control (e.g., when they are required in an emergency or until repairs can be undertaken).

- Provide specifications for:

  service connections:

  - freeze protection and thawing measures (e.g., heat trace, circulation pump, pitorifice);

  - materials (e.g., pipe, insulation);

  - above-ground service pipe requirements (e.g., structural integrity, foundation, etc.)

  Bleeders (water wasting):

- design, position, location of water wasting device/system;

- conditions for wasting water (e.g., time limit, charges).

  water conservation:

- fixture flow rates (maximum).

In cold regions the utility regulations for vehicle-haul services should:

- define the method of service (e.g., on-demand, fixed schedule);

- specify the frequency of service or schedule (for areas or building types);

- specify the policy, process and charges for unscheduled service;

- regulate the disposal of greywater and excreta (honey bags);

- provide specifications for building tanks:

  - connections (access, size, location),

  - capacity (minimum size),

  - design and materials (structural, foundation),

  - controls (high/low fluid indicator, water cutoff when wastewater tank is full); and

- provide specifications for water conservation:

  - fixture flow rates (maximum).

2.11 References

Adam, K.M., R.F. Piotrowski, J.M. Collins, and B.T. Silver. 1984. Snow roads for pipeline installation on the Arctic pilot project. *Proceedings, 3rd Specialty Conference on Cold Regions Engineering,* Canadian Society for Civil Engineering, Montreal, Quebec, 603-618.

Alter, A.J. 1972. Arctic environmental health problems. *CRC Critical Reviews in Environmental Control,* January, 459-515.

American Water Works Association. 1980. *Water Utility Management.* AWWA Manual 5, American Water Works Association, Denver, Colorado.

Bond, R.W. 1985. *Sewage Disposal Alternatives to Honey Bucket Bunker Disposal Systems.* Environmental Health Branch, Alaska Area

Native Health Services, U.S. Public Health Service, Anchorage, Alaska.

Brocklehurst, C. and G. W. Heinke. 1985. *The Effect of Water Supply and Sanitation on Health on Indian Reserves in Manitoba.* Department of Civil Engineering, University of Toronto, Toronto, Ontario.

Cairncross, S., et al. 1980. *Evaluation for Village Water Supply Planning.* John Wiley and Sons, New York.

Cameron, J.J. 1979. *Northwest Territories Water and Sanitation Systems Analysis Computer Program.* Vol. 1 and 2. Department of Local Government, Government of the Northwest Territories, Yellowknife, NWT.

Cameron, J.J. 1982. Economic analysis of water and sanitation alternatives for communities in northern Canada. In: *Utilities Delivery in Cold Regions,* Environment Canada, EPS 3-WP-82-6, Ottawa, Ontario, 77-101.

Cameron, J.J. 1985. *Culture and Change in the Northwest Territories: Implications for Community Infrastructure Planning.* Master's thesis, School of Community and Regional Planning, University of British Columbia, Vancouver, B.C.

Cameron, J.J. and B.C. Armstrong. 1980. Water and energy conservation alternatives for the north. In: *Utilities Delivery in Northern Regions,* Environment Canada, EPS 3-WP-80-5, Ottawa, Ontario, 47-88.

Campbell, M. 1980. *Appropriate Water and Sanitation Technologies for Indian Communities in Northern Ontario.* Royal Commission on the Northern Environment, Toronto, Ontario.

Clark, E., et al. 1973. *Expedient Snow Airstrip Construction Technique.* U.S. Army Corps of Engineers, Cold Regions Research and Engineering Laboratory Special Report No. 198. Hanover, New Hampshire.

Communications. 1974. *Seeding Recommendations for Revegetation of Arctic and Subarctic Soils.* U.S. Dept. of Agriculture, Soil Conservation Service, Anchorage, Alaska; Institute of Agriculture, University of Alaska, Palmer, Alaska; and Alaska State Division of Aviation, Anchorage, Alaska.

Department of Public Works and Services. 1993. *Design Standards and Guidelines for New Public Buildings.* Government of the Northwest Territories, Yellowknife, NWT.

Feachem, R.G., D.J. Bradley, H. Garelick, and D.D. Mara. 1983. *Sanitation and Disease: Health Aspects of Excreta and Wastewater Management.* John Wiley and Sons, Toronto, 501 pp.

Feilden, R.E.K., J. Van Praet, E. Payne, J.J. Cameron, and M. Michael. 1988. Economics of piped services in Northwest Territories hamlets. Presented at Canadian Society of Civil Engineers Annual Conference, Calgary, Alberta.

Flanders, S.N., J.S. Buska, and S. Barrett. 1981. Window performance in extreme cold. In: *Proceedings, Specialty Conference on the Northern Community.* American Society of Civil Engineering, New York, 396-408.

Government of the Northwest Territories. 1986. *General Terms of Reference for a Community Water and Sanitation Services Study.* Community Works and Capital Planning Division, Municipal and Community Affairs, Government of the Northwest Territories, Yellowknife, NWT.

Grey, D.M. and D.H. Male (editors). 1981. *Handbook of Snow: Principles, Processes, Management and Use.* Pergamon Press, Toronto, Ontario.

Heinke, G.W. 1984. *Report on Public Health Effects of Municipal Services in the Northwest Territories.* Department of Local Government, Government of the Northwest Territories, Yellowknife, 45 pp.

Heinke, G.W. and D. Prasad. 1976. *Disposal of Human Waste in Northern Areas. Some Problems of Solid and Liquid Waste Disposal in the Northern Environment.* Environment Canada, Ottawa, EPS-4-NW-76-2, 87-140.

Hutcheon, N.B. and G.O.P. Handegord. 1983. *Building Science for a Cold Climate.* John Wiley and Sons, Toronto (Also: Construction Technology Centre Atlantic Inc., Fredericton, N.B.).

James F. MacLaren Ltd. 1980. *Guidelines for Municipal Type Wastewater Discharges in the Northwest Territories: Background Report.* Prepared for the Northwest Territories Water Board, Yellowknife, NWT.

Johnston, G.H. (ed). 1981. *Permafrost, Engineering Design and Construction.* John Wiley and

Sons, Toronto, Ontario, 173-218.

Kalbermatten, J.M., J. De Annes, D.D. Mara, C.G. Gunnerson. 1980. *Appropriate Technology for Water Supply and Sanitation: A Planner's Guide*. World Bank, Washington, DC.

Lapinleimu, K., M. Stevik and L. Soininen. 1984. Virus isolations from sewage in Finland. *Circumpolar Health*, 84: 213-216.

Martin, J.D. 1982. The impact of housing and sanitation on communicable disease in the Northwest Territories. In: *Utilities Delivery in Cold Regions*, Economic and Technical Review Report, EPS 3-WP-82-6, Environment Canada, 204-215.

McFadden, T.T. and F.L. Bennett. 1991. *Construction in Cold Regions - A Guide for Planners, Engineers, Contractors, and Managers*. John Wiley and Sons, New York.

McGary, M.G., T. Jackson, W. Rybczynsk, A.V. Whyte, and A.P. Zimmerman. 1980. Appropriate technology for water supplies and sanitation in northern communities. In: *Utilities Delivery in Northern Regions*. Environment Canada, EPS 3-WP-80-5, Ottawa, Ontario, 121-135.

Michael, M. 1984. *Effects of Municipal Services and Housing on Public Health in the Northwest Territories*. Municipal and Community Affairs, Government of the Northwest Territories, Yellowknife, NWT.

Municipal and Community Affairs. 1991a. *Community Works Management System*. Government of the Northwest Territories, Yellowknife, NWT.

Municipal and Community Affairs. 1991b. *Water and Sewage Services By-laws Handbook*. Municipal and Community Affairs, Government of the Northwest Territories, Yellowknife, NWT (Draft).

Rees, W.E. 1986. *Stable Community Development in the North: Properties and Requirements - An Econo-Ecological Approach*. School of Community and Regional Planning, University of British Columbia, Vancouver, B.C., U.B.C. Planning Papers, Studies in Northern Development, #13.

Reinhard, K.R. 1976. Resource exploitation and the health of western arctic man. *Circumpolar Health*, University of Toronto Press, Toronto, Ontario, 615-627.

Robinson, B.A. and G.W. Heinke. 1990. *The Effect of Municipal Service on Public Health in the Northwest Territories*. Municipal and Community Affairs, Government of the Northwest Territories, Yellowknife, NWT.

Robinson, B.A. and G.W. Heinke. 1991. *The Effect of Municipal Service Improvements on Public Health in the Northwest Territories*. Municipal and Community Affairs, Government of the Northwest Territories, Yellowknife, NWT.

Sanger, F.J. 1969. *Foundations of Structures in Cold Regions*. U.S. Army, Cold Regions Research and Engineering Laboratory, Monograph III-C4, Hanover, NH.

Saunders, R.J. and J.J. Warford. 1976. *Village Water Supply*. The John Hopkins University Press, Baltimore, Maryland.

Schliessmann, D.J., F.O. Atchley, M.J. Wilcomb, and S.F. Welch. 1958. *Relationship of Environmental Factors to the Occurrence of Enteric Diseases in Areas of Eastern Kentucky*. U.S. Public Health Monograph No. 54, Washington, DC.

Simon, J.C., R.R. Forster, T. Alcose, E.A. Brabec and F. Ndubisi. 1984. *A Culturally Sensitive Approach to Planning and Design With Native Canadians*. Prepared for Canada Mortgage and Housing Corporation, Ottawa, Ontario.

Sims, M.J. 1982. Acceptance of treated water by northern community residents. In: *Utilities Delivery in Cold Regions*, Environment Canada, Economic and Technical Review Report, EPS 3-WP-82-6, 140-156.

Squires, A.D. 1977. Preparation of an operations and maintenance manual. In: *Proceedings, Symposium on Utilities Delivery in Arctic Regions*, Environmental Protection Service Rep. No. EPS 3-WP-77-1, Ottawa, Ontario.

Tester, F.J. 1976. Community attitudes and perception of waste management problems at Resolute, Northwest Territories. *Circumpolar Health* 76: 635-642.

Tobiassion, W. 1971. Deterioration of structures in cold regions. In: *Proceedings, Symposium on Cold Regions Engineering*, University of Alaska, College, Alaska.

Weller, G.R. and P. Manga. 1987. The politics of health in the circumpolar north. *Arctic Medical Research*, 46(2): 52-63.

World Health Organization. 1982. *Environmental Health Problems in Arctic and Subarctic Areas*. Regional Office for Europe, World Health Organization, Copenhagen.

Young, T.K. 1982. The Canadian North and the Third World: Is the analogy appropriate? Presented at the 73rd Annual Conference of Canadian Public Health Association, Yellowknife, NWT, June 22, 1982.

SECTION 3

GEOTECHNICAL CONSIDERATIONS

3rd Edition Steering Committee Coordinators

William L. Ryan

3rd Edition Principal Author

R.L. (Buzz) Scher

Special Contributions

Alan J. Hanna

William L. Ryan

Section 3   Table of Contents

Section 3   List of Figures

Section 3   List of Tables

3 GEOTECHNICAL CONSIDERATIONS

This section provides a review of the basic geotechnical engineering necessary for the planning, design, and construction of utilities in cold climates. Illustrative examples are presented, where possible, for clarification. The reader is referred to the referenced literature and bibliography for in-depth treatments of cold-region geotechnical engineering problems and solutions. In many cases assistance from qualified, registered professionals will be required. Many technical terms are used throughout this section that may be unfamiliar to a sanitary or utility engineer; most are defined in the Glossary.

Utility systems in cold regions must function under severe climatic conditions. Understanding the climatic and geotechnical site conditions is critical in designing for freeze protection, foundation stability, thermal stresses, and economy. The limited data typically available for the project area must usually be modified either to estimate the extreme climatic conditions or to allow for highly variable surface changes within a small area. Site-specific surveys consistent with the thermal and structural design considerations are imperative. Of primary concern is movement and the possible structural damage due to the freezing and thawing of the soil. Therefore, the maximum thickness of the active layer, and the thermal properties, thaw stability and frost susceptibility of the soils must be determined. In permafrost areas, particularly in soils with a high ice content, the subsurface explorations must extend to the maximum range of anticipated major thermal effects, and surface conditions and drainage patterns must be noted.

3.1 Cold Regions Characteristics

3.1.1 Climate. The general climatic conditions, past, present and future (global changes), need to be estimated for facility design and interpretation of subsurface conditions; this includes daily temperature (mean, maximum, and minimum), wind speed and direction, precipitation, and solar radiation. These data are usually available for selected stations from national weather departments, and some have been summarized in published charts and reports (National Research Council of Canada, 1970; Canadian Department of Transport, 1967; Hartman and Johnson, 1978). However, the local microclimate can significantly modify these values for other sites even though they may be nearby. From records of the climatic conditions, one can predict seasonal

freeze and thaw penetration depths, and model the ground thermal response to the project.

The few permanent weather stations covering the vastness of the cold regions may not yield the information required for a project site. Depending on the size, significance, and risks to a project, temporary weather stations may have to be set up on-site to collect relevant short-term data.

There has been much debate, especially in the past 10 years, on the effect of increased "greenhouse gases" in the atmosphere and changes in the ozone layer. It is apparent that in some areas, the most recent 20 to 30 year average annual air temperatures have changed from averages published 10 to 20 years ago. In some areas, the temperature has been warming while in other areas it has been cooling (Lachenbruch and Marshall, 1986; Etkin, 1989). Various predictions have been made (Etkin, 1989) which include warming scenarios between 0.06 and 0.8°C per decade. The arctic designer should be aware of changes and trends in the air temperatures in the location of specific projects. In some areas where a warming trend is evident, it may be appropriate to consider an increase of 0.5 to 1.0°C over the life of the project (Esch, 1988). For critical structures, more rigorous attention should be given to potential warming in the specific area.

Freeze, Thaw and Heating Indices. In most cold regions, the annual air temperature variation can be modelled as a sinusoidal function, characterized by a mean annual temperature (T_{ma}), seasonal variation (amplitude, A_o), and seasonal phase lag (t_{lag}, days), as illustrated in Figure 3-1 and the following equations.

$$T(t) = T_{ma} - A_o \cdot \cos[2\pi/365 \cdot (t - t_{lag})]$$

$$t_1 = 365/2\pi \cdot \cos^{-1}[(T_{ma} - T_f)/A_o] + t_{lag}$$

$$t_2 = 365/2\pi \cdot [2p - \cos^{-1}[(T_{ma} - T_f)/A_o]] + t_{lag}$$

where:

t = time in days from January 1

t_1 = beginning of thaw season

t_2 = end of thaw season

T_f = freeze/thaw temperature (typically 0°C).

The annual temperature conditions in cold-region engineering are typically characterized using three indices: freeze, thaw and heating. The air freezing

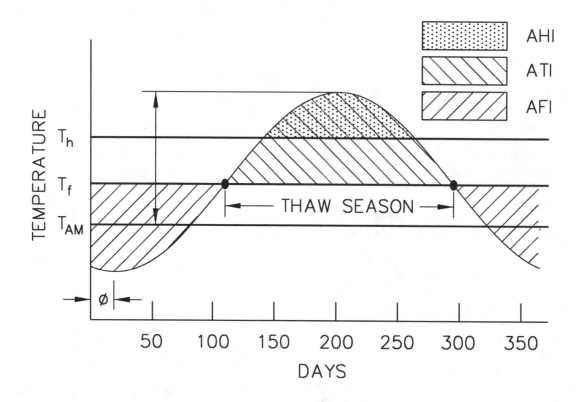

FIGURE 3-1   SINUSOIDAL ANNUAL AIR TEMPERATURE MODEL

index (AFI) is the annual summation of the freezing degree days. A freezing degree day is defined as the freezing temperature (T_f) minus the mean daily temperature multiplied by one day, when the mean daily temperature is below freezing. The air thawing index (ATI) is the annual summation of the thawing degree days. A thawing degree day is defined as the mean daily temperature minus T_f multiplied by one day, when the mean daily temperature is above freezing. The air heating index (AHI) is the annual summation of the heating degree days. A heating degree day is defined as a standard (T_h, typically 18°C) minus the mean daily temperature multiplied by one day, when the mean daily temperature is below the standard. These indices can be determined by integrating over the area bounded by the sinusoidal and appropriate reference temperature curves (Figure 3-1).

The design freezing (or thawing) index is typically determined as the average AFI (or ATI) of the three coldest winters (or warmest summers) during the most recent 30 years on record. If fewer than 30 years are on record, select the value carefully. In some cases the air freezing (or thawing) index for the coldest winter (or warmest summer) during the most recent 10 year period should be selected. If records for fewer than 10 years are available, then relationships such as those shown in Figure 3-2 could be used to predict a design (maximum) index based on the mean index from the general region. When design indices are based on less than 30 years of climate data, the lack of confidence should be reflected in the selection of a corresponding safety factor. The safety factor should be larger if the range of the air freezing (or thawing) index is large. (See Appendix E for climatic data for the North-American cold regions. The curves shown for the freezing, thawing, and heating indexes are average values for the period of record.)

Other Temperature Characteristics.   Daily cold temperatures often fluctuate about freezing, especially during fall and spring, and winter in more moderate cold regions. The freeze and thaw indices cannot be used to quantify these fluctuations. Wexler (1983) presented simple regression equations for qualifying the annual and monthly number of these fluctuations, based on mean temperatures.

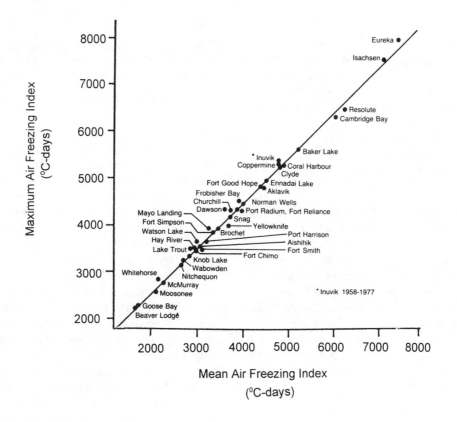

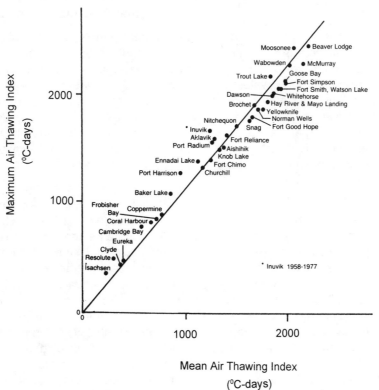

FIGURE 3-2   RELATION BETWEEN MAXIMUM AND MEAN AIR
FREEZING AND THAWING INDICES FOR NORTH-
ERN CANADA DURING THE PERIOD 1949 TO 1959
(after Johnston, 1981)

Three cold-day conditions are defined; a "frost day" is a day with a minimum temperature below 0°C, an "ice day" is a day with a maximum temperature below 0°C, and a "freeze-thaw day" is a day with a minimum temperature below 0°C and the maximum above 0°C. The number of freeze-thaw days are a measure of the daily temperature fluctuations about freezing, which are of particular use in rationalizing a safety factor for the design air freezing index, and in pavement design.

3.1.2 Permafrost. The term "permafrost" only describes the condition in which any material stays below 0°C for two or more years, without consideration of material type, ground ice distribution, or thermal stability. Solid rock could be permafrost even though thawing would generally not affect its performance as a construction material. Ice could also be permafrost, but with considerable adverse effects on overlying structures if it thaws. Permafrost may also include soil that is below 0°C but with the pore fluids in a liquid state because of salinity and other dissolved substances.

The mention of permafrost often strikes fear into the hearts of engineers and construction workers. Whether or not it is actually a detriment depends on the materials, ice distribution, the proposed facility, and the desired construction season. Sometimes, sands and gravels can be easier to excavate when frozen than when thawed because groundwater is frozen and less shoring may be needed. Also, many organic surfaces and materials are easier to excavate when frozen. In bog or muskeg areas, for example, excavation could be difficult during the summer due to an inadequate bearing surface for the equipment, but simpler in the winter when the surface is frozen.

Mean annual permafrost temperatures near 0°C (characterized as "warm" if above -4°C) are found near the southern limits of the continuous permafrost region. In far northern areas, the mean annual temperature can be -10°C or lower. Marginal or warm permafrost is much more difficult to design for than cold permafrost, because any small physical disturbance has a much greater thermal effect on the depth of the active layer and material strength and creep behavior. The equilibrium between freezing or thawing is much more delicate in the warm permafrost areas.

The moisture (ice) content, soil gradation and consistency of permafrost are also important considerations in qualifying stability. Thaw settlements are generally directly proportional to the frozen mois-

ture content. Dense, coarse-grained soils typically exhibit very little strain upon thaw, while that in ice-rich, fine-grained soils can be quite significant.

Distribution of Permafrost. The approximate distribution of permafrost in the Northern Hemisphere is shown in Figures 3-3, 3-4, and 3-5. Permafrost underlies about 20 percent of the land area of the earth. This includes about half of Russia and Canada, and 80 percent of Alaska. Brown and Péwé (1973) provide a thorough review of the permafrost distribution and environmental relationships in North America. The southern limit of the continuous permafrost zone coincides roughly with the -9°C mean annual air temperature isotherm. The southern limit of scattered discontinuous permafrost seems to coincide approximately with the -1°C mean annual air temperature isotherm. Permafrost can also be found south of this limit in shaded areas on north-facing slopes, and at high elevations. Permafrost also extends seaward from the north coast below the Arctic Ocean (Lewellen, 1973; Sellmann, 1980; Hunter, 1988). For example, permafrost is predicted to be about 60 m thick for a distance of 1,000 m offshore in areas with a stable (noneroding) shoreline along the North American north coast.

With present climates and heat flow outward from the center of the earth, the maximum permafrost thickness in Alaska is about 700 m, which occurs in the vicinity of Prudhoe Bay. Thicknesses of 1,700 m have been measured in Siberia and up to 720 m in Canada. A maximum thickness of about 1,000 m has been predicted for northern Canada (Judge, 1973). Some typical regional permafrost thicknesses are listed in Table 3-1.

Permafrost is usually absent under large lakes and rivers that do not freeze to the bottom during winter. The maximum ice thickness on a lake in Canada or the United States is about 2.5 m; a lake deeper than this probably contains an unfrozen bulb. The depth of this unfrozen bulb for smaller lakes has been found to be about equal to the minimum width of the lake. Thus, if the minimum horizontal dimension of the lake is greater than the thickness of the permafrost, there is probably no permafrost under the center of the lake. This rule of thumb can be used when looking for a source of water in a permafrost area. Permafrost under rivers or streams is much harder to predict. Usually, if the river flows significantly year-round, no permafrost is under the river.

River meandering, which is typical of northern rivers, can greatly influence the location of the thaw bulb under the river (Figure 3-6). The effect on per-

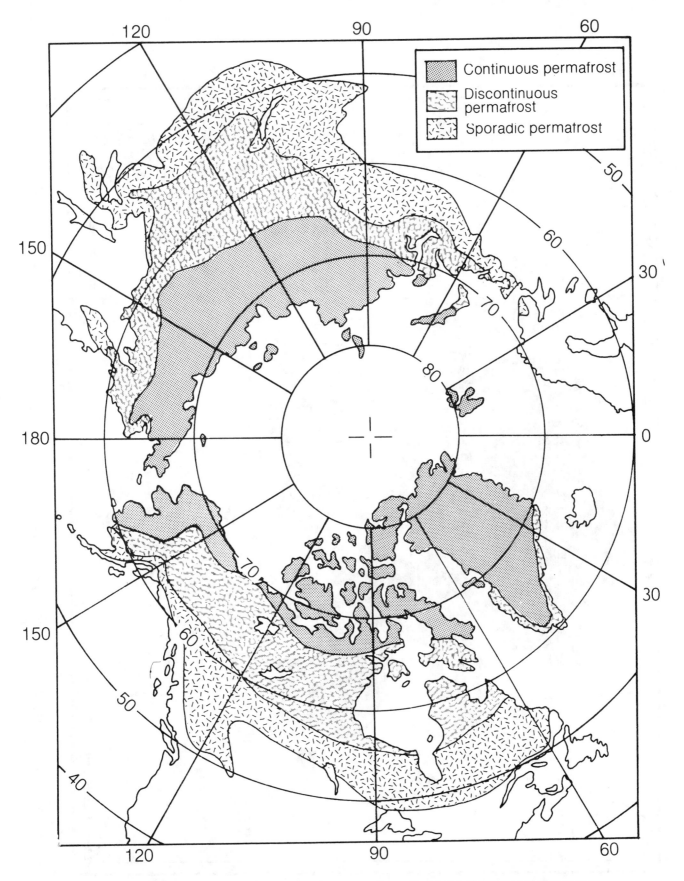

FIGURE 3-3   AREAL DISTRIBUTION OF PERMAFROST IN THE NORTHERN HEMISPHERE

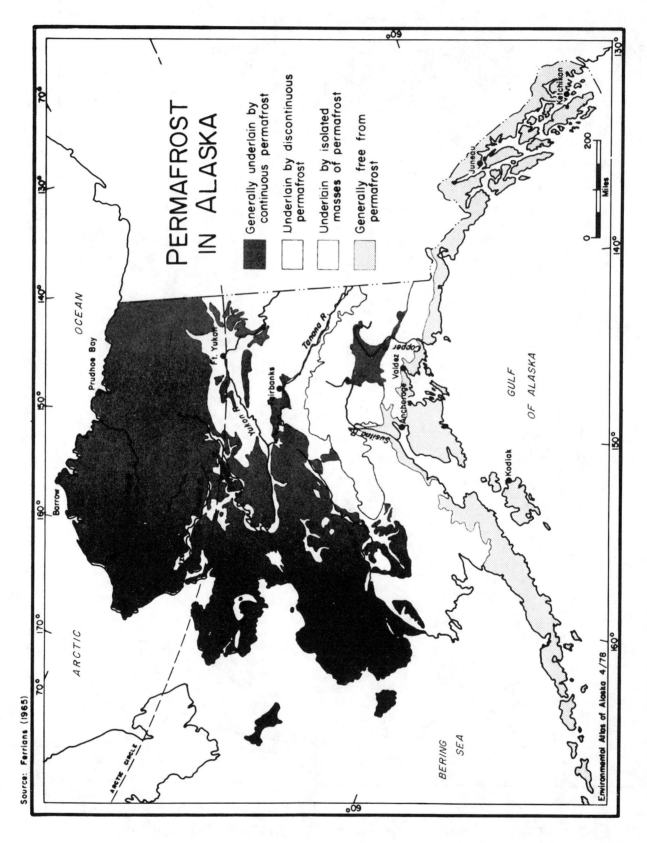

FIGURE 3-4 PERMAFROST IN ALASKA (Hartman and Johnson, 1978)

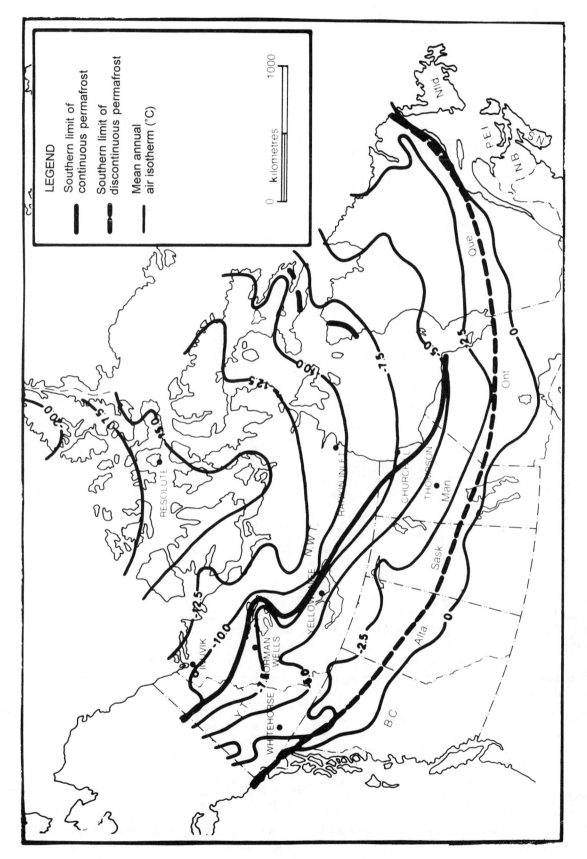

FIGURE 3-5   DISTRIBUTION OF PERMAFROST AND GROUND TEMPERATURE OBSERVATION SITES IN CANADA

TABLE 3-1   *EXAMPLES OF PERMAFROST THICKNESS*

Location	Thickness (m)
United States (Alaska)	
Barrow	600
Bethel	130
Dillingham	30
Fairbanks	90
Prudhoe	700
Anchorage (sporadic)	10
Nome	110
Canada	
Churchill, Manitoba	60
Dawson, YT	60
Fort Simpson, NWT	15
Inuvik, NWT	120
Kelsey, Manitoba	15
Norman Wells, NWT	55
Rankin Inlet, NWT	350
Resolute, NWT	400
Yellowknife, NWT	100
Winter Harbor, NWT	50
Cameron Island	720

mafrost by typical meandering rivers is shown in Figures 3-7 and 3-8 taken at Bethel and Noatak, Alaska. For example, the Kuskokwim River is continually moving northward at an average rate of 10 to 15 m a year due to the eroding permafrost at the top of the meander. This is forcing the City of Bethel to relocate farther to the north.

Surface Conditions.   The conditions of the ground surface, in terms of vegetation, snow cover, and patterned features are useful in detecting permafrost and the presence of ground ice. Permafrost near the surface tends to retard the growth of trees with tap roots, such as birch. In interior Alaska, the boundary between slopes underlain by permafrost and slopes that are free of permafrost is often delineated by a change in vegetation (Péwé, 1982). Stunted black spruce forests typically grow on poorly draining, gentle slopes underlain by shallow permafrost as close as 300 mm below the surface. White spruce, birch, aspen, and alder are typically found on better drained soils where permafrost is absent or only present at great depths. In areas where tundra fires or land clearing have destroyed the surface insulation, sink holes or thermokarsts form when the ice wedges and lenses in the permafrost melt. Figure 3-9 shows the effects of vegetation removal over a 26-year period on a test plot near Fairbanks, Alaska underlain by permafrost. Additionally, cut banks along streams and rivers may expose permafrost ice features (Figure 3-10).

Vegetation and snow cover have a substantial influence on the depth of the active layer and thus the

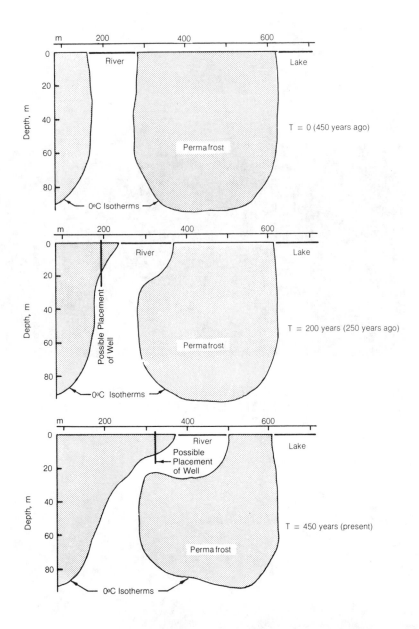

FIGURE 3-6   COMPUTER SIMULATION OF THE EFFECT
OF A RIVER ON PERMAFROST (after Smith
and Hwang, 1973)

upper level of the permafrost. Snow accumulating in the fall can reduce freezing depths in the winter, but snow falling in the late winter and spring can provide insulation against thawing in the spring. Vegetation and snow cover affects the radiant and convective heat transfer between the air and the underlying inorganic surface; thus directly affecting the correlation between the air and ground surface temperatures. Dry mosses and lichens can provide an insulating value comparable to that of fiberglass batt insulation. Large trees tend to shade the ground surface. Lower brush shades the ground and also tends to trap snow in the winter. Trees and brush both reduce wind velocities near the ground surface which, in turn, reduces the amount of heat lost from the ground. Mosses not only insulate the surface but also protect the ground from erosion by drainage water. Flowing water can moderate ground temperatures enough to cause thawing. The thermal conductivity of water and thus its heat transfer

FIGURE 3-7   FRONT STREET EROSION AT BETHEL, ALASKA,
               DUE TO THE MEANDERING OF THE KUSKOKWIM
               RIVER

FIGURE 3-8   RIVER ERODING PERMAFROST AT NOATAK, ALASKA
               (Sewage lagoon drained when dike eroded. Also thawed ice
               wedges are visible in lagoon bottom.)

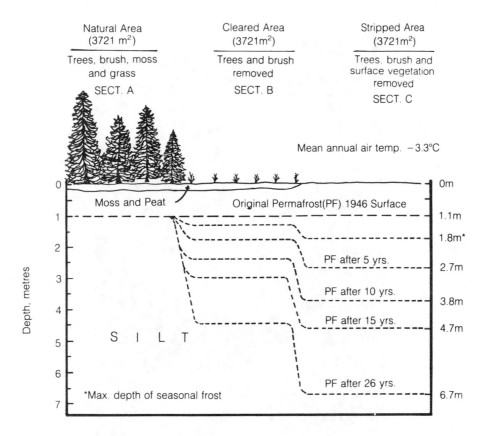

Natural Area
(3721 m²)

Trees, brush, moss
and grass
SECT. A

Cleared Area
(3721m²)

Trees and brush
removed
SECT. B

Stripped Area
(3721m²)

Trees. brush and
surface vegetation
removed
SECT. C

Mean annual air temp. −3.3°C

Depth, metres

0

1

2

3

4

5

6

7

Moss and Peat

S I L T

*Max. depth of seasonal frost

Original Permafrost(PF) 1946 Surface

PF after 5 yrs.

PF after 10 yrs.

PF after 15 yrs.

PF after 26 yrs.

0m

1.1m

1.8m*

2.7m

3.8m

4.7m

6.7m

FIGURE 3-9   PERMAFROST DEGRADATION WITH VEGETATION REMOVAL
(Linell, 1973)

properties are approximately 16 times greater than those of air. Once the ground has thawed, the water quickly erodes it to expose deeper soils to thawing.

Patterned ground features in the form of frost polygons (Figures 3-11 and 3-12) indicate the presence of permafrost and, more importantly, ice that forms at the upper surface of the permafrost downward in the shape of a wedge (Figure 3-13). Under certain conditions, the upper surface of the permafrost contracts with cooling in the winter and cracks. These small vertical cracks fill with water the following summer when the active layer thaws. This water freezes against the permafrost. This process is repeated each year, and the ice wedge "grows".

Beaded streams (Figure 3-14) also indicate the existence of ice wedges, even though they may not be evident from the surface. The flowing water melts the ice wedges as it crosses them creating an obvious "bead" or pool in the stream.

Another surface feature unique to permafrost areas is a pingo (Figure 3-15), not to be confused with a seasonal frost mound. Pingos are created when hydrostatic pressure builds up as the remnant thaw bulb of a past lake freezes, resulting in vertical surface movements. Pingos can be over 30 m high and 500 m in diameter.

3.1.3 Ground Temperatures. The subsurface soil mass temperature varies with time and depth. The temperatures are a function of the deep geothermal temperatures, the ground surface conditions, soil thermal properties, and groundwater movements (convective heat transfer). The magnitude and depth of ground temperature fluctuations are often characterized by the so-called "trumpet curve" (Figure 3-16). This envelope defines the range of periodic ground temperatures caused by annual surface temperature variations. The depth at which no periodic change occurs is usually about 10 to 15 m. The interception of the freezing temperature and the trumpet curve estimates the permafrost table depth and the active layer thickness.

FIGURE 3-10   PERMAFROST EXPOSED ON A
"CUT BANK" OF KUSKOKWIN
RIVER

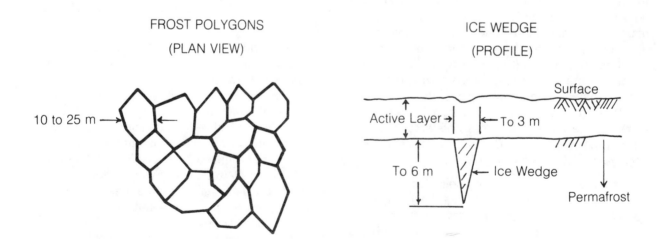

FIGURE 3-11   TYPICAL FROST POLYGONS AND ICE WEDGES

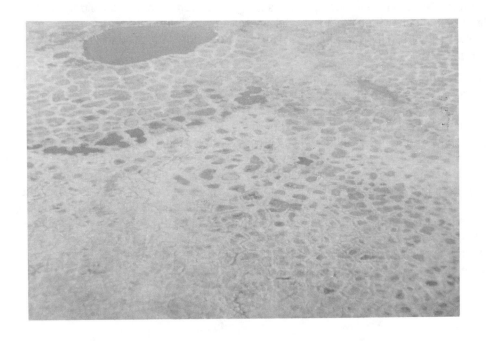

FIGURE 3-12   FROST POLYGONS ON ALASKA'S NORTH SLOPE

FIGURE 3-13   ICE WEDGE (Ryan, 1981)

Also important is the concept of freeze point depression and zones of subzero temperatures that are unfrozen (noncryotic) with no ice forming in the soil. This condition is particularly relevant to coastal, saline permafrost regions (Hivon and Sego, 1991; Hanna, 1989).

Geothermal Gradient. The geothermal gradient is the observed increase in temperature with depth. It has been reported (Carslaw and Jaeger, 1959) to vary from 0.01 to 0.05°C/m on land; 0.03°C/m is generally used. The geothermal gradient can be projected to the surface for an estimate of mean annual surface temperature. However, accurate ground temperature readings to depths of 30 m or more are required. A characteristic uniform geothermal gradient breaks down in areas of volcanic or geothermal activity. Changes in slope or bends in the geothermal gradient can also indicate past changes in the surface climate (Lachenbruch and Marshall,

1986). For example, the mean surface temperature at Barrow is currently -8°C, but ground temperature measurements at considerable depth indicate that a few hundred years ago the mean surface temperature was around -12°C.

Periodic Surface Temperatures. The ground surface temperatures in cold regions generally vary sinusoidally, similar to the air temperature variation. Long-term periodic surface temperature changes tend to establish a similar though delayed and dampened response in ground temperatures. Equations to model air-to-ground, and subsurface temperatures are provided in Section 4 (Thermal Considerations). Arnold (1978) also presents a model to estimate ground surface temperatures in permafrost areas based on relative humidity and dewpoint temperature.

FIGURE 3-14   "BEADED" STREAM (Ryan, 1981)

FIGURE 3-15   PINGO NEAR TUKTOYAKTUT, NWT

3.2   Subsurface Investigations

A thorough investigation of a proposed construction site should be undertaken prior to design to determine the site conditions, including subsurface soils and ground ice, surface and ground temperatures, and groundwater. Such investigations typically involve two or three levels of work including reconnaissance, test borings and pits, and/or geophysical surveys. In addition to defining the subsurface soil conditions, the primary objectives of soil explorations in arctic and subarctic regions are to qualify and delineate:

- the boundaries of the frozen and thawed zone within the depths to be influenced by the proposed construction;

- the amount and structure of interstitial, segregated, and massive ice in the frozen soil; and

- the composition and properties of the soil itself.

3.2.1   Reconnaissance.   All available information on site conditions should first be researched. Such existing information could include regional profiles, geologic reports, maps, and surveys, aerial photography, and investigations from other projects in the general vicinity. The objectives during this level of investigation are to qualify the surface conditions and shallow geology in order to plan for more definitive investigations including scope, drilling/excavating equipment, sampling tools, sample handling/storage/shipping needs, logistics and personnel requirements. For preliminary work, visual observations, both on-site and from photographs, can be very helpful in determining what type of soil might be encountered.

Remote Sensing and Terrain Unit Mapping.   In permafrost regions, the presence of ground ice can often be characterized at the surface by the distribution of vegetation, terrain relief, and patterned features. Aerial photographs can be used to initially estimate ground ice distribution as well as to delineate general terrain soil units, especially when cali-

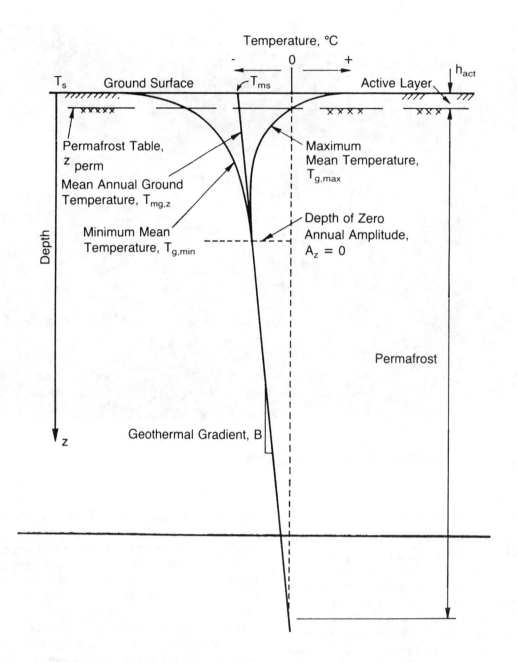

FIGURE 3-16   SCHEMATIC GROUND TEMPERATURE "TRUMPET" CURVE

brated with some test borings and geophysical surveys (Kreig and Regar, 1982; Ferrians and Hobson, 1973). Thematic Mapper satellite data have also been used to predict the occurrence of permafrost in discontinuous zones (Morrissey, 1988).

3.2.2 Test Borings and Pits.   The most common type of subsurface investigation includes drilling test borings or excavating shallow pits at the project site. The type and number of test borings or pits and soil

samples needed depends on the type and size of facility, expected variability of the materials encountered both vertically and horizontally, equipment availability, time, and the degree of risk acceptable due to unknown conditions encountered during construction. A balance must be maintained between the available equipment, the quality of information and risk desired, and the transportation costs when planning to obtain soil information. The smaller the equipment, the more time and personnel required to obtain the samples. The primary difference between drilling and sampling in cold regions versus more temperate climates is maintaining the thermal integrity of the samples between the time of taking and testing.

Recent general reviews of drilling and sampling in permafrost are provided by Reimers (1980), and Riddle and Hardcastle (1991). Methods of drilling and sampling frozen ground are generally similar to those for unfrozen ground, except for some significant particulars. Frozen soils are generally more difficult to penetrate, and the samples taken are usually disturbed thermally as well as mechanically, which can affect some of the tests. Generally, the larger the hole and sampler diameter, the less physical and thermal disturbance will occur.

It is important to maintain detailed and accurate records of the progress of the drilling and sampling. This record should include bit type, penetration rate, type of circulation media, temperature of circulation media, bit wear, and any other data useful to the designer. A sample test boring log is shown in Figure 3-17.

A common mistake in sampling is to collect a sample from one depth or location and assume it is representative of a greater thickness or area than appropriate. Therefore, it is very important to obtain a detailed field log. Also, while in the field, it is far better to collect too many samples than too few. As a rule of thumb, the subsurface should be explored and sampled to a depth equal to the width of the structure or at least to the expected depth of pile foundations or influence of the structure. This is based on the requirement that any soil subject to thawing during the existence of the structure should be investigated. It is also highly desirable that boreholes be drilled deep enough so that ground temperatures can be obtained near the depth of no seasonal change (10 to 15 m).

The most reliable method of sampling and observing the existing, near surface soil conditions is to excavate a hole or trench large enough so that the walls can be visually inspected and samples taken by hand. This method disturbs the samples the least. Excavated pits is the preferred method for granular borrow source investigations, so that a good log of the stratification and lensing can be obtained. However, it is sometimes not possible to excavate pits

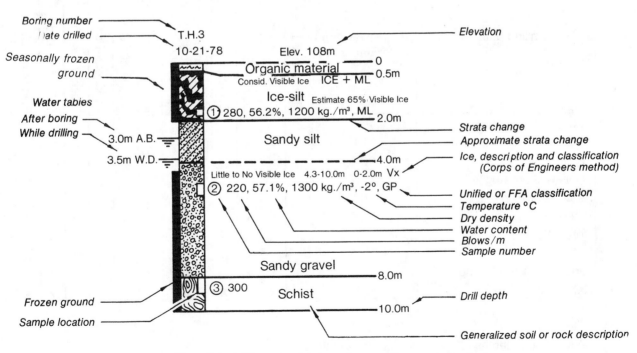

Also list drilling method and sampling tools/technique.

FIGURE 3-17   SAMPLE BORING LOG

because of the large equipment necessary, especially when the ground is frozen.

Drilling Methods. Drilling methods that have been used in permafrost investigations have included auger, percussive hammer, vibratory, rotary, and wash boring equipment. Depending on the site access and area studied, this drilling equipment has been mounted on wheeled and track carriers, skids for pulling, or comprised of helicopter transportable units. Table 3-2 summarizes the appropriate drilling method versus type of soil, sampler tool and power unit carrier.

The use of continuous-flight augers is an effective drilling method in frozen soils. With proper, but different bit selection, auger drilling can be performed in both coarse-grained (gravels and coarse sands) and fine-grained (fine sand, silt and clay) frozen soils.

Inappropriate bits will tend to slowly melt the soil, and may eventually ball up around the bit and reduce the drilling rate. Limited-flight or bucket auger drills have been used for some shallow soil investigations. However, these rigs lack a cased hole which may be a problem in poorly bonded materials, or when groundwater is encountered; thus they are primarily used for pile and utility pole installations.

The stem of continuous-flight augers can be either solid or hollow. Solid-stem augers are generally smaller in diameter and require a smaller power unit. They can penetrate frozen soils faster and at less cost than hollow-stem augers. However, samples are disturbed and usually mixed with materials from other depths. Therefore, solid-stem augers are typically used only for borrow material source investigations, or to economically supplement other more

TABLE 3-2   DRILLING METHODS FOR PERMAFROST *(after Riddle and Hardcastle, 1991)*

Drilling Method	Typical[a] Application Depth in Permafrost (m)	Subsurface Materials				Sampling[d] Techniques				Applicable Carrier			
		Fine-Grained Ice-Rich	Fine-Grained Ice-Poor	Coarse-Grained	Bedrock	Hand Sample	Split Spoon	Modified Shelby Tube	Continuous Core	Helicopter Portable	Truck Mounted	Track Mounted	Skid Mounted
Augers:													
Bucket or Highway Auger	6.1	x	x	x		x					x	x	
Solid Flight Auger	30.5	x	x	x		x					x	x	
Hollow Stem Auger	45.7	x	x	x		x	x	x	x	x	x	x	x
Percussive Hammer	30.5			x		x	x	x			x	x	
Vibratory	22.9		x	x					x	x	x	x	
Rotary Drilling	91.5[b]	x	x	x	x		x	x		x	x	x	x
Rotary Percussion	91.5			x	x		x	x		x	x	x	x
Jet-Drive (Wash-Boring)	61		x	x			x	x		x	x	x	x
Refrigerated Coring[c,d]	30.5[b]	x	x	x	x				x	x	x	x	x

Notes:
(a) Maximum depths affected by numerous factors.
(b) Depths in bedrock may be considerably greater.
(c) Refrigerated coring is also considered as a sampling method.
(d) See Table 3-3.

expensive drilling methods. The use of hollow-stem augers is likely the most widespread and available drilling method in North America. Most types of sampling are possible through the center of the augers, which provide a cased hole as the boring advances to counter most of the detrimental effects associated with drilling in variable soil and groundwater conditions. However, due to the larger diameter of some hollow-stem augers (typically 200 mm OD), larger and heavier power units (more expensive to operate and transport) are required for drilling. Additionally, all auger drilling is typically very difficult in soils containing cobbles and boulders, and in very dense frozen materials such as till.

Percussive hammer drilling generally consists of advancing a double-walled casing into the ground with a pile driving drop hammer. Air is injected down the annulus between the casings, which then carries the soil cuttings to the surface up the inside casing. This system is especially effective in dense, coarse-grained frozen materials, but will not perform well in frozen fine-grained soils (tends to plug), materials containing boulders, or in bedrock. The system provides a continuous sample which is totally disturbed, and likely thawed and segregated by particle size.

Vibratory or sonic drilling involves using an oscillating motion at the surface to drive a hollow drill string or casing into the ground. A continuous disturbed core sample can be obtained, but the drill string must be withdrawn from the ground to extrude the sample. Depending on the resistance to drilling, there can be extreme heat generated causing total thermal disturbance. The system appears to work well in completely frozen, finer-grained materials and bedrock (by adding a rotary motion). Sonic drilling equipment is relatively expensive and maintenance requirements can be significant. An advantage of this system is that the drilling components can typically be broken into relatively small loads for transportation and helicoptor transportable units have been used.

Rotary drilling and wash boring involve pumping a fluid, generally water or mud, down the drill string to a bit to aid in penetration by cooling the bit and removing cuttings. Casing is typically advanced behind the drill bit using a drop hammer for hole stability. With casing, the drill string can also be periodically withdrawn to obtain bulk soil samples with down-hole tools. These systems generally result in high drilling rates and require relatively lightweight power units, compared with continuous-flight auger

equipment. Disadvantages with rotary and wash systems are associated with the thermal disturbance and possible contamination of the sample by the drilling fluid. Chilling units can be used to minimize the thermal disturbance caused by the drilling mud (Isaacs and Code, 1972). Additionally, equipment may be required to clean and recycle the drilling fluid, depending on the location, proximity to water, and environmental concerns. Wash boring is also less effective in ice-rich, fine-grained soils, and in bedrock. Additionally, a chop-type bit is usually required in frozen coarse-grained soils.

In well-bonded frozen ground, rotary drilling with compressed air circulation is common in Canada. The cuttings used for logging are of course, completely disturbed and some thermal disturbance also occurs. More detailed sampling is undertaken at regular intervals using the methods outlined below.

Sampling. The type of sampler to use will be a function of the desired thermal state of the recovered material, disturbance, and volume of sample required. Depending on the size of the project and design risks, disturbed samples are acceptable for laboratory index tests necessary for classification and empirical correlation with mechanical parameters. However, if the project requires laboratory testing of mechanical properties for design it is critical that undisturbed samples be obtained and maintained at their in-situ temperature. Table 3-3 summarizes the applicable sampling methods (tools) for permafrost versus soil type, disturbance, and category of laboratory testing.

Grabbing bulk samples from the auger flights or cuttings returns is one of the simplest and least expensive methods. However, the samples are completely disturbed and may be mixed with soils from other depths. Additionally, the ice structure, as well as some of the frozen state will likely be destroyed, although this may not be a concern, especially when simply logging general soil units (by gradation), and in borrow source investigations.

Driven samplers are a commonly used tool in association with hollow-stem auger and rotary drilling. Drive samplers include a cutting shoe attached to either a split-spoon or solid barrel. Both are typically advanced using a drop hammer at the surface. In thawed soils, and some finer-grained, poorly bonded frozen materials, the sampler penetration rate can be used to indicate in-place relative density or consistency of the materials. In well bonded frozen soils, the sampler penetration rate may give an indication of temperature sensitivity of the mate-

TABLE 3-3   SAMPLING METHODS FOR PERMAFROST
(after Riddle and Hardcastle, 1991)

Sampling Methods	Subsurface Materials				Example Lab Tests	
	Fine-Grained Ice-Rich	Fine-Grained Ice-Poor	Coarse-Grained	Bedrock	Index[c]	Engineering Properties[d]
Hand Sample[a]	x	x	x	x	x	
Split Spoon[a]	x	x	x		x	
Modified Shelby Tube[b]	x	x			x	x
Continuous Auger Core[b]	x	x	x		x	x
CRREL Barrel[b]	x	x			x	x
Refrigerated Core[b]	x	x	x	x	x	x

Notes:
(a) Disturbed sample.
(b) Undisturbed sample.
(c) Such as grain size distribution, plasticity, classification, moisture and salinity content.
(d) Such as unit weight, thaw consolidation, shear strength, creep behavior.

rials. Split-spoon samplers are widely used in coarse-grained frozen soils and some weathered bedrock. Generally, larger diameter sampler barrels are used, up to 300 mm OD, to maximize the material recovery. In most frozen soils, especially coarse-grained, the sample driving action often completely destroys the soil structure, as well as causing some melting.

Modified shelby tubes are probably the most widely used tool in Alaska for sampling fine-grained frozen soils. This tool is especially effective in ice-rich materials, and through ice lenses. This method utilizes a special cutting head with tungsten carbide teeth, attached to either a solid or split-barrel sampler. The tool is rotated into the soil under a downward pressure; in effect, coring the soil. Modified shelby tubes up to 450 mm in diameter have been used. There typically is some thermal and physical disturbance to the outer surface of the sample depending on the

soil gradation, ice content, temperature, and sampler rotating rate. Faster rotating and advancement commonly thaws up to about 10 to 12 mm of frozen fine-grained material around the outside of the sample.

More expensive sampling tools for permafrost investigations include continuous-auger core barrels, and using refrigerated fluids while coring to minimize thermal disturbance. The continuous-auger core barrels or "CRREL" barrels, typically 80 to 130 mm in diameter, have been used very successfully to obtain large samples of frozen fine- and coarse-grained soils. However, this tool can considerably disturb more gravelly materials, depending on the gradation and ice bonding. Additionally, modifications to the drilling rig may be necessary before using this type of core barrel. Refrigerated coring is very expensive, requires very experienced drilling personnel, and has not proven entirely successful in obtaining thermally undisturbed samples in all types of frozen soils. However, it may be the only method currently available to obtain relatively undisturbed samples of frozen, very coarse gravels and bedrock. The most success with this method has been achieved using refrigerated petroleum-based fluids. However, these type of products are generally no longer environmentally acceptable.

Care of Samples. The necessity for obtaining high-quality samples cannot be overstressed, because the designer must accurately know the properties of the soils at the site. For fine-grained soils scheduled for shear strength, creep or thaw-strain/consolidation testing, samples of permafrost should be kept frozen at their original temperature until the tests can be run in the laboratory. If this is not possible, at least the in-situ undisturbed temperature should be recorded in the field. The samples should not be refrozen in the laboratory as most frozen soils may react with considerable difference on successive freezings. The preparation and transportation of undisturbed frozen samples from the field to the laboratory is a difficult and costly operation. They must be shipped and stored in refrigerated containers or insulated boxes. Packing with dry ice is a good means to control temperatures for shipping. All samples, frozen or thawed, must be tightly sealed in waterproof and airtight containers immediately after collection because of the importance of accurate moisture content measurements. Moisture content substantially affects the strength and other properties of soils, including thermal properties.

3.2.3 Geophysical Exploration. The use of geophysical exploration methods has proven to be a successful approach to supplementing test boring programs, and expanding the understanding of the permafrost conditions over a larger area. However, these methods can be expensive and subject to multiple interpretations. In-depth treatments of geophysical exploration methods applicable to permafrost studies are provided by Andersland and Anderson (1978), Arcone et al. (1978), Barnes (1963), Hoekstra (1978), Hoekstra and McNeill (1973), Judge et al. (1991) and Osterkamp and Jurick (1980).

In general, when interstitial water in rocks and soils freezes, the physical properties that change the most include the electrical conductivity and elastic modulus. Therefore, electrical and shallow seismic geophysical exploration methods are considered the most useful for permafrost studies. However, gravity survey methods may also be applicable for iden-

tifying massive buried ice features such as lenses and wedges. Ground-penetrating radar has also proven successful in detecting massive ice.

Electrical Resistivity Surveys. Electrical resistivity measurements can be obtained using both ground contact (e.g., galvanic resistivity) or noncontact methods (e.g., magnetic induction and surface impedance). The apparent resistivity values measured in the field using these methods generally increase substantially where the permafrost contains segregated or massive ice. Table 3-4 summarizes ranges of field measurements of apparent resistivity in various rock and soil types for direct current methods. Caution must be used when interpreting survey results in low-moisture or high-saline soils.

Based on a case study in interior Alaska (Osterkamp et al., 1980), changes in apparent resistivity were found to accurately delineate ice-rich soils, but it was not possible to distinguish between segregated and massive ice. Additionally, the authors found the apparent resistivity values varied seasonally and spatially over similar soil/ground-ice conditions. They recommended the most favorable time to conduct these type of surveys in relatively warm, ice-rich, fine-grained soils would be when the active layer was completely frozen in mid- to late winter and early spring. Finally, the authors concluded the magnetic induction method appeared to be the fastest and simplest procedure under field conditions.

Other recent electrical resistivity survey case studies in permafrost areas have been reported by Morack and Rogers (1978), Lafleche et al. (1988), Washburn and Phukan (1988), and Sinha (1988). Arcone and Delaney (1988) reported a case study using geophysical surveys down boreholes for measuring electrical properties of frozen silt in Alaska, and for assessing the volumetric ice contents.

Seismic Refraction. Permafrost increases the velocity of the compressional wave. Experience has shown that a velocity change occurs near the freezing point of water. This velocity change causes a time delay in thawed areas (near freezing), which can be mistaken for frozen ground. The time delay somewhat resembles a fault on seismic records. Results of the seismic method can also be misleading in both dry areas and saline water areas. Table 3-5 summarizes ranges of seismic compression wave velocities measured in various permafrost soils and rock.

Compressional waves are also influenced by soil properties and temperatures. Low velocities are re-

TABLE 3-4   CHARACTERISTIC DIRECT CURRENT FIELD
MEASUREMENTS OF FROZEN SOIL AND ROCK
RESISTIVITY (after Barnes, 1963)

Material	Resistivities Ohm-cm x 10^{-4}	Approximate Temperature, °C
Silt and organic matter[a]	20 to 80	-1
Sand and fine gravel[a]	63 to 240	-1
Gravel[a]	78 to 410	-1
Conglomerate[b]	120 to 160	-1
Glacial drift[b]	30 to 400	-11
Gneiss[b]	730 to 1370	-11
Basaltic debris[c]	1.5 to 15	-21
Marine gravel[c]	0.32 to 4.0	?

Notes:

(a) Interior Alaska
(b) Greenland
(c) Antarctica

corded when the interstitial water is saline and only partially frozen. Lower velocities also occur in fine-grained soils when the freezing point is lowered by interfacial forces. Waves show slightly higher velocities with lower ground temperatures and in older, compact, less porous rock (Barnes, 1963). Wave velocities are also higher when travelling horizontally through soil formations than when they are travelling vertically.

The active layer can also influence seismic surveys. If seismic reflection methods are to be used, the active layer should be frozen so that it does not dissipate the energy. However, if seismic refraction methods are used, the active layer should be completely thawed, so there is no chance of the overlying layer (active layer) showing higher velocities than the permafrost.

Ground Penetrating Radar.   There have been significant advances in the development of ground-penetrating radar (GPR) capabilities for arctic subsurface surveys in recent years (Judge et al., 1991; Robinson et al., 1993). GPR has been used for mapping or delineating permafrost, bedrock profiles and structure, and buried objects such as pipelines and tanks. In permafrost applications it has been successfully used for mapping both frozen and un-

frozen interfaces as well as identifying zones of massive ice.

GPR is similar to seismic refraction, except it consists of an electromagnetic pulse of energy that is reflected back to the surface from interfaces with changes in dielectric properties. Thus, depending on the strength of the received signals it is possible to identify strata with differing dielectric constants. Since frozen soil has a significantly lower dielectric constant than unfrozen soil, the interface can usually be detected fairly clearly. Pure ice has an even lower dielectric constant and massive ice has been successfully mapped in granular deposits in the Mackenzie Delta region.

3.2.4 Measuring Ground Temperatures. Local ground temperatures will typically not be available, and extrapolations are not advisable since they can vary significantly with the air temperature, snow cover, vegetation, drainage, topography, and soil properties. By far the most reliable approach is to obtain actual site measurements over a long term or at a sufficient depth (10 to 15 m) to obtain the mean annual ground temperature. Various instruments can be used to measure ground temperatures, including steel probes, frost tubes, thermometers, thermistors, and thermocouples. The method used depends on the accuracy needed, the equipment available for installation, the importance and cost of the facilities to be constructed, and the frequency of readings. An accuracy of ±0.5°C is usually sufficient for engineering purposes (Johnston, 1981). Where temperatures are near 0°C, however, greater accuracy may be required. In-depth discussions of ground temperature monitoring and thermometry are presented in Krzewinski and Tart (1985).

Steel Rod Probes and Frost Tubes. Sometimes the only information required is whether the ground at a given level is frozen or thawed. A steel rod, approximately 25 mm in diameter, can be driven or twisted into the ground at different times of the year to determine by probing the upper surface of the frozen ground. This method usually works satisfactorily in fine-grained soils at depths of no more than 3 m. However, in gravelly soil or dry (unbonded) soil or soil containing occasional rocks, this method is not reliable. To determine the maximum thickness of the active layer, it would be best to probe late in fall or early in winter. It should be understood that the bottom of the active layer can still be thawing even after freezing has started at the ground surface, especially in areas with an active layer over 2 m thick. The active layer may freeze back from the bottom up, so thaw depths from fall measurements can be misleading.

Frost tubes have been used to measure the depth and rate of freeze-thaw penetration below the surface. These tubes consist of an outer casing, and an inner casing which can be removed. The inner casing is transparent and is filled with an indicator fluid, such as methylene blue, that changes color upon freezing. The tubes are placed deeper than the active layer. The outer casing may also have to be anchored to offset frost-heave forces when used in fine-grained soils. Frost tubes can be readily fabricated and easily read. They are inexpensive and accurate to about 50 mm. A schematic frost tube design is provided in Andersland and Anderson (1978; pg. 468).

Thermometers. The simplest, and probably least expensive, method of measuring shallow soil temperatures is to use either a bimetallic or a long-stemmed, mercury-filled thermometer. The cost of these thermometers varies directly with their accuracy. They are seldom used for temperature readings at depths greater than one metre.

Thermistors and Thermocouples. Electric resistance thermometers are most often used for

TABLE 3-5   CHARACTERISTIC FIELD-MEASURED SEISMIC COMPRESSION WAVE VELOCITIES IN PERMAFROST (after Barnes, 1963)

Material	Seismic Velocity (km/sec)	Estimated Ground Temperature (°C)
Soils		
Fine-grained and organic[a]	1.5 to 3.7	-1 to -9
Coarse-grained		
Aeolian and alluvium[a]	2.4 to 4.6	-1 to -3
Glacial till and outwash	2.3 to 4.7	-2 to -11
Bedrock		
Mesozoic sediments[a,b]	2.5 to 4.3	-5 to -10
Paleozoic and older sediments[c]	4.1 to 5.9	-11
Metamorphic[a,c]	4.0 to 6.1	-1 to ?

Notes:
(a) Alaska
(b) Canada
(c) Greenland

ground temperature measurements. They operate on the principle that the electric resistance of most metals increases with increasing temperatures; nickel wire (or sometimes platinum) is commonly used. These thermometers are highly sensitive and accurate, even more so than thermistors. They are very stable over a period of years. A field accuracy of ±0.05°C can be obtained. Because of the size of the sensors and their high cost, electrical resistance thermometers are not used extensively. Ice-bath calibration is required for both thermistors and thermocouples.

Thermistors are thermally sensitive resistors that exhibit a change in electrical resistance with a change in temperature; they are becoming more popular as a method of measuring ground temperatures. The change in resistance versus temperature in thermistors is nonlinear, but very large for small temperature ranges. They are highly sensitive to small temperature changes and accurate to ±0.01°C. Two problems encountered with thermistors are their sensitivity to pressure and their instability in moist conditions. The use of epoxy resins to seal connections can overcome the moisture problem.

Thermocouples are also commonly used for measuring ground temperatures for engineering purposes. They consist of two dissimilar metal wires joined at the point where the temperature is to be read. Different types of metal combinations have different temperature ranges through which they can be used. For ground temperature readings, one wire is usually made of copper and the other of constantan (55 percent copper and 45 percent nickel). A current is generated proportional to the temperature at the joined end. Thermocouples can be obtained with different accuracies. A potentiometer (Wheatstone bridge) is used to measure the temperature, which varies linearly with changes in resistance. The potentiometer must be kept in an area where the temperature is above 0°C while readings are being taken. It must also be stabilized with its surrounding temperature, i.e., readings should not be taken until the potentiometer has reached the temperature of its surroundings.

Procedure for Installing Thermocouples or Thermistors. When measuring ground temperatures, prefabricated strings (cables) of thermocouples or thermistors at specified intervals are usually installed in a predrilled hole. However, they can be laid directly in the ground, insulation, gravel pads, or fills, or placed directly into flowing fluids.

For temporary locations, the temperature instrumentation is typically buried in place in a test hole. It may be desirable to tape them to a wood dowel or some other low-heat-conducting material to the string to ensure the temperature sensors are correctly positioned in the hole. To minimize damage to the strings, the drilled hole should be backfilled with finely ground soil cuttings or sand. The backfill material should be compacted as closely as possible to the original soil density. The backfill should preferably be dry so that temperatures will stabilize to equilibrium with the surrounding ground as soon as possible. The surface end of the string must be fully protected from moisture, animals, and construction activities.

For an alternate method, the drill hole could be cased with a plastic pipe (or some other low-heat-conducting material). The annulus between the hole and casing should be completely backfilled with finely ground cuttings or sand, and compacted as closely as possible to the original soil density. After placing the string, the pipe should be filled with a mixture of nonfreezing fluid. Propylene glycol or silicon oil is commonly used to fill the casing. In the past, mixtures of about 80 percent diesel oil and 20 percent wax were also used. This configuration allows the wires to be removed when no longer required, or repaired or replaced in the event of a malfunction. The casing should be as small in diameter as possible, and the fluid fill as dense or viscous as possible in order to minimize convective currents. The surface end of the string is typically routed to an instrument panel for permanent installations.

The act of drilling the hole and installing the wires disturbs the normal temperature for a period. This disturbance is especially critical when measuring geothermal gradients deep in the ground. Depending on the drilling operation, the depth at which the readings are taken, and the required accuracy, it may take several weeks before an adequate equilibrium is reached. In relatively shallow installations (not over 2 to 3 m deep), for an accuracy of 0.1 to 0.5°C, and if care was taken during installation, a week should be sufficient to dissipate thermal disturbances caused by drilling and installation. Figure 3-18 shows a typical thermistor string installation for obtaining ground temperature measurements over several years. Discussions of typical thermistor and thermocouple installation methods and results are presented in Krzewinski and Tart (1985) and Klein et al. (1986).

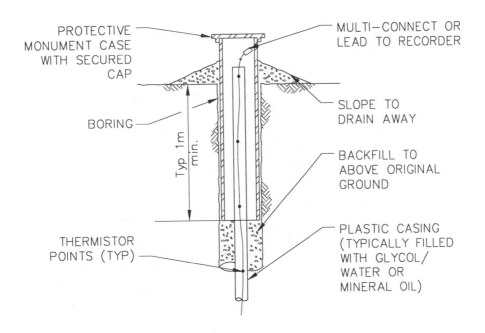

FIGURE 3-18   SCHEMATIC THERMISTOR STRING INSTALLATION

3.3   Soil Mechanics

Knowledge of the mechanical properties of soils is essential for foundation design and construction in all climates. Key are the index parameters and consistency/density, as well as the shear strength and compressibility properties. In cold regions, additional parameters must be considered, including the soil frost susceptibility and porewater chemistry, and in permafrost the creep temperature, adfreeze, and thaw-strain relationships. An in-depth review of the physics, chemistry, and mechanics of frozen soils is presented in Anderson and Morgenstern (1973) and Andersland and Anderson (1978).

Based on the subsurface investigation findings, the soils encountered should be grouped into general units based on visual classification, depositional environment, consistency/density, thermal condition, etc. A number of soil samples from each general unit should then be tested in the laboratory to qualify the characteristic geotechnical parameters and properties for design. A detailed case study of a laboratory testing program to measure some frozen and thawed properties of permafrost soils is presented by Watson et al. (1973).

3.3.1 Index Testing. The index properties typically define the soil classification and in-situ condition. In cold regions the key index properties include grain-size distribution, natural moisture content, consistency/density, plasticity, and porewater salinity. Depending on the scope of the facility, and knowledge of the soil conditions, it is often enough to simply perform laboratory index tests and determine the other mechanical parameters (shear strength, compressibility, etc.) based on published empirical correlations.

The methods of conducting soil tests in the field or the laboratory are presented in most textbooks about soils and foundations, and are not covered here (Andersland and Anderson, 1978; Taylor, 1967; Shuster, 1970). Tests should be performed according to standardized specifications whenever possible, such as the American Society of Testing and Materials (ASTM, 1994). The U.S. Army Cold Regions Research and Engineering Laboratory (CRREL) in Hanover, New Hampshire has also done considerable work on the testing and analysis of frozen soils and has publications available on the subject (Chamberlain, 1981). The index test results should be included on the test hole logs (Figure 3-17).

Total Moisture Content. One of the most important index tests that must be performed is to determine the total gravimetric moisture content (W_{tot}). This value is helpful in classifying soils based on visual inspection, and interpretation of other index

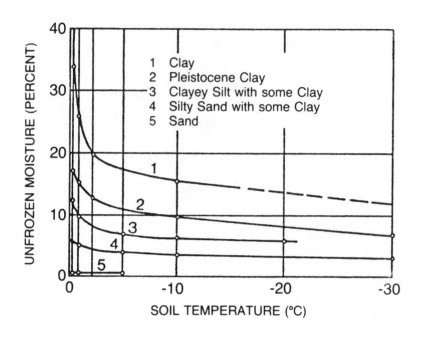

FIGURE 3-19   UNFROZEN MOISTURE CONTENT IN SOILS
(Tsytovich, 1975)

parameters such as density and saturation, as well as for estimating some mechanical properties. This test is most often determined by oven drying the sample to a constant weight. This test can be performed on disturbed as well as undisturbed soil samples.

In frozen soils, it is important to qualify the laboratory moisture content with respect to the in-situ ice distribution of the tested sample, including the visible volume and condition (see Section 3.3.3, Classification). Also of particular importance in cold regions is the fact that the measured total moisture of a frozen soil is actually present in-situ in two phases, ice and water. This condition significantly affects the thermal, hydraulic and mechanical properties including creep, shear strength, and bearing properties of frozen soil, as well as complicating frost-heave and thermal modelling. The water to ice ratio depends primarily on the soil temperature, grain size, pressure, and the mineral content of the water. This ratio is higher at any temperature for soils with small grain sizes such as clays. The ratio decreases as the temperature decreases (Figure 3-19). Tsytovich (1975) points out that tests have shown that essentially all pore water is frozen at -70°C. The gravimetric unfrozen water content (W_{uf}) at any temperature can be represented by:

$$W_{uf} = A \cdot (-T)^B$$

where:

A, B =   constants dependent on the soil type

T   =   temperature (°C)

Table 3-6 summarizes ranges of the A and B constants characteristic of fine-grained soils.

Bulk Density.   The in-situ bulk density can be determined by weighing the sample and computing the volume on the basis of its average dimensions. The volume of frozen cores can also be determined in a cold room using a displacement technique, by immersion in a liquid such as kerosene. Note that this test can only be performed on undisturbed and intact soil samples. From the moisture content and the bulk density, the dry density can then be computed. In frozen soils, it is important to note the in-situ ice distribution (see Section 3.3.3, Classification) in the sample tested. For example, what volume of the tested sample was segregated or massive ice.

Particle Size Distribution.   Another important test is the distribution or gradation of particle sizes. The grain size distribution is typically reported as a percentage by weight smaller than a standard reference. This test is performed by mechanically shaking the sample through a series of standard sieves. The distribution of silt- and clay-sized particles (less than 0.074 mm) is typically measured using a hydrometer. These tests can be performed on disturbed as well as undisturbed soil samples. Note particularly the percentage of the sample (by weight) smaller than 0.074 mm (#200 U.S. sieve size) and 0.02 mm. These sizes are typically used as an index to express the susceptibility of the soil to frost heave (Chamberlain, 1981).

Soil Plasticity. The Atterberg limits include the liquid limit, plastic limit, and plasticity index and are measures of the soil plasticity. The plasticity of a soil is important for classification, characterization and empirical correlation with shear strength and bearing capacity. These tests can be performed on disturbed or undisturbed soil samples.

3.3.2 Salinity. There are many arctic regions where the porewater (or pore-ice) is saline because of the marine influence during soil deposition or subsequent reworking due to inundation in geological time (Hivon and Sego, 1991; Miller and Johnson, 1990). Marine salinity is typically in the order of 30 to 35 ppt. Porewater salinities as high as 46 ppt have been reported in the Baffin Island region (Hanna, 1989). It is thought that high "salinity" encountered in the lower Mackenzie Valley (Hivon and Sego, 1991) can be attributed to local bedrock (limestone and possibly gypsum) rather than marine salinity. The salinity is often determined on the basis of electrical conductivity, and the porewater chemistry is not always analyzed. Nixon (1988) presented a simple laboratory test procedure to measure salinity. The permafrost design engineer must establish if there is any soil chemistry that might significantly alter the soil thermal and physical behavior.

Salinity in the porewater depresses the freeze point by as much as 1.8°C for normal seawater concentrations of 30 ppt. Therefore, saline soil does not start to freeze until nearly -2°C.

As freezing occurs in saline soils, ice crystals form in the voids and they contain almost fresh water as the salts are expelled upon freezing. As the temperature becomes colder, the ice crystals grow such that the crystals and the soil particles are surrounded by a highly concentrated brine solution. It is well known that all medium- to fine-grained soils will have a portion of the porewater remain unfrozen (Figure 3-19), even at temperatures of -5 to -10°C, due primarily to surface tension effects. In saline soils the amount of unfrozen porewater at any given freezing temperature is considerably greater due to the resulting high salt concentrations. It is very important that the very different unfrozen moisture content curve be accurately represented in any geothermal analyses.

The impact of this phenomenon on the engineering properties of saline soils has been shown by Nixon and Lem (1984). Because there is effectively less "bonding" within the frozen saline soil mass compared to a comparable fresh water soil, the strength and the creep deformation characteristics are considerably less favorable.

3.3.3 Classification. A number of classification systems are used to define the soil phase and frozen mass.

Soil Phase. In the United States, the soil phase is typically classified by one of two systems for civil engineering projects. The Unified Soil Classification System (Figure 3-20) is generally used for earthworks and foundation designs associated with buildings, utilities and dams. For roads, the American Association of State Highway Officials (AASHTO, 1990) classification system is generally used. In Canada, the Modified Unified Soil Classification System has introduced an intermediate plastic category, CI, between liquid limits of 30 and 50 percent.

TABLE 3-6   CHARACTERISTIC UNFROZEN WATER CONTENT PARAMETERS (after Nixon, 1991)

Soil	Specific Surface	A	B
Silts	20 to 60±	0.04 to 0.06	-0.30 to -0.52
Clays	> 60±	0.09 to >0.20	-0.25 to -0.55

Coarse-grained soils, more than half of material larger than no. 200 sieve size

Major Division			Group Symbol	Typical Name	Field Identification Procedure	Information Required for Describing Soils	Laboratory Classification Procedure
Gravels, more than half of coarse fraction larger than no. 4 sieve size	Clean gravels (little or no fines)		GW	Well-graded, gravel-sand mixtures, little or no fines	Wide range in grain sizes and substantial amounts of intermediate particle sizes	For undisturbed soils add information on stratification, degree of compactness, cementation, moisture conditions, and drainage characteristics.	$C_u = D_{60}/D_{10} > 4$; $C_c = D_{30}^2/D_{10}D_{60}$ (>1, <3)
			GP	Poorly graded gravels or gravel-sand mixtures, little or no fines	Predominantly one size or a range of sizes with some intermediate sizes missing		Not meeting all gradation requirements for SW
	Gravels with fines (appreciable amount of fines)		GM	Silty gravels, gravel-sand-silt mixtures	Nonplastic fines or fines with low plasticity (for identification procedures see ML)	Give typical name; indicate approximate percentages of sand and gravel; maximum size; angularity, surface condition, and hardness of the coarse grains; local or geologic name and other pertinent descriptive information; and symbol in parentheses.	Atterberg limits below A line or PI<4 — Above A line with 4<PI<7 are borderline cases requiring use of dual symbols
			GC	Clayey gravels, gravel-sand-clay mixtures	Plastic fines (for identification procedures see CL)		Atterberg limits above A line with PI>7
Sands, more than half of coarse fraction smaller than no. 4 sieve size	Clean sands (little or no fines)		SW	Well-graded sands, gravelly sands, little or no fines	Wide range in grain size and substantial amounts of all intermediate particle sizes	Example: Silty sand, gravelly; about 20% hard, angular gravel particles 10 mm maximum size; rounded and subangular sand grains, coarse to fine; about 15% nonplastic fines with low dry strength; well compacted and moist in place; alluvial sand; (SM).	$C_u = D_{60}/D_{10} > 6$; $C_c = D_{30}^2/D_{10}D_{60}$ (>1, <3)
			SP	Poorly-graded sands or gravelly sands, little or no fines	Predominantly one size or a range of sizes with some intermediate sizes missing		Not meeting all gradation requirements for SW
	Sands with fines (appreciable amount of fines)		SM	Silty sands, sand-silt mixtures	Nonplastic fines or fines with low plasticity (for identification procedures see ML)		Atterberg limits below A line or PI<4 — Above A line with 4<PI<7 are borderline cases requiring use of dual symbols
			SC	Clayey sands, sand-clay mixtures	Plastic fines (for identification procedures see CL)		Atterberg limits above A line with PI>7

Determine percentages of gravel and sand from grain-size curve, depending on percentage of fines (fraction smaller than no. 200 sieve size) coarse-grained soils are classified as follows:

Less than 5% = GW, GP, SW, SP
More than 12% = GM, GC, SM, SC
5 to 12% = borderline cases requiring use of dual symbols

Fine-grained soils, more than half of material smaller than no. 200 sieve size

Identification Procedures on Fraction Smaller Than No. 40 Sieve Size

Major Division	Group Symbol	Typical Name	Dry Strength (crushing characteristics)	Dilatancy (reaction to shaking)	Toughness (consistency near PL)	Information Required for Describing Soils
Silts and clays, liquid limit < 50	ML	Inorganic silts and very fine sands, rock flour, silty or clayey fine sands or clayey silts with slight plasticity	None to slight	Quick to slow	None	For undisturbed soils add information on structure, stratification, consistency in undisturbed and remolded states, moisture and drainage conditions.
	CL	Inorganic clays of low to medium plasticity, gravelly clays, sandy clays, silty clays, lean clays	Medium to high	None to very slow	Medium	Give typical name; indicate degree and character of plasticity; amount and maximum size of coarse grains; color in wet condition; odor, if any; local or geologic name and other pertinent information; and symbol in parentheses.
	OL	Organic silts and organic silty clays of low plasticity	Slight to medium	Slow	Slight	Example: Clayey silt, brown; slightly plastic; small percentage of fine sand; numerous vertical root holes; firm and dry in place; loess; (ML).
Silts and clays, liquid limit > 50	MH	Inorganic silts, micaceous or diatomaceous fine sandy or silty soils, elastic silts	Slight to medium	Slow to none	Slight to medium	
	CH	Inorganic clays of high plasticity, fat clays	High to very high	None	High	
	OH	Organic clays of medium to high plasticity, organic silts	Medium to high	None to very slow	Slight to medium	
Highly organic soils	Pt	Peat and other highly organic soils	Readily identified by color, odor, spongy feel, and frequently by fibrous texture			

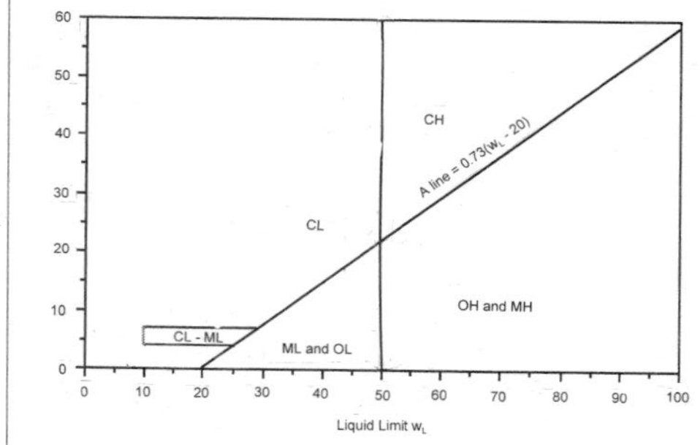

Plasticity Chart for Laboratory Classification of Fine-Grained Soils

Plasticity Chart (A line = 0.73(w$_L$ – 20)): axes Plasticity Index (0–60) vs Liquid Limit w$_L$ (0–100); regions CL-ML, CL, CH, ML and OL, OH and MH.

1. Boundary classifications: soils possessing characteristics of two groups are designated by combinations of group symbols. For example, GW-GC, well-graded gravel-sand mixture with clay binder.
2. All sieve sizes on this chart are United States standard.

FIELD IDENTIFICATION PROCEDURES FOR FINE-GRAINED SOILS OR FRACTIONS

These procedures are to be performed on the minus no. 40 sieve-size particles, approximately 0.4 mm. For field classification purposes, screening is not intended; simply remove by hand the coarse particles that interfere with the tests.

Dilatancy (reaction to shaking)

After removing particles larger than no. 40 sieve size, prepare a pat of moist soil with a volume of about 160 mm³. If necessary, add enough water to make the soil soft but not sticky.

Place the pat in the open palm of one hand and shake horizontally, striking vigorously against the other hand several times. A positive reaction consists of the appearance of water on the surface of the pat, which changes to a livery consistency and becomes glossy. When the sample is squeezed between the fingers, the water and gloss disappear from the surface and the pat stiffens finally it cracks or crumbles. The rapidity of appearance of water during shaking and of its disappearance during squeezing assists in identifying the character of the fines in a soil.

Very fine clean sands give the quickest and most distinct reaction whereas a plastic clay has no reaction. Inorganic silts, such as a typical rock flour, show a moderately quick reaction.

Dry Strength (crushing characteristics)

After removing particles larger than no. 40 sieve size, mold a pat of soil to the consistency of putty, adding water if necessary. Allow the pat to dry completely by oven, sun, or air drying and then test its strength by breaking and crumbling between the fingers. This strength is a measure of the character and quantity of the colloidal fraction contained in the soil. The dry strength increases with increasing strength.

High dry strength is characteristic for clays of the CH group. A typical inorganic silt possesses only very slight dry strength. Silty fine sands and silts have about the same slight dry strength but can be distinguished by the feel when powdering the dried specimen. Fine sand feels gritty whereas a typical silt has the smooth feel of flour.

Toughness (consistency near plastic limit)

After particles larger than the no. 40 sieve size are removed, a specimen of soil about 160 mm³ in size is molded to the consistency of putty. If too dry, water must be added, and if sticky, the specimen should be spread out by hand on a thin layer and allowed to lose some moisture by evaporation. Then the specimen is rolled out by hand on a smooth surface or between the palms into a thread about 3 mm in diameter. The thread is then folded and rerolled repeatedly. During this manipulation the moisture content is gradually reduced and the specimen stiffens, finally loses its plasticity, and crumbles when the plastic limit is reached.

After the thread crumbles, the pieces should be lumped together and a slight kneading action continued until the lump crumbles.

The tougher the thread near the plastic limit and the stiffer the lump when it finally crumbles, the more potent the colloidal clay fraction in the soil. Weakness of the thread at the plastic limit and quick loss of coherence of the lump below the plastic limit indicate either inorganic clay of low plasticity or materials such as kaolin-type clays and organic clays which occur below the A line.

Highly organic clays have a very weak and spongy feel at the plastic limit.

FIGURE 3-20   UNIFIED SOIL CLASSIFICATION (developed from Linell and Kaplar, 1963)

Frozen Soil Mass. In frozen materials, the thawed soil phase is typically classified following the Unified System (Figure 3-20). The description of the in-situ frozen mass is then annotated to the thawed classification, based on the visible ice distribution and bonding (Figure 3-21). The Canadian system for ground ice classification is very similar (Pihlainen and Johnston, 1963).

3.3.4 Frost Susceptibility. The frost susceptibility of a soil is a function of many factors, particularly the grain-size distribution, particle surface area, porewater chemistry, temperature gradient and stress condition. Additional insight into the frost susceptibility of soils relative to texture is presented in Chamberlain (1981), Jones and Lomas (1983) and Rieke et al. (1983) and for saline soils in Chamberlain (1983).

For many civil engineering projects, the frost susceptibility of a soil is simply characterized by the classification and percent of the particles, by weight, smaller than 0.02 mm ($P_{0.02}$), such as the U.S. Army Corp of Engineers Frost Design Soil Classification System illustrated on Figure 3-22. This classification system is based on an extensive series of one-dimensional laboratory frost-heave tests (Kaplar, 1974). The frost susceptibility criteria was characterized based on the measured heave rates during these tests. Note that the test results (Figure 3-22) illustrate a wide range of heave in all types of soil, even those classified as "non-frost-susceptible". Other frost susceptibility factors and index parameters are presented below in Section 3.3.8.

The frost classification in Figure 3-22 is sometimes modified to further characterize the frost susceptibility of pavement structure fill materials. The U.S. Department of the Army (1985) defined two additional frost classification groups; "possibly frost susceptible" (PFS) and "very low to medium frost susceptibility (S). Gravels with $P_{0.02}$ between 1.5 and 3 percent, and sands with $P_{0.02}$ between 3 and 10 percent are grouped as PFS, which require laboratory testing to determine frost design classification. The S group, considered suitable for subbase materials, included gravels (S1) with $P_{0.02}$ between 3 and 6 percent, and sands (S2) with $P_{0.02}$ between 3 and 6 percent.

3.3.5 Hydraulic Conductivity. Mass transfer of water through frozen soils has been demonstrated in the laboratory (Harlan, 1973; Loch and Kay, 1978; Perfect and Williams, 1980; Oliphant et al., 1983) and observed in the field (Mackay et al., 1979; Smith, 1985; Harris, 1988). Generally, the hydraulic force,

or soil-water potential in shallow nonsaline frozen soils is primarily a function of the thermal gradient. As such, a thermal gradient in frozen soil could establish the conditions under which water would move (in the direction of the colder temperature), at a rate dependent on the conductivity, in turn dependent upon the soil type, temperature, degree of saturation, and unfrozen moisture content. Thermally induced water movements in frozen soils could produce long-term heave problems associated with foundation subgrade cooling systems (Section 3.4.3) and buried chilled pipelines (natural gas) in frozen ground (Nixon, 1987a). Nixon (1991) characterized the hydraulic conductivity of a frozen soil (k_f) using a power law of the form:

$$k_f = k_o/(-T)^\alpha$$

where:

k_o = hydraulic conductivity at -1°C

T = temperature (°C)

α = slope of the relationship between k and T on a log-log plot.

The ranges in hydraulic conductivity versus temperature reported for a number of frozen fine-grained soils is illustrated in Figure 3-23. Nixon (1991) suggested the parameter α can be estimated as roughly equal to -5 times the unfrozen parameter B in Table 3-6.

3.3.6 Mechanical Properties. The mechanical properties of frozen soils of primary interest when designing foundations and earthworks for utility systems include strength, creep, and consolidation if allowed to thaw. These properties in frozen soils are complex, interrelated functions of the soil structure, total moisture and ice content, saturation, temperature, stress, and loading rate. The strength and creep deformation behavior in a given frozen soil will be dominated by temperature, ice content, and loading/strain rate, while consolidation deformations will be dominated by the total moisture content, soil particle structure, and load.

Ice has a high creep rate under a given load compared to soil, and even shows some creep under very small loads. Thus, the more ice in the soil, the higher the creep rate. Furthermore, unfrozen water in frozen soils increases creep and reduces the long-term strength. "Warmer" frozen soils have lower strengths and higher creep rates than soils at temperatures considerably below freezing. Short-term tests should not be used for designing foundations, because they do not measure creep. That is, short-

Classify Soil Phase by the Unified Soil Classification System:

PART 1: Description of Soil Phase (Independent of Frozen State)	Major Group — Description	Major Group — Designation	Subgroup — Description	Subgroup — Designation	Field Identification	Pertinent Properties of Frozen Materials Which Can Be Measured by Physical Tests to Supplement Field Identification	Thaw Characteristics	Guide for Construction on Soils Subject to Freezing and Thawing — Criteria
	Segregated ice not visible by eye	N	Poorly bonded or friable	Nf	Identify by visual examination; to determine presence of excess ice, use procedure under note (3) and hand magnifying lens as necessary; for soils not fully saturated estimate degree of ice saturation (medium, low); note presence of crystals or of ice coatings around larger particles	In-place temperature; Density and void ratio — a. In frozen state, b. After thawing in place; Water content (total H₂O, including ice) — a. Average, b. Distribution	Usually thaw-stable	The potential intensity of ice segregation in a soil is dependent to a large degree on its void sizes and for pavement design purposes may be expressed as an empirical function of grain size as follows: Most inorganic soils containing 3 percent or more of grains finer than 0.02 mm in diameter by weight are frost-susceptible for pavement design purposes. Gravels, well-graded sands and silty sands, especially those approaching the theoretical maximum density curve, which contain 1.5 to 3 percent finer by weight than 0.02 mm size should be considered as possibly frost-susceptible and should be subjected to a standard laboratory frost susceptibility test to evaluate actual behavior during freezing. Uniform sandy soils may have as high as 10 percent of grains finer than 0.02 mm by weight without being frost-susceptible. However, their tendency to occur interbedded with other soils usually makes it impractical to consider them separately. Soils classed as frost-susceptible under the above pavement design criteria are likely to develop significant ice segregation and frost heave if frozen at normal rates with free water readily available. Soils so frozen will fall into the thaw unstable category. However, they may also be classed as thaw stable if frozen with insufficient water to permit ice segregation. Soils classed as non-frost-susceptible under the above criteria usually occur without significant ice segregation and are usually thaw stable for pavement applications. However, the criteria are not exact and may be inadequate for some structure applications; exceptions may also result from minor soil variations.
			Well bonded — No excess ice	Nb, n				
			Well bonded — Excess ice	e				
Part II: Description of Frozen Soil	Segregated ice visible by eye (ice 25 mm or less thick)	V	Individual ice crystals or inclusions	Vx	For ice phase, record the following as applicable: Location, Orientation, Thickness, Length, Spacing, Hardness, Structure } per part III below; Size, Shape, Pattern of arrangement; Color. Estimate volume of visible segregated ice present as percent of total sample volume	Strength — a. Compressive, b. Tensile, c. Shear, d. Adfreeze; Elastic properties; Plastic properties; Thermal properties; Ice crystal structure (using optical instruments) — a. Orientation of axes, b. Crystal size, c. Crystal shape, d. Pattern of arrangement	Usually thaw-unstable	In permafrost areas, ice wedges, pockets, veins, or other ice bodies may be found whose mode of origin is different from that described above. Such ice may be the result of long-time surface expansion and contraction phenomena or may be glacial or other ice which has been buried under a protective earth cover.
			Ice coatings on particles	Vc				
			Random or irregularly oriented ice formations	Vr				
			Stratified or distinctly oriented ice formations	Vs				
Part III: Description of Substantial Ice Strata	Ice greater than 25 mm thick	ICE	Ice with soil inclusions	ICE + soil type	Designate material as ice and use descriptive terms as follows, usually one item from each group, as applicable: Hardness: hard, soft (of mass, not of individual crystals); Structure: clear, cloudy, porous, candled, granular, stratified; Color: colorless, gray, blue; Admixtures: contains few thin silt inclusions	Same as part II above, as applicable, with special emphasis on ice crystal structure		
			Ice without soil inclusions	ICE				

DEFINITIONS

Ice coatings on particles are discernible layers of ice found on or below the larger soil particles in a frozen soil mass. They are sometimes associated with hoarfrost crystals, which have grown into voids produced by the freezing action.

Ice crystal is a very small individual ice particle visible in the face of a soil mass. Crystals may be present alone or in combination with other ice formations.

Clear ice is transparent and contains only a moderate number of air bubbles.

Cloudy ice is translucent but essentially sound and nonpervious.

Porous ice contains numerous voids, usually interconnected and resulting (1) from melting at air bubbles or along crystal interfaces from presence of salt or other materials in the water or (2) from the freezing of saturated snow. Though porous, the mass retains its structural unity.

Candled ice is ice which has rotted or otherwise formed into long columnar crystals, very loosely bonded together.

Ice lenses are lenticular ice formations in soil occurring essentially parallel to each other, generally normal to the direction of heat loss and commonly in repeated layers.

Ice segregation is the growth of ice as distinct lenses, layers, veins, and masses in soils, commonly but not always oriented normal to direction of heat loss.

Well bonded signifies that the soil particles are strongly held together by the ice and that the frozen soil possesses relatively high resistance to chipping or breaking.

Poorly bonded signifies that the soil particles are weakly held together by the ice and that the frozen soil consequently has poor resistance to chipping or breaking.

Friable denotes a condition in which material is easily broken up under light to moderate pressure.

Thaw-stable frozen soils do not, on thawing, show loss of strength below normal long-time thawed values or produce detrimental settlement.

Thaw-unstable frozen soils show, on thawing, significant loss of strength below normal long-time thawed values and/or significant settlement, as a direct result of the melting of the excess ice in the soil.

NOTES

1. When rock is encountered, standard rock-classification terminology should be used.
2. Frozen soils in the N group may, on close examination, indicate presence of ice within the voids of the material by crystalline reflections or by a sheen on fractured or trimmed surfaces. However, the impression to the unaided eye is that none of the frozen water occupies space in excess of the original voids in the soil. The opposite is true of frozen soils in the V group.
3. When visual methods may be inadequate, a simple field test to aid evaluation of volume of excess ice can be made by placing some frozen soil in a small jar, allowing it to melt and observing the quantity of supernatant water as a percent of total volume.
4. Where special forms of ice, such as hoarfrost, can be distinguished, more explicit description should be given.
5. The observer should be careful to avoid being misled by surface scratches or frost coating on the ice.
6. The letter symbols shown are to be affixed to the Unified Soil Classification letter designations or may be used in conjunction with graphic symbols, in exploration logs or geologic profiles. For example, a lean clay with essentially horizontal ice lenses:

CL- / Vs   or   Vs

The descriptive name of the frozen soil type and a complete description of the frozen material are the fundamental elements of this classification scheme. Additional descriptive data should be added where necessary. The letter symbols are secondary and are intended only for convenience in preparing graphical presentations. Since it is frequently impractical to describe ice formations in frozen soils by words alone, sketches and photographs should be used where appropriate to supplement descriptions.

FIGURE 3-21   DESCRIPTION AND CLASSIFICATION OF FROZEN SOILS (developed from Linell and Kaplar, 1963)

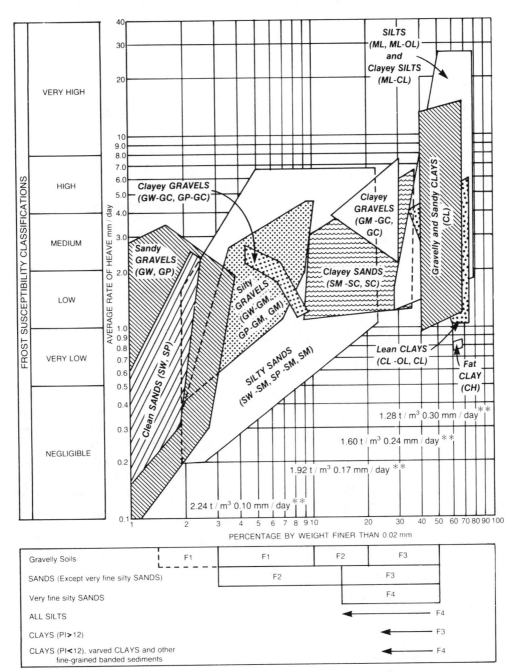

FIGURE 3-22   RANGE IN THE DEGREE OF FROST SUSCEPTIBILITY OF SOILS
(Kaplar, 1974)

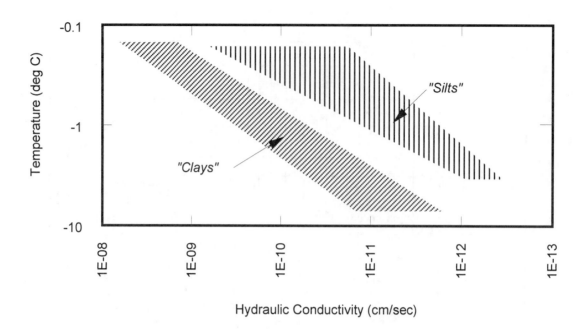

FIGURE 3-23   CHARACTERISTIC HYDRAULIC CONDUCTIVITY OF FROZEN
FINE-GRAINED SOILS (after Nixon, 1991)

term strengths could indicate considerably higher loads than the soil can actually handle in the long term. Loads applied over extended periods create additional stresses between the soil grains and ice in the soil. These stresses can cause pressure melting of the ice and plastic deformation or creep of the soil. Foundation design must take into account the deformation and resulting settlement caused by creep.

Elastic Properties.   The elastic properties of frozen soils can be determined in the laboratory. Several publications provide data for various soils (Kaplar, 1963; Shuster, 1970; Taylor, 1967; Tsytovich, 1975; Johnson, 1981). Frozen soils are more brittle, stiffer (higher modulus of elasticity), and have less damping capacity than unfrozen soils, when it comes to the sort of dynamic loading that would be caused by an earthquake. These properties vary widely with temperature, ice content, soil type, and load characteristics. Even rock is stiffer when frozen. Stress wave velocities are higher in frozen soils. Young's modulus of elasticity (E) may be several times greater for frozen soils than for unfrozen soils. All earthquake design techniques for thawed soils are applicable to frozen soils but the effects may be more pronounced because of the brittleness, greater stiffness, and rock-like behavior.

Strength.   The compressive, shear, and tensile strengths are of interest in design. In thawed soils, strength is primarily a function of the cohesion and friction between the soil particles. Further, the strength in thawed soils is generally not time or load-rate dependant (to a point). However, in frozen soils, the presence of ice will greatly affect the strength-deformation behavior. Ice is a non-Newtonian material, exhibiting strong time-dependent plastic deformation, For a given loading rate and temperature, the strength deformation behavior in frozen soils depends on the relative volumes of ice and soil particles (Goughnour and Andersland, 1968).

In general, the strengths of frozen soils will vary directly with loading rate, and inversely with temperature. In other words, the strengths will increase at faster loading rates, and/or with decreasing temperatures. Reported laboratory data indicate the strength of frozen soils does not approach elastic behavior except at very rapid strain rates (>1.0 sec^{-1}) and very low temperatures ($<-40°C$), neither of which could be expected under field conditions.

It is also very important to note that because of the time dependant behavior, the "long-term" strengths of frozen soils to be used for design will generally be much smaller than the ultimate or short-term strength. Laboratory strength tests are rarely extended long enough to measure anything other than short-term behavior. Additionally, similar to thawed soils, some magnitude of strain is required to mobilize the strength in frozen materials. The design strength value selected must also be based on an acceptable magnitude of strain.

The strength of frozen soils is a combination of the time dependant ice matrix strength, and the time independent strength associated with friction between the soil particles. The shear strength of frozen soils can be modelled as:

$$\tau = c_f + \sigma_n \tan\phi_f$$

where:

c_f = the frozen "cohesion", dominated by the presence of ice and generally taken as the unconfined compressive strength at a strain rate and temperature similar to field conditions;

σ_n = the normal stress on the shear plane; and,

ϕ_f = the angle of internal friction for the soil structure.

For most civil engineering designs, it is practical to characterize frozen soils into two categories, ice-rich/fine-grained and ice-poor, for assessing strength behavior. Ice-rich/fine-grained soils are defined as having a bulk density less than about 1,000 kg/m^3 (Phukan, 1985) to 1,700 kg/m^3 (Weaver and Morgenstern, 1981). The strength behavior of ice-rich soils is dominated by the interstitial ice, and is generally modelled as purely cohesive or nonfrictional. McRoberts (1982) summarized long-term shear strengths for ice-rich soils ranging from about 60 to 175 kPa at -1°C, up to 100 to 300 kPa at -4°C.

Ice-poor soils generally include coarse-grained and dense fine-grained materials. The shear strength of ice-poor soils includes a long-term frictional component, and a short-term time dependent cohesive contribution. The friction angles of some ice-poor frozen soils have been reported to be less than the same soil thawed, under rapid short-term strain rates (Andersland and Anderson, 1978). However, for bearing capacity design, it is generally appropriate to model the long-term shear strength of ice-poor soils using only the friction angle of a similar thawed material.

Summaries of frozen soil strengths versus the interactive relationships with ice content, temperature, strain rate, and confining pressure are provided in Andersland and Anderson (1978), Kaplar (1974), and Tsytovich (1975).

Creep. Creep, as referred to herein, is a time-dependent deformation behavior of great importance when designing foundations in frozen soils. Creep is defined as slow, progressive movement of a loaded frozen soil without thawing or consolidation (no total volume change). Movement is in the shearing mode; the soil "flows" rather than "settles". Creep is caused by pressure-melting of ice in the soil at points of soil grain contact, migration of unfrozen water to regions of lower stress, breakdown and plastic deformation of the pore ice, and readjustment in particle arrangement. Frozen soils exhibit creep under long-term loads as low as five to ten percent of their rupture strength. Creep must be reduced to that value which the structure can withstand throughout its existence. Moisture or ice content influences the creep rate for frozen soils. In fact, for ice-rich soils low in concentration of solids, the long-term strength can approach zero, if no movement can be tolerated. The creep rate is also greatly influenced by the frozen soil temperature, since the soil temperature influences the amount of unfrozen water in the soil.

Under a constant load, creep may ideally be divided into three periods (Figure 3-24), more thoroughly discussed in Andersland and Anderson (1978) and Ladanyi (1972):

- The initial period (primary creep) occurs immediately after the application of load and results in a high, but decreasing or attenuating creep rate.

- The second period (secondary creep) is long in comparison with the period of primary creep, and the cumulative strains developed with time can be large. The strain rate is roughly constant in this period, and the strain vs time curve can be approximated by a straight line. Once creep has entered the secondary stage, failure will eventually occur (unless the curve is horizontal) either through excessive deformation or tertiary creep (the third period).

- Tertiary creep can be very short and ends with failure of the soil. This period is characterized by accelerating strain rates. While not ultimate, the soil is typically considered to be at failure at the time tertiary creep begins.

The ultimate or instantaneous bearing capacity or instantaneous shear strength should not be used for design purposes, unless the design calls for a very short term of loading, because creep deforma-

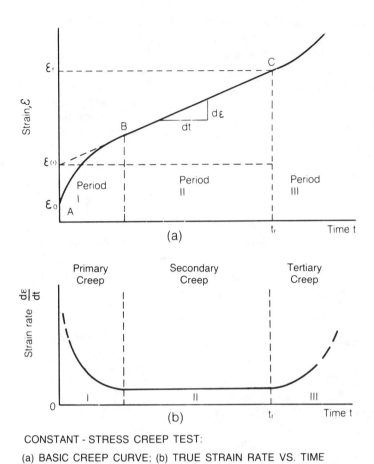

CONSTANT - STRESS CREEP TEST:

(a) BASIC CREEP CURVE; (b) TRUE STRAIN RATE VS. TIME

FIGURE 3-24   *CONSTANT STRESS CREEP TEST*
(Andersland and Anderson, 1978)

tion will undoubtedly occur. Ice-poor soils generally exhibit only primary creep behavior when tested at less than ultimate loads. In ice-rich soils, long-term deformations under a constant load are typically modelled only assuming secondary creep, as the primary period contribution is small relative to time. Long-term designs in ice-rich soils are usually based on loads for which the magnitude of creep over the facility life would be negligible.

Adfreeze Strengths.   A very important soil strength that must be considered when designing foundations is the adfreeze strength or bond between the soil and the surface of a pile, post, foundation, etc. Adfreeze bonds determine a part of the load-carrying capacity of most foundations, especially piling. Creep occurs in the adfreeze bond at the contact surface between the pile and the surrounding soil. The adfreeze strength is primarily temperature-de-

pendent, but it also varies with soil type, moisture content, and surface characteristics of the piling or foundation. Adfreeze bonds are also important when assessing "frost jacking" (Section 3.3.8).

3.3.7 Organic Soils. Muskeg is a term used for organic terrain which comprises a surface layer of living plants underlain by peat or mineral soil. The surface vegetation is composed of mosses, lichens, sedges, grasses, or a combination of all these, with or without tree and shrub growth. The next layer is peat, usually saturated with water. Peat is a highly compressible organic material. Peat is further defined (MacFarlane, 1969) as an accumulation of fragmented organic material derived from past vegetation, but now chemically changed and fossilized. The peat is underlain by the native mineral soil. Peat is by far the most critical element of muskeg terrain. It is primarily the peat layer that determines design and affects the methods of construction. Muskeg or peat bog areas normally have poor surface drainage and a high water table. Peat is also a very effective insulator, especially when dry. For this reason isolated masses of permafrost have been found below elevated, drier, peat bog areas, even in subarctic areas such as Anchorage, Alaska and Fort McMurray, Alberta.

For most civil engineering purposes, peat is characterized based on the structure (fibrous or amorphous), and organic (or ash) content tested by incineration in the laboratory. An in-depth review of the peat classification and engineering properties is presented in MacFarlane (1969) and Burwash and Wiesner (1984).

The permeability of peat ranges from about 1×10^{-7} to 4×10^{-3} mm/s, with the horizontal permeability larger than vertical permeability. The void ratio varies from 2 (dense) to 25 (fibrous). Compression greatly reduces the permeability. The specific gravity varies between 1.1 and 2.68; 1.5 is considered the average. If the specific gravity is greater than 2, the peat contains granular or mineral material. The density varies from 80 to 320 kg/m³ (dry) to 320 to 1,200 kg/m³ (in the natural state). Air drying can shrink peat by 10 to 50 percent of the original volume. Peat is normally very acid because of the carbon dioxide and humic acid given off during decay of the vegetation. In some peats, pH values as low as 3.5 have been measured. Corrosion can be a definite problem for metals (such as pipelines) buried in peat.

As the moisture content decreases (usually accomplished by dewatering or drainage), the void ratio decreases and permeability decreases; it shrinks and becomes more dense. The volume, however, cannot be reduced below a limiting value by drainage alone. Peat can be stabilized to some extent by drainage, but this is usually a slow and expensive process. In muskeg areas, a common solution to construction problems associated with soft, organic deposits is to construct in the winter, when the surface is frozen and can support transport and construction equipment (MacFarlane, 1969).

The shear strength of unfrozen peat can be relatively high, but the material is also extremely compressive. Shallow, near-vertical temporary cuts can be made in very fibrous peats for utility installations. The strength of peat increases dramatically with consolidation. The shear strength of undisturbed peat usually varies between 2 and 30 kPa and varies inversely with the water content (Burwash and Wiesner, 1984). The compressive strength of frozen peat at -9.5°C has been measured at 3.9 MPa at a strain rate of 1 percent/min, and 4.6 MPa with a strain rate of 2 percent/min. MacFarlane (1968) presents what little information has been published in English regarding the strength of frozen peat.

The bearing capacity (P_{ult}) and maximum thickness of fill (H_{ult}) placed over unfrozen peat can generally be estimated as:

$$P_{ult} = (5 \text{ to } 6) \cdot \tau$$

$$H_{ult} = (5.5 \text{ to } 6) \cdot \tau/\gamma$$

where:

τ = the shear strength of peat (often taken as the field vane shear test value)

γ = the unit weight of the fill.

Muskeg areas are often preconsolidated with a load in excess of the design load being applied. Under this increased load, settlement is allowed to continue to the value predicted for the design load and then the surcharge is removed. Figure 3-25 is an example of the analysis for the surcharging method of consolidation. In addition to compression, lateral shear failures can occur under the fill if care is not observed during placement.

3.3.8 Frost Heave. Frost heave and frost jacking are probably the most troublesome consequences of soil freezing. Frost heaving can take place in all freezing soils; thus it occurs in permafrost or nonpermafrost areas. If heave occurred uniformly over a large area it might not be as serious a problem. However, heave is generally not uniform and varies considerably over distances of only a few metres. Frost heave is made up of two components. The first and major component is due to the migration of moisture to the freezing front, and the associated growth of segregated ice lenses perpendicular to the direction of heat flow. The second component is due to the volumetric expansion associated with the phase change of the interstitial and migrating water as it freezes.

Generally, three "ingredients" – winter, water, "wick" (3Ws) – must be present for frost heave to occur. If any of these ingredients are removed or avoided, the magnitude of heaving can be effectively reduced:

- winter for freezing temperatures in the soil, (or artificial such as ice rinks, bulk freezers, etc.);

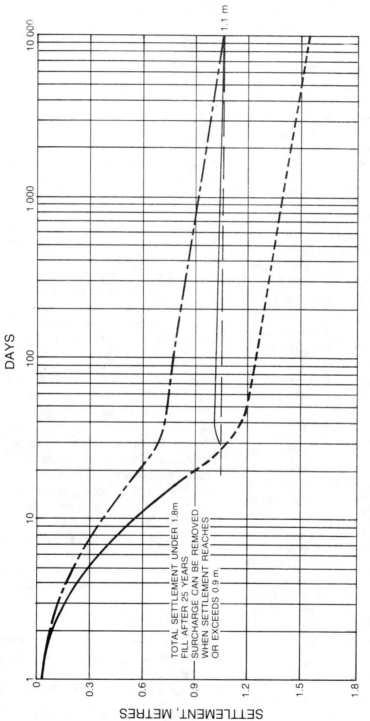

FIGURE 3-25   EXAMPLE OF SURCHARGING MUSKEG FOR CONSOLIDATION (MacFarlane, 1969)

- a source of water, such as the groundwater table – the shallower the water table, the higher the potential for heaving; and

- a frost-susceptible soil (Section 3.3.4).

The mechanisms associated with frost heave are extremely complex. A soil does not necessarily have to be saturated to exhibit some frost heave. The lateral extent of an individual lens may not depend on the uniformity of the soil, water supply, or temperature gradient. The thickness of the lens will vary between soils, and always depends on the balance between moisture supply and heat flow. The larger the heat flow or the faster the freezing front moves downward, the less heaving occurs. Additionally, the magnitude of heave decreases with increasing overburden or confining pressure. One-dimensional laboratory heave tests can be performed on soil samples from a project site. However, the results of such tests can only be used to assess the potential for frost heave rather than for predicting magnitudes of heave, unless the boundary conditions, rate of freeze, moisture availability, and overburden stress are accurately accounted for. Very sophisticated equipment and procedures are required to accurately test all these factors in the laboratory.

Generally speaking, the amount of heave that a given soil undergoes increases with an increasing amount of fines. Heave also seems to become less pronounced in a given soil after the first freeze-thaw cycle. This is probably due to the structural reorganization of the soil grains with the first freeze-thaw cycle. Results of many frost-heave tests are illustrated in Figure 3-22, and in Sherif et al. (1977).

Theories of ice segregation and frost-heave processes in saline soils are reviewed by Chamberlain (1983).

Frost-Heave Models. While the theory of frost heave relative to the general mechanisms occurring at the freeze front are fairly well understood, models to predict the potential and magnitude of heave for a specific set of conditions are quite limited. An ideal frost-heave model would have to

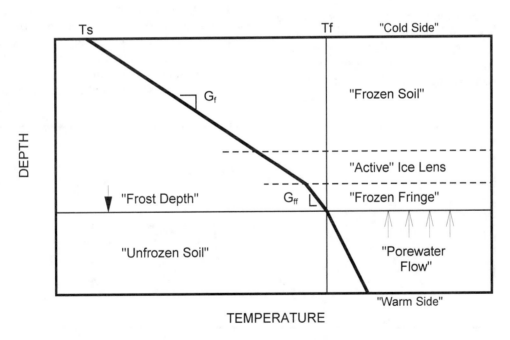

G_f = Temperature gradient in frozen soil

G_{ff} = Temperature gradient in frozen fringe

FIGURE 3-26   SCHEMATIC TEMPERATURE PROFILE AT AN ACTIVE ICE LENS (after Nixon, 1987b)

couple the mass flow of water to the freeze front with conductive and convective heat transfer. Additionally, functions would be required to account for the critical relationships between hydraulic conductivity and ice content with temperature, and increasing overburden pressures as the frost front advances. Figure 3-26 illustrates a simplified model of the thermal conditions, temperature gradients and water movements below and through an advancing freeze front.

Konrad and Morgenstern (1980; 1981) presented a fairly practical frost-heave prediction model. In this model, the heave rate sensitivity to hydraulic conductivity and other poorly defined factors is defined by a "segregation potential" (SP) parameter. The SP parameter is not unique for a soil and cannot be predicted based on other more standard soil index properties. Rather, the SP parameter is more of a constant of proportionality between the rate water moves to the freeze front, and the temperature gradient behind the freeze front. It is determined based on observations from laboratory frost-heave tests; thus is sensitive to the freeze rate and overburden stress conditions tested. Figure 3-27 summarizes the characteristic range of reported SP data of fine-grained soils as a function of overburden pressure. The SP (mm^2/(s•˚C)) can be an effective parameter for characterizing the frost susceptibility of a material (Konrad and Morgenstern, 1983), as it depends

on the composition of the total soil groundwater system and the overburden loads.

Following Konrad and Morgenstern (1982), the magnitude of frost heave (h) can be predicted as a function of time (t) by integrating:

$$dh/dt = 1.09•SP•G_f + 0.09•n•dX/dt$$

where:

SP = segregation potential, mm^2/(s•˚C)

G_f = the temperature gradient through the frozen soil, ˚C/mm

n = the porosity of the soil, reduced to take account of the interstitial water that will not freeze

X = the frost depth, mm.

This model is very sensitive to the SP and temperature gradient functions. The SP parameter must be based on the anticipated field freeze rates and overburden conditions. Referring to Figure 3-26, Nixon (1987b) demonstrated that the temperature gradient in the above equation should actually be that through the frozen fringe, but is often taken as the average gradient through the bottom of the frozen soil. The temperature gradient at the bottom of the frozen soil can be predicted with a fair degree of accuracy, subject to the accuracy of the boundary conditions and soil properties, using analytical or

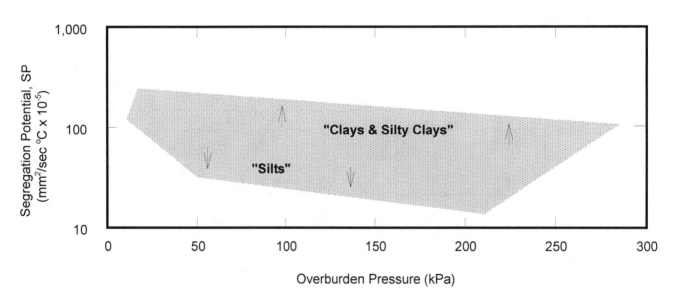

FIGURE 3-27   CHARACTERISTIC SEGREGATION POTENTIAL PARAMETER IN FINE-GRAINED SOILS (after Nixon, 1987b)

computer solutions presented in Section 4 (Thermal Considerations). The above equation, coupled with computer-generated thermal gradients, has been used to reasonably model the frost heave at controlled and instrumented test sites (Nixon, 1982; 1986).

The SP method for evaluating frost heave is semi-empirical and does not explicitly model the relationships between heave rate, temperature gradient and other more fundamental soil properties (Nixon, 1991). Nixon (1991) presented a frost heave theory that predicts discrete ice lens locations and effects of overburden pressure on predicted heave rate. This approach relies on the unfrozen water content and frozen hydraulic conductivity relationships discussed above, as functions of temperature.

Frost-Heave Pressures. Frost heave can impart enormous lifting forces on shallow and deep foundation systems, as well as on buried utilities transitioning between frost-susceptible and nonsusceptible soils. Field and laboratory investigations of frost-heave forces on horizontal and vertical surfaces are reported in Buska and Johnson (1988), Domaschuk (1982), Penner (1969; 1974), Penner and Gold (1971), and Tsytovich (1975).

Based on the above references, the following generalizations have been reported. Adfreeze stresses are affected by the soil type, soil temperature, and rate and magnitude of heave at the surface. Basal heave pressures are also affected by the depth of frost penetration. Peak adfreeze stresses generally occur early in the freeze period when heave rates and soil temperature differentials are greatest. However, maximum uplift forces often occur much later near the time of maximum frost penetration. Adfreeze stresses may be highest against steel, which is a better thermal conductor relative to soil, wood or concrete; in other words the temperatures across the bond would be coldest along buried steel components. However, the surface roughness of the foundation material is also an important factor contributing to the magnitude of adfreeze bond. Adfreeze stresses along piling may be somewhat inversely proportional to the diameter. And, adfreeze stresses in saturated, very cold, coarse-grained soils can be extremely high. As such, saturated coarse-grained soils at the surface over heaving soils can act in effect as a vice, greatly increasing the uplift forces on buried foundation components.

Domaschuk (1982) presents summaries of the maximum measured basal heave and adfreeze stresses reported, and notes that the maximum pressures appear to be affected more by the test method (lab or field, member size, frost penetration rates, etc.) than by soil type. Following Domaschuk (1982), maximum basal heave pressures ranging from 2,000 to 3,000 kPa were measured against steel plates during field tests over highly frost-susceptible soils. Maximum heave/jacking adfreeze stresses up to roughly 2,700 kPa have been measured against small-scale wood piles tested in the laboratory. Maximum heave/jacking adfreeze stresses ranging from roughly 100 to 300 kPa have been measured during larger-scale field tests on a variety of pile materials and concrete block walls.

Frost Jacking. The consequence of foundation components, especially piles and posts, buried in or penetrating heaving soils bonded to the foundation is referred to as "frost jacking". The process is as follows. During the freezing season, the ground heaves, lifting the object because of the adfreeze stresses between the soil and the object. In the spring, when the ground thaws and settles back, the object may not return to its original position or depth because soil and water may have filled the void underneath it or there is insufficient load to overcome soil friction. This process can continue to occur annually. Figure 3-28 illustrates schematically the frost-jacking cycle. Jacking displacements of 450 mm in one season have occurred.

For a pole or post to be jacked, the upward force must be greater than the total downward forces. In nonpermafrost areas, the downward force is provided by friction from the underlying unfrozen soil and the load. In permafrost areas, the downward force is supplied by the soil adfreezing to the length of the pole embedded in the underlying permafrost and the structural dead load. Unless the load is significant and constant, it should be ignored in the prediction of jacking. Adfreeze strengths are very temperature dependent, and when the active layer is applying the maximum upward force (in early winter), the reactive force downward in the permafrost is at a minimum because the warmest temperatures from the previous summer are still affecting those depths. Therefore, knowing the critical temperatures is very important when analyzing frost jacking.

Examples of frost heaving and jacking are numerous in Arctic and subarctic areas and common in the continental United States and southern Canada. The problem of frost jacking along with the formation of frost boils is a major cause of road and airfield deterioration in cold regions. In cold regions, load limits are imposed on most highways during

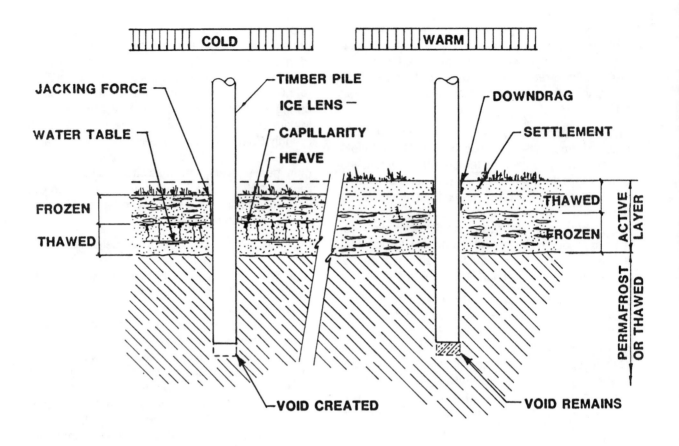

FIGURE 3-28   FROST ACTION AND PILE JACKING (Nottingham and Christopherson, 1983)

spring to minimize the damage caused by the deterioration in subgrade support. Furthermore, the differential heaving of road surfaces can create travel hazards, often indicated by the "bump" and "dip" signs along the highways.

Pile-supported railroad bridges and bridge approaches along the Alaska Railroad have undergone movements that have resulted in the uncoupling of railways cars. An example of jacking of a pile-supported bridge approach is shown in Figure 3-29. A temporary solution is to saw off the piling and relevel

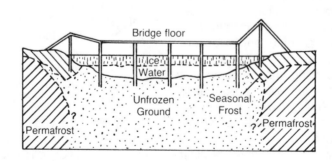

FIGURE 3-29   FROST JACKING OF BRIDGE PILING

*FIGURE 3-30   FROST HEAVE OF POWER PILE AND DIFFERENTIAL
SETTLEMENT IN YAKUTSK, RUSSIA*

the track each spring. After a few years, however, the piling must be replaced.

Power poles are common victims of frost jacking (Figure 3-30). If left alone, the poles slowly jack out of the ground until they lose support and either topple over or hang on the wires.

Caskets have been known to "jack" out of the ground. A good example of this can be found in Barrow, Alaska. After the measles epidemic in the late 1930s, many people were buried in wooden caskets in shallow graves within the active layer because of the difficulty of digging the frozen ground with the tools available. Since that time the caskets have slowly heaved and look like peat mounds in the area of the Esatkuat Lagoon.

There are also numerous examples of differential heave of house and building foundations and floor slabs. Considerable damage can result, because the heaving is seldom uniform.

Methods to mitigate or design foundations to resist frost-heave forces and jacking are presented in Section 3.4.5.

Frost Loads around Buried Utilities.   The concept of frost loads on buried utilities in cold-region municipalities has been identified for some time (Smith, 1976; Fielding and Cohen, 1988). Problems have been encountered in some arctic communities in a similar, yet more severe manner, where very high freezeback pressures in the active layer have caused collapse of buried HDPE sewer pipes (Hanna and Cucheran, 1991; 1994). More details on these problems and suggested means to mitigate the situation are provided in Section 9.

3.3.9   Thaw Deformation.   Thaw deformation refers to the total movement associated with thawing frozen soils. The respective deformation components of a thawing soil can include the phase change from thawing segregated and interstitial ice, the collapse of the soil matrix once the ice bonding is lost, elastic compressibility of the soil matrix under an overburden pressure, and the classical consolidation of the soil matrix associated with the drainage of pore water under external stress. The total thaw deformation is referred to herein as thaw settlement. The component associated with the drainage of pore water versus rate of thaw can be a critical factor,

FIGURE 3-31   EXAMPLE OF THAW SETTLEMENT IN YAKUTSK, RUSSIA

especially in fine-grained soils, and is referred herein alone as the thaw consolidation.

Thaw deformations can be detrimental to foundations (Figure 3-31) and pavement performance. Thaw consolidation can have adverse consequences when the shear resistance of the soil is at issue, such as for slope stability. If the rate of water generation during thawing exceeds the rate at which the water can drain from the soil, excess pore pressures will develop, with a corresponding reduction in soil matrix effective stresses. This condition can lead to the failure of foundations or slopes, discussed further in Section 3.5. Additionally, the bearing capacity of the thawing, poorly draining materials above a frozen mass can drop significantly, and a "frost boil" results. Some of the potholes found in many road surfaces during late spring are visible results of this.

Total Thaw Settlement.   The total thaw settlement is most often tested in the field or laboratory by simply allowing a confined frozen sample to thaw one-dimensionally under a fixed surcharge load. Subsequent load increments are usually applied to include the effective stress of interest. The results are typi-

cally reported as a total thaw strain, and a compressibility for the applied effective stress range.

The results of many laboratory thaw-settlement tests on undisturbed, native frozen soil samples from Alaska and western Canada are reported by Hanna et al. (1983) and Nelson et al. (1983). The results of laboratory thaw-settlement tests on samples of compacted frozen fill materials are reported by Kinney and Troost (1984), Mahmood et al. (1984), and Scher (1982). In general, the data from all of these references, summarized on Figure 3-32, indicate a direct correlation between the thaw strain and frozen dry density at the time of thawing. In turn, the frozen dry density is a direct function of the material, saturation in undisturbed soils, and in disturbed compacted fills, total moisture content and compactive effort/energy (see Section 3.6.2).

Thaw settlement prediction can be based on the density variation upon thawing as determined by simple laboratory tests. A more thorough treatment of the prediction of thaw settlements can be found in Andersland and Anderson (1978), Crory (1973), Hanna et al. (1990), Scher (1982), and Tsytovich (1975).

Thaw Consolidation. The theory and modelling of the thaw-consolidation condition are presented for single and multiple layered soils by Morgenstern and Nixon (1971) and Nixon (1973). This theory combined a model of the heat conduction problem, defining the movement of the thaw plane, with the conventional relationships of porewater drainage from a thawed material under stress. This theory demonstrates the potential for generating excess pore pressures when the rate of thaw exceeds the rate of drainage, which can have adverse affects on bearing capacity and slope stability. Morgenstern and Nixon (1971) define a thaw consolidation ratio, "R", as a measure of the relative rates of generation and drainage of excess pore fluids. Conditions presenting an R parameter greater than unity would suggest danger of generating substantial pore pressures at the thaw plane, and the subsequent possibility of instability (Morgenstern and Nixon, 1971).

Laboratory testing of thaw consolidation requires much more control and apparatus than simple total thaw-settlement tests. Laboratory thaw-consolidation tests must measure the rate of thaw and the rate and volume of pore water ejected from the sample under a given load, and the resultant settlement as the sample is thawed in a controlled one-dimensional setting. Thaw-consolidation test apparatus and procedures are presented in Nixon and Morgenstern (1974).

Usually, permeable, coarse-grained soils consolidate (and the settlement takes place) in direct relationship to the advance of the thaw plane. Fine-grained soils, however, show only a limited amount of consolidation while thawing, and large pore pressures build up because of low permeability. An example of a major arctic project that required the application of thaw consolidation was the design for

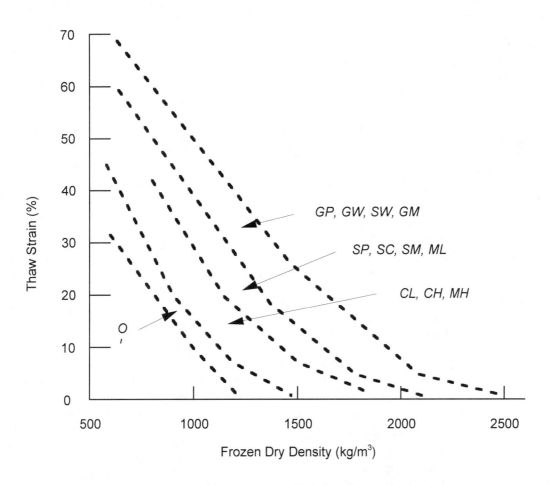

FIGURE 3-32 CHARACTERISTIC THAW STRAIN RELATIONSHIPS IN UNDIS-TURBED FROZEN SOILS (after Hanna et al., 1983; Nelson et al., 1983)

thawing ice-rich slopes for the Norman Wells oil pipeline (Hanna and McRoberts, 1988).

3.4   Foundations

In nonpermafrost cold regions, foundation designs must satisfy both conventional geotechnical requirements for bearing capacity and settlement, as well as ensuring against adverse effects of seasonal frost such as heaving or jacking and reduced strength during thaw. Shallow and deep foundations typically bear on thawed strata below the depth of seasonal frost, and means are provided to isolate or limit frost jacking.

In permafrost regions, foundations must also address the thermal impact from construction and long-term facility operations on the frozen soils, as well as conventional requirements and seasonal frost effects (active layer). The critical thermal impacts could include thaw deformation, reduced strength, and creep. As a bearing material, frozen soils are typically very strong, and bearing capacities are not usually controlled by classical punching or general shear failures. Rather, design capacities are typically based on limiting facility movements (heave, thaw settlement, and/or creep) to tolerable levels. Utility buildings and structures such as treatment plants and storage tanks can be very heavy, and settlement limits may be very small because of interconnecting piping and equipment. On the other hand, small noncritical structures may tolerate greater settlements. In permafrost regions, the use of shallow versus deep foundations is a function of the frozen soil thermal stability and creep characteristics, as well as building/utility loads, function and construction costs. A schematic flow chart of the foundation selection and design process is illustrated on Figure 3-33. The first step is to define the site conditions and facility criteria. Key site conditions include climate, soils, permafrost and ground temperatures. Based on the soils data, primary geotechnical parameters are assigned, such as unit weights, shear strength, and creep and thaw strain parameters. Key facility criteria include loads, desired performance, settlement tolerances, acceptable maintenance levels and design life.

An initial foundation scheme is then selected as a function of the site conditions and facility criteria. Generally, a shallow foundation could be more appropriate for unheated and heated structures over cold, thaw stable permafrost, or for buildings with very high floor loadings such as warehouses. Deep foundations could be more appropriate for structures with low to moderate floor loads elevated over warm, thaw unstable permafrost.

Shallow foundation systems are often set on-grade or elevated over a gravel pad. The gravel pad can include a layer of insulation and/or air ventilation ducts to reduce the thermal disturbance in the underlying permafrost, and to reduce the gravel fill quantities (Figure 3-34). In the Canadian Arctic, shallow "Greenland" foundations are most common, comprised of footings placed directly on the permafrost, usually with insulation to reduce the required depth of burial. Deep foundations include a variety of wood, steel and concrete piles, installed following a number of different methods.

After a schematic foundation scheme is designed, the thermal response of the underlying permafrost must be evaluated. Most strength, creep and frost heave/jacking characteristics of frozen soils are strongly dependent upon the temperature. Therefore, foundations cannot be designed in cold regions without a thorough analysis of the seasonal and subsurface thermal conditions both prior to construction and during long-term facility operation. If the thermal analyses indicate detrimental effects to soil properties or facility performance (such as excessive settlements or heaving), the foundation scheme can be modified or an alternate system could be evaluated. Foundation modifications include earthwork, increasing the thermal resistance under the structure or adding a subgrade cooling system below the facility. Earthwork modifications may include overexcavating undesirable materials and replacing with engineered fill, or prethawing and consolidating unstable materials. The thermal resistance under the building could be improved by elevating the structure to form an air gap, adding insulation and/or an air ventilation system. Subgrade cooling systems include passive and mechanical (active) circulating units designed to remove the heat generated from the building and to further cool the permafrost temperatures.

Once the appropriate foundation scheme is determined, construction and long-term cost estimates should be prepared. Subject to these costs, it may again be appropriate to re-evaluate an alternate foundation scheme. Final design of the most economical, yet functional foundation scheme would then proceed. Obviously the magnitude of the facility in terms of cost and function will control the engineering time available for each of these general design steps.

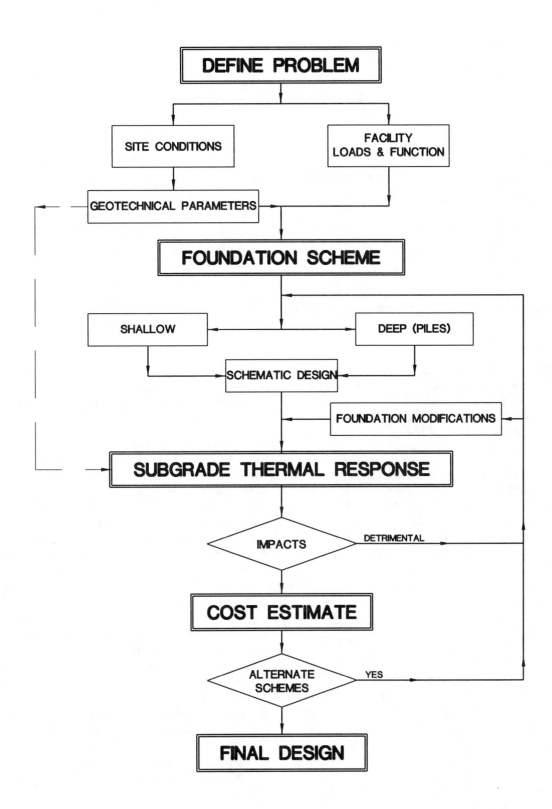

FIGURE 3-33   SCHEMATIC PERMAFROST FOUNDATION EVALUATION PROCESS

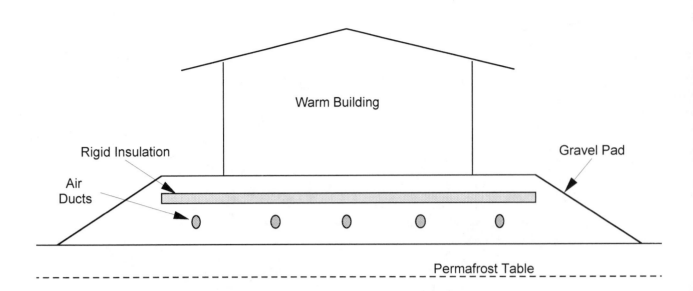

FIGURE 3-34   SCHEMATIC (ON-GRADE) SHALLOW FOUNDATION

The following discussions focus on the current geotechnical design factors and methods for building and pipeline foundations and earth retaining structures. Retherford (1983) provides additional information regarding overhead power line foundations in subarctic and arctic regions.

The reader is referred to the bibliography at the end of this section for in-depth references covering foundation design in cold regions. These noted references include numerous design examples. The important thermal factors and analysis methods and models for foundation design are covered in Section 4 (Thermal Considerations).

Much of the North American Arctic is classified as moderately to highly seismic (as defined by the Canadian National Building Code or the Uniform Building Code). All structures and foundations should be designed for the specific conditions defined in the codes.

3.4.1 Shallow Foundation Systems.   Generally, shallow foundations, such as footings and on-grade slabs, elevated on gravel pads are the least expensive system in permafrost regions, if sufficient non-frost-susceptible fill is available. In permafrost areas, the pad must be designed with an adequate thickness to protect the underlying permafrost from thawing. In a nonpermafrost area, the pad must be designed to prevent significant freezing of any underlying frost-susceptible material to minimize frost

heave. The use of insulation or cooling pipes to reduce the required thickness is discussed later. The pad must, of course, be designed to carry the proposed structure.

Soil borings indicate whether the underlying ground is ice-rich and thus whether significant settlement can result from its thawing. Borings should be taken at all four corners of a small structure as a minimum and at intermediate points for larger structures to determine whether the expected settlements would be uniform or not. For some uses relatively small magnitudes of uniform settlement can be tolerated. If the sampling indicates that thawing of the permafrost will cause tolerable settlement, the gravel pad would be designed as if it were in a temperate climate, simply for the dead and live loads of the structure. However, in permafrost terrain and especially in a discontinuous or sporadic permafrost area, the soil properties (including whether the soil is frozen or not) can vary substantially within a distance of 5 to 10 metres. In this case, the structure should be placed on a gravel pad of sufficient thickness so that the resultant thaw bulb will not extend into the frost-susceptible material. The depth of thaw under a slab placed on a gravel pad can be estimated conservatively using the modified Berggren method (Section 4).

For most heated structures constructed on-grade, the thickness of gravel fill (without insulation) required to fully protect the underlying permafrost from thawing will be uneconomical. Even in the construction of roads and airfields, up to several metres of gravel may be required just to overcome the reduced insulating value of the surface organic mat which becomes compressed during construction or by the gravel fill itself.

Board insulation, air ventilation ducts, or subgrade cooling systems are typically incorporated into or below the fill pad to reduce thermal disturbance and fill quantities. Additionally, elevating the building over the pad to introduce a space for air circulation below the heated building floor is an efficient way of providing the necessary insulating capacity. The height of air space needed depends on such factors as the magnitude of the wind, and orientation and size of the building. Generally for pumphouses or residences an air space of 0.6 to 1.2 m is adequate. The sides of the open space must not be

blocked with snow or other objects in the winter. Any skirting would have to have a open area equivalent to 0.6 to 1.2 m.

Insulation. Board insulation is commonly buried under and around shallow foundations in cold regions. In general, insulated foundations can be used very economically for both heated and unheated structures. The compressive strength of the insulation must be considered when buried below bearing foundation components. Additionally, the effects of moisture absorption (McFadden, 1989) and UV radiation on the thermal and strength properties of the insulation must be considered. Due to its high compressive strength and low water absorption characteristics, extruded polystyrene insulation boards are more commonly used in buried applications. Expanded styrene "bead boards" have also

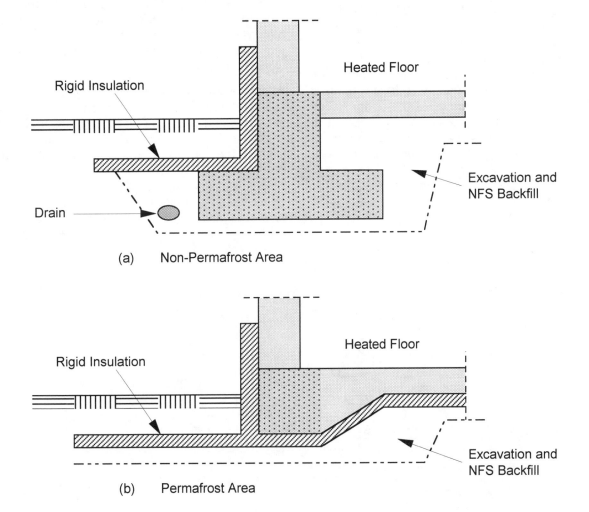

(a)   Non-Permafrost Area

(b)   Permafrost Area

FIGURE 3-35   SCHEMATIC INSULATED SHALLOW FOUNDATIONS

been used, however, this material has inferior insulation values when it absorbs water.

In nonpermafrost areas, board insulation is commonly buried around or under shallow foundations to reduce the depth of freeze for frost-heave protection, reduce building heat losses, and reduce foundation construction costs (Figure 3-35). Robinsky and Bespflug (1973) present procedures and examples for designing insulated foundations for seasonal frost concerns, based on the air freezing index, building temperature, and foundation soil type. Robinsky and Bespflug (1973) concluded that a considerable reduction could be achieved in building heat loss and construction costs if the insulation outside the foundation wall is extended above the ground surface about 0.3 metres.

In permafrost areas, insulation is commonly buried in elevated fill pads to reduce the depth of thaw, and reduce fill quantities (Figures 3-34 and 3-35). A general form analytical solution for designing insulated foundations to prevent thaw below heated, on-grade structures over frozen soils is presented in Nixon (1983). Based on a parametric evaluation of the analytical solution performed by Nixon (1983), it would appear that insulation alone under floors heated between 10 and 20°C (T_s) over saturated fine-grained permafrost may only be economical to prevent thaw below the building when ground temperatures are less than roughly -10°C. For warmer ground conditions, the thickness of insulation required to prevent thaw could well be economically and practically excessive.

Air Ventilation Ducts. For larger structures with high loadings (warehouses, firehalls, storage tanks, etc.) construction of a heavy structural floor above an open air space is not economical and on-grade construction is used. In such cases, it may be most economical to incorporate either a thick layer of insulation or a thinner layer of insulation and an air ventilation system by means of ducts buried within the pad to reduce thermal disturbance. Passive or mechanical subgrade cooling systems can also be placed in the pads to remove heat, and are discussed in Section 3.4.3.

The ventilation ducts are located in the gravel below the insulation. There the ducts remove heat as it travels downward through floor and insulation. Duct foundations are sometimes more expensive than either piling or posts and pads. However, with larger buildings and heavy floor loading it is often necessary to use an on-grade foundation with ducts. A building with a ventilated foundation may experience a slightly greater heat loss than a conventional building, because under the floor the temperature is closer to ambient.

There are three common methods for providing circulation through the ducts in the pad. The "brute force" method consists of installing fans that blow cold air through the ducts in the winter. In the summer the fans have to be turned off and the ducts closed at the ends. This method provides the best control over the movement of heat, but it does require a source of electricity, which can be expensive. A second method is to use the natural "chimney" effect by creating a draft through the ducts. This is achieved by connecting the ducts through a

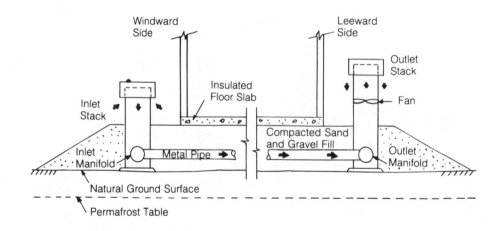

FIGURE 3-36   THULE HANGAR FOUNDATION (Andersland and Anderson, 1978)

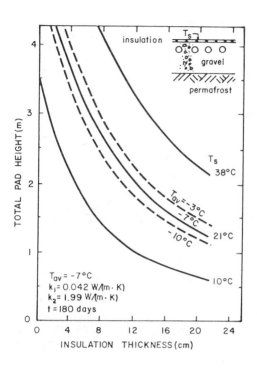

FIGURE 3-37   VENTILATED PAD THICKNESS MODEL
FOR SUMMER PERIOD WITH AIR
DUCTS CLOSED (Nixon, 1978b)

manifold to a stack higher on the leeward side (outlet) than on the windward side (inlet). This approach was used for ventilating the pad of a large hangar in Greenland (Figure 3-36). The third approach is to orient the ducts or collection headers, or both, in the direction of the prevailing wind. This type of foundation was tried in Greenland and Inuvik. The problem with the last two methods is that they rely on the wind, which can be unpredictable. Also, with the older method of open ducts, there was the potential for blockage by snow drifts or other debris.

Air ventilation systems are typically sized based on the building heat load. Linell and Lobacz (1980), Nixon (1978a; 1978b) and Krzewinski and Tart (1985) present analytical solutions for designing the ventilation system for a large building on an insulated gravel pad whereby the thaw bulb is contained within the gravel pad. Figure 3-37 provides a solution for estimating the gravel fill thickness versus insulation thickness for ventilated pads during the summer when the air ducts are closed as a function of the floor temperature (T_s) and mean annual air duct temperature (T_{av}). This figure is based on typi-

cal thermal properties for gravel at 2.5 percent moisture content, and for board insulation. Note the analytical solution used to generate Figure 3-37 neglects the three-dimensional effects of the environment around the building. These effects can be significant in terms of reducing required insulation and pad thicknesses for smaller buildings (Nixon, 1983).

Figure 3-38 illustrates a solution for estimating the ventilation capacity or flow rate under a heated floor as a function of insulation thickness and temperatures T_s and T_{av}. This figure is based on the same thermal properties as for Figure 3-37, and assumes the annual temperature rise through the ventilation ducts is limited to about 3°C. Actual system design should be based on a more rigorous thermal solution, as provided in the above references.

In addition to thermal design, a number of other factors must be considered in designing air ventilation systems. The ducts must be sloped so that any surface water, groundwater, or condensation can drain to a low point for removal in the summer. Condensation or melting of ice crystals in the ducts can result if moist air is allowed to enter the system. This

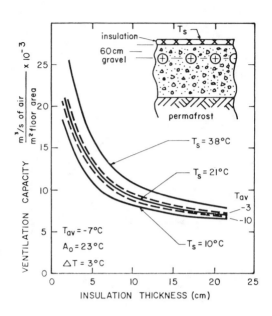

FIGURE 3-38   COOLING DUCT VENTILATION REQUIRE-
MENTS FOR WINTER PERIOD (Nixon, 1978b)

can be a problem if the ducts are kept in operation when the outside temperatures are higher than the duct wall temperatures. The ducts should also be above the surrounding ground or pad level to facilitate maintenance, inspection and air flow. If they are placed below grade they, or at least the manifolds on the ends, should be sized large enough to permit access by maintenance crews for inspection and cleaning.

The use of blowers for air circulation in the duct system has both advantages and disadvantages. They provide more air at a steady rate and allow smaller ducts to be used; however, blowers use more energy and create more maintenance problems. It is also critical that they be turned off and on at the right time.

Snowdrifting can create blockage problems at duct or plenum entrances. Building orientation and location can reduce these problems. It is usually best to orient buildings so that the shortest dimension of the foundation is at right angles to the prevailing wind. More information on snowdrifting and suggested methods of causing the drifts to form at a distance from the building, instead of against it, can be found in Mellor (1965) and Adam and Piotrowski (1980).

If possible, systems should be designed so that the air circulation is automatically turned on in early winter and off in spring. Automation is considered desirable, because operators tend to forget operations that require attention only twice a year. Automatic operation can be controlled by thermostats. An alarm system should be set up to warn of excessive heat penetration beneath the structure.

In-depth, recent case studies of shallow foundations over permafrost incorporating insulation and air ventilation systems are presented in Auld et al. (1978), Odom (1983) and Smith et al. (1991).

Footings. A footing is an enlargement of a column or wall to distribute concentrated loads over sufficient area so that the allowable bearing capacity on the underlying soil will not be exceeded. Linell and Lobacz (1980) provide an in-depth discussion of possible types of footings that should be used under various conditions. For small buildings the most common type of "footing" foundation is "posts and pads" on gravel (Figure 3-39). This type of foundation is commonly used for residences in remote areas. Some movement can be expected, but not enough to damage the structure. Many times wedges are installed under the post that can be driven in to correct for differential settlement. In other cases, connections between the post and floor joists have been slotted to allow the use of jacks and spacers or shims to be inserted. The clear air space between the building and the gravel pad should be large enough to allow free air circulation under the building.

In Canada today, many small buildings are constructed on pile foundations. However, for those with surface footings on a gravel pad, it is most common to place the gravel pad directly on the natural surface.

Footing designs for larger buildings depend on the load that must be transmitted to the soil. The footings should extend below the level of deepest thaw or freeze penetration for the life of the structure. It is important that heat lost through the floor of the building is removed by providing for air circulation. Access to the air space must be provided for foundation adjustment (jacking or shimming) and for inspection and repair of utilities. If this is impossible, ventilated pads or heat extractors should be considered. The effect of shading of the building assists in reducing the surface thawing index, but movement of air must be assured. The south and west faces of buildings are subjected to considerable warming due to direct and reflected sunlight. Structures should

be cantilevered out over the outside piles or footings, if possible, to provide some shading which allows an n value of 1.0 to be used for the thaw season and possibly permits the permafrost to build up, instead of degrade, under the structure. Alternatively, some insulation buried around the perimeter could be used to reduce the ground warming.

The thaw depth under a structure elevated on piles or footings and incorporating a free air circulation space can be computed using the Neumann, Stefan, or modified Berggren equations. An n factor of 1.0 should be used to determine the surface thawing index of the shaded area under the building. If the circulation is not adequate, the air thawing index for the area may have to be increased. The amount of such adjustment would have to be on the basis of the specifics of the site and structure.

If the base of footings is not placed into the permafrost, the ground between the bottom of the footing and the upper surface of the permafrost should be removed and replaced with engineered fill (Figure 3-40). Depending on the soil conditions, it may even be necessary to excavate into the permafrost to reach soil with an adequate bearing capacity. In

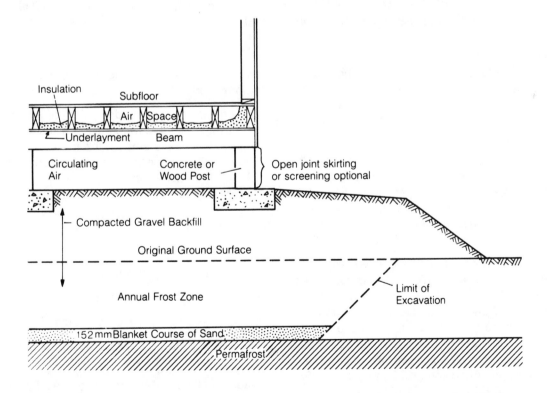

FIGURE 3-39   TYPICAL "POST AND PAD" DESIGN FOR LIGHT STRUCTURE WITH
AIR SPACE AND GRAVEL MAT (Linell and Lobacz, 1980)

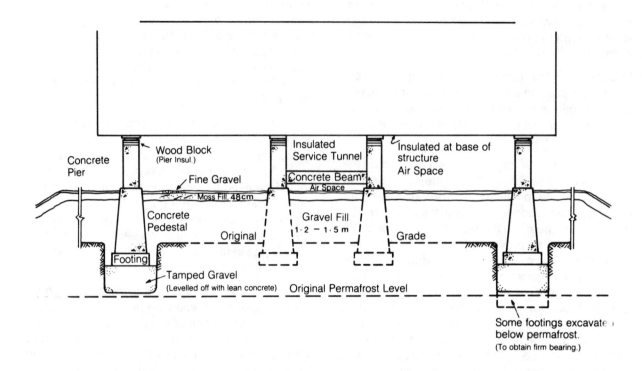

FIGURE 3-40   LARGE TWO-STORY STEEL FRAME BUILDING ON FOOTINGS AND PIERS AT
CHURCHILL, MANITOBA (Subgrade is sand and silt interspersed with gravel and
large boulders.) (Linell and Lobacz, 1980)

doing this, however, extreme care must be taken to prevent thaw in the remaining permafrost near where the footing is to be placed. The adfreeze bond must be broken in the active layer on the riser or post (Section 3.4.5).

In nonpermafrost areas, footings should be founded below the depth of seasonal frost penetration, and should be sized following conventional thawed soils bearing capacity theory in standard engineering texts. However, the detrimental effects of frost heave and jacking must be considered (Section 3.4.5).

In permafrost areas, footings are typically sized based on the allowable settlements associated with long-term creep in the frozen foundation soils, in turn dependent on the ground temperature and thermal disturbance due to construction and facility operation. In-depth design procedures for sizing footings over permafrost for creep are presented in Andersland and Anderson (1978), Ladanyi (1983), McRoberts (1982), Nixon (1978b), and Phukan (1985). Linell and Lobacz (1980) also provide empirical equations derived from laboratory tests on

certain soils, which can be used to estimate creep deformation.

The primary problem in designing for creep settlements under shallow foundations is associated with defining the stress state in the frozen soils, which behave as nonlinear, viscous materials (Nixon 1978b). In general form, the creep settlement rate ($\dot{s}$, mm/yr) below footings can be estimated using:

$$\dot{s} = I \cdot a \cdot B \cdot p^n$$

where:

I   =   the stress distribution influence factor

a   =   the footing radius or half width, mm

B, n   =   secondary creep constants, B, $kPa^{-n} \cdot yr^{-1}$

p   =   applied (increased) design bearing pressure, kPa.

Figure 3-41 illustrates a model of the influence factor under a circular footing elevated on a gravel pad. This solution is based on a linear elastic stress distribution below the load. Based on a review of creep

two points should be considered when using the influence factor model in Figure 3-41. First, while ice is a nonlinear, viscous material, the influence factor under a footing bearing directly over frozen ground (no gravel pad) based on linear elastic stresses reasonably matches the influence factor predicted following a nonlinear expanding cavity model and a nonlinear computer analysis, when the creep exponent (n) is less than about 3. Second, the influence factor under a strip footing predicted by the linear elastic and nonlinear expanding cavity models is on the order of two to three times larger than under a circular footing, again when the creep exponent (n) is less than about 3.

The reader is referred to Ladanyi (1983) for an in-depth presentation of the secondary creep parameters and appropriate influence factor model as a function of footing shape and soil condition.

3.4.2 Piling Foundation Systems. Piles are probably the most successful and accepted method of providing a foundation for buildings in permafrost. Piles transmit the structural loads to depths where the supporting strength of the soil remains relatively stable throughout the life of the structure. They also must isolate the structure from the seasonal heave

and settlement in the active layer. Wood and especially steel piling are commonly used in cold regions; both are commonly installed in pre-augered holes. Steel piles are also typically installed by driving.

Pile design in frozen soils in most cases is based on limiting long-term creep settlements. Creep settlements are a function of the pile-soil adfreeze stress and temperature. In addition to compressive loads, design of all piling in nonpermafrost and permafrost areas must also consider the detrimental seasonal frost effects associated with heave and jacking. Subgrade cooling systems (Section 3.4.3) are often incorporated with or integral to the piles to further improve load capacity and reduce thermal disturbance.

In general, piling should be installed with as little disturbance as possible to the surface of the ground and the temperature regime of the permafrost. The correct design of a piling foundation depends on predicting accurately the depth of the active layer and long-term ground temperatures, after construction has been completed. After installation, cold air should be allowed to circulate freely around the piling. A ventilation opening of 0.6 to 1.2 m is usually adequate for circulation unless the smallest dimen-

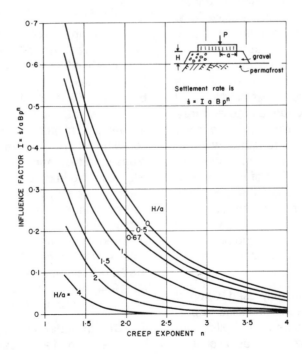

FIGURE 3-41   SETTLEMENT RATE FOR CIRCULAR
FOOTING BASED ON LINEAR ELASTIC
STRESSES (Nixon, 1978b)

sion of the building is greater than 9 m. Time must be allowed after installation for the surrounding soil or slurry to freeze (adhere) to the pile itself.

The following discussions concentrate on piles loaded vertically in compression. Tension-loaded or lateral-loaded piles react differently (see Foriero and Ladanyi, 1991; Neukirchner and Nixon, 1987a; Nixon, 1984), and require more rigorous design and often on-site testing because of the lack of soil-pile behavior information available to the designer.

Types of Piling. Several materials have been used for piles in cold regions; timber and steel pipe or H-piles are the most common. Concrete piles are not usually suited to cold-regions work. Under frost-heaving conditions tensile forces could develop in the pile, which would crack the concrete and expose the reinforcing. Also, cast-in-place concrete piles using normal concrete mixes cannot be formed in permafrost because of the possibility of freezing the concrete before it sets, or because extensive thawing of the permafrost could occur from the heat of hydration of the concrete. Fast-setting grouts containing accelerators and water-reducing agents are used commonly in the Canadian Arctic.

Timber piles are usually less expensive and easier to handle than steel. They are available in lengths up to 21 m and, if they are treated in the active layer portion and frozen into saturated permafrost, should last indefinitely. Also they are not subject to corrosion and have a lower thermal conductivity than steel. It is recommended that the upper end be painted if pressure treatment is impossible. Note that preservative treatments may reduce the adfreeze bond strength, so should be avoided on the lower length of pile if possible. The strength of wood at low temperatures is influenced by its moisture content. Figure 3-42 shows the effect of various temperatures and moisture contents on the strength of wood. Unlike wood at warmer temperatures, wood at cold temperatures fails suddenly and violently without the usual warning cracks. Wood piles generally cannot be driven into permafrost by hammering.

Both pipe or H-piles are commonly used. Capped (at the bottom) pipe piles can be installed in pre-drilled holes. Open-ended pipe and H-piles can sometimes be driven into frozen silts as discussed below. An advantage of driven piles over drilled and slurried piles is that the freezeback, and thus loading, is essentially immediate, although the ultimate capacity may be less. Steel piles may also have to be protected by a corrosion protection system, if the

surface water and soils are corrosive. Low temperature grade steel should be used in situations where the piles will be subject to impact loadings or fatigue cycles at temperatures as low as -45°C.

Rooney et al. (1977) found that H-piles without web reinforcement are susceptible to deformation during driving into dense soils. Drive shoes to maintain tip shape on pipe and H-piles should not be larger than the outside diameter of the pile itself. Otherwise, the skin friction adfreeze bond strengths can be significantly reduced because of the void created. For H-piles installed by driving, the web can be reinforced with angle iron (Figure 3-43). This modification also provides an annulus for installing a temperature monitoring thermistor string (during freezeback or to monitor facility thermal disturbance), or a thermosyphon at a later date if warranted.

In many remote communities in the Canadian Arctic the pile installation equipment is often limited to an Air Trak drilling rig, which is generally more practical because of the relatively shallow igneous bedrock. Steel piles are most typically 114 and 141 mm in diameter. In the past the piles have typically been water pipe, however, the lacquered finish is detrimental to a good adfreeze bond. Nowadays, structural (HSS) steel is more commonly specified.

The axial capacity of pipe piles in frozen ground can be improved by adding lugs (Andersland and Alwahhab, 1983), rings or helices (Figure 3-44) (Domaschuk et al., 1994; Long, 1973) or expanded bases (Domaschuk, 1984) to the pile through the load transfer section. Piles with such modifications are installed in predrilled holes and backfilled with a soil-water slurry. These modifications effectively increase the side and end-bearing areas of the pipe pile, resulting in greater load capacity. Additionally, the side capacity becomes more a function of the soil shearing at the outside of the modifications, versus adfreeze along the pile skin. Long (1973) reported shear strengths through frozen soils at the outside of the rings at least three times greater than the adfreeze strength along the pile. Domaschuk et al. (1994) discuss the load transfer along pipe pile modified with rings or helices.

Pile Design Axial Capacity. The capacity of piles in nonpermafrost areas can be calculated using conventional thawed-ground solutions found in standard engineering texts. In permafrost areas, the capacity of the piling is most often controlled by limiting either the long-term adfreeze strength or creep displacements (settlement) along the pile-soil interface.

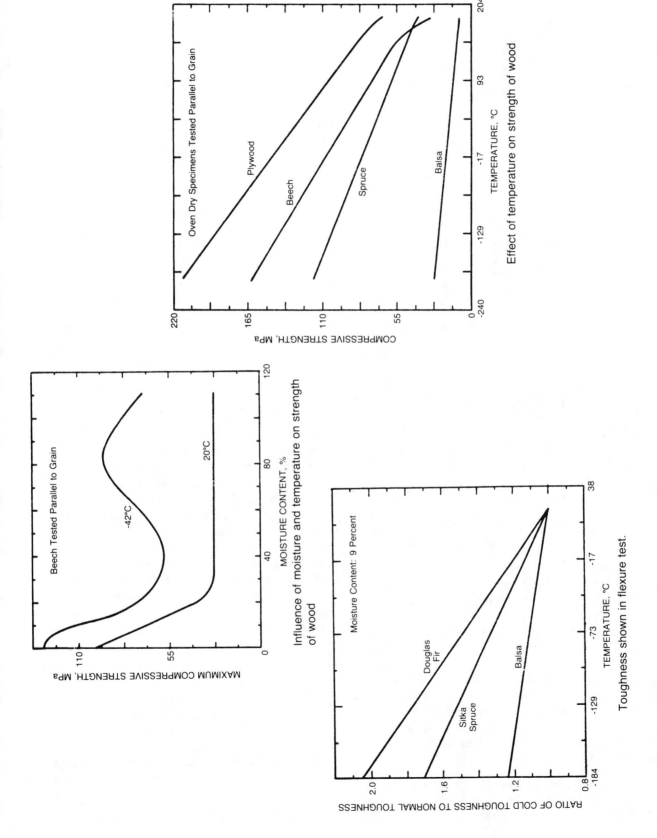

FIGURE 3-42   PROPERTIES OF WOOD (Linell and Lobacz, 1980)

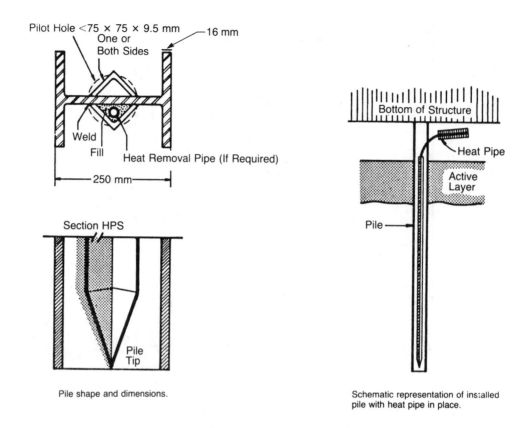

Pile shape and dimensions.

Schematic representation of installed pile with heat pipe in place.

FIGURE 3-43   H-PILE REINFORCEMENT (Rooney et al., 1977)

The creep rate is a function of the pile material and size, soil type, and temperature. In general, the capacity of piles in frozen soils, especially ice-rich materials, can be improved by either increasing embedment, increasing pile diameter, or lowering the ground temperature such as by using a thermopile or thermosyphon (Long, 1963; Nixon, 1978b).

The basic design approach includes first defining the tolerable pile displacements as well as the soil conditions and warmest average ground temperature profile at the site. With this information, appropriate allowable adfreeze stresses can be selected and the allowable pile capacity versus embedment length determined. In addition to the compressive loads from the structure, the piles must also be designed to withstand downdrag from settlements due to thaw or consolidation, as well as provide sufficient anchorage and tensile strength to prevent upward displacement due to frost heave in the active layer.

Piles in more ice-rich materials are often designed based only on creep-limited shear stress on the side of the pile. The end-bearing capacity in ice-rich soils is usually ignored in piling design in permafrost, and can be relatively insignificant in ice-poor soils. For most pile installations in permafrost, the fraction of the pile load carried in end bearing is only about five to eight percent (Nixon and McRoberts, 1976; Weaver and Morgenstern, 1981). Additionally, full end-bearing and full skin friction may not be mobilized simultaneously due to different shear-strain characteristics and should be neglected (Weaver and Morgenstern, 1981). However, end-bearing capacity can be recognized when the tip diameter is large and is in contact with bedrock, hard clay, dense sand or gravel (see below).

The ability of a pile to use the shear strength of the soil through the use of rings, lugs (Andersland and Alwahhab, 1983), helical fins, bottom plates, or in-place corrugations can greatly increase the allowable load. Shear strengths of a minimum of three times the adfreeze strength were reported by Long (1973). The shear planes, where the strength is computed, is the outer edge of the rings or protrusions.

General procedures and examples for designing piles in all types of permafrost soils are provided in

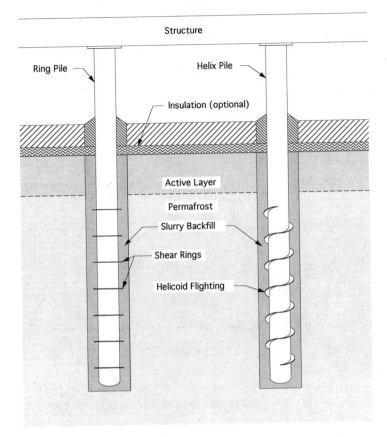

FIGURE 3-44   HELICAL FINS OR RINGS ON PIPE
PILES (Arctic Foundations, Inc.)

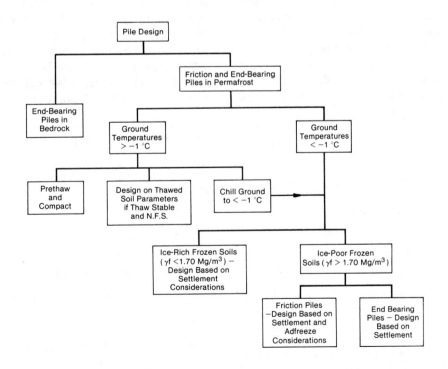

FIGURE 3-45   PROPOSED PILE DESIGN PROCEDURE (Weaver
and Morgenstern, 1981)

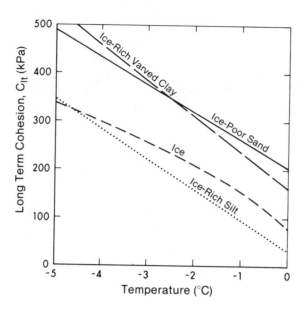

FIGURE 3-46   LONG-TERM COHESION OF FROZEN
SOILS (Weaver and Morgenstern, 1981)

Crory (1963; 1982), Nixon and McRoberts (1976), Nottingham and Christopherson (1983), Weaver and Morgenstern (1981), Neukirchner (1985) and Nixon (1988). Figure 3-45 illustrates a schematic flow chart for designing piles in permafrost as a function of bearing material, ground temperature, and soil ice conditions. In general, piles should be designed based on the lesser of long-term adfreeze strength, or creep settlement limits. The long-term adfreeze strength (t_{lt}) can be estimated as:

$$t_{lt} = m \cdot c_{lt}$$

where:

m = the pile material roughness coefficient:

   = 0.6 for steel and concrete

   = 0.7 for uncreosoted wood

   = 1.0 for corrugated steel pipe (Weaver and Morgenstern, 1981)

c_{lt} = the long-term cohesion, kPa, of the frozen soil at the design ground temperature (Figure 3-46).

The creep settlements of piles in frozen soils is a function of the time, pile diameter, and soil type and temperature. Weaver and Morgenstern (1981) developed a series of equations for predicting allowable shear stress and pile capacity versus creep

settlement. The equations for ice-rich soils were based on an established flow law model for steady-state (secondary) creep in polycrystalline ice. The equations for ice-poor soils were based on an established primary creep law model. The following general equations for predicting the load capacity of friction piles in ice-rich (P_{IR}) and ice-poor (P_{IP}) soils are based on the respective expanded forms presented in Weaver and Morgenstern (1981).

$$P_{IR} = 2\pi a L \left(\frac{u_a(n-1)}{atB3^{(n+1)/2}} \right)^{1/n}$$

$$P_{IP} = 2\pi a L \left(\frac{u_a(c-1)}{at^b 3^{(c+1)/2}} \right)^{1/c} w(\theta+1)^k$$

where:

a   = pile radius, m

L   = pile length in frozen soil, m

u   = pile displacement/settlement, mm/hr

t   = time, hr

n, B   = secondary creep equation constants, B, kPa^{-n}/yr

c, b, w, k = primary creep equation constants, (w, MPa•hr$^{b/c}$/°C^k)

θ = the absolute temperature below the freezing point of water, °C.

The secondary creep constants (for ice-rich soils) and primary creep constants (for ice-poor soils) are functions of the soil type and temperature. Examples of these constants are provided in Weaver and Morgenstern (1981).

The salinity of the soil also greatly affects the creep rates (Biggar et al., 1989; Miller and Johnson, 1990; Nixon, 1988). Figure 3-47 illustrates the predicted creep settlement rate versus applied load and soil salinity.

As noted in Section 3.1.1, the designer must also consider recent changes and trends in average air temperatures and determine what temperature is appropriate for the design life of the project.

If the adfreeze bond is ruptured because of excessive load, the pile essentially turns into a conventional friction pile with a considerably lower load capacity. The adfreeze bond does not mend until the complete pile is put through a thaw-freeze cycle. Long (1973) reports that loading after failure produced strengths of about 46 percent of the initial strength.

When steel pipe piles are driven into permafrost using conventional hammers, the load capacity should be reduced to about 75 percent of that for the same pile slurried in with a sand slurry, if the natural soil is not as strong as the sand slurry. This reduction is not as great when comparing driven H-piles with those slurried in place, because of the larger surface area of the H-pile compared to a pipe pile. The reduction for H-piles has been measured at only about 10 percent.

The permafrost temperature annually varies with depth and time. For pile design, the warmest temperature profile, based on "average" surface temperature fluctuations, along the embedded length are typically assumed for selecting allowable adfreeze stresses. Depending on climate and soil conditions, this procedure could cause considerable overdesign. If the permafrost temperatures vary considerably over the year, the warmest permafrost temperatures could be used to obtain the pile displacements for half of the year with no pile settle-

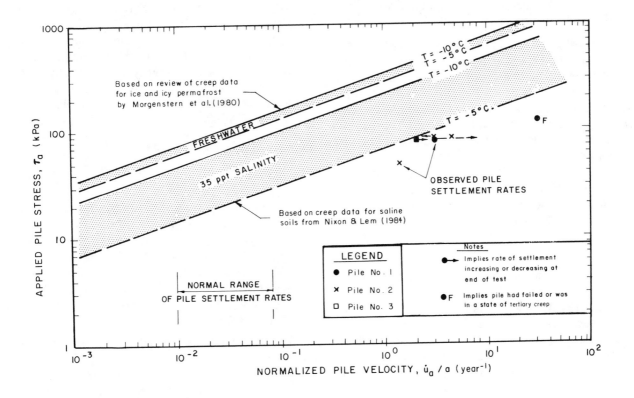

FIGURE 3-47   PREDICTION FOR CREEP SETTLEMENT FOR FRESHWATER AND SALINE SOILS (Nixon, 1988)

ments assumed for the coldest half of the year (Nixon and McRoberts, 1976).

End-Bearing Piles. It is sometimes possible to design piles in the Arctic on the basis of end bearing, which typically can provide much greater pile capacities than adfreeze piles, especially in ice-rich, saline soils. The conditions required for end-bearing piles are usually, sound ice-poor bedrock or competent unfrozen soils at depth beneath a relatively thin surface layer of icy permafrost.

For piles installed into bedrock it is common to drill a socket of at least 2 m depth into the rock. This allows a reasonable confirmation of bedrock, as opposed to large boulders, and provides a good depth for grouting. The grout, which should be an acceptable arctic grade grout, serves two purposes – to distribute the normal load and to provide some resistance to frost-jacking forces.

If the piles are to be founded on underlying unfrozen ground, the basic design should be as for regular unfrozen design principles. If the overlying soil is ice-rich and is allowed to thaw, considerable "downdrag" forces can develop as a result of thaw settlement. The extra load caused by the downdrag will have to be taken into account as a design load on the piles (Hanna et al., 1990).

Installation of Piling. It is best to use the installation method causing the least thermal and physical disturbance to the existing ground. All piles can be installed in predrilled, oversized holes backfilled with a soil/water slurry which must be allowed time to freeze back before loading the pile. This method can produce a lot of thermal disturbance, especially in warm permafrost. Alternatively, steel H-piles can be driven into warm permafrost, or cold permafrost with a small predrilled pilot hole. Steel pipe piles can also be driven into permafrost, but commonly require a pilot hole closer to the pile diameter. Driving piles produces relatively less thermal disturbance.

The driving resistance of piles in permafrost is dependant on a number of factors. Following Davidson et al. (1978) driving resistance increases with increased mean grain size (fine to coarse) and pile cross-sectional area. The driving resistance decreases with an increased ice content (low to high concentration), increased permafrost temperature (cold to warm), increased hammer energy, and increased pilot hole size.

A number of factors should be considered when predrilling pilot holes for driving or oversized holes for the slurry method. The hole should be kept clean of cuttings and slough. Use of drilling fluids should be avoided as they can cause greater thermal disturbance – unless warming is desired to "soften" the soil for easier driving. The hole should also be protected from warm air and surface/melt water.

Probably the most common method of pile installation currently in use is to drill an oversized hole, set the piling, and place a slurry in the annular space. The slurry is then frozen, naturally or mechanically. The soil/water mixture (slurry) is very important in this method. A dense slurry is most efficient at transferring the load from the pile into the surrounding soil and has higher adfreezing strengths than the native soil. The temperature and the quantity of the slurry must be strictly controlled. The heat it gives off in freezing must be absorbed by the surrounding permafrost without causing further degradation. The hole is typically drilled 100 mm to 150 mm greater in diameter than the pile to be placed into it. A minimum clearance of 50 mm around the pile is needed to place and compact the slurry. In Alaska the hole may be overdrilled by about 150 mm in depth so that dry sand can be added to the required depth before pile installation. The slurry temperature should be at or less than 4.5°C. It should have the consistency of 150 mm slump concrete (approximately 30 percent moisture content). During placement it should be vibrated with a small-diameter spud vibrator and continuously rodded. Another procedure is to partially fill the hole with slurry and then vibrate the pile into the slurry while adding more slurry as the pile advances. Saturated, medium-grained sands produce the highest adfreeze strength. The slurry should be made up, if possible, of material with the gradation shown in Table 3-7.

Sands with the optimum gradation may have to be imported. Alternatively, the cuttings from the hole may be used, but adjustments in the design adfreeze bond should be made according to its makeup. Clays or organics should not be used. Gravel also should not be used, except possibly in the active layer to reduce the frost-jacking potential if desired, or as a lower-water content, high-shear slurry for placing helical piles in oversized holes.

Timber or closed-end pipe piles can float out of the hole during slurry placement. One way to prevent this is to place a small amount of slurry in the hole and allow it to freeze back before filling the annular space to ground level. A load of some type could also be placed to hold the piling down until freezeback is sufficient.

TABLE 3-7   TYPICAL SAND SLURRY GRADATION (after Sanger, 1969)

Sieve Size (mm)	Percent Passing by Weight
10	100
5	93 to 100
2	70 to 100
0.4	15 to 57
0.074	0 to 17

Freezeback of Piling.   Freezeback of driven piling in frozen ground can be relatively quick, unless a large amount of heat was generated during the driving, or the permafrost is warm. Partial loading of driven piles can often begin soon after installation. Freezeback of slurried piling depends on permafrost temperatures, the properties and amount of slurry used, and the thermal degradation caused by the drilling and installation. It can vary from a few days to over a year. Before a load is put on the pile, freezeback must have taken place to the extent necessary to support the load. Thermistors should be installed on selected piling at several depths to monitor the progress of the freezeback. Insufficient freezeback before loads are applied has caused many failures of pile foundations in arctic areas. Also, slurries containing higher concentrations of fine materials do not freeze completely at 0˚C. This is another advantage of using sand slurries over silt slurries. In warm permafrost (-1˚C) there may not be enough "cold" to freeze back the piling, because the freezing temperature of the slurry could be lower than the permafrost temperature. It would then be necessary, as it would where loading must proceed before natural freezeback can take place, to use a mechanical or forced freezeback method. The higher the moisture content of the slurry, the larger the amount of heat that must be removed. Thus, it is best to keep the moisture content as low as possible, yet maintain adequate workability.

Natural Freezeback.   Natural freezeback relies on the cold reservoir in the permafrost to extract heat from the slurry. Freezeback proceeds from the outside of the hole towards the pile. In winter, freezing can proceed from the pile outward also, if the pile is hollow. The amount of heat that must be removed is governed by the slurry volume, water content, and dry density. Figure 3-48 presents the equation for calculating natural freezeback times. The permafrost temperature variation with depth must be known to predict the freezeback time.

Knowing the temperature profile at different times of the year also allows one to select the time of year for the fastest freezeback. If a temperature profile is unavailable, then at least the maximum expected permafrost temperature throughout the pile depth should be used. Under most conditions, drilling or augering frozen soils is not significantly more difficult, and usually the drill bits clean more easily when the soil is cold. Winter installation also reduces the possibility of surface water entering the holes, and the holes do not slough as easily. Equipment is more mobile during winter over many soil surface conditions. These advantages indicate that late winter is often the best time to install piling.

Figure 3-49 shows the actual heat paths and temperature profiles at different times for the natural freezing of slurry placed at 0˚C. Tobiasson and Johnson (1978) present some actual measured freezeback curves for piling in Barter Island, Alaska, for permafrost temperatures between -3˚C and -9˚C.

Artificial Freezeback.   Artificial refrigeration must be used when permafrost temperatures are relatively high, the amount of heat introduced by the slurry is large, or the pile must be loaded before natural freezeback can take place. This is usually accomplished by circulating a fluid through longitudinal or spiral steel or copper tubing in or near the pile. Thermal piles or heat extractors can be used during the winter. If artificial refrigeration is to be used only to freezeback the slurry, the lines should be filled with oil and sealed after the initial freezeback has been completed. If it is ever necessary to restart the refrigeration or refreeze at a later date, it can be done easily. For a small installation, the piling can be filled

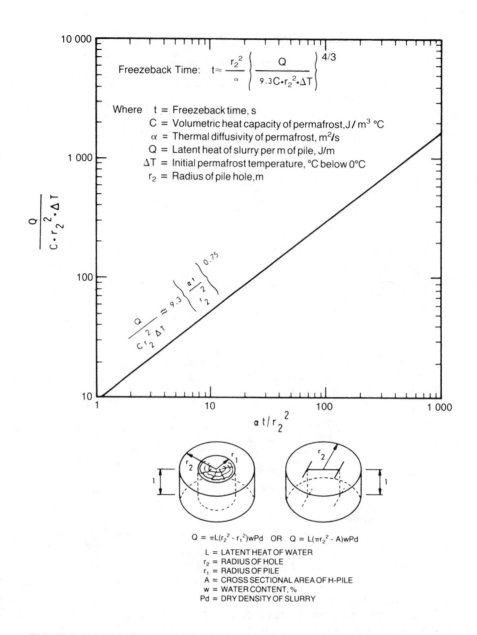

FIGURE 3-48 GENERAL SOLUTION OF PILE SLURRY FREEZEBACK
(Johnston, 1981)

with dry ice to speed freezeback. This could be very expensive where a large number of piles are involved.

Brine or glycol solutions or just cold winter air have been used as the circulating medium for artificial freezeback, but propane or refrigerants of similar characteristics have been found more efficient and economical with most installations.

Alternate Pile Backfill Materials. A pile backfill that was developed and tested for the recent upgrading of the Canadian portion of the North Warning System installations, was a cement grout. Arctic grouts had been in use for well casings and rock-socketed piles or anchors for a number of years. The unpublished results of 1987 tests on some arctic grouts conducted at the University of Alberta indicated that the adfreeze bonds developed in these grouts were substantially less than expected, espe-

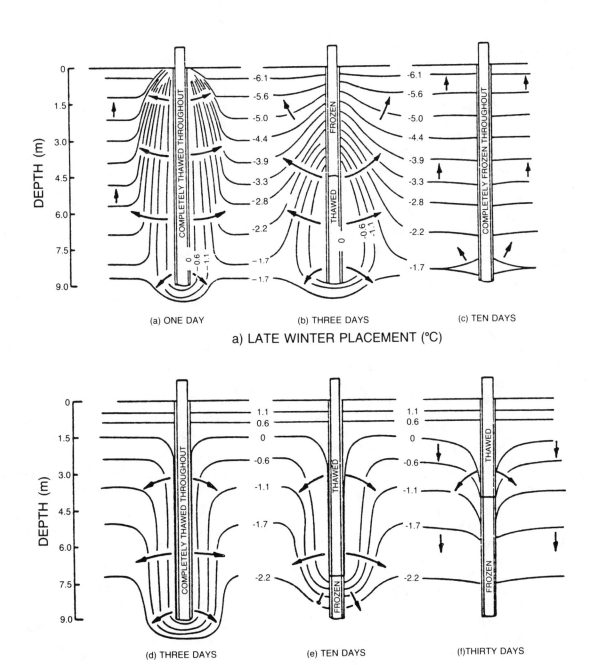

FIGURE 3-49   NATURAL FREEZEBACK OF PILES IN PERMAFROST DURING WINTER
AND SUMMER (Arrows represent heat flow) (Linell and Lobacz, 1980)

cially in a saline permafrost environment. Partly as a result of such tests, other grouts were utilized such as SET-45 (Master Builders) or new grout mixtures were developed such as Sika Grout Arctic 100.

The grouted pile concept was tested by Holubec and Brzezinski (1989) and Biggar and Sego (1989). Additional grout test results are provided by Biggar et al. (1991). The grouts used are capable of properly curing in ground as cold as -10˚C provided the manufacturer's instructions are followed. The main advantage to using grout as the backfill is that the pile diameter effectively becomes the hole diameter. In addition, the referenced tests showed even greater relative increases in strength, probably due to an irregular grout/native soil interface that tends to mobilize the shear strength in the native soil rather than rely strictly on the adfreeze bond.

Pile Load Tests. Pile tests at the actual construction site are recommended, especially at new remote sites. This is especially important for larger projects because of the variability of the soil and its properties from one location to the next. Pile tests are also more important in marginal (warm) permafrost areas. Pile load tests should be mandatory for heavy loads (such as water storage tanks) in marginal permafrost areas. On-site testing is expensive but not nearly as expensive as repairs due to overloaded piling. Pile testing should be conducted in late fall or early to mid-winter when ground temperatures are at their warmest. Several piles must be installed for the tests so that a stable reference exists for displacement measurements. Three or more piles are required to provide sufficient capacity to test the pile, because piles in tension can be significantly weaker than those in compression. Load increments should be small and held for several days instead of the method recommended in ASTM for pile tests in thawed ground. Small increments, held for several days, yield load settlement data, which is useful to assess the ultimate and working stress levels, but only long-term tests (a year or more) can truly define creep behavior. During the tests a maximum pile displacement of around 50 mm should be reached. The adfreeze bond is usually considered ruptured after a displacement of approximately 12 mm or less (Linell and Lobacz, 1980).

Discussions of methods and recommended pile load test procedures in permafrost are provided by Manikan (1983) and Neukirchner (1988). Methods of analyzing creep rates based on pile load tests are presented by Neukirchner and Nyman (1985). Case studies of pile load tests in saline soils are presented by Biggar et al. (1994), Miller and Johnson (1990) and Nixon (1988).

3.4.3 Subgrade Cooling Systems. Subgrade cooling systems refer to rather specialized passive (natural convective heat transfer) or mechanical (energy input required to drive system) heat transfer units incorporated into the facility foundation section. These units may include thermal piles (Long, 1963), thermosyphons, or refrigerant circulation loops. Subgrade cooling systems are often used to counter the thermal disturbance below heated structures, especially over warm or ice-rich, thaw unstable permafrost. Additionally, these systems can be used to improve the strength and reduce creep in some frozen soils by further cooling the ground temperatures. Thermal piles can be used as a combination of a structural member and heat extractor, or just as a heat extractor. If they are to be used as structural members (piles), they must be designed to support the load. Thermal piles transfer considerably more heat out of the ground than could occur by conduction through the soil alone. Seasonal heat transfer can be up to 10.5 million kJ per thermal pile depending on the pile construction and installation, soil type and properties, and air temperature.

There are three general types of subgrade cooling systems, illustrated in pile configurations on Figure 3-50:

- two-phase extractors, which operate on a passive evaporation-condensation cycle;

- single-phase extractors, which rely on passive convection currents within the liquid inside the extractor; and

- mechanical systems, where a convective heat transfer cooling fluid is circulated through the extractor by a pump or fan.

The first two types are referred to as passive or self-operating systems. They automatically start transferring heat out of the ground when the surface temperature drops to a point below the temperature in the ground, and stop when the surface temperature warms above that point. The mechanical systems, however, must be physically stopped when the surface temperature is above ground temperature or, if allowed to continue, they transfer heat into the ground from the surface and raise the ground temperature.

The two-phase cooling system units include thermal piles and thermosyphons, charged with a refrigerant that evaporates at the temperature in the

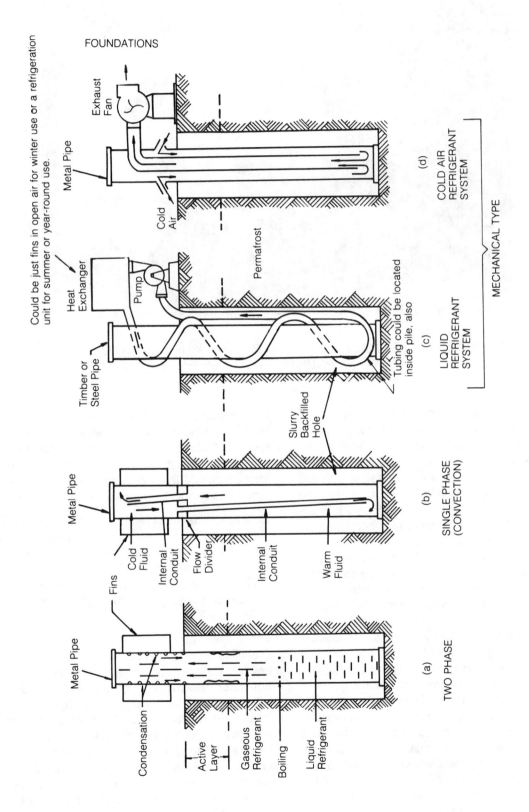

FIGURE 3-50   THERMAL PILE TYPES (Johnston, 1981)

FIGURE 3-51   THERMAL PILE FOUNDATION AT AKIUK MEMORIAL
SCHOOL – KIPNUK, ALASKA (Arctic Foundations, Inc.)

ground, rises to the top of the unit as a vapor, condenses there due to the colder temperatures at the surface and, under gravity, runs back down the inside of the unit. The cycle is then repeated. At the top, it gives up the heat of condensation which is transferred through the pile wall and fins (on the outside) to the atmosphere. The refrigerant is selected on the basis of its evaporation-condensation temperature at different pressures, its density, its ability to flow, and its lack of corrosiveness. Propane, carbon dioxide, and ammonia are commonly used. Propane was used in the past, primarily because of its lower pressure requirements, but recently carbon dioxide has been used because it is more efficient. A leak under any pressure will eventually stop the operation of the unit. The units are sealed after the refrigerant has been added. Leakage could be a fire hazard with some types of subgrade cooling systems. Therefore they must be pressure tested at frequent intervals. They must be constructed of corrosion resistant materials because of the danger and the expense of repairing them, once they have been installed. The two-phase systems transfer considerably more heat than the single-phase systems. Self-refrigerating thermal piles typically affect an area in a 2.4 to 3.0 m radius

around themselves. As with the ventilated pads, the heat extractors only work in winter. They must establish a "cold reservoir" that absorbs any heat that reaches the pad throughout the year. The evaporator sections of thermosyphon units placed under structures should be installed at a minor angle.

Zarling and Haynes (1985b) provide an in-depth review of the history, mechanics, composition, and operation of two-phase thermosyphons. A typical thermal-pile installation is illustrated in Figure 3-51. Typical inclined thermosyphon installations under shallow, on-grade building foundations are illustrated in Figure 3-52.

The single-phase, convective systems work in vertical piles or horizontal circulating loops under shallow foundations, on the same principal as gravity hot-water systems. The system is completely filled with a suitable nonfreezing fluid characterized by large density gradients such as air, glycol, and oil, and sealed (but it does not have to be pressurized). Heat is moved upward out of the ground by natural circulation induced by a density gradient of fluid resulting from the temperature difference between the exposed top and the embedded bottom. Leakage can be a problem, because the fluid could cause environmental damage or destroy the adfreeze bond or both, if it infiltrates the soil. Disadvantages of the single-phase systems include a possibility of some downward heat flow in the summer by conduction, and they require a larger initial temperature difference within the coolant to initiate circulation in the winter without mechanical input. An example of a single-phase subgrade cooling system is illustrated in Figure 3-53.

The condensation segments (above-ground portion) of subgrade cooling units are sometimes provided with fins to increase the surface area and rate of heat transfer (dissipation). With no wind this is a critical point in the removal of heat.

Horizontal two-phase thermosystems have become more popular in recent years beneath heated, on-grade structures in northern Canada. More active systems have also been used such as refrigeration plants at hangars in Inuvik and Resolute Bay and a community center in Inuvik. Also, heat pumps have been successfully operated at two Yukon facilities (Goodrich and Plunkett, 1990).

Design. Subgrade cooling systems are typically sized to remove all the heat load generated from the structure floor, plus remove the latent heat associated with refreezing the soils below the structure thawed in the summer. Procedures for designing subgrade cooling systems require fairly rigorous thermal analyses. Recommended design procedures for two-phase thermal-pile and thermosyphon cooling systems are presented in Hayley (1982), Haynes and Zarling (1988), Krzewinski and Tart (1985), Long (1963) and Zarling and Haynes (1985a; b). Domaschuk et al. (1994) discuss the load transfer along thermopiles in frozen soils. The design procedures for single-phase natural convection piles and subgrade circulating loops are provided in Cronin (1977; 1983) and Bhargava et al. (1991).

The amount of exposed surface area, and exposure to wind, are very important factors in determining the heat removed by a subgrade cooling system. Thus, the above-ground condenser (or radiator) sections must be kept clear of snow and wind shading to the maximum extent possible. It can also be important to paint the condenser section surface with a light-colored paint to reduce absorption of incident solar radiation.

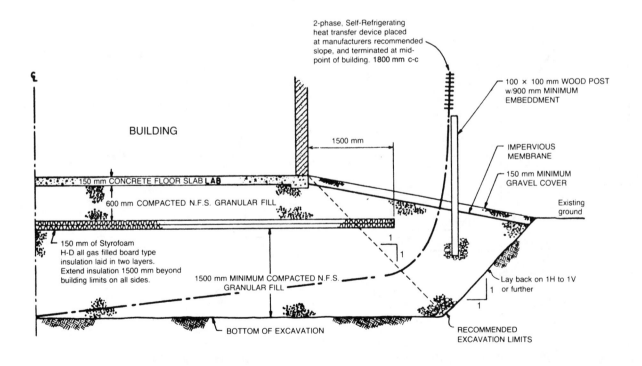

FIGURE 3-52   *TWO-PHASE HEAT EXTRACTOR FOR REMOVING HEAT FROM UNDER BUILDING (Phukan et al., 1978)*

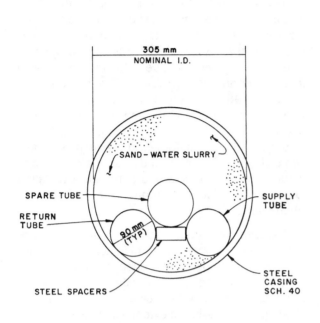

(a) CASING SECTION

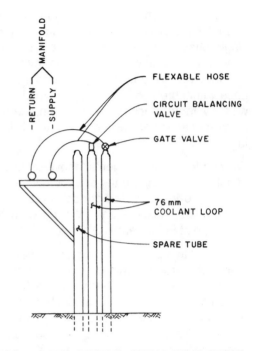

(b) ABOVE GROUND CIRCULATING TUBES

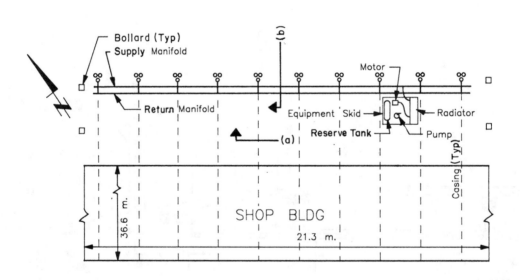

(c) SUBGRADE COOLING SYSTEM LAYOUT

*FIGURE 3-53   SINGLE-PHASE MECHANICAL SUBGRADE COOLING SYSTEM, ADOT MAINTE-
NANCE BUILDING – JIM RIVER, ALASKA (after Rooney et al., 1991)*

Performance. Data on the predicted performance of various thermal-pile and thermosyphon units are available from the manufacturers. Laboratory testing of thermosyphon performance relative to evaporator slope and wind speed across the condenser are reported for for inclined evaporator sections by Haynes and Zarling (1988) and Zarling and Haynes (1985a) and for horizontal evaporator sections in Haynes et al. (1992) and Haynes and Zarling (1988). In-depth case studies of the field performance of thermosyphons are provided in Hayley (1982) and Zarling et al. (1990). An in-depth case study of the field performance of single-phase passive subgrade cooling system is provided in Cronin (1983).

Rooney et al. (1991) and Scher (1991) present a performance case study of a mechanical, single-phase subgrade cooling system installed after original construction to mitigate thaw settlement problems at a transportation maintenance facility in interior Alaska. Of special interest in this case was the apparent heave induced by operation of the subgrade cooling system, even when the surrounding soils remained frozen year-round. Floor heave was recorded for a number of years, despite evidence the cooling system was operating properly, and that the ground temperatures were below freezing. It was proposed that unfrozen moisture in the warm, fine-grained permafrost was attracted towards the cooling loops in response to thermal gradients established when the system was operated in the winter. Scher (1991) provides a method to evaluate the potential for this situation when designing all subgrade cooling systems.

Monitoring. Subgrade cooling systems are typically designed to require minimal maintenance after construction. However, it is very important that the systems be inspected periodically to ensure satisfactory operation. Periodic monitoring should include inspection of:

- the above-ground components for indications of corrosion, wear, or other physical damage;

- the charging medium (fluid or gas) volumes and purity;

- operating performance criteria such as the pressure of the charging medium in two-phase systems, and the circulating rates and in-out temperature differences in single-phase units; and,

- the condition of moving parts in mechanical systems such as pumps and fans.

In-depth discussion of techniques to monitor thermosyphons is presented by McFadden (1987) and Yarmak and Long (1986). Suggested monitoring techniques include visual observation, infrared observation, measurement of condenser temperature, evaporator section and soil temperature measurements, and internal pressure readings.

3.4.4 Retaining Walls. Utility facilities may include earth retaining walls in structure foundations, to support terraced fills and pads, or for protection along drainages. The conventional design of retaining walls for earth pressures is covered in standard engineering texts.

In cold regions, special considerations must also be given to the detrimental effects of heaving and jacking in seasonally freezing soils. If the soils behind earth retaining structures do not drain well, and are protected from becoming saturated, enormous horizontal and vertical forces could be imparted, which would certainly move or damage the wall. Additionally, for cantilever and gravity buttress-type walls founded over ice-rich, frozen soils, the base dimensions must be sized considering the creep deformations in response to the lateral sliding and overturning loads, as with shallow foundations. Soo and Muvdi (1992) present a method for designing a frozen soil retaining wall.

3.4.5 Heave and Jacking Mitigation. There has been much effort and money expended on methods for reducing or eliminating frost-heaving and keep jacking forces on buried foundation components. Generally, deep foundations are designed to provide enough uplift capacity to resist frost-jacking forces. However, shallow foundations often are not anchored, nor is the foundation dead weight sufficient to counter the basal heave forces. Following are some of the more economical methods used to mitigate heave and jacking forces.

Overexcavation. With shallow foundations, it is typical to overexcavate frost-susceptible soils from below the footings, and backfill with non-frost-susceptible materials. This same approach can be applied to piling or shallow posts. The frost-susceptible soils are overexcavated from around the piles, through the active layer, and backfilled with non-frost-susceptible materials. Another approach is to use board insulation to reduce the thickness of the thaw-freeze zone.

Wrapping or Coating the Piles. Probably the most successful method at the lowest cost consists of wrapping the pole or pile, within the active layer,

(a)   *Wrapping Pile Plastic Wrap*

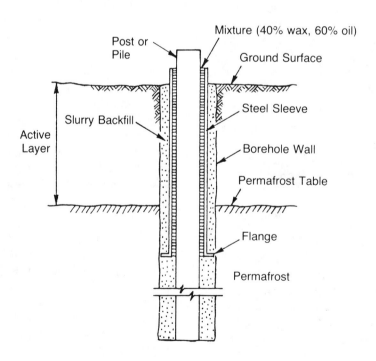

(b)   *Oil-Wax Method for Preventing Frost Jacking*

FIGURE 3-54   METHODS OF PREVENTING FROST JACKING

with two or three layers of plastic wrap. As a further precaution, grease can be applied to the pile and the plastic sheet. A more recent selection in the Canadian Arctic has been to use a loosely fitting shrink sleeve over the greased pile. Another method is to pour an oil-paraffin mixture between the piling and a surrounding paper or plastic tube (Figure 3-54). A typical mixture is Standard Oil's Microval No. 1650 Wax and Mentor No. 28 oil in a 2:3 ratio.

As an alternative, an epoxy or vinyl surface coating in the active layer region not only adds corrosion protection, but also reduces the adfreeze resistance to about 10 percent of the uncoated value. Even coating wood posts or piling with preservative in the active layer reduces the tendency to frost-jack.

Increasing Resistance to Uplift. There are many methods of increasing the resistance to uplift within the soil below the active layer. Commonly, wood piles are placed with the butt (large end) down, so it tapers upward. Wood piles can also be notched, or cross members can be added. Steel piles can have expanded bases (Domaschuk, 1984) or fins or rings can be fastened to the outside of each pile (Figure 3-44) (Long, 1973; Domaschuk et al., 1994), as discussed above. The purpose behind using the fins or rings on pipe piling is to use the usually higher shear strength of the soil instead of its normally lower adfreeze strength to counteract the upward force in frost jacking.

"Corrugating" the Piles. Pipe piles have also been "corrugated" in place after being driven into an undersized hole. The corrugations are positioned close enough to mobilize the full shear strength of the cylinder of soil defined by the vertical projection of the finished corrugations (Long, 1973). In stronger soils, the allowable pile load is increased by the mobilized shear strength of the soil plus the increased diameter of the pile. In low-shear-strength soils, the increase is due only to the increased diameter.

Surcharge and Drainage. The magnitude of potential frost heave around foundations can sometimes be reduced by adding surcharge over the site and grading to drain away from the structure. The heave rate is in part a function of the overburden stress at the freeze front (Section 3.3.8). Figure 3-27 illustrates how the segregation potential, SP, as a parameter to quantify frost heave, decreases as the overburden pressure increases. For example, assume a typical frost depth of about 2 metres in silty soils with an overburden pressure of about 35 to 40 kPa at that depth. Referring to Figure 3-27,

the SP parameter could be reduced by roughly half if enough surcharge were added to more or less double the overburden stress.

Similar to designs in more temperate regions, the ground surface around all foundations subject to frost action should be graded to drain surface water away from the structure. The object is to keep the shallow soils around the buried foundation elements as dry as possible, thereby ideally reducing the frozen adfreeze bond and uplift forces.

Chemically Treating the Soil. Chemical treatment of soils has been demonstrated in laboratory investigations, with variable success, to reduce the frost susceptibility of fine-grained soils. These investigations have involved additives such as void fillers, cementing agents, aggregants, salts (freeze point depression), water-proofers and dispersants. Lambe and Kaplar (1971) and Lambe et al. (1971) report results from heave tests with 52 additives in 25 different soils ranging from plastic clays to sandy gravels. They concluded ferric chlorides, at less than about 0.5 percent weight, and phosphate based dispersants, at 0.1 to 0.5 percent weight, showed the most promise as potential low cost soil additives for reducing frost heave effects. However, further studies are required to evaluate the long-term stability in the materials (versus leaching during thaw cycles) as well as the improved load bearing capacity of treated materials during thaw.

Kozisek and Rooney (1986) and Vita et al. (1986) report results from laboratory and field tests using low concentrations of portland cement mixed with native soils to stabilize the surfacing and bearing characteristics. While expensive to mix and place (US$ 21/m^3, 1986), the cement treated silty soils were demonstrated to provide a more economical base course versus transporting a clean sandy gravel to the site. However, it was concluded that additional types of chemical treatments may be required where frost heave is severe and cannot be tolerated, such as for airfields.

Preventing the Flow of Capillary Water. In theory, frost heaving could be reduced by placing an impermeable barrier or "capillary break" such as plastic sheets, asphalt membranes or thick non-woven geotextiles (Henry, 1988) along the subgrade to prevent the flow of capillary water to the freezing front. However, the barrier must be placed below the seasonal frost penetration depth which could require a great deal of earthwork. The membranes must also be sloped so that water infiltrating from above drains to the side.

FIGURE 3-55   RUSSIAN "PIPE STAND" FOR POWER POLE

Pontoon Supports. Frost jacking of power and telephone utility poles is a major problem in cold regions. Often the cost and time are prohibitive to incorporate any of the above mitigating measures. In such cases, the utility poles must be periodically reset or shored. Figures 3-55 and 3-56 illustrate fairly inexpensive methods of shoring utility poles. Additional information on overhead utility foundations is provided in Retherford (1983).

3.4.6 Safety Factors. The safety factor (F) used should be based on the reliability of the data used in design, the serviceability and economy of the structure, and the probability and consequences of failure.

Unlike thawed soils, the strength and behavior of frozen soils are strongly dependent on the magnitude and rate of loading, soil texture, porewater chemistry and thermal condition. Field sampling and laboratory testing of frozen samples is often very expensive. The volume of such existing data for fro-

zen soils is much less than for thawed soils. Therefore, depending on the scale of the project, the geotechnical data used during design of foundations in frozen ground can often be less reliable than for design in thawed materials. Further, the probability and consequences of foundation failure are higher in frozen ground that is not thaw stable or could exhibit excessive creep behavior if warmed. As such, the design of foundations in frozen soils should be based on conservative soil parameters, conditions and safety factors.

The magnitude of safety factors used to design shallow and deep foundations in frozen ground generally range between roughly 2 and 4, but sometimes as high as 6. Further, it can sometimes be uneconomical to design a foundation to support the full live, snow and transient (wind, seismic, etc.) loads, when the capacity is based on a limiting creep settlement over a long time. Therefore, several design capacities are sometimes determined using differ-

FIGURE 3-56   "TRIPOD" POWER POLE TO REDUCE JACKING
PROBLEM

ent safety factors to reflect different load combinations. Foundation design capacities based on ultimate frozen soil shear or adfreeze strengths may be warrant a relatively high safety factor. For example, a safety factor between 2 and 3 could be considered in thaw stable course-grained soils or for short-term load combinations, and from 3 to 6 in thaw unstable fine-grained materials or for sustained (long-term) loads. Alternatively, foundation design capacities based on limiting creep settlements over a long time (typically 20+ years) may justify a lower safety factor. For example, a safety factor between 2 and 3 could be used for sustained load combinations, if the capacity is determined using conservative creep strengths and thermal conditions.

A minimum safety factor of about 2 to 2.5 is typically assumed for resisting frost-jacking forces, when using the average sustainable adfreeze strength. This safety factor should include the dead load, if it is acting continuously and has no live load. There is some danger in including dead load, in that the pile could be vulnerable to frost heave during construction, especially if the unloaded pile will be sitting through a winter.

3.5 Slope Stability

The problems of slope stability are covered only briefly here; the reader is referred to the various references cited for a more thorough treatment. Slopes in the frozen state are generally very stable, but if the thermal equilibrium is disturbed, slopes as shallow as 5° can become unstable if the materials are thaw unstable or exhibit high creep rates. Slope instability is characterized by the down and outward movement of soil or rock. Slope failures are generally related to inadequate shear resistance or excessive creep deformation.

Shallow slope failures are usually classified as either a skin-flow or slide. With a flow, the failed mass moves down the slope with the characteristics of a viscous fluid. With a slide failure, essentially intact soil masses move down the slope as a unit(s). Flows can occur in any cold region, but they are more common in permafrost regions where they occur with fine-grained soils of the active layer. There are also skin-flows in which only the surface vegetation (mosses, brush, etc.) starts sliding, picking up mineral soil in the active layer and, in turn, causing it to flow. Large landslides seldom occur in permafrost areas. Where they do occur, a layer of permafrost, composed of sands and gravels, overlies a layer of permafrost composed of clays. The failure plane is between the granular soils and the clay.

In-depth reviews of documented slope failures and causes in permafrost are reported in Donovan and Krzewinski (1978) and McRoberts and Morgenstern (1974b). McRoberts (1975) presents a model and parametric study of permafrost slope deformations associated with secondary creep.

3.5.1 Shallow Thawing-Slope Failures.

In permafrost regions, skin-flows associated with the thawing of the active layer on slopes are the more common form of landslide. The stability along a slope surface depends upon the shear strength of the thawed soil. The shear strength is a function of the effective normal stress. Effective normal stress, in this situation, is the stress created by the overburden pressure minus the pore pressure created during the thawing. The pore pressure may consist of a hydrostatic component and an excess pressure component created by thawing in the presence of excess ice and the unconsolidated nature of the soil. The magnitude of the excess pore pressure generated is a function of the thaw-consolidation proper-ties (Section 3.3.9); typically characterized by the thaw-consolidation ratio (R). This is a measure of the rate of thaw versus the consolidation coefficient (function of permeability) for the soil (Morgenstern and Nixon, 1971). Pore pressures are generally higher at the thaw interface, and decrease with distance above the thaw front. An increase in pore pressure causes a decrease in the effective normal stress and thus the effective shear strength. An increased pore pressure is the reason why most failures occur at the thaw interface. Basically, shallow flow slope failures are most common in soils where conditions result in a thaw-consolidation ratio greater than about 0.3. Thaw-consolidation ratios equal to or greater than about 0.3 could be expected in frozen, saturated fine-grained clayey silts and clays subject to moderate to rapid thawing.

The stability of shallow thaw failures along slopes is typically analyzed using an infinite-slope model (Figure 3-57), and the results of laboratory thaw-consolidation tests. In-depth reviews of infinite-slope stability analysis procedures are provided in McRoberts and Morgenstern (1974a) and McRoberts and Nixon (1977). An in-depth case study

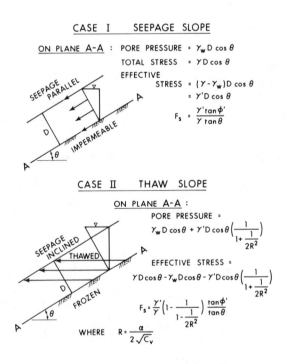

FIGURE 3-57   INFINITE SLOPE ANALYSIS FORMULATION (McRoberts and Morgenstern, 1974a)

of permafrost slope design for a buried oil pipeline is presented by Hanna and McRoberts (1988).

3.5.2 Thermal and Physical Erosion. In addition to failure through the thawing active layer, several other factors associated with thermal and/or physical erosion can contribute to slope instability. Rivers and streams are a major natural factor contributing to landslides when they meander and undercut slopes (Figures 3-7, 3-8, and 3-10). The instability is caused by thermal as well as physical erosion. The correction of this problem requires adequate bank protection, which can be very expensive.

The vegetation cover forms an insulating layer to protect the underlying frozen soil, with root systems also providing a small degree of physical stability. However, the surface vegetation is often damaged by tundra fires. Care should be taken not to destroy its insulating value during construction. In fact, the clearing of brush and trees, if absolutely necessary, should be done by hand. Cuts along roadways are better left alone to heal by themselves. Along the Alaska oil pipeline, tests were conducted to replace the natural insulating value of the tundra mat in a cut using foamed insulation (Donavan and Krzewinski, 1978). The results were conclusive; it is better to let the nearly vertical cut banks heal themselves by melting and flowing as shown in Figure 3-58. The mat at the top of the slope bends down to cover and protect the slope as the soil sloughs into the roadside ditches. This natural healing is much more successful than the application of artificial insulation. It was necessary to cut the banks at a slope steeper than normal, with a wide ditch which can be cleaned out occasionally as the slope melts and flows to its equilibrium position.

Thermal erosion associated with seepage along buried utility trenches crossing slopes is also a very important design consideration (Vita and Rooney, 1978). The potential for seepage-induced erosion along utility trenches may be greatest in more temperate permafrost zones, with fine-grained frozen soils susceptible to piping. Vita and Rooney (1978) present trench geometry and backfill modifications to reduce the potential for thermal erosion along buried utilities crossing slopes.

3.5.3 Slope Surface Treatments. Many researchers have demonstrated improvements to slope stability, for cases of surface thawing or thaw erosion, by covering with either an insulative layer or gravity buttress. Analyses of such improvements

are presented by Mageau and Rooney (1984) and McRoberts and Nixon (1977). General conclusions drawn from these authors include:

- Slope surface treatments designed to allow some thawing of the underlying slope will reduce thermal erosion and improve stability, but they may be short-term solutions. Without additional measures, some degree of slope failure could still be expected with time.

- Slope surface treatments designed to insulate against thawing are more effective. However, long-term slope movements could again be expected, associated with the gradual deteriration of the covering materials.

- Surface treatments designed simply to increase the reflectivity to reduce thaw, are generally ineffective.

- Revegetating generally can be a successful surface treatment for ice-rich slopes. However, this treatment usually requires flatter slopes and higher moisture content soils for better overall growing conditions. The presence of the other surface treatment materials may inhibit the revegetation growth rate.

McRoberts and Nixon (1977) demonstrated that an unprotected slope with a thaw-consolidation ratio of 0.6 would be expected to fail (shallow thaw-flow) above an angle of about 8°. However, this slope covered with a gravel blanket would remain stable up to about 11°, as long as the depth of thaw below the gravel did not exceed twice the cover thickness. The addition of 40 to 60 mm of board insulation would further improve the stability of this example slope significantly.

For the Alaska oil pipeline, slopes were covered with a free-draining material to improve stability. These materials consisted of gravel to coarse sand (150 mm to 2 mm), containing less than five percent (by weight) passing the 2 mm sieve, and exhibiting an angle of internal friction between 40° and 45°.

Careful drainage of slopes to reduce the pore pressures can help prevent slides. Underdrains across the slope near the bottom of the active layer can be effective. Revegetation of the slopes helps; the natural mosses and lichens are extremely slow to grow back.

Buttresses as shown in Figure 3-59 were used with some success on cuts along the Alaska oil pipeline.

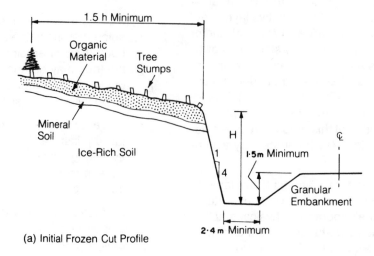

(a) Initial Frozen Cut Profile

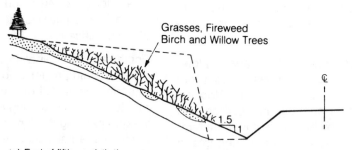

(b) End of first thaw season. Slope is mostly unstable and very unsightly; ditch will require cleaning if massive ice is present.

(c) End of fifth or sixth thaw season.

Slope stabilizes with reduced thaw and vegetation established.

FIGURE 3-58   CUT STABILITY PROGRESSION IN ICE-RICH
PERMAFROST (Johnston, 1981)

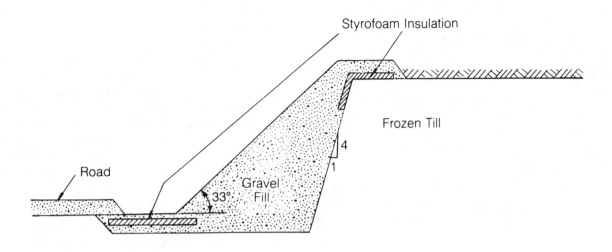

FIGURE 3-59   EXAMPLE OF BUTTRESS ALONG THE ALASKA PIPELINE

Insulation in the form of wood chips was selected to reduce the rate of thaw on numerous ice-rich slopes on the Norman Wells pipeline (McRoberts et al., 1985; Hanna and McRoberts, 1988). An initial performance of the pipeline slopes is provided by Hanna et al. (1994).

3.6   Earthworks

Historically, heavy construction has been a seasonal activity in arctic and even subarctic regions. Winter construction has been thought to be too expensive. But with the advent of oil development on the North Slope of Alaska and in northern Canada, more and more contractors were forced into winter construction not only because of the sheer magnitude of the work but also because much of the construction has been taking place on the environmentally fragile tundra areas, where most surface travel is forbidden during the summer. Roberts et al. (1978) provides a discussion of predicting the feasibility of cold weather earthwork.

Earthworks can be classified into two basic activities: excavation, and placement (fill and compaction). These activities result in a number of problems unique to cold regions and influence the stability of the soils involved. The construction of dams and the related influence of the resulting impoundment are special cases. Roadway and runway construction are also special applications of earthwork activities; they can have a profound effect on the ground environment.

3.6.1   Excavation.   Excavation, either to remove unwanted material or material needed elsewhere, or to acquire fill material of suitable quality, is common to most construction projects. An early method of excavating (still used in gold dredging and, occasionally, for piling) was to use steam or hot water probes to thaw ground so it could be excavated using conventional methods. The use of cool water (>10˚C) was demonstrated to be the most energy efficient. Nearly all the heat required to thaw frozen ground is taken up in converting ice to water, thus the ice content of the ground determines the fuel requirements for thawing. The primary disadvantage of steam- or water-thawing is that the amount of thaw cannot be precisely controlled and usually more ground is thawed than absolutely necessary. Not knowing exactly how much is thawed makes it almost impossible to predict freezeback times. In "warm" permafrost areas, the ground, thermally disturbed by steam- or water-thawing, may never freeze back naturally. Steam- or water-thawing is not recommended where freezeback of the excavation (such as piling) is necessary.

Large equipment, such as backhoes, with reinforced rock buckets and rock saws have been used successfully to excavate frozen ground. For excavation of large areas, heavy dozers with rippers can be quite effective. Depending on the surface conditions, it often proves easier and cheaper than summer excavation.

In areas of high groundwater, freezing before excavation can also reduce the need for dewatering and the need for shoring. The timing of the construction is very critical when using this method. This method was used in the construction of a 35-story office tower in Calgary. The foundation had to be constructed close to the foundation of another tall building in an area with a high water-table. Steel pipes driven into the soil on the perimeter of the area to be excavated carried a brine solution for freezing. After the excavation, a monolithic concrete slab was cast into place.

Trench excavation in frozen ground with a backhoe can be very slow and expensive work. Several methods have been used to loosen the material ahead of the backhoe.

Kudoyarov et al. (1978) provides a summary of Russian earth and rock excavation techniques for construction in permafrost regions.

The excavation of granular borrow materials is conducted in either summer or winter. In summer, the materials are usually stripped off in thin layers as they become thawed and then stockpiled for drainage. As many sources are remote from the development site the borrow is usually then transported during the "winter haul." Alternatively, the excavation can be conducted in winter, usually with ripper assist, and the material hauled to the site or community for stockpiling or winter placement.

Rock Sawing and Blasting. A rocksaw is sometimes used to saw the trench sides, and then after the material is blasted between the cuts, a conventional backhoe can easily and speedily remove it (Cerutti et al., 1982). CRREL has gathered information on the use of liquid explosives for blasting frozen ground (Mellor, 1973). Liquid explosives are more efficient than solid ones. Frozen soil can absorb considerable energy, therefore more explosive must be used than when blasting rock. After using a rocksaw to cut horizontal lines (the trench sides) Alyeska used 0.36 kg of dynamite per m^3 of frozen ground. The Public Health Service used between 0.35 and 0.60 kg of Atlas 40 percent special Gelatin per m^3 of frozen ground to blast utilidor trenches in Nome, Alaska. The Nome trenches were not presawn.

Drilling and Blasting. The Russians also use a method of drilling or "punching" holes ahead of the backhoe. A method that has proven to be quite satisfactory is that of drilling and blasting ahead of the backhoe. However, in some areas such as a town, blasting is unsafe.

Trenching. Several wheel-type, track-mounted trenchers were tested before construction of the Alaska oil pipeline. The larger ones did well in frozen silts, but in frozen gravel they were continually breaking down. Maintenance of the buckets and the carbide teeth on the cutting bar is expensive. The vast majority of the excavation for the Norman Wells pipeline was successfully performed in winter using arctic wheel ditchers that had been developed for the Arctic gas pipeline. In some very strongly bonded silty sands, ripper assist was used. The ditchers were also used on many of the slopes (up to about 17°).

Dredging. Dredging is being used more and more extensively in arctic areas as a way of moving large quantities of sand and silt during the short summer construction season. Dredging is not significantly different in cold climates from that in more temperate ones; in fact, it is well adapted to cold regions. The U.S. Public Health Service constructed the water supply dam in Barrow, Alaska, using a combination of gravel hauling by truck and dredging.

For certain projects, dredging may be the more economical excavation method. A major deepening of the harbor at Tuktoyaktuk, NWT, was undertaken by dredging. Dredges were already in the area for the construction of artificial islands to serve as drilling platforms. The 80,000 m^3 of dredged material was used to fill a low-lying area for a site for a housing development. Dredging was estimated to save over 50 percent in cost compared to the cost of hauling and placement by conventional methods (trucks and scrapers). In Tuktoyaktuk and many communities in northern and western Alaska, buildable land is in very short supply. Dredging in low-lying areas and shallow ponds is a less expensive method of providing additional building area. The low-lying areas can be filled to a height sufficient to place the structures above flooding levels. The dredged fill was placed at a cost of only about \$5/m^3 (Cdn, 1980) instead of the estimated \$16/m^3 (Cdn, 1980) that trucked fill would have cost. The material readily drains during placement, therefore only minor settlement and heave problems are expected.

Based on the success of fill-dredging, a circular water storage impoundment was also constructed by dredging in Tuktoyaktuk in summer 1981 (Figure 3-60). The impoundment, 300 m in diameter and 12 m high, is constructed from sand dredged from the harbor during a two-week period and worked with earth-moving equipment.

FIGURE 3-60   TUKTOYAKTUK WATER SUPPLY IMPOUNDMENT

The impoundment was designed to store a year's supply of water from surface sources for the community. After treatment, water is distributed by truck to the homes. The reservoir is lined with an unreinforced liner material, because it has better elongation properties to accommodate the expected differential settlements in the impoundment floor. Settlements occur when the few ice lenses present thaw, after the reservoir has been put into operation. The inside slopes were 14° and the outside slope was 18°. Approximately 1.3 m of dredged sand was placed over the liner to protect it from erosion by wave and ice action.

Dredging can cause potential environmental problems which must be considered:

- Significant localized increases in turbidity and suspended solids at the source.

- Direct entrainment of fish at the dredge cutter head results in their death. These mortalities are not usually considered significant enough to affect the large range and distribution of the fish species.

- Virtually all clams and benthic populations will be destroyed in the area dredged. Recolonization of the disturbed area by organisms from the undisturbed areas may occur quickly, if the disturbed area is not too large. However, recurring dredge operations such as harbor maintenance may prevent recolonization of the disturbed area.

- Dredging should normally be suspended during critical migratory or spawning seasons. If it continues, the migratory patterns of fish are interrupted.

3.6.2   Fill and Compaction.   Preferably, all engineered fills should be constructed with thawed, classified materials. However, fills can and have frequently been constructed under winter conditions with frozen materials, or with materials chemically modified to depress the freeze point. However, fills placed under freezing conditions or with frozen materials generally cannot be economically placed and compacted to a relatively stable condition. As such, earthworks with frozen materials typically present problems during thaw (at least the first thaw cycle) associated with thaw settlement and reduced bearing capacity and stability.

Compaction of fill materials during construction is usually gauged relative to a standardized laboratory fill unit weight, such the Proctor or relative density values. In thawed soils, the compacted unit weight is a direct function of the moisture content of the soil during placement and compactive effort. Each soil has an optimum moisture content for a particular method of compaction. The optimum moisture content allows the maximum unit weight to be attained for a given compactive effort.

In compacting frozen ground, the clumps of soil particles held together by ice must be broken up. As one would expect, the optimum moisture content is much smaller for soil when it is frozen than when it is thawed. In fact, as the temperature drops below freezing, the optimum moisture content approaches the air-dry state. Ice crystals occupy space that would otherwise be filled by soil particles. Additionally, ice bonds between the soil particles and fro-

zen clumps result in a "looser" soil structure (i.e., higher void ratios), versus thawed fills with the same grain size distribution.

A typical laboratory compaction curve for a thawed and frozen sandy gravel fill is illustrated in Figure 3-61. It is clear that the inclusion of any frozen moisture in a fill will greatly reduce the probable compacted unit weight. Recall that the magnitude of settlements in fills compacted in a frozen condition is directly related (inversely) to the frozen dry unit weight (see Section 3.3.9 and Figure 3-32).

Temperature will also affect the compacted frozen dry unit weight, as illustrated in Figure 3-62. With knowledge of compaction costs of the same, or similar fill when thawed, Figure 3-63 can be used to estimate compaction costs of a frozen fill.

More in-depth discussions of compaction of frozen coarse-grained fills, and the potential problems resulting from poor compaction are provided in Bernell (1965), Brooker (1992), Haas and Barker (1989), Kinney and Troost (1984), Linell and Kaplar (1963),

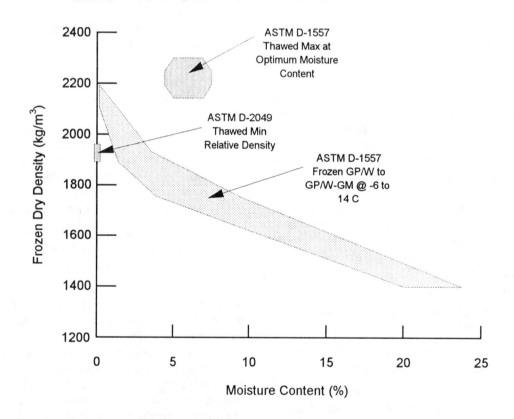

FIGURE 3-61   CHARACTERISTIC LABORATORY COMPACTION CURVES FOR FROZEN SANDY GRAVEL FILL (after Haas and Barker, 1989; Kinney and Troost, 1984; Mahmood et al., 1984; Scher, unpublished data)

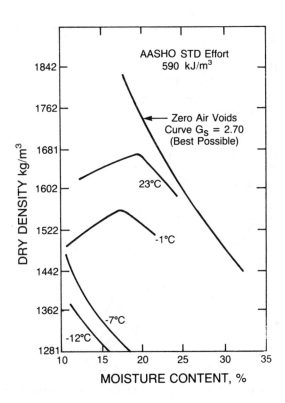

*FIGURE 3-62   EFFECT OF FREEZING TEMPERATURES
ON COMPACTION OF A SILTY FINE
SAND (Linell and Kaplar, 1963)*

Mahmood et al. (1984), Schlegel and Stangl (1990), and Tart and Luscher (1981). A discussion of compacting cohesive soils in winter treated with freeze-point depressant agents is provided by Bathenchuk et al. (1972).

3.6.3 Geosynthetics. Geosynthetics in geotechnical and construction engineering typically refer to manufactured grids, textiles or fabrics, or membranes incorporated in or below engineered fills and foundations. The applications of geosynthetics in thawed soils include reinforcement, material separation, cushioning, filtration, transmission, and/or isolation such as liners in liquid reservoirs or under fuel storage facilities. In cold regions, geosynthetics are often used for all these applications, but in association with the presence with thawed soils.

The design for these type of applications can be found in the manufacturer's literature and in recent texts, not reviewed herein (see Canadian Geotechnical Society, 1992; Holtz, 1988). Attention should

be given in design to the cold temperature properties.

To date, geosynthetics have been used for only a few applications somewhat specific to cold regions and permafrost. These applications are generally in association with embankment construction for roads and airfields (Section 3.7.2). Smith (1978) reported on use of geomembranes to encapsulate less than desirable fill soils in embankments, rendering these soil materials usable. Henry (1988) reported on laboratory experiments to test the effectiveness of geotextiles as capillary breaks or barriers in embankments to mitigate frost heave. Kinney and Abbott (1984), and Kinney and Connor (1987; 1990) report on the use of high-modulus geotextiles and geogrids in road embankments to bridge underlying voids created in the natural foundation soils by the thaw of more massive ice features. Kinney and Savage (1989) and Savage (1991) report on using high-modulus geotextiles and geogrids to reinforce the

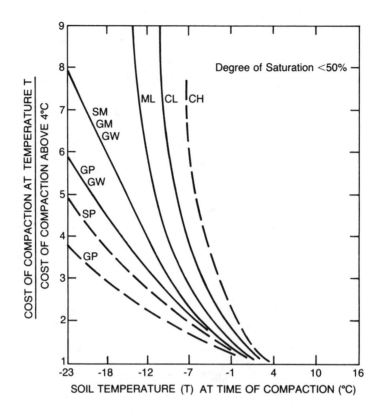

FIGURE 3-63   APPROXIMATE RATIO OF COST OF SOIL
COMPACTION AT SUBFREEZING TEM-
PERATURES (Linell and Kaplar, 1963)

pavement structures over road embankments experiencing lateral spreading or fault displacements associated with uneven thawing in the foundation soils, especially under the side slopes. These applications have been used to construct a few test sections along roads in Alaska, and the reader is referred to the noted references for design and performance details.

3.6.4 Earth Dams. Earth dams are commonly used in permafrost regions for water supply reservoirs and sewage lagoons. In constructing a dam the thermal regime under the impoundment, and possibly downstream, will be changed. A case study of the changes to the thermal regime under an earth-fill dam in Greenland is reported by Fulwider (1973). The large mass of water in the reservoir absorbs considerable heat in the summer. This heat is lost to the air and soil around the impoundment and dam. Depending on the size of the impoundment, this heat loss to the air can change the climate in the area. Average annual temperatures and precipitation in-

crease. The heat lost to the soil thaws the permafrost under the dam and impoundment. The thawing causes erosion problems around the waterline of the impoundment, and what is worse, increases percolation of water around and under the dam. The percolation then causes additional thawing which could lead to failure of the dam. The amount of percolation depends on the permeability of the foundation material in the thawed state. If the soil is ice-rich, the thawing could result in serious settlements below the dam. Even dams founded on bedrock can experience critical percolation problems due to cracks that were ice filled, and essentially impermeable before the dam was constructed.

Construction Methods.   There are two primary methods used in dam design and construction in permafrost areas: the warm method and the cold method (Figure 3-64). The feasibility of each method is primarily dictated by the local climate conditions.

The warm construction method, most feasible in the discontinuous permafrost region, attempts to keep the mass of the dam in a thawed state. Percolation must be controlled with an impervious core, and water migration from the upstream face to the downstream face must be designed as a dam would be designed for a warm climate. Figure 3-64(a) indicates the steady-state condition for a "warm" dam. Settlement can be a problem in underlying ice-rich soils and must be considered before dam construction. A possible method would be to prethaw the permafrost where the proposed dam is to be built. This would also allow a more accurate determination of future percolation rates. The warm method of dam construction can usually be successfully used, provided the frozen foundation material is not ice-rich (and thus not subject to uneven settlement), has sufficient bearing capacity when thawed, and maintains its permeability when thawed. If these criteria cannot be met, the cold construction method should be used.

The cold construction method, which is more practical in the continuous permafrost region, maintains the frozen condition of the underlying ground and most of the dam itself Figure 3-64(b). This ensures the stability of the dam and essentially, prevents percolation. The cold method requires a larger dam mass than would be needed according to design criteria for warmer climates. Flat slopes, especially on the upstream face, are necessary to protect the frozen core of the dam. The upstream toe of the dam must be designed to take any settlement that will occur when the permafrost beneath it thaws. This could mean yearly filling work to replace the settled earth until equilibrium has been reached. Keeping the downstream half of the dam frozen prevents nearly all percolation. If this cannot be accomplished with the natural freezing temperatures in winter and by actually constructing the dam in the winter (allowing it to freeze as it is built), heat extractors or refrigeration coils may have to be added to ensure the core remains frozen (Figure 3-64(c)). If large unfrozen areas (Figure 3-64(b)) exist in the permafrost below the proposed dam, they must be frozen before construction starts, or the dam must be moved to a site where no thawed areas exist. If they are allowed to remain thawed, they can provide a path for percolating water. Frozen areas under the upstream toe of the dam should probably be thawed before construction to lessen the possible settlement at the toe. If the cold method is used several banks of thermocouples or thermisters should be installed to monitor the frozen state of the dam and foundation closely. If conditions are thought to be marginal,

some pipe for subsequent cooling systems could be installed during construction.

Causes of Dam Failures: Spillways and Water Outlets. Spillways and water outlets have caused more dam failures than any other single cause. Spillways must be placed outside the dam itself, not through or over it on piling, so that it has as little effect on the frozen ground as possible. Stilling wells and other depressions should not be allowed at the discharged end of the spillway. Siphon spillways are sometimes used over the center of the dam for the day to day flows. Conventional spillways are located at the side to handle only large, short-term breakup or flood flows. Large dams (over 9 m) are not as well understood, and the problem needs special attention. Those that have been built have nearly always been founded on rock where the existence or nonexistence of permafrost really did not have a great effect. Or they have been constructed with mechanical cooling coils, so that their frozen state can be maintained.

Dam Case Studies. Russian engineers have constructed dams in permafrost regions for many years (Gromov, 1974; Maslovski et al.,1978; Savarenskii, 1960; Tsvetkova, 1960; Tsytovich, 1975). Most of these dams have been constructed following the cold method, and incorporated thermal piles or refrigerating coils to ensure the frozen condition. Not all of these dams have been successful (Tsvetkova, 1960).

One of the oldest Russian dams over permafrost was built in 1792 in Petrovsk-Zabaykalskiy. Most of the larger dams in Russia are founded on bedrock. Figure 3-65 is of the Chernyshevskiy hydropower dam on the Vilyuy River in Siberia. The dam is rockfill on bedrock. It is 75 m high and 600 m long. It produces 650 MW of power and backs up an impoundment 80 km wide and 322 km long. The dam is founded on permafrost. As the impoundment caused the foundation to thaw, they have had to grout the cracks in the bedrock to control the seepage. They are also closely monitoring the thaw and erosion under and around the impoundment. In the first three years after the impoundment was filled, 2.7 m of thaw had been measured in the permafrost under the impoundment.

The water supply dam for the town of Mirnyy is a cold-method dam. Mechanical cold-air fans maintain the dam and its foundation in a frozen condition. The 21 m high dam is located on the Irelyakh River. It was constructed in the winter, thereby assuring that the core started out frozen. Mechanical

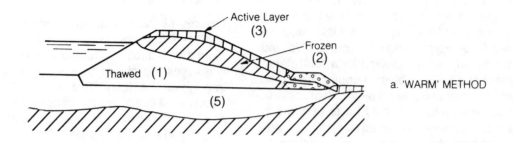

a. 'WARM' METHOD

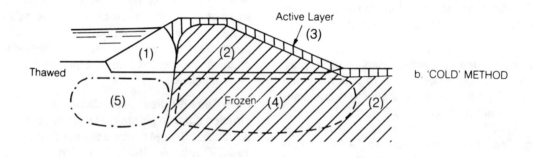

b. 'COLD' METHOD

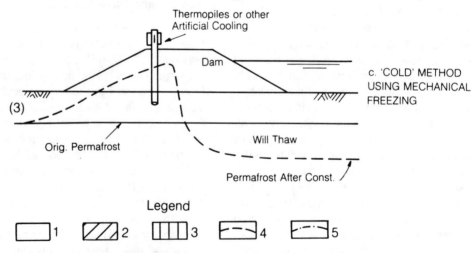

c. 'COLD' METHOD USING MECHANICAL FREEZING

Legend

□ 1   ▨ 2   ⊞ 3   ⌇ 4   ⌐ 5

1–zone of permanently thawed soil; 2–zone of permanently frozen soil; 3–zone of alternating thawing and freezing; 4–zone in which it is necessary to freeze naturally unfrozen ground; and 5–zone in which it is desirable to thaw previously frozen ground.

FIGURE 3-64   METHODS OF DAM CONSTRUCTION ON PERMAFROST

FIGURE 3-65   CHERNYSHEVSKIY DAM IN RUSSIA

circulating thermal piles were added to the dam after construction. They extend to a depth of about 30 m and are operated when the temperature is below -15°C. The temperatures of the core and foundation are continuously monitored with thermocouples. Settlement problems were experienced in the concrete-lined spillway when it was used continuously. To solve this problem, two large pipe siphons, extending over the top of the dam, bypass sufficient water for normal operation. Flood flows are still taken through the conventional spillway, because they are of short enough duration not to cause significant thawing. The thermal piles are spaced at 1.5 m and they have been operated only intermittently in the last few years.

In North America, probably the largest dam system in the permafrost region is the Kelsey Power Generating Facilities on the Nelson River, 644 km north of Winnipeg (Brown and Johnston, 1970; MacDonald, 1963; Savarenskii, 1960). The main dam is 37 m high rock-fill with the spillway cut through solid rock. The maximum generating capacity is 313 MW. It is a warm-method dam; the permafrost below the dikes was allowed to thaw. The settlement, 1.2 to 1.5 m in some areas, was replaced, as it occurred, with fill material.

In 1967-68, the U.S. Public Health Service constructed a cold-method dam, 9 m high, to store water for domestic use in Kotzebue, Alaska. It was constructed in the winter and thermocouples were installed to monitor the frozen condition of the base and foundation. The soil temperatures have stayed very near and a little below freezing in and below the dam itself, even though several metres of thaw have occurred under the impoundment itself. There has been some settlement on the upstream face, but nothing that routine maintenance cannot handle.

The U.S. Public Health Service also constructed a water supply dam in Barrow, Alaska, using about 76,340 m³ of gravel collected on the beach and several hundred thousand cubic metres of material dredged out of the reservoir itself (Figure 3-66). The dam is approximately 300 m long and 10 m wide at the crest. It splits the Esatkuat lagoon, which was originally brackish due to storm waves from the Arctic Ocean. The maximum height is about 6 m, but an elaborate, impervious core was not provided because there was only 1.5 m of head differential between the upstream and downstream water levels. The foundation for the dam consists of the same material that was dredged out of the impoundment: silts and sands ranging from loose to very dense in

FIGURE 3-66   BARROW WATER SUPPLY DAM – OUTLET STRUCTURE

both the frozen and unfrozen state (but usually below 0˚C). The material was not ice-rich, and little settlement was experienced. Some settlement occurred in the abutment areas where ground ice melted because of the increase in the upstream water level.

The dredged material provided the binder (fines) needed to control the seepage through the well-graded beach gravel, and to provide erosion protection for the dam faces. Its removal from the impoundment also increased the fresh-water storage capacity. The construction was accomplished in four stages.

- The gravel was placed by end-dumping with trucks to form the "core".

- Material was dredged from the future impoundment up onto the sides of the gravel core and shaped with a bulldozer. The dredge can be seen operating in the distance.

- The sides were then sandbagged with a mixture of cement and sand. These were for temporary erosion control and long-term stability.

- More material (sand and silt) was then dredged up over the sand bags of the upstream and downstream faces, and allowed to flow to its normal angle of repose (probably 7 to 8˚). The area around the intake structure and the overflow tubes (Figure 3-66) was left as shown with the sand-cement bags and not covered by additional dredged material.

Many other smaller dams have been constructed in the permafrost areas throughout Alaska and Canada (Smith et al., 1984; 1989) primarily for water supplies. The U.S. Army Corps of Engineers have constructed and monitored small dams at Unalakleet, Alaska, and Thule, Greenland (Gromov, 1974). There are several embankments and dikes that have been constructed around Yellowknife, NWT, to contain and decant mine tailing wastes.

3.7   Roads and Airfields

Roads and airfields in cold regions range from temporary trails made of snow and ice, to long-term paved travelways. A primary difference between roads and runways in cold regions and those in warmer regions is the higher construction and maintenance costs. Construction costs in cold regions are generally impacted by unfavorable climate, physical hazards such as wetlands and thaw unstable foundation soils, and often limited quantities of desirable materials. Maintenance costs are often higher because of differential frost heave and settlement. Operations are also impacted. Load limits must typically be lowered in the spring when the subgrade is thawing and weakened due to reduced effective stresses (thaw consolidation; see Section 3.3.9).

Most roads and airfields in cold regions are constructed with conventional gravel or paved surfaces over earth-fill embankments. Two major gravel roads have been constructed in permafrost areas in recent years. These are the Alaska oil pipeline haul road that extends north from the Yukon River to Prudhoe Bay (Berg et al., 1978), and the Dempster highway in Canada that starts near Dawson, YT and ends at Inuvik, NWT (CARC, 1979; Huculak et al., 1978). The Mackenzie Highway has been completed as far north as Wrigley, NWT and many sections of the Northwest Territories and Yukon Territory highways are now paved.

Temporary, seasonal snow and ice roads and airfields are also commonly used as both an economical and sometimes environmentally preferred construction alternative to earth fills.

Bridge and pier design will not be discussed in this monograph. In-depth discussion of this subject in cold regions is provided in Johnston (1981), Montgomery et al. (1984), and Caldwell and Crissman (1983).

In-depth reviews of the unique design, construction, and operation considerations of airfields in the arctic regions is provided in Crory (1988) and Crory et al. (1978).

3.7.1 Route Selection. Proper route selection is an important consideration in minimizing the construction and maintenance problems of roads. Careful selection of the route can also considerably reduce maintenance problems. Remember, the shortest distance is usually not the cheapest and most satisfactory solution. Aerial photographs and topographic mapping can be used for preliminary route selection. Additionally, an on-site reconnaissance is highly recommended. Some pointers that help in selecting the route:

1. Pick areas where borrow materials are available.

2. Stay away from unstable slopes and slide areas.

3. Avoid residual snow patch areas and areas with icing problems (see Section 3.7.3).

4. Avoid areas requiring substantial cuts and fills. However, if cuts are necessary:

   a. Avoid north-facing slopes.

   b. Make nearly vertical back-slopes on cuts in frozen, ice-rich soil; let the ground thaw and slough find its own final position (Figure 3-58).

   c. Provide a wide ditch at the base of a cut to allow removal of sloughing material as necessary.

   d. Clear trees and brush, but minimize any harm to the organic mat from the top of the slope, back a distance about equal to the height of the slope.

   e. Do not attempt to reseed the slope until it has stabilized.

5. Avoid deep valleys and depressions likely to harbor snow for prolonged periods.

6. Avoid marshes, springs, and other wet and/or muskeg areas.

7. Avoid areas with patterned ground relief indicating tension cracks and ice wedges (Figures 3-11 and 3-12).

8. Stay on ridges or small, rounded hilltops where soil conditions are better and snow will blow free, thereby reducing drifting problems.

9. Cross rivers and streams in straight stretches where there are gravel bars on both sides. Avoid meanders with cut banks on either side of the river.

10. Avoid isolated ice-rich permafrost if practical.

11. Avoid aligning roads along contours across flat slopes or drainages to minimize ponding of surface runoff. Any changes in the amount of surface drainage or the drainage patterns can have a significant impact on the underlying permafrost.

3.7.2 Design and Construction. Road and airfield designs require knowledge of the vehicle loads and repetitions, and desired trafficability. In cold regions, a thorough understanding of the local climate and subsurface conditions is also imperative, especially air temperatures, soil types, groundwater and ice. For roads and airfields, the alignment/ geometrics (not covered herein), embankment, and

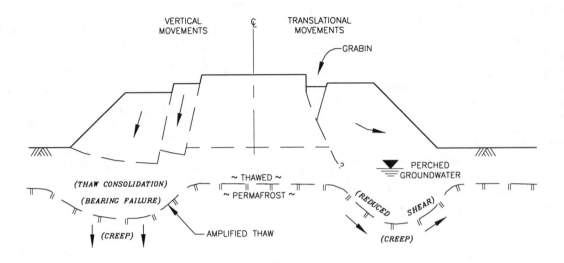

FIGURE 3-67   LONGITUDINAL CRACKING IN AN EMBANKMENT OVER THAW
UNSTABLE PERMAFROST (after Scher, 1996)

pavement surface are typically designed in separate steps. In-depth reviews of these design procedures will not be presented here, but appropriate references are provided. The following discussions focus on the more unique aspects of road and airfield design in cold regions.

Embankments.   In nonpermafrost cold regions, earth-fill embankments are designed similar to temperate climates, with attention given to alignment, grades, and drainage. Seasonal frost effects are also accounted for in terms of frost heave and thaw weakening potential (see pavement structure below).

In permafrost areas, the stability of earth-fill embankments is directly related to the thermal stability of the natural foundation soils. Thermal instability in foundation soils under embankments over permafrost may result from:

- a loss of the natural ground cover's insulative characteristics;

- significant change in surface temperature conditions;

- water ponding next to the embankment which then warms the natural ground;

- potential snow drifting or snow ploughed wind row; or,

- uneven thawing under the embankment, which can lead to both lateral spreading and

differential fault or graben displacements in the pavement surface (Figure 3-67), especially over ice-rich foundation soils.

Placing fill on top of the organic ground cover, muskeg or peat compresses these materials, which significantly reduces their insulating value. Additionally, when break-up occurs in the spring, the ground under the surface mat is still frozen so very little of the snow melt or rain infiltrates the ground. Nearly all of it must run off as surface drainage. Culverts for summer drainage are usually still plugged with snow and ice. They are buried in the fill material, and therefore thaw after the surface runoff has stopped. Most road subgrades, where culverts are usually located, are designed so that their lower portions remain frozen or thaw late in the spring. The culverts can be physically thawed ahead of break-up using steam points or other means, but this is expensive and not usually done.

Figures 3-68 through 3-70 show the progressive deterioration of an earth-fill embankment road in Bethel, Alaska, caused by deeper seasonal thawing from about 0.5 to 1.2 m. This deeper thawing was because of the standing water ponded behind the road in the spring, and because the 450 mm of sand fill did not nearly replace the insulating value of the 100 to 125 mm of the natural organic ground cover.

The "amplified" thaw under the embankment side slopes (Figure 3-67) is in response to the warmer

FIGURE 3-68   PONDING BEHIND ROADS IN BETHEL, ALASKA –
SPRING, 1970 (Ryan, 1981)

FIGURE 3-69   ROAD CONSTRUCTION ACROSS SLOPE IN BETHEL, ALASKA
– SUMMER, 1969 (Ryan, 1981)

surface conditions compared to those on the adjacent pavement and natural ground surfaces. In the winter, these side slopes are generally covered with snow cleared from the pavement or drifted against the embankment, which effectively insulate the slopes from the cold winter temperatures (Esch, 1983a). As a result, the mean annual surface temperatures on the side slopes have been found to sometimes be several degrees Celsius warmer than the mean temperature of the pavement surface (Johnson, 1988; Zarling and Braley, 1987). Also, the ratio of fill thickness to original peat insulation thickness is lowest in the sides of the embankment.

Embankment movements and deformation over warm permafrost has also been attributed to secondary creep in underlying ice-rich materials. Case studies and analytical models to evaluate embankment movements due to creep are presented in McHattie and Esch (1988) and Phukan (1983).

A number of methods have been tested and used in road and airfield projects in permafrost to mitigate the detrimental effects of thaw unstable foundation soils. These include insulation, geosynthetics, embankment toe/stabilization berms, air ventilation and subgrade cooling systems, and artificial side slope coverings, summarized in Esch (1983b; 1988) and Johnson (1988). McHattie and Esch (1983) discuss the beneficial effects of buried peat under roads over permafrost. The success of these various methods has been mixed, and very site-specific. To date, only the first two methods (insulation and geosynthetics, discussed below) have found common use; the others have not been significantly effective in mitigating the problems associated with thaw unstable foundation soils below an embankment over permafrost. Esch (1988) provides an overview of the design methods and features field tested to mitigate thaw unstable foundation soil conditions, including post-construction performance, for a number of roads and railroads in Alaska and Canada. To date, no economical solution has been found for constructing stable embankments over thaw unstable, ice-rich permafrost.

The most common method to mitigate thaw instability problems is to include board insulation in the embankment to reduce the thermal disturbance to the underlying permafrost. Figure 3-71 illustrates the theoretical effects of uninsulated and insulated road embankments on the underlying permafrost table. However, when the mean annual surface tempera-

FIGURE 3-70   DETERIORATION (WASHOUT) OF ROADS SHOWN IN FIGURES
3-68 AND 3-69 – SUMMER, 1970 (Ryan, 1981)

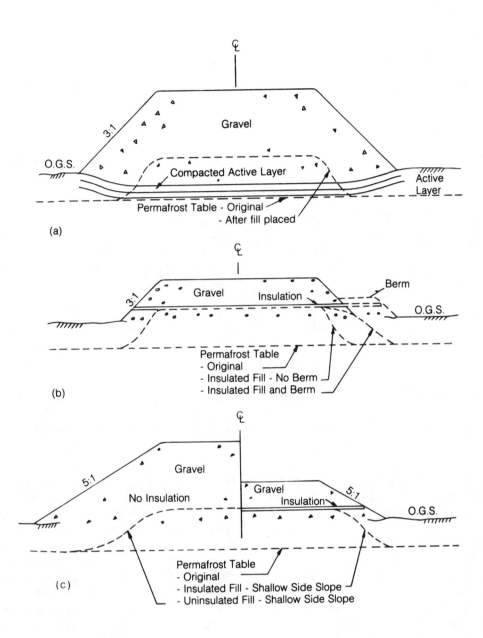

*FIGURE 3-71   EFFECT OF UNINSULATED AND INSULATED ROAD
EMBANKMENTS (Johnston, 1981)*

tures are close to or above freezing, the use of insulation generally only slows the rate of long-term thawing under the embankment, and may have little impact on the amplified thawing under the side slopes (Figure 3-72). This use of insulation is similar to that for shallow building foundations (Section 3.4.1). In addition to the above references, case studies of the design, construction, and performance of insulated road and airfield embankments are presented in Berg (1974), Esch (1973; 1984; 1986),

Johnston (1983), Johnson and Bradley (1988) and Olsen (1984). A simplified approach to the design and construction of insulated gravel work pads and roads is provided in Wellman et al. (1977).

Test sections of several roads in Alaska have incorporated geosythetics in the embankments. In these applications, the geosynthetics were not used to reduce the thermal disturbance, but rather to improve the physical stability of the embankment and

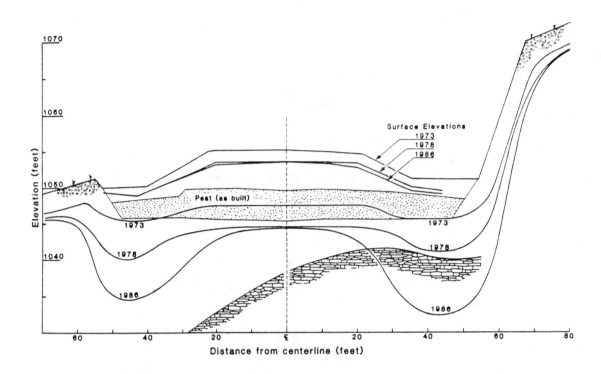

FIGURE 3-72   AMPLIFIED EMBANKMENT SIDE-SLOPE THAW DEPTHS (Reckard et al., 1988)

increase the safety of the pavement surface. The performance of these test sections has been encouraging to date, and several recent highway projects in Alaska have included longer sections of use spanning several kilometres of road. Figure 3-73 illustrates three of the major geosynthetic applications in road embankments over permafrost, including encapsulation, bridging voids, and reinforcing the pavement structure over spreading or differential faulting (see also Section 3.6.3).

The use of air ventilation or subgrade cooling systems may be practical to improve conditions over very thaw unstable permafrost. A review of the design and performance of an air ventilation system in a highway embankment over permafrost in Alaska is provided in Zarling et al. (1983). The design and performance of two-phase thermosyphons installed in an attempt to stabilize a section of an airfield in western Alaska is reported by McFadden (1985) and McFadden and Siebe (1986). Similar stabilization of a Canadian railway embankment was achieved by thermosyphons (Hayley, 1988).

Pavement Section.   The pavement section refers to the materials within a depth influenced by the surface load, and typically includes the surface course, base and subbase materials. The materials immediately below the pavement structure are referred to herein as the subgrade; this could refer to the engineered embankment fill or in-situ natural foundation soils. Gravel-surfaced roads and airfields are probably the most economical in cold regions, especially over permafrost subject to moderate or low traffic volumes. This type of surfacing is often the least expensive to construct and maintain compared to flexible or rigid pavements. Flexible pavements (asphalt concrete) are also used on surfaces subjected to large traffic volumes, and in some areas for public convenience to minimize broken windshields. However, paved roads typically cost much more to construct and maintain, especially in nonpermafrost areas over frost-susceptible subgrades, and in warm permafrost areas over thaw unstable permafrost. An example of an insulated, asphalt paved road is illustrated in Figure 3-74. It must be recognized that the dark asphalt surface will absorb more heat in summer, thus accentuating the road section instability problems discussed above.

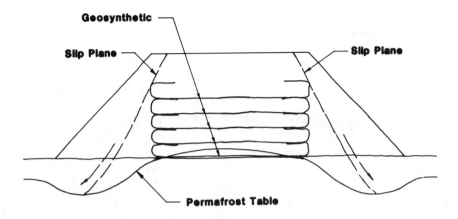

(a)   *Reinforced Embankment Core*

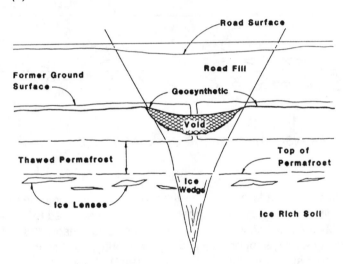

(b)   *Bridging Thermokarst Voids*

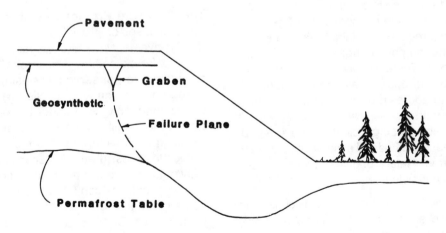

(c)   *Bridging Fault Cracking*

FIGURE 3-73   *GEOSYNTHETICS APPLICATIONS IN EMBANKMENTS*
                OVER PERMAFROST (after Johnson, 1988; Kinney, 1984)

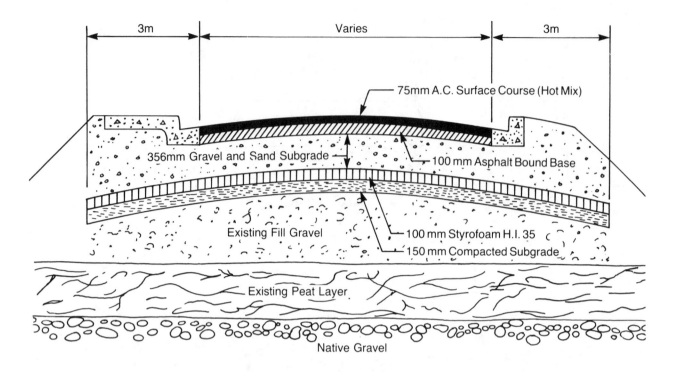

FIGURE 3-74   TYPICAL INUVIK INSULATED PAVING CROSS SECTION

The design of pavement structures for roads and airfields in cold regions is typically complicated by two significant problems. Probably the most critical problem is the reduction or loss of bearing support in the subgrade because of its frost susceptibility, compressibility during thaw and weakening due to excess pore pressures during thaw. The loss of support causes the surface to deteriorate. The second problem stems from the consequence of generally poor quality aggregates available near the project for constructing the pavement structure or for mixing with asphalt. There are five general methods commonly used to design pavement structures in cold regions to address the subgrade frost-heave and thaw-strength issues, each based at least in part on the vehicle loadings, climate, and subgrade conditions. These methods all assume the subgrade is stable, in terms of lateral or differential movements associated with thawing.

The earliest methods include three discussed in-depth by Linell et al. (1963) and Lobacz et al. (1973): complete protection, limited-subgrade protection, and reduced-subgrade strength. More recently, the Alaska Department of Transportation and Public Facilities (1982) developed the "excess-fines"

method, and the now common mechanistic design method has been adopted to cold-regions application (Mahoney and Vinson 1983). Design of airport pavement overlays is discussed in Anderson and Heimark (1984). Stabilization of the pavement section using soil-cement mixtures is presented in Vinson et al. (1984).

The **Complete Protection Method** (Linell et al., 1963) is based on providing a non-frost-susceptible (NFS) pavement structure deeper than the expected depth of freeze or thaw. This freeze/thaw depth can be estimated conservatively using the modified Berggren equation (see Section 4). Board insulation is commonly used under the pavement structure to reduce fill quantities. Use of high-moisture-retaining NFS sand can also reduce the required section thickness (better frozen to unfrozen conductivity ratio and greater latent heat effects). When operating conditions can tolerate some surface heaving, slightly to moderately frost-susceptible soils, such as S1 or S2 (Section 3.3.4) can be used in the subbase.

The **Limited-Subgrade Protection Method** (Linell et al., 1963) allows some frost penetration or thaw into the subgrade based on assumed acceptable small deformations at the surface due to heave or settlement. This method is appropriate for uniform subgrades consisting of silts, very fine silty sands with more than 15 percent less than 0.02 mm by weight, clays with plastic indices less than about 12, or varved clays and other banded fine-grained sediments. Figure 3-75 provides a graph to determine the minimum pavement structure (as "base") for limited-subgrade protection as a function of the fill/ subgrade moistures and maximum depth of freeze.

The **Reduced-Subgrade Strength Method** (Linell et al., 1963) neglects seasonal heave, and is based on providing the minimum pavement structure thickness for bearing. This method has been used for pavement design when the subgrade is fairly uniform, consists of F1, F2 or F3 soils (Figure 3-22), and the traffic volumes are relatively low. Figure 3-75 provides a graph to determine the total thickness of NFS pavement structure as a function of the subgrade frost classification and vehicle design index. The design index (Table 3-8) is a measure of the vehicle weights and loadings. For low volume roads, a design index of 4 to 6 may be appropriate; for heavy vehicles, a value of 6 to 8 should be used.

The **Excess-Fines Method** (Alaska Department of Transportation and Public Facilities, 1982) is based on a minimum load bearing, asphalt pavement structure comprised of some frost-susceptible materials. Frost heave is neglected with this method. The method is based on a statistical review of the pavement deflection response (using Falling Weight Deflectometer (FWD) data) versus the gradation of the pavement structure. This method can be used for new pavements or overlays. Briefly, the method allows the designer to determine the required asphalt thickness as a function of the vehicle loadings and distribution of fines (less than 0.074 mm) with depth in the pavement structure.

The **Mechanistic Approach** to designing pavements is covered in standard engineering texts. This is a very flexible method, which allows the designer to determine the most economical pavement structure as a function of specific vehicle loads and material properties. The disadvantage of this method is that it requires knowledge of the modulus of elasticity and Poisson's ratio for each material type. Laboratory tests to measure these values are very expensive. Additionally, these properties are extremely sensitive to the loading (strain) rate and material temperature (especially frozen or thawed). As such, most applications utilize values reported from other studies on similar materials, or from val-

ues back-calculated from field FWD surveys (Mahoney and Vinson, 1983).

Snow and Ice Roads and Airfields. Design criteria for snow and ice roads and airfields over ground and lake/river/sea ice are presented in Adam (1978), Adam et al. (1984), Albele (1990), Barthelemy (1975; 1992), Clark (1973), Hayley and Valeriote (1994), Keyes (1977), Mellor (1993), and Tomayko (1974). The reader is referred to these publications for details on the construction and maintenance of snow and ice roads and runways.

3.7.3 Operations Hazards and Considerations.

Operations on cold-regions roads and airfields typically include greater maintenance due to the unique hazards associated with reduced subgrade strength during thaw, embankment failures, icing in culverts, and cracking due to thermal contraction of the pavement structure and/or embankment in the winter.

Reduced Subgrade Support. Load restrictions are commonly placed on roads in the spring when the subgrade bearing support is at a minimum. The restrictions are usually based on deflection tests under given loads.

Embankment Failures. Embankments over ice-rich permafrost are prone to significant lateral spreading or differential fault/graben cracking in response to the uneven foundation soil thawing, as described in 3.7.2 (see Scher, 1996). Figure 3-76 is an example of such lateral spreading and differential cracking along an interior Alaska highway. Sloughing in the side slopes is especially common. These surface cracks can create a very serious safety condition, posing danger to wheeled vehicles, especially motorcycles. Once these problems begin to occur, regular maintenance will be required to patch and relevel the affected area. In areas prone to this type of condition, gravel-surfaced roads are obviously preferable.

Culvert and Drainage Icings. Culverts can create major maintenance problems. Large culverts allow the circulation of warm air during the summer which can cause undesired thawing in the subgrade. Flexible covers have been proposed for the ends of large culverts to reduce the circulation of warm air in the summer. Another possibility is to insulate them so that the warm air passing through cannot thaw the permafrost below them. In permafrost areas, culverts typically protrude from the subgrade at an upward angle on each end, because they have caused thawing in the center of the road and settled, causing a reverse grade that leads to ponding.

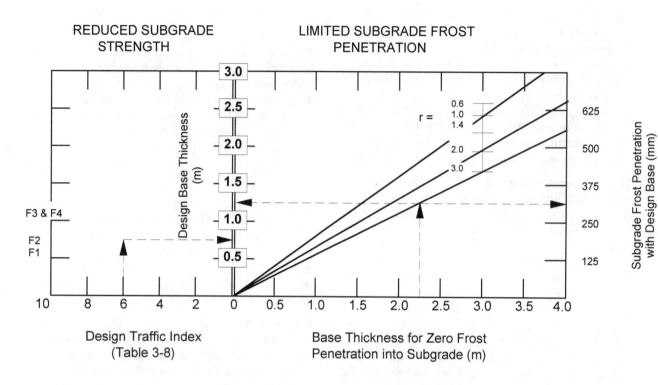

FIGURE 3-75   *DESIGN PAVEMENT STRUCTURE THICKNESS OVER FROST-SUSCEPTIBLE SUBGRADE SOILS (after Lobacz et al., 1973)*

TABLE 3-8   *VEHICULAR TRAFFIC DESIGN INDEX (after U.S. Navy, 1979)*

Design Index	Traffic Characteristics				
	Percent Vehicle Group			Vehicles Per	Approximate
	1	2 + 3	3	Day	EAL
1	100				1 to 5
2	>90	<10	0	<1000	6 to 20
3	>90	<10	<1	<3000	21 to 75
4	>85	<15	<10	<6000	76 to 250
5	>75	<25	<15	<6000	251 to 900
6		>25		>6000	901 to 3000

Group 1: Passenger cars and panel and pickup trucks
Group 2: Two-axle trucks
Group 3: Trucks with ≥ three axles
EAL: Equivalent 80 kN single-axle loads

FIGURE 3-76   LATERAL SPREADING AND DIFFERENTIAL CRACKING,
GOLDSTREAM ROAD – FAIRBANKS, ALASKA (Scher, 1996)

Natural icings can also occur on side slopes (Figure 3-77) and at stream crossings. They can grow to the point where they endanger bridges, roads, and even buildings or entire towns. In a typical stream in the Fairbanks, Alaska area, icings (or overflow) consisted of about four percent of the yearly runoff volume and 40 percent of the winter stream flow. Stream icings are an important method of water storage because they melt much more slowly than the surrounding snow and augment streamflow after the peak snowmelt runoff. Stream icings can create ice thicknesses well over five times the normal water depth in the stream. Rivers or streams that derive a significant portion of their flow from groundwater are more prone to icing than those deriving their flow mostly from surface water. Such rivers or streams usually have higher flows in the winter because of:

- considerable rain just before freeze-up;

- low air temperature with little snow cover during the first half of the winter. A heavy snow insulates and reduces the rate of frost penetration and thus the possibility of icings;

- the close proximity (shallow) of an impervious layer such as permafrost or bedrock.

Man-made icings are caused by uneven freezing of the active layer due to physical snow removal or shading under a structure. The migrating or flowing groundwater is forced to the surface through the path of least resistance where it spreads out and freezes in uneven layers. Figure 3-78 shows how the natural ice sheet can thicken under a bridge (shaded from snow cover), causing an icing. Figure 3-79 indicates how a road or railroad constructed on a cross slope can promote icings. In both cases, snow had been plowed off the road and piled in the ditches on either side of the road. The snow removal results in excess frost penetration beneath the road, thus restricting the groundwater flow. On the other hand, the snow helps insulate the active layer. Under snow, the active layer is thinner and thus becomes the weak point where the seeping groundwater (under pressure) can break through to the surface and freeze. Because of the active layer freezing downward, significant pressures are built up forcing the water to break through the surface at the upstream weak point.

Studies in Alaska of actual road-caused icings, and evaluation of possible solutions are provided in Johnson and Esch (1977) and Livingston and Johnson (1978). Methods used currently to mitigate

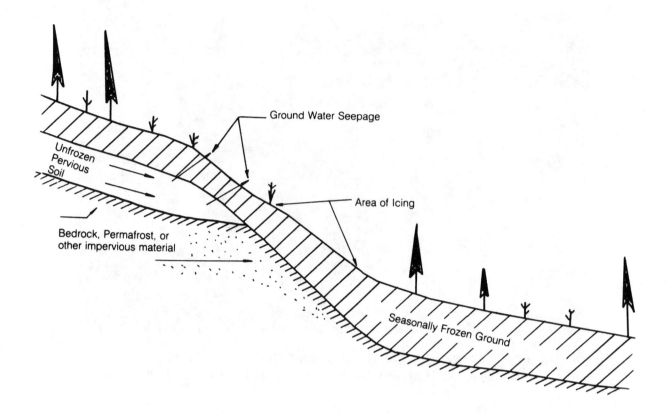

FIGURE 3-77   CONDITIONS FOR THE DEVELOPMENT OF NATURAL ICING

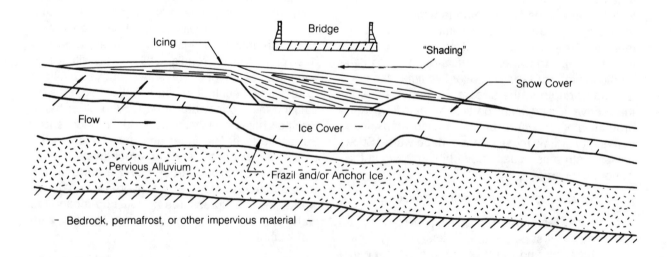

FIGURE 3-78   STREAM ICING CAUSED BY AN ELEVATED ROAD BRIDGE

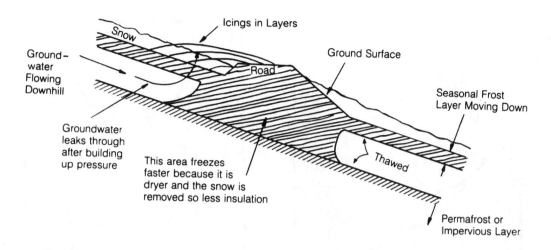

FIGURE 3-79   ROAD-CAUSED ICING

icing problems include improved open-ditch drainages to avoid culverts where possible, insulating culverts and groundwater to limit winter freezing, creating a frost belt above the culvert to block winter flow (Figure 3-80), and incorporating a steam-circulating or electric wire circuit through the culvert to facilitate thawing.

Pavement Cracking. An additional maintenance problem especially common in cold regions is pavement cracking. These cracks can be in the pavement only, or in extremely cold areas they can extend through the subgrade. The deeper cracks require filling and resurfacing. Differential heaving can also cause cracks (Scher, 1996). Excluding cracks associated with embankment failures, pavement cracking can be divided into two general categories, based on the cause: fatigue or thermal (McHattie, 1980).

Fatigue cracking and rutting is a cyclical process where the damage done depends on traffic loading

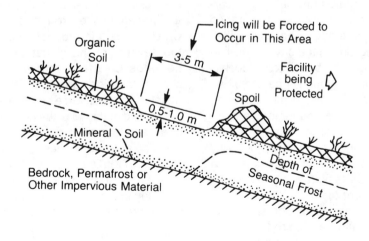

FIGURE 3-80   TYPICAL CROSS SECTION OF A FROST
BELT INSTALLATION

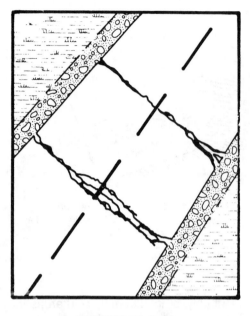 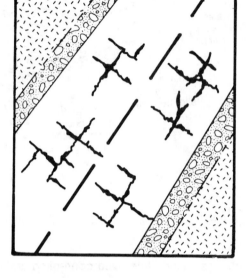

a. TRANSVERSE CRACKING   b. MAP CRACKING

FIGURE 3-81   PAVEMENT THERMAL CRACKING (McHattie, 1980)

and local climate. The climate strongly affects the ability of the pavement structure to support traffic loads, but where high-quality (low fines content) materials have been used, the climate usually has little effect. Fatigue cracking begins at the base of the pavement and propagates upward to the surface, primarily with traffic action. The appearance of fatigue cracks at the pavement surface coincides with the end of the design life of the pavement. The crack propagation upward can be critically accelerated by the loss of subgrade support under the pavement during spring thawing, in turn related to the gradation of the materials. The greater the percentage of fines (silts and clays) or the shallower their occurrence, the sooner and greater the probability of fatigue cracking. Wheel path rutting results from repeated loadings and abrasion (usually caused by studded tires). It is thought that only 20 percent of the total rut depth in Alaska is due to abrasion. Ruts deeper than 13 mm should be repaired, because they can create water ponding and thus hydroplaning, which lessens steering control.

Thermal cracking appears as transverse cracks or map cracks (Figure 3-81). Although it is widespread and very noticeable, it is less serious than the fatigue cracking. Major transverse cracks ("tire thumpers") usually cross the entire roadway. Spacing and severity of transverse cracks seem to be most closely related to temperature changes, thermal coefficients of expansion of all materials in the road, and frozen tensile strength of the upper soil layers. Crack spacing is probably more closely related to the severity of the climate, whereas crack width depends on the month to month temperature changes. Major transverse cracks are a particular maintenance nuisance because of their cyclic movements and because of their frequent penetration into deep soil layers. In-depth reviews of the thermal cracking and magnitudes are presented in Hajek and Haas (1972), Osterkamp (1986) and Rix (1969). A case study of attempts to seal and control thermal cracking in the Fairbanks (Alaska) International Airport runway is provided in Ecsh and Franklin (1989).

With time, the thermal cracks will evolve into a "map" or blocky pattern (Figure 3-81(b)). It is almost directly related to low temperature asphalt stiffness. Map cracking consists of randomly oriented small cracks affecting only the pavement.

FIGURE 3-82   TYPICAL MEANDERING RIVER IN ARCTIC ALASKA

3.8 Hydrological Hazards

Consideration of the hydrologic factors influencing utilities in cold regions is discussed in-depth later in this monograph (Section 5). The following provides limited discussion of several general hydrologic factors related in part to geotechnical considerations.

3.8.1 Dams. Dams on rivers affect up- and downstream flows and velocities. Rivers in cold regions usually experience greater flow variations during the year than those in warmer climates. Ratios of maximum to minimum flows of 50 to 100 are common. Dams tend to lower the average level of the river downstream as well as the bed load it normally carries. They also provide a measure of flood protection to the area. Other thermal impacts and design approaches have been discussed in Section 3.6.4.

3.8.2 Erosion and Sediment Transport. The magnitude of the movement of meanders and channel shifting on alluvial fans can have a considerable influence on the design of pipeline or road crossings. Erosion and meandering of rivers in cold regions was studied carefully in connection with the design and route selection of the Alaska oil pipeline. Rivers tend to meander to a great extent in the arctic areas of Alaska, possibly because of the flat-

ness of the terrain (Figures 3-7 and 3-8). In most valleys, several old river channels are usually visible from the air (Figure 3-82). Ox-bow lakes are often evident where the river or stream has cut through bends tens to thousands of years ago. When designing a road pipeline crossing in the vicinity of such an ox-bow, keep in mind that erosion upstream probably increases considerably when the meander "cuts through", because the length of the river will be shortened. Riddle et al. (1988) present a case study of methods to stabilize thermal and physical bank erosion along the Yukon River in Alaska. Veldman and Yaremko (1978) provide a summary of the design and construction of river "training" structures to counter erosion.

As the water temperature decreases, the viscosity increases and so does the sediment carrying capacity of the river or stream. Clark et al. (1988) present a general review of suspended sediment transport in arctic rivers. Glacier-fed streams are numerous in Alaska. They are milky in color because of the very fine soil particles suspended in the water. Glacier-fed streams or rivers can have sediment concentrations as high as 2,000 mg/L. This is 10 to 20 times greater than nonglacial streams in cold regions which themselves usually have sediment

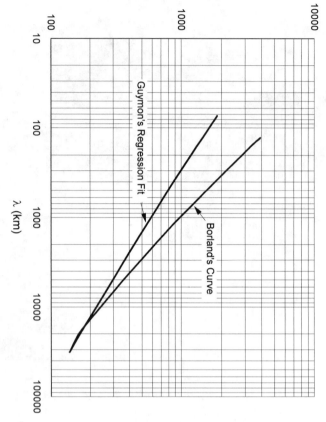

Avg. Annual Suspended Sediment Yield
(tonnes/km²/yr)

where:

$$\lambda = \left(\frac{A_t}{A_g} \right) \bullet L_g$$

A_t = total catchment area, km²
A_g = area of glacier, km²
L_g = length of stream below the glacier, km

FIGURE 3-83 *SUSPENDED SEDIMENT YIELD OF GLACIER-FED STREAMS (developed from Guymon, 1974)*

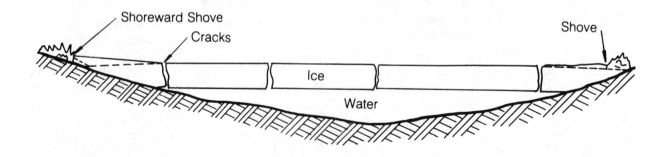

FIGURE 3-84 *SHEET ICE EXPANSION AND CONTRACTION*

concentrations greater than rivers in warmer climates (for at least a portion of the year) because of the colder water temperatures. Guymon (1974) combined the data from several studies to produce Figure 3-83 which can be used to estimate the sediment yield from glacier-fed streams.

3.8.3 Ice Forces. In addition to ice flows driven by currents or wind, ice, after it forms on an impoundment or lake, can cause severe forces on dams or shore structures. It contracts when cooled and expands when warmed like most other materials. When an ice sheet contracts, it usually cracks. The cracks immediately fill with water, which usually freeze. Then, when the ice warms, it expands and thrusts shoreward (Figure 3-84). This is a kind of ratchet action all in one direction, shoreward. The thrust exerts large forces on both dam face and shore structures. The amount of movement, and thus the lateral thrust, varies with the thickness of the ice sheet, the temperature change, and the solar radiation. An in-depth review of causes, mechanics, and magnitudes of ice forces on structures is provided in Caldwell and Crissman (1983).

Studies in France (Drouin, 1974) concluded that ice thrust does not significantly increase with ice sheets greater than 300 mm, because the ice below this depth is not exposed to the large temperature fluctuations to which the surface is subjected. Snow cover also reduces the magnitude of these forces considerably by reducing the effect of sudden air temperature changes on the ice temperature. Ice thrust loading on a dam can be reduced considerably by sloping the upstream face.

3.9 References

Abele, G. 1990. *Snow Roads and Runways.* U.S. Army Corps of Engineers, CRREL Monograph 90-3.

Adam, K.M. 1978. *Building and Operating Winter Roads in Canada and Alaska - Environmental Studies No.4.* Department of Indian and Northern Affairs, Ottawa, Ontario.

Adam, K.M. and R. Piotrowski. 1980. Solving snow drifting problems at Baker Lake, NWT, using snow modelling techniques. In: *Proceedings Utilities Delivery in Northern Regions,* Ottawa, Ontario, 394-408.

Adam, K.M., R.F. Piotrowski, J.M. Collins, and B.T. Silver. 1984. Snow roads for pipeline installation on the arctic pilot project. In: *Proceedings 3rd International Conference on Cold-Regions Engineering,* Edmonton, Alberta, 603-618.

Alaska [State of] Department of Transportation and Public Facilities (ADOT&PF). 1982. *Guide for Flexible Pavement - Design and Evaluation.* Juneau, Alaska. 27 p.

American Association of State Highway and Transportation Officials (AASHTO). 1990. *Standard Specifications for Transportation Materials and Methods of Sampling and Testing (Parts I & II),* Washington, D.C.

American Society for Testing and Materials (ASTM). 1994. *Annual Book of ASTM Standards, Volume 04.08, Soil and Rock.* Philadelphia, Pennsylvania.

Andersland, O.B. and M.R.M. Alwahhab. 1983. Lug behaviour for model steel piles in frozen sand. In: *Proceedings: 4th International Conference on Permafrost,* Fairbanks, Alaska, 16-21.

Andersland, O.B. and D.M. Anderson. (Editors) 1978. *Geotechnical Engineering for Cold Regions.* McGraw-Hill Inc., New York, 566 p.

Anderson, K.O. and H. Heimark. 1984. Airport pavement overlays in cold regions. In: *3rd International Conference on Cold Regions Engineering,* Edmonton, Alberta, 637-657.

Anderson, K.O. and N.R. Morgenstern. 1973. Physics, chemistry, and mechanics of frozen ground: A review. In: *North American Contribution 2nd International Permafrost Conference,* Yakutsk, USSR, 257-288.

Arcone, S.A. and A.J. Delaney. 1988. Borehole investigations of the electrical properties of frozen silt. In: *Proceedings 5th International Permafrost Conference,* Trondheim, Norway, 910-915.

Arcone, S.A., P.V. Sellmann, and A.J. Delaney. 1978. Shallow electromagnetic geophysical investigations of permafrost. In: *3rd International Permafrost Conference,* Edmonton, Alberta, 502-507.

Arnold, C.L. 1978. Ground surface temperatures in permafrost areas. In: *Proceedings Applied Techniques for Cold Environments,* Anchorage, Alaska, 112-121.

Auld, R.G., R.J. Robbins, L.W. Rosenegger, and R.H.B. Sangster. 1978. Pad foundation design and performance of surface facilities in the Mackenzie Delta. In: *Proceedings 3rd International Permafrost Conference*, Edmonton, Alberta, 765-771.

Barnes, D.F. 1963. Geophysical methods for delineating permafrost. In: *Proceedings, 1st International Permafrost Conference*, Lafayette, Indiana, 349-355.

Barthelemy, J.L. 1975. *Snow-Road Construction – A Summary of Technology from Past to Present*. Naval Facilities Engineering Command, Civil Engineering Lab. Technical Report R 831. 17 p.

Barthelemy, J.L. 1992. Nomographs for operating wheeled aircraft on sea-ice runways: McMurdo Station, Antarctica. In: *11th International Offshore Mechanics and Arctic Engineering Symposium*, Calgary, Alberta, 27-33.

Bathenchuk, E.N., et al. (translated by H.R. Hayes, 1972). *Placing Cohesive Soils in Winter in the Far North*. National Research Council of Canada, Technical Translation 1501, Ottawa. 119 p.

Berg, R. 1974. *The Use of Thermal Insulating Materials in Highway Construction in U.S.*, U.S. Army. CRREL, Report MP 539, Hanover, N.H.

Berg, R., Brown, J. and Haugen, R.K. 1978. Thaw penetration and permafrost conditions associated with the Livengood to Prudhoe Bay road, Alaska. In: *Proceedings 3rd International Permafrost Conference*, Edmonton, Alberta, 615-621.

Bernell, L. 1965. Properties of frozen granular soils and their use in dam construction. In: *Proceedings 6th International Conference on SMFE*, Vol. 2, 451-455.

Bhargava, R., J. Patterson and D.J. Bassler. 1991. A passive radiator for a refrigerated foundation in permafrost. *ASHRAE Transactions*, Vol 97.

Biggar, K.W. and D.C. Sego. 1989. Field load testing of various pile configurations in saline permafrost and seasonally frozen rock. In: *Proceedings 42nd Canadian Geotechnical Conference*, Winnipeg, Manitoba, 304-312.

Biggar, K.W., D.C. Sego and M.M. Noel. 1991. Laboratory and field performance of light alumina cement based grout for piling in permafrost. In: *Proceedings 44th Canadian Geotechnical Conference*, Calgary, Alberta, Vol. 1, Paper 42.

Biggar, K.W., D.C. Sego and R.P. Stahl. 1994. The performance of a long-term pile load testing system in saline and ice-rich permafrost. In: *47th Canadian Geotechnical Conference*, Halifax, Nova Scotia, 441-450.

Brooker, E.W. 1992. Abuse of engineering principles, negligence and litigation. In: *Proceedings of the 45th Canadian Geotechnical Conference*, Toronto, Ontario, Paper 107.

Brown, R.J.E. and T.L. Péwé. 1973. Distribution of permafrost in North America and its relationship to the environment: A review, 1963-1973. In: *North American Contribution 2nd International Permafrost Conference*, Yakutsk, USSR, 71-100.

Brown, W.G. and G.H. Johnston. 1970. Dikes on permafrost: predicting thaw and settlement. *Canadian Geotechnical Journal*, 7(4): 365-371.

Burwash, A.L. and W.R. Wiesner. 1984. Classification of peats for geotechnical engineering purposes. In: *Proceedings, 3rd International Specialty Conference on Cold Regions Engineering*, Edmonton, Alberta, 979-998.

Buska, J.S. and J.B. Johnson. 1988. Frost heave forces on H and pipe foundation piles. In: *Proceedings 5th International Permafrost Conference*, Trondheim, Norway, 1039-1044.

Caldwell, S.R. and R.D. Crissman. (editors) 1983. *Design for Ice Forces*. American Society of Civil Engineers (ASCE), New York. 218 p.

Canada Department of Transport (CDT). 1967. *The Climate of the Canadian Arctic*. Meteorological Branch, Air Services, Toronto, Ontario.

Canadian Arctic Resources Committee (CARC). 1979. The Dempster highway (road to resources). *Northern Perspectives*, 7(1).

Canadian Geotechnical Society. 1992. *Canadian Foundation Engineering Manual,* 3rd Edition. BiTech Publishers, Inc. Vancouver, B.C. 512 p.

Carslaw, H.S. and J.C. Jaeger. 1959. *Conduction of Heat in Solids*, 2nd edition, Oxford University Press. 510 p.

Cerutti, J.L., et al. 1982. Underground utilities in Barrow, Alaska. In: *Proceedings Symposium* on *Utilities Delivery in Cold Regions,* Edmonton, Alberta.

Chamberlain, E.J. 1981. *Frost Susceptibility of Soil.* U.S. Army CRREL, Monograph 81-2, Hanover, New Hampshire, 110 p.

Chamberlain, E.J. 1983. Frost heave of saline soils. In: *Proceedings 4th International Permafrost Conference,* Fairbanks, Alaska, 121-126.

Clark, E. 1973. *Expedient Show Airstrip. Construction Technique.* U.S. Army CRREL, Special Report No. 198, Hanover, N.H.

Clark, M.J., A.M. Gurnell, and J.L. Threlfall. 1988. Suspended sediment transport in arctic rivers. In: *Proceedings 5th International Permafrost Conference,* Trondheim, Norway, 558-563.

Cronin, J.E. 1977. A liquid natural convection concept for building subgrade cooling. In: *Proceedings 2nd International Symposium on Cold Regions Engineering,* Fairbanks, Alaska, 26-41.

Cronin, J.E. 1983. Design and performance of a liquid natural convection subgrade cooling system for construction on ice-rich permafrost. In: *Proceedings 4th International Permafrost Conference (1976),* Fairbanks, Alaska, 198-203.

Crory, F.E. 1963. Pile foundations in permafrost. In: *Proceedings 1st International Permafrost Conference,* Lafayette, Indiana, 467-476.

Crory, F.E. 1973. Settlement associated with thawing of permafrost. In: *North American Contribution 2nd International Permafrost Conference,* Yakutsk, USSR, 599-607.

Crory, F.E. 1982. Piling in frozen ground. *Journal of the Technical Councils ASCE,* 108: 112-124.

Crory, F.E. 1988. Airfields in Arctic Alaska. In: *Proceedings 5th International Permafrost Conference,* Trondheim, Norway, 49-55.

Crory, F.E., R.L. Berg, C.D. Burns, and R. Kachadoorian. 1978. Design considerations for airfields in NPRA. In: *Proceedings Applied Techniques for Cold Environments,* Anchorage, Alaska, 441-458.

Davidson, B.E., J.W. Rooney, and D.E. Bruggers. 1978. Design variables influencing piles driven in permafrost. In: *Proceedings Applied*

Techniques for Cold Environments (1981), Anchorage, Alaska, 307-318.

Domaschuk, L. 1982. Frost heave forces on embedded structural units. In: *Proceedings 4th Canadian Permafrost Conference,* Calgary, Alberta, 487-486.

Domaschuk, L. 1984. Frost heave resistance of pipe piles with expanded bases. In: *Proceedings, 3rd International Offshore Mechanical and Arctic Engineering Symposium,* New Orleans, Louisiana, Vol. III, 58-63.

Domaschuk, L., D. Dubois, and D.H. Shields. 1994. Load transfer from thermopiles to frozen soil. In: *47th Canadian Geotechnical Conference,* Halifax, Nova Scotia, 451-460.

Donavan, N.C. and T.G. Krzewinski. 1978. Slope stability studies in a cold environment. In: *Proceedings Applied Techniques for Cold Environments,* Anchorage, Alaska, 840-851.

Drouin, M. and B. Michel. 1974. *Pressures of Thermal Origin Exerted By Ice Sheets Upon Hydraulic Structures.* U.S. Army, CRREL, Translation No. 427, Hanover, N.H.

Esch, D.C. 1973. Control of permafrost degradation beneath a roadway by subgrade insulation. In: *North American Contribution 2nd Internatinal Permafrost Conference,* Yakutsk, USSR, 608-622.

Esch, D.C. 1983a. Design and performance of road and railway embankments on permafrost. In: *Final Proceedings 4th International Permafrost Conference,* Fairbanks, Alaska, 25-30.

Esch, D.C. 1983b. Evaluation of experiemental design features for roadway construction over permanent. In: *Proceedings 4th International Permafrost Conference,* Fairbanks, Alaska, 283-288.

Esch, D.C. 1984. Surface modifications for thawing of permafrost. In: *Proceedings 3rd International Conference Cold Regions Engineering,* Edmonton, Alberta, 711-725.

Esch, D.C. 1986. Insulation performance beneath roads and airfields in Alaska. In: *Proceedings 4th International Conference Cold Regions Engineering,* Anchorage, Alaska, 713-722.

Esch, D.C. 1988. Roadway embankments on warm permafrost, problems and remedial treatments. In: *Proceedings 5th International Per-*

mafrost Conference, Trondheim, Norway, 1223-1228.

Esch, D.C. and D. Franklin. 1989. Asphalt pavement crack control at Fairbanks International Airport. In: *Proceedings 5th International Conference Cold Regions Engineering,* St. Paul, Minnesota, 59-69.

Etkin, D.A. 1989. Greenhouse warming: Consequences for arctic climate. In: *Proceedings Workshop on Climate Changes and Permafrost: Significance to Science and Engineering.* St. Paul, Minnesota.

Ferrians, O.J., Jr. and G.D. Hobson. 1973. Mapping and predicting permafrost in North America: a review, 1963-1973. In: *North American Contribution 2nd International Permafrost Conference*, Yakutsk, USSR, 479-498.

Fielding, M.B. and A. Cohen. 1988. Predicting pipeline frost load. *Journal of the American Water Works Association,* 80(11): 62-69.

Foriero, A. and B. Ladanyi. 1991. Design of piles in permafrost under combined lateral and axial loads. *Journal of Cold Regions Engineering*, 5(3): 89-105.

Fulwider, C.W. 1973. Thermal regime in arctic earthfill dam. In: *North American Contribution 2nd International Permafrost Conference*, Yakutsk, USSR, 622-628.

Goodrich, L.E. and J.C. Plunkett. 1990. Performance of heat pump chilled foundations. In: *Proceedings 5th Canadian Permafrost Conference*, Quebec City, Quebec.

Goughnour, R.R. and O.B. Andersland. 1968. Mechanical properties of a sand-ice system. *Journal of the Soil Mechanics Foundation Division, American Society of Civil Engineers*, 94 (SM4): 923-950.

Gromov, A. 1974. *Design and Construction of Hydraulic Structures on Permafrost*. U.S. Army CRREL, Translation No. 416, Hanover, N.H.

Guymon, G.L. 1974. Regional sediment yield analysis of Alaska streams. *Journal of the Hydraulics Division*, 100(HY1): 41-51.

Haas, W.M. and A.E. Barker. 1989. Frozen gravel: a study of compaction and thaw-settlement behavior. In: *Proceedings 5th International Conference Cold Regions Engineering*, St. Paul, Minnesota, 308-319.

Hajek, J.J. and R.C.G. Haas. 1972. Predicting low-temperature cracking frequency of asphalt concrete pavements. *Highway Research Record No. 407*, 39-54.

Hanna, A.J. 1989. Baffin region salinity, pile design basis and recent pile modifications. In: *Saline Permafrost Workshop*, Winnipeg, Manitoba.

Hanna, A.J. and J. Cucheran. 1991. Problems with buried sewers in the Town of Iqaluit, NWT. In: *Proceedings 10th Canadian Hydrotechnical Conference/Engineering Mechanics Symposium,* Vancouver, B.C., Vol. IV, 354-363.

Hanna, A.J. and J. Cucheran. 1994. Monitoring and remediation of problems with the buried sewer system in the Town of Iqualuit, NWT. In: *Proceedings 7th International Cold Regions Engineering Specialty Conference*, Edmonton, Alberta, 569-577.

Hanna, A.J., R.J. Forsyth, and D. Garvin. 1990. Thaw settlement around a building on warm ice-rich permafrost. In: *Proceedings 5th Canadian Permafrost Conference*, Quebec City, Quebec, 419-424.

Hanna, A.J. and E.C. McRoberts. 1988. Permafrost slope design for a buried oil pipeline. In: *Proceedings 5th International Permafrost Conference* Trondheim, Norway, 1247-1252.

Hanna, A.J., J.M. Oswell, E.C. McRoberts, J.D. Smith, and T.W. Fridel. 1994. Initial performance of permafrost slopes: Norman Wells pipeline project, Canada. In: *Proceedings 7th International Cold Regions Engineering Specialty Conference* ASCE/CSCE, Edmonton, Alberta, 369-396.

Hanna, A.J., R.F. Saunders, G.N. Lem, and L.E. Carlson. 1983. Alaska Highway gas pipeline project (Yukon section) – thaw settlement design approach. In: *Proceedings 4th International Conference on Permafrost*, Fairbanks, Alaska, 439-444.

Harlan, R. 1973. Analysis of coupled heat-fluid transport in partially frozen soil. *Water Resources Research*, 9: 1314-1322.

Harris, S.A. 1988. Observations on the redistribution of moisture in the active layer and permafrost. In: *Proceedings 5th International Permafrost Conference*, Trondheim, Norway, 364-369.

Hartman, C.W. and P.R. Johnson. 1978. *Environmental Atlas of Alaska*. Institute of Water Resources. University of Alaska, Fairbanks, 95 p.

Hayley, D.W. 1982. Application of heat pipes to design of shallow foundations on permafrost. In: *Proceedings 4th Canadian Permafrost Conference (1981)*, Calgary, Alberta, 535-544.

Hayley, D.W. 1988. Maintenance of a railway grade over permafrost in Canada. In: *Proceedings 5th International Conference on Permafrost*, Trondheim, Norway, 3, 43-48.

Hayley, D.W. and M.A. Valeriote. 1994. Engineering a 300 km winter haul road in Canada's Arctic. In: *7th International Cold Regions Engineering Specialty Conference*, Edmonton, Alberta, 413-425.

Haynes, F.D. and J.P. Zarling. 1988. Thermosyphons and foundation design in cold regions. *Cold Regions Science and Technology*, 15: 251-259.

Haynes, F.D., J.P. Zarling, and G.E. Gooch. 1992. Performance of a thermosyphon with a 37-meter-long, horizontal evaporator. *Cold Regions Science and Technology*, 20: 261-269.

Henry, K. 1988. Use of geotextiles to mitigate frost heave in soils. In: *Proceedings 5th International Permafrost Conference*, Trondheim, Norway, 1096-1101.

Hivon, E.G. and D.C. Sego. 1991. Distribution of saline permafrost in the Northwest Territories. In: *Proceedings 44th Canadian Geotechnical Conference*, Calgary, Alberta, Vol. I, Paper 38.

Hoekstra, P. 1978. Electromagnetic methods for mapping shallow permafrost. *Geophysics*, 43(4): 782-787.

Hoekstra, P. and D. McNeill. 1973. Electromagnetic probing of permafrost. In: *North American Contribution 2nd International Permafrost Conference*, Yakutsk, USSR, 517-526.

Holubec, I. and L.S. Brzezinski. 1989. Arctic grout pile. In: *Proceedings 8th International Conference on Offshore Mechanics and Arctic Engineering*, The Hague, 731.

Holtz, R.D. (editor) 1988. Geosynthetics for soil improvement. In: *Geotechnical Special Publication No. 18*, ASCE Geotechnical Engineering Division, New York, NY.

Huculak, N.A., J.W. Twach, R.S. Thomson, and R.D. Cook. 1978. Development of the Dempster highway north of the Arctic Circle. In: *Proceedings 3rd International Conference on Permafrost*, Edmonton, Alberta, 799-805.

Hunter, J.A. 1988. Permafrost aggradation and degradation on Arctic coasts of North America. In: *Proceedings 5th International Permafrost Conference*, Trondheim, Norway, Vol 3, 27-34.

Isaacs, R.M. and J.A. Code. 1972. Problems in engineering geology related to pipeline construction. In: *Proceedings Canadian Northern Pipeline Research Conference*. NRC Technical Memorandum 104.

Johnson, E.G. (editor). 1988. *Embankment Design and Construction in Cold Regions*. ASCE, New York. 173 p.

Johnson, E.G. and G.P. Bradley. 1988. Protection of warm permafrost using controlled subsidence at Nunapitchuk Airport. In: *Proceedings 5th International Permafrost Conference*, Trondheim, Norway, 1256-1261.

Johnson, E.G. and Esch, D.C. 1977. Investigation and Analysis of the Paxson roadway icing. In: *Proceedings 2nd International Symposium on Cold Regions Engineering (1976)*, Fairbanks, Alaska, 100-126.

Johnston, G.H. 1981. *Permafrost, Engineering Design and Construction*. John Wiley and Sons, Toronto, Ontario, 540 p.

Johnston, G.H. 1983. Performance of an insulated roadway on permafrost, Inuvik, NWT. In: *Proceedings 4th International Conference on Permafrost*, Fairbanks, Alaska, 548-553.

Jones, R.H. and K.J. Lomas. 1983. The frost susceptibility of granular materials. In: *Proceedings 4th International Permafrost Conference*, Fairbanks, Alaska, 554-559.

Judge, A.S. 1973. The prediction of permafrost thickness. *Canadian Geotechnical Journal*, 10(1): 1-11.

Judge, A.S., C.M. Tucker, J.A. Pilon, and B.J. Moorman. 1991. Remote sensing of permafrost by ground penetratiing radar at two airports in Arctic Canada. *Arctic,* 44, Suppl. 1: 40-48.

Kaplar, C.W. 1963. Laboratory determination of the dynamic moduli of frozen soils and of ice. In: *Proceedings 1st International Permafrost Conference,* Lafayette, Indiana, 293-300.

Kaplar, C.W. 1974. *Freezing Test for Evaluating Reactive Frost Susceptibility of Various Soils.* U.S. Army CRREL, Technical Report No. 250, Hanover, N.H.

Keyes, D.E. 1977. Ice and snow construction: workpads, roads, airfields and bridges. In: *Proceedings 2nd International Symposium on Cold Regions Engineering (1976),* Fairbanks, Alaska, 369-382.

Kinney, T.C. and R. Abbott. 1984. Geotextiles used to reinforce roads over voids. In: *Proceedings 3rd International Conference on Cold Regions Engineering,* Edmonton, Alberta, 493-505.

Kinney, T.C. and B. Connor. 1987. Geosynthetics supporting embankments over voids. *Journal of Cold Regions Engineering,* 1(4): 158-170.

Kinney, T.C. and B. Connor. 1990. Geosynthetic reinforcement of paved road embankments on polygonal ground. *Journal of Cold Regions Engineering,* 4(2): 102-112.

Kinney, T.C. and B.M. Savage. 1989. Using geosynthetics to control lateral spreading of pavements in Alaska – preliminary results. In: *Proceedings Geosynthetics '89 Conference,* San Diego, California, 324-333.

Kinney, T.C. and K.A. Troost. 1984. Thaw strain of laboratory compacted frozen gravel. In: *Proceedings 3rd International Conference on Cold Regions Engineering,* Edmonton, Alberta, 811-822.

Klein, C.A., C.R. Wilson, B.D. Benson, and G.W. Carpenter. 1986. Installation of thermistor strings in test borings: a comparison of methods and results. In: *Proceedings 4th International Conference Cold Regions Engineering,* Anchorage, Alaska, 200-206.

Konrad, J.M. and N.R. Morgenstern. 1980. A mechanistic theory of ice lens formation in fine-grained soils. *Canadian Geotechnical Journal,* 17(4): 473-486.

Konrad, J.M. and N.R. Morgenstern. 1981. The segregation potential of a freezing soil. *Canadian Geotechnical Journal,* 18(4): 482-491.

Konrad, J.M. and N. Morgenstern. 1982. Effects of applied pressure on freezing soils. *Canadian Geotechnical Journal,* 19: 494-505.

Konrad, J.M. and N.R. Morgenstern. 1983. Frost susceptibility of soils in terms of their segregation potential. In: *Proceedings 4th International Permafrost Conference,* Fairbanks, Alaska, 660-665.

Kozisek, L. and J.W. Rooney. 1986. *Soil Stabilization Test Strips, Bethel, Alaska.* Alaska DOT&PF Report AK-RD-86-27. 74 p.

Kreig, R.A. and R.D. Regar. 1982. *Air-Photo Analysis and Summary of Landform Soil Properties Along the Route of the Trans-Alaska Pipeline System.* Division of Geological and Geophysical Surveys, Report 66, Department of Natural Resources, State of Alaska. 149 p.

Krzewinski, T.G. and R.G. Tart, Jr. (editors). 1985. *Thermal Design Considerations in Frozen Ground Engineer.* ASCE, New York. 277 p.

Kudoyarov, V.I., M.P. Pavchich, and V.G. Radchenko. 1978. Earth and rock excavation technique for construction of hydraulic structures in regions of permafrost. In: *Proceedings Applied Techniques in Cold Environments,* Anchorage, Alaska, 201-215.

Lachenbruch, A.H. and B.V. Marshall. 1986. Changing climate: geothermal evidence from permafrost in the Alaskan Arctic. *Science,* 234(Sept.): 689-696.

Ladanyi, B. 1972. An engineering theory of creep of frozen soils. *Canadian Geotechnical Journal,* 9: 63-80.

Ladanyi, B. 1983. Shallow foundations on frozen soil: creep settlement. *Journal of Geotechnical Engineering,* 109(11): 1434-1448.

Lafleche, P.T., A.S. Judge and J.A. Pilon. 1988. The use of ground probing radar in the design and monitoring of water retaining embankments in permafrost. In: *Proceedings 5th International Permafrost Conference,* Trondheim, Norway, 971-976.

Lambe, T.W. and C.W. Kaplar. 1971. *Additives for Modifying the Frost Susceptibility of Soils,*

Part I. US Army COE CRREL TR 123 I, Hanover, New Hampshire. 44 p.

Lambe, T.W., C.W. Kaplar, and T.J. Lambie. 1971. *Additives for Modifying the Frost Susceptibility of Soils, Part II*. US Army COE CRREL TR 123 II, Hanover, New Hampshire. 44 p.

Lewellen, R.I. 1973. The occurrence and characteristics of nearshore permafrost, northern Alaska. In: *North American Contribution 2nd International Permafrost Conference*, Yakutsk, USSR, 131-136.

Linell, K.A. 1973. Long term effects of vegetative cover on permafrost stability in an area of discontinuous permafrost. In: *North American Contribution 2nd International Permafrost Conference*, Yakutsk, USSR, 688-693.

Linell, K.A., F.G. Hennion, and E.F. Lobacz. 1963. *Corps of Engineers Pavement Design in Areas of Seasonal Frost*. U.S. Highway Research Board, Record 33, Washington, D.C.

Linell, K.A. and C.W. Kaplar. 1963. Description and classification of frozen soils. In: *Proceedings 1st International Permafrost Conference*, Lafayette, Indiana, 481-486.

Linell, K.A. and E.F. Lobacz. 1980. *Design and Construction of Foundations in Areas of Deep Seasonal Frost and Permafrost*. Special Report No. 80-34, U.S. Army CRREL, Hanover, N.H.

Livingston, H. and E. Johnson. 1978. Insulated roadway subdrains in the subarctic for the prevention of spring icings. In: *Proceedings Applied Techniques in Cold Environments*, Anchorage, Alaska, 513-521.

Lobacz, E.F., G.D. Gilman, and F.B. Hennion. 1973. Corps of Engineers design of highway pavements in areas of seasonal frost. In: *Proceedings Synposium on Frost Action on Roads*, Oslo, Norway,142-152.

Loch, J. and B. Kay. 1978. Water redistribution in partially frozen, saturated silts under several temperature gradients and overburden loads. *Soil Science Society of America Proceedings*, 42: 400-406.

Long, E.L. 1963. The long thermopile. In: *Proceedings 1st International Permafrost Conference*, Lafayette, Indiana, 487-491.

Long, E.L. 1973. Designing friction piles for increased stability at lower installed cost in permafrost. In: *North American Contribution 2nd International Permafrost Conference*, Yakutsk, USSR, 693-699.

MacDonald, D. 1963. Design of Kelsey dykes. In: *Proceedings 1st International Permafrost Conference*, Lafayette, Indiana, 492-496.

MacFarlane, I.C. 1968. Strength and deformation tests on frozen peat. In: *Proceedings 3rd International Peat Congress*, Quebec.

MacFarlane, I.C. (editor) 1969. *Muskeg Engineering Handbook*. University of Toronto Press, Ontario, 320 p.

Mackay, J.R., J. Ostrick, C.P. Lewis, and D.K. Mackay. 1979. Frost heave at ground temperatures below 0°C, Inuvik, Northwest Territories. *Current Research, Part A; Geologic Survey of Canada*, Paper 79-1A, 403-405.

Mageau, D.W. and Rooney, J.W. 1984. *Thermal Erosion of Cut Slopes in Ice-Rich Soil*. ADOT&PF, Report No. FHWA-AK-RD-85-02. 94 p.

Mahmood, A., M.G. Schleger, and S.C. Shrestha. 1984. Geologic origin and fill properties of the Barrow unit materials. In: *Proceedings 3rd International Conference Cold Regions Engineering*, Edmonton, Alberta, 1045-1058.

Manhoney, J.P. and T.S. Vinson. 1983. A mechanistic approach to pavement design in cold regions. In: *Proceedings 4th International Permafrost Conference*, Fairbanks, Alaska, 779-784.

Manikian, V. 1983. Pile driving and load tests in permafrost for the Kuparuk pipeline system. In: *Proceedings 4th International Permafrost Conference*, Fairbanks, Alaska, 804-880.

Maslovski, G.F., L.N. Toropov, A.F. Vasiliev, L.K. Domanski and A.A. Kusnetsov. 1978. Hydraulic structures in the north regions of the USSR. In: *Proceedings Applied Techniques in Cold Regions*, Anchorage, Alaska, 522-537.

McFadden, T. 1985. *Performance of the Thermotube Permafrost Stabilization System in the Airport Runway at Bethel, Alaska*. ADOT&PF, Report AK-RD-86-20. 46 p.

McFadden, T. 1987. Using soil temperatures to monitor thermosyphon performance. *Journal of Cold Regions Engineering,* Vol.1(4), 145-157.

McFadden, T. 1989. Thermal performance degradation of wet insulations in cold regions. *Journal of Cold Regions Engineering,* 2(1): 25-34.

McFadden, T. and C. Siebe. 1986. Stabilization of a permafrost subsidence in the airport runway at Bethel, Alaska. In: *Proceedings 4th International Conference Cold Regions Engineering*, Anchorage, Alaska, 118-133.

McHattie, R. 1980. Highway pavement cracks: An Alaskan overview. *The Northern Engineer*, 12(4): 17-21.

McHattie, R.L. and D.C. Esch. 1983. Benefits of a peat underlay used in road construction on permafrost. In: *Proceedings 4th International Permafrost Conference*, Fairbanks, Alaska, 826-831.

McHattie, R. and Esch, D. 1988. Embankment failure from creep of permafrost foundation soils. a case history. In: *Proceedings 5th International Permafrost Conference*, Trondheim, Norway, 1292-1297.

McRoberts, E.C. 1975. Some aspects of a simple secondary creep model for deformations in permafrost slopes. *Canadian Geotechnical Journal*, 12: 98-105.

McRoberts, E.C. 1982. Shallow foundations in cold-regions: design. *Journal of Geotechnical Engineering Division, ASCE*, 108(GT10): 1338-1349.

McRoberts, E.C. and N.R. Morgenstern. 1974a. The stability of thawing slopes. *Canadian Geotechnical Journal*, 11: 447-469.

McRoberts, E.C. and N.R. Morgenstern. 1974b. Stability of slopes in frozen soil, Mackenzie Valley, NWT. *Canadian Geotechnical Journal*, 11: 554-573.

McRoberts, E.C. and J.F. Nixon. 1977. Extensions to thawing slope stability theory. In: *Proceedings 2nd International Symposium on Cold Regions Engineering (1976)*, Fairbanks, Alaska, 262-276.

McRoberts, E.C., J.F. Nixon, A.J. Hanna, and A.R. Pick. 1985. Geothermal considerations for wood chips used as permafrost slope insula-tion. In: *Proceedings 4th International Symposium on Ground Freezing*, Sapporo, Japan.

Mellor, M. 1965. *Blowing Snow*. U.S. Army, CRREL, Monograph III A3C, Hanover, N.H.

Mellor, M. 1973. *Use of Liquid Explosives for Excavation of Frozen Ground*. U.S. Army, CRREL, Report No. MP 600, Hanover, N.H.

Mellor, M. 1993. *Notes on Antarctic Aviation*. U.S. Army Corps of Engineers, CRREL Report 93-14.

Miller, D.E. and L.A. Johnson. 1990. Pile settlement in saline permafrost – a case history. In: *Proceedings 5th Canadian Permafrost Conference*, Quebec City, Quebec, 371-378.

Montgomery, C.J., R. Gerard, W.J. Huiskamp, and R.W. Kornelsen. 1984. Application of ice engineering to bridge design standards. In: *Proceedings 3rd International Conference Cold Regions Engineering*, Edmonton, Alberta, 795-810.

Morack, J.L. and J.C. Rogers. 1978. Extending foundation studies in sporadic permafrost. In: *Proceedings Applied Techniques for Cold Environments*, Anchorage, Alaska, 51-58.

Morgenstern, N.R. and J.F. Nixon. 1971. One-dimensional consolidation of thawing soils. *Canadian Geotechnical Journal*, 8: 558-565.

Morrissey, L.A. 1988. Predicting the occurrence of permafrost in the Alaskan discontinuous zone with satellite data. In: *Proceedings 5th International Permafrost Conference*, Trondheim, Norway, 213-217.

National Research Council of Canada (NRCC). 1970. *Climatic Information for Building Design in Canada*. NRC No. 11153, Ottawa, Ontario.

Nelson, R.A., U. Lusher, J.W. Rooney, and A.A. Stramler. 1983. Thaw strain data and thaw settlement predictions for Alaskan soils. In: *Proceedings 4th International Permafrost Conference*, Fairbanks, Alaska, 912-917.

Neukirchner, R.J. 1985. Pile creep designs for frozen layered profiles. In: *Proceedings Arctic Offshore Engineering Conference*, San Francisco, California, 1103-1111.

Neukirchner, R.J. 1988. Standard method for pile load tests in permafrost. In: *Proceedings 5th*

International Permafrost Conference, Trondheim, Norway, 1147-1151.

Neukirchner, R.J. and J.F. Nixon. 1987. Behavior of laterally loaded piles in permafrost. *Journal of Geotechnical Engineering (ASCE)*, 113(1): 1-4.

Neukirchner, R.J. and K.J. Nyman. 1985. Creep rate analysis of pile load test data. In: *Proceedings Arctic Offshore Engineering*, San Francisco, California, 1112-1121.

Nixon, J.F. 1973. Thaw-consolidation of some layered systems. *Canadian Geotechnical Journal*, 10: 617-631.

Nixon, J.F. 1978a. Geothermal aspects of ventilated pad design. In: *Proceedings 3rd International Permafrost Conference*, Edmonton, Alberta, 840-846.

Nixon, J.F. 1978b. First Canadian geotechnical colloquium: foundation design approaches in permafrost areas. *Canadian Geotechnical Journal*, 15: 96-112.

Nixon, J.F. 1982. Field frost heave predictions using the segregation potential concept. *Canadian Geotechnical Journal*, 19: 526-529.

Nixon, J.F. 1983. Geothermal design of insulated foundations for thaw prevention. In: *Proceedings 4th International Permafrost Conference*, Fairbanks, Alaska, 924-927.

Nixon, J.F. 1984. Laterally loaded piles in permafrost. *Canadian Geotechnical Journal*, 21(3): 431-438.

Nixon, J.F. 1986. Pipeline frost heave predictions using a 2-D thermal model. In: *Proceedings ASCE Convention*, Boston, Massachusetts, 67-82.

Nixon, J.F. 1987a. Thermally induced heave beneath chilled pipelines in frozen ground. *Canadian Geotechnical Journal*, 24: 260-266.

Nixon, J.F. 1987b. Ground freezing and frost heave: a review. *The Northern Engineer*, 19(3&4): 8-18.

Nixon, J.F. 1988. Pile load tests in saline permafrost at Clyde River, Northwest Territories. *Canadian Geotechnical Journal*, 25(1): 24-32.

Nixon, J.F. 1991. Discrete ice lens theory for frost heave in soils. *Canadian Geotechnical Manual,* 28(6): 843-859.

Nixon, J.F. and G. Lem. 1984. Creep and strength testing of frozen saline fine-grained soils. *Canadian Geotechnical Journal*, 21: 518-529.

Nixon, J.F. and E. McRoberts. 1976. A design approach for pile foundations in permafrost. *Canadian Geotechnical Journal*, 13: 40-57.

Nixon, J.F. and N.R. Morgenstern. 1974. Thaw-consolidation tests on undisturbed fine-grained permafrost. *Canadian Geotechnical Journal*, 11: 202-214.

Nottingham, D. and A.B. Christopherson. 1983. *Design Criteria For Driven Piles in Permafrost.* ADOT&PF, Report No. AK-RD-83-19, 33 p.

Odom, W.B. 1983. Practical application of underslab ventilation system: Prudhoe Bay case study. In: *Proceedings 4th International Permafrost Conference*, Fairbanks, Alaska, 940-944.

Oliphant, J.L., A.R. Tice, and Y. Nakano. 1983. Water migration due to a temperature gradient in frozen soil. In: *Proceedings 4th International Permafrost Conference*, Fairbanks, Alaska, 951-956.

Olsen, M.E. 1984. Synthetic insulation in arctic roadway embankments. In: *Proceedings 3rd International Conference Cold Regions Engineering*, Edmonton, Alberta, 739-752.

Osterkamp, T.E. and R.W. Jurick. 1980. Detecting massive ground ice in permafrost by geophysical methods. *The Northern Engineer,* 12(4): 27-30.

Osterkamp, T.E., R.W. Jurick, G.A. Gislason, and S.I. Akasofu. 1980. Electrical resistivity measurements in permafrost terrain at the Engineer Creek road cut, Fairbanks, Alaska. *Cold Regions Science and Technology*, 3: 277-286.

Osterkamp, T.E., et al. 1986. *Low Temperature Traverse Cracks in Asphalt Pavements in Interior Alaska.* ADOT&PF, Report No. AK-RD-86-26, 60 p.

Penner, E. 1969. *Particle Size as a Basis for Predicting Frost Action in Soils of Building Research*, Paper No. 406, Ottawa, Ontario.

Penner, E. 1974. Uplift forces on foundations in frost heaving soils. *Canadian Geotechnical Journal*, 11: 323-338.

Penner, E. and L.W. Gold. 1971. Transfer of heaving forces by adfreezing to columns and foundation walls in frost-susceptible soils. *Canadian Geotechnical Journal*, 8: 514-526.

Perfect, E. and P.J. Williams. 1980. Thermally induced water migration in frozen soils. *Cold Regions Science and Technology*, 3: 101-109.

Péwé,T.L. 1982. *Geological Hazards of the Fairbanks Area, Alaska*. Alaska Division of Geological and Geophysical Surveys Special Report 15. 109 p.

Phukan, A. 1983. Long-term creep deformation of roadway embankment on ice-rich permafrost. In: *Proceedings 4th International Permafrost Conference*, Fairbanks, Alaska, 994-999.

Phukan. A. 1985. *Frozen Ground Engineering*. Prentice-Hall International Series in Civil Engineering Mechanics.

Phukan, A., R.D. Abbott, and J.E. Cronin. 1978. Self-refrigerated gravel pad foundations in frozen soils. In: *Proceedings Applied Techniques for Cold Environments*, Anchorage, Alaska, 1003-1016.

Pihlainen, J.A. and G.H. Johnston. 1963. *Guide to a Field Description of Permafrost*. Canada, National Research Council, Associate Committee on Soil and Snow Mechanics, Technical Memorandum 79, 23 p.

Reckard, M., D. Esch and R. McHattie. 1988. *Peat Used as Roadway Insulation over Permafrost*. Alaska DOT&PF Report No. AK-RD-88-11, 28 p.

Reimers, S. 1980. Drilling and sampling in frozen ground. *The Northern Engineer*, 12: 13-17.

Retherford, R.W. 1983. Power lines in the Arctic and subarctic – experience in Alaska. In: *Proceedings, 4th International Permafrost Conference,* Fairbanks, Alaska, 1060-1065.

Riddle, C.H. and P.K. Hardcastle. 1991. Drilling and sampling of permafrost for site investigation purposes: a review. In: *Proceedings International Arctic Technology Conference*, Anchorage, Alaska, 611-620.

Riddle, C.H., J.W. Rooney, and S.r. Bredthauer. 1988. Yukon River bank stabilization: a case study. In: *Proceedings 5th International Permafrost Conference*, Trondheim, Norway, 1312-1317.

Rieke, R.D., T.S. Vinson, and D.W. Mageau. 1983. The role of specific surface area and related index properties in the frost heave susceptibility of soils. In: *Proceedings, 4th International Permafrost Conference*, Fairbanks, Alaska, 1066-1071.

Rix, H.H. 1969. Vertical movements and crack-width changes on highway pavement surfaces. *Canadian Geotechnical Journal,* 6: 253-270.

Roberts, W.S., C.W. Lovell, and T.R. West. 1978. Predicting feasibility of cold weather earthwork. In: *Proceedings Applied Techniques for Cold Environments,* Anchorage, Alaska, 960-972.

Robinsky, E.I. and K.E. Bespflug. 1973. Design of insulated foundations. *Journal of the Soil Mechanics and Foundations Division*, 99(9): 649-667.

Robinson, S.D., B.J. Moorman, A.S. Judge, and S.R. Dallimore. 1993. The characterization of massive ground ice at Yaya Lake, Northwest Territories using radar stratigraphy techniques. In: *Current Research, Part B: Geological Survey of Canada*, Paper 93-1B, 23-32.

Rooney, J.W., D. Nottingham, and B.E. Davison. 1977. Driven H-pile foundations in frozen sands and gravels. In: *Proceedings 2nd International Symposium on Cold Regions Engineering (1976)*, Fairbanks, Alaska, 169-188.

Rooney, J.W., C.H. Riddle, and R.L. Scher. 1991. Foundation rehabilitation at ADOT/PF Jim River maintenance camp. In: *Proceedings International Arctic Technology Conference*, Anchorage, Alaska, 829-840.

Ryan, W.L. 1981. Permafrost's impact on utility design. In: *Design of Water and Wastewater Services for Cold Climate Communities.* Pergamon Press, Oxford, England, 37-53.

Sanger, F.J. 1969. *Foundations of Structures in Cold Regions.* U.S. Army, CRREL, Monograph III-C4, Hanover, N.H., 91 p.

Savage, B.M. 1991. *Use of Geogrids for Limiting Longitudinal Cracking in Roads on Permafrost.* ADOT&PF Statewide Research AK-RD-91-12, 107 p.

Savarenskii, F.P. 1960. *Dams in Permafrost Regions.* U.S. Army, CRREL, Translation No. 29, Hanover, N.H.

Scher, R.L. 1982. Thaw settlement prediction in drill sites built with frozen gravels. In: *Proceed-*

ings 14th Offshore Technology Conference, Houston, Texas, 449-458.

Scher, R.L. 1991. Foundation heave induced by a subgrade cooling system over permafrost. In: *Proceedings International Arctic Technology Conference,* Anchorage, Alaska, 483-492.

Scher, R.L. 1996. Environmental-induced longitudinal cracking in cold regions pavement. In: *Proceedings 8th International Cold Regions Engineering Conference,* Fairbanks, Alaska [In press].

Schlegel, M.G. and K.O. Stangl. 1990. Use of weathered rock for engineered fill in permafrost regions of Alaska. In: *Proceedings 5th Canadian Permafrost Conference,* Quebec, 379-388.

Sellmann, P.V. 1980. *Regional Distribution and Characteristics of Bottom Sediments in Arctic Coastal Waters of Alaska.* U.S. Army CRREL, Special Report 80-15, Hanover, N.H.

Sherif, M.A., I. Ishibashi, and W. Ding. 1977. Frost-heave potential of silty sands. In: *Proceedings 2nd International Symposium on Cold Regions Engineering (1976),* Fairbanks, Alaska, 239-251.

Shuster, J.A. 1970. Laboratory testing and characterization of permafrost for foundation uses. In: *Proceedings Symposium on Cold Regions Engineering,* Fairbanks, Alaska, 73-118.

Sinha, A.K. 1988. EM soundings for mapping complex geology in the permafrost terrain of northern Canada. In: *Proceedings 5th International Permafrost Conference,* Trondheim, Norway, 994-999.

Smith, L.B., J.P. Graham, J.F. Nixon, and A.S. Washuta. 1991. Thermal analysis of forced-air and thermosyphon cooling systems for Inuvik airport expansion. *Canadian Geotechnical Journal,* 28: 399-409.

Smith, L.B., W.G. Notenboom, M. Campbell, S. Cheema, and T. Smyth. 1989. Pangnirtung water reservoir: geotechnical aspects. *Canadian Geotechnical Journal,* 26(3): 335-347.

Smith, L.B., A. Shevkenek, and R. Milburn. 1984. Design and performance of earthworks water reservoirs in the Northwest Territories. In: *Proceedings 3rd International Specialty Conference on Cold Regions Engineering,* Edmonton, Alberta, 323-342.

Smith, M.W. 1985. Observations of soil freezing and frost heave at Inuvik, Northwest Territories, Canada. *Canadian Journal of Earth Science,* 22: 283-290.

Smith, M.W. and C.T. Hwang. 1973. Thermal disturbance due to channel shifting, Mackenzie Delta, NWT, Canada. In: *North American Contribution 2nd International Permafrost Conference,* Yakutsk, USSR, 51-60.

Smith, N. 1978. Techniques for using membrane encapsulated soil layers in roads and airfields in cold regions. In: *Proceedings Applied Techniques for Cold Regions,* Anchorage, Alaska, 560-570.

Smith, W.H. 1976. Frost loadings on underground pipe. *Journal of the American Water Works Association,* 68(12): 673-674.

Soo, S. and B.B. Muvdi. 1992. Design method for frozen-soil retaining wall. *Journal of Cold Regions Engineering,* 6(2): 73-89.

Tart, R.G., Jr. and U. Luscher. 1981. Construction and performance of frozen gravel fills. In: *Proceedings Specialist Conference Northern Community Environments,* Seattle, Washington, 693-704.

Taylor, D.W. 1967. *Fundamentals of Soil Mechanics.* John Wiley & Sons, New York.

Tobiasson, W. and P. Johnson. 1978. The details behind a typical Alaskan pile foundation. In: *Proceedings 3rd International Permafrost Conference,* Edmonton, Alberta, 891-897.

Tomayko, D.J. 1974. Elevated snow roads in the Antarctic. *The Military Engineer* No. 429.

Tsvetkova, S.G. 1960. *Experiences in Dam Construction in Permafrost Regions.* Publishing House of Academy of Sciences, USSR, Moscow.

Tsytovich, N.A. 1975. *The Mechanics of Frozen Ground.* McGraw-Hill, New York. 426 p.

U.S. Department of the Army. 1985. *Pavement Design for Seasonal Frost Conditions.* Technical Manual TM 5-818-2, Washington, D.C.

U.S. Department of the Navy. 1979. *Civil Engineering, Pavement.* NAVFAC Design Manual 5.4, Alexandria, Virginia.

Veldman, W.M. and E.K. Yaremko. 1978. Design and construction of river training structures. In:

Proceedings Applied Techniques for Cold Environments, Anchorage, Alaska, 852-863.

Vinson, T.S., J.P. Mahoney, and M.J. Kaminski. 1984. Cement stabilization for road construction in cold regions. In: *Proceedings 3rd International Conference Cold Regions Engineering,* Edmonton, Alberta, 619-636.

Vita, C.L. and J.W. Rooney. 1978. Seepage-induced erosion along buried pipelines. In: *Proceedings Applied Techniques in Cold Environments,* Anchorage, Alaska, 864-874.

Vita, C.L., T.S. Vinson, and J.W. Rooney. 1986. *Bethel Airport CTB-AC Pavement Performance Analysis.* Alaska DOT&PF Report AK-RD-86-31, 74 p.

Washburn, D.S. and A. Phukan. 1988. Discontinuous permafrost mapping using the EM-31. In: *Proceedings 5th International Permafrost Conference,* Trondheim, Norway, 1018-1023.

Watson, G.H., W.A. Slusarchuk, and R.K. Rowley. 1973. Determination of some frozen and thawed properties of permafrost soils. *Canadian Geotechnical Journal,* 10: 592-606.

Weaver, J.S. and N.R. Morgenstern. 1981. Pile design in permafrost. *Canadian Geotechnical Journal,* 18: 357-370.

Wellman, J.H., E.S. Clark, and A.C. Condo. 1977. Design and construction of synthetically insulated gravel pads in the Alaskan Arctic. In: *Proceedings 2nd International Symposium on Cold Regions Engineering,* Fairbanks, Alaska, 62-85.

Wexler, R.L. 1983. Diurnal freeze-thaw frequencies in the high latitudes: a climatological guide. In: *Proceedings 4th International Permafrost Conference,* Fairbanks, Alaska, 1390-1395.

Yarmak, E., Jr. and E.L. Long. 1986. Monitoring techniques for thermosyphons. In: *Proceedings 4th International Conference Cold Regions Engineering,* Anchorage, Alaska, 207-219.

Zarling, J.P. and A.W. Braley. 1987. *Thaw Stabilization of Roadway Embankments Constructed over Permafrost.* ADOT&PF, Report No. FHWA-AK-RD-87-20, 34 p.

Zarling, J.P., B. Connor, and D.J. Goering. 1983. Air duct systems for roadway stabilization over permafrost areas. In: *Proceedings 4th*

International Permafrost Conference, Fairbanks, Alaska, 1463-1468.

Zarling, J.P. and F.D. Haynes. 1985a. Laboratory tests and analysis of thermosyphons with inclined evaporator sections. In: *Proceedings 4th International Offshore Mechanics and Arctic Engineering Symposium,* Dallas, Texas, 31-37.

Zarling, J.P., and F.D. Haynes. 1985b. *Thermosyphon Devices and Slab-on-Grade Foundation Design.* ADOT&PF, Report No. AK-RD-86-16. 63 p.

Zarling, J.P., P. Hansen, and L. Kozisek. 1990. Design and performance experience of foundations stabilized with thermosyphons. In: *Proceedings 5th Canadian Permafrost Conference,* Quebec, 365-370.

3.10 Bibliography: Key References for Cold Regions Geotechnical Engineering

Andersland, O.B. and D.M. Anderson. (editors)1978. *Geotechnical Engineering for Cold Regions.* McGraw-Hill, Inc., New York. 566 p.

Canadian Geotechnical Society. 1992. *Canadian Foundation Engineering Manual,* 3rd Edition. BiTech Publishers, Inc. Vancouver, B.C. 512 p.

Johnson, E.G. (editor). 1988. *Embankment Design and Construction in Cold Regions.* Technical Council in Cold Regions Engineering Monograph, American Society of Civil Engineers, New York, NY. 173 p.

Johnston, G.H. 1981. *Permafrost, Engineering Design and Construction.* John Wiley and Sons, Toronto, Ontario. 540 p.

Jordan, D.F. and G.N. McDonald. (editors). 1983. *Cold Regions Construction.* Technical Council in Cold Regions Engineering Monograph, American Society of Civil Engineers, New York, NY. 121 p.

Krewinski, T.G. and R.G. Tart, Jr. (editors). 1985. *Thermal Design Considerations in Frozen Ground Engineering.* Technical Council in Cold Regions Engineering Monograph, American Society of Civil Engineers, New York, NY. 277 p.

McFadden, T.T. and F.L. Bennett. 1991. *Construction in Cold Regions.* John Wiley & Sons, Inc. 615 p.

Nixon, J.F. 1978. First Canadian geotechnical colloquium: foundation design approaches in permafrost areas. *Canadian Geotechnical Journal*, 15: 96-112.

Phukan, A. 1985. *Frozen Ground Engineering*. Prentice-Hall, Inc. 336 p.

Tsytovich, N.A. 1975. *The Mechanics of Frozen Ground*. (edited by G.K. Swinzow). McGraw-Hill, New York. 426 p.

Weaver, J.S. and N.R. Morgenstern. 1981. Pile design in permafrost. *Canadian Geotechnical Journal*, 18: 357-370.

SECTION 4

THERMAL CONSIDERATIONS

3rd Edition Steering Committee Coordinator

Daniel W. Smith

3rd Edition Principal Author

Daniel W. Smith

Section 4   Table of Contents

Section 4 List of Figures

Section 4 List of Tables

4 THERMAL CONSIDERATIONS

4.1 Introduction

Utility systems in cold regions must function under severe climatic conditions. Thermal considerations are at least as important as the hydraulics and structural features of utility systems and must be included in the selection and design of materials, components and processes. Thermal analyses are necessary to design for freeze protection, foundation stability, thermal stress, and economy. In this section, the major emphasis is on the thermal design of piped water and sewer systems. Thermal aspects are discussed, and simple equations and illustrative examples are presented for solving some of the thermal problems encountered in utility system design.

4.1.1 Site Considerations.
Climatic and geotechnical site information is necessary for thermal analysis and design. Adequate air temperature data, including the range, mean, and various indices are usually available from nearby weather stations or from published charts and reports (Canadian Department of Transport,1967; National Research Council, 1970; Hartman and Johnson, 1978) but the local microclimate can modify these values (see Section 17). Local ground temperatures are often not available, and extrapolations are inadvisable since they can vary significantly with air temperature, snow cover, vegetation, drainage, topography, and soil properties. The most reliable approach is to obtain actual long-term site measurements. Limited data must be modified to estimate the extreme climatic conditions or used with allowances for surface changes resulting from construction and development.

Geotechnical conditions are frequently highly variable within a small area. Site-specific surveys consistent with the thermal and structural design considerations are imperative. Of primary concern are movement and possible failure of structures due to the freezing and thawing of the ground. Therefore, the maximum thickness of the active layer, the soil thermal properties, and frost susceptibility must be determined. In permafrost areas, particularly in soils with a high ice content, the soil survey must extend to the maximum range of anticipated thermal effects. Surface conditions and drainage patterns must also be noted.

4.1.2 Design Considerations.
The primary areas of concern in the design of utilities in cold regions are failure of pipes due to freezing of water,

thaw-settlement or heaving of foundation soil, thermal strain and associated stress, and economical operation.

Utility systems in cold regions are thermally designed with a conservative safety factor, and often the worst conditions that could occur simultaneously are considered. This is justified by:

- simplifications and assumptions within the thermal equations and models;

- limited data and random nature of the climatic and physical site conditions;

- variations and assumptions in the physical and thermal properties of materials such as insulation, soil and pipes; and

- the extreme consequences of system failure.

Although thermal characteristics may be precisely defined, the application of the principles and the control of heat movement are complicated. In practice, it is often the unexpected or unforeseen conditions that result in damage or failure. Therefore thermal design must encompass more than precise thermal analysis.

4.2 Thermal Considerations of Ground

Ground temperatures can be predicted knowing the soil properties and the climatic data for a given location. But the analytical methods presented here must not be substituted for actual ground temperature measurements at the site, if actual readings can be obtained. Also, the longer the period of record the more confidence one can have in their applicability to the proposed construction. The formulas presented have predicted ground temperatures and freezing or thawing depths to within 90 percent of those measured. Most of the formulas, however, are based on observations of homogeneous and isotropic soil, which is seldom found. All thermal nomenclature and units are in accordance with standards adopted by the Canadian Geotechnical Society (Barsvary, 1980).

4.2.1 Thermal Properties of the Soil.
Thermal properties of the soil are extremely important to the design engineer, because they help determine the amount of thawing or freezing that will result from changes at the surface and subsurface of the ground. Many experiments and field tests have been carried out to determine the thermal properties of

the soil once its physical properties such as dry density and moisture content are known (see Figure 3-9). As stated earlier, it is best to actually measure the soil properties in the field at the proposed location of the construction.

Thermal Conductivity. Thermal conductivity is a measure of the rate at which heat moves through a material to bring about a one-degree temperature difference. Thermal conductivity (k) generally varies directly with soil moisture and density. It increases when soil freezes. Thermal probes have been developed which can be used in the field to measure thermal conductivity (Penner, 1970; Slusarchuk and Foulger, 1973). In this time-consuming process extremely sensitive thermisters are used. Accuracies are in the order of ±4 percent.

Kersten (1949; 1963) developed an approach for estimating the soil thermal conductivity, provided the dry density and moisture content of the soil are known. His curves are widely used and have proven reasonably accurate, even though they were based on laboratory tests carried out on remoulded samples. Figures 4-1 through 4-3 were drawn by Andersland and Anderson (1978) according to Kersten's formulas. The data for heat conduction through frozen soil (in the process of thawing) or thawed soil (in the process of freezing) are given for three soil types. The following are guidelines for determining the three soil types:

1) Silt and clay soils are defined as those containing more than 50 percent silt and clay.

2) Sandy soils consist of 70 percent or more gravel and sand.

3) If a soil contains 30 to 50 percent clay and silt or sandy loam, use a combination of Figure 4-1 and 4-2.

Note: Gravels are not covered by these curves. See Andersland and Anderson (1978).

The thermal conductivity of thawed soil is much more sensitive to the composition of the soil than it is for the same, but frozen, soil. Therefore, more care should be taken in determining the thermal conductivity of unfrozen soils. The thermal conductivity of peat, for example, is more than 10 times greater in the winter than in the summer. Therefore, a peat deposit acts as a "heat pump", conducting heat rapidly out of the ground in the winter, and preventing its re-entry in the summer. This helps to preserve the permafrost. In fact, if the peat is dry in the summer, it can be nearly as good an insulator as fiberglass (k = 0.07W/(m• °C)).

Specific Heat Capacity. The specific volumetric heat capacity (C) of a substance is the heat absorbed by a unit volume to raise its temperature by 1°C.

The specific heat capacity (c) is the amount of heat required to raise a unit mass 1°C.

These two properties are related by:

$$C = c\rho_d$$

where:

$C = kJ/(m^3 \cdot °C)$

$c = kJ/(kg \cdot °C)$

$\rho_d = kg/m^3$

The specific heat capacity of soils does not vary greatly.

For dry soils,　　$c = 0.710$ kJ/(kg•°C)

and for water,　　$c_w = 4.187$ kJ/(kg•°C)

Ice has half the specific heat capacity of water. For thawed soils with water content, w (%):

For frozen soil:

$$c_f = 0.710 + \frac{0.5 \bullet c_w \bullet w}{100} (\frac{kJ}{kg \bullet °C})$$

Thermal Diffusivity. Thermal diffusivity (α) is an index of the rate at which a material undergoes a temperature change. Practically, this is a measure of the rate at which heat moves into soil in the spring and summer.

Thawed soils:　　$\alpha_u = k_u/C_u$

Frozen soils:　　$\alpha_f = k_f/C_f$

where:

k_u, k_f = thermal conductivity, W/(m•°C)

C_u, C_f = specific volumetric heat capacity, (J/(m³•°C))

α_u, α_f = thermal diffusivity (m²/s)

Soil with high diffusivity heats or cools faster than one with low diffusivity. Wet soil has a lower thermal conductivity and a higher heat capacity than dry soil, and therefore a lower diffusivity. This accounts for considerably deeper freeze and thaw depths in dry gravels and sands than in wet peat or silt areas.

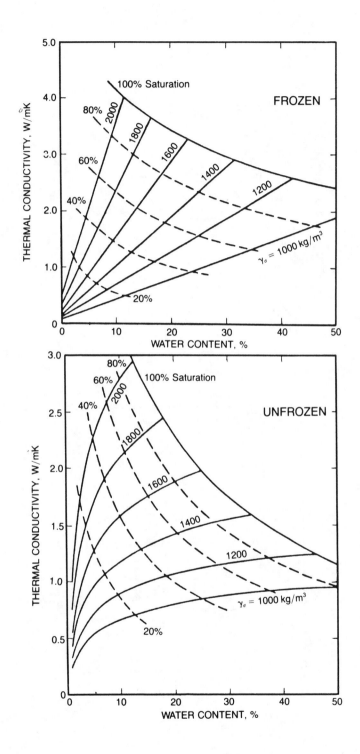

FIGURE 4-1   THERMAL CONDUCTIVITY OF SANDY
SOILS (after Kersten, 1949)

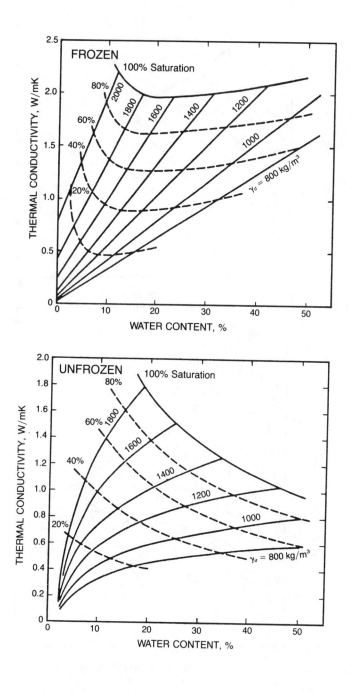

FIGURE 4-2   THERMAL CONDUCTIVITY OF SILT
AND CLAY SOILS (after Kersten, 1949)

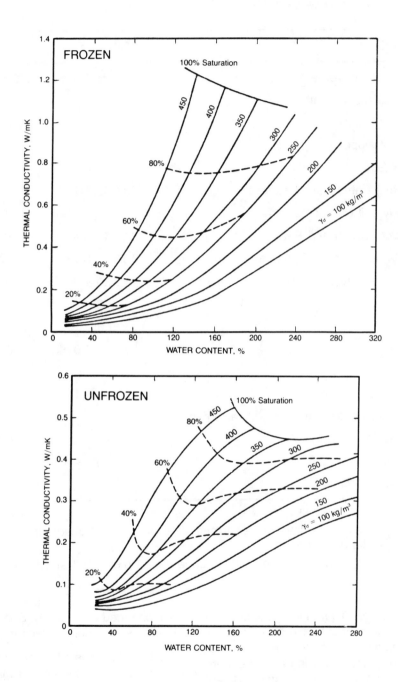

FIGURE 4-3   THERMAL CONDUCTIVITY OF PEAT (Kersten, 1949)

TABLE 4-1   n-FACTORS FOR RELATING AIR TEMPERATURE TO SURFACE TEMPERATURES

Surface	Thawing**	Freezing**
Concrete pavement*	1.5 to 2.1	0.6 to 0.8
Asphalt pavement*	1.75 to 1.9	0.7 to 0.9
Gravel road*	1.4 to 1.9	0.75 to 0.9
Gravel slope*	1.4	0.8
Dark Gravel*	1.3 to 1.7	–
White painted paved surface*	0.8 to 1.2	–
Sandy soil with some snow	–	0.5
Spruce, Brush and Peat	0.35 to 0.45	0.3 to 0.5
a. Same cleared of spruce	0.73 to 0.78	0.3 to 0.4
b. Same cleared to mineral soil surface	1.2 to 1.6	0.3 to 0.7
Willows	0.82	–
Weeds	0.86	–
Grass (turf)	0.8 to 1.0	0.5
Snow	–	0.8 to 1.0
Peat bales on road	1.4 to 2.1	–
Beneath heated building with air space		
a. Unskirted	1.0	0.9
b. Skirted	0.7	0.5
Shaded surface	1.0	0.9

*   Free of snow and ice

**   These values have all been measured by different observers in Arctic Alaska and Canada.

Ref: (Ryan, 1981; Johnston, 1981; Lunardini, 1978; Linell and Lobacz, 1980)

Latent Heat of Fusion.   The volumetric latent heat of fusion (L) is the amount of heat required to thaw water (or heat released when freezing water) per unit volume of soil without causing a temperature change. L depends only on the moisture content (w, %) of the material with dry density, ρ_d:

$$L = \left(334 \frac{kJ}{kg}\right)\left(\frac{w \bullet \rho_d}{100}\right)$$

When working in saline soils it should be remembered that L is reduced to 225 kJ/kg and the equation changed accordingly. Sea water (35,000 mg/L total dissolved solids) freezes at -1.91°C.

4.2.2 Heat Transfer at the Ground Surface. The complex energy exchange regime operative at the ground surface is influenced by many factors. The surface characteristics of the ground (color, vegetation, snow, etc.) have a significant effect on both the amount of solar heat actually penetrating the surface to thaw the ground in the summer, and the rate and amount of heat lost back to the atmosphere in the winter causing the ground to freeze. Freezing and thawing indexes, mean annual temperatures,

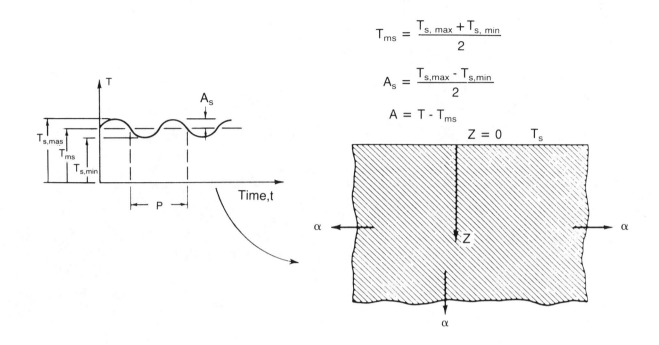

$$T_{ms} = \frac{T_{s,\,max} + T_{s,\,min}}{2}$$

$$A_s = \frac{T_{s,max} - T_{s,min}}{2}$$

$$A = T - T_{ms}$$

Assumptions
1. No latent heat of fusion effects. See (Lachenbruch, 1959) for estimate of effects.
2. Asymptotic points do not change with time. Pseudo-steady-state conditions exist.
3. Homogeneous soil mass. See (Lachenbruch, 1959) for layered systems.

(a)   Periodic temperature variation at the surface of a semi-infinite solid
        (after Chapman, 1967; Lachenbruch, 1959)

FIGURE 4-4   PERIODIC GROUND TEMPERATURE FLUCTUATIONS
              (after Chapman, 1967; Lachenbruch, 1959)

and other climatic data available from weather agencies are values measured in the air at standard weather-measuring stations located about 1.4 m above the ground surface. The mean annual soil surface temperature (T_{ms}) is 3 to 6°C warmer than the mean annual air temperature (T_{ma}). However, to determine what effects the climate has on the ground under the surface, the published values must be adapted to the actual values at the ground surface. Several studies (Ryan, 1981; Lunardini, 1978; Linell and Lobacz, 1980) have tried to determine the relationship between air and surface temperature. The multiplication factors used to convert the air tem-

peratures to surface temperatures are listed in Table 4-1, according to which the surface temperature equals the n factor times the air temperature.

Many variables influence the magnitude of n. At a given location it can vary with time because of precipitation, surface temperatures, season, cloud cover, humidity, albedo, wind speed, drainage, snow cover, etc. For a given surface it also increases with latitude and wind for the freezing season, and decreases for the thawing season. Within Alaska or Northern Canada, however, the variation of n with latitude is not significant. The wind can have a sig-

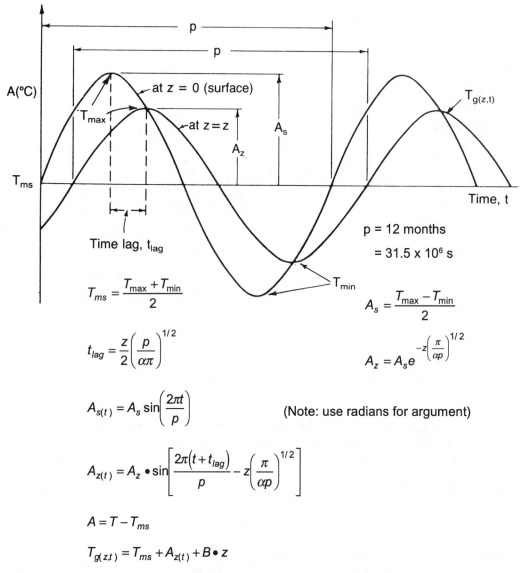

$$T_{ms} = \frac{T_{max} + T_{min}}{2}$$

$$A_s = \frac{T_{max} - T_{min}}{2}$$

$$t_{lag} = \frac{z}{2}\left(\frac{p}{\alpha\pi}\right)^{1/2}$$

$$A_z = A_s e^{-z\left(\frac{\pi}{\alpha p}\right)^{1/2}}$$

$$A_{s(t)} = A_s \sin\left(\frac{2\pi t}{p}\right)$$

(Note: use radians for argument)

$$A_{z(t)} = A_z \bullet \sin\left[\frac{2\pi\left(t + t_{lag}\right)}{p} - z\left(\frac{\pi}{\alpha p}\right)^{1/2}\right]$$

$$A = T - T_{ms}$$

$$T_{g(z,t)} = T_{ms} + A_{z(t)} + B \bullet z$$

(b) Temperature distribution at depth z in semi-infinite soil mass subjected to periodic surface temperature (after Chapman, 1967)

FIGURE 4-4   PERIODIC GROUND TEMPERATURE FLUCTUATIONS (after Chapman, 1967; Lachenbruch, 1959)

nificant effect on n. For example, the variation with wind on a concrete surface would be n = 1.8 for thawing with no wind, but about 1.2 with a 16 km/h wind.

There is considerable variation in the measured n values. It is suggested that the averages be used but weighted according to the latitude and wind comparisons given above.

4.2.3 Ground Temperature Variation.

The soil mass temperature is subjected to geothermal and cyclic climatic influences. Geothermal influences are usually considered in terms of the geothermal gradient. Periodic changes in surface tem-

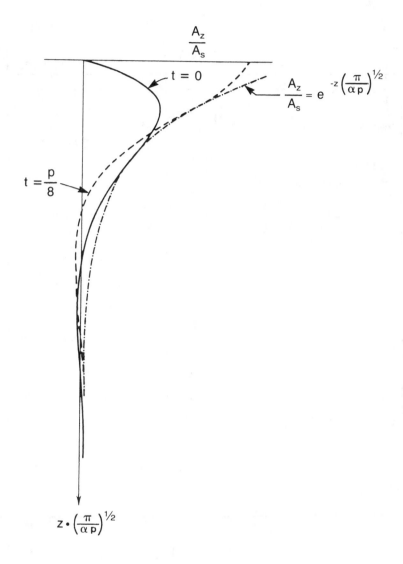

(c)   Alternation of temperature amplitude as a function of depth
       and time of year

FIGURE 4-4   *PERIODIC GROUND TEMPERATURE*
              FLUCTUATIONS (after Chapman, 1967;
              Lachenbruch, 1959)

perature greatly influence ground temperatures. The magnitude and depth of influence are a function of the thermal properties of the ground and the characteristics of the cyclic climatic influences.

Geothermal Gradient.   The geothermal gradient is the observed increase in temperature with depth. It has been reported (Carslaw and Jaeger, 1959) to vary from 0.01 to 0.05 °C/m on land. Generally, 0.03 °C/m is used. In Figure 4-4(d), the geothermal gradient is projected to the surface for an estimate of

mean annual surface temperature, T_{ms}. This method is more accurate than averaging the mean annual air temperature, T_{ma}, and correcting it with n for T_{ms}. However, accurate ground temperature readings at depths of 30 m or more are required.

Periodic Surface Temperatures.   The sinusoidal variation in surface temperatures is well established (see Figure 4-4(e)). Long-term periodic surface temperature changes tend to establish a similar though delayed and dampened response in ground

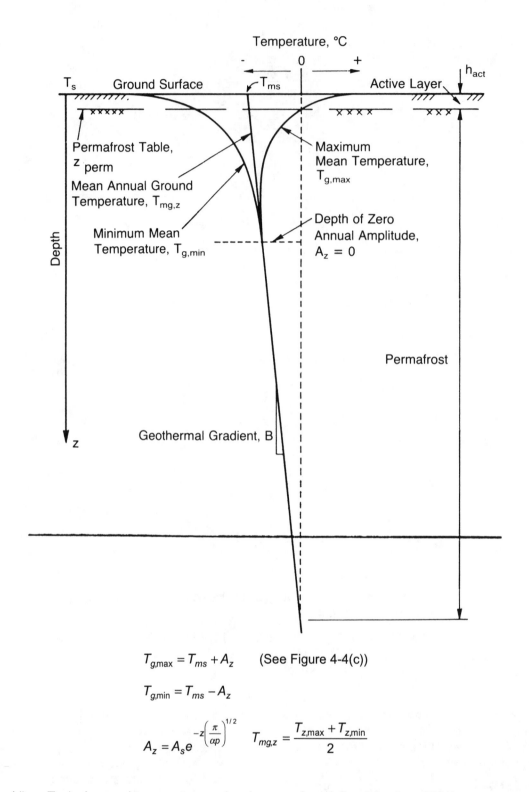

$$T_{g,max} = T_{ms} + A_z \quad \text{(See Figure 4-4(c))}$$

$$T_{g,min} = T_{ms} - A_z$$

$$A_z = A_s e^{-z\left(\frac{\pi}{\alpha p}\right)^{1/2}} \quad T_{mg,z} = \frac{T_{z,max} + T_{z,min}}{2}$$

(d)　Typical ground temperature regime in permafrost (after Johnston, 1981)

FIGURE 4-4　*PERIODIC GROUND TEMPERATURE FLUCTUATIONS (after Chapman, 1967; Lachenbruch, 1959)*

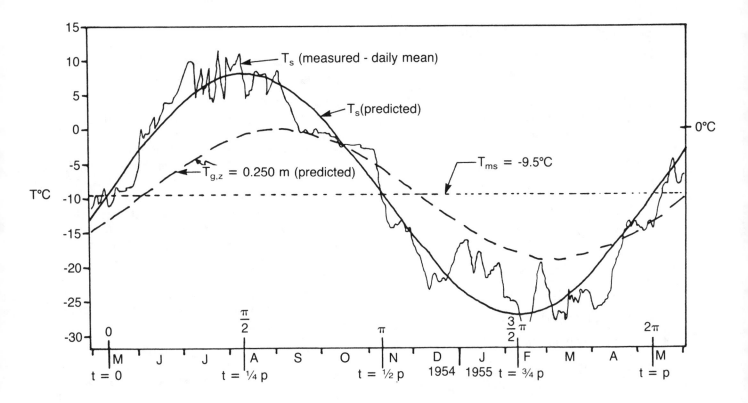

$T_{g(z,t)}$ for z = 0.25 m from Figure 4-4(b)

and $T_s = T_{g(z,t)}$ for z = 0.0 m

(e)   Example of surface and ground temperature variation near Barrow, AK (after Lachenbruch, 1959)

FIGURE 4-4   *PERIODIC GROUND TEMPERATURE FLUCTUATIONS (after Chapman, 1967; Lachenbruch, 1959)*

temperatures. Figure 4-4(a) and (b) describe the effect of periodic surface temperatures ($T_{g(z,t)}$, where z = 0).

Figure 4-4(d) describes the so-called "trumpet curve". This envelope of temperatures defines the zone of periodic ground temperature variation caused by surface temperature variation. The depth at which no periodic change occurs is usually 10 to 15 m. The interception of the freezing temperature, T_0, and the trumpet curve estimates the permafrost table depth, z_{perm}, and the active layer thickness, h_{act}. Equations mathematically describing Figure 4-4(a)-(d) accompany the figure. Figure 4-4(c) illustrates the attenuation of cyclic ground temperature variation as a function of depth or time of year.

Periodic Ground Temperatures.   Figure 4-4(b) shows that as z increases, the ground temperature amplitude, A_z is attenuated and there is a time lag, t_{lag}, introduced between the surface temperature cycle and the ground temperature cycle. The period remains the same. The ground temperature at depth (z) and time (t) is $T_{g(z,t)}$ and is described by the equation in Figure 4-4(b). This equation tends to overestimate the temperature since the tempering effect of the latent heat of fusion, L, is not accounted for in the equation. Also, the effects of stratification of the soil mass or a layer of snow or other material are not considered. Lachenbrach (1959) examines these problems at length.

The mean annual surface temperature, T_{ms}, is important for establishing time, t = 0 and t = p as shown in Figure 4-4(e). The record in Figure 4-4(e) for Barrow, Alaska demonstrates the use of these equations for predicting T_s and T_g throughout the year.

Anomalies. The equations and curves presented so far assume homogeneous, isotropic, dry soil, and no outside influences. If the soil properties change with depth, the curves are not smooth; they will have changes in slope corresponding to points where the soil properties change.

Nomenclature.

A_a = air temperature amplitude, °C

A_s = surface temperature amplitude = $n \cdot A_a$, °C

A_z = ground temperature amplitude at depth z, °C

A_0 = minimum readable temperature $\cong 0.15$ °C

a = thermal diffusivity, m^2/d

B = geothermal gradient, °C/m ($\cong 0.03$ °C/m)

p = period, in time units (often 365 d)

t = time, d

$T_{s(t)}$ = surface temperature, °C, at time, t

$T_{g(t)}$ = ground temperature, °C, at depth z, at time, t

$T_{g(z)}$ = ground temperature at depth z, °C

T_{ma} = mean annual air temperature, °C

T_{ms} = mean annual surface temperature, °C

     = $n \cdot T_{ma}$

T_{mg} = mean annual ground temperature at depth, °C

z_{active} = depth of active layer, m

$z_{permafrost}$ = depth of permafrost, m

z = depth, from surface, m

p = 3.14159

The whole concept of a constant geothermal gradient breaks down in areas of volcanic or geothermal activity. Significant movement of groundwater or large bodies of surface water can also affect ground temperatures significantly. Figure 4-5 indicates what effect damming a major tributary of the Lena River in Siberia had on existing ground temperatures. The remaining permafrost (in 1973) will thaw in time as heat is flowing into it from both top and bottom. The effect this thawing will have on the surrounding terrain and the dam itself is being carefully monitored.

Changes in slope or bends in the geothermal gradient can also indicate past changes in the surface climate. For example, the T_{ms} at Barrow is currently -8°C, but ground temperature measurements at considerable depth indicate that a couple of hundred years ago the T_{ms} was around -12°C.

4.2.4 Predicting Freezing and Thawing Depths.

The equations describing heat conduction are second-order differential equations which have a limited number of "closed form" solutions that generally relate to geometrically simple boundaries, homogeneous materials, and steady-state conditions. These solutions are useful since their explicitness allows for relatively easy numerical computations. There are also computer programs available (Kent and Hwang, 1980; Sheppard, et al., 1978; Aitken and Berg, 1968) for solving these equations. Complex problems involving nonisotropic, nonhomogeneous conditions can be solved using finite difference and finite element techniques. The reader should seek out qualified professionals in these cases.

Several equations have been developed from the original work of Neumann who studied the problem of movement of a thawing front and the associated ground temperatures (U.S. Department of the Army, 1966; Moulton, 1969; Nixon and McRoberts, 1973; Aldrich and Paynter, 1953; 1966). His solution is the basis for all other methods, which are simplifications. Nixon and McRoberts (1973) provided an excellent discussion of the Neumann method and the related simplifications. Figure 4-6 illustrates the problem defined by Neumann. The movement of the thawing interface is defined by:

$$z = C(t)^{1/2}$$

where:

z = depth of thaw (m)

t = thaw duration (d)

C = constant

     = $f(k_u, k_f, C_u, C_f, T_g, T_s, L)$

Figure 4-6(b) (Moulton, 1969) provides an exact graphical solution to the Neumann equation. The solution is insensitive to the ratio α_u/α_f which has been used as 0.7 in Figure 4-4(b).

Example 4.1

How much heat flows to the surface of the earth from the centre?

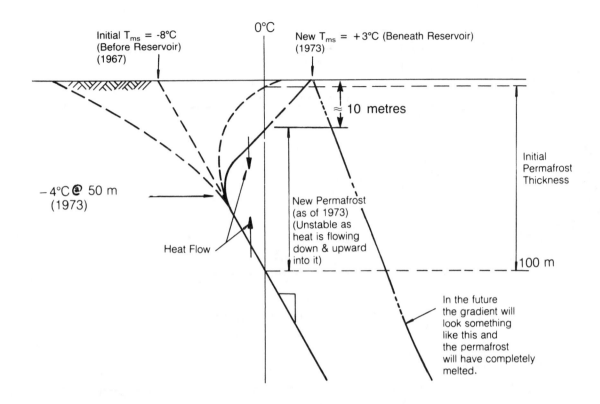

Initial T_{ms} = -8°C
(Before Reservoir)
(1967)

0°C

New T_{ms} = +3°C (Beneath Reservoir)
(1973)

≈ 10 metres

Initial Permafrost Thickness

−4°C @ 50 m
(1973)

New Permafrost
(as of 1973)
(Unstable as
heat is flowing
down & upward
into it)

Heat Flow

100 m

In the future
the gradient will
look something
like this and
the permafrost
will have completely
melted.

FIGURE 4-5   ACTUAL TEMPERATURE MEASUREMENTS IN SIBERIA BEFORE AND
AFTER THE CONSTRUCTION OF A DAM ON THE LENA RIVER

Assume: The thermal conductivity (k) of the earth averages 1.73 W/(m•°C).

Solution:

Then heat flow per unit area $= k \dfrac{\Delta T}{\Delta z}$

where:

$\dfrac{\Delta T}{\Delta z}$ is the geothermal gradient at 0.03 °C/m

Thus, heat flow $= \dfrac{1.73W}{m \cdot °C} \bullet \dfrac{0.03°C}{m}$

$= 0.0519 \ W/m^2$

of heat flows to the surface.

This value has actually been measured on Alaska's North Slope, where it equals about 0.0568 W/m².

Example 4.2

If the water temperature at the bottom of Cook Inlet (near Anchorage) is 0°C and the oil from a well on a platform in the inlet is 95°C, what is the depth of the field?

Solution:

$\Delta T = 95°C - 0° = 95°C$

assume $\dfrac{\Delta T}{\Delta z} = 0.03°$ C/m

then z $= 95°C \bullet \dfrac{m}{0.03°C}$

$= 3{,}170$ m beneath the inlet.

This agrees closely with the depth from which the oil is pumped, considering that there is some heat loss in the oil as it is pumped to the surface.

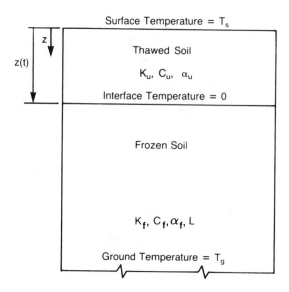

FIGURE 4-6(a) THE NEUMANN PROBLEM
(after Moulton, 1969)

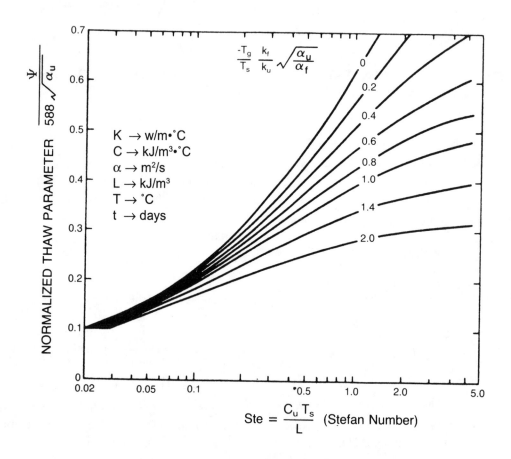

FIGURE 4-6(b) GRAPHICAL SOLUTION OF THE NEUMANN EQUATION

Simplifications of this exact solution include the Stefan and Berggren equations. Although these equations are redundant given the simplicity of obtaining the exact solution by Figure 4-6(b), they are commonly used in practice. The Stefan equation, used for roughly estimating the thaw or freezing depth, has the form:

$$z = 13.15 \left[\frac{k_u \bullet n \bullet TI}{L} \right]^{0.5}$$

where:

z   =   thaw (or freezing) depth (m)

k_u =   unfrozen thermal conductivity W/(m•°C)

TI =   air thawing index (°C•d) (or use of the freezing index, FI)

n   =   surface correction factor for TI

L   =   specific volumetric latent heat of fusion (kJ/m³)

The modified Berggren has the form:

$$z = 13.15 \bullet \lambda \left[\frac{k_u \bullet n \bullet TI}{L} \right]^{0.5}$$

where:

k, TI, L are common to the Stefan equation

λ = correction coefficient (see Figure 4-7)

A special thermal problem, the thawing of a two-layer system, can be solved using a modified Stefan equation.

$$z = \frac{k_2}{k_1} \bullet h^2 + \left[\frac{172.8 \bullet k_2 \bullet n \bullet TI}{L_2} \right]^{0.5} - \left[\frac{k_2}{k_1} - 1 \right] \bullet h$$

where:

z   =   thaw depth (m)

h   =   thickness of surface layer (m)

k_1 =   thermal conductivity of surface layer (W/(m•°C))

k_2 =   thermal conductivity of subsoil mass (W/(m•°C))

TI =   air thawing index (°C•d)

n   =   surface correction factor for TI

L_2 =   specific volumetric latent heat of fusion for subsoil, (kJ/m³)

Care should be exercised in the units of measure used in these equations. Units other than those noted can lead to incorrect solutions.

Example 4.3 (Thawing Problem)

A frozen, saturated, silty clay is exposed to a thawing front. Determine the depth of thaw using the Neumann, Stefan and modified Berggren equations if the soil has the following properties.

Given:

T_{ma} =   -1°C

T_g =   -3°C

w   =   43%

ρ_d =   1176 kg/m³

TI =   1095°C•d

n   =   0.73

t   =   80 d

Solution:

From Figure 4-2

k_u =   1.0 W/(m•°C) and

k_f =   2.2 W/(m•°C)

$$C_u = \left(0.710 + \frac{4.187 \bullet w}{100} \right) \bullet \rho_d$$

$$C_u = \left(0.710 + 4.187 \left(\frac{43}{100} \right) \right) \bullet 1176 \frac{kg}{m^3}$$

$$= 2950 \text{ kJ/(m}^3 \bullet °C)$$

$$C_f = \left(0.710 + (4.187 \bullet 0.5w) \right) \bullet r_d$$

$$C_f = \left[0.710 + 4.187 \bullet 0.5 \bullet \left(\frac{43}{100} \right) \right] \bullet 1176 \frac{kg}{m^3}$$

$$= 1895 \text{ kJ/(m}^3 \bullet °C)$$

$$L = \left(334 \frac{kJ}{kg} \right) \left(\frac{w \bullet \rho_d}{100} \right)$$

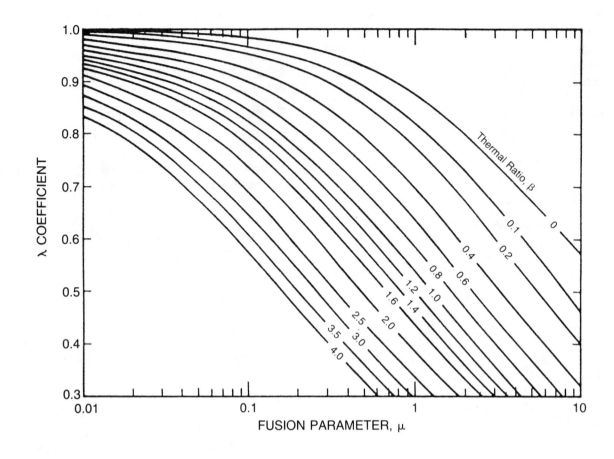

Freezing

$$\beta = \frac{(T_{ma} - T_o) \bullet t}{(n) \bullet (FI)}$$

$$\mu = \frac{(C_f) \bullet (n) \bullet (FI)}{(L) \bullet (t)}$$

Thawing

$$\beta = \frac{(T_o - T_{ma}) \bullet t}{(n) \bullet (TI)}$$

$$\mu = \frac{(C_u) \bullet (n) \bullet (TI)}{(L) \bullet (t)}$$

FIGURE 4-7   CORRECTION COEFFICIENT λ FOR MODIFIED BERGGREN EQUATION
(Johnston, 1981)

$$L = \left(334\frac{kJ}{kg}\right)\left(\frac{43 \bullet 1176\frac{kg}{m^3}}{100}\right)$$

$$= 169,000 \text{ kJ/m}^3$$

$$\alpha_u = \frac{k_u}{1000 \bullet C_u} = \frac{1.0\frac{W}{m \bullet °C}}{1000 \bullet 2950\frac{kJ}{m^3 \bullet °C}}$$

$$= 3.39 \bullet 10^{-7}\frac{m^2}{s}$$

$$\alpha_f = \frac{k_f}{1000 \bullet C_f} = \frac{2.2\dfrac{W}{m \bullet °C}}{1000 \bullet 1895\dfrac{kJ}{m^3 \bullet °C}}$$

$$= 1.16 \bullet 10^{-6}\frac{m^2}{s}$$

Neumann

$$Ste = \frac{C_u \bullet T_s}{L} = \frac{C_u \bullet n \bullet Tl}{L \bullet t}$$

$$= \frac{2950\dfrac{kJ}{m^3 \bullet °C} \bullet 0.73 \bullet (1095°C \bullet d)}{169000\dfrac{kJ}{m^3} \bullet 80d} = 0.17$$

$$\left(\frac{-T_g}{T_s}\right) \bullet \left(\frac{k_f}{k_u}\right) \bullet \left(\frac{\alpha_u}{\alpha_f}\right)^{0.5} = \left(\frac{3}{0.73 \bullet \dfrac{1095°C \bullet d}{80d}}\right)$$

$$\bullet \left(\frac{2.2\dfrac{W}{m \bullet °C}}{1.0\dfrac{W}{m \bullet °C}}\right) \bullet \left(\frac{3.39 \bullet 10^{-7}\dfrac{m^2}{s}}{1.16 \bullet 10^{-6}\dfrac{m^2}{s}}\right)^{0.5}$$

$$= +0.36$$

from Figure 4-6(b)

$$\frac{\psi}{588 \bullet (\alpha_u)^{0.5}} = 0.25$$

$$\psi = \left[(0.28) \bullet (588) \bullet (3.39 \bullet 10^{-7})\right]^{0.5} = 0.086$$

$$z = \psi(t)^{0.5} = (0.086) \bullet (80d)^{0.5} = 0.77m$$

Stefan

$$z = 13.15\left[\frac{k_u \bullet n \bullet Tl}{L}\right]^{0.5}$$

$$z = 13.15\left[\frac{\left(1.0\dfrac{W}{m \bullet °C}\right) \bullet (0.73) \bullet (1095°C \bullet d)}{169000\dfrac{kJ}{m^3}}\right]^{0.5}$$

$$= 0.90 \text{ m}$$

Modified Berggren

$$z = 13.15 \bullet \lambda\left[\frac{k_u \bullet n \bullet Tl}{L}\right]^{0.5}$$

from Figure 4-7

$$\mu = \frac{C_u \bullet n \bullet Tl}{L \bullet t}$$

$$\mu = \frac{\left(2950\dfrac{kJ}{m^3 \bullet °C}\right) \bullet (0.73) \bullet (1095°C \bullet d)}{169000\dfrac{kJ}{m^3} \bullet (80d)} = 0.17$$

$$\beta = \frac{(0°C - T_{ma}) \bullet t}{(n) \bullet (Fl)}$$

$$\beta = \frac{(0°C - (-1°C)) \bullet (80d)}{(0.73) \bullet (1095°C \bullet d)} = 0.10 \quad \text{and therefore}$$

$$\lambda = 0.95$$

Substituting:

$$z = 13.15 \bullet (0.95) \bullet$$

$$\left[\frac{\left(1.0\dfrac{W}{m \bullet °C}\right) \bullet (0.73) \bullet (1095°C \bullet d)}{\left(169000\dfrac{kJ}{m^3}\right)}\right]^{0.5}$$

$$= 0.86 \text{ m}$$

Example 4.4 (Freezing Problem)

Determine the frost depth using the Neumann and Stefan equations for a gravel material with the following properties:

Given:

T_g = 0°C

w = 3%

ρ_d = 2000 kg/m³

FI = 500°C·d

n = 0.9

t = 150 d

Solution:

From Figure 4-1

k_u = 1.7 W/(m·°C)

k_f = 1.3 W/(m·°C), and

$$C_u = \left(0.710 + 4.187 \cdot \frac{3}{100}\right) \cdot 2000\frac{kg}{m^3}$$

$$= 1670\frac{kJ}{m^3 \cdot {}^\circ C}$$

$$C_f = \left(0.710 + 4.187 \cdot 0.5 \cdot \frac{3}{100}\right) \cdot 2000\frac{kg}{m^3}$$

$$= 1545\frac{kJ}{m^3 \cdot {}^\circ C}$$

$$L = \left(334\frac{kJ}{kg}\right)\left(\frac{3 \cdot 2000\frac{kg}{m^3}}{100}\right) = 20000\frac{kJ}{m^3}$$

$$\alpha_u = \frac{k_u}{1000 \cdot C_u} = \frac{1.7\dfrac{W}{m \cdot {}^\circ C}}{1000 \cdot 1670\dfrac{kJ}{m^3 \cdot {}^\circ C}}$$

$$= 1.02 \cdot 10^{-6}\frac{m^2}{s}$$

$$\alpha_f = \frac{k_f}{1000 \cdot C_f} = \frac{1.3\dfrac{W}{m \cdot {}^\circ C}}{1000 \cdot 1545\dfrac{kJ}{m^3 \cdot {}^\circ C}}$$

$$= 8.41 \cdot 10^{-7}\frac{m^2}{s}$$

Neumann

$$Ste = \frac{C_u \cdot T_s}{L} = \frac{C_u \cdot n \cdot TI}{L \cdot t}$$

$$= \frac{1545\dfrac{kJ}{m^3 \cdot {}^\circ C} \cdot 0.9 \cdot (500{}^\circ C \cdot d)}{20000\dfrac{kJ}{m^3} \cdot 150d} = 0.23$$

$$\left(\frac{-T_g}{T_s}\right) \cdot \left(\frac{k_f}{k_u}\right) \cdot \left(\frac{\alpha_u}{\alpha_f}\right)^{0.5} = \left(\frac{(0{}^\circ C)}{0.9 \cdot \dfrac{500{}^\circ C \cdot d}{150d}}\right) \cdot$$

$$\left(\frac{1.3\dfrac{W}{m \cdot {}^\circ C}}{1.7\dfrac{W}{m \cdot {}^\circ C}}\right) \cdot \left(\frac{1.02 \cdot 10^{-6}\dfrac{m^2}{s}}{8.41 \cdot 10^{-7}\dfrac{m^2}{s}}\right)^{0.5} = 0$$

from Figure 4-6(b)

$$\frac{\psi}{588 \cdot (\alpha_u)^{0.5}} = 0.33$$

$$\psi = (0.33) \cdot (588) \cdot (8.41 \cdot 10^{-7})^{0.5} = 0.18$$

$$z = \psi \cdot (t)^{0.5}$$

z = 2.2m

Stefan

$$z = 13.15 \cdot \lambda\left[\frac{k_u \cdot n \cdot TI}{L}\right]^{0.5}$$

$$z = 13.15 \left[\frac{2.2 \frac{W}{m \cdot °C} \cdot 0.9 \cdot 450°C \cdot d}{20000 \frac{kJ}{m^3}} \right]^{0.5}$$

$$= 2.8 \text{ m}$$

The discussion and examples so far in this section have focused on the thawing of soil. The same equations are applicable for freezing problems, provided the frozen and unfrozen properties as well as the FI and TI values are switched in the equations.

Ground thermal analysis can become a very complex problem in real applications. The foregone discussion was simplified to explain the various factors affecting thawing rate and depths. Moisture content, freezing fronts, and heterogeneous soil conditions may greatly alter the applicability of the equations. Sophisticated numerical methods may be required to analyze ground thermal conditions properly. A specialist should be consulted.

4.3 Heat Loss From Pipes

The thermal design must result in a system which will prevent the freezing of water and wastewater within pipes or tanks that are located in environments below 0° C, and provide for the most economical operation possible. The primary method of preventing freezing is the reduction of heat loss complemented by the replacement of heat lost. Heat loss is proportional to the difference between the temperature of the fluid and the ambient external temperature, and the thermal resistance of the intervening materials. Measures that provide a more favorable environment, such as buried pipes, and measures that increase the thermal resistance, such as insulation, reduce the rate of heat loss. These methods however, do not eliminate the possibility of freezing, therefore, heat losses must be replaced by removing the cooled water before it freezes or by heating the fluid or the pipe surface.

Estimates of the maximum rate of heat loss are required to determine the freeze-up time and accordingly design heaters, circulation pumps, and heat-tracing systems. Annual heat loss estimates are necessary not only to determine total energy requirements but also to assess methods of reducing these. Example solutions to these thermal problems are given in Section 4.7.

4.3.1 Pipe Environment.
The freeze-up time and the total heat loss depend on the temperatures encountered in the pipe environment. Above-ground piping systems must be thermally designed for the lowest expected air temperatures, perhaps -40 to -60°C. For reliable and economical operation during winter, exposed pipes must be completely insulated and usually a flow through the pipes must be maintained. Heating of the water may also be necessary or economical.

Extremely low air temperatures are significantly moderated by the ground surface conditions, primarily snow cover and vegetation, and the thermal properties of the soil. Of importance to heat loss and the design of buried pipes are the minimum ground temperatures and the maximum depth of freezing or thawing. Surface temperature variations are attenuated with depth, depending upon the thermal diffusibility (thermal inertia) and latent heat (related to moisture content) of the soil. While air temperatures may have an annual range of 90°C, the temperature at a depth of two metres may vary from 2°C in saturated organic soils in undisturbed areas to as much as 25°C in exposed dry soils or rock. At a depth of 10 m, seasonal temperature fluctuations are usually negligible. Daily air temperature fluctuations are insignificant below 0.5 to 1.0 m of bare soil or ice; below 0.5 m, if an undisturbed snow cover is present. Air temperature fluctuations are attenuated with depth, so that ground temperature fluctuations are not felt until some time later. The lag time depends on the surface conditions and thermal properties of the soil: at a depth of two metres the lag time may be from a few weeks to five months or more. Therefore, minimum ground temperatures and maximum frost penetration may occur when the extreme winter temperatures have passed. Frost penetration is greatest in rock or bare, dry soils. An undisturbed snow cover may reduce the depth of frost penetration by an amount equal to its own thickness (Legget and Crawford, 1952). Placing utility lines away from traveled, plowed areas – at the back of lots rather than under roadways – may significantly reduce heat loss from the soil and increase the time before freeze-up. But years with little snowfall as well as snowdrifting and community induced changes must be considered.

Conventional municipal piping can be installed below the maximum depth of seasonal frost. In cold regions, the frost penetration is often greater than the common pipe depths of two to three metres; in fact, frost penetration may be six metres or more in exposed dry soil or rock. Deep frost penetration, high groundwater, hilly terrain, rock or other factors may make it more practical and economical to install all

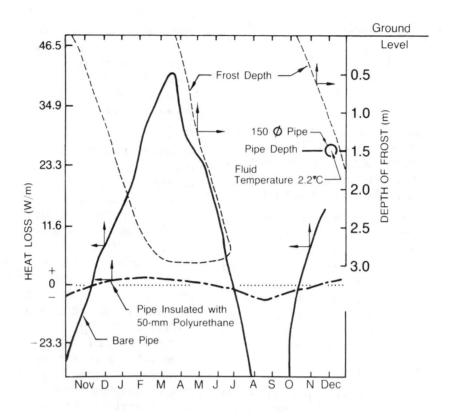

FIGURE 4-8   FROST PENETRATION AND HEAT LOSS FROM BARE
AND INSULATED PIPES WITHIN SEASONAL FROST
(Gunderson, 1975)

or portions of the utility system within the frost zone (James, 1980; Whyman, 1980; Nyman, 1965; Janson, 1966). In these cases, the degree of freeze protection necessary depends on the ground temperatures at pipe depth. Where pipes are only intermittently or periodically within frost zones, conventional bare pipes may be adequate, provided a minimum flow can be maintained by pumping, bleeding, or consumption to replace heat loss to the ground. Frost-proof appurtenances, stable backfill, and some heating may also be necessary. Even in these marginal cases, insulating pipes in shallow or low-temperature areas is highly desirable because it greatly increases the reliability of the system. Heat loss and freeze danger are significantly reduced by insulating the pipes (see Figure 4-8).

Insulated pipes can be installed in shallow trenches or within berms at the ground surface. In these cases, there is often little thermal advantage in deep bury, but the minimum desirable cover (0.5 to 1.0 m

for exposed surfaces) should be such that daily temperature fluctuations have no influence. The minimum depth of cover is also governed by the ability of the insulation and pipe to withstand anticipated traffic loads. This depth is usually one metre or more. Other factors which influence the average and local depth of bury are:

- the pipe grade and terrain;

- frost-heaving problems, which are greater for shallow pipes; and

- access for maintenance.

The reduction in excavation cost can sometimes be balanced with the cost of insulation and other freeze-protection measures necessary for shallow buried pipes within the seasonal frost zone (James, 1980; Whyman, 1980; Nyman, 1965; Janson, 1966).

In seasonal frost zones, buried pipes can be pre-insulated usually with polyurethane. Alternatively, a layer of insulating board, such as polystyrene, can be placed above the pipe (Figure 4-9). Although insulating board is often less expensive than polyurethane, the installation cost is higher, and because this insulating method is used with bare pipes and fittings, the effectiveness of the insulation is lower than direct insulation of the pipes. Since frost penetration below the pipes can be prevented by an insulating board, the board method has been used where the soils underlying the pipe are frost-susceptible. The necessary thickness and width of the board varies indirectly with the depth to which pipes are buried and directly with the frost penetration of the soil. Generally, the insulation should be at least 1.2 m wide for a single pipe. In terms of reducing frost penetration, polystyrene foam insulation 50 mm thick (k = 0.035•W/m•°C) is roughly equivalent to 1.2 m of sand or silt or 1.0 m of clay cover over the pipe (Gerriutsen, 1977). The heat loss and trench width can be reduced by placing the insulation in an inverted U-shape. Proper backfilling is difficult when using insulating boards.

The economics for freeze protection, compared to separately insulated pipes, is improved when water pipes and warm sewer or central-heating lines share a common trench under a board. Design information on this method is available from descriptions of its use in Scandinavia (Gunderson, 1975; 1978). Extreme caution must be applied with this approach as failure or leakage of a sewer pipe could contaminate a long section of the trench. Also repairs may be very difficult.

4.3.2 Freezing of Pipes. Freeze damage to containers of fluid, including pipes, occurs because of the expansion of water when it changes to ice (McFadden, 1977). This imposes a pressure on the still unfrozen liquid that can reach thousands of kilopascals. Failure is caused by hydrostatic pressure, not by the ice expanding directly on the walls (Houk, 1974). The freezing of quiescent water in pipes occurs in stages as shown in Figure 4-10 (Gilpin, 1977a; 1977b). Water must always supercool, a state typically reached at -3° to -7°C in qui-

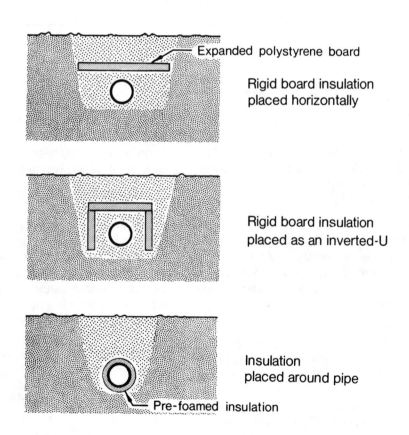

FIGURE 4-9   METHODS OF INSULATING BURIED PIPES

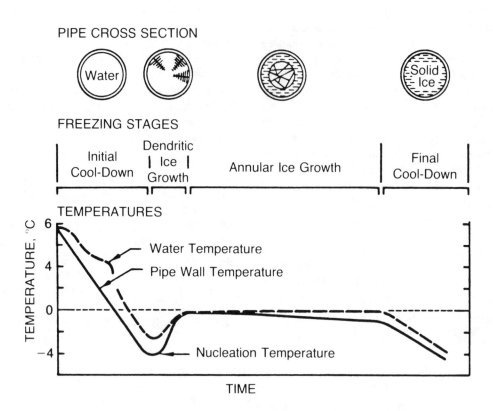

PIPE CROSS SECTION

FREEZING STAGES

TEMPERATURES

FIGURE 4-10   STAGES IN THE FREEZING OF A QUIESCENT WATER PIPE
(Gilpin, 1977a)

escent pipes during slow cooling, before nucleation and freezing produces dendritic ice growth. Further cooling results in the growth of an annulus of ice inward from the pipe walls.

Dendritic ice formation increases the start-up pressure required to reinstate flow, particularly in small-diameter pipes and at a slow fluid cooling rate, and can effectively block the pipe in much less time than that required for the pipe to become blocked by annular growth of ice (Gilpin, 1977a).

The freezing process for fluid flowing in pipes is not simply by annular growth; such an oversimplification can lead to false conclusions and errors in design. Gilpin (1977b; 1979; 1981) observed that for flowing pipes ice does not grow as a uniform thickness along the pipe, but occurs as cyclical ice bands of a tapered flow passage (Figure 4-11). Further cooling freezes the narrow neck closed and stops the flow. Subsequent freezing can result in pipe breaks between each ice band. No ice formation in

pipes can be safely tolerated and fluid temperatures should not be allowed to drop below 0.5°C.

As soon as the water temperature drops to freezing, ice may start to form somewhere in the system, usually on metal valves and fittings. Ice plugs formed in this manner can prevent start up or draining of the pipes long before the entire system freezes. The recommended design freeze-up time is the time available before water in an inoperative system reaches the freezing point. This time period must be sufficient to permit repairs or drain the system. The length of time allowed depends upon the ease of maintenance and the availability of maintenance personnel and equipment. The design freeze-up time may have to be several days in small communities, whereas in larger centres the time may, with caution, be reduced to less than one day. The shortest freeze-up time and hence the largest number of freeze-ups are associated with small-diameter service lines. The maximum freeze-up time is the time necessary for the fluid to drop to -3°C, the nucle-

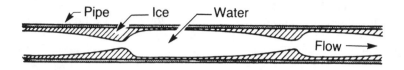

FIGURE 4-11   ICE BANDS FORMED DURING THE FREEZING
OF FLOWING PIPE (Gilpin 1977b, 1979, 1981)

ation temperature. No portion of the latent heat for complete pipe freezing should be included in freeze-up calculations. (See Figure 4-8, Example 4.13).

Gravity pipelines or open channels may also ice up. The initial filling of water is often a critical period with respect to freezing in this manner. Sewer pipes may "glacier" causing the pipe to slowly fill with ice. This type of icing usually does not stress the pipe to failure. Furthermore, frosting may occur in the crown of sewer lines when at below-freezing temperature. Icing may occur, if the pipe environment is below freezing and the heat input is lower than the heat outflow. Crown ice and glacier ice may melt when the flow is increased.

4.3.3 Freeze Protection.

The design freeze-up time can be increased with higher operating temperatures, increased insulation, and locating pipes where ambient temperatures are warmer. Excess temperature increases heat loss, introduces inefficiency, and is expensive. Utility systems that are highly insulated are preferable, since they require less heating and provide a longer freeze-up period. Insulation only retards heat loss, and whenever ambient temperatures are below freezing, heat loss from pipelines must be replaced. The length of time each year that freeze-up protection is required for freeze-up prevention is related to the flow and temperature of the water. A similar argument applies to buried pipes.

Small diameter pipes, such as service connections, have less specific heat capacity and latent heat of fusion available, and the design freeze-up time is usually only a matter of minutes to a few hours in a below-freezing environment. Freezing usually occurs in small-diameter pipes first because they are particularly vulnerable to any interruptions in flow. Thawing capability is therefore mandatory for small-diameter pipes.

Properly designed, buried, insulated piping does not usually affect the ground temperatures significantly,

but poorly insulated or bare pipes have a warmed and thawed zone around them. This should not be relied upon to increase the freeze-up time, since soil texture and moisture content vary significantly, and the freezing temperature of moisture in soils is often below 0°C, particularly in cohesive soils (Andersland and Anderson, 1978). Therefore, the deliberate thawing of ground is not recommended.

Some flexible pipe materials, such as polyethylene, special polyvinyl chloride mixes, and small-diameter copper lines, may not rupture upon freezing but their expansion depends upon the manner of freezing. Various methods to prevent or limit the freeze damage to pipelines have been proposed (McFadden, 1977). These methods provide additional thermal resistance at convenient, pre-determined locations where the hydrostatic pressure that occurs with freezing can be lowered, be it by a pressure relief valve, bursting diaphragm, expandable section, or some other device; or else, a short section of pipe is allowed to be sacrificed.

4.3.4 Thawing of Frozen Pipes.

Providing a means for thawing all water and sewer pipes and wells that may freeze is recommended and often mandatory. Remote electrical thawing methods, which must be incorporated into the original design, include skin effect, impedance, and various resistance wire and commercial heating cable systems. Frozen wells have been thawed by applying a low voltage from a transformer to a copper wire located inside the drop pipe (Eaton, 1964). Once a small annulus of ice has been melted, the flow is restored in the course of which the remaining ice in the pipe is thawed. A common approach to thawing frozen, small-diameter metal pipes is to pass an electrical current from a welder or transformer through the pipe (Bohlander, 1963). This operation is greatly facilitated if readily accessible thaw wire connections are provided. A frozen pipe can also be thawed by passing warm water through a small-diameter pipe which is installed on the exterior surface of the pipe. This thaw tube system can also be used as a carrier pipe for electric heat tracing, a return line for recirculation, or a bypass of the main line. The capital cost of this system is much lower than electric heat tracing but it requires manual operation. Utilidors that incorporate pipe heat tracing can be thawed, provided that these heating lines use an antifreeze solution or that they are drained before freeze-up. Exposed pipes may also be thawed with warm air. Injection thawing with steam or hot water can be used to thaw water and sewer lines if adequate access and room for hoses, equipment, and personnel is provided in manholes. Small-diameter water lines of any material can also be quickly thawed by pumping warm water into a smaller diameter plastic hose that has been inserted into the frozen pipe (Nelson, 1976: Currey, 1980). These thawing methods are detailed in Appendix D.

The energy required to thaw a frozen pipe is largely the heat necessary to melt ice. Knowing the heat that must be supplied and the energy output of a thawing system, the time to melt the ice can be calculated. In practice, an opening that permits a flow to commence may be all that is necessary to thaw the remaining ice, provided that the flushing water is warm. Long lengths may be thawed this way in stages. In many cases, heat-tracing systems are sized not just to supply the heat loss necessary to prevent freezing but to thaw or reopen the line in reasonable time.

4.4   Replacement of Lost Heat

Heat loss cannot be completely prevented. If ambient temperatures are below freezing, it is simply a matter of time before freezing occurs, unless either heat is added to the fluid or the cold fluid is replaced with warmer fluid. Heat can be added either along the length of the pipe or at point sources.

4.4.1   Fluid Replacement.   Freezing does not occur when the residence time of the liquid in the pipeline is less than the time necessary for it to cool to the freezing point. The quantity and temperature of the replacement water; i.e., the total heat available, must be sufficient, and the flow must be reliable. Operation without additional heating is restricted to situations where relatively warm water supplies, such as groundwater, are used or where the flow rate is reliable and high, as it is in some water supply pipelines or trunk mains. Bleeding of water has been used to maintain or enhance the flow in service lines, dead ends, and intermittently flowing pipelines, but the wasting of large quantities of water can be inefficient and lead to water

supply and wastewater treatment problems. Recirculation maintains flow and uniform temperature within the system, and prevents premature freezing at locations with lower than average ambient temperature or at poorly insulated sections. The water temperature declines, however, unless warmer water is added or the recirculating water is heated.

4.4.2   Point Sources of Heat.   Water may be heated at the source, treatment plant, pumping stations, along the pipeline, or within distribution systems as required. Heat is commonly obtained from oil- or gas-fired boilers, however, simple electric water heaters have been used where the heat requirements are very low. The heating of water can be a cost efficient use for low-temperature waste heat from sources such as electric power generation. Where the raw water temperature is too low, it must be raised to a specified operating temperature. The heating capacity required to do that is determined by the maximum flow rate; the annual energy requirements are determined by the total water demand during the heating period.

There must be sufficient flow within the piping system to distribute the added heat. If the normal water demand is too low or intermittent, then bleeding or recirculation is necessary. A minimum water temperature can be maintained within the piping system by increasing either the flow rate or the input water temperature while keeping the other parameters constant, or by adjusting them simultaneously. High temperatures enhance heat loss and introduce inefficiency. As a general rule, the temperature drop along a pipeline should always be kept to less than 5°C, preferably less than 2.5°C, by insulation, high flow rates, or intermediate heating along the pipeline or within the system. Little reduction in heat loss results from limiting the temperature drop to less than 0.5°C, yet the capital and operating costs rise substantially. Velocities greater than 0.1 m/s for 150 mm pipes and 0.5 m/s for 50mm pipes are of little benefit in reducing total energy input to maintain a specified minimum water temperature. Higher velocities – pitorifice systems, for example, require a minimum velocity of 0.6 m/s – must be balanced with the electrical energy requirements for pumping (see Example 4.10).

Sewer lines are generally warmer than water mains, but freezing can occur where flows are intermittent or slow. In these cases, a flush tank can be used to discharge wastewater or warm water periodically into the sewer line (see Section 9). Bleeding water directly into the sewer lines also adds heat, but this practice is usually only recommended as a temporary measure. Direct heating of wastewater is not practiced.

4.4.3 Pipe Heat-Tracing Systems.

Heating requirements to replace heat losses in order to maintain a minimum temperature to prevent freezing can be supplied by pipe heat-tracing systems. Such systems are more commonly used in multipipe utilidors but have been used for single pipes. Heat is provided by warming air spaces by convection, and in some cases by conduction. The pipe heat-tracing system may either be part of a central heating system or designed to heat only a utilidor. If part of a central heating system in above-ground utilidors, some of the high-temperature pipe insulation may have to be removed to provide enough heat to prevent freezing of adjacent pipes during the period of lowest ambient temperatures (Hull, 1980). The heat source cannot be manipulated and when the ambient temperature increases, inefficient and undesirable overheating of the utilidor and water main occur (see Example 4.9). Freezing of pipes can result from thermal stratification in large utilidors or from the shielding of small diameter heat-tracing pipes (Reed, 1977).

For pipe heat tracing, low-temperature fluids, generally between 80°C and 98°C, are much simpler to use than either steam or high-temperature water, and the fluid mixture can be adjusted to depress the freezing point to the lowest expected ambient temperatures. The use of an antifreeze solution protects the heat-trace piping, allows start up during winter and provides a means of thawing frozen pipes. The viscosity of glycol/water solutions is greater than that of water alone, therefore the required pumping capacity and friction losses are higher. The heat transfer characteristics are also poorer than for water alone. For example, a 50 percent solution of glycol and water would require a 14 percent increase in flow rate to achieve the same heat transfer. Design information on these heating systems is available from the American Society of Heating, Refrigeration and Air Conditioning Engineers (ASHRAE, 1981a; 1981b). It should be remembered that ethylene glycol is toxic and, in general, should not be used in water supply systems. Alternatively, propylene glycol, which is nontoxic, can be used cautiously. These solutions are corrosive to zinc, and they can leak through joints and pump seals through which water at the same pressures does not leak. Mechanical seal pumps should be used. Some boiler manufacturers nullify their warranty if glycol solutions are used. Special organic fluids can also be used instead of water/glycol solutions. They must be checked for toxicity before they are used.

4.4.4 Electric Heat-Tracing Systems.

Electric heat-tracing systems are relatively easily installed and controlled. They are appropriately used where the freeze-up time is short and where the lines cannot be blown clear or drained. They may be installed either continuously on water and sewer pipelines, or only at freeze-susceptible or critical locations, such as road crossings, service connections, or appurtenances such as fire hydrants. Because of the relatively high cost of electrical energy, these systems are usually installed for freeze prevention in the event of a failure in operation, such as a prolonged no-flow condition, rather than as the primary method of maintaining a minimum operating temperature or to heat up the fluid. The capability and ease of remote thawing is also important in many situations. The designer must weigh these advantages against the risks, costs of electric heat-tracing systems, and alternative freeze-protection methods.

A variety of electric heat-tracing systems and products are available. These are discussed in detail in Appendix D.

4.4.5 Friction.

The amount of heat generated by friction depends on the flow rate, fluid viscosity, and the pipe size and roughness. Friction heating is negligible for smooth (new) pipes with fluid velocities of less than 2 m/s, which is about the desirable upper limit for flow in pipes, but for high fluid velocities it may be significant. Since this energy is supplied by pumping, deliberately increasing the velocity (friction) is a very inefficient method of heating. Equations for heat input by friction are presented in Figure 4-8.

4.5 Physical Methods of Reducing Heat Loss

The primary physical method of reducing heat loss is insulation. Since the absence of moisture content is a key factor in determining the thermal performance of insulations, only near-hydrophobic insulations should be used. Even these insulations usually require some moisture protection to prevent ground moisture, humidity, or water from pipe failures from reaching the insulation.

Systems that are liberally insulated are generally preferred since they require minimum heating and circulation, provide a longer freeze-up period, and have less influence on the thermal regime of the ground. An economic analysis to balance heating

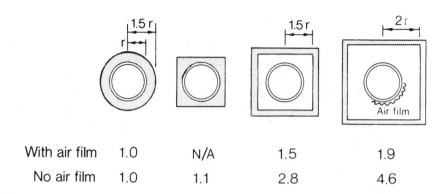

With air film	1.0	N/A	1.5	1.9
No air film	1.0	1.1	2.8	4.6

FIGURE 4-12   RELATIVE HEAT LOSS FROM SINGLE PIPES
INSULATED WITH SAME VOLUME OF INSULATION

and insulating costs should be performed to determine the most economical amount of insulation (see Example 4.11). Other factors, such as the freeze-up time, the maximum rate of heat loss, and practical dimensional considerations must also be considered in the selection of insulation thickness.

Heat loss estimates for piped systems must adequately allow for poorly insulated or exposed sections of pipes, joints and appurtenances, and thermal breaks such as at pipe anchors. A 150 mm gate valve has a surface area equivalent to one metre of bare pipe. If this valve were left exposed it would lose as much heat as about 60 m of 150 mm pipe insulated with 50 mm thick polyurethane insulation; freezing occurs at the valve first. Therefore, the thermal resistance around appurtenances should be about 1.5 times that required around connecting pipe lengths. This illustrates not only the danger of restricting thermal analysis and design to straight sections of piping but also the importance of fully insulating the piping system.

Heat loss and the volume of materials can be reduced by minimizing the perimeter and exposed surface area. This is most important for above-ground pipes and facilities. Insulation is most effective when it is placed directly around the source of heat. The relative insulative ability of various simple shapes and locations is shown in Figures 4-12 and 4-13. In an air space the thermal resistance of the air film around the pipe can be quite significant and must be considered. For a single pipe, insulation is best applied in an annulus directly around the pipe. Heat loss from pipes in a utilidor is less than that from separate pipes insulated with the same total

volume of insulation, if the utilidor is compact. This is often not possible. The spacing requirements for pipes, appurtenances, installation and maintenance, particularly for central heating lines, often requires that utilidors be designed with large air spaces. These utilidors are thermally less efficient than separately insulated pipes. Heat supplied from warm sewer pipes, central heating lines, or heat tracing favours utilidors in terms of thermal efficiency. Pipes within utilidors may also have a longer freeze-up time because of the larger heat source available compared to separate pipes. This is particularly important for small-diameter pipes. Heat loss can also be reduced and freeze protection improved by installing one water pipe inside a larger one, rather than using two separate pipes. This technique is applicable for freeze protection of small-diameter recirculation pipes used to maintain a flow in supply lines or in dead ends within a water distribution system.

Heat loss can also be reduced by lowering the operating temperature of water pipelines, but the benefits of this are only significant where the external ambient temperatures are just slightly below freezing. The heat loss from insulated storage tanks and pipes is often small compared to the energy requirements to raise the source water to the operating temperature of the system. Therefore, lower operating temperatures may significantly reduce energy requirements for preheating. Because of the greater relative reduction in heating requirements, utility systems in the subarctic are more often operated nearer freezing than those in arctic regions.

4.5.1 Insulation Techniques. The appropriate type and thickness for the insulation of piping and structures in cold regions must be selected. The thickness may be determined from economic analysis (see Example 4.11), or other considerations such as freeze-up time or building comfort. Common insulating materials are plastics, minerals and natural fibers, or composite materials. For design purposes, the structural and thermal properties for the worst conditions should be used. These conditions occur after the insulation has undergone aging, compaction, saturation, and freeze-thaw cycles. Other selection considerations are ease of installation, vapor transmission, burning characteristics, and susceptibility to damage by vandals, animals, chemicals, and the environment.

The insulating value of a material depends more or less directly on the volume of entrapped gas in the material. If the material becomes wet and the voids fill with water, the insulating properties are lost since the thermal resistance of air is about 25 times that of water and 100 times that of ice. In the past, the lack of a near-hydrophobic insulation made the design of piping in moist environments very difficult (Churakov, 1959) and is a major reason for the development of above-ground utilidors (Johnston, 1981). The availability now of rigid closed-cell plastic foam insulations with low thermal conductivity and high resistance to water absorption has drastically influenced the design of utility systems in cold regions. However, they have limitations, and knowledge of their properties is essential.

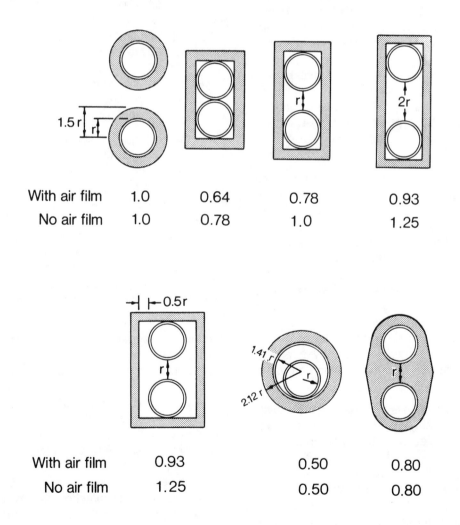

FIGURE 4-13   RELATIVE HEAT LOSS FROM TWO PIPES INSULATED WITH SAME VOLUME OF INSULATION

Polyurethane foam is used extensively in cold regions to insulate pipes, storage tanks, and some buildings and foundations. Urethane bonds to most materials. Piping or other components can be preinsulated or polyurethane can be prepared on-site from the raw chemicals (which are about 1/30th the final volume) and then applied. Field applications are restricted by climatic conditions. The density and thermal conductivity is often poorer than values attainable under factory conditions. The foam must be protected from ultraviolet radiation during shipping and use. Only a metal skin has proven effective to prevent "aging", the loss of entrapped heavy gas. Entrapped heavy gas increases the thermal conductivity by about 30 percent above the theoretical minimum value (Shirtliffe, 1977). Depending upon the formulation, urethanes can have a higher flame-spread rating than other building materials. They are combustible, and a flame-protective barrier is usually required by building insurers and some building codes. If ignited, plastic foams release smoke and toxic gases. Foams with densities over 100 kg/m^3 are essentially impermeable, but lighter foams, which are better insulators, require coatings to prevent water absorption, since freeze-thaw cycles of the moisture can lead to deterioration of the insulation.

Extruded polystyrene, particularly the high-density products (50 kg/m^3), deteriorates least from moisture absorption and freeze-thaw cycles (Kaplar, 1974), but the outer 5 mm of unprotected buried insulation should be disregarded in thermal analyses. Molded polystyrene absorbs some moisture and should not be used in moist conditions. Polystyrene is available in board stock or beads. Board stock has been extensively used to reduce frost penetration. Beads are useful for filling voids in utilidors for maintaining easy access to pipes. Although the thermal conductivity of polystyrene is higher than polyurethanes, the volumetric cost is usually less (Table 4-2).

Glass fiber batt insulation is the most common building insulation, primarily because it is fire resistant and relatively inexpensive. But glass fiber is not water resistant. Wetness reduces its insulating value significantly; by half, if eight percent by volume is water, for example. For this reason, glass fiber should not be used underground but may be considered wherever dry conditions persist. Cellular glass is very water resistant, but is seldom used because it is brittle, difficult to work with, and deteriorates with freeze-thaw cycles. Lightweight insulating concrete made with polystyrene beads, pum-

ice or expanded shale can be formulated for relatively high strength and thermal resistance. It can be poured into place around piping but should be protected from moisture to prevent freeze-thaw deterioration.

Many other insulating materials, including new products, such as sulphur foam and urea-formaldehyde, may also find specific applications in cold-regions engineering.

4.6 Thermal Calculations

Unfortunately, many of the thermal problems that are encountered in practice either do not have exact mathematical solutions, or other complexities make precise solutions impossible. The second-order differential equation, which describes the conduction of heat, has a limited number of closed-form solutions, and these generally relate to geometrically simple boundaries, relatively homogeneous materials, and steady-state conditions (Carslaw and Jaeger, 1959). However, such solutions are useful, since their explicit nature allows relatively easy numerical computation and encourages quantitative insight into thermal problems.

The analytical thermal equations presented in this section imply important idealizations. The analyst must assess their applicability for particular problems and use the results as a guide to engineering design. The analyst is advised to consider various models and a range of values for physical and thermal conditions. More accurate results are obtained when reliable data are available. The analyst must often be content with approximate solutions, and it will be necessary to assume either a large safety factor or the worst conditions to arrive at conservative estimates.

Boundary temperatures in the field vary continuously with both random and periodic components and are often a result of very complex heat exchange effects. The materials encountered in practice frequently represent rather poor approximations of homogeneous isotropic media. Soil, for example, is a complex, multiphase, heterogeneous medium, the behavior of which is further complicated by the water component, which undergoes phase transition in the temperature regime of concern. Some of these physical and thermal complexities may be taken into account by using strictly numerical techniques to solve the appropriate differential equations (Kent and Hwang, 1980; Goodrich, 1978).

Equations to calculate ground temperatures and depth of freezing and thawing are presented in Sec-

TABLE 4-2   COMPARISON OF INSULATION PROPERTIES OF VARIOUS MATERIALS

Material	Thermal Conductivity w/(m·°C) (k)	Density (kg/m³) (ρd)	Compressive Strength at 5% Deflection (Pa)	Effect of Moisture Content (w) on Thermal Conductivity (k)	Relative Volumetric Cost
Polyurethane					
New	0.016	30	0.002	Negligible	4 to 6.5
Aged	0.023	30	0.002	Negligible	
High Strength	0.081	65	0.007	Nil	
Expanded Polystyrene	0.029 to 0.035	15 to 45	0.002 to 0.003	Negligible	3.5
Molded Polystyrene	0.030 to 0.038	15 to 30	0.007 to 0.0018	Absorption, w = 3% max.	2
Sulfur Foam	0.42	175	0.0032	Absorption, w = 2% max.	
Insulating Concrete	0.105 to 0.604	300 to 1500	0.007 to 0.060	Nonrepellent	
Glass Fiber	0.035 to 0.052	25 to 55		Deterioration at w = 8%, k = 0.06	1
Glass Foam	0.058	150 to 200	0.007 (Ultimate)	Nil	
Vermiculite	0.070	200		Deteriorates	
Sawdust (dry)	0.058 to 0.09	150 to 250		Deteriorates	
Fine-Grained Soil					
Moisture 10% (Frozen)	1.05 (1.05)	1600		Increases	
Moisture 30% (Frozen)	1.74 (2.32)	1600		Increases	
Coarse-Grained Soil					
Moisture 5% (Frozen)	1.63 (1.16)	1750		Increases	
Moisture 15% (Frozen)	2.09 (3.02)	1750			
Water	0.58	1000			
Ice	2.21	900			

TABLE 4-3   LIST OF SYMBOLS

A	=	area, m^2
a	=	thaw factor; $= T'$ arccosh (z_p/r_p), dimensionless
B	=	geothermal gradient, °C/m
b	=	breadth, m
C	=	specific volumetric heat capacity, kJ/$(m^3 \cdot °C)$
c	=	specific heat capacity, kJ/(kg·°C)
D	=	diameter, m
e	=	$z_p^2 - r_p^2$, m
FI	=	air freezing index, °C·d
f	=	correction modulus for emission of radiated heat, dimensionless
G	=	$Q_v \cdot C \cdot R$, m (if C is J/$(m^3 \cdot °C)$)
HI	=	heating index, °C·d
h	=	height or thickness, m
H	=	hydraulic head, m
k	=	thermal conductivity, W/(m·°C)
L	=	volumetric latent heat of fusion, kJ/m^3
l	=	length, m
N	=	1.12 W/$(m^{7/4} \cdot °C^{5/4})$
n	=	$\dfrac{SFI}{FI}$ or $\dfrac{STI}{TI}$
P	=	perimeter, m
p	=	period; d, h, s
Q	=	quantity of heat, J
Q_v	=	volumetric flow rate, m^3/s
q	=	rate of heat transfer, W
R	=	thermal resistance, m·°C/W
r	=	radius, m
S	=	Stefan-Boltzman constant, W/$(m^2 \cdot °K^4)$
s	=	constant, dimensionless
SFI	=	surface freezing index, °C·d
STI	=	surface thawing index, °C·d
Ste	=	Stefan number, dimensionless
TI	=	air thawing index, °C·d
T	=	temperature, °C = °K − 273.15
t	=	time; d, h, s
u	=	coeff. of thermal expansion, m/(m·°C)
v	=	velocity, m/s
W	=	0.56 v + 1
w	=	water content, %
x	=	direction vector
z	=	depth below surface, m
α	=	thermal diffusivity, m^2/s

β	=	thermal ratio, modified Berggren equation, dimensionless
Δ	=	shift in permafrost table, m
δ	=	absolute value of $\dfrac{e + z_{active} + \delta_{j=1}}{e - z_{active} - \delta_{j=1}}$
λ	=	correction coefficient, modified Berggren equation, dimensionless
μ	=	fusion parameter for modified Berggren equation, dimensionless
μ	=	near surface conduction thermal transfer coef., W/$(m^2 \cdot °C)$
π	=	3.14159
ρ	=	density, kg/m^3
ψ	=	constant in Neumann equation, dimensionless

Subscript Symbols

A	=	ambient
a	=	air
af	=	air film
C	=	conduit
c	=	conduction
d	=	convection
E	=	exterior casing, utilidor
F	=	friction
f	=	frozen
g	=	ground
I	=	insulation
i	=	interior conduit
J	=	subscript number, general case
j	=	subscript number, general case
L	=	thermal lining, utilidor
ma	=	mean annual air temperature
ms	=	mean annual surface temperature
o	=	freezing
p	=	pipe
r	=	radiation
SF	=	safety factor
S	=	surface
T	=	total
t	=	thawed
U	=	utilidor
u	=	unfrozen
w	=	water
z	=	direction
χ	=	infinity

TABLE 4-4   *THERMAL CONDUCTIVITIES OF COMMON MATERIALS*

Material	Dry Density kg/m³ (ρ_d)	Specific Heat Capacity kJ/(kg•°C) (c)	Thermal Conductivity W/(m•°C) (k)
Air, no convection (0°C)		1.00	0.024
Air film, outside, 24 km/h wind (per air film)			0.86
Air film, inside (per air film)			0.24
Polyurethane foam	32	1.67	0.024
Polystyrene foam	30	1.26	0.036
Rock wool, glass wool	55	0.84	0.040
Snow, new loose	85	2.09	0.08
Snow, on ground	300	2.09	0.23
Snow, drifted and compacted	500	2.09	0.7
Ice at -40°C	900	2.09	2.66
Ice at 0°C	900	2.09	2.21
Water (0°C)	1000	4.19	0.58
Peat, dry	250	2.09	0.07
Peat, thawed, 80% moisture	250	1.34	0.14
Peat, frozen, 80% ice	250	0.92	1.73
Peat, pressed, moist	1140	1.67	0.70
Clay, dry	1700	0.92	0.9
Clay, thawed, saturated (20%)	1700	1.76	1.6
Clay, frozen, saturated (20%)	1700	1.34	2.1
Sand, dry	2000	0.80	1.1
Sand, thawed, saturated (10%)	2000	1.21	3.2
Sand, frozen, saturated (10%)	2000	0.88	4.1
Rock typical	2500	0.84	2.2
Wood, plywood, dry	600	2.72	0.17
Wood, fir or pine, dry	500	2.51	0.12
Wood, maple or oak, dry	700	2.09	0.17
Insulating concrete (varies)	200 to 1500		0.07 to 0.60
Concrete	2500	0.67	1.7
Asphalt	2000	1.67	0.72
Polyethelene, high density	950	2.26	0.36
Polyvinyl chloride, PVC	1400	1.05	0.19
Asbestos cement	1900		0.65
Wood stave (varies)			0.26
Steel	7500	0.50	43
Ductile iron	7500		50
Aluminum	2700	0.88	200
Copper	8800	0.42	375

*   Values are representative of materials but most materials have a variation in thermal properties.

tion 4.2. Portions of this section, including time-independent, steady-state problems of heat flow, are adapted from Thornton (1977). The symbols used are defined in Table 4-3, and the thermal conductivity of some common materials is presented in Table 4-4. Solutions to a number of utility system problems are presented to illustrate the computational procedures and typical results.

4.6.1 Basic Heat-Loss Relationships. Heat is a form of energy distinguished by a temperature differential. Whenever there is a difference in temperature between two points, energy is transferred from one point to the other in an attempt to achieve equilibrium. Thermodynamics deals with energy transfer and the prediction of equilibrium conditions. Heat transfer provides methods of analysis for the prediction of rates of energy transfer. Heat transfer can involve three different mechanisms: conduction, convection, and radiation. Conduction and convection require a material medium. Heat transfer by radiation can occur through a vacuum. Conduction is the only mode of heat transfer for an opaque solid. In the steady-state, one-dimensional case without heat generation, the basic equation takes the form:

$$q_c = -kA\frac{dT}{dx} = \frac{kA(T_1 - T_2)}{L}$$

where:

q_c = rate of heat flow by conduction, W

k = thermal conductivity, W/(m•°K)

A = area through which heat is transferred, m²

T = temperature, °K = °C + 273.15

x = distance in direction of heat flow, m

L = thickness of material between points 1 and 2, m

Subscripts 1, 2 = reference points in the direction of heat flow

The emission of radiant energy from a perfect blackbody surface follows the relationship:

$$q_r = f_{1-2} \bullet SA\left(T_1^4 - T_2^4\right)$$

where:

q_r = rate of heat flow by radiation, W

f_{1-2} = a modulus which modifies the equation to account for emissions and relative geometries of the actual bodies.

(=1.0 for black bodies)

S = Stefan-Boltzman constant

   = $5.67 \times 10^{-8} \dfrac{W}{m^2 \bullet K^4}$

A = surface area, m²

T = absolute temperature, °K = °C + 273.15

Pore spaces and open areas around pipes can allow convection to occur. The rate of convection heat transfer, q_d, can be written as:

$$q_d = \mu_d A(T_s - T_{w,\infty})$$

where:

μ_d = average convection heat transfer coefficient at the fluid-to-solid interface, W/(m²•°K)

A = surface area in contact with fluid, m²

T_s = surface temperature, °K

$T_{w,\infty}$ = temperature of undisturbed fluid far away from heat transfer surface, °K

These simple equations represent the basic one-dimensional relationships. Normally heat loss from utility pipe systems can be predicted by a second-order differential equation called the diffusion or Fourier equation. Assuming geometrically simple boundaries and constant physical properties, closed-form solutions are possible (Carslaw and Jaeger, 1957; Kreith and Black, 1980). Steady-state solutions (time independent) describe the long-term (infinite) heat flow. This approach was followed by Thornton (1977) in assembling the relationships presented in this section. The cylindrical coordinate system has been used when discussing heat loss from pipes.

4.6.2 Steady-State Pipeline Solutions. Figure 4-14 deals with heat flow from: a bare pipe, an insulated pipe, a single pipe in an insulated box, and a utilidor carrying multiple pipes. In each case, some of the major approximations in addition to the implied time-independent, steady-state assumptions, are indicated. Some comments intended to facilitate application of the formulas are also included. Where applicable, expressions are presented for relevant thermal resistance, rates of heat flow, and insulation thicknesses.

	(a) BARE PIPE	(b) INSULATED PIPE	(c) SINGLE PIPE IN A BOX	(d) MULTIPLE PIPE UTILIDOR
Sketch	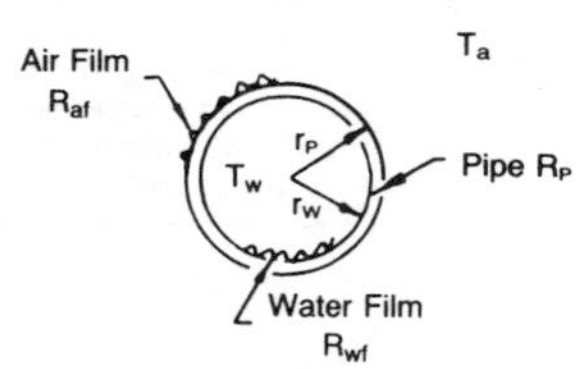 Air Film R_{af}; T_a; Pipe R_p; r_p, r_w, T_w, t_w; Water Film R_{wf}	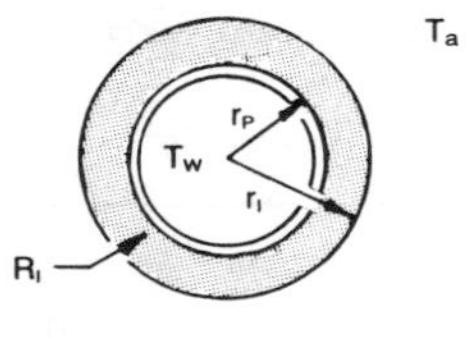 T_a; r_p, T_w, r_I; R_I	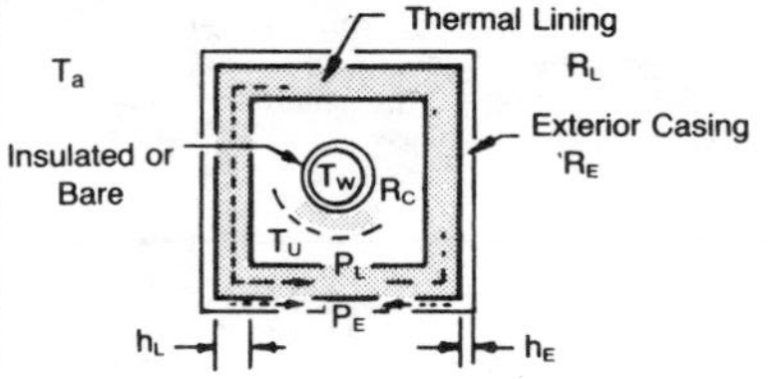 T_a; Thermal Lining R_L; Exterior Casing R_E; Insulated or Bare; T_w R_C; T_U P_L; P_E; h_E; h_L	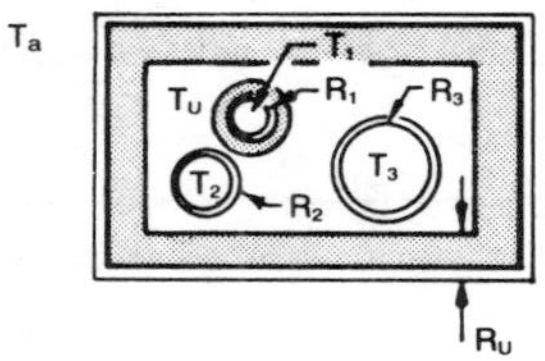 T_a; R_3, R_1, T_3; T_1; T_U, R_U; T_2 R_2
Assumptions	Thin-walled pipe (i.e. $r \approx r_w$). R_{wf} is negligible. $R_p \le R_{af}$.	All thermal resistances but that of the insulation are neglected.	Convection ensures the temperature inside the utilidor, T_U, is uniform. Utilidor air films are neglected.	Same as (c).
Thermal Resistance	$R_p = (r_p - r_w)/(r_p + r_w)\pi k_p$ $\mu_a = N\left(\dfrac{T_w - T_a}{r_p}\right)^{0.25} \cdot W$ $R_{af} = (2\pi r_p \mu_a)^{-1}$ $R_C = R_p + R_{af}$ $N = 1.12 \text{ J/s} \cdot m^{7/4} \cdot °C^{5/4}$ $W = \sqrt{0.56v + 1}$ for $v = m/s$	$R_C = R_I = \dfrac{\ln(r_I/r_p)}{2\pi k_I}$	Calculate R_C, the thermal resistance of the interior conduit by: using (b) if insulated or using (a) if bare and replacing T_a in the formula for μ_a by an estimate for $T_U (\le T_w)$. $R_L = h_L/p_L k_L$ $R_E = h_E/p_E k_E$ $R_U = R_L + R_E$ $R_T = R_C + R_U$ $T_U = \dfrac{(T_w/R_C) + (T_a/R_U)}{(1/R_C) + (1/R_U)}$ If bare pipe, iterate T_U	Calculate R for each pipe as in (c) to get R_J, where $J = 1,2,3,\dots$ Calculate R_U as in (c) $T_U = \dfrac{\sum_J (T_J/R_J) + (T_a/R_U)}{\sum_J (1/R_J) + (1/R_U)}$ If bare pipes present, iterate T_U
Rate of Heat Loss	$q = (T_w - T_a)/R_C$	$q = (T_w - T_a)/R_I$	$q = (T_w - T_a)/R_T$	$q_U = (T_J - T_U)/R_J$ (per pipe) $q_U = \sum_J q_J = (T_U - T_a)/R_U$
Insulation Thickness (given Q)	N/A	$r_I - r_p =$ $r_p \{\exp[2\pi k_I (T_w - T_a)/q] - 1\}$	Obtain R_E and R_C as above $h_L = p_L k_L\left[\left(\dfrac{T_w - T_a}{q}\right)R_E - R_C\right]$ If bare interior pipe, iterate T_U, R_C and hence h_L	Given acceptable q_L, calculate R_J as above and evaluate $T_U = T_J - R_J q_J$ for each pipe for which q_J is known. Using the maximum T_U found, calculate new q_J as above. Using these q_J and the same T_U, evaluate $h_L = p_L k_L\left[\dfrac{T_U - T_a}{\sum q_J} - R_E\right]$ If bare pipes present, iterate T_U, R_J and hence h_L as (c).
Comments	Often, for metal pipes, R_p may be neglected. If R_p is significant, the expression above for μ_a will generate an overestimate of q. If $T_a > T_w$ switch T_a and T_w in the expression for μ_a.	The neglected thermal resistances given in (a) may be included if desired. Estimate a value for the insulation surface temperature and calculate μ_a and R_{af} iterate.	The value of μ_a, and hence R_{af}, is fairly insensitive to the choice of T_U, and so one iteration on T_U is usually sufficient. Often R_e may be neglected. Similar calculational procedure may be performed for pipes and utilidors of different cross-section.	If it is clear that one pipe dominates the heat loss process, (c) may be used to estimate T_U. It is wise to consider the heat loss from the various pipes if certain other pipes cease to function.

FIGURE 4-14   STEADY-STATE THERMAL EQUATIONS FOR ABOVE-SURFACE PIPES (adapted from Thornton, 1977)

	(a) BARE, NO THAW	(b) BARE, WITH THAW ZONE	(c) INSULATED, NO THAW	(d) INSULATED, WITH THAW ZONE
Sketch	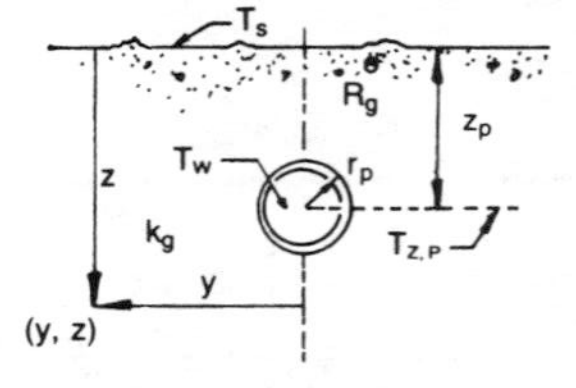	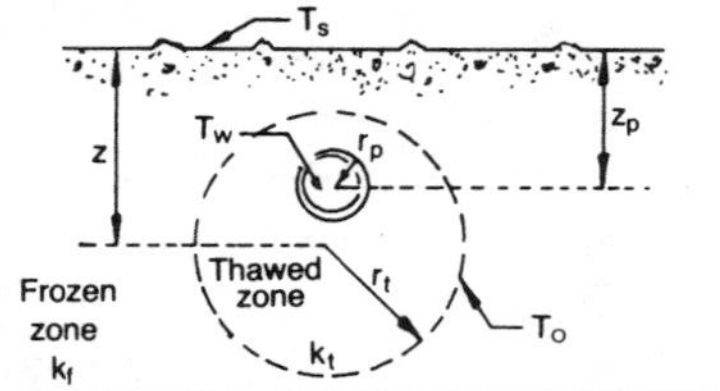	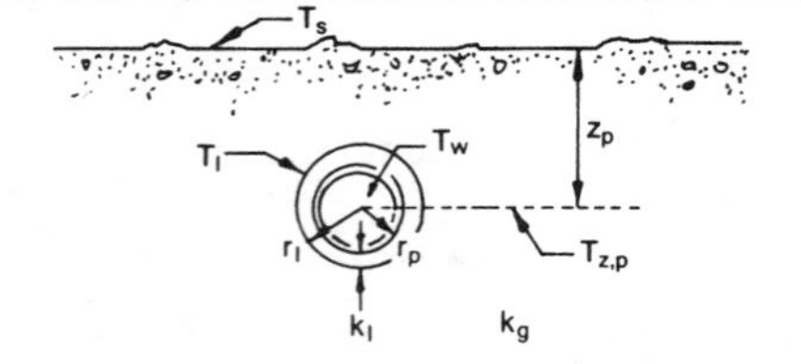	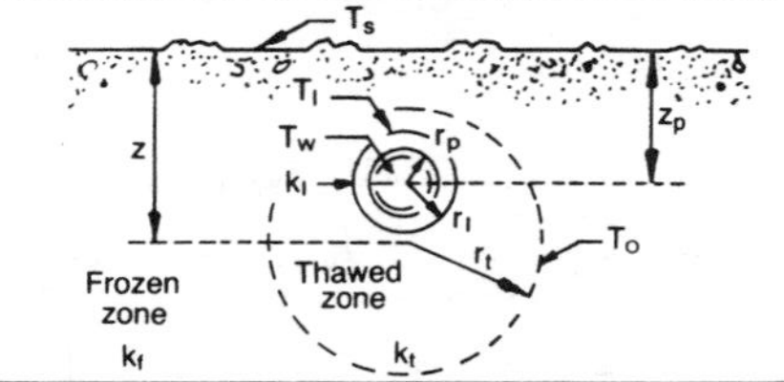
Assumptions	Neglect all thermal resistances except that of the soil. Pipe is in zone of influence from surface temperatures.	Same as (a), but accounting for the different conductivities of thawed and frozen soil.	Neglecting all thermal resistances except those of the soil and insulation. Outer surface of insulation assumed to be isothermal. $r_l - r_p \ll z_p$. Pipe is in zone of influence from surface temperature.	Same as (c) but accounting for the different conductivities of thawed and frozen soil.
Thermal Resistance and Thaw Zone Parameters	$$R_g = \frac{\text{arccos h}\left(\frac{z_p}{r_p}\right)}{2\pi k_g}$$	$$T_w = \frac{k_t}{k_f}(T_w - T_o) + T_o$$ $$T' = \frac{T_o - T_s}{T_w - T_s}$$ $$e = \sqrt{z_p^2 - r_p^2}$$ $$a = T' \text{ arccosh}(z_p/r_p) \qquad r_t = e \text{ csch } a$$ $$z = e \text{ coth } a$$ R_t, R_f and $R_g (= R_f + R_t)$ as given in (d), but with r_l replaced by r_p.	R_l as given in Figure 4-14(b) R_g as given in (a), but with r_p replaced by r_l $$T_l = T_w - \frac{R_l(T_w - T_s)}{R_g + R_l}$$ For known T_w, T_s, and R_g, the minimum insulation thickness to prevent thaw (ie. $T_l = T_o$) is given by: $$R_l' = \frac{T_w - T_o}{T_o + T_s} R_g$$	R_l as given in Figure 4-14(b) T_w, T', e, z, r_t and R_g as in (b) but with r_p replaced by r_l and using: $$a = T' [\text{arccosh}(z_p/r_z) + 2\pi k_l R_l]$$ $$T_l = T_w - \frac{R_l(T_w - T_s)}{R_g + (k_f/k_t)R_l}$$ Also: $$R_t = [\text{arccosh}(z_p/r_l) - \text{arccosh}(Z/r_l)]/2\pi k_t$$ $$R_f = [\text{arccosh}(z/r_l)]/2\pi k_f$$ $$R_g = R_f + R_t$$
Rate of Heat Loss	$$q = \frac{T_w - T_s}{R_g}$$	$$q = \frac{T_w - T_s}{R_g'} \text{ where } R_g' = \frac{\text{arccosh}(z_p/r_p)}{2\pi k_t}$$	$$q = \frac{T_w - T_s}{R_l + R_g}$$	$$q = \frac{T_w - T_s}{R_g' + (k_f/k_t)R_l}$$
Insulation Thickness	N/A	N/A	For no thawing outside the insulation the minimum insulation thickness is given by: $$r_l - r_p = r_p[\exp(2\pi k_l R_l) - 1]$$	$$R_l = [(a/T') + \text{arccosh}(z_p/r_l)]/2\pi k_l$$ $r_l - r_p$ as in (c) but with R_l replaced by R_l from above.
Comments	For calculations of heat loss when there is a temperature gradient in the soil and $z_p > 2r_p$, T_w may be replaced by T_{zp}, the undisturbed ground temperature at the pipe axis depth. For an upper limit on heat loss use $k_g = k_f$ otherwise use $k_g = (k_f + k_t)/2$	The thawed zone is a circle in cross section.	May be used to approximate (d) if $k_t = k_f$ and/or $r_l = r_t$, and thaw zone parameters are not required. Use $k_g = k_f$ or $k_g = (k_f + k_t)/2$ as in (a).	Often the above expressions for R_t, R_f and R_g are not required.

FIGURE 4-15   STEADY-STATE THERMAL EQUATIONS FOR BELOW-SURFACE PIPES (adapted from Thornton, 1977)

CONDITIONS	SKETCH	THERMAL RESISTANCE
Square Insulation	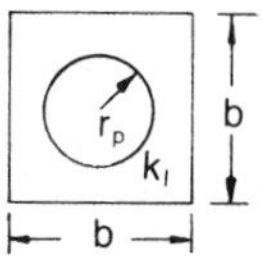	$R = \dfrac{1}{2\pi k_I} \cdot \ln 1.08 \dfrac{b}{2r_p}$
Rectangular Insulation	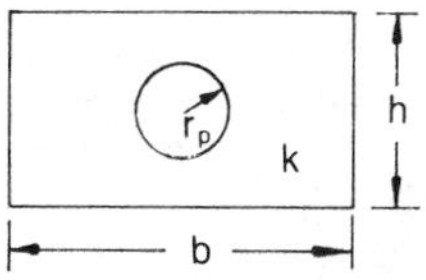	$R = \dfrac{1}{2\pi k} \cdot \ln\left(\dfrac{4h}{\pi r_p} + 2s\right)$ b/h s b/h s b/h s 1.00 0.08290 2.00 0.00373 4.00 6.97×10^{6} 1.25 0.03963 2.25 0.00170 5.00 3.01×10^{7} 1.50 0.01781 2.50 0.00078 ⋮ ⋮ 1.75 0.00816 3.00 0.00016 ∞ 0
Eccentric Cylindrical Insulation	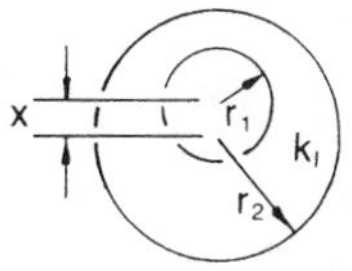	$R = \dfrac{1}{2\pi k} \cdot \ln \dfrac{\sqrt{(r_2 + r_1)^2 - x^2} + \sqrt{(r_2 - r_1)^2 - x^2}}{\sqrt{(r_2 + r_1)^2 - x^2} - \sqrt{(r_2 - r_1)^2 - x^2}}$
Two Buried Bare Pipes	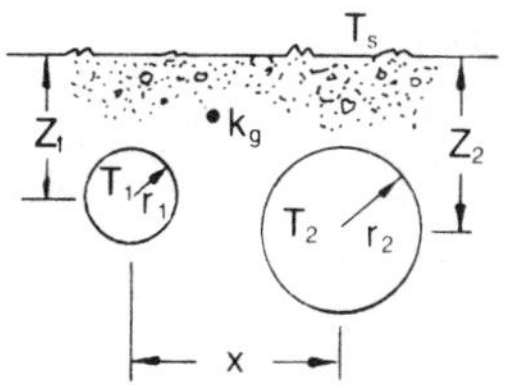	$R_1 = \dfrac{1}{2\pi k_g} \cdot \dfrac{\left[\ln \dfrac{2Z_1}{r_1} \cdot \ln \dfrac{2Z_2}{r_2}\right] + \left[\ln \sqrt{\dfrac{(Z_1 + Z_2)^2 + x^2}{(Z_1 - Z_2)^2 + x^2}}\right]^2}{\left[\ln \dfrac{2Z_2}{r_2}\right] - \left[\left(\dfrac{T_2 - T_s}{T_1 - T_s}\right) \cdot \left(\ln \sqrt{\dfrac{(Z_1 + Z_2)^2 + x^2}{(Z_1 - Z_2)^2 + x^2}}\right)\right]}$ where $4\mu < Z_1 \le Z_2$ for R_2, interchange subscripts 1 and 2
Buried Rectangular Duct	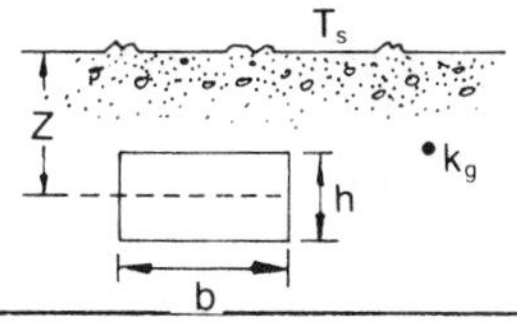	$R = \dfrac{1}{k_g\left(5.7 + \dfrac{b}{2h}\right)} \ln \dfrac{3.5Z}{b^{0.25} \cdot h^{0.75}}$
Surface Thermal Resistance	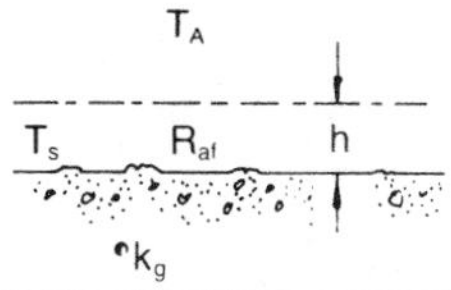	Surface thermal resistance between ground and air can be approximated as the equivalent thickness of the underlying soil equal to where $\quad h = \dfrac{m^2 \cdot {}^\circ C}{W} \qquad\qquad R_{af} = \dfrac{k_g}{R_{af}}$
Composite Wall	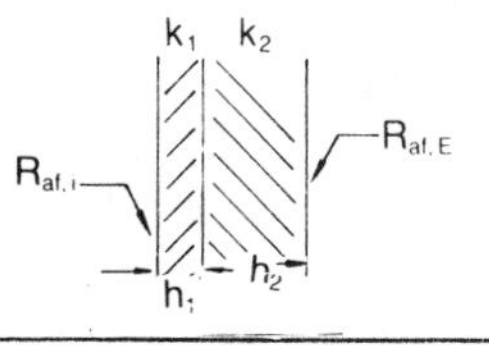	$R_T = R_{af,i} + R_{af,E} + \dfrac{h_1}{k_1} + \dfrac{h_2}{k_2}$ where $R_{af} = \dfrac{m^2 \cdot {}^\circ C}{W}$

FIGURE 4-16 STEADY-STATE THERMAL RESISTANCE OF VARIOUS SHAPES AND BODIES (after Kutateladze, 1963)

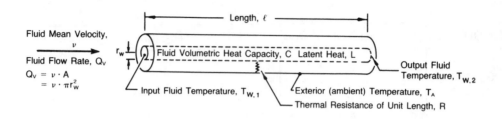

Comments: The above sketch is schematic. R and T_A appearing in these equations can be replaced by the thermal resistance and corresponding exterior temperature for any shape or configuration.

Heat Loss and Temperature Drop in a Fluid Flowing Through a Pipe

$$G = Q_v \cdot C \cdot R \left(\frac{m^3}{s} \cdot \frac{J}{m^3 \cdot °C} \cdot \frac{s \cdot m \cdot °C}{J} \right)$$

Given: R, T_A, $T_{W,1}$ or $T_{W,2}$

Find: $T_{W,1}$ or $T_{W,2}$

$$T_{W,1} = T_A + \frac{(T_{W,2} - T_A)}{\exp(-\ell/G)}$$

or $\quad T_{W,2} = T_A + (T_{W,1} - T_A) \cdot \exp(-\ell/G)$

Given: $T_{W,1}$, $T_{W,2}$, T_A

Find: R

$$R = -\ell \cdot \left(Q_v \cdot C \cdot \ln \left[\frac{(T_{W,2} - T_A)}{(T_{W,1} - T_A)} \right] \right)^{-1}$$

Given: $T_{W,1}$, $T_{W,2}$, T_A, R

Find: Fluid Mean Velocity, v

$$v = -\ell \cdot \left(A \cdot R \cdot C \cdot \ln \left[\frac{(T_{W,2} - T_A)}{(T_{W,1} - T_A)} \right] \right)^{-1}$$

Given: $T_{W,1}$ or $T_{W,2}$, T_A, v, R

Find: Heat Loss Rate, q

$$q = \frac{G}{R} \cdot (T_{W,2} - T_A) \cdot \left[\exp\left(\frac{\ell}{G}\right) - 1 \right]$$

$$= \frac{G}{R} \cdot (T_{W,1} - T_A) \cdot \left[1 - \left(\exp\left(\frac{-\ell}{G}\right)\right) \right]$$

Given: v, H_F (headloss due to friction, m/m)

Find: Friction Heat Gain, q_F

$$q_F = 3.074 \times 10^4 \frac{J}{m^4} \cdot Q_v \cdot H_F$$

q_F is negligible for $v < 2$ m/s

Freeze-Up Time For a Full Pipe Under No-Flow Conditions ($v = 0$)

Given: R, $T_{W,1}$, T_A

Find: Freeze-up Time, t_f

$\quad t_f$ = time for liquid temperature to drop to freezing point

Design Time

$$t_f = A \cdot R \cdot C \cdot \ln \left[\frac{(T_{W,1} - T_A)}{(T_0 - T_A)} \right]$$

Safety Factor Time

$\quad t_{f,SF}$ = time for fluid temperature to drop to nucleation temperature

$\quad t_{f,SF} = t_f \quad$ where $T_0 = -3°C$ for water.

Complete Freezing Time

$\quad t_{f,T}$ = time for liquid 0°C to freeze solid.

$$t_{f,T} = \frac{A \cdot R \cdot L}{(T_0 - T_A)} \qquad \text{(s)}$$

Given: duration of flow stoppage, t

Find: R

$$R = \frac{t}{A \cdot C} \left[\ln \left(\frac{T_{W,1} - T_A}{T_0 - T_A} \right) \right]^{-1}$$

$R_{SF} = R \quad$ where $T_0 = -3°C$ for water

FIGURE 4-17   TEMPERATURE DROP AND FREEZE-UP TIME IN PIPES (adapted from Thornton, 1977)

Figure 4-15 gives similar information for uninsulated and insulated buried pipes. In each of these two cases, the presence of thawed ground around the pipe is considered, and formulas are included which indicate the dimensions of the resulting thaw cylinder.

Figure 4-16 contains expressions for the thermal resistance of various shapes and bodies from which heat loss can be calculated. Formulas are given in Figure 4-17 for estimating the temperature drop (or gain) along a pipeline system, and simple expressions relating to freeze-up times under no-flow conditions are included.

Steady-state thermal influences in isotropic, homogeneous soils can be summed and geometric modifications and approximations made to these basic equations. For example, a layered soil can be represented by an "effective" soil thickness with the same total thermal resistance as the layered soil.

When pipes are buried below the influence of short-term air temperature fluctuations, the ground temperatures around the pipeline resemble a slowly changing series of steady-state conditions (Prokhaer, 1959). This is illustrated in Figure 4-18. The heat loss from deeply buried pipes can be calculated from steady-state equations for a cylinder of material around a pipe, if the fluid temperature

and the soil temperature at a known distance from the pipe are measured, and the soil and insulation thermal conductivities are known. Heat loss from deep pipes can also be conveniently estimated by replacing the ground surface temperature in the steady-state equations with the ground temperature at the pipe depth (Janson, 1966) (see Example 4.7).

Steady-state temperatures around a pipe (real or equivalent) can be easily determined from equations. In permafrost soils, the maximum thermal influence of the pipe can be estimated by simply adding the steady-state pipe temperatures to the maximum ground temperatures expected in the permafrost when no pipe is present. The solution can be further simplified if only the temperatures below the pipe (i.e., Y = 0) or only the maximum thaw are required. For the given conditions, these formulas overestimate thawing, but possible transient thermal effects of the actual installation, and other factors such as subsurface water flow along the trench are not considered. See Figure 4-19.

4.7 Example Problems

The following problems are presented to illustrate the application of thermal equations to solve some of the thermal problems encountered in utility system design. Various units of measure are still in common usage and this can lead to confusion and er-

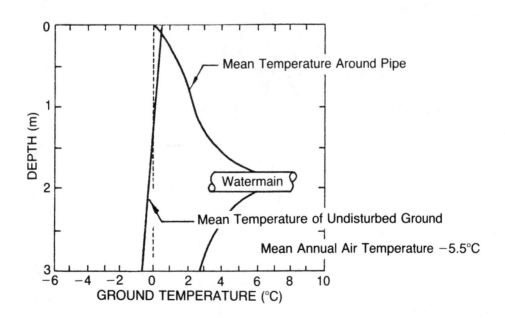

FIGURE 4-18   GROUND TEMPERATURES AROUND BURIED WATER PIPE
AT YELLOWKNIFE, NWT, CANADA (Prokhaer, 1959)

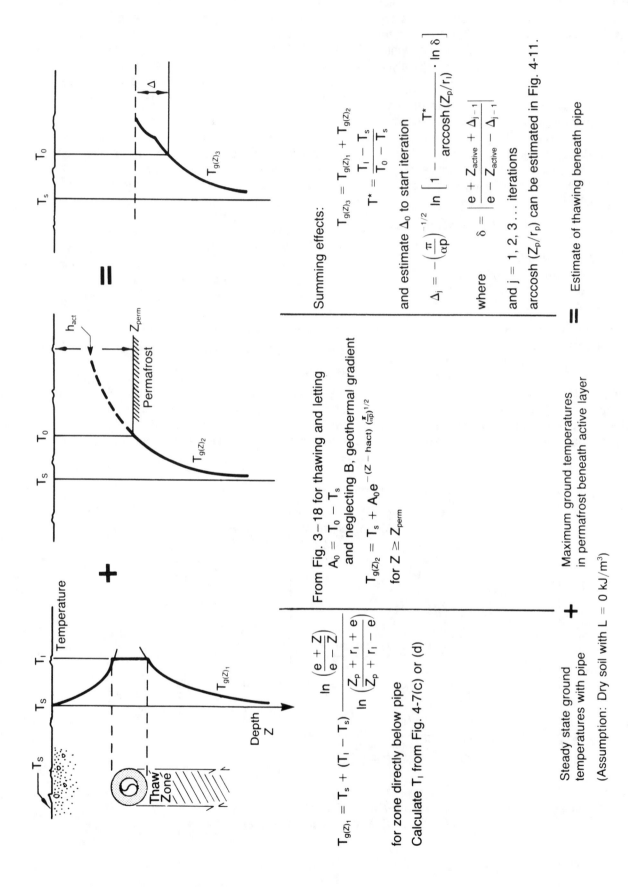

$$T_{g(Z)_1} = T_s + (T_l - T_s) \frac{\ln\left(\frac{e + Z}{e - Z}\right)}{\ln\left(\frac{Z_p + r_l + e}{Z_p + r_l - e}\right)}$$

for zone directly below pipe
Calculate T_l from Fig. 4-7(c) or (d)

Steady state ground
temperatures with pipe

From Fig. 3—18 for thawing and letting
$$A_0 = T_0 - T_s$$
and neglecting B, geothermal gradient
$$T_{g(Z)_2} = T_s + A_0 e^{-(Z - hact)\left(\frac{\pi}{\alpha p}\right)^{1/2}}$$
for $Z \geq Z_{perm}$

Maximum ground temperatures
in permafrost beneath active layer

Summing effects:
$$T_{g(Z)_3} = T_{g(Z)_1} + T_{g(Z)_2}$$
$$T^* = \frac{T_l - T_s}{T_0 - T_s}$$
and estimate Δ_0 to start iteration
$$\Delta_j = -\left(\frac{\pi}{\alpha p}\right)^{-1/2} \ln\left[1 - \frac{T^*}{\text{arccosh}(Z_p/r_l)} \cdot \ln \delta\right]$$
where $\quad \delta = \left|\frac{e + Z_{active} + \Delta_{j-1}}{e - Z_{active} - \Delta_{j-1}}\right|$
and j = 1, 2, 3 . . . iterations
arccosh (Z_p/r_p) can be estimated in Fig. 4-11.

Estimate of thawing beneath pipe

(Assumption: Dry soil with L = 0 kJ/m³)

FIGURE 4-19   GROUND TEMPERATURES AND THAWING AROUND BURIED PIPES IN PERMAFROST

rors. Most of the equations presented in this section can be utilized with any system of units as long as the units are consistent. The analyst is encouraged to work in SI units. Conversion factors to SI units are given after the appendices.

Example 4.5

Calculate the rate of heat loss per unit length from an above-ground, 136 mm in diameter, plastic pipe with outside diameter of 166 mm. The thermal conductivity k_p is 0.36W/(m• °C). The pipe is enclosed in 50 mm of polyurethane foam, k_l = 0.023 W/(m•°C). Water temperature is 5°C, and air temperature drops to -40°C. Wind speed is estimated at 25 km/h.

Solution:

Problem resembles that of Figure 4-14(b).

$$r_p = \frac{166mm}{2} = 83 \text{ mm or } 0.083 \text{ m}$$

$$r_l = r_p + 50 \text{ mm} = 133 \text{ mm or } 0.133 \text{ m}$$

$$r_l/r_p = 1.6 \text{ and } k_l = 0.023 \text{ W/(m•°C)}$$

Therefore

$$R_c = R_l = \frac{\ln\left(\dfrac{r_l}{r_p}\right)}{2\pi \bullet k_l} = \frac{\ln(1.6)}{2\pi \bullet 0.023} = 3.3 \frac{m \bullet °C}{W}$$

Assume R_p and R_{af} are negligible.

$$q = \frac{T_w - T_a}{R_l} = \frac{(5°C) - (-40°C)}{\dfrac{3.3 m \bullet °C}{W}} = 14 \frac{W}{m}$$

Include neglected thermal resistances as a check: T_s, the insulation surface temperature is estimated as 35°C, a value close to T_a. From Figure 4-14(a),

$$R_p = \frac{r_p - r_w}{(r_p - r_w)\pi \bullet k_p}$$

$$= \frac{0.083m - 0.068m}{(0.083m + 0.068m)\pi \bullet \dfrac{0.36W}{m \bullet °C}}$$

$$= \frac{0.088 m \bullet °C}{W}$$

$$R_{af} = (2\pi r_l m_a)^{-1}$$

$$\mu_a = N\left(\frac{T_s - T_a}{r_l}\right)^{0.25} \bullet W_w$$

where:

$$N = \frac{1.12W}{m^{\frac{7}{4}} \bullet °C^{\frac{5}{4}}}$$

$$W = (0.56v + 1)^{0.5}$$

where v = 25 km/h = 6.9 m/s

$$W_w = \left(0.56\left(6.9\frac{m}{s}\right) + 1\right)^{0.5} = 2.2$$

$$\mu_a = \frac{1.12W}{m^{\frac{7}{4}} \bullet °C^{\frac{5}{4}}} \bullet \left(\frac{-35°C - (-40°C)}{0.133m}\right)^{0.25} \bullet (2.2)$$

$$= 6.1 \frac{W}{m^2 \bullet °C}$$

$$R_{af} = \left[2\pi(0.133m) \bullet \left(\frac{6.1W}{m^2 \bullet °C}\right)\right]^{-1} = 0.20 \frac{m \bullet °C}{W}$$

$$T_s = T_a + \frac{R_{af}(T_w - T_a)}{R_{af} + R_l + R_p}$$

$$T_s = -40°C + \frac{\left(0.20\dfrac{m \bullet °C}{W}\right) \bullet (5°C - (-40°C))}{(0.20 + 3.3 + 0.088)\dfrac{m \bullet °C}{W}}$$

$$= -37°C$$

Try another iteration with T_s = -37°C.

This results in

T_s = -37°C and

R_{af} = 0.23m•°C/W

Therefore,

$R_c = R_{af} + R_p + R_l$

$= 0.23 + 0.088 + 3.3 = 3.6 \ m\cdot°C/W$

and

$$q = \frac{5°C - (-40°C)}{3.6 \dfrac{m\cdot°C}{W}} = 13 \frac{W}{m}$$

Note: If there is no wind, $T_s = -35°C$, $R_{af} = 0.43$ $m\cdot°C/W$ and $q = 12 \ W/m$

Example 4.6

Calculate the rate of heat loss per unit length of pipe from Example 4.5 if the pipe is buried to a depth of 1.22 m in a soil with thermal conductivity, $k_g = 2.0$ $W/(m\cdot°C)$ when the water temperature is 5°C and ground surface temperature is -40°C.

Solution:

This problem is similar to the one in Figure 4-15(c).

From Example 4.5,

$R_p = 0.088 \ m\cdot°C/W$

$R_l = 3.3 \ m\cdot°C/W$

$$R_g = \frac{\text{arccos}h\left(\dfrac{z_p}{r_l}\right)}{2\pi\cdot k_g}$$

$z_p = 1.22 \ m$

$r_l = 0.133 \ m$

$k_g = 2.0 \ W/(m\cdot°C)$

$$R_g = \frac{\text{arccos}h\left(\dfrac{1.22}{0.133}\right)}{2\pi\cdot(2.0)} = 0.23 \frac{m\cdot°C}{W}$$

Therefore

$R_c = R_g + R_l + R_p$

$\quad = 0.23 + 3.3 + 0.088 = 3.6 \ m\cdot°C/W$

and

$$q = \frac{5°C - (-40°C)}{3.6 \dfrac{m\cdot°C}{W}} = 13 \frac{W}{m}$$

Example 4.7

For the pipes in Examples 4.5 and 4.6, calculate the annual heat loss and maximum rate of heat loss per unit length when the ambient temperatures are:

Mean Monthly Temperatures (°C)

	J	F	M	A	M	J
Air	-28.4	-24.9	-24.0	-16.8	-11.7	-1.2
Depth of 1.22 m	-14.0	-15.0	-15.0	-13.8	-10.4	-8.0

	J	A	S	O	N	D
Air	4.5	2.6	-0.2	-8.8	-15.2	-21.4
Depth of 1.22 m	-4.9	-2.8	-2.8	-3.1	-6.7	-10.0

Mean annual air temperature = -12.8˚C

Mean annual ground temperature at 1.22 m = -9.7˚C

Minimum air temperature = -50.0°C

Minimum mean daily ground temperature at 1.22 m = -17.8°C

From Example 4.5:

$R_c = 3.6 \ m\cdot°C/W$

$T_w = 5°C$

From Example 4.6:

$R_c = 3.6 \ m\cdot°C/W$

$T_w = 5°C$

Solution:

Calculate the water freezing index for time period when $T_a \leqq T_o$ i.e. $\Sigma\,(T_w - T_a)$ on a monthly basis.

Above-ground pipe:

$\Sigma\,(T_w - T_a) = (5°C - (-28.4°C))\cdot31d +$
$\qquad\qquad\qquad (5°C - (-24.9°C))\cdot28d + ...$

$\qquad\quad = 6227 \ °C\cdot d$

Buried pipe:

$\Sigma\,(T_w - T_g) = (5°C - (-14.0°C))\cdot31d +$
$\qquad\qquad\qquad (5°C - (-15.0°C))\cdot28d + ...$

$\qquad\quad = 5050 \ °C\cdot d$

Calculate annual heat loss and maximum rate of heat loss.

Above-ground pipe: (Example 4.5)

$$Q = \frac{(6227^{\circ}C \cdot d) \cdot \left(24\frac{h}{d}\right) \cdot \left(3600\frac{s}{h}\right)}{R_c}$$

$$Q = \frac{\left(538 \cdot 10^6 \, ^{\circ}C \cdot s\right)}{3.6\frac{m \cdot s \cdot ^{\circ}C}{J}} = 150\frac{MJ}{m}$$

and

$$q = \frac{5^{\circ}C - (-50^{\circ}C)}{3.6\frac{m \cdot ^{\circ}C}{W}} = 15\frac{W}{m}$$

Buried pipe: (Example 4.6)

$$Q = \frac{(5050^{\circ}C \cdot d) \cdot \left(24\frac{h}{d}\right) \cdot \left(3600\frac{s}{h}\right)}{3.6\frac{m \cdot s \cdot ^{\circ}C}{J}} = 121\frac{MJ}{m}$$

and

$$q = \frac{5^{\circ}C - (-17.8^{\circ}C)}{3.6\frac{m \cdot ^{\circ}C}{W}} = 6.3\frac{W}{m}$$

Note: The annual heat loss from the above-ground pipe is only slightly higher than a pipe buried in permafrost; however, the maximum rate of heat loss (and freeze-up risk) is much higher. The relative differences become greater in warmer areas, particularly in non-permafrost conditions. In the extreme case, the pipe may be located below the maximum frost penetration therefore heating and freezing risk are nil.

Example 4.8

Calculate the interior temperature and rate of heat loss per unit length of the plywood box shown below which contains a 150 mm, Class 1500 asbestos cement pipe. Assume free convection, radiation is negligible, and air leakage from the box.

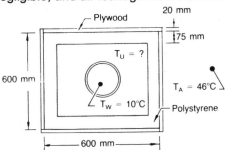

Solution:

This problem resembles that of Figure 4-14(c) with an uninsulated pipe. Estimate T_u as 2°C and use method of Figure 4-14(a).

Parameters describing box and pipe:

D_p = 150 mm or 0.15 m (inside)

D_b = 180 mm or 0.18 m (outside)

r_p = 0.090 m

r_w = 0.075 m

From Table 4-4

k_p = 0.65 W/(m•°C)

k_E = 0.17 W/(m•°C)

k_L = 0.036 W/(m•°C)

From Figure 4-14(a):

$$R_p = \frac{(0.090m - 0.075m)}{(0.090m + 0.075m)\pi\left(0.65\frac{W}{m \cdot ^{\circ}C}\right)}$$

$$= 0.045\frac{m \cdot ^{\circ}C}{W}$$

$$R_{af} = (2\pi r_p m_a)^{-1}$$

and

$$\mu_a = 1.12\left(\frac{10^{\circ}C - 2^{\circ}C}{0.090m}\right)^{0.25} = 3.4$$

$$R_{af} = \left[2\pi \cdot (0.090)(3.4)\right]^{-1} = 0.51\frac{m \cdot ^{\circ}C}{W}$$

R_c = R_{af} + R_p = 0.56 m•°C/W

P_E = plywood center line perimeter, m

P_L = insulation center line perimeter, m

$$P_E = 4\left(0.6m + \frac{0.020m}{2}\right) = 2.44m$$

$$P_L = 4\left(0.6m + \frac{0.075m}{2}\right) = 2.25m$$

$$R_E = \frac{\text{thickness of plywood}}{(\text{center line perimeter}) \cdot K_E}$$

$$R_L = \frac{\text{thickness of plywood}}{(\text{center line perimeter}) \cdot K_L}$$

From Figure 4-14(c):

$$R_E = \frac{0.020}{(2.4) \cdot (0.17)} = 0.048 \frac{m \cdot {}^\circ C}{W}$$

$$R_L = \frac{0.075}{(2.3) \cdot (0.036)} = 0.93 \frac{m \cdot {}^\circ C}{W}$$

$R_u = R_E + R_L$

$R_u = 0.049 + 0.91 = 0.95 \ m \cdot {}^\circ C/W$

$$T_u = \frac{\dfrac{T_w}{R_c} + \dfrac{T_a}{R_u}}{\dfrac{1}{R_c} + \dfrac{1}{R_u}}$$

$$T_u = \frac{\dfrac{10.}{0.56} + \dfrac{-46}{0.95}}{\dfrac{1}{0.56} + \dfrac{1}{0.95}} = -10.4{}^\circ C$$

Try new estimate of T_u, say $T_u = -7{}^\circ C$

$$\mu_a = 1.12 \left(\frac{10{}^\circ C - (-7{}^\circ C)}{0.090 m} \right)^{0.25} = 4.2$$

$$R_{af} = \left[2\pi \cdot (0.090)(4.2) \right]^{-1} = 0.43 \frac{m \cdot {}^\circ C}{W} \ , \ \text{and}$$

$R_c = 0.045 + 0.43 = 0.47 \ m \cdot {}^\circ C/W$

$$T_u = \frac{\dfrac{10.}{0.47} + \dfrac{-46}{0.95}}{\dfrac{1}{0.47} + \dfrac{1}{0.95}} = -8.5{}^\circ C$$

Try another iteration, $T_u = -8.5{}^\circ C$

$$\mu_a = 1.12 \left(\frac{10{}^\circ C - (-8.5{}^\circ C)}{0.090 m} \right)^{0.25} = 4.2$$

$$R_{af} = \left[2\pi \cdot (0.090)(4.2) \right]^{-1} = 0.43 \frac{m \cdot {}^\circ C}{W}$$

$$T_u = \frac{\dfrac{10.}{0.43} + \dfrac{-46}{0.95}}{\dfrac{1}{0.43} + \dfrac{1}{0.95}} = -8.1{}^\circ C$$

$R_T = R_p + R_{af} + R_u$

$\quad = 0.045 + 0.43 + 0.95$

$\quad = 1.4 \ m \cdot {}^\circ C/W, \text{ and}$

$$q = \frac{10{}^\circ C - (-46{}^\circ C)}{1.4 \dfrac{m \cdot {}^\circ C}{W}} = 39 \frac{W}{m}$$

Example 4.9

A utilidor having the same configuration as Example 4.8 but including a high temperature (120°C) water (HTW) pipe, 50 mm in outside diameter covered with 40 mm of asbestos fibre, $k_i = 0.09$ W/(m•°C). Find the interior temperature of the utilidor and the rate of heat loss per unit length.

From Example 4.8:

$T_a = -46{}^\circ C$

$T_{w,1} = 10{}^\circ C$

$R_u = 0.95 \ m \cdot {}^\circ C/W$

$R_{p,1} = 0.045 \ m \cdot {}^\circ C/W$

$r_{p,1} = 0.090 \ m$

Solution:

Call the water pipe, 1 and HTW pipe, 2. Neglect the thermal resistance of the HTW pipe and air film. Calculate $R_{l,2}$ from Figure 4-14(d) which resembles this problem.

$r_{l,2} = 25 \ mm + 40 \ mm = 65 \ mm \ \text{or} \ 0.065 \ m$

$r_{p,2} = 25 \ mm \ \text{or} \ 0.025 \ m$

$$R_{c,2} = R_{l,2} = \frac{\ln\left(\dfrac{0.065 mm}{0.025 mm} \right)}{2\pi \cdot \left(\dfrac{0.09 W}{m \cdot {}^\circ C} \right)} = 1.7 \frac{m \cdot {}^\circ C}{W}$$

Now estimate T_u. Say $T_u = 5{}^\circ C$

$R_{c,1} = R_{p,1} + R_{af,1}, \text{ and}$

$R_{af,1} = (2\pi \cdot r_{p,1} \cdot m_a)^{-1}$

where

$$\mu_a = 1.12 \left(\frac{10°C - (5°C)}{0.090m} \right)^{0.25} = 3.1$$

$R_{af,1} = (2\pi (3.1)(0.090))^{-1} = 0.58$ m•°C/W

$R_{c,1} = 0.045 + 0.58 = 0.62$ m•°C/W

$$T_u = \frac{\dfrac{10}{0.62} + \dfrac{120}{1.7} + \dfrac{-46}{0.95}}{\dfrac{1}{0.62} + \dfrac{1}{1.7} + \dfrac{1}{0.95}} = 12°C$$

Try another iteration, using $T_u = 12°C$

$$\mu_a = 1.12 \left(\frac{12°C - (10°C)}{0.090m} \right)^{0.25} = 2.4$$

Therefore

$R_{af,1} = (2\pi (0.090)(2.4))^{-1} = 0.73$ m•°C/W

$R_{c,1} = 0.045 + 0.73 = 0.77$ m•°C/W

$$T_u = \frac{\dfrac{10}{0.62} + \dfrac{120}{1.7} + \dfrac{-46}{0.95}}{\dfrac{1}{0.62} + \dfrac{1}{1.7} + \dfrac{1}{0.95}} = 12°C$$

which closes.

Rate of heat loss:

$$q_2 = \frac{120°C - 12°C}{1.7 \dfrac{m•°C}{W}} = 64 \frac{W}{m}$$

For water pipe:

$$q_2 = \frac{10°C - 12°C}{0.77 \dfrac{m•°C}{W}} = 12.6 \frac{W}{m} \quad \text{say } 0.0 \frac{W}{m}$$

For the utilidor:

$$q_2 = \frac{12°C - (-46°C)}{0.95 \dfrac{m•°C}{W}} = 61 \frac{W}{m}$$

Example 4.10

A buried, 150 mm outside-diameter bare metal pipe lies 1.2 m below grade in a clay soil. The soil has thawed and frozen thermal conductivities of 1.6 and 2.1 W/(m•°C), respectively. Calculate the mean size of the thawed zone and the average rate of heat loss per unit length, if the mean surface temperature is -2.5°C and the water in the pipe is 7°C.

Solution:

This problem resembles that of Figure 4-15(b).

r_p = 150 mm/2 = 75 mm or 0.075 m

z_p = 1.2 m

T_w = 7°C

T_s = -2.5°C

T_o = 0°C

k_t = 1.6 W/m•°C

k_f = 2.16 W/m•°C

$$T_w' = \frac{1.6°C}{2.1} (7°C - 0°C) = 5.3°C$$

$$T' = \frac{0°C - (-2.5°C)}{5.3°C - (-2.5°C)} = 0.32°C$$

e = $(z_p^2 - r_p^2)^{1/2}$

= $((1.2)^2 - (0.075)^2)^{1/2}$ = 1.2 m

a = T' arccosh (z_p/r_p)

= 0.32 arccosh (1.2/0.075) = 1.1

and

r_t = thaw radius, m

= e csch a = 1.2 csch (1.1) = (1.2)(0.75)

= 0.90 m

z = centre of thaw radius

= e coth a = 1.2 coth (1.1)

= (1.2) • (1.3)

= 1.5 m

Hence, under steady-state conditions, a thawed zone will be present within a cylinder parallel to the pipe, of radius 0.90 m and approximately 0.30 m below the axis of the pipe.

Heat loss calculations:

$$R'_g = \frac{arccosh\left(\dfrac{z_p}{r_p}\right)}{2\pi \bullet k_f} = \frac{arccosh\left(\dfrac{1.2}{0.075}\right)}{2\pi \bullet (2.1)}$$

$$= 0.26 \frac{m \bullet {}^\circ C}{W}$$

$$q = \frac{5.3^\circ C - (-2.5^\circ C)}{0.26 \dfrac{m \bullet {}^\circ C}{W}} = 30 \frac{W}{m}$$

Example 4.11

A metal pipe of external diameter 152 mm is buried in permafrost (kg = 1.73W/(m• °C)) with its axis 1.22 m below the ground surface whose mean temperature is -2.5°C. What is the minimum thickness of polyurethane foam insulation (k = 0.024W/(m• °C)) which will maintain the soil in a frozen state (on average) if water flowing through the pipe has a mean temperature of 7.2°C? What is the average rate of heat loss if this insulation thickness is used?

Solution:

Given parameters are:

r_p = 152 mm/2 = 76 mm or 0.076 m

z_p = 1.22 m

z_p/r_p = 1.22 m/0.076 m = 16

k_g = 1.73 W/(m•°C)

T_I = T_o = 0°C

T_s = -2.5°C

T_w = 7.2°C

k_I = 0.024 W/(m•°C)

$$R_g = \frac{arccosh\left(\dfrac{z_p}{r_p}\right)}{2\pi \bullet k_f} = \frac{arccosh(16)}{2\pi \bullet (1.73)} = 0.32 \frac{m \bullet {}^\circ C}{W}$$

$$R'_I = \frac{7.2^\circ C - 0^\circ C}{0^\circ C - (-2.5^\circ C)} \bullet \frac{0.32 m \bullet {}^\circ C}{W} = 0.92 \frac{m \bullet {}^\circ C}{W}$$

Minimum insulation thickness, $r_I - r_p$:

$r_I - r_p$ = r_p (exp (2 π k_I R'_I) -1)

= 0.011 m or 11 mm

For insulation thickness of 11 mm:

From Figure 4-14(b)

$$R_I = \frac{\ln\left(\dfrac{r_I}{r_p}\right)}{2\pi \bullet k_I} = \frac{\ln\left[\dfrac{(0.076+0.011)}{0.076}\right]}{2\pi\left[\dfrac{0.024W}{m \bullet {}^\circ C}\right]} = 0.92 \frac{m \bullet {}^\circ C}{W}$$

Therefore

$$q = \frac{7.2^\circ C - (-2.5^\circ C)}{(0.92+0.32)\dfrac{m \bullet {}^\circ C}{W}} = 7.8 \frac{W}{m}$$

C = 4.187 MJ/(m³•°C)

From Figure 4-4(b):

p = 3600 s/h • 24 h/d • 365 d

= 31.5 x 10⁶ s

e = (z_p^2 - r_I^2)^0.5 = (1.22² - 0.087²)^0.5

= 1.22 m

z_p/r_I = 14

$$T^* = \frac{0.0^\circ C - (-2.5^\circ C)}{0.0^\circ C - (-2.5^\circ C)} = 1.0$$

and estimate Δ_o = 1.0 m

Therefore

$$\delta = \frac{1.22+1.2+1.0}{1.22-1.2-1.0} = 3.49 \text{ m}$$

Solution:

1. Steel pipe:

r_p = 0.084 m

r_w = 0.077 m

$$u = 3.5 \bullet 10^{-6} \frac{m}{m \bullet {}^\circ C}$$

E = 206,000 MPa

Change in length if unrestrained

$$\Delta l = 100m \bullet \left(3.5*10^{-6}\ \frac{m}{m \bullet {}^\circ C}\right)(50{}^\circ C)$$

$$= 0.018\ m\ \ or\ \ 18\ mm.$$

Example 4.12

For the buried insulated pipe in Example 4.11, estimate the maximum thawing under the pipe (autumn) if the maximum thaw depth (active layer) is 1.2 m and the volumetric heat capacity of the soil is 1800 kJ/(m³•°C).

Given from Example 4.11:

k_g = 1.73 W/(m • °C)

r_l = 0.087 m

r_p = 0.076 m

r_l = r_p + 0.011 m = 0.087 m

T_l = 0.0°C

T_s = -2.5°C

R_g = 0.32 m•°C/W

z_p = 1.22 m

Solution:

Resembles problem of Figure 4-18.

Given:

z_{active} = 1.2 m

C = 1800 kJ/(m³ • °C)

$$\alpha = \frac{k_g}{C} = 1.73\frac{J}{s \bullet m \bullet {}^\circ C} \bullet \frac{m^3 \bullet {}^\circ C}{1800kJ} \bullet \frac{kJ}{1000J}$$

$$= 0.961 \bullet 10^{-6}\ \frac{m^2}{s}$$

First iteration, j = 1

$$\Delta_1 = -\left(\frac{\pi}{\left(0.961 \bullet 10^{-6}\right) \bullet \left(31.5 \bullet 10^6\right)}\right)^{-0.5}$$

$$\bullet \ln\left(1 - \frac{T^*}{arccos\, h(14)} \bullet \ln(3.49)\right)$$

$$= -3.10 \bullet (-0.470)\ =\ 1.46\ m$$

Second iteration,

Δ_{j-1} = 1.46 m

δ = 2.69 m

Δ_2 = -3.10 • (-0.353) = 1.09 m

Third iteration,

Δ_{j-1} = 1.09 m

δ = 3.29 m

Δ_3 = -3.10 • (-0.442) = 1.37 m

Fourth iteration,

Δ_{j-1} = 1.37 m

δ = 2.81 m

Δ_4 = -3.10 • (-0.371) = 1.15 m

Fifth iteration,

Δ_{j-1} = 1.15 m

δ = 3.16 m

Δ_5 = -3.10 • (-0.424) = 1.31 m

... and so on until convergence. Given the limitations of the ground thermal equation (Figure 4-4), Δ is probably close enough at this point, i.e. $\Delta \approx 1.3$ m.

Example 4.13

Calculate the design freeze-up time, safety factor time, and time for complete freeze-up for the insulated above-ground pipe in Example 4.5.

Given:

R_c = 3.6 m•°C/W

T_a = -40°C

T_w = 5°C

r_w = 0.068 m

L = 334 MJ/m³

Solution:

From Figure 4-16:

Design time,

$$t_f = \pi r_w^2 \bullet R_c \bullet C \bullet \ln\left(\frac{T_w - T_a}{T_o - T_a}\right)$$

$$t_f = \pi(0.068m)^2 \bullet \left(3.6\frac{m \bullet {}^\circ C}{W}\right)$$

$$\bullet\left(4.187\bullet10^6\ \frac{J}{m^3\bullet{}^\circ C}\right)\bullet\ln\left(\frac{5^\circ C-\left(-40^\circ C\right)}{0.0^\circ C-\left(-40^\circ C\right)}\right)$$

$$= 25.8\times10^3\ s\ =\ 7.2\ h$$

Safety factor time, $t_{f,sf}$

$t_{f,sf}\ =\ t_f$ where $T_o\ =\ -3^\circ C$

$t_{f,sf}\ =\ 42.9\times10^3\ s\ =\ 12\ h$

Complete freezing time, $t_{f,T}$

$$t_{f,T}=\frac{A\bullet R_c\bullet L}{T_o-T_a}$$

$$=\frac{\pi\left(0.068m\right)^2\bullet\left(3.6\frac{m\bullet{}^\circ C}{W}\right)\bullet\left(334\bullet10^6\ \frac{J}{m^3}\right)}{0.0^\circ C-\left(-40^\circ C\right)}$$

$$=\ 437\bullet10^3\ s\ =\ 121\ h\ or\ 5\ d$$

Example 4.14

Calculate the input temperature and the rate of heat loss (heat input) for a 3,000 m recirculating water system for various flow rates if the buried pipe is the same as Example 4.6, and the water temperature is to be maintained at a minimum of 5°C when the ground temperature at the pipe depth is -10°C.

Given from Example 4.6:

$r_w\ =\ 0.068$ m

$R_c\ =\ 3.6\ m\bullet{}^\circ C/W\ =\ 3.6\ (m\bullet s\bullet{}^\circ C/J)$

$C\ =\ 4.187\ MJ/(m^3\bullet{}^\circ C)$

Solution:

$I\ =\ 3000$ m

$T_{w,2}=\ 5^\circ C$

$T_A\ =\ -10^\circ C$

From Equations in Figure 4-17:

$$G=\pi\bullet\left(0.068m\right)^2\bullet v\bullet\left(4.187\bullet10^6\ \frac{J}{m^3\bullet{}^\circ C}\right)$$

$$\bullet\left(3.6\frac{m\bullet s\bullet{}^\circ C}{J}\right)$$

$$=\ (219000\bullet v)\ s$$

$$T_{w,1}=-10^\circ C+\frac{5-\left(-10^\circ C\right)}{\exp\left(\dfrac{-3000}{219000v}\right)}$$

and

$$q=\frac{219000v}{3.6}\left(5^\circ C-\left(-10^\circ C\right)\right)$$

$$\bullet\left(\exp\left(\dfrac{3000m}{219000v}\right)s\right)-1$$

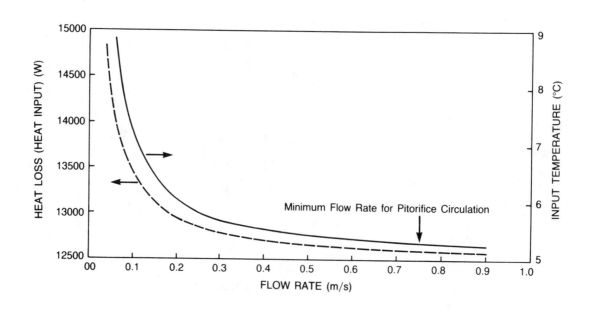

Thus $T_{w,1}$ and q can be evaluated for various values of v in m/s. These are plotted on page 4-46.

Note: Heat loss and input temperature are lower at high flow rates, however, there is little benefit from flow rates greater than 0.1 m/s. A pitorifice system would require a minimum flow rate of approximately 0.75 m/s.

Example 4.15

Determine the economical thickness of insulation for an above-ground, 168 mm outside-diameter steel water pipe which is maintained at 5°C for temperature conditions of Example 4.7. Fuel oil contains 39 MJ/L and costs $0.24/L. The heating plant is 85 percent efficient. Installed cost of polyurethane foam (k_l = 0.024W/(m• °C)) for various thicknesses is given below. The economic life is 20 years and the net discount rate is 8 percent/year.

From Example 4.7, the water freezing index is 6100 °C•d.

Solution:

1. Cost of heat $= \left(\dfrac{\$0.20}{L}\right) \bullet \left(\dfrac{L}{39M}\right) \bullet \left(\dfrac{1}{0.85}\right)$

$$= \frac{\$0.006}{MJ}$$

2. Thermal resistance (neglect all except insulation):

$$R_c = R_l = \frac{\ln\left(\dfrac{r_l}{0.084}\right)}{2\pi\left(0.024\right)}$$

Nominal (m)	Insulation Thickness r_l (m)	Installed Cost ($/m)	PV of Annual Heating Cost ($/m)	Total PV ($/m)
0.025	0.109	11.80	17.96	29.76
0.050	0.134	15.40	10.02	25.42
0.075	0.159	17.70	7.33	25.03
0.100	0.184	20.30	5.97	26.27

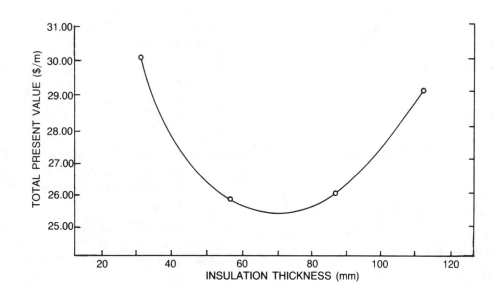

$$= 6.63 \bullet \ln\left(\frac{r_I}{0.084}\right)\frac{s \bullet m \bullet {}^\circ C}{J}$$

3. From financial tables available elsewhere:

Present valve (PV) factor for 20 years at 8 percent rate = 9.8181

PV annual heating cost

$$= \frac{6100{}^\circ C \bullet d}{6.63\frac{s \bullet m \bullet {}^\circ C}{J} \bullet \ln\left(\frac{r_I}{0.084m}\right)} \bullet \left(\frac{86400s}{d}\right)$$

$$\bullet \left(\frac{MJ}{10^6 J}\right) \bullet \left(\frac{\$0.006}{MJ}\right) \bullet 9.8181$$

$$= \frac{\$4.68}{m \bullet \ln\left(\frac{r_I}{0.084m}\right)}$$

for the following available insulation thicknesses and costs:

These results, which have been plotted on the next page, indicate that the most economically attractive thickness of insulation is approximately 70 mm.

Note: Other factors such as the freeze-up time, the maximum rate of heat loss (heating system capacity), and practical dimensions must also be considered in the selection of insulation thickness. The sensitivity of results to the assumptions, in cluding the discount rate and energy costs, should also be checked.

Example 4.16

Calculate the expansion and load for 100 m lengths of 150 mm diameter steel and polyethelene pipes which undergo a temperature change of 50°C.

Load if restrained

$$P = \pi\left(0.084^2 - 0.077^2\right)m^2 \bullet \left(206000\frac{MN}{m^2}\right)$$

$$\bullet \left(3.5 \bullet 10^{-6}\frac{m}{m \bullet {}^\circ C}\right) \bullet \left(50{}^\circ C\right)$$

$$= 0.128 \text{ MN or } 128 \text{ kN}$$

2. Polyethylene pipe (high density):

r_p = 0.133 m

r_w = 0.083 m

$$u = 4.0 \bullet 10^{-5}\frac{m}{m \bullet {}^\circ C}$$

E = 412 MPa

Change in length if unrestrained

$$\Delta l = 100m \bullet \left(4.0 \bullet 10^{-5}\frac{m}{m \bullet {}^\circ C}\right) \bullet \left(50{}^\circ C\right)$$

$$= 0.200 \text{ m or } 200 \text{ mm}$$

Load is restrained

$$P = \pi\left(0.133^2 - 0.083^2\right)m^2 \bullet \left(412\frac{MN}{m^2}\right)$$

$$\bullet \left(4.0 \bullet 10^{-5}\frac{m}{m \bullet {}^\circ C}\right) \bullet \left(50{}^\circ C\right)$$

$$= 28 \bullet 10^{-3} \text{ MN or } 28 \text{ kN}$$

Note: Although the thermal expansion of the plastic pipe is much more than the metal pipe, the load to restrain thermal expansion is considerably less.

4.8 References

Aitken, G. and R. Berg. 1968. *Digital Solution of the Modified Berggren Equations*. U.S. Army Cold Regions Research and Engineering Laboratory, Report No. SR-122, Hanover, N.H.

Aldrich, H.P. and H.M. Paynter. 1953. *Analytical Studies of Freezing and Thawing of Soils*. U.S. Army, Arctic Construction and Frost Effects Laboratory, Technical Report No. 42, Hanover, N.H.

Aldrich, H.P. and H.M. Paynter. 1966. *Depth of Frost Penetration in Non-Unifrom Soil*. U.S. Army Cold Regions Research and Engineering Laboratory, Monograph 111-C5b, Hanover, N.H.

Andersland, O.B. and D.M. Anderson. 1978. *Geotechnical Engineering for ColdRegions*. McGraw-Hill Book Company, New York, 566 p.

ASHRAE. 1981a. *ASHRAE Guide and Data Book*. American Society of Heating, Refrigeration and Air Conditioning Engineers Inc.

ASHRAE. 1981b. *ASHRAE Handbook of Fundamentals*. American Society of Heating, Refrigeration and Air Conditioning Engineers Inc.

Barsvary, A.K. 1980. List of terms, symbols, and recommended SI units and multiples for geotechnical engineering. *Canadian Geotechnical Journal*, 17: 89-96.

Bohlander, T.W. 1963. Electrical methods for thawing frozen pipes. *Journal AmericalWater Works Association*, 55: 602-608.

Canada Department of Transport. 1967. *The Climate of the Canadian Arctic*. Meteorological Branch, Toronto, Ontario.

Carslaw, H.S. and J.C. Jaeger. 1959. *Conduction of Heat in Solids*. Oxford University Press, England.

Chapman, A.J. 1967. *Heat Transfer*. Second Edition. MacMillan, London. 617 p.

Churakov, B. H. 1959. *Installation of Sanitary Engineering Utilities in Permafrost*. National Science Library, Ottawa, Ontario.

Currey, J.R. 1980. Thawing of frozen service lines. In: *Proceedings, Utilities Delivery in Northern Regions*. Environmental Protection Service Report No. EPS 3-WP-80-5, Ottawa, Ontario, 310-313.

Eaton, E.R. 1964. Thawing of Wells in Frozen Ground by Electrical Means. *Waterand Sewage Works,* 3(8): 350-353.

Gerriutsen, E. D. 1977. Northern Ontario water distribution systems. *Journal AmericalWater Works Association*, 69: 242-244.

Gilpin, R.R. 1977a. A study of pipe freezing mechanisms. In: *Proceedings, Utilities Delivery in Arctic Regions*. Environmental Protection Service Rep. No. EPS 3-WP-77-1, Ottawa, Ontario, 207-220.

Gilpin, R.R. 1977b. The effects of dendritic ice formation in water pipes. *International Journal of Heat and Mass Transfer,* 20: 693.

Gilpin, R.R. 1981. Ice formation in a pipe containing flows in the transition and turbulent regimes. *ASME Journal of Heat Transfer,* 103: 363-368.

Gilpin, R.R. 1979. The morphology of ice structure in a pipe at or near transition Reynolds numbers. American Institute of Chemical Engineers Symposium 189, *Heat Transfer,* 75: 89-94.

Goodrich, L.E. 1978. Some results of a numerical study of ground thermal regimes. In: *Proceedings, Third International Conference on Permafrost*. National Research Council, Ottawa, Ontario, 29-34.

Gunderson, P. 1975. *Frostproofing of Pipes*. U.S. Army Cold Regions Research and Engineering Laboratory, Draft Translation TL 497, Hanover, N.H.

Gunderson, P. 1978. *Frost Protection of Buried Water and Sewer Pipes*. U.S. Army Cold Regions Research and Engineering Laboratory, Draft Translation TL 666, Hanover, N.H.

Hartman, C.W. and P.R. Johnson. 1978. *Environment Atlas of Alaska*. Institute of Water Research, University of Alaska, Fairbanks, 95 p.

Houk, J. 1974. Freeze damage in water containers. *The Northern Engineer*, 6(2): 4-6.

Hull, J.A. 1980. *Thermodynamic Analysis of the Water Distribution System in Inuvik, NWT*. Environmental Protection Service Report No. EPS 3-WP-80-5, Ottawa, Ontario, 332-346.

James, F. 1980. Critical evaluation of insulated shallow buried pipe systems in the Northwest Territories. In: *Proceedings, Utilities Delivery in Northern Regions*. Environmental Protection Service Report No. EPS 3-WP-80-5, Ottawa, Ontario, 150-186.

Janson, L.R. 1966. Water supply systems in frozen ground. *Proceedings International Permafrost Conference*. National Academy of Science, Washington, D.C., 403-433.

Johnston, G. H. (editor) 1981. *Permafrost Engineering Design and Construction*. John Wiley & Sons, Toronto, Ontario, 540 p.

Kaplar, C.W. 1974. *Moisture and Freeze-Thaw Effects on Rigid Thermal Insulations*. U.S. Army Cold Regions Research and Engineering Laboratory TR 249, Hanover, N. H.

Kent, D. and C.T. Hwang. 1980. Use of a geotechnical model in northern municipal projects. *Proceedings, Utilities Delivery in Northern Regions*. Environmental Protection Service Report No. EPS 3-WP-80-5, 317-331.

Kersten, M.S. 1949. *Thermal Properties of Soils*. Engineering Experiment Station Bulletin 28,

University of Minnesota, 227 p.

Kersten, M.S. 1963. Thermal properties of frozen ground. In: *Proceedings, 1st International Permafrost Conference*, Purdue University, West Lafayette, Indiana.

Kreith, F. and W.Z. Black. 1980. *Basic Heat Transfer.* Harper & Row, New York, 556 p.

Kutateladze, S. S. 1963. *Fundamentals of Heat Transfer.* Academic Press, New York, 485 p.

Lachenbruch, A.H. 1959. Periodic heat flow in a stratified medium with application to permafrost problems. *U.S. Geological Survey Bulletin*, 1083: 1-36.

Legget, R.F. and C.B. Crawford. 1952. Soil temperatures in water works practice. *Journal of the American Water Works Association*, 44: 923-939.

Linell, K.A. 1973. Long-term effects of vegetative cover on permafrost stability in an area of discontinuous permafrost. In: *North American Contribution to Permafrost, Second International Conference*, National Academy of Science, Washington, D.C., 688.

Linell, K.A. and E.F. Lobacz. 1980. *Design and Construction of Foundations in Areas of Deep Seasonal Frost and Permafrost.* Special Report No. 80-34, U.S. Army Cold Regions Research and Engineering Laboratory, Hanover, N.H.

Lunardini, V.J. 1978. A correlation of n-factors. In: *Proceedings, Conference on Applied Techniques for Cold Environments.* American Society of Civil Engineers, New York, 233-244.

McFadden, T. 1977. Freeze damage prevention in utility distribution lines. In: *Utilities Delivery in Arctic Regions.* Environmental Protection Service Report No. EPS 3-WP-77-1, Ottawa, Ontario, 221-234.

Moulton, L.K. 1969. *Prediction of the Depth of Frost Penetration: A Review of Literature.* Report No. 5, West Virginia University, Morgantown.

National Research Council. 1970. *Climatic Information for Building Design in Canada.* NRC No. 11153, Ottawa, Ontario.

Nelson, L. M. 1976. Frozen water services. *Journal of the American Water Works Association*, 68: 12-14.

Nixon, J.F. and E.C. McRoberts. 1973. A study of some factors affecting the thawing of frozen soils. *Canadian Geotechnical Journal,* 10: 439-452.

Nyman, F. 1965. Insulation vs. deep trenching for Alaskan water mains. *Civil Engineering*, 35: 40-41.

Penner, E. 1970. *Measuring Thermal Conductivity of Frozen Soils.* Research Paper No. 27, National Research Council, Division of Building Research, Ottawa, Ontario

Prokhaer, G. V. 1959. *Underground Utility Lines.* National Research Council, Technical Translation TT-1221, Ottawa, Ontario.

Reed, S. 1977. Field performance of a sub-arctic utilidor. In: *Proceedings, Utilities Delivery in Arctic Regions.* Environmental Protection Service Report No. EPS 3-WP-77-1, Ottawa, Ontario, 448-468.

Ryan, W.L. 1981. Permafrost's impact on utility design. In: *Design of Water and Wastewater Services for Cold Climate Communities.* Pergamon Press, Oxford, England, 37-53.

Sheppard, M.I., B.D. Kay, and J.P.G. Loch. 1978. Development and testing of a computer model for heat and mass flow in freezing soils. In: *Proceedings, Third International Conference on Permafrost.* National Research Council of Canada, Ottawa, Ontario, 75-81.

Shirtliffe, C. J. 1977. *Polyurethane Foam as a Thermal Insulation: A Critical Review.* Division of Building Research, Note No. 124, National Research Council of Canada, Ottawa, Ontario.

Slusarchuk, W.A. and P.H. Foulger. 1973. *Development and Calibration of a Thermal Conductivity Probe Apparatus for Use in the Field and Lab.* Technical Paper No. 388, National Research Council, Division of Building Research, Ottawa, Ontario.

Thornton, D.E. 1977. Calculation of heat loss from pipes. In: *Proceedings, Utilities Delivery in Arctic Regions.* Environmental Protection Service Report No. EPS 3-WP-77-1, 131-150.

U.S. Department of the Army. 1966. *Calculation Methods for Determination of Depths of Freeze and Thaw in Soils.* Technical Manual No. 5-852-6, Washington, D.C.

Whyman, A.D. 1980. Pre-insulated high-density polyethylene piping - the evolution of a northern achievement. In: *Proceedings, Utilities Delivery in Northern Regions.* Environmental Protection Service Report No. EPS 3-WP-80-5, Ottawa, Ontario, 187-220.

4.9 Bibliography

Aldrich, H.P. and H.M. Paynter. 1953. *Analytical Studies of Freezing and Thawing of Soils.* U.S. Army, Arctic Construction and Frost Effects Lab, Technical Report No. 42, Hanover, N.H.

Aldrich, H.P. and H.M. Paynter, H.M. 1966. *Depth of Frost Penetration in Non-Uniform Soil.* U.S. Army, Cold Regions Research and Engineering Laboratory, Monograph 111-C5b, Hanover, N.H.

Aldworth, G.A. and A.W. Petrie. 1977. Some recent developments in shallow-buried water and sewage systems. Presented at: Water Works Association, Atlantic Canada Section Conference.

Alter, A.J. 1969. *Sewage and Sewage Disposal in Cold Regions.* U.S. Army, Cold Regions Research and Engineering Laboratory, Monograph 111-C5a, Hanover, N.H.

Alter, A.J. 1969. *Water Supply in Cold Regions.* U.S. Army, Cold Regions Research and Engineering Laboratory, Monograph 111-C5a, Hanover, N.H.

Anon. 1966. Plastic foam insulates Yukon waterline. *Engineering and Contract Record*, March, 29.

Associated Engineering Services Ltd. 1978. *Freeze Protection Analysis for The Water Distribution System in Buffalo Narrows.* Prepared for: Department of Northern Saskatchewan, Prince Albert, Saskatchewan.

Brown, W. G. 1963. *Graphical Determination of Temperature Under Heated or Cooled Areas on the Ground Surface.* Division of Building Research, Paper No. 163, National Research Council of Canada, Ottawa, Ontario.

Cameron, J.J. 1976. Waste impounding embankments in permafrost regions: the sewage lagoon embankment, Inuvik, N.W.T. In: *Some Problems of Solid and Liquid Waste Disposal in the Northern Environment.* Northwest Region, Environment Canada, EPS 4-NW-76-2, 141-230.

Cameron, J.J. 1977. Buried utilities in permafrost regions. In: *Proceedings, Utilities Delivery in Arctic Regions.* Environmental Protection Service Report No. EPS 3-WP-77-1, Ottawa, Ontario, 151-200.

Carson, N.B. 1976. Pipeline heating with electrical skin current. *Engineering Journal,* 43(9): 61-65, 1960.

Chemelex Division, Raychem Corporation. 1978. *Thermal Design Guide.* Redwood City, California

Chapman, F.S. and F.A. Holland. 1966. Keeping piping hot - part II, by heating. *Chemical Engineering,* January 17, 133-144.

Cheriton, W.R. 1966. Electrical heating of a water supply pipeline under arctic conditions. *Engineering Journal,* 49(9): 31-35.

Copp, S.S., C.B. Crawford, and J.W. Grainge. 1956. Protection of utilities against permafrost in northern Canada. *Journal of the American Water Works Association,* 48(9): 1115-1166.

Cory, F. 1973. Settlement associated with the thawing of permafrost. In: *North American Contribution Permafrost Second International Conference.* National Academy of Sciences, Washington, D.C.

Fukasako, S. and N. Seki. 1988. Freezing and melting characteristics in internal flow. *Northern Engineer,* 20(1): 4-18.

Giles, S. 1956. *A Proposed System of Utility Piping Installation in Show, Ice, and Permafrost.* U.S. Naval Civil Engineering Research and Evaluation Laboratory. Technical Note N-261, Port Hueneme, California.

Gilpin, R.R. and M.G. Faulkner. 1977. Expansion joints for low-temperature above-ground water piping systems. In: *Proceedings, Utilities Delivery in Arctic Regions.* Environmental Protection Service Report No. EPS 3-WP-77-1, Ottawa, Ontario, 346-363.

Gilpin, R.R. 1977. Ice formation in pipes. In: *Proceedings, 2nd International Symposium on Cold Regions Engineering.* Cold Regions Engineers Professional Association, Fairbanks, Alaska, 4-11.

Gilpin, R.R. 1978. A study of factors affecting the ice nucleation temperature in a domestic water supply. *Canadian Journal of Chemical Engineering,* 56: 466-471.

Gold, L.W. and A.H. Lachenbruch. 1973. Thermal conditions in permafrost: a review of North American literature. In: *North American Contribution Permafrost Second International Conference.* National Academy of Sciences, Washington, D.C., 3-23.

Goodrich, L.E. 1973. Computer simulations. In: *North American Contribution Permafrost Second International Conference.* National Academy of Science, Washington, D.C.,23-25.

Greenland Technical Organization. *Guide for Carrying Out Electrical Pipe Heating Works.* Copenhagen, Denmark, (In Danish).

Hsin, J. 1978. *A Study of the Auto-Trace Heaters Application in the Sclaircor Piping System.* Chemelex Division, Raychem Corporation, Redwood City, California.

Hull, J. 1980. Thermodynamic analysis of water distribution system in Inuvik, NWT. In: *Proceedings, Utilities Delivery in Northern Regions.* Environmental Protection Service Rep. No. EPS 3-WP-80-5, Ottawa, Ontario.

Jahns, H.O., T.W. Miller, L.D. Power, W.P. Rickey, T.P. Taylor, and J.A. Wheeler. 1973. Permafrost protection for pipelines. In: *North American Contribution Permafrost Second International Conference,* National Academy of Science, Washington, D.C. 673-684.

James, F.W. 1976. Buried pipe systems in Canada's Arctic. *The Northern Engineer,* 8(1): 4-12.

Janson, L.E. 1964. *Frost Penetration in Sandy Soil.* Royal Institute of Technology Transaction 231, Stockholm, Sweden.

Johnston, G.H. 1973. Ground temperature measurements using thermocouples. In: *Proceedings, Seminar on the Thermal Regime and Measurements in Permafrost,* National Research Council of Canada, Ottawa, Technical Memorandum 108, 1-12.

Judge, A. S. 1973. Ground temperature measurements using thermistors. In: *Proceedings, Seminar on the Thermal Regime and Measurements in Permafrost.* National Research Council, Technical Memorandum 108, Ottawa, Ontario, 3-25.

Jumikis, A.R.1966. *Thermal Soil Mechanics.* Rutgers University Press, New Brunswick, New Jersey.

Jumikis, A.R. 1977. *Thermal Geotechnics.* Rutgers University Press, New Brunswick, New Jersey.

Jumikis, A.R. 1978. Graphs for disturbance - temperature distribution in permafrost under heated rectangular structures. In: *Proceedings, Third International Conference on Permafrost,* National Research Council, Ottawa, Ontario, 589-596.

Irwin, W.W. 1979. New approaches to services in permafrost areas - Norman Wells, NWT. In: *Proceedings, Utilities Delivery in Northern Regions.* Environmental Protection Service Rep. No. EPS 3-WP-80-5, Ottawa, Ontario, 507-542.

Kardymon, V.F. and V.P. Stegantsev. 1972. *The Positioning of Heating Cable for Protecting Water Supply Lines from Freezing.* Translated from Russian for: Northern Technology Unit, Environmental Protection Service, Ottawa, Ontario.

Kent, D. and C.T. Hwang. 1980. Use of a geothermal model in northern municipal projects. In: *Proceedings, Utilities Delivery in Northern Regions.* Environmental Protection Service Rep. No. EPS-3-WP-80-5, Environment Canada, Ottawa, Ontario.

Kersten, M.S. 1949. *Thermal Properties of Soils.* University of Minnesota Bulletin No. 28, Minneapolis, Minnesota.

Klassen, H.P. 1960. Water supply and sewerage system at Uranium City. *Engineering Journal,* 43(9): 61-65.

Klassen, H.P. 1965. Public utilities problems in the discontinuous permafrost areas, In: *Proceedings of the Canadian Regional Permafrost Conference,* National Research Council of Canada, Technical Memo 86, Ottawa, Ontario, 106-118.

Linell, K.A. and G.H. Johnston. 1973. Engineering design and construction in permafrost regions: a review. In: *Northern American Contribution Permafrost Second International Conference.* National Academy of Sciences, Washington, D.C., 553-575.

Lunardini, V.J. 1978. A correlation of n-factors. In: *Proceedings, Applied Techniques for Cold Environments*, American Society of Civil Engineers, New York, 233-244.

Monie, W.D. and C.M. Clark. 1974. Loads on underground pipe due to frost penetration. *Journal of the American Water Works Association*, 66(6): 353-358.

Morgenstern, N.R. and J.F. Nixon. 1971. One-dimensional consolidation of thawing soil. *Canadian Geotechnical Journal*, 8(4): 558-565.

Moulton, L.K. 1969. *Prediction of the Depth of Frost Penetration: A Review of Literature*. West Virginia University, Report No. 5, Morgantown, West Virginia.

Nixon, J.F. and E.C. McRoberts. 1973. A study of some factors affecting the thawing of frozen soils. *Canadian Geotechnical Journal*, 10: 439-452.

O'Brien, E.T. and A. Whyman. 1977. Insulated and heat-traced polyethylene piping aystems: a unique approach for remote cold regions. In: *Proceedings, Utilities Delivery in Arctic Regions*, Environmental Protection Service, Rep. No. EPS 3-WP-77-1, Ottawa, Ontario, 309-339

Okada, A. 1973. A rough estimation method of heat loss in buried underground pipes. *Society of Heating, Air Conditioning and Sanitary Engineers, Japan (SHASA), Transactions*, II: 46-52.

Orlov, V.A. 1966. *The problems of Heat Supply in Settlements in the Permafrost Regions*. Problems of the North, No. 10, National Research Council of Canada, Ottawa, Ontario.

Page, W.B. 1955. Arctic sewer and soil temperatures. *Water and Sewage Works*, 102(8): 304-308.

Page, W.B. 1956. Heat loss from underground pipelines. In: *Fourth Alaskan Science Conference, Juneau, Alaska*, Alaska Division, American Association for the Advancement of Science, 41-46.

Ryan, W.L. 1977. Design guidelines for piping systems - panel discussion. In: *Utilities Delivery in Arctic Regions*, Environmental Protection Service, Rep. No. EPS 3-WP-77-1, Ottawa, Ontario, 243-255.

Saltykov, N.I. 1944. *Sewage Disposal in Permafrost in the North of the European Portion of the USSR*. Academy of Sciences, Moscow, U.S.S.R., 1944, Translated for St. Paul District, Corps of Engineers, U.S. Army.

Smith, W.H. 1976. Frost loadings of underground pipe. *Water Technology DistributionJournal*, December: 673-674.

Squires, A.D. 1977. Preparation of an operations and maintenance manual. In: *Utilities Delivery in Arctic Regions*, Environmental Protection Service, Rep. No. EPS 3-WP-77-1, Ottawa, Ontario, 256-265.

Srouji, G.A. 1978. Thermal analysis of heated recirculating water distribution systems in northern regions. In: *Proceedings, Applied Techniques for Cold Environments*, American Society of Civil Engineers, New York.

Stanley, D.R. 1965. Water and sewage problems in discontinuous permafrost regions. In: *Proceedings, Canadian Regional Permafrost Conference*, National Research Council, Technical Memo 86, Ottawa, Ontario, 93-105.

Tsytovich, N.A. 1975. *The Mechanics of Frozen Ground*. Translation from Russian, Edited by G. K. Swinzow, McGraw-Hill Book Company, New York, New York.

U.S. Department of the Army. 1966. *Calculation Methods for Determination of Depths of Freeze and Thaw in Soils*. Department of the Army, Technical Manual No. 5-852-6, Washington, D.C.

Vyalov, S.S. and G.V. Porkhaev (eds). 1975. *Handbook for the Design of Bases and Foundations of Buildings and Other Structures on Permafrost*. National Research Council, Technical Translation NRC/CNR TT-1865, Ottawa, Ontario.

Yastrebov, A.L. 1972. *Engineering Utility and Sewage Lines in Permafrost Soil*, U.S. Army Foreign Science and Technology Center, Rep. No. FSTC-HT-1392-73, Charlottesville, Virginia.

Zarling, J.P. 1978. Growth rates of ice. In: *Proceedings, Applied Techniques for Cold Environments*, American Society of Civil Engineers, New York, 100-111.

Zenger, N.N. 1965. *Ways in Which to Improve the Economics of Water Supply in the Northern*

Regions. Problems of the North, No. 9, National Research Council of Canada, Ottawa, Ontario.

SECTION 5

WATER SOURCE DEVELOPMENT

3rd Edition Steering Committee Coordinator

William L. Ryan

3rd Edition Principal Author

William L. Ryan

Special Contributions

Ron Kent

Section 5   Table of Contents

Section 5   List of Figures

Section 5   List of Tables

5   WATER SOURCE DEVELOPMENT

5.1   Introduction

Selecting and developing water sources for cold-climate conditions requires special attention to a number of details usually not considered in warmer locations. Hydrologic conditions differ in several important aspects. The amount of winter precipitation is low in most areas and frozen soil conditions can result in a high runoff during spring break-up (Ryan and Crissman, 1990).

Ground conditions can influence surface and groundwater movement. In continuous permafrost areas year-round groundwater movement is restricted to subpermafrost zones. In summer and fall some suprapermafrost water movement can occur. In intermittent permafrost areas, groundwater recharge and movement is influenced by frozen areas. In regions of deep seasonal frost, winter snow and ice may melt and run off before soils are thawed.

There are many small, shallow lakes and ponds in northern regions. Again, the thermal conditions of the ground will influence the water movement. The presence and thickness of ice cover varies according to local conditions, but generally lasts from five to ten months. Constructed reservoirs and impoundments have been built to store water. They can be structured to minimize the shortcomings of shallow natural lakes.

Hydrologic data on northern lakes, streams, and groundwater are scarce and typically cover periods of short duration. This makes it difficult to predict reliable yields for water supply purposes.

Regardless of the apparent merits of the source under consideration, there is no substitute for detailed preliminary engineering studies and direct observation of local conditions prior to final selection. Some type of water source may be developed in virtually any part of the cold regions of North America, but the physical difficulty and cost of operation and maintenance or the associated construction costs may be highly unattractive.

Environment Canada produces annual and historical surface water data on CD-ROM which lists stream flows at selected stations in the territories and all the provinces. In addition to these data, special reports may be available on particular locations, dealing with short-term data, groundwater, or special water quality studies. Weather information as well as other climatological factors such as ice thickness may also be available for selected locations. For this information, the following office should be contacted.

Atmospheric Environment Branch
Environment Canada
Yellowknife, NT X1A 2R2
Phone: 403-920-8500
Fax: 403-873-2970

The annual U.S. Government publication, "Water Resources Data for Alaska," includes information on stream flows and water quality for selected streams. This publication is available from:

U.S. Geological Survey
Water Resources Division
4230 University Drive
Anchorage, Alaska 99503
Phone: 907-561-1181

Other information is available from local U.S. Geological Survey offices or state geological survey offices.

In addition to these publications, special reports may be available on particular locations, dealing with short term data, groundwater, or special water quality studies. Both nations maintain agencies concerned with weather and other climatological factors and regularly publish their data. Examples include ice thickness data and weather information. For this information the following offices should be contacted:

Atmospheric Environment Service
Environment Canada
4905 Dufferin Street
Downsview, Ontario K1A 0E7
Phone: (416) 739-4939

National Weather Service Forecast Office
636 6th Avenue
Anchorage, Alaska 99501
Phone: 907-936-2525

Environment and Natural Resources Institute
University of Alaska
707 'A' Street
Anchorage, Alaska 99513
Phone: 907-257-2733

5.2   Water Sources

The basic water sources in cold regions are the same as those in temperate regions except for snow and ice. The extreme cold however, results in some peculiarities in surface and groundwater hydrology.

5.2.1 Surface Water.

Surface water sources suitable for continuous year-round supply are rivers and larger lakes. During the winter, shallow lakes and small streams cannot serve as a continuous supply source for water because they are mostly frozen. In this case one may have to resort to various constructed reservoirs or impoundments which increase winter storage. Larger streams and deep lakes may remain usable, but flows and volumes are reduced since there is little or no precipitation to replenish the water used.

Rivers. Rivers are usually an excellent water source in winter because sediment transport is minimal and overland flows, which tend to lower water quality, do not occur. The disadvantages with rivers as a supply source include low water temperatures and flowing ice during freeze-up and break-up period, which may damage or destroy water intake structures. Also, in alluvial streams it can be difficult to locate a permanent channel, especially under the ice.

Seasonal water quality changes may be significant for cold-region rivers. Winter flows consist of water from interflow, groundwater, and spring water. Although the sediment content may be low, mineral and organic contents can be higher than in the summer.

Summer river flows, which tend to be much greater than winter flows, may contain sediments or glacial rock flour from overland runoff or melting glaciers which increase treatment costs. Figure 5-1 shows hydrographs for typical arctic rivers and streams.

Streams that flow all winter or nearly all winter may experience water pressures great enough to force water through cracks to the surface, where it then freezes. This frozen water is called "aufeis." Aufeis is essentially winter flow water that is stored until melting. In some locations the large ice mass formed by repeated overflows may have the following effects:

- restrict the flow in the stream channel during break-up resulting in some or more out-of-bank flow;

- thicker than usual or calculated ice masses in the stream during break-up; and

- more flow in the stream after break-up flows have subsided due to the slower melting of the ice.

Knowledge of the potential for upstream aufeis formation is therefore important for the design of the stream water sources.

Just as aufeis modifies the runoff pattern during spring break-up, so do large snowdrifts. Usually the larger cold thermal mass requires longer to melt, therefore extending the period of melt-related flows.

Erosion also contributes to sediment loads during the summer flows. Erosion problems along river banks can be much more serious in permafrost regions. Figure 5-2 shows examples of bank failure mechanisms associated with ground ice presence (Lawson, 1983).

Lakes. Lakes can be a good continuous source of water, but this depends upon the size (area and volume) of the lake and the severity of the climate. Shallow lakes may freeze to the bottom in arctic regions or may freeze-concentrate impurities below the ice to the extent that the water is no longer of acceptable quality (Ryan and Crissman, 1990). Many lakes, however, which are not suitable as a continuous water source, may be used at selected times of the year to fill water storage tanks.

Reservoirs or impoundments are essentially constructed lakes. They can be formed by berming or diking an area, damming a stream, or excavating a hole. The shortcomings of natural shallow lakes can be reduced because they can be constructed with a much greater depth to surface area ratio. On the North Slope of Alaska these reservoirs usually consist of an old meander bend in a river out of which gravel has been mined for roads and building pads. These pits are usually 10 to 30 metres deep and provide a good source of water even with the 3 to 4 metres of ice that forms on the surface in winter. The reservoirs are usually recharged by underground flow from the river even in the winter. Artificial or constructed impoundments are also used in Chesterfield Inlet, Pangrirtung, and Tuktoyaktuk. The circular reservoir at Tuktoyaktuk was formed using dredged material and a liner and is filled by pumping from surrounding shallow lakes during the summer (Figure 3-60).

Deep pools that do not freeze to the bottom in rivers or lakes can be critical habitat for overwintering fish, especially in rivers that cease to flow in the winter (Ryan and Crissman, 1990). Withdrawing water from these pools can adversely affect the fish and can be deemed unacceptable to regulatory agencies.

Two important factors should be determined for lake or pond source use. The first is to determine water

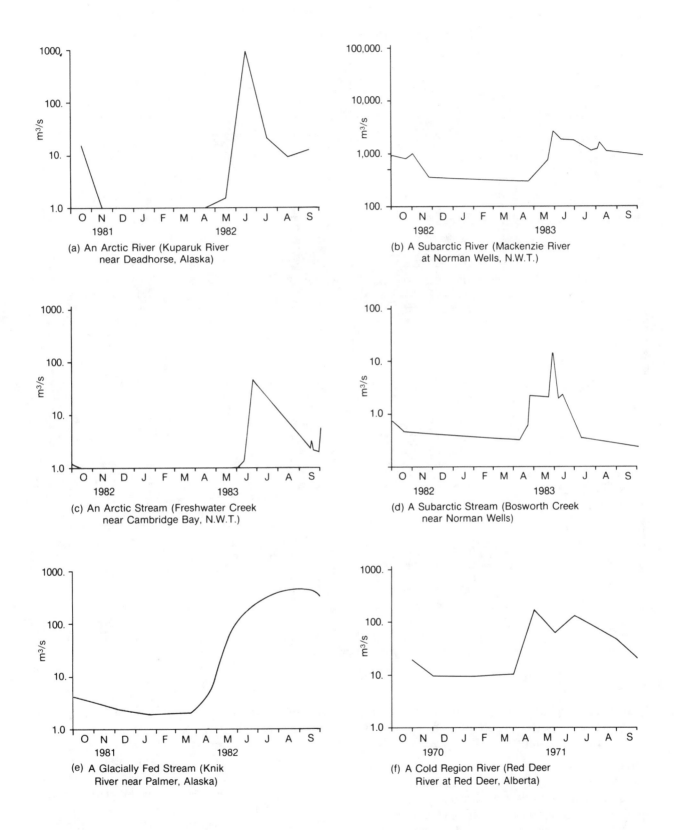

FIGURE 5-1   HYDROGRAPHS FOR RIVERS IN COLD REGIONS

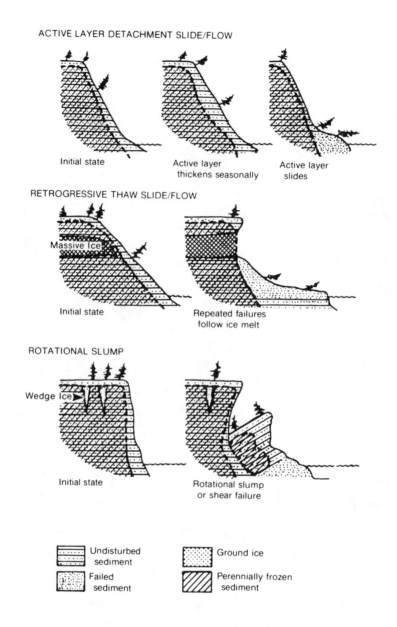

ACTIVE LAYER DETACHMENT SLIDE/FLOW

Initial state

Active layer
thickens seasonally

Active layer
slides

RETROGRESSIVE THAW SLIDE/FLOW

Massive Ice

Initial state

Repeated failures
follow ice melt

ROTATIONAL SLUMP

Wedge Ice

Initial state

Rotational slump
or shear failure

Undisturbed
sediment

Failed
sediment

Ground ice

Perennially frozen
sediment

FIGURE 5-2   BANK FAILURE TYPES WHERE GROUND ICE
IS PRESENT (Shafer, 1981)

yield in comparison to water need. The second factor is to determine winter storage capacity while considering winter water use and ice growth.

Water Yields.   Several methods are available to estimate water yields (Ryan and Crissman, 1990). The most common include comparison of precipitation with runoff through field observation, regression analysis and basin analogy, and by application of mass balance principles.

Aerial photographs and topographic information are normally used to delineate watershed boundaries. In many cases, because of flat terrain and subtle relief, this is difficult. Because significant runoff occurs during break-up, when the ground is frozen, confirmation of boundaries and watershed yield can be observed and measured. Historic aerial photographs may be used to assess water courses, drainage basins, or the lateral extent of small ponds or lakes at different times of the year.

Once an estimate of the watershed boundary has been determined, traditional hydrologic runoff methods can be used to estimate water contributions. Several factors need to be included in the mass balance equations. These include:

- inflow – rain, snow accumulation (drifting), condensation, stream inflow;

- outflow – stream outflow, use, evaporation, evapotranspiration, sublimation,

- storage.

Information is usually very limited. Historical rainfall, snowfall, and temperature information may be available for some areas, but it is usually of limited duration. Assumptions are usually required to evaluate the influence of all parameters. Snow accumulation can be either positive or negative depending on the topography, vegetation, wind direction, duration, and magnitude (Ryan and Crissman, 1990). A conservative estimate should be made for a given source. Because of relatively cool evenings, condensation on vegetation can occur. In coastal (higher humidity) areas, the assumption that condensation is approximately equal to net evapotranspiration has been used. Sublimation is the direct phase change of snow to vapor. It has had limited study, but values of up to 15% of the water content of snow have been used.

Continued water demand, loss of volume due to ice growth, and lack of recharge results in a steady drawdown in lakes or impoundments during the winter months. In order to analyze the situation and ensure a sufficient quantity and quality of water will be available at the end of the season, an estimate of

population growth, water use, and rate of ice growth must be made (Ryan and Crissman, 1990). Population projections can be obtained from a number of sources or generated by including estimates based on historic information, mathematical projections, planning studies, or demographic models. Water use can be based on historic water use records, or assumed average values such as in Tables 5-2 and 5-3. Ice growth can be calculated using historic mean monthly temperatures. Rates of ice growth should be estimated for the minimum monthly temperature of record.

A general expression for ice thickness is called Stefan's equation:

$$h = \sqrt{\frac{2 \bullet (k) \bullet (SFI)}{L_v}}$$

Simplifying this equation (Zarling, 1978):

$$h = \varepsilon \bullet \sqrt{SFI}$$

where,

h   =   ice thickness, m

SFI   =   freezing index at ground surface, °C•s

L_v =   volumetric latent heat of fusion of water, J/m^3

k   =   thermal conductivity of ice, W/(m•°C)

ε =   coefficient of proportionality, m/(°C$^{1/2}$•s$^{1/2}$) (see Table 5-1)

Snow cover has a significant insulating effect and can reduce the maximum ice thickness. The total

TABLE 5-1   SOME EXAMPLES OF ε-FACTORS FOR ICE THICKNESS
(from Zarling, 1978)

ε-Factor $\dfrac{m}{°C^{1/2} \bullet s^{1/2}}$ x 10^{-5}	Conditions
10.4 – 11.0	Practical maximum for ice not covered with snow
9.3	Windy lakes with no snow
8.1 – 9.3	Medium-sized lakes with moderate snow cover
6.7 – 7.5	Rivers with moderate flow
4.6 – 5.8	River with snow
2.3 – 4.6	Small river with rapid flow

ice thickness can also be greater than calculated. For instance, if the weight of snow or the lowering of the water level causes cracks in the ice, water overflows onto the surface, where this water is then drawn into the snow by capillary action. The resulting slush freezes and bonds to the original ice. This "snow ice" appears white, whereas pure water ice appears clear (or black if looking toward the bottom of a lake).

Impurities, such as most salts and dissolved organics, are rejected from freezing water, making the ice relatively pure. The more slowly ice is formed, the more efficient the rejection process becomes. The effect of this process is that impurities become concentrated in water under the ice. As the liquid content of a lake or pond approaches 50 percent reduction by ice formation, dissolved salt concentrations double.

Example 5.1

Calculate the expected maximum ice thickness for a medium-sized lake with moderate snow cover and no snow cover when the annual surface freezing index is 3750°C·d.

Solution (a): Moderate snow cover.

From Table 5-1:

average ε-factor = $8.7 \times 10^{-5} \dfrac{m}{°C^{1/2} \cdot s^{1/2}}$

(Range of $\varepsilon = 8.1 \times 10^{-5}$ to 9.3×10^{-5})

SFI = 3750°C·d = 3.24×10^{8} °C·s

Ice thickness, h = $\varepsilon(SFI)^{1/2}$

$$= \left(8.7 \times 10^{-5} \frac{m}{°C^{1/2} \cdot s^{1/2}}\right) \cdot \left(3.24 \times 10^{8} °C \cdot s\right)^{1/2}$$

= 1.57 m~1.6 m

(Range of h = 1.46 to 1.67 m)

Solution (b): No snow cover (using Eq. 5-1).

$$h = \sqrt{\frac{2(k)(SFI)}{L_v}}$$

k_{ice} = $2.21 \dfrac{W}{m \cdot °C}$

L_v = 3.3472×10^{8} J/m³

SFI = 3750°C·d = 3.24×10^{8} °C·s

$$h = \sqrt{\frac{2\left(2.21\dfrac{W}{m \cdot °C}\right)\left(3.24 \times 10^{8} °C \cdot s\right)}{3.3472 \times 10^{8} \dfrac{J}{m^3}}}$$

h = 2.07 m ~ 2.1 m

Examples incorporating the effects of wind are available in Zarling (1978). Complete analytical models for examining the effects of snow cover on ice growth are presented in Stanley and Smith (1990).

Thermal and chemical analyses are required to identify a lake or pond that may freeze deep enough to create problems with high dissolved solids concentrations. A mass balance can be used to estimate concentrations that may be present in the remaining water. Figure 5-3 shows the effect of ice growth on the concentration of some ions (Boyd, 1959).

The design or analysis process can be summarized as follows:

1. Water demand is calculated by combining the population with per capita usage values. This is converted to m³. A stage vs. volume curve is developed. The relationship between water level and volume is used to determine water available for use. Beginning at the start of the critical time period, usually September or October, the monthly water use is subtracted from the assumed reservoir full volume, to estimate the end-of-month volume remaining.

2. Since water volume and surface area varies with depth, a new estimate of the water surface elevation is made; volume and surface area is then calculated. The ice thickness must also be calculated. This volume of ice is subtracted from the end-of-month volume to determine useable volume. The procedure is repeated for the winter months to ensure that sufficient volume remains available at the end of spring.

Enhanced Lake Storage. Since smaller natural lakes in the continuous permafrost regions of North America are often not deep enough to provide adequate storage of unfrozen water to meet supply requirements through a winter, they must be modified if additional storage is not provided. One method of modification is lake deepening by dredging. Caution must be used when considering this approach since disturbance of the thermal regime of and around the lake may cause melting and lake enlargement. There are also equipment mobilization

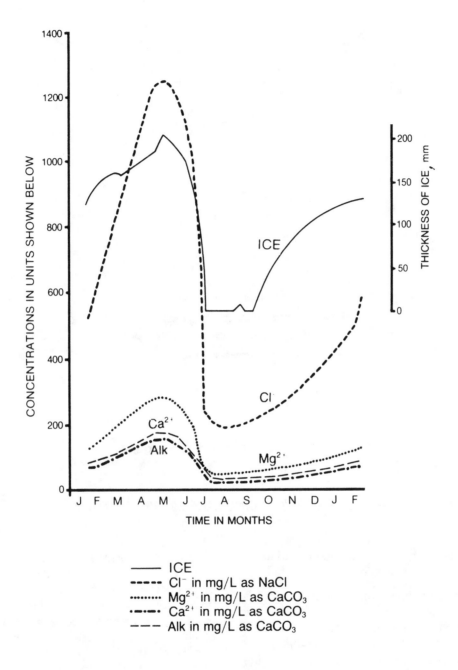

FIGURE 5-3   EFFECT OF ICE FORMATION ON SOME IONS IN
LAKE WATER (after Boyd, 1959)

problems, dredge spoil disposal problems, and an alternate source could be needed for at least a year. The newly exposed (unweathered) sediments can have quite an effect on water quality, particularly for an aggressive water. Figure 5-4(a) shows lake deepening as a method of enhancing lake storage of

water (Shafer, 1981). This method was also used at Devil's Lake in Kotzebue, Alaska (Figure 5-5).

Another method of improving lake capacity is by constructing a diked reservoir within an existing lake (Figure 5-4(b)). The internal reservoir must be lined to allow the water level to be raised. Also, hydro-

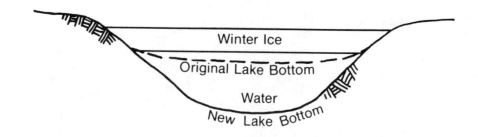

(a) Lake deepening

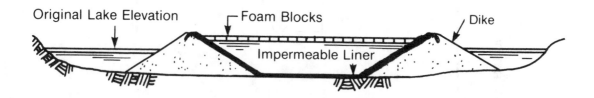

(b) Reservoir within existing lake

FIGURE 5-4   METHODS OF ENHANCING LAKE STORAGE (Shafer, 1981)

phobic, closed-cell insulation has been used to reduce surface heat loss (Shafer, 1981).

In some cases air bubbles may allow heat from the deeper portions of the lake to be used to offset the heat being transferred through the ice (Ashton, 1974; Carey, 1983). This technique results in greater total heat loss from the water body, but may reduce the amount of ice formation. Factors, such as the mixing of water with low dissolved oxygen concentrations and higher dissolved organics and inorganics from the bottom throughout the water column, must be evaluated for each case.

An effective method of reducing ice growth would decrease the quantity of storage capacity lost to ice and therefore reduce the total volume of a new reservoir. For existing reservoirs a reduction in ice growth would increase the volume of useable water and hence increase the service life of the reservoir.

A Canadian study recently evaluated available methods of reducing ice growth in the design of new reservoirs or to increase the service life of existing reservoirs (Stanley and Smith, 1990). Methods were evaluated for their effectiveness at reducing the loss of storage capacity to ice, their economic efficiency and their general acceptability of being used on a potable water source. Methods evaluated include: geometrical optimization of the design of the reservoir; reducing ice thickness; and shortening the period that storage is required. The study found that the geometry of a reservoir can definitely affect the quantity of water lost to ice. Of the factors considered, the depth was found to be the most important. The greater the depth, the greater the utilization of the total reservoir volume. To reduce ice thickness, the study considered methods that altered the heat exchange at the top surface of the ice, methods that altered the heat exchange at the bottom surface of

FIGURE 5-5   WATER SOURCE DREDGING – KOTZEBUE, ALASKA

FIGURE 5-6   CONSTRUCTED WATER STORAGE POND – MEKORYUK, ALASKA

FIGURE 5-7   SNOW FENCE AND WATER STORAGE POND – SHISMAREF, ALASKA

the ice and methods that altered the chemical properties of the ice. Of the methods considered only two proved to be both effective and economically viable; accumulating snow on the surface with the use of snow fences and the placement of insulation on the ice surface (Slaughter et al., 1975). It is also possible to accelerate ice decay in the spring which will quicken the release of water that is tied up as ice. Figure 5-6 shows a constructed water storage reservoir in Mekoryuk, Alaska.

The ice growth evaluations were based on freezing theory and verified with field measurements. The feasibility of the various methods of ice thickness reduction by snow deposition has been studied in field installations and proven effective. Several of the other methods were found to be uneconomical except in very specialized cases. Little information is available on operational difficulties associated with them.

Watershed Yield Improvement.   The amount of water supplied by a given watershed is related to the precipitation received by the watershed. Other influences on yield are evaporation to the atmosphere and transpiration by vegetation.

Records of precipitation, evaporation, and temperature are needed to estimate runoff or yield (Ryan and Crissman, 1990). The agencies listed in Section 5.1 maintain current data on many locations. Some agencies calculate and publish watershed yields along specific streams. When no other information is available, these data should be used as a guide. For large projects where accurate information is critical, data should be collected for several years before final design to confirm that the required yield will, in fact, be available.

In cold regions, particularly in the Arctic, a significant proportion of the precipitation is in the form of snow. Due to winds, however, it does not necessarily remain where it falls initially. This can actually be turned to advantage by inducing the occurrence and growth of snowdrifts to increase the annual water yield of a small watershed (Tabler, 1980; Ryan and Crissman, 1990). Figure 5-7 shows the snow fence constructed in Shismaref to increase watershed yield (Wheaton, 1980; Farmwald and Crum, 1986).

Design criteria for snow fences have been derived from several different projects.

Location	Yield m^3/lin. m (thousand gal./lin. ft.)	Height of Fence m (ft)
Barrow, Alaska	103.0 (8.3)	1.21 (4)
Wainwright, Alaska	146.5 (11.8)	2.4 (8)
Shishmaref, Alaska	31.0 (2.5)	2.4 (8)
Kotzebue, Alaska (design study only)	45.9 (3.7)	3.6 (12)
Kipnuk, Alaska	?	2.4 (8)
Baker Lake, NWT	0.3	5.5 (18)

An analysis of the Baker Lake snow fences provided the following cost information. The 5.5 m (18 ft) high snow fence cost $398 per metre (Cdn. 1993) to construct. This included material and labor including piling foundation and wind bracing. The materials for the fence were purchased in 1990/1991.

Off-stream storage may be an economic means for reducing problems of quantity or quality. Water storage in a lake was used in Inuvik, NWT to meet community water needs during freeze-up and break-up. This method can also be used to meet summer requirements when concentrations of sediment or other contaminants are high. (See Section 7 for more information on impoundments.)

5.2.2 Groundwater. Groundwater is usually the most desirable source of water in cold regions for several reasons:

- Normally, groundwater temperature is nearly constant and warmer than surface water in the winter.

- Mineral quality of groundwater is more constant than surface water.

- Groundwater from under the permafrost is almost always a year-round source of supply so that alternate or dual-source systems are often not needed.

- There is much less chance of contamination from surface activities than with surface water (especially deep groundwater).

However, the cost of exploring, drilling, developing, and maintaining wells in cold, remote areas can be high (Ryan and Crissman, 1990).

Groundwater in Permafrost Areas. Groundwater in areas of continuous permafrost may be found in three general locations;

- above permafrost, within the active layer (suprapermafrost);

- within permafrost in thawed areas (intrapermafrost); and

- under permafrost (subpermafrost).

Waters found in the active layer above the permafrost are generally not potable unless treated extensively. Such water is usually found within one to two metres of the surface and frequently has a high mineral or organic content, or both. Because they are shallow, such aquifers are also subject to contamination from privies, septic tanks, and animals. The quantity of water from this source is often small and unreliable and not available in the winter. A suprapermafrost water collection system developed for Point Hope, Alaska (McFadden and Collins, 1978) is shown in Figure 5-8.

Intrapermafrost water is quite rare and usually highly mineralized. Such water must contain high concentrations of impurities to depress the freezing point below that of the surrounding permafrost. There is no reliable method to locate pockets of interpermafrost water with present state-of-the-art techniques (Smith et al., 1979). For these reasons it is not normally a suitable water source.

Subpermafrost groundwater is the most reliable and satisfactory groundwater source in permafrost regions. Recharge of subpermafrost aquifers occurs beneath large rivers and lakes where there is no permafrost. When fine-grained soils are frozen, the downward movement of water to the groundwater is effectively prevented. Satisfactory wells have been located in the thawed areas beside or under rivers or large lakes, since the ground in these areas may not freeze.

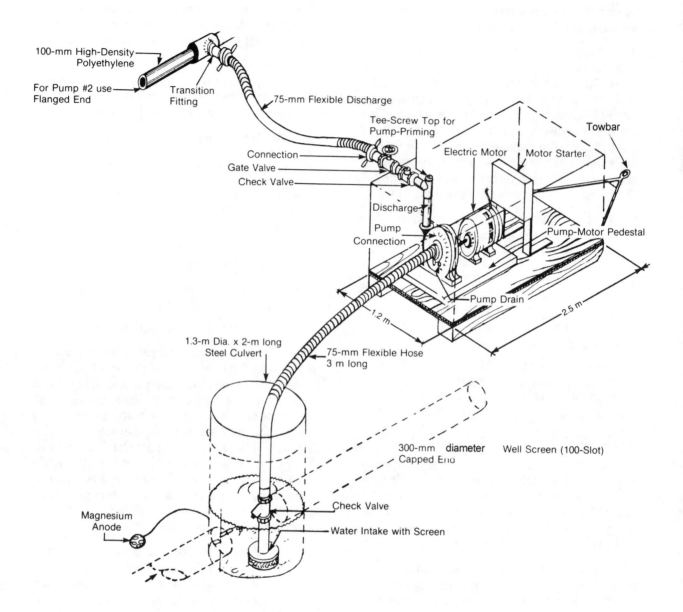

FIGURE 5-8   SUPRAPERMAFROST WATER SOURCE PUMP SYSTEM AND INFILTRATION GALLERY

Subpermafrost groundwater is generally deficient in dissolved oxygen. As a result, high concentrations of some minerals, such as iron and manganese, which are soluble under these conditions, are present. Higher concentrations of polyvalent cations (hardness) are also common in subpermafrost groundwater. Occasionally this groundwater contains dissolved organic substances as well.

Costs for drilling a well and its maintenance are higher in permafrost areas. The water must be pro-

tected from the cold permafrost and the permafrost needs to be shielded from the heat of the water. This often requires special well casings, grouting methods and heat-traced water lines, all contributing to the cost.

Groundwater in Intermittent Permafrost Areas.
The construction and operation of wells in intermittent permafrost areas requires special consideration of the thermal conditions. In such regions the permafrost is generally warmer than in the continuous

region. Thermal disturbance of the surface by drilling a well or by the well itself can result in thawing and, perhaps, well failure (Linell, 1973).

In these regions the potential for successful well development is also much higher. Thawed "windows" permit more water to move from the surface to the groundwater. This can also result in a better quality of water. In areas where buried or surficial organics are slowly decomposing, the quality of the groundwater can be poor.

Groundwater in Deep Seasonal Frost Areas. The volume of groundwater in these areas is seriously influenced by frost conditions. The amount of water moving from the surface to the water table is restricted during the period of frozen ground. If the percolation rate is low and the water pumping rate is high, water mining may occur.

Estimation of Groundwater Yield. The predictable yield of groundwater from a watershed may be estimated in a manner similar to that in warmer regions where the permafrost layer essentially acts as a confining layer. Once a proper well has been installed, well testing to determine the safe yield follows normal practice. The point of discharge must be selected so that the pumped water has the least possible effect on the surrounding ground.

Because of the effects of permafrost as a confining layer, and the limited recharge during the winter months, longer periods of test pumping are recommended to assist in obtaining information on long-term yields and quality. Variations in recharge have also led to seasonal changes in water quality, which can be assessed by sampling at different time periods.

Well Construction and Design. Location, siting and design of water wells in cold regions follow the same procedures and principles as in other locations but additional considerations are needed (Ryan and Crissman, 1990). Critical factors include:

- Considering bulk fuel storage and fuel fill line locations to minimize the potential for contamination.

- Drop pipes should be provided with drainback valves at least to a level below the deepest expected frost penetration.

- Heat tracing should be used if no drainback is used or for wells which penetrate permafrost to prevent freezing, however overheating must be avoided.

- The top of the casing should be insulated, and consideration should be given to providing a building at the well head for location of controls and other components and to assist in the operation and maintenance of the facility under adverse weather conditions.

- Seasonal frost can damage the well casing through frost jacking. Bentonite is usually used to grout the annular shape around the casing in the seasonal frost region. Concrete grout, as used in warmer climates, bonds tightly to the steel casing and the frost heaving can pull the casing apart, ruining the well.

- The increasing depth of the active layer during the winter can increase the pressure on the groundwater beneath the advancing freezing front. A well using this source can become artesian or even flowing artesian as winter progresses. There are several recorded instances where this pressure has caused large amounts of water to flow out of the well causing considerable damage to structures. This process is similar to the method of icing (aufeis) formation by streams.

5.2.3 Other Water Sources. Snow, ice and rain catchments are potential water sources which may be considered for small or temporary establishments. See Section 14 for detailed information.

Seawater. Desalinated seawater has been used for domestic supplies but the associated operation and maintenance problems can be considerable. Intakes in the ocean or on the beach are subject to ice forces of great magnitude. Ice scour during fall and spring can be serious. Along the Arctic Ocean, scour can occur at any time. During the winter months, shore-fast ice and frozen beaches pose special problems.

Brackish Water. Brackish water with total dissolved solids (TDS) of 10,000 mg/L or less is occasionally the only source available. Such waters may be treated by reverse osmosis or distillation, but significant problems must be anticipated with small installations. Treatment methods are discussed in Section 6.

Harvesting Snow and Ice. Snow and ice can be melted for water. Most larger communities and camps in the North American Arctic have developed groundwater or surface water sources and snow or ice is only used in isolated homes or temporary hunting, trapping, or oil exploration camps. In Antarctica, however, major settlements melt snow for their water supply. Snow is usually gathered with front-end loaders and transported to melters which operate on fuel oil and/or waste heat collected from the engines used for electrical power generation. Because of the possible exhaust contamination from the generators and heating boilers, the snow to be collected must be carefully selected, usually upwind from the camp. The soot from the engines tends to give the melted water an oily taste. Antarctica experience indicated the snow often contained volcanic ash which created turbidity and taste problems. Thus, the melted snow needed to be thoroughly filtered and disinfected. It takes over one litre of fuel to produce the heat to melt 100 litres of ice (more if melting snow), therefore, snow and ice are usually a last resort. On the Greenland icecap, a cavern is often formed in the ice cap by injecting steam into a drilled hole around 45 m deep. The thawed water is then pumped out for use in the camp. This method has produced better quality water than melting snow, but is still very expensive.

Water Reuse. In the absence of ample supplies of fresh water, water reuse may be considered. In Alaska, bath and laundry wastewaters are often reclaimed and used at several locations for toilet flushing and other nonpotable purposes. Reclamation of wastewater for conversion to potable water is not a commonly used technology in northern regions. Yet the effluent from existing secondary or tertiary sewage treatment plants is nearly free from suspended materials and should be more economical to treat than seawater. Seawater contains approximately 35 times more dissolved solids than domestic sewage.

5.3 Water Requirements

Determining the quantity of water needed and rate of flow is an initial step in the design of analysis of a water supply. Data on existing usage and projections of future needs are necessary.

5.3.1 Variations in Water Use. Short-term and seasonal variations in flow include daily, weekly, and monthly fluctuations. Average daily rates of water use can be determined from utility records, with at least two years of data reviewed. Alternatively, typical values depending on geographic location and type of system can be used.

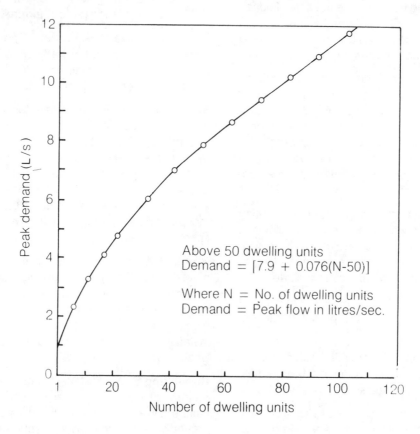

Above 50 dwelling units
Demand = [7.9 + 0.076(N-50)]

Where N = No. of dwelling units
Demand = Peak flow in litres/sec.

FIGURE 5-9   HOURLY PEAK WATER DEMAND IN SMALL COLD-CLIMATE COMMUNITIES

TABLE 5-2   RECOMMENDED DESIGN WATER DEMAND

	Water Consumption, L/(p•d)	
	Average	Normal Range
Households		
1. Self-haul from watering point	10	5 to 25
2. Truck system		
a. Nonpressure water tank, bucket toilet and central	10	5 to 25
b. Nonpressure water tank, bucket toilets and no central	25	10 to 50
c. Nonpressure water tank, waste holding tank	45	20 to 70
d. Pressure water system[1], waste holding tank, and normal	90	40 to 250
e. Pressure water system, waste holding tank, low flush	60	
3. Piped system[2] (gravity sewers)	225	100 to 400
Piped water supply and trucked sewer pumpout	110	
4. Piped system (vacuum or pressure sewers)	145	60 to 250
5. Well and septic tank and tilefield	160	80 to 250
Trucked water delivery and individual septic tanks	100	
Institutions (piped system)		
Day student school	10	2 to 18
Boarding school student	100	100 to 400
Nursing station or hospital (per bed)	100	
Hotels per bed	100	
Restaurants, bars, per customer	5	
Offices	10	
Showers (two per person per week)	10	
Laundry (two loads per family per week)	7	
Camps		
Base camp	200	
Drilling pad	130	
Temporary, short-duration camps	100	
Greenland Guidelines for New Water Works[3]		
Water collected in buckets		30 to 80
Water supplied by trucks or piped to nonsewered houses		80 to 150
Piped to sewered houses		150 to 200
USSR Guidelines for Systems in Permafrost Regions[4]		
Building with hydrant columns only		30 to 50
Building with running water and sewers, no baths		125 to 160
Building same with baths and local water heaters		160 to 230
Building same with central hot-water supply		250 to 350

Notes:

[1] Conventional flush toilets are strongly discouraged with truck system in NWT

[2] The figures for piped systems do not make allowances for the practice of letting fixtures remain open in cold weather to provide continuous flow in water mains to prevent freezing. Water use under that practice used only in older systems may be as high as several 1,000 L/(p•d).

[3] Rauschenberger (1983)

[4] Fedorov and Zaborshchikov (1979)

Demand factors are needed to design sources and storage facilities. Maximum daily demand should be computed at 2.3 times the average daily demand. Maximum hourly demand should be computed at 4.5 times the annual average daily demand. Figure 5-9 is presented for estimating hourly peak water demand in small cold-climate communities.

5.3.2 Water Usage. The amount of water used by inhabitants of northern communities depends on several factors. Cultural background is particularly important, since many cold-climate communities in Alaska and Canada are populated by indigenous people. Traditionally, these people have not had access to large quantities of water and therefore tend to use water conservatively. As more water becomes available, however, it is used more freely. For information on water use in individual dwellings and construction camps see Sections 13 and 14. Water usage can also increase considerably with higher pressures in the delivery system. Pressures should range between 100 and 350 kPa and no higher.

The minimum amount of water considered adequate for drinking, cooking, bathing and laundry is 60 litres per person per day (L/(p•d)). Even this may be difficult to achieve where piped delivery is not practical or possible. In communities without residential piped water distribution systems, and which have honey-bag waste systems, water consumption is between 4 and 12 L/(p•d).

Analysis of data collected at Wainwright, Alaska, indicates that when more water was made available, water use in homes rose from about 2 L/(p•d) to about 5.5 L/(p•d) in a three-year period. Wainwright, however, has a central facility that provides for bathing and laundry away from the home. This water use has not been included in these figures. In spite of these apparently low quantities for household use, significant health benefits were observed in the village from improved water supply and sanitation facilities. This is consistent with information in Figure 2-1.

Table 5-2 lists water use factors for various types of communities in cold regions. General design information is also available for Greenland (Rauschenberger, 1983) and the USSR (Fedorov and Zaborshchikov, 1979).

Nonresidential Use. The total community water use varies with the extent of nonresidential activities (i.e., commercial, institutional and industrial) in the community. The nonresidential activities tend to

increase in proportion to the population of the community. Unless specific data are available to estimate nonresidential water use in a community, for design purposes, the total water use per capita for both trucked and piped systems is calculated using the following equations in the NWT:

Total Community Population	Total Water Use Per Capita
0 to 2,000	RWU • [1.0 + (0.00023 • Population)]
2,000 to 10,000	RWU • [-1.0 + (0.323 • Ln (Population)]
>10,000	RWU • [2.0]

where,

RWU = Residential Water Use

Ln = natural logarithm

Water Conservation. Even though these recommended quantities are recognized as adequate, many communities in Alaska and Canada consume large quantities of water of which much is wasted. Table 5-3 gives examples of water use. In some cases water consumption is excessive. In many communities this is the result of water bleeding (see Section 8).

System designers must be alert to the possibilities for conserving water and potential reasons why systems sometimes encourage waste. Water and its treatment costs money, as do sewers and sewage treatment which must handle the hydraulic loads. Section 14.3 summarizes the various household water conservation fixtures, including those for toilet systems.

The City of Yellowknife, NWT, undertook initiatives to conserve water over the past ten years. Measures such as infrastructure upgrade, bleeder detection and elimination, water metering and leak detection surveys have resulted in substantial decreases in consumption and considerable monetary savings to the municipality. The City now estimates that actual use is about 280 L/(p•d), a figure that has stayed relatively constant since 1980. At one time, the City distributed approximately 560 L/(p•d). Water losses have been reduced 25% since they started their conservation program, a savings of nearly $860,000 (Cdn) annually.

Bleeding. Bleeding is the practice of allowing water to run to prevent freeze-up of water service lines. There could be isolated instances where bleeding

TABLE 5-3   *EXAMPLES OF ACTUAL WATER USE*

City	Water Consumption (L/(p•d))	Approximate Number of People	Type of System
Alaska			
Allakaket (1993)	15	200	Watering point
Anchorage	890*	220,000	Conventional water and gravity sewers
Bethel	270	1,200	Circulating water and gravity sewers
Brevig Mission (1993)	50	137	Washeteria
Dillingham	2,300*		Conventional water and gravity sewer
Dot Lake	190	50	Central heat-conventional water (utilidors), individual sewer
Elim	170	196	Circulating piped; gravity sewer
Fairbanks	650*	25,000	Circulating water and gravity sewer
Galena (1980)	12	200	Truck delivery only. Showers and laundromat available elsewhere.
Golovin (1993)	70	146	Truck delivery; septic tanks
Gulkana (1993)	110	103	Circulating piped; gravity sewer
Holy Cross (1993)	280	275	Circulating piped
Homer	1,630*		Conventional water and gravity sewers
Kenai	380		Conventional water and gravity sewer
Kiana (1993)	170	410	Circulating piped; gravity sewer
Kotzebue (1980)	246	2,544	Circulating distribution system; gravity sewer
Kotzebue (1993)	280	3,500	Circulating piped; gravity sewer
Little Diomede (1993)	23	180	Washeteria
Metlakatla (1993)	1,000	900	Conventional piped; gravity sewer
Minto	190	180	Circulating water-gravity sewers
Nunapitchuk (1979-80)	17	312	Washeteria consumption
Palmer	760*		Conventional water and gravity sewer
Saxman (1993)	1,000	380	Conventional piped; gravity sewer
Seldovia	680*		Conventional water and gravity sewer

continued

TABLE 5-3   *continued*

City	Water Consumption (L/(p•d))	Approximate Number of People	Type of System
Seward	4,500**		Conventional water and gravity sewers (winter)
	2,300		(summer)
Shaktookik (1979-80)	20	160	Self-haul
	38		Washeteria plus self-haul
	68		Total summer use, summer distribution line; cluster septic tank
Shaktoolik (1993)	200	200	Circulating piped
Shishmaref (1993)	38	466	Washeteria
St. Michael (1993)	15	330	Watering point
Toksook Bay (1978-79)	136	317	Circulating piped
Toksook Bay (1993)	170	510	Circulating piped; gravity sewer
Tununak (1980)	22	283	Washeteria consumption
Unalakleet (1978)	352	600	Circulating water-gravity sewers
Wales (1993)	38	147	Washeteria
Alberta			
Edmonton	236	500,000	Conventional, residential only
Denmark			
Copenhagen (1972-73	115-253		Range for 33 multistory apartment building***
Greenland			
Godthab (1981)	123	9,423	Piped water and sewer
Julianehaab (1981)	232	2,574	Part piped, part self-haul
	340		Piped house only
	35		Self-haul only
Sukkertoppen (1981)	145	3,013	Piped
Northwest Territories			
Aklavik	32	797	Trucked delivery
	63		Summer piped system
Fort Franklin (1978)	45	472	Truck delivery
Fort Good Hope (1976-77)	32	440	Truck delivery
Fort McPherson	250	850	Piped portion of community; remainder trucker

continued

TABLE 5-3   *continued*

City	Water Consumption (L/(p•d))	Approximate Number of People	Type of System
Inuvik (1970)	20	1,300	Trucked water, honey bags
Inuvik (1976)	485 to 550	3,500	Circulating water and gravity sewer
Resolute Bay (1970)	163		Circulating water
Resolute Bay (1977)	23	160	Trucked water, honey bags
Sacks Harbour (1977)	22	180	Truck delivery
Yellowknife	485 to 560	10,000	Piped water, gravity sewer
	90		Trucked water, sewage pumpout
Yukon			
Clinton Creek	1,140*	381	Circulating water and gravity sewer
Dawson City	6,400*	745	Circulating water and gravity sewer
Faro	1,140*		Circulating water and gravity sewer
Mayo	1,700*	462	Circulating water and gravity sewer
Whitehorse	1,680*	11,217	Circulating water and gravity sewer

Notes:
*   Some water bleeding to prevent freezing of service line.
**   Leakage in old water pipes.
***   Rauschenberger (1983)

is economical, however, users must be discouraged from doing this where it is not required. This type of wastage is most common in the spring when frost penetration is greatest. To compound the problem, the amount of water available is lowest in early spring. Subarctic communities seem to be more prone to this situation, probably because service lines in the Arctic are designed for low temperatures and are usually heated or recirculated. Remedies for this problem are to educate water users, meter all service connections, provide inexpensive and quick methods of thawing frozen service lines, and construct service lines that are less apt to freeze. Constructing service lines that are less apt to freeze means:

- burying lines below frost line until within the thaw bulb of the house;

- insulating lines where the surrounding ground may freeze;

- recirculating the water in service lines;

- providing heat tapes on lines in the frost zone; and

- heating water in the distribution system.

Leakage. A lot of water is wasted not only because of leakage from old or broken service lines and mains but also because of poorly-maintained plumbing within buildings. Possible methods of minimizing losses of this kind are to:

- maintain pressure in mains at the lowest pressure necessary (approximately 170 kPa);

- promptly repair all leaks in mains and service lines;

- check the system for leaks frequently by isolating sections and pressure testing;

- inform users about the causes of leaks, and train them to repair leaking fixtures such as faucets and toilets; and

- install water meters on all services.

5.3.3 Water Quality. The concern for the quality of the water source is based primarily on the ease with which the water may eventually be treated to make it potable, and the cost of the required treatment. Reliability in quality is as important as reliability in quantity.

Surface Water Quality. Surface waters are more readily polluted by people and animals; thus emphasis should be placed on bacteriological and biological quality of the water and watershed. Cysts, bacteria, and viruses live for long periods in cold waters, and pose a potential health problem for long distances downstream from their entry point.

Water sources should be selected and the watershed protected in a manner acceptable in any climate.

Group I   water may be used as public water supplies without treatment (deep groundwater supplies) (disinfection in the distribution system is always required);

Group II   water may be used after disinfection only (groundwater under the influence of surface water); and

Group III   waters require either complete conventional treatment (including coagulation, sedimentation, filtration and disinfection) or direct filtration and disinfection (Malcolm Pirnie, Inc. and HDR Engineering, Inc., 1990).

Because of the high probability of contamination of surface water by animals harboring various helminth eggs and protozoan cysts, all surface waters must now be coagulated, filtered and disinfected before use in the public water supply in the United States (see Section 6).

Surface water sampled for quality during warm weather may yield misleading values. The sun's ultraviolet light can reduce concentrations of microbial constituents in the water. Freeze rejection of minerals and other impurities during ice formation causes remaining liquid to be of significantly poorer quality. Lakes in cold regions also "turn over" in the fall (Ryan and Crissman, 1990). This process is caused by the decreasing temperature of the lake's surface with the start of winter and the fact that the density of fresh water is greatest at 4°C. Sudden increases in the total dissolved solids and suspended solids will result. Also, during the summer, runoff filters through the mosses and lichens collecting color and organics. These can increase water treatment needs, especially if trihalomethanes are formed during disinfection.

Groundwater Quality. Suprapermafrost water must be considered of questionable quality, since contamination by pit privies, septic systems, and animals can easily occur. Subpermafrost waters are generally unpolluted, but may contain high concentrations of minerals such as iron (as high as 175 mg/L), manganese, magnesium, and calcium as well as organics. Concentrations of iron below 7 mg/L and hardness below 150 mg/L are reasonably easy to reduce by treatment and do not significantly detract from the value of the source. In the highly mineralized areas of Alaska, some groundwaters have been found to contain unacceptably high quantities of arsenic. High concentrations of nitrates have been observed in other groundwaters near Fairbanks, Alaska, and on Nunivak Island.

Quality Improvement of Lake Water. The quality of water in a small saline pond or lake can often be improved by pumping out the concentrated brines which remain under the ice near the end of winter and allowing fresh runoff to replace it. Repeated one or more times, this method may permit the use of an initially unacceptable water body as a source of supply. The U.S. Public Health Service, in developing an improved water source for Barrow, Alaska, used this method with good results. Total dissolved solids concentration in the pond, when the ice cover had fully developed, was about 7,000 mg/L. The range of total dissolved solids in the pond is shown in Figure 5-10. Soil salinity and brine pockets in the soil beneath the impoundment may limit the amount of improvement which can be realized from this technique.

5.4   Structures

Structures relating to water supplies range from a simple temporary intake on river ice to a complex dam on permafrost with a year-round intake and pumping station. Wells and their appurtenances are also considered supply structures.

This discussion does not intend to provide a guide for detailed design of any facility but rather to point out features that may require special attention in cold climates. Designs should be prepared by experienced engineers, qualified to work in cold regions.

5.4.1   River Intakes. Intake structures may be either temporary or permanent. Permanent structures are the most desirable, because they permit a certain freedom from attention during such critical times

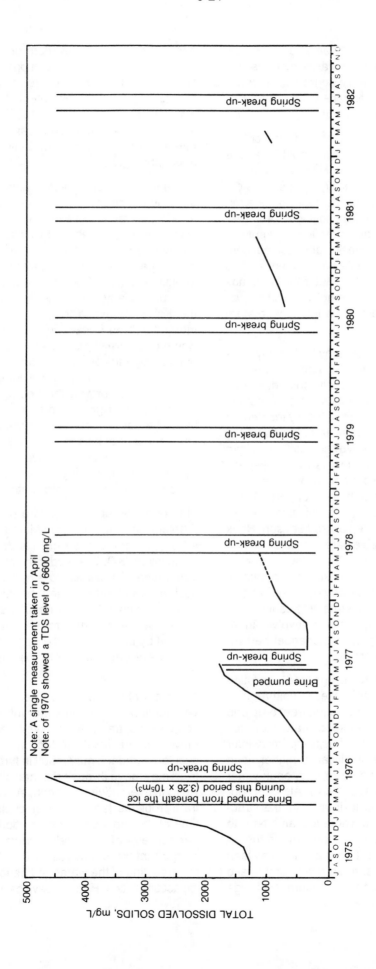

FIGURE 5-10   WATER QUALITY IN UPPER ESATKUAT LAGOON – BARROW, ALASKA, 1975-82

as freeze-up and break-up (Ryan and Crissman, 1990). On the other hand, temporary structures may be less expensive and permit a degree of latitude in operation not afforded by permanent structures (Figure 5-11, 5-12 and 5-13).

Temporary facilities require attention during critical times of the year, and the operator must be aware of existing or potential problems. However, depending upon the overall system design and source capabilities, intakes may be required for only a short period each year. Under some conditions a protected pump on the river shore or on the ice may suffice. This requires neither much design attention nor continuous operator attention, unlike permanent intakes (Figures 5-14, 5-15 and 5-16). But if the demands are such that they cannot be met by providing water storage, then more elaborate intake works are required.

Intake Types.   River intakes may be broadly classified into two main categories: bank intakes and bed intakes. Both types are usually comprised of a river intake structure and a pumping station (located on the bank). Bank intakes are typically located on the steep banks along the outside of bends where secondary flows naturally move sediment away from the inlet ports and the channel is at its deepest. Multiple intake ports are common, allowing for sediment-free withdrawals from high-level ports during summer floods and effective withdrawal from lower ports during the low flow/ice covered season. Bank intakes require a high, stable bank for the combined pumphouse/intake structure. On shallow and/or laterally unstable streams, bed intakes may provide a viable alternative. They may be comprised of one or more permanent pier-type structures constructed on the channel bed, or may simply involve an intake pipe or screen lying on the channel bed (or supported on a trestle or cradle).

In addition to the standard design and operation considerations, open-water intakes in cold regions are jeopardized by flowing ice during freeze-up and break-up and must be given special attention. Ice jams, which may form during freeze-up, midwinter or break-up, may cause significantly higher flood stages, flow velocities, and scour than open water events (Ryan and Crissman, 1990). Also, the potential for impact damage to in-river intake structures is greater in reaches where ice jams tend to form. Frazil ice is an important concern, as the inherently adhesive nature of frazil particles may lead to clogged intake screens, trash racks, and intake pipes. The effects of ice on a minimum discharge

and minimum and maximum stage must also be considered in the design and operation of water intakes in cold regions. Intake location and design should also consider the potential of scour and erosion (Ryan and Crissman, 1990). Usually historic erosion rates and directions can be measured thus future rates and directions estimated using past aerial photographs.

Frazil Ice.   Frazil particles, (small discs of ice 1 to 3 mm in diameter) form in supercooled water (water at temperatures of a few hundredths of a degree less than 0˚C) (Ryan and Crissman, 1990). Frazil particles are extremely sticky and will adhere not only to each other (forming frazil slush), but to any foreign material that has a favorable crystalline structure, particularly steel, cast iron, and copper. The frazil slush floats to the surface and the unsubmerged portion freezes into the familiar "pan ice." Some of the frazil particles may adhere to the bed before accumulating in sufficient quantities to float to the surface. When the frazil adheres to very large gravel or boulders on the bed it is often buoyant enough to float gravel and small boulders off the bed. The development of a surface ice cover prevents rapid heat loss from the water and thus local frazil production stops. However steep reaches, such as rapids sections, may remain open continuing to generate frazil ice well into winter.

The formation of frazil ice is a particularly serious problem and must be considered in the design of supply systems (Tsang, 1982; Foulds, 1974; Henshaw, 1887). The predominant problems are the adherence of frazil ice to structures in the water, such as water intakes, and the constriction of flow by the accumulations of frazil under an ice cover. Intake screens, trash racks, etc. may become choked by frazil in a matter of hours. Even lake intakes have been known to experience frazil clogging.

A number of methods have been employed to prevent frazil problems on water intake structures. In some cases, trash racks have been completely removed before frazil production begins. However, such an approach may result in further problems as frazil may move through the intake lines into the wet well. An alternative approach, suitable only for mild frazil problems, is to employ mechanical ice removal. One of the most common solutions is to heat the bar screens of the intake works and other submerged intake structures to 0.1˚C. This prevents supercooling of the water and hence prevents frazil ice accumulation. Heating devices must be activated

Winter Pumping Mode

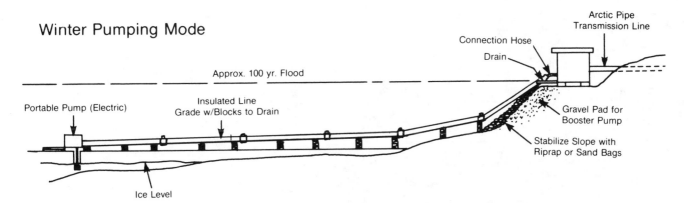

Above depicts wintertime pumping mode. River pump is disconnected, drained, and stored in booster station after each tank filling.

In the summer the pump in placed on the beach with intake extending into river, held off bottom with a float.

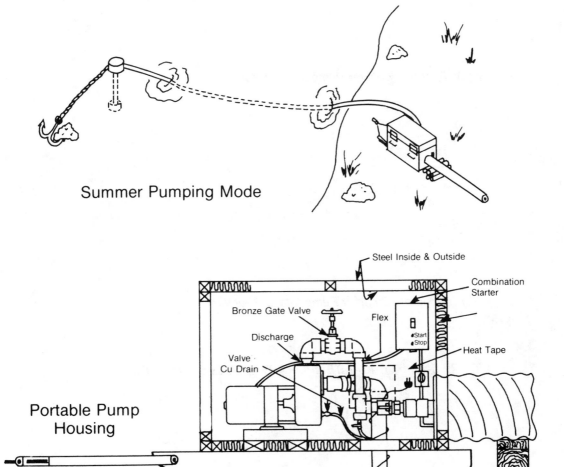

Summer Pumping Mode

Portable Pump Housing

FIGURE 5-11   TEMPORARY WATER INTAKE STRUCTURE

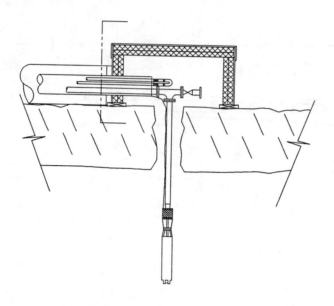

(a) Temporary Winter Pump Intake Structure

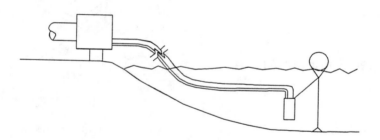

(b) Temporary Summer Pump Intake Structure

FIGURE 5-12   TYPICAL TEMPORARY INTAKE STRUCTURES

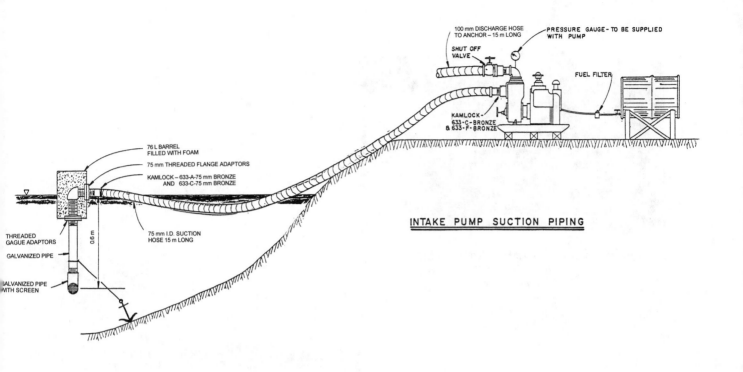

100 mm DISCHARGE HOSE
TO ANCHOR - 15 m LONG

PRESSURE GAUGE- TO BE SUPPLIED
WITH PUMP

SHUT OFF
VALVE

FUEL FILTER

KAMLOCK-
633-C-BRONZE
& 633-F-BRONZE

76 L BARREL
FILLED WITH FOAM

75 mm THREADED FLANGE ADAPTORS

KAMLOCK – 633-A-75 mm BRONZE
AND 633-C-75 mm BRONZE

THREADED
GAGUE ADAPTORS

GALVANIZED PIPE

GALVANIZED PIPE
WITH SCREEN

0.6 m

75 mm I.D. SUCTION
HOSE 15 m LONG

INTAKE PUMP SUCTION PIPING

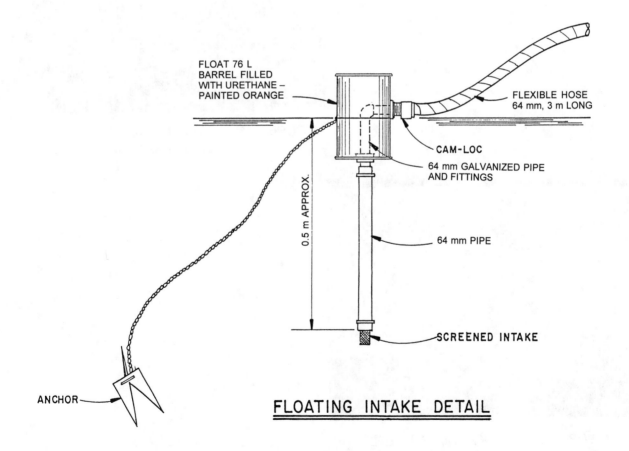

FLOAT 76 L
BARREL FILLED
WITH URETHANE –
PAINTED ORANGE

FLEXIBLE HOSE
64 mm, 3 m LONG

CAM-LOC

64 mm GALVANIZED PIPE
AND FITTINGS

0.5 m APPROX.

64 mm PIPE

SCREENED INTAKE

ANCHOR

FLOATING INTAKE DETAIL

FIGURE 5-13   TYPICAL FLOATING SUMMER INTAKE STRUCTURES

FIGURE 5-14   WATER TRUCK FILLING AT
WATER SUPPLY LAKE IN
CAMBRIDGE BAY, NWT

FIGURE 5-15   WATER INTAKE AT FORT
MCPERHSON, NWT

FIGURE 5-16   INTAKE HOUSE IN PEEL
CHANNEL, MACKENZIE RIVER –
INUVIK, NWT

before the problem develops, as it is not feasible to melt the ice by heating the bars. Trash racks should also be completely submerged, as heat loss from the bars to the atmosphere may increase the possibility of frazil development and accumulation. Other alternative measures to minimize frazil accumulation include substituting wood, plastic, fiberglass or other materials for metal, and through the use of chemical coatings. These have been employed with varying success. The simplest solution to frazil problems is to locate the intake in a reach of river where surface ice forms quickly, for example, in a long, calm reach. However, caution should be taken to ensure that the site is not downstream of an open reach where frazil will continually be generated.

Frazil can also cause indirect problems for river intakes. For example, when frazil pans or anchor ice become entrained in the flow, impact on trash racks can cause severe damage, particularly when the anchor ice contains gravel and small boulders. The accumulation of frazil ice on the bed in front of a water intake can also be a major problem and reaches prone to such accumulations are unsuitable intake sites.

Tidal Influence. Many rivers in cold regions have very flat gradients near their mouths where they flow into salt water. Intake placement must consider salt water intrusion and tidal salinity at lower tides. For example, the Kuskokwim River at Bethel, Alaska has considerable tidal fluctuation even though it is over 50 river miles from the Bering Sea. The heavier saline water wedge can travel upriver many miles at high tides and low river flows. To obtain usable water it may be necessary to pump from near the surface of the river or to pump only at low tides and high river flows. A sampling program is extremely important to determine the true quality of the water available in these circumstances. A critical time to sample (at depth as well as at the river surface) will be during the winter when most Arctic stream flows are at the lowest and during high tides.

Low Flows and Stages. Although discharges may be at their lowest during the midwinter period, minimum stages often tend to occur in the late fall. This is because the additional resistance afforded by the ice cover reduces the conveyance capacity of a channel, causing higher stages than that which occurs for the same discharge without an ice cover (Ryan and Crissman, 1990). Unless off-stream storage is available, water intake design must account for the possibility of extremely low flows or stages, as well as the potential for frazil slush accumulation

under the ice cover, both of which would limit effective withdrawals. In addition, the intake must be located low enough to avoid becoming frozen into the ice cover.

Use of multiple intakes is a recommended approach which may enhance system reliability. A system in which multiple intakes remain functional throughout the winter can provide continuous circulation of water. This may be effective in minimizing the thickness of the ice cover, thus preventing freezing of the intake system.

Ice Jams. Ice jams may occur during freeze-up, midwinter or spring break-up (Ryan and Crissman, 1990). Formation of an ice jam can lead to a significant increase in upstream water levels (which may threaten the intake pumphouse), and reduced discharges and water levels downstream. The dynamic release of an ice jam can result in flood discharges and water levels far in excess of open water events. The resulting high stages and velocities can lead to pumphouse inundation, ice impacts on structures and scour/deposition patterns which may adversely affect intake operation.

Intake Examples. Numerous arrangements and configuration of river intakes have been designed with varying degrees of success. Figure 5-17 shows the piping schematic for a matched pair of intakes. Such designs are continually evolving to make use of more sophisticated concepts and materials (Wahanik, 1978).

Figure 5-18 shows a lake water intake. Note that the intake line is installed to allow water to flow to the wet well by gravity so that even if the intake itself was damaged, water would remain available at the pump. Note also the insulation, wet-well heater, heat trace for intake line, and recirculation line for the townsite.

Figures 5-19 and 5-20 show different approaches to the design of water intakes for impoundments. Figure 5-19 shows the incorporation of water filtering capability in the collector. Figure 5-20 shows the use of a steel hydraulically-cleaned intake screen.

5.4.2 Infiltration Galleries. Infiltration galleries may be constructed parallel to the water sources, across the water course, either vertical or radially. Schematics of such systems are shown in Figure 5-21.

Infiltration galleries offer some advantages over conventional river or lake intakes. The most obvious advantage is that they are located away from

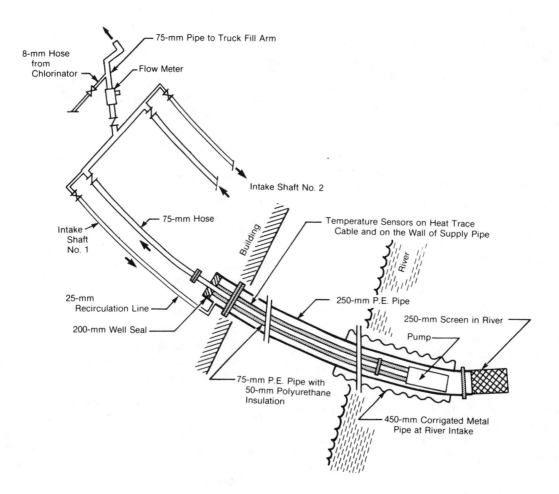

75-mm Pipe to Truck Fill Arm

8-mm Hose from Chlorinator

Flow Meter

Intake Shaft No. 2

Intake Shaft No. 1

75-mm Hose

Building

River

Temperature Sensors on Heat Trace Cable and on the Wall of Supply Pipe

25-mm Recirculation Line

200-mm Well Seal

250-mm P.E. Pipe

250-mm Screen in River

Pump

75-mm P.E. Pipe with 50-mm Polyurethane Insulation

450-mm Corrigated Metal Pipe at River Intake

FIGURE 5-17   PIPING SCHEMATIC FOR RIVER WATER INTAKE AT FORT NORMAN, NWT

the river and thus the hazards imposed by ice during freeze-up and break-up. Infiltration galleries may be placed in the thaw bulb of streams in permafrost areas where they collect water even when the streams appear solidly frozen. Usually some flow of water occurs within the stream bed itself, particularly when the bed material is relatively coarse.

A second advantage offered by infiltration galleries is the filtration of water by the materials surrounding the collectors. This may be a very significant advantage in streams which carry a load of suspended material such as silt or glacial rock flour. They can provide "filtration credits" toward U.S. EPA surface water treatment requirements.

Galleries must be protected against freezing, especially in permafrost areas. Some sort of heating system is usually installed during construction. Both electric and steam heating systems have been suc-

cessfully used. Usually heat lines are placed on the upper surface of the lateral and a second heat line is installed 0.4 to 0.6 m above the lateral. Insulation with snow is another way to reduce frost penetration. For the snow cover to be effective insulation the area should receive no traffic so that the snow remains uncompacted (Feulner, 1964). Figures 5-22 and 5-23 show actual infiltration galleries in Grayling and Shungnak, Alaska.

Periodic gallery cleaning may be necessary to remove silt and other sediments that enter the laterals and sumps. The use of geotextiles may reduce this requirement and improve overall system performance.

Springs can be developed by installing horizontal infiltration galleries in the aquifer. This is generally an approved method since it reduces the possibility

FIGURE 5-18   INTAKE STRUCTURE AT DEVIL'S LAKE IN KOTZEBUE, ALASKA

of water contamination at the point of collection (Section 14).

5.4.3 Wells. A producing well is most successfully located through detailed examination of geological conditions: type and permeabilities of soils and rocks; position of strata; character of cracks, fissures, and other large openings; and a study of performance records of other wells in the area (Driscoll, 1986). Especially in the arctic regions, professional hydrogeologists familiar with permafrost should be consulted during the preliminary stages when groundwater is being considered for a water supply source.

In all cases, wells must be located a safe distance from potential sources of pollution. Local health departments usually have specific requirements, but in the absence of other guides, wells should be at least 35 metres from the nearest source of pollution. Another important consideration is the direction of groundwater movement and the effect of ground freezing on that movement.

Well Drilling. There are several opinions on the best method for drilling water wells. Location, accessibility, size of well required and other factors

influence the methods used. Due to the transportability of equipment, cable tool drilling is popular in remote areas. "Jetting" of smaller wells has the advantage of relatively low cost, and the machinery is easier to move into remote areas. It can have adverse effects, however, on fine-grained, ice-rich permafrost. Cable tool systems require less water for the drilling operation than the jetting method, although most water used in drilling can be reused. Rotary drilling machinery can be used in all types of geological formations, but is larger and more expensive.

In some cases, for example, when drilling through frozen soil, the drilling fluid may have to be heated during the operation. The heated fluid prevents the mud from freezing in the permafrost and aids in thawing the ice-rich soil as the drill penetrates.

Well Seals. Sanitary well seals in the top of well casings prevent contamination by surface sources, yet permit easy removal of the pump when necessary. If pumps and pipes to the building can be installed below the seasonal frost line, a single pipe from the well suffices. However, if the well is installed in permafrost or pipes cannot be installed below the

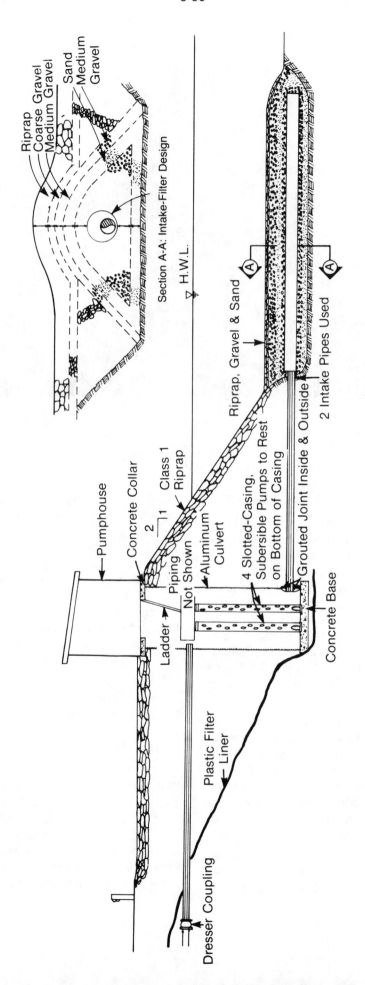

Riprap
Coarse Gravel
Medium Gravel
Sand
Medium Gravel

Section A-A: Intake-Filter Design

H.W.L.

Riprap, Gravel & Sand

2 Intake Pipes Used

Grouted Joint Inside & Outside

4 Slotted-Casing, Subersible Pumps to Rest on Bottom of Casing

Aluminum Culvert

Piping Not Shown

Ladder

Class 1 Riprap

Concrete Collar

Pumphouse

Concrete Base

Plastic Filter Liner

Dresser Coupling

FIGURE 5-19   WATER SUPPLY DAM BED INTAKE STRUCTURE – SAND POINT, ALASKA

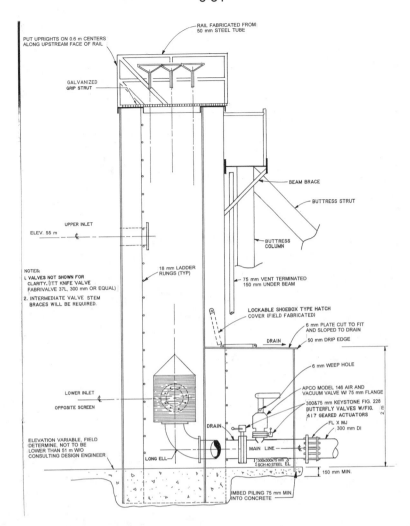

PUT UPRIGHTS ON 0.6 m CENTERS
ALONG UPSTREAM FACE OF RAIL

RAIL FABRICATED FROM
50 mm STEEL TUBE

GALVANIZED
GRIP STRUT

BEAM BRACE

BUTTRESS STRUT

UPPER INLET

ELEV. 55 m

BUTTRESS
COLUMN

NOTES:
1. VALVES NOT SHOWN FOR
CLARITY. (ITT KNIFE VALVE
FABRIVALVE 37L, 300 mm OR EQUAL)

2. INTERMEDIATE VALVE STEM
BRACES WILL BE REQUIRED.

18 mm LADDER
RUNGS (TYP)

75 mm VENT TERMINATED
150 mm UNDER BEAM

LOCKABLE SHOEBOX TYPE HATCH
COVER (FIELD FABRICATED)

6 mm PLATE CUT TO FIT
AND SLOPED TO DRAIN

DRAIN

50 mm DRIP EDGE

6 mm WEEP HOLE

APCO MODEL 146 AIR AND
VACUUM VALVE W/ 75 mm FLANGE

LOWER INLET

OPPOSITE SCREEN

300&75 mm KEYSTONE FIG. 228
BUTTERFLY VALVES W/FIG.
417 GEARED ACTUATORS

FL X MJ
300 mm DI

DRAIN

MAIN LINE

ELEVATION VARIABLE, FIELD
DETERMINE, NOT TO BE
LOWER THAN 51 m W/O
CONSULTING DESIGN ENGINEER

LONG ELL

300x300x75 mm
SCH 40 STEEL EL

150 mm MIN.

IMBED PILING 75 mm MIN.
INTO CONCRETE

FIGURE 5-20   SHEET PILE DAM AT PORT GRAHAM, ALASKA

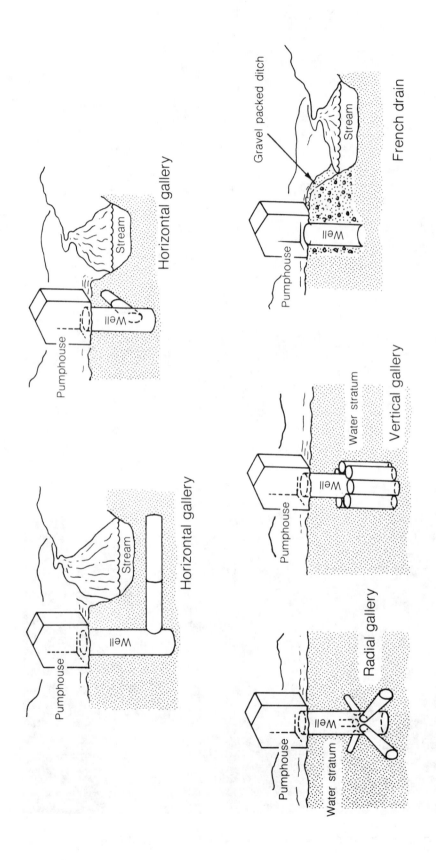

FIGURE 5-21   INFILTRATION GALLERIES (Feulner, 1964)

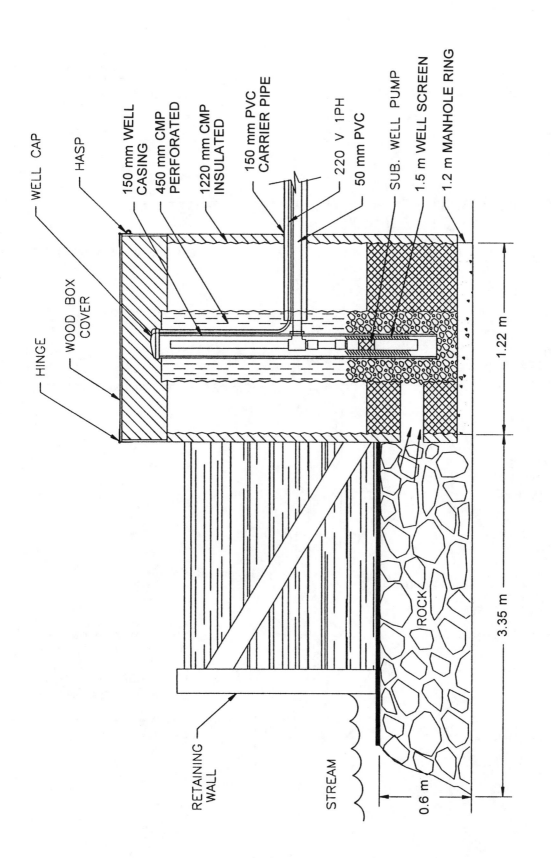

WELL CAP

HASP

150 mm WELL CASING

450 mm CMP PERFORATED

1220 mm CMP INSULATED

150 mm PVC CARRIER PIPE

220 V 1PH

50 mm PVC

SUB. WELL PUMP

1.5 m WELL SCREEN

1.2 m MANHOLE RING

WOOD BOX COVER

HINGE

RETAINING WALL

STREAM

ROCK

1.22 m

3.35 m

0.6 m

FIGURE 5-22   GRAYLING INFILTRATION GALLARY

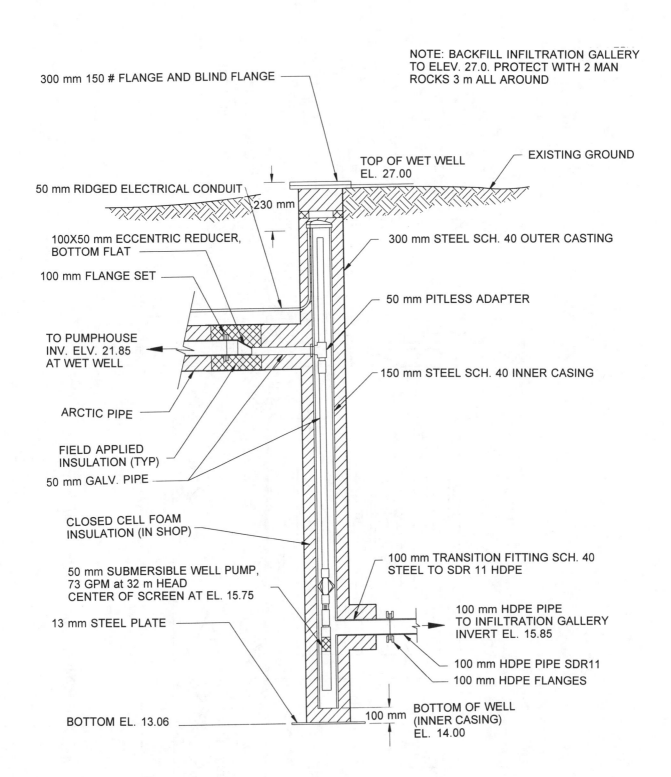

300 mm 150 # FLANGE AND BLIND FLANGE

NOTE: BACKFILL INFILTRATION GALLERY
TO ELEV. 27.0. PROTECT WITH 2 MAN
ROCKS 3 m ALL AROUND

EXISTING GROUND

TOP OF WET WELL
EL. 27.00

50 mm RIDGED ELECTRICAL CONDUIT

230 mm

100X50 mm ECCENTRIC REDUCER,
BOTTOM FLAT

300 mm STEEL SCH. 40 OUTER CASTING

100 mm FLANGE SET

50 mm PITLESS ADAPTER

TO PUMPHOUSE
INV. ELV. 21.85
AT WET WELL

150 mm STEEL SCH. 40 INNER CASING

ARCTIC PIPE

FIELD APPLIED
INSULATION (TYP)

50 mm GALV. PIPE

CLOSED CELL FOAM
INSULATION (IN SHOP)

100 mm TRANSITION FITTING SCH. 40
STEEL TO SDR 11 HDPE

50 mm SUBMERSIBLE WELL PUMP,
73 GPM at 32 m HEAD
CENTER OF SCREEN AT EL. 15.75

100 mm HDPE PIPE
TO INFILTRATION GALLERY
INVERT EL. 15.85

13 mm STEEL PLATE

100 mm HDPE PIPE SDR11

100 mm HDPE FLANGES

BOTTOM EL. 13.06

100 mm

BOTTOM OF WELL
(INNER CASING)
EL. 14.00

FIGURE 5-23   INFILTRATION GALLERY IN SHUNGNAK, ALASKA

seasonal frost line, some method is required to prevent freezing of the well.

The use of bentonite grout instead of cement provides an adequate seal and reduces the possibility of damage to the well casing due to frost heave.

5.4.4 Pumping Stations. Pumphouses can provide shelter for pumping equipment controls, boilers, treatment equipment, and maintenance personnel who must operate and service the facility. The structural design depends on the requirements of each location, and must be considered individually. The type of equipment housed within the shelter also depends on the individual system and may vary from a simple pump to a complex system with boilers for heat addition, standby power, and alarm systems to alert operators of malfunction. Any system must provide the degree of redundancy and other safeguards required by the nature of the operation and location.

All pumphouses should be designed with moisture-proof floors, since the floors get wet frequently. Pumphouses must be large enough to accommodate additional equipment such as heaters and their controls. Oversizing the original pumphouse at an installation should be considered carefully in relation to design life and the accuracy of demand predictions.

Most pumphouses or treatment buildings include redundant pumps, standby power sources and alarm systems. Voltage control devices are recommended to protect electrical equipment where power is of questionable consistency or dependability.

Heat Addition. Heat addition at the source of the water supply is usually required, since the water is very cold, often approaching 0˚C. Protection from freezing is the overriding reason for heating water at the source. Enough heat must be added to compensate at least for heat loss during transmission. It is generally accepted that water in transmission lines should be at least 4˚C to provide an adequate margin for heat loss, should a pump fail. However, thermal analysis should be made in detailed design. The treatment of raw water is much more efficient at warmer temperatures. Thus most of the heat that is needed for distribution and storage should be added before treatment if possible.

Disinfection. Additional disinfection may need to be provided at pump stations and storage facilities, particularly where water warming or long periods of storage may encourage growth of various types of bacteria, such as iron-oxidizing bacteria where iron pipes and fittings are used or health-related bacteria in any installation.

5.4.5 Protection of Transmission Lines. Pipelines carrying water from the intake and pumphouse to the storage reservoir may be either buried or laid above ground, depending upon local conditions. In general, buried lines are preferred to reduce maintenance and heat loss. Surface or elevated lines must have additional insulation and should be protected with some form of heat tracing or thawing method. For more detailed design information see Sections 4 and 8.

Pumps and transmission lines should be provided with drains, preferably automatic, to evacuate the water in case of power loss or other long-term failure, and thus prevent rupture due to freezing. These essential provisions may be as simple as the elimination of check valves and providing positive gradient so the line drains back to the source after pumping. Other solutions that have been used include using HDPE pipe which will not rupture if it freezes or using sacrificial sections in the line so the damage can be limited to sections where repairs can easily be made. The sacrificial sections are created by adding insulation which ensures that they will be the last place to freeze and the location of the damage.

5.4.6 Corrosion Control. The treatment and control of internal corrosion for small water systems can be difficult and expensive. The problems with corrosive water are unique to each system. Several options are normally available for the modificiation of water qualtiy to reduce the potential for internal corrosion, and include the following:

1. pH and alkalinity adjustment – This is one of the most common parameters to modify in water treatment practice. Soda ash and sodium bicarbonate are generally used for small water systems due to ease and safety in handling. Soda ash was added in Gulkana, Alaska following reverse osmosis treatment. For waters low in alkalinity and hardness, such adjustment may be insufficient to inhibit corrosion through calcium carbonate film deposition.

   For waters high in carbon dioxide, removal through air stripping has been used. Removal results in increased pH. Air stripping of dissolved gas was used in Kotzebue and Galena, Alaska as a pretreatment process.

2. Inhibitors and sequestering chemicals – The addition of chemicals to form protective barriers that

isolate the water from the piping materials can effectively reduce corrosion. Several common additives including various forms of phosphates and silicates have been used. Facilities to inject sequestering agents have been installed in several communities including St. Michael and Golovin, Alaska.

3. Passivation – Calcium carbonate addition through the use of limestone contactors has been used in a number of small communities in Alaska. Water is passed through a bed of crushed limestone contained in a pressure vessel, and dissolves calcium carbonate. Systems have been installed in Kwigillingok, Little Diomede, and Koliganak, Alaska.

4. Material replacent – Under many circumstances, it is possible to use piping materials that are unaffected by internal corrosion. In particular, house plumbing with polybutalene materials in lieu of copper pipe can be used. Such an approach was taken for the new water system in Kokhanok, Alaska in which all plastic pipe material was specified.

Several recent publications are available that describe theory, experimental procedures, and recommendations (AWWA, 1994; AWWA Research Foundation, 1985; 1990; Benjamin et al., 1992; Letterman, 1993; U.S. EPA, 1995). Due to the complexity and variability of water quality, bench-scale and field-scale evaluations are recommended.

5.5 References

Ashton, G.D. 1974. *Air Bubbler Systems to Suppress Ice*. U.S. Army Cold Regions Engineering and Research Laboratory Special Report 210, Hanover, New Hampshire.

AWWA. 1994. *Desktop Studies for Corrosion Control*. American Water Works Association, Denver, Co.

AWWA Research Foundation. 1985. *Internal Corrosion of Water Distribution Systems*. American Water Works Association, Denver, Co.

AWWA Research Foundation. 1990. *Chemistry of Corrosion Inhibitors in Potable Water*. American Water Works Association, Denver, Co.

Benjamin, L., et al. 1992. Pilot testing a limestone contactor in British Columbia. *Journal of the American Water Works Association*, May.

Boyd, W.L. 1959. Limnology of selected Arctic lakes in relation to water supply problems. *Ecology*, 40(1): 49-54.

Carey, K.L. 1983. *Melting Ice with Air Bubblers*. U.S. Army Cold Regions Engineering and Research Laboratory, Cold Regions Technical Digest, No. 83-1, Hanover, New Hampshire.

Driscoll, F. 1986. *Groundwater and Wells*. UOP Johnson Division, Saint Paul, Minnesota.

Farmwald, J. and J.A. Crum. 1986. Developing a community water system for Shishmaref, Alaska. *Proceedings, Cold Regions Engineering Specialty Conference*, American Society of Civil Engineers, New York.

Fedorov, N.F. and O.V. Zaborshchikov. 1979. *Manual for Designing Water Supply and Sewage Systems in Permafrost Soil Regions*. Stroyisdat, Leningrad, USSR. Translation No. 14012 by U.S. Army Cold Regions Research and Engineering Laboratory, Hanover, New Hampshire. 153 p.

Feulner, A.J. 1964. *Galleries and Their Use for Development of Shallow Ground Water Supplies with Special Reference to Alaska*. U.S. Geological Survey Water Supply Paper 1809-E, Washington, D.C. 16 p.

Foulds, D.M. 1974. Ice problems at water intakes. *Canadian Journal of Civil Engineering*, 1: 137-140.

Henshaw, G.H. 1887. Frazil ice: on its nature, and the prevention of its action in causing floods. *Transactions, Canadian Society of Civil Engineers*, 1: 1-23.

Lawson, D.E. 1983. *Erosion of Perennially Frozen Streambanks*. U.S. Army Cold Regions Research and Engineering Laboratory, Rep. 83-29, Hanover, New Hampshire. 22 p.

Letterman, R. 1993. Discussion: Pilot Testing a Limestone Contactor in British Columbia. *Journal of the American Water Works Association*, February.

Linell, K.A. 1973. Risk of uncontrolled flow from wells through permafrost. *Permafrost, Second International Conference*, National Academy of Sciences, Washington, D.C. 462-468.

Malcolm Pirnie, Inc. and HDR Engineering, Inc. 1990. *Guidance Manual for Compliance with the Filtration and Disinfection Requirements for Public Water Systems using Surface Wa-*

ter Sources. U.S. Environmental Protection Agency, Contract No. 68-01-6989.

McFadden, T. and C. Collins. 1978. Case study of water supply for coastal villages surrounded by salt water. *Proceedings, Applied Techniques for Cold Environments*, American Society of Civil Engineers, New York, 1029-1040.

Rauschenberger, K. 1983. Water consumption in Greenland. *Proceedings, Cold Regions Environmental Engineering Conference*, Department of Civil Engineering, University of Alaska, Fairbanks, 112-121.

Ryan, W.L. and R.D. Crissman. 1990. *Cold Regions Hydrology and Hydraulics.* American Society of Civil Engineeers, New York, New York. 823 p.

Shafer, R.V. 1981. *Water Sources for Industrial Activities on the Arctic Coastal Plane.* SOHIO Alaska Petroleum Company, Anchorage. 11 p.

Slaughter, C.W., M. Mellor, P.V. Sellmann, J. Brown, and L. Brown. 1975. *Accumulating Snow to Augment the Fresh Water Supply at Barrow, Alaska.* U.S. Army Cold Regions Engineering and Research Laboratory Special Report 217, Hanover, New Hampshire.

Smith, D.W., G.A. Smith, J.M. Brown, J.M., R.L. Schraeder, and L. Kosikowski. 1979. *Rapid Detection of Water Sources in Cold Regions, A Selected Bibliography of Potential Techniques.* U.S. Army Cold Regions Engineering and Research Laboratory, Cold Regions Technical Digest, No. 79-10, Hanover, New Hampshire. 75 p.

Stanley, S.J. and D.W. Smith, 1990. *Water Reservoir Ice Thickness Reduction Study.* Community Works and Capital Planning, Government of the Northwest Territory, Yellowknife, NWT.

Tabler, R. 1980. Geometry and density of snow drifts formed by snow fences. *Journal of Glaciology*, 26(94).

Tsang, G. 1982. *Frazil and Anchor Ice. A Monograph.* NRC Subcommittee on Hydraulics of Ice Covered Rivers, Ottawa, Ontario. 93 p.

US EPA. 1995. *Effect of pH, DIC, Orthophosphate and Sulfate on Drinking Water Cuprosolvency.* EPA/600/R-95/085.

Wahanik, R.J. 1978. Influence of ice formation in the design of intakes. *Proceedings, Applied Techniques for Cold Environments*, American Society of Civil Engineers, New York, 582-597.

Wheaton, S. 1980. *Ponds as Potable Water Sources at Shismaref, Alaska.* Alaska Area Native Health Service, Office of Environmental Health and Engineering, Anchorage, Alaska.

Zarling, J.P. 1978. Growth rates of ice. *Proceedings, Applied Techniques for Cold Environments*, American Society of Civil Engineers, New York, 100-111.

SECTION 6

WATER TREATMENT

3rd Edition Steering Committee Coordinator

Daniel W. Smith

3rd Edition Principal Authors

Steven J. Stanley

Daniel W. Smith

Section 6   Table of Contents

Section 6   List of Figures

Section 6   List of Tables

6   WATER TREATMENT

6.1   Introduction

The supply of a good, safe drinking water is essential for sustaining a good quality of life and protection of public health. Generally, treatment processes used in cold regions are the same as those used in more temperate climates. However, there are a few unique circumstances found in cold regions that must be considered in the design and operation of these treatment facilities. Many of these can be related to cold water and air temperatures. In addition, many facilities are located in small, isolated communities which presents additional challenges for the design and operation of the facilities.

The quality of treated drinking water is dependent on both the quality of raw water and the performance of the treatment process. In all cases, the best quality of raw water possible should be used to decrease the reliance on treatment processes. Once the source has been selected, it is also important that actions are taken to protect source water quality to

ensure the continued production of a good potable water. As outlined in Figure 6-1, the performance of a treatment facility is dependent on having both a plant which is capable of proper treatment and operation of the plant such that the capability of the plant is met. The capability of the plant is dependent on: the treatment processes used, their proper design, and the maintenance and administration of the facility to continue this capability. A significant challenge facing small facilities is to properly operate the facility in order to fully realize its capabilities.

Cold water and air temperatures present an additional challenge for water treatment in cold regions. Protective measures must be designed into the system to prevent failure and damage due to freezing. In addition, for long periods of the year raw water temperatures can be at or near 0°C. Cold water temperatures tend to reduce reaction kinetics for many treatment processes, reduce sedimentation veloci-

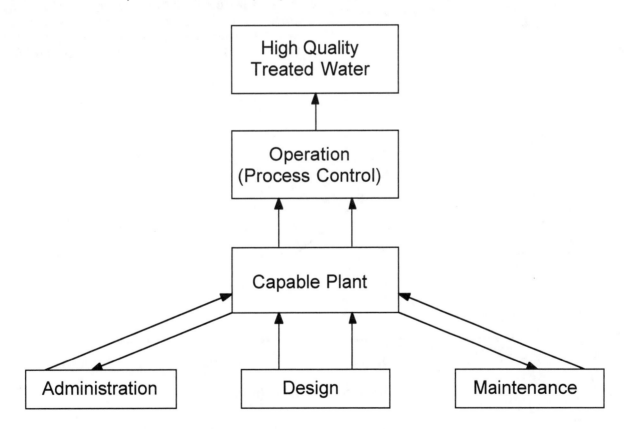

FIGURE 6-1   CAPABLE WATER TREATMENT PLANT MODEL

ties, as well as impact the performance of other treatment processes. Although consideration of these factors in the design and operation of treatment facilities in cold regions may not result in dramatic changes from that done in more temperate climates, the failure to consider them may result in a treatment system unable to meet the desired objectives.

The goal of water treatment is to produce a water which is "safe" to drink. The safety of drinking water is normally assessed by comparison with national or international quality standards and guidelines such as those established by the World Health Organization, and the governments of Canada and the United States. However, there may also be local or regional requirements which are more restrictive.

6.2 Water Quality Characteristics

Water quality parameters in cold regions vary widely depending on their source, location, time of season and as well as many other factors. Presented below is a general discussion of the physical, chemical and biological characteristics of water found in cold regions. It should be emphasized however, the first step in the design or assessment of any water treatment facility is a detailed, site-specific characterization of water quality.

6.2.1 Physical Characteristics. Cold climates impact a number of physical water quality characteristics. Of importance to drinking water treatment are water temperature, turbidity, color, and taste and odor. These parameters can affect the quality of drinking water, the efficiency of the treatment processes and aesthetics.

Water temperature is directly influenced by the natural environment. Variations in temperature affect the chemical, biological and physical characteristics of the water. Temperature affects the absolute viscosity of the water inversely, i.e., as water temperature decreases there is an increase in viscosity, as shown in Figure 6-2. Viscosity in turn affects the settling velocity of particles as well as energy requirements for mixing.

Hydrology of many northern watersheds can result in substantial changes in water quality over the year. During winter and ice cover conditions turbidity and color tend to be low and stable. With spring melt and break-up, both the turbidity and color can increase substantially. This is especially true for northern rivers. Increased sediment loads occur due to increased velocities in rivers, scour from ice during break-up, and the surface runoff from snowmelt and rain. Both Milburn and Prowse (1994) and Ferric and Weyrick (1994) have shown that sediment flux in-

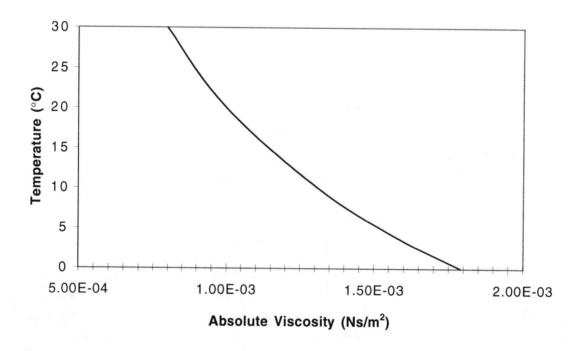

FIGURE 6-2   DYNAMIC VISCOSITY OF WATER AT ATMOSPHERIC PRESSURE

creases dramatically during the break-up event. The relationship between snowmelt and water quality, however, can be complex (Cheng et al., 1993). For example, in a study of suspended solids released during snowmelt on Gleen Creek, Alaska (Chacho, 1990), it was discovered that the major release of these solids occurred well after the initial snowmelt release.

Ice cover and cold water temperatures can also affect taste and odor of water. Ice cover on rivers limits the volatilization of many taste and odor compounds and as a result they may persist much longer than would be the case under open water conditions. A study by Kenefick et al. (1994) found that in the Athabasca River in northern Alberta, odors orginating from an industrial discharge were noticeable some 1,000 km downstream of the source under ice-covered conditions. In addition, for small lakes and raw water storage facilities, an ice cover can result in anoxic conditions which may also negatively impact taste and odor of the water.

6.2.2 Chemical Characteristics. Both inorganic and organic chemical characteristics of water can be influenced by conditions found in cold regions. Low temperatures reduce the rate of chemical reactions as well as solubility. This affects both the fate of chemicals in natural waters as well as treatment performance. In addition, as described earlier, ice covers on rivers and lakes can limit the loss of chemical constituents through volatilization as well as limit other chemical processes such as photolysis.

The natural hydrological cycle also affects the chemical characteristics of water. During winter, a greater portion of river flow is attributed to groundwater sources which can result in increased concentrations of ions (Cheng et al., 1993). However, the concentrations of these chemicals depend on the geology of the area and vary remarkably even in undisturbed aquatic systems. Ionic concentrations also increase with the onset of snowmelt (Molot et al., 1989).

Ice formation can also cause an increase in chemical concentrations; during the ice formation process most chemicals are excluded from crystal structure and are concentrated in the remaining unfrozen water. This is most significant in small lakes and raw water storage facilities where the ice formed during winter can occupy a significant portion of the total volume. A study by Stanley and Smith (1991) found that in the Northwest Territories a number of reservoirs lost up to 50% of the total reservoir vol-

ume to ice. They discussed methods available to reduce ice thickness on reservoirs to improve the storage capacity as well as reduce the concentration of contaminants in the stored water.

As in more temperate climates there are a number of inorganic and organic chemicals, often associated with human activity, that are of concern in drinking water. Monitoring of these chemicals in both the raw and treated water supplies is required to ensure concentrations are within prescribed limits. However, a few chemicals are of particular concern in cold regions. Various metals have been found in elevated concentrations in many northern raw waters. Iron and manganese, although normally not at concentrations high enough to cause health concerns, often cause aesthetic problems. Due to sewage treatment and disposal practices in many small cold-region communities, nitrite and nitrate concentrations may exceed recommend levels for the protection of public health. High nitrate and nitrite concentrations can result from poor placement or performance of septic fields, animal rearing facilities, natural organic deposits and leakage from lagoon systems. It should be noted that movement of contaminants such as nitrite and nitrate in the soil is influenced by frozen ground and they may not behave as expected in unfrozen ground conditions.

The high organic content of soil in many permafrost locations can result in high color and organic content in raw water. Although these organics, normally made up of fulvic and humic acids, pose more of an aesthetic than a health concern, upon chlorination they can form chlorinated organics byproducts which do pose a health concern. Because of these chlorinated byproducts more stringent guidelines have been developed. In order to meet these guidelines, areas that have high natural organic concentrations in their raw water source are required to reduce organics prior to chlorine addition. It should be stated however, that although the formation of disinfection byproducts should be minimized, the risks from these are small compared with the risks associated with inadequate disinfection. It is important that disinfection should not be compromised in attempting to control such byproducts (WHO, 1993)

6.2.3 Biological Characteristics. As stated by the World Health Organization, "Infectious diseases caused by pathogenic bacteria, viruses and protozoa or by parasites are the most common and widespread health risk associated with drinking water." Although many of the concerns with microbial contaminants are similar to those in temperate climates, studies have found that microorganisms may persist longer in cold regions. These studies (Gordon, 1972; Davenport et al., 1976; Putz et al., 1984; Stanley et al., 1992) have found that the combination of cold temperatures and ice cover results in increased microorganism survival. The cold water retards respiration and predation, while the ice sheet eliminates the lethal effects of sunlight (Putz et al., 1984).

A number of studies have found that disease rates associated with waterborne microbial contaminants tend to be higher in some northern communities (Hrudey and Raniga, 1981; Heinke, 1984; Brocklehurst et al., 1985; Robinson and Heinke, 1990). Brocklehurst et al. found that the level of the servicing of the water supply system could be related to community health. Robinson and Heinke found that risk of diarrhea was related to low water consumption.

As is the case in most other locations including cold regions, one of the major challenges in water treatment is associated with the inactivation of *Giardia* and *Cryptosporidium* in the treatment system. Both of these protozoan cysts have been found to be resistant to chlorine disinfection (AWWA, 1990). In addition, these pathogenic microorganisms are not well represented by traditional microbial indicators such as total and fecal coliforms (WHO, 1993). New strategies are evolving to control risks due to these waterborne pathogens. The Surface Water Treatment Rule as set out by the US EPA requires that all surface waters be filtered, has introduced the concept of CT (concentration of disinfectant x contact time) and reduced turbidity standards (AWWA, 1989). The importance of turbidity as a parameter to indicate the microbial quality of drinking water is evident by the US EPA using turbidity to justify pathogen removal credits in their most recent standards (Letterman, 1994). In these standards maximum credits are earned with turbidity of $\leq$ 0.5 NTU 95% of the time.

6.3 Process Design

Cold water temperatures, unique water quality characteristics and the often remote and isolated location of water treatment facilities results in the need for additional considerations in process design and the assessment of water treatment processes in cold regions. These special considerations will be discussed below. However, the well established principles of water treatment common to all locations should also be employed.

6.3.1 Heat Addition.

Nearly all cold region water systems require heat addition. If possible this should be done prior to other water treatment efforts. Cold water temperature affects treatment processes in numerous ways, such as:

- reduce settling and filtration rates;
- reduce chemical reaction rates;
- reduce efficiency of disinfectants;
- increase mixing requirements; and
- increase energy for pumping.

The two major options in the design of treatment facilities treating cold waters are to account for the impact that cold temperatures have on treatment processes and design accordingly, or add sufficient heat to the cold water such that these effects are mitigated. Heating the water will allow the use of standard design in areas where cold water temperatures are found. This is especially important when package-type treatment systems are used, as most are designed based on water temperatures found in more temperate climates. Raw-water temperatures must be raised to meet design specifications for the treatment unit.

There are a number of methods available for heating water supplies and all should function satisfactorily if well designed. For each of the methods it is important to consider the effect raising the water temperature may have on water quality parameters. Of most importance is its effect on the solubility of dissolved gases (decreases with increasing temperature), and on the solubility of dissolved solids (may cause precipitation). Possible impacts of these changes will be discussed below.

Generally, heat made be added to the raw water by a direct-fired boiler or by blending. A direct-fired boiler uses an oil, gas or coal furnace. Three basic types of heat exchangers are used:

- water tube;
- fire tube; and
- cast-iron water jacket.

The boiler must be operated in a manner that prevents the cooling of exhaust gases to below the dew point. This helps to prevent pitting-type corrosion in the boiler.

Often water is used as the hot fluid in the system and the furnace is operated to maintain water temperatures below boiling. Antifreeze may be used in the hot fluid, however as some antifreezes are toxic caution must be exercised to prevent the possibility of leaks and cross-connections.

Water may also be heated by blending. Occasionally a source of hot water is available that can be blended with cold water to achieve the desired temperature. Fairbanks Alaska, successfully used this system to warm cold well water. In Whitehorse, Yukon, geothermal water is added to raw water and provides substantial savings in the cost of water heating. Other sources of heat may be water from a central heating system or the cooling water from an engine. It is important that the quality of water used in blending be investigated to ensure it is of acceptable quality.

6.3.2 Corrosion Potential. In all water systems it is important that the treated water is not excessively corrosive for both protecting public health and maintaining the performance and service life of the system. Corrosion can lead to elevated concentrations of metal compounds in drinking water; some have serious health concerns while others cause aesthetic problems (Kirmeyer and Logsdon, 1983). Internal corrosion also results in high costs due to the replacement and repair of prematurely deteriorated pipes. Reduction in hydraulic capacity of pipes due to corrosion will result in increased pumping costs. Microbial agents often associated with corrosion can also pose health and aesthetic concerns. Due to the water characteristics found in many locations in cold regions and the high costs associated with building and maintaining the systems, protection against corrosion may be more significant than in more temperate climates. In water supply systems in cold regions five causes of corrosion are of primary importance:

- dissolved corrosive gases;
- solubility;
- biological agents;
- electrochemical cells; and
- erosion.

It is important to note that many of the processes used to treat the water can increase the corrosivity of water. For example, the addition of heat, chlorine, and hydroflorosilicic acid was found by Facey and Smith (1995) to increase the corrosivity of a soft, poorly buffered and slightly acidic water found in northern Canada by between 40% and 65%.

Dissolved Corrosive Gases. Dissolved oxygen (DO) and carbon dioxide (CO_2) are the principal corrosive gases of concern in water supplies. Both can be responsible for pitting-type corrosion.

The solubility of oxygen in water varies inversely with temperature. As water which has DO concentrations at or near saturation is warmed, oxygen will come out of solution. This molecular oxygen can then react with iron and other metals causing pipes, pumps and tanks to corrode. Consideration of this phenomenon is important if the water is to be heated. The system must be designed such that oxygen is released at a controlled location.

Solubility. The solubility of many materials is dependent on pH. Changes in pH results in the concentration of dissolved materials either greater than solubility limits causing precipitation or less than solubility limits, increasing the potential for the dissolution of solid materials the water comes in contact with. A study of water quality in the NWT by Facey and Smith (1993) found that water in many locations in the NWT was very soft, poorly buffered and slightly acidic. These types of waters would be especially prone to wide fluctuations in pH. Care must be exercised to ensure that the addition of treatment chemicals does not change the pH dramatically. Corrosion by soft, low alkalinity water becomes significant through the use of acidic water treatment chemicals and by the acidification of surface waters by acid precipitation (AWWA, 1985). It also should be noted that in some locations in the NWT it was found that the water was very hard, highly buffered and slightly alkaline. This highlights the need to perform detailed water quality analyses prior to making decisions on the water treatment system.

The corrosivity of water in terms of solubility is mostly assessed based on corrosion indices. Corrosion indices used are divided into two classes: calcium carbonate saturation indices, and indices based on other solution properties. The Langelier Saturation Index (LSI), the Aggressive Index (AI) and the Caldwell-Lawerence Diagrams are calcium carbonate based indices. The indices all predict whether water will precipitate or dissolve calcium carbonate. Waters which are supersaturated with respect to calcium carbonate tend to be noncorrosive; this is attributed to the deposition of a protective layer of calcium carbonate on the inside of the pipe. However, these indices do not provide a direct measurement of the corrosivity and should only be used to provide a general indication of corrosion potential (AWWA, 1985).

The Larson Ratio is a corrosion index that is based on water quality parameters other than calcium carbonate. This index is based on the relative corrosive behavior of chloride and sulfate ions and the protective properties of bicarbonate.

The indices above do not properly consider the temperature of water. Schock (1984) investigated temperature and ionic strength corrections for the Langelier Index.

Biological Agents. There are a number of naturally occurring bacteria which are capable of mediating corrosion in water treatment distribution systems. These bacteria are referred to as either corrosion-causing or corrosion-intensifying bacteria. Corrosion-causing microorganisms are those bacteria whose activity directly causes corrosion. Corrosion-intensifying microorganisms are those bacteria whose activity contributes to corrosion. Examples of corrosion-causing microorganisms include heterotrophic iron-, sulfate-, sulfite- and thiosulfate-reducing bacteria. Examples of corrosion-intensifying microorganisms include sulfur- and iron-oxidizing bacteria. Together these bacteria are known to cause or intensify corrosion by the following mechanisms (Facey and Smith, 1993):

- the production of a metabolic byproduct corrosive to metals (e.g., H_2S);

- the creation of biofilms that favor corrosion; and

- the presence of growths or deposits on the metal surface encourages the formation of differential aeration or concentration cells, causing localized corrosion.

The biofilm formed in pipes may also harbor microorganisms of concern to public health. A study of a northern Canadian distribution system found that treated water from the system met all regulatory requirements for total coliforms however, tubercles removed from the system contained up to 5×10^4 total coliforms per gram of tubercle (Emde et al., 1991).

Problems related to microbial corrosion can be greatly reduced by proper disinfection of water prior to distribution, maintenance of a disinfectant residual through the system and incorporation of a maintenance program to remove biofilms from the pipe.

Electrochemical Cells. Whenever there is an electropotential between two locations due to different metals or environmental conditions, and there is a means of electron transfer, corrosion can occur. This type of corrosion can occur naturally or can be induced by grounding.

The grounding of electrical systems to water distribution piping has been practiced widely in the North. Part of the reason is that the large metal pipe system provides a good conductor to the earth. Grounding of systems in permafrost areas is very difficult due to the lower electrical conductivity of the frozen ground. Although grounding to the distribution system is economical, it is discouraged due to the possible acceleration of corrosion and for safety concerns.

Corrosion by electrochemical cells can be prevented by cathodic protection techniques.

Erosion Corrosion. Erosion corrosion, also called impingement attack, is the result of excessive flow velocities (>1.2 m/s). At one time it was thought to be purely mechanical in nature. However, it has been found that high velocities disrupt the formation of protective films, which allows other types of corrosion such as electrochemical attack, to occur at a higher rate.

Impingement attack is characterized by rough surfaces and horseshoe or U-shaped pits. This type of corrosion occurs most severely in areas of high turbulent energy such as downstream of fittings and in recirculating systems. Control is accomplished by limiting pipe velocities.

6.3.3 Coagulation and Flocculation. The coagulation and flocculation process is used in water treatment systems to change particle characteristics to improve removal in the subsequent separation processes. Most commonly the separation process used is sedimentation; the goal of coagulation and flocculation is to form larger, more settleable flocs. However, sometimes other separation processes such as direct filtration or membrane processes may be used, and the goal of coagulation and flocculation may be slightly different.

Coagulation is a chemical process involving the destabilization of colloids. Once these particles are destabilized they can be flocculated, which involves mixing to promote particle-particle contact to form larger more settleable agglomerates. Although the primary goal of the coagulation and flocculation process is the removal of particulate matter, it is also effective at the removal of color and natural organic matter (NOM). The importance of coagulation and flocculation in removal of organic contaminants is also evident in the US EPA's (1994) decision to select enhanced coagulation as the "best available technology" for the meeting of new disinfection byproduct regulations. Enhanced coagulation involves the optimization of the process for the removal of organic material which act as precursors to disinfection byproducts (Cheng et al., 1995).

It has been suggested that water temperature changes, from 20˚C to near freezing, could affect the flocculation process in the following areas (Hanson and Cleasby, 1990):

- abrupt changes in the physical chemistry of water as the solid/liquid phase boundary is approached;

- changes in the repulsive and attractive particle forces which may affect particle-particle interactions;

- changes in the surface chemistry of clay;

- changes in the structure of the turbulent flow field due to viscosity changes; and

- changes in the chemistry of the system, both the water and the coagulant.

Of these factors, Hanson and Cleasby (1990) found that effects on mixing energy due to changes in viscosity and changes in system chemistry impact flocculation kinetics the most. Other factors listed above could not explain difference in performance at different temperatures. The overall conclusion of their study was that system chemistry is more important on flocculation kinetics at different water temperatures than the choice of energy input. With appropriate system chemistry, i.e., using ferric sulfate or cationic polymer as the primary coagulant, the effect of cold water temperatures on flocculation kinetics was eliminated.

In other studies of the effect of temperature on coagulation and flocculation, Mohtadi and Rao (1973) found that lower water temperatures necessitated increased coagulant (alum) to achieve the same degree of flocculation. They suggested that discrepancies in published information was related to failures in keeping the pH at optimum. The optimum pH varies inversely with temperature.

Morris and Knocke (1984) reported that with either alum coagulation or ferric chloride coagulation, the flocs at 1˚C are generally smaller than those formed at 20˚C (Figure 6-3 (a) and (b)). They also found that low temperature has a significant effect on turbidity removal for both coagulants, as shown in Figure 6-3 (c) and (d), although the effect was not as severe for ferric chloride as for aluminum sulfate.

Hanson and Cleasby (1990) and Morris and Knocke (1984) found that iron-based coagulants and polymers appear to perform better in cold water conditions.

6.3.4 Mixing. Mixing is an important function in water treatment. It is required for distributing chemicals, for flocculation, and for dissolving solids. Mixing is strongly dependent on temperature because of changes in the viscosity of the liquid. Figure 6-4 can be used to make the necessary adjustments in design criteria for temperature-induced viscosity changes. It is plotted with 20˚C as the base level. The power input for mechanical flocculation is directly dependent on fluid viscosity, as defined by:

$$P = G^2 V \mu$$

where

P = power input, W

G = root mean square velocity gradient, s^{-1}

V = flocculator tank volume, m^3

μ = dynamic fluid viscosity, Pa•s

To maintain the same velocity gradient in the tank as the liquid temperature decreases, the 20˚C power requirement has to be adjusted by the multiplier derived from Figure 6-4. This relationship is valid for any type of mechanical mixing where power is directly related to viscosity.

Mixing requirements are usually selected according to the product of the root mean square velocity gradient and mixing time (Gt). It is influenced by the time required for desired reactions to occur and is often arbitrarily based on the successful performance of similar units. Recommended detention times for flocculation in water range from 15 to 30 minutes. Increasing this detention time compensates for lower water temperatures, if the power input is not changed. The multipliers from Figure 6-4 can also be used for this purpose. Multiple basins in series is the most effective way to increase detention time, provided some basins can be bypassed during periods when the water is warmer.

One alternative to extended flocculation time is the use of higher chemical doses. Another is to adjust pH to the optimum for the temperature of the water being treated. Optimum pH varies inversely with water temperature. It is advisable to evaluate each alternative, since one may be more economical than another.

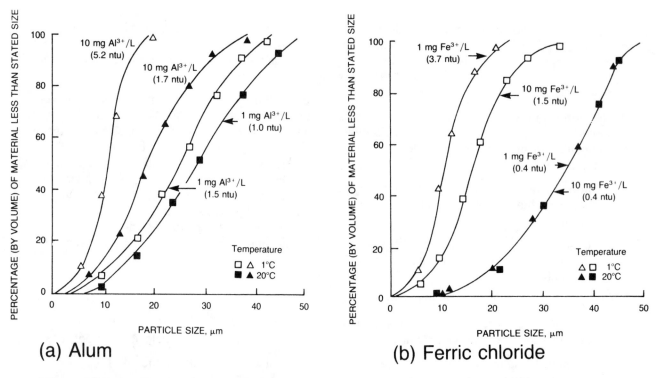

Effect of Solution Temperature and Dosage on Particle Size Distribution and Residual Turbidity.

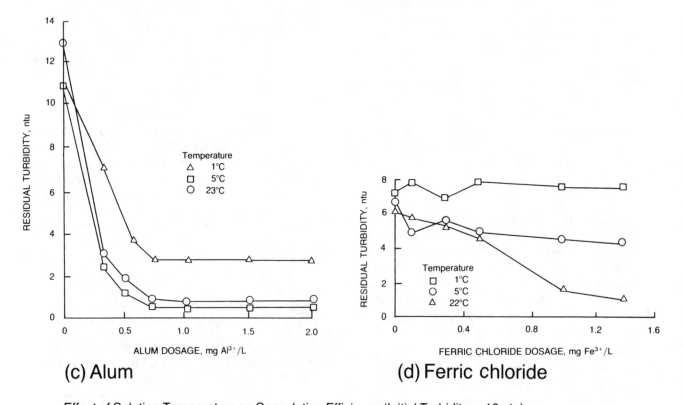

Effect of Solution Temperature on Coagulation Efficiency (Initial Turbidity = 10 ntu)

FIGURE 6-3   EFFECTS OF TEMPERATURE ON COAGULATION

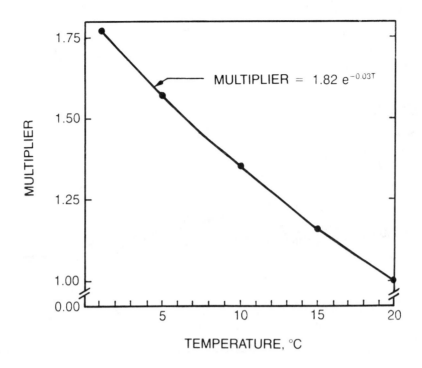

FIGURE 6-4   CORRECTION FACTOR FOR THE EFFECT OF
TEMPERATURE ON VISCOSITY-DOMINATED
PROCESSES

6.3.5 Sedimentation and Upflow Clarifiers.
The influence of temperature on the settling rate can be described for Type 1 (discrete, nonflocculating particles in a dilute solution), and Type 2 (discrete, flocculating particles in a dilute solution). Type 1 particles act according to Stokes' Law and changes in settling velocity can be explained by changes in viscosity. With Type 2 sedimentation, adsorption and physical attachment factors vary with temperature.

The effect of temperature on Type 3 (zone-settling) is less quantifiable. Figure 6-5 provides a correction factor for zone settling detention times at various temperatures (Reed and Murphy, 1969).

Upflow and sludge blanket clarifiers may not be as sensitive to low temperature as conventional clarifiers. However, temperature variations cause density differences and thermal currents, which may reduce efficiency. Sludge blanket and upflow clarifiers should be operated at nearly constant temperatures.

6.3.6 Filtration.
Filtration is affected by low water temperature to the extent that head losses through the filter are directly proportional to viscosity. The relative head loss changes 2.5 to 3.5 percent for each degree of temperature change. Rapid sand filter hydraulic loading rates vary widely and are largely dependent on the quality of influent water to the filter and filter design. Single media filter load rates vary from 4 to 13 m/h while multimedia filters have been reported to operate from 4.8 to 25 m/h. As a result multimedia filter beds often provide more efficient use of space in cold-region facilities. The multiplier values from Figure 6-4 should be used to adjust filtration head loss. For example, if the initial design head loss is 1 m at 20°C, it is about 1.5 m at 5°C.

Depending on the levels of particulate matter in the untreated water, filters may be used either following sedimentation, if high levels of particulate matter are found in the raw water, or without sedimentation when the raw water has low levels of particulate matter. The latter type of filtration is termed direct filtration. For both types of filtration it is important that the coagulation and flocculation process precedes the filtration step so that the process destabilizes colloidal particles and improves their removal characteristics within the filter.

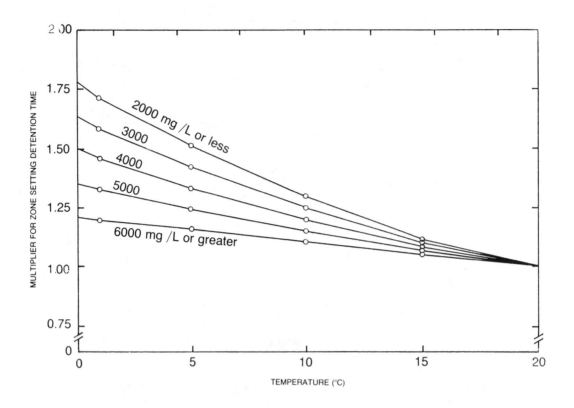

FIGURE 6-5   ZONE SETTLING DETENTION TIME VERSUS TEMPERATURE

Backwashing of filters is also affected. Power for pumping varies with viscosity changes as shown in Figure 6-4. Adjustments for filtration and backwashing are based on viscosity changes. However, the minimum upflow velocities or wash rates to fluidize and clean filter media are reduced because of increased fluid density. For example, if it takes a velocity of 3.24 m/h to fluidize a sand bed at 20°C, it requires only 2.16 m/h at 5°C.

Work by Valencia and Cleasby (1979) found that the efficiency of the backwashing process is more closely related to the velocity gradient (G) than the backwash velocity. A summary of their analysis is shown in Figure 6-6 for sand 0.5 mm in diameter. The G value increases with increased water temperature at the rate of 5.5/(s•°C). This was calculated assuming the same approach velocity. This suggests that during the winter when G is smaller, the expansion is greater and the interstitial velocity is less, resulting in less scouring of the sand grains. It may be possible to obtain the same degree of cleaning with longer backwash periods at the cost of using more product water.

Successful direct multimedia (coal, sand, and garnet) filtration of water high in glacial silt has been reported (Ross et al., 1982), although more development work has been suggested.

If water is heated in the treatment system, dissolved gases coming out of solution can result in air binding of the filters. This problem is made worse if negative pressures occur in the filter due to excessive headlosses.

6.3.7 Adsorption. The adsorption process using either granular activated carbon or powdered activated carbon can be an effective method to remove organics from water. Granular activated carbon has the advantage that the carbon can be regenerated. However, for small facilities costs associated with this could be prohibitive. For facilities that experience periodic episodes of high organics, color, or taste and odor in their raw water, powdered activated carbon may be an effective choice due to the lower capital costs associated with it.

Adsorption is an exothermic process and in theory, lower water temperature should have a slight positive effect on the extent of adsorption (Environmental Health Directorate, 1993). However, studies have shown that adsorption capacity and rates are inhibited at low temperatures (Magsood and Benedek, 1974; 1977). This may be because of limitations in mass transport at colder temperatures.

6.3.8 Disinfection. Disinfection is the process of destroying or inactivating disease-causing organisms in the water. Traditionally, disinfection efficiency has been monitored by the reduction or elimination of indicator bacteria. With the recognition that indicator bacteria may not be representative of pathogenic organisms such as protozoan cysts (*Giardia*) and viruses, the CT concept has been developed to provide additional assessment of the disinfection process. The CT concept sets required CT values (concentration of disinfectant x contact time) for various disinfectants, required removals and water quality characteristics. As part of the Surface Water Treatment Rule (SWTR) the US EPA has set CT values for a number of disinfectants based on log reductions for *Giardia* and viruses (AWWA, 1989). The required log reduction for a treatment system is dependent on concentration of Giardia cysts in the raw water. Removal credits are given for conventional treatment with the remaining required reduction being met by disinfection. For pipelines the value used for contact time "T" in the SWTR is the theoretical detention time (volume of the pipe divided by the flow rate). For mixing basins, storage reservoirs, and other process units the SWTR de-

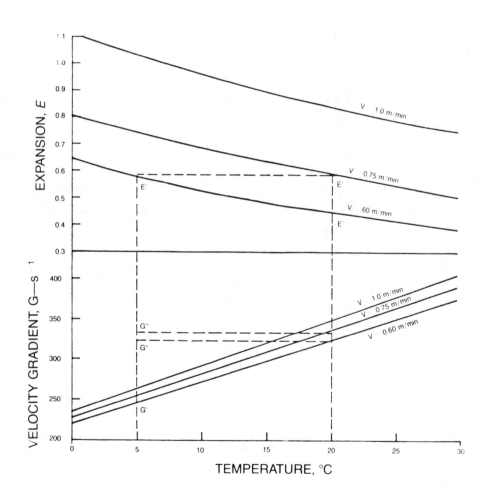

FIGURE 6-6   *VELOCITY GRADIENTS AND EXPANSION FOR DIFFERENT TEMPERATURES (Uniform, 0.5 mm diameter sand bed with initial porosity of 0.43) (Valencia and Cleasby, 1979)*

TABLE 6-1   *CT VALUES FOR DIFFERENT DISINFECTANTS (pH 6 to 9 except where noted for free chlorine)*

CT Values for Inactivation of *Giardia* (mg/L•min)

Disinfectant	Inactivation	Temperature		
		≤ 1°C	10°C	20°C
Ozone	0.5 log	0.48	0.23	0.12
	1 log	0.97	0.48	0.24
	2 log	1.9	0.95	0.48
	3 log	2.9	1.43	0.72
Chlorine dioxide	0.5 log	10	4	2.5
	1 log	21	7.7	5
	2 log	42	15	10
	3 log	63	23	15
Chloramine	0.5 log	635	310	185
	1 log	1270	615	370
	2 log	2535	1230	735
	3 log	3800	1850	1100
Free chlorine (pH = 7.0) (C = 2 mg/L)	0.5 log	39	21	10
	1 log	79	41	21
	2 log	157	83	41
	3 log	236	124	62

CT Values for Inactivation of Viruses (mg/L•min)

Disinfectant	Inactivation	Temperature		
		≤ 1°C	10°C	20°C
Ozone	2 log	0.9	0.5	0.25
	3 log	1.4	0.8	0.4
	4 log	1.8	1.0	0.5
Chlorine dioxide	2 log	8.4	4.2	2.1
	3 log	25.6	12.8	6.4
	4 log	50.1	25.1	12.5
Chloramine	2 log	1243	643	321
	3 log	2063	1067	534
	4 log	2883	1491	746
Free chlorine (pH = 6 to 9)	2 log	6	3	1
	3 log	9	4	2
	4 log	12	6	3

fines the contact time as the T10 (time for 10% of the water to pass through the system) rather than the theoretical detention time. The T10 must normally be determined by tracer techniques. Table 6-1 gives a few select CT values for illustrative purposes. The greater the CT value the less effective the disinfectant. It is important to note in Table 6-1 the effect water temperature has on the CT value. Table 6.1 can also be used to compare the effectiveness of the various disinfectants.

Chlorine. Chlorine is widely used as a disinfectant for potable waters. The greatest advantages of chlorine are:

- the ability to maintain and monitor residuals;

- low chemical and reactor costs; and

- relatively low dosage requirements.

Also, the availability of chlorine as Cl_2 gas, calcium hypochlorite $Ca(OCl)_2$, and sodium hypochlorite (NaOCl) enables safe handling of the disinfectant under various conditions. Logistics and operator qualifications should guide the designer in selecting the chlorine source.

Solubility of chlorine in cold water is reduced. However, chlorine is never added in quantities which approach solubility limits.

As discussed, in recent years there has been concern over the products formed by using chlorine as a disinfectant in waters containing organic compounds (Peters and Perry, 1981). Therefore prechlorination should not be used in the production of potable water. Furthermore, the precursors to the formation of trihalomethanes should be removed before chlorination, if trihalomethane concentrations in the product are likely to exceed recommended limits. The US EPA (1994) recommends enhanced coagulation as the best available technology for this task.

As indicated in Table 6-1, chlorine disinfection is hindered by cold water. At the same chlorine residual concentration the required contact time has to increase almost four times to get the same log reduction in *Giardia*.

If ammonia is present in the water, free chlorine will react with it to produce chloramines. As indicated in the Table 6-1 the effectiveness of chloramines against *Giardia* and viruses is substantially less than free chlorine. In some instances it may be desirable to form chloramines by the addition of ammonia. Although chloramines are less effective than chlo-

rine they have the advantage of being able to better retain a residual in distribution systems.

Ozone. As indicated in Table 6-1, ozone is the strongest disinfectant used in water treatment. Ozone also has other possible uses due to its strong oxidation properties. Ozone is only slightly temperature sensitive and no significant modification of standard techniques is required other than a slight increase in contact time.

There is little question of the effectiveness of ozone as a disinfectant. However, it is generally expensive to install and operate, although costs are decreasing relative to alternative methods. Where costs are not as important or are offset by logistic considerations, the use of ozone may provide an advantage, as it is generated from air using electricity. Its use is unaffected by problems of resupply in remote areas, but it does require a reliable electricity supply.

A major drawback remains that no residual remains in the treated water to protect against post contamination. As a result chlorine addition is required for distribution system protection.

Chlorine Dioxide. Chlorine dioxide (ClO_2) has been used successfully as a disinfectant and oxidant. But on-site generation requirements, chemical transportation problems, and process complexity all lead to the conclusion that it is unlikely that its use will be adapted to remote locations. In addition there is some concern with the production of chlorite and chlorate which may pose a health concern.

6.3.9 Fluoridation. Fluoridation of water supplies to an optimum concentration of about 1.2 to 1.4 mg/L in northern communities is considered to be an effective practice to improve dental health. The recommended concentration in cold regions is higher than normal due to the lower per capita consumption of drinking water. However, care must be taken not to overdose fluoride, as doses greater than 4 mg/L may produce skeletal fluorosis and greater than 2 mg/L may produce dental fluorosis (AWWA, 1990). The method of addition is governed largely by the sophistication of the water supply system. In communities with a conventional water treatment plant, addition can be implemented using conventional technologies. In communities that do not have a conventional treatment plant nor a year-round raw water storage, it is recommended the fluoridation be done in the holding tank. It is not recommended that fluoride be added during the filling of the water trucks as a reliable dosage within concentration limits required cannot be achieved (Marianayagam, 1986). In communities with year-round reservoirs it has been found that the safest and most effective means of fluoridating these water supplies is to add the fluoride to the reservoir (through the influent line) during filling (Marianayagam, 1986). However, there is a concern with freeze concentrations, as ice can take up over 50% of the reservoir volume (Stanley and Smith, 1991). To prevent high fluoride levels at the end of winter in these reservoirs the volume of ice must be considered in determining the amount of fluoride to add.

6.3.10 Water Softening. Water softening may be required in cold-region communities where the water supply contains very high concentrations of calcium, magnesium, or other polyvalent cations. Water softening is normally done for aesthetic reasons. However, excessively hard water can result in the precipitation of solids in the system, which can limit its performance and service life.

Two methods of softening are well developed: ion exchange and chemical precipitation.

Ion Exchange. Low water temperatures influence the rate of flow through the exchange media because of higher viscosities. The ion exchange rate itself is rapid enough to not limit softening at low temperatures. Note that most ion exchangers are replacing the polyvalent cations with monovalent sodium ions. High sodium concentrations may not be desirable.

Water from some subpermafrost aquifers, lakes which are anoxic in the lower levels, and other oxygen-free sources may contain relatively high concentrations of soluble forms of iron. If they are oxidized and precipitate in the ion exchange resin, they can reduce the exchange capacity or "foul" the resin. Therefore, if possible, iron should be removed before ion-exchange treatment. At low iron concentrations, the iron can be complexed with sodium hexametaphosphate.

Another method of controlling soluble ferrous ions is to prevent any exposure to oxygen, even during regeneration and backwash. For example, small amounts of sodium sulphite (Na_2SO_3) may have to be added to the solution to remove any oxygen present.

Chemical Precipitation. Chemical precipitation can be used where:

- the water to be treated is turbid;

- the volumes are large enough to make the process more economical;

- the addition of sodium ions is not desired or acceptable; or

- other chemical softening benefits are expected.

This process is affected by low water temperatures since it involves mixing, chemical reactions, flocculation, sedimentation, filtration, and sludge-handling. The softening process can be applied for carbonate hardness only (using lime) or both carbonate and noncarbonate hardness (using lime-soda ash). Chemical dosages should be established by jar testing.

In Minneapolis (Cold, 1976), a series of tests softening cold water showed that at 1°C, hardness removal was slower, and the distribution of resulting particles shifted to larger sizes (Figure 6-7).

6.3.11 Desalination. In coastal areas of the Arctic and in some areas where groundwater sources are highly mineralized, desalination may be required. Several desalination methods including distillation, reverse osmosis (RO), and freeze treatment are available.

Distillation. Distillation is the best known and most highly developed means of removing dissolved materials from water. Cold water increases the operational costs of distillation slightly. The relatively high skill requirement for operators makes this an undesirable process in remote areas. Small stills (about 5.6 L/s) require about 1 L of diesel fuel for 175 L of distilled water.

Reverse Osmosis. Reverse osmosis (RO) uses mechanical energy to drive water through a semipermeable membrane. RO effectively removes most organic, bacterial, and particulate matter from water (AWWA, 1990). RO is by far, the most used of the membrane processes. In 1993 it was reported that greater than 190 membrane plants had been constructed in North America with 80% being RO type facilities (Morin, 1994). By proper membrane selection, potable water may be produced. Morin (1994) also reported that feedwaters with TDS concentrations up to 6,500 mg/L are being treated to drinking water standards with these units. RO is temperature-sensitive; the best results are obtained when water temperatures are in the range of 20°C to 30°C.

The cost of RO is relatively high due to equipment operation and maintenance requirements. Small RO

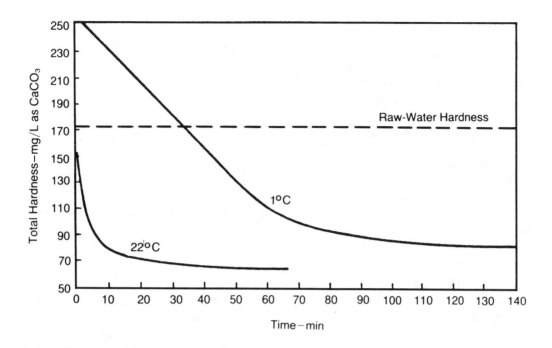

FIGURE 6-7   LIME SOFTENING RATE AT 1°C AND 22°C (Cold, 1976)

units are available in sizes from 3.5 m³/d to 3,500 m³/d. Units have been built treating up to 50 ML/d. Morin (1994) found RO requires about 2.0 kWh/m³ of water treated. The ratio of costs to benefit for RO desalination is lowered by the following factors:

- higher water temperatures;

- industrial/commercial water users; and

- sewers for brine disposal.

The ratio of costs to benefits is increased by the following factors:

- high land costs;

- high electrical power costs;

- high population density per dwelling unit; and

- high interest costs.

RO treatment systems must be protected from freezing at all times. Membranes that have been frozen are unreliable even if freezing has occurred during shipping prior to installation.

Freezing. Treatment by freezing is a system that may be practical in cold regions, where the cold is used as a resource. This method is based upon the fact that impurities are slowly "rejected" or "salted out" as water freezes, and the ice contains only pure water. It must be recognized however, that freezing does not kill all bacteria, viruses and protozoan cysts. Three types of freezing processes have been used successfully in pilot projects:

(a) The reservoir process involves freezing a large volume of water. When ice containing the desired volume of water has been formed, the brine below the ice is withdrawn. The ice, upon melting, is purified water (see Figure 5-3).

(b) Layer-freezing involves the freezing of brackish water in successive sheets. The first water melted in warm weather is discarded, because it contains most of the impurities.

(c) Spray-freezing involves spraying brackish water through a modified lawn sprinkler to form a cone of ice. Pure water is frozen more quickly, and brine drains away continuously throughout the winter. In a pilot-scale test in Saskatchewan, chloride content was reduced from 2,000 mg/L to 500 mg/L for 75 percent of the brackish water sprayed (Spyker and Husband, 1973). Estimated total costs for the process are about $0.22/m³ (Cdn) of purified water.

6.3.12 Membrane Processes. With the improvement of membrane processes, they are not only applicable for desalination as described above but are also being used to produce potable water from a variety of raw water sources. Processes being used include, microfiltration (MF), ultrafiltration (UF), nanofiltration (NF) and reverse osmosis (RO). A study by Wiesner et al. (1994) found that low pressure membrane filtration (only UF and NF were investigated) appears to be a cost-effective option for small facilities (18,900 m^3/d) that need to remove particles, DOC, or both from raw waters with low to moderate turbidities and concentrations of DOC. They found the cost of membrane filtration is largely a function of the permeate flux. It was also found that permeate flux varies as a function of raw-water quality as well as membrane type. Although membrane processes show promise in water treatment, more work is required to assess long-term operating costs, reliability and performance.

6.3.13 Ozonation. Ozone as a disinfectant was briefly covered in Section 6.3.8. However, experience with ozone in water treatment has shown numerous benefits in addition to disinfection.

Iron and manganese are effectively removed by oxidation with ozone, followed by separation. Applied early in the treatment chain, ozone rapidly oxidizes iron and manganese and aids in flocculation.

Oxidation of organics can be accomplished by ozone addition in the middle of the overall process. If followed by treatment with granular activated carbon, it is very effective in removing organics and produced none of the side effects on taste and odor associated with chlorination.

6.4 Specific Contaminant Removal

6.4.1 Iron Removal. Iron is of considerable concern in waters in cold regions. It may exist in any of nine valence states, but Fe^{II} and Fe^{III} are the most important in water. Not only elemental iron, but also organic iron complexes present problems (Gjessing, 1981).

Limiting iron in water supplies is more for the sake of visual appeal and economic concerns than for health reasons. Concentrations of less than 0.3 mg/L is a suggested objective, since higher concentrations can be detected visually and may stain clothes and plumbing fixtures.

High iron concentrations in cold-region waters may be explained as follows. Groundwaters and waters under the ice in ponds and shallow lakes can become deficient in dissolved oxygen. The dissolution of iron results in a highly soluble complex. In surface waters, where sufficient oxygen exists to oxidize the iron, much less soluble species are formed. The species formed are a function of pH.

Treatment of waters containing excessive ferrous ions requires oxidation, which may be accomplished by aeration or by the addition of chlorine, ozone or potassium permanganate. Simple aeration is not effective for removing organic iron complexes. Low temperatures restrict the aeration process. Although oxygen is more soluble at low temperatures, overall gas transfer and reaction rates are lower. The gas transfer rates depend also on the type of aeration system used.

Coarse bubble diffusers tend to be more maintenance-free than other types of aerators. Aeration can be accomplished by cascade or "waterfall" systems, but these systems are associated with humidity problems in the treatment plant.

Aeration tanks should have a width to depth ratio of 2 to 1 or greater to promote good mixing. Detention time is about 10 to 30 minutes, and the air volume range required is 0.05 to 1.25 m^3/m^3 of water treated.

6.4.2 Color Removal. Objectionable color is frequently found in water originating in tundra regions where organics are leached from decaying vegetative matter. Color may be reduced or removed by chemical coagulation using alum, ferric compounds, or aluminum ions (Vik, 1972). As mentioned, enhanced coagulation has been recommended by the US EPA for organic removal. It can also be removed by carbon adsorption, or by chemical oxidation with chlorine or ozone, or both. Testing of particular water samples is recommended.

6.4.3 Organics Removal. Organic materials in water, much like color, can be partly removed by coagulation and sedimentation. Removal by coagulation is optimized by careful control of the pH during the process. More complete removal requires carbon adsorption, ozone treatment, or both.

Activated carbon is effective for removal of organics, but when granular carbon is used, there is a strong potential for enhancing bacterial growth in the carbon bed. Organics removed from the water become food for microorganisms which may eventually be washed through the carbon bed and into the treated water. Disinfection after organics removal is required in such situations.

The use of both ozonation and carbon adsorption has been found extremely effective in organics removal and in treating iron-organic complexes. Ozone destroys microorganisms and breaks down many organics. The remaining organics are absorbed on the carbon beds.

6.5 Plant Design

There are several important aspects to water treatment plant design in cold regions. Nearly all functions of water treatment must be housed for protection as well as for ease of maintenance and operation of equipment. Certain process units require heated shelter, while for others an unheated shelter suffices. Generally, processes which include equipment such as pumps and exposed piping must be housed and heated to prevent damage from freezing. Water temperature, not air temperature, determines process efficiency.

6.5.1 Buildings.
Combining different functions under one roof rather than in a group of smaller buildings reduces surface area and heating requirements. Piping and electrical runs may be shorter and less expensive to install and possibly easier to maintain. Possible expansion in the future should be considered when selecting the single-building concept. The floor plan of cold-region systems can be critical to building efficiency. Designers should place areas which require stable heat in the interior of the building. Rooms for storage and other less vital functions can be allotted the space along the outer walls. In this manner, heat lost from the interior is used to heat other space before escaping outdoors. Building shape can be optimized to reduce surface area and oriented to take advantage of sunshine. Placement of some functions below ground and banking buildings with earth on outside walls reduces heat loss.

Vapor barriers are very important in cold climates and more so in buildings such as treatment plants where moist air is prevalent. Poor vapor barriers permit moisture to penetrate into wall and roof insulation and reduce its effectiveness.

Dehumidifiers may be required, particularly in areas where electrical or electronic equipment is located.

6.5.2 Ventilation.
Ventilation is important both for human health and comfort and for building economy. Moist air in indoor process areas must be expelled to prevent condensation and related problems. Air may be reused from one type of space to another to economize on the amount of warmup required. For example, air may be moved from office space to lab space before exhaust, or from office to process areas. If dehumidification equipment is installed, air from the process area can be reused in either office or lab area.

An alternative to this direct reuse of air is to extract heat from warm, moist exhaust air to preheat incoming fresh air. Several fairly efficient devices are available to perform this function.

Continuous ventilation may not be necessary in all spaces. To force-ventilate offices and labs when they are not occupied is unnecessary and wastes heat.

6.5.3 Lighting.
In cold northern climates, adequate lighting is important both inside and outside treatment plants to compensate for reduced winter daylight. Controls for lighting circuits should be designed so that minimum light is provided in unoccupied areas with supplementary lights available as needed. Unless relatively small lighting circuits are provided, appreciable power may be used for a function not actually needed.

6.5.4 Controls.
Process controls are sensitive devices which may be adversely affected by low temperature or high humidity. Plant design should provide controls on interior walls or in special cabinets to be protected from moisture and low temperature. Ventilation should be provided to reduce the possibility of damage by atmospheric changes.

6.5.5 Standby Equipment.
In cold regions, equipment is subjected to rigorous conditions. As a result, equipment failure may be more frequent. Providing redundant equipment may increase system reliability.

The designer should not place total reliance on one item, if that reliance could as easily be placed on two smaller items. For example, rather than one large pump, two or three smaller ones are generally more desirable. This also permits periodic maintenance without suspending operation entirely.

6.6 Miscellaneous Considerations

6.6.1 Drainage.
Whenever possible, plants should be designed and built so that they can be drained by gravity, if necessary. This is particularly true in remote areas and at smaller installations.

6.6.2 Auxiliary Power.
Auxiliary power supply should be provided to support minimum operation of the treatment plant and distribution system.

6.6.3 Space/Process Trade-Offs.
Under some circumstances the designer may wish to evaluate the overall benefits of energy-intensive processes, which are space efficient. For example, pressure filters or precoat filters, which require pumping, may

have overall advantages compared to standard or multimedia gravity filters. Or in another case a centrifuge may be preferable to a standard clarifier.

In cold regions these trade-offs must be evaluated carefully, because heating requirements are significant. Where plant expansion is contemplated, these considerations may be particularly valuable, because they can lead to additional capacity within existing space.

6.6.4 Replacement Parts. Plant management should maintain a stock of replacement parts for equipment subject to failure or wear. This is particularly important in remote areas where much time may be lost in shipment of parts from sources of supply.

6.7 Operation

As shown in Figure 6.1, for a plant to produce an acceptable quality of water, the plant must not only be designed so it is capable of meeting this goal but also operated in a manner such that this capability is met. A study by Smith et al. (1995) in Northern Alberta found that many small facilities have difficulty meeting set guidelines. Of communities serving less than 500 people approximately 30 to 45 percent of them did not meet microbiological requirements. In addition as many as 60 percent of them did not meet guidelines for THMs. Site visits to a number of these facilities found that much of the difficulty could be attributed to poor operation of the facility. Similar findings have also been reported in the United States (Goodrich et al., 1992). Difficulties result because many systems serving populations fewer than 500 are in trouble in terms of the financial and technical expertise needed to manage their drinking water problems (Prendergast, 1993).

In many of these small systems Prince et al. (1995) found that on-line monitoring of turbidity and chlorine residual would provide the additional information needed for the operator to make process decisions.

6.8 References

AWWA, 1985. *Internal Corrosion of Water Distribution Systems*. American Water Works Association Research Foundation, Denver, Colorado, 690 p.

AWWA. 1989. *Guidance Manual for Compliance with the Filtration and Disinfection Requirements for Public Water Systems using Surface Water Sources*. American Water Works Association, Denver, Colorado.

AWWA. 1990. *Water Quality and Treatment*. Fourth Edition, American Water Works Association. McGraw-Hill, Inc., Toronto, Ontario, 1194 p.

Brocklehurst, C., G.W. Heinke, and H. Hodes. 1985. *The Effect of Water Supply and Sanitation on Health on Indian Reserves in Manitoba*. University of Toronto, Department of Civil Engineering. 15 pp.

Chacho, E.F. 1990. Water and suspended solids discharge during snowmelt in a discontinuous permafrost basin. In: *Fifth Canadian Permafrost Conference*, Quebec City, Quebec, 167-173.

Cheng, H., S. Leppinen, and G. Whitley. 1993. Effects of river ice on chemical processes. In: *Environmental Aspects of River Ice*, T.D. Prowse and N.C. Gridley, (Eds.), Environment Canada, NHRI Science Report #5, 75-96.

Cheng, R.C., S.W. Krasner, J.F., Green, and K.L. Wattier. 1995. Enhanced coagulation: a preliminary evaluation. *Journal American Water Works Association*, 87(2): 91-103.

Cold, L.D. 1976. Surface water treatment in cold weather. *Journal American Water Works Association,* 68(1): 22-25.

Davenport, C.V., E.B. Sparrow, and R.C. Gordon. 1976. Fecal indicator bacteria persistence under natural conditions in an ice-covered river. *Applied and Environmental Microbiology*, 32(4): 527-536.

Emde, K.M.E., D.W. Smith, and R.M. Facey. 1991. Initial investigation of microbially-mediated corrosion in a low temperature water distribution system. *Water Research*, 26: 169-175.

Environmental Health Directorate. 1993. *Water Treatment Principles and Applications - A Manual for the Production of Drinking Water*. Department of National Health and Welfare, Canadian Water and Wastewater Association. 365 p.

Facey, R.,M. and D.W. Smith. 1993. *Northwest Territories Water Quality Study 1990 - 1993 Summary Report*. Prepared by Daniel W. Smith & Associates for Department of Municipal and Community Affairs, Government of Northwest Territories, Yellowknife, NWT. 72 p.

Facey, R.M. and D.W. Smith. 1995. Soft, low-temperature water-distribution corrosion: Yellowknife, NWT. *Journal of Cold Regions Engineering*, 9(1): 23-40.

Ferrick M.G. and P.B. Weyrick. 1994. On the sediment transport capacity of rivers during breakup. In: *Proceedings of the Workshop on Environmental Aspects of River Ice*, T.D. Prowse (Editor), National Hydrology Research Institute, Saskatoon, Saskatchewan, NHRI Symposium Series, No 12, 161-175.

Gjessing, E.T. 1981. Water Treatment Considerations - Aquatic Humus. In: *Design of Water and Wastewater Services for Cold Climate Communities*, Pergamon Press, Oxford, 95-101

Goodrich, A.G., J.Q. Adams, B.W. Lykins Jr., and R.M. Clark. 1992. Safe drinking water from small systems: treatment options. *Journal American Water Works Association*, 84(5): 49-55

Gordon, R.C. 1972. *Winter Survival of Fecal Indicator Bacteria in a Subarctic Alaskan River*. Report No. EPA-R2-72-013, U.S. Environmental Protection Agency, Washington, D.C.

Hanson, A.T. and J.L. Cleasby. 1990. The effects of temperature on turbulent flocculation: fluid dynamics and chemistry. *Journal American Water Works Association*, 82(11): 56-73.

Heinke, G.W. 1984. *Report on Public Health Effects of Municipal Service in the Northwest Territories*. Department of Local Government, Government of the Northwest Territories, Yellowknife, 45 p.

Hrudey, S.E. and S. Raniga. 1981. Greywater characteristics, health concerns and treatmen technology. In: *Design of Water and Wastewater Services for Cold Climate Communities*. Pergamon Press, Oxford, 137-154.

Kenefick, S., B. Brownlee, E. Hrudey, L. Gammie, L. and S. Hrudey, S. 1994. *Water Odor Athabasca River, February and March, 1993*. Northern River Basin Study Prject Report No. 42. Edmonton, Alberta.

Kirmeyer, G.J. and G.S. Logsdon. 1983. Principles of internal corrosion and corrosion monitoring. *Journal American Water Works Association*, 75: 1433-1443.

Letterman, R.D. 1994. What turbidity measurements can tell us. *Opflow*, 20: 8.

Magsood, R. and A. Benedek. 1977. Low-temperature organic removal and denitrification in activated carbon columns. *Journal Water Pollution Control Federation*, 49(10): 2107-2117.

Magsood, R. and R. Benedek. 1974. The feasibility of the physical-chemical treatment of sewage at low temperatures. In: *Symposium on Wastewater Treatment inCold Climates*, Environmental Protection Service Report No. EPS 3-WP-74-3, Ottawa, Onxtario, 523-548.

Marianayagam, C.C. 1986. *Water Fluoridation in the Northwest Territories*. M.A.Sc. Thesis. Department of Civil Engineering, University of Toronto.

Milburn, D. and T.D. Prowse. 1994. Observations of sediment-chemistry interactions during northern river break-up. In: *Proceedings of the Workshop on Environmental Aspects of River Ice*, T.D. Prowse (Editor), National Hydrology Research Institute, Saskatoon, Saskatchewan, NHRI Symposium Series, No 12, 21-41.

Mohtadi, J.F. and P.N. Rao. 1973. Effects of temperature on flocculation of aqueous dispersions. *Water Research*, 7: 747.

Molot, L.A., P.J. Dillon, and B.D. LaZerte. 1989. Factors affecting alkalinity concentrations of streamwater during snowmelt in central Ontario. *Canadian Journal of Fisheries and Aquatic Science*, 46: 1658-1666.

Morin, O.J. 1994. Membrane plants in North America. *Journal American Water Works Association*, 87(12): 42-54.

Morris, J.K. and W.R. Knocke. 1984. Temperature effects on the use of metal-ion coagulants for water treatment. *Journal American Water Works Association*, 76: 74-79.

Pendergast, J. 1993. Fit to drink. *Civil Engineering*, New York, 63(5): 52-55.

Peters, C.J. and R. Perry. 1981. The formation and control of trihalomethanes in water treatment processes. In: *Design of Water and Wastewater Services for Cold Climate Communities*. Pergamon Press, Oxford, 103-123.

Prince, D.S., S.J. Stanley, and D.W. Smith. 1995. Performance assessment and control in small and medium sized water treatment plants with on-line monitoring. In: *Proceedings of the Ca-*

nadian Water and Wastewater Conference, Edmonton, Alberta.

Putz, G., D.W. Smith, and R. Gerard. 1984. Microorganism survival in an ice-covered river. *Canadian Journal of Civil Engineering*, 11(2): 177-186.

Reed, S.C. and R.S. Murphy. 1969. Low temperature activated sludge settling. *Journal Sanitary Engineering Division, ASCE*, 95(SA4): 747-767.

Robinson, B.A. and G.W. Heinke. 1990. *The Effect of Municipal Services on Public Health in the Northwest Territories*. Department of Municipal and Community Affairs, Government of the Northwest Territories. 81 p.

Ross, M.D., R.A. Lowman, and R.S. Sletten. 1982. *Direct Filtration of Streamborne Glacial Silt*. U.S. Army Cold Regions Research and Engineering Laboratory Rep. No. 82-23, Hanover, New Hampshire, 17 p.

Schock, M.R. 1984. Temperature and ionic strength correction to the Langelier Index-Revisited. *Journal American Water Works Association*, 76: 72-76.

Smith, D.W., S.J. Stanley, and D.S. Prince. 1995. Drinking water quality of communities in northern Alberta. In: *Proceedings of the Canadian Water and Wastewater Conference*, Edmonton, Alberta.

Spyker, J.W. and W.H.W. Husband. 1973. *Desalination of Brackish Water by Spray Freezing*. Saskatchewan Research Countil Report No. EDC-73-CIV3, Saskatoon.

Stanley S.J., D.W. Smith, and G.D. Milne. 1992. Microorganism survival in ice-covered marine environments. *Journal of Cold Regions Engineering*, 6(2): 58-72.

Stanley, S.J. and D.W. Smith, D.W. 1991. Reduction of ice thickness on northern water reservoirs. *Journal of Cold Regions Engineering*, 5(3): 106-124.

US EPA (United States Environmental Protection Agency) 1994. National primary drinking water regulations; Disinfectants and disinfection byproducts; Proposed rule. *Federal Register* 59(145):38668.

Valencia, J.A. and J.L. Cleasby. 1979. Velocity Gradients in granular filter backwashing. *Journal*

American Water Works Association, 71: 732-738.

Vik, E.A. 1972. *Treatment of Potable Water Containing Humus by Electrolytic Addition of Aluminum*. Dissertation, Department of Civil Engineering, University of Washington, Seattle, 297 p.

Weisner, M.R., J. Hackney, S. Sethi, J.G. Jacanglo, and J. Laine. 1994. Cost estimates for membrane filtration and conventional treatment. *Journal American Water Works Association*, 87(12): 33-41.

WHO, 1993. *Guidelines for Drinking-Water Quality*. Second Edition, World Health Organization, Geneva.

SECTION 7

WATER STORAGE

3rd Edition Steering Committee Coordinator

Vern Christensen

James A. Crum

3rd Edition Principal Author

Richard Feilden

Special Contributions

Kurt Egelhofer

Dan Schubert

Al Shevkenek

John Warren

Section 7   Table of Contents

Section 7　List of Figures

Section 7   List of Tables

7   WATER STORAGE

7.1   Introduction

Community water storage facilities fall into two general classes: raw water reservoirs and treated water reservoirs.

Raw water reservoirs are most often used to provide continuity of supply:

- through seasons when natural sources are frozen;

- through predictable time periods when source water quality is so poor that treatment is not practical or economical; and

- during temporary supply interruptions due, for instance, to weather.

Other possible uses are:

- to improve water quality through sedimentation, or through mixing and averaging out of fluctuation in quality; and

- to provide constant, equalized flow to a water treatment plant.

In many arctic locations, good natural water sources are available only during a very brief summer season. In such cases, raw water reservoirs are commonly sized for a full year's consumption based on design level of service and planning horizon.

Owing to their large size, raw water reservoirs usually are outdoor impoundments formed by dikes or dams. In favorable settings, basins have been cut into solid rock. Tankage has also been used where favored by small population, low water use rate and absence of a more economical option. Earthen impoundments are often made watertight using a membrane liner, which involves careful detailing in the design and skillful installation in the field.

Treated water reservoirs are used to:

- level the demand peaks that occur during the operating day,

- provide dedicated reserves of water for firefighting, and

- provide additional dedicated reserves against possible interruption in supply.

Relative to community water demands, treated water reservoirs typically range in volume from less than one full day's consumption to several days'.

Treated water reservoirs are tanks built of steel, concrete or timber. In thaw-sensitive permafrost settings, a permafrost-protective style of foundation normally is needed, and as a consequence the entire reservoir is above ground. Insulation, the protective outer jacket which covers the insulation, and any tank liner needed (to defend against corrosion or leakage, or both) are important and sometimes difficult aspects of the design.

In this section, capacity considerations are discussed first, earthen and rock reservoirs next, and tankage of various designs last. Topics covered include:

- functional analysis (predesign);

- siting considerations;

- design options and details;

- construction constraints to be accommodated;

- operation and maintenance considerations; and

- other information derived from experience in Alaska and Canada.

Related topics not discussed in this section include:

- the design of dams or weirs across active stream courses – a specialized field addressed in Section 3; and

- water storage tanks which serve a single building or building group, rather than a community or district. Building tanks are dealt with in Section 14.

7.2   Capacity

7.2.1   Water Demands and Design Horizon. Forecasting community water demands is discussed in Section 2. Because both raw water and treated water reservoirs tend to be quite large and expensive, a fairly distant planning horizon is recommended to avoid the premature need for expansion.

7.2.2   Raw Water Reservoir Capacity. The required net ("live") capacity of a raw water reservoir is the product of daily demand during the period of use times the longest anticipated duration, plus reserve for safety, less any ongoing replenishment. In cold regions, many communities depend on raw water reservoirs as the sole water source from freeze-up to break-up, more or less.

"Total" storage includes usable live storage, plus any capacity lost to ice in winter, plus unusable ("dead") storage below outlet level. In an open reservoir that thaws reliably before its normal refilling schedule begins, some of the maximum ice volume may actually be used as live storage.

Of course, there must be a suitable water source, and a sufficiently long window of availability for re-filling.

7.2.3 Treated Water Reservoir Capacity. Community water utilities need reserves of treated water to:

- cover short-term demand peaks that exceed the water supply or treatment rate;

- provide large, short-term flows for fire-fighting;

- cover water system shutdowns of reasonable duration, whether scheduled or unexpected; and

- provide water for winter period.

Storage volume allocated to one purpose is usually considered not to be available for another, because demands are likely to be concurrent. The total storage requirement is taken to be the sum of the separate peaking, fire and outage requirements.

For design purposes, fire flows are considered to act concurrently with the peak day water demand. The peak day demand normally is near the design limit of the water supply and treatment facilities, and for this reason, as well as for reliability, it is usual practice to assume that fire demands are met entirely from storage. With trucked water distribution, the fire storage requirement is limited by low delivery rate. Fire protection is discussed in Section 15.

It is usually possible to design tankage to minimize inaccessible "dead" storage. As well, tankage volume is not usually lost to ice. Generally, freezing has to be prevented in tanks to avoid ice damage to the fittings, structure, or anticorrosion liner.

7.3 Earthen Reservoirs

7.3.1 Overview. This section deals with the design of earth-diked impoundments used to store raw water. The discussion is partially applicable to projects of a similar nature, such as the enlargement or creation of storage by raising natural lakes or by diking natural gullies.

Earth dikes and the reservoir floor are made (reasonably) watertight by a core or liner of clay, a core of frozen soil, or by a manufactured membrane liner. Clay and frozen soil designs often are not feasible, leaving membrane liners the most common design solution. However, membranes can involve difficult design and installation problems, and are relatively vulnerable to damage in service. Groundwater pressures or gas developed by biologic activity in the underlying thawed organic soils have floated liners entirely off their beds, underlining the need for design to remove the liquid or gas that develops under the liner.

In northern settings, evaporation losses tend to be balanced roughly by precipitation gains. Leakage losses can significantly reduce useful capacity.

Ice formation in outdoor impoundments resembles ice growth in natural water bodies. The pattern of change in ice growth rate as the winter season progresses, and the estimation of ice thickness were described in detail by Stanley and Smith (1991). In a cold location, ice thickness may reach two metres. It is recognized that reduction of ice growth, for instance through trapping snow on the ice surface as insulation, could significantly improve reservoir efficiency.

The inner faces of earthen dikes usually have fairly flat slopes, often 1:3 to 1:4. In reservoirs which are filled annually in summer and then drawn down over the entire winter, the water surface area at the freezing plane diminishes progressively as the water level declines. As a result, the total volume of ice formed over the entire winter is somewhat less than the volume defined by considering maximum ice thickness to occur at the high water level. Note too, that 1.1 m of ice is equivalent to 1.0 m of water. A time-step calculation that models drawdown progress and concurrent ice growth through the winter yields the best estimate of useful volume lost to ice.

"Dead storage" is the water volume below the outlet level, which cannot be drawn out of the reservoir. It is not always desirable to reduce dead storage volume to as little as possible. The amount of water that remains in an ice-covered impoundment towards the end of winter must be great enough to buffer against freeze-concentration effects, and any other water quality problems likely to develop.

The reservoir's large mass of permanently unfrozen water is a heat source which creates a thaw bulb in subsoils. Thaw-subsidence may occur over a long period of time.

TABLE 7-1   COVER MATERIAL AND LINER INTERFACE

Liner Material	Average Residual Angle of Friction (degrees)	Factor of Safety		
		3H:1V	3.5H:1V	4H:1V
HDPE	21.3	1.17	1.36	1.56
Reinforced CPE	24.6	1.37	1.60	1.83
Unreinforced CPE	24.1	1.34	1.56	1.78
Hypalon	23.7	1.32	1.53	1.76

Seepage losses carry heat. Surface and subsurface water flows can disturb the thermal regime in, for instance, dike foundation soils (Cameron, 1976).

7.3.2 Geometric Considerations. The design of reservoir dike sections depends on the geotechnical properties of available materials. The low coefficients of friction at the faces of a smooth membrane liner may govern the slope of the inner face. Scour and erosion of inner side slopes of reservoirs by ice and waves may not be wholly avoidable. Table 7-1 shows some average dimensions for reasonably stable materials.

Balancing cut and fill, the usual objective in southern locations, is often not feasible in a lined reservoir in a permafrost location. Local soils often are ice-rich, and would need to be thawed and consolidated before being used in construction of dikes. The remaining option is to build dikes using imported, preferably granular soils, selected for thaw (and re-freeze) stability.

A second problem with the concept of excavating is that the thaw bulb that grows below and around the reservoir tends to become a water-filled basin within the surrounding soils which are rendered impervious by permafrost. If the reservoir floor elevation is below the local groundwater table, usually ground surface, there will be corresponding uplift on the underside of the liner. Pumping to keep the groundwater table below a certain level is likely to be expensive and not necessarily reliable.

Two indicators of reservoir efficiency are the ratio of useful live storage to total storage or "utilization", and the ratio of useful live storage to dike earth volume. Capital cost is quite closely related to dike earth volume. Reservoir efficiency is highly sensitive to geometry.

Reservoir depth has the greatest influence on storage utilization. An 8 m deep square reservoir with net storage of 33,500 m³ and 2 m of ice has a utilization of about 70 percent. At the same depth, 8 m, a change in inner side slope makes surprisingly little change in utilization, but obviously it has a very large and direct effect on dike soil volume. On a level surface, round and square outlines are about equally efficient. Efficiency declines in rectangular reservoirs as the ratio of length to width becomes larger, but the longer straight side of a rectangular outline may facilitate future expansion by twinning (Stanley and Smith, 1991).

Slope over the site dramatically increases the ratio of dike soil volume to useful impoundment volume. Even on a level site, the volume of soil needed for dike construction is likely to be considerably greater than the net storage provided. Efforts aimed at optimizing the choice of site, and at optimizing the dike section and layout geometry will be worthwhile.

7.3.3 Water Quality Considerations. In summer, raw water quality may be improved in a reservoir by sedimentation, and may be degraded by the effect of waves and wind currents in suspending (or re-suspending) fine material along the dike shoreline. Biological processes quite often have little noticeable effect in summer in cold regions. Regimes in natural lakes may be a useful indicator. In general, use of the reservoir by waterbirds, etc. has no greater health or quality significance than similar use of the source water by the same populations. On the other hand, the reservoir needs to be protected from contamination by humans, dogs, vehicles, fuel spills and wildlife that may carry human pathogens i.e., giardia.

Raw water kept in a lined reservoir during winter is affected by sedimentation, biological activity, freeze concentration, and thermal stratification. Death and decay of biota, and freeze-concentration of dissolved materials can cause significant taste and odor problems in earth-diked raw water reservoirs as the winter progresses.

Sedimentation removes settleable turbidity, but not colloidal material causing color. Sedimentation proceeds rapidly as soon as the first skim of ice cover puts an end to wave action and wind currents.

Ice cover prevents oxygen transfer from the air, and also reduces production of oxygen by algae; snow cover will stop oxygen production. If food and nutrients exist to support bacteria – usually the case, especially with a tundra pond source – then aerobic metabolism will soon use up the dissolved oxygen reserves. Anoxic conditions are likely to produce taste and odor problems, often through the reduction of sulphate to produce hydrogen sulphide. A slow rate of aeration is one possible defense, but it is likely to stir up any turbidity present. Another is air stripping of dissolved gases in the water treatment scheme.

Freeze concentration occurs and should be considered. Dissolved material tends to be left at increasing concentration in the unfrozen water fraction. If a lake has no inflow or outflow during winter, and its bathymetry is well known, then estimates of freeze concentration can be made for different ice depths simply by considering relative volumes of frozen water and unfrozen water. In a water supply reservoir, the ongoing consumption of the unfrozen fraction must also be considered. A time-step calculation is needed that takes into account the geometry of the reservoir, decline in water level, growth of ice, and decline in the unfrozen water fraction.

Stagnation effects due to thermal stratification may be avoided by a slow rate of aeration which will keep the reservoir contents gently agitated.

In an unlined reservoir, bottom (and dike) soils containing organic matter such as grass, detritus and root mat will adversely affect water quality. Leaching column tests will provide some indication of possible uptake of dissolved oxygen, of leaching of color, organic carbon and ammonia, and of effect on pH.

If an unlined reservoir is to be built, for example, by diking off an existing valley or raising the level of an existing lake, the question arises as to how much surface preparation needs to be done before flooding vegetated land. At five diverse sites in Alaska, the costs of various degrees of advance clearing were compared to the ten-year life cycle costs of treating the water as would be needed if the vegetation were left in place. At four sites, stripping appeared to be the most cost effective option (Smith and Justice, 1975; 1976).

Finally, seasonal turnover in raw water impoundments can cause abrupt changes in water quality.

7.3.4 Site Requirements. Principal criteria include possible locational advantages, topography, and geotechnical requirements.

Location. Preferred locational advantages include:

- easy access to the community without undue exposure to contamination;

- easy access to the source;

- short haul distance from borrow areas;

- in a piped system, an economical pumping arrangement.

In a trucked system, the road to the community needs to be aligned to minimize snowdrifting problems.

Topography. It may be possible to use natural topographic features to advantage. More often, however, side slopes will tend to increase the volume of dike material needed to achieve a certain net volume of useful storage, giving the advantage to a level site. Room for expansion is often a consideration. In making economic comparisons between sites, include capital and operating costs of roads and the fill line, and the effects of change in supply distance to the community.

Hillsides above a site may lead to "icing" difficulties during the winter. Groundwater may be confined in the seasonal talik between impermeable surface frost and impermeable permafrost, so that downslope drainage creates artesian pressure. The thaw bulb below the reservoir intersects the talik, creating a natural relief point, possibly leading to liner failure. The design of an underdrain system that will relieve groundwater pressure reliably through the winter may be a difficult problem.

Geotechnical. Foundation soils need sufficient strength and stability to carry dike loads without undue deformation. In permafrost locations, possible thaw-subsidence and thaw-consolidation effects need to be assessed. Active measures to prevent a thaw bulb from developing beneath the reservoir are not likely to be practical, owing to the large area involved.

The design of dikes (interior and exterior side slopes in particular), and membrane liner bedding and protection will depend on the characteristics and quantities of available materials, and on feasible construction techniques, including quality control. Assurance of good compaction may allow use of steeper slopes in the design. Processing of aggregate for specific purposes may be necessary. A reservoir construction project usually is large enough to warrant the import of specific items of construction equipment, if needed.

Geotechnical Investigation. In broad terms, the geotechnical investigation for an earth-diked reservoir will need to determine:

- foundation stability, including thaw related issues;

- availability, volumes and properties of borrow available for building dikes and roads, including natural liner soils (if used) or membrane bedding and cover materials; and

- equipment availability, practical construction techniques, unfulfilled equipment requirements, approximate construction season, probable rate of construction progress.

Typically, a fairly extensive subsurface program will be needed, which in turn may involve mobilization of drilling equipment to the location.

7.3.5 Design Details. Design details are in many respects governed by site-specific geotechnical parameters. Dike liner failures due to dike settlement, uplift, erosion, and defects in installation are common, indicating a need for careful design and detailing.

Underdrains. The need for subdrains and their design is highly site specific. Reliance on pumping systems is not appropriate in many locations. If drainage must operate in winter, heat tracing may be needed to keep discharge points open – which introduces significant operating expense as well as the additional risk of failure.

Bringing upstream ends of subdrains to the surface can help inspection and cleaning.

Dikes. The dike crest of a reservoir provides room for equipment mobility for maintenance, for a fence (if provided), and minor reserve against erosion loss. Wind erosion may need to be considered if dikes are of fine noncohesive soil.

Freeboard provides safety against overtopping by waves, and includes some allowance for dike settlement.

Dike slopes are designed for geotechnical stability, considering specific material properties, moisture regime, and the degree of compaction expected. In a membrane-lined reservoir, the coefficients of friction between the liner and its bedding and cover soils are likely to govern the inside slope, as discussed later.

Clay Core or Liner. Impervious soils suitable for use as impervious cores of dikes and reservoir floors often are not readily available. Where they are, geotechnical analysis and design are site-specific. Possible effects of thaw subsidence and of heave on refreezing need to be considered.

Admix liners, like bentonite clay, are similar to those of compacted native clays. The bentonite liner is subject to variations in placement content, incomplete mixing, and damage due to the thin nature of the liner, usually 50 to 150 mm. Bentonite liners have a lower inherent permeability and are more likely to swell and self-heal than most naturally occurring clay soils. Asphalts and soil cements are subject to shrinkage and tension cracks, and inconsistent application (Thornton and Blackall, 1976; Folkes, 1981; Kirby, 1979).

Frozen Core and Subfloor. In very cold settings, it may be possible to make dikes and the floor impermeable by having them freeze (or maintaining an existing frozen condition) permanently, through all seasons. Cooling ducts have been used to develop and maintain ice cores in dams in Siberia. Thermopiles, such as the "cryo-anchor", might be used in the same way. Reservoir dikes at Pond Inlet, NWT, have been designed to freeze naturally.

Membrane Liner. Flexible membrane liners can be used effectively to limit seepage losses through permeable dikes and reservoir floors (see Figure 7-1). Materials most often used include:

- chlorinated polyethylene (CPE), reinforced;

- chlorosulphonated polyethylene (Hypalon), reinforced;

- high density polyethylene (HDPE), not reinforced.

The liner material and any solvents used in seaming must have FDA or AWWA approval for use in a public water system.

The inclusion of reinforcement in CPE and Hypalon liners improves tear resistance at the expense of

FIGURE 7-1   MEMBRANE PLACEMENT – SHISHMAREF, ALASKA

reducing allowable elongation. HDPE currently cannot be obtained reinforced.

Folds and pleats can be left in the liner to accommodate expected soil subsidence.

Material thicknesses considered for the 1988 Arviat, NWT reservoir were 0.914 mm (36 mils) for reinforced CPE and reinforced Hypalon, with protective soil cover, 2.03 mm (80 mils) for unreinforced HDPE with protective soil cover, and 2.54 mm (100 mils) for unreinforced HDPE without soil cover.

Liners must be protected against overstretching due to ground movements and against mechanical damage due to ice loads. They need to be anchored effectively against down-slope creep, and against being lifted by wind.

Liner preparations are potential trouble spots. On the other hand, burying an intake pipe below the liner protects the pipe from ice loads in the shoreline zone.

Membrane Liner Bedding.   Membrane liners are placed on a protective, stable bedding of clean sand, maximum grain size 5 mm (or less), not more than five-percent silt and clay sizes. It is worthwhile to

send samples of bedding, along with grain size distribution curves, to prospective liner suppliers, to obtain confirmation of suitability.

If no soil cover is used over the liner, then guidelines set by the liner manufacturer probably will govern the dike's inside slopes. The factors which govern the need for a soil cover over the liner include installation method, liner tensile strength and stretching, and potential ice action. For liners without soil cover, problems may develop with underlying gas formation floating the liner.

Membrane Liner Cover.   If a protective soil cover is placed over a membrane liner, then friction at the liner-cover interface must be considered and likely will govern the dike's inside slope. Laboratory tests may be needed to determine a value usable for design unless the manufacturer is able to supply the information. Take into account the effects of reservoir filling and drawdown on soil stability. Data in Table 7-1 provide some general indication of ranges, but it is emphasized that these are specific to particular soils and materials and are not for direct application elsewhere.

The liner protection recommended for the Arviat reservoir was clean sand, followed by pit run gravel, followed by rip-rap. The sand thickness recommended was 500 mm (measured perpendicular to the liner face) if placed using a light rubber-tired bobcat, less if hand placed.

At Tuktoyalatuk, NWT, geoweb has been used to reduce wave erosion of the liner soil cover (Crist, 1993).

Spillway. Accidental overfilling of the reservoir could lead to dike instability, and almost certainly to erosion failure upon overtopping. A positive safeguard is needed, for instance, overflow culverts through dikes, which discharge into an erosion-protected drainage route.

Intake. Inclined shaft intakes (discussed in another section) have worked well installed beneath membrane liners, where they are well protected from ice damage. The intake shaft elbows up through the liner at its lower end, and terminates in a well screen. A bolted flange is used to form a compression seal between the pipe and the liner.

7.3.6 Construction Aspects Affecting Design. Various earthwork operations warrant consideration owing to large size or to specialization. Liner installation is a difficult and critical operation.

Earthwork Capability. As mentioned earlier, the equipment capacity and human skills available to undertake a large and somewhat demanding earthwork project need to be assessed carefully. Achievable quality in such aspects as aggregate processing, moisture control, and compaction need to be recognized in the design. The overall project plan may involve purchasing or leasing certain items of equipment and moving them to the site. In off-road locations, the usual poor fit between construction window and transportation window may tie up specialized equipment at the site for far longer than it is actually needed to do the work.

The need to manufacture or process aggregates for specialized uses (such as in subdrain systems and liner protection) and equipment needs, warrants careful investigation. In this respect, as in others, a thorough geotechnical assessment of both the site and potential borrow sources is essential to effective planning and execution of the project, to realistic forecasting of costs and to overall project economy.

Membrane Liners. CPE and Hypalon liners are seamed by solvent welding. Careful quality control

is needed. HDPE liners are seamed by heat fusion, which involves significant operator skill, specialized equipment, and careful quality control. Successful experience with some minimum number of similar installations is sometimes set out in bid documents as a prerequisite qualification, in order to bid supply and/or installation of a membrane liner.

Because of the subtle differences between sheet-material formulations, the material manufacturer is sometimes required to provide an affidavit attesting to the suitability for the purpose and full compliance with specifications. The presence of the manufacturer (or representative) on-site throughout installation is sometimes specified.

Shop drawings prepared by the liner manufacturer need to show the layout of panels (code number each corner), anchoring details, jointing details, location of all shop-made and field-made seams. Shop drawings can be updated in the field to show any changes and repairs.

The liner material needs to arrive on-site undamaged, which requires proper packaging for the mode of transport and for any periods of storage.

Continuous inspection of all liner work is recommended. Seams need to be tested for watertightness, by vacuum box or spark test methods.

Light, rubber-tired equipment may be needed to place and spread cover soils.

7.3.7 Operational Considerations Affecting Design. The reservoir needs to be protected adequately from contamination caused, for instance, by fuel spills and snowmobile traffic. Fencing may be needed.

A membrane liner left uncovered may present a serious safety hazard. Under frosty conditions, it may be so slippery that a worker could slide down into the reservoir and be unable to get back out.

Solvent-welded liners are easier to repair in the field than heat-fused ones. Specialized equipment is not needed.

Provide a liner repair kit to maintenance personnel, and training in its use.

Significant leakage through tears, poor seams and other defects in soil-covered membrane liners may run undetected. If leakage begins to create problems, leak locations may be quite difficult to find.

Emptying only one cell of a two- (or more) cell system may lead to uplift on the empty cell's liner, owing to groundwater pressure and movement through a common dike. Subdrains need to be designed for the maximum possible inequality in water levels between cells.

7.3.8 Maintenance Considerations Affecting Design.

The potential for dikes and the reservoir floor to settle due to thaw subsidence, thaw consolidation and other factors needs to be evaluated in the course of the geotechnical investigation, and taken into account in design. If settlement is anticipated, the design needs to include provision for restoration of dikes, liner, fencing and structures.

Scour and erosion of inner side slopes by wind, waves and ice cannot be avoided entirely. Periodic repair may be needed. If so, there may be an ongoing need for particular equipment, which the project may have to provide.

Membrane liners are sensitive to damage. Provision must be made for repair if needed. Leakage losses can substantially reduce useful capacity. Note that the liner cover on one hand protects the liner, but on the other makes leaks much harder to locate.

There may be a need to clean accumulated sediment from a raw water reservoir at infrequent intervals.

7.3.9 Experience.

The Government of the Northwest Territories has constructed earth-diked reservoirs in a number of locations. Key data are shown in Table 7-2.

A winter-source raw water reservoir was successfully constructed at Pangnirtung, NWT, in 1985-87 by cut and fill, on a site having a seven-percent crossfall. The liner is HDPE, with a cover of sand and gravel (Smith et al., 1989).

Pangnirtung is within a region of continuous, deep permafrost. The subsoil at the reservoir site is a glacial till, of dense silty sand containing occasional cobbles and boulders. The till is overlain by thin peat deposits. Soils are extremely ice-rich near the surface, saturated in the upper few metres, and drier at increasing depth. Vertical ice lenses up to one metre wide were encountered below the surface. The site is poorly drained and very wet in summer.

The crossfall made it possible to install gravity flow subdrains, which relieve groundwater pressures beneath the liner. As well, a deep interceptor ditch along the reservoir's uphill side directs surface water, and mobile active-layer groundwater around one of the reservoir's ends.

Because the surface soils will not carry traffic when thawed, most of the earthwork was done in winter. Of a total excavation volume of 140,000 m³, 90,000 m³ needed to be blasted. The dikes were built of "boulders" of excavated frozen material. The more ice-rich material was placed on the outer embankment slope, and the drier material on the inner one. Where reservoir side slopes were formed in cut, ice-rich material was subexcavated to a depth of 3 m and replaced with drier material. All new embankments were allowed to thaw, settle and drain naturally during the following summer. Inner and outer side slopes were built at 1:4, due to the expected instability of the wet, thawing material.

Sand for the bedding and liner cover was obtained by screening pit run material from a local source.

7.3.10 Hydraulically Dredged Reservoirs.

Floating hydraulic dredges are used in the Canadian and Alaskan arctic to modify existing ponds and impoundments for water reservoirs. The dredging process economically moves large quantities of semi-liquid material to build a reservoir dike or structure or to create additional lake volume for storage. The use of salty soils in the dredging process may require a lining or other treatment of the water.

7.3.11 Cost Information.

Capital cost data are presented in Table 7-2.

7.4 Rock Reservoirs

A raw water reservoir can be cut into bedrock where a suitable rock formation exists. Uses of rock reservoirs and functional design considerations closely follow the earlier discussion of earth-diked reservoirs, except with respect to geotechnical aspects and liners.

A rock reservoir usually has vertical sides, which in comparison to earth-diked reservoirs reduces the proportion of total storage volume lost to ice. Note, however, that rock is a fairly good conductor of heat. Some additional allowance must be made in volume calculations for growth of ice against the upper parts of reservoir walls.

Water quality considerations closely follow the earlier discussion in Section 7.3. Freeze-concentration may be less (for reservoirs of similar net storage), if there is a more favorable ice:water ratio.

Even relatively sound rock has joints and fissures. Construction of a rock reservoir will involve sealing of cracks, either by injection of grouting or by a surface treatment such as gunite. The cost of sealing may be a decisive factor in weighing the cost effec-

TABLE 7-2   SUMMARY OF EARTHEN RESERVOIRS – NORTHWEST TERRITORIES

Reservoir Location	Year Built	Max. Reservoir Capacity (m³)	Usable Volume Under Ice (m³)	Max. Water Depth (m)	Design Ice Thick-ness (m)	Dead Storage Depth (m)	Freeboard (m)	Water Surface Dimensions (Reservoir Full)	Inside Slopes
1. Pangnirtung	1967/68	18,000	8,000	6.5	2.0	0.6	0.6	45 m x 167 m 1 Cell	2.8:1 to 6.4:1
2. Arviat	1976	23,800	9,100	4.5	1.8	0.9	0.9	57 m x 122 m 1 Cell	2.1:1 to 3:1
3. Pine Point	1977/78	464,800	336,600	8.2	1.2	0.9	0.9	180 m x 218 m 2 Cells	3:1
4. Broughton Island	1978	31,500	22,000	9.0	1.5	0.0	1.0	84 m x 84 m 1 Cell	3:1
5. Fort Good Hope	1978/79	26,400	12,800	5.3	1.8	0.6	1.2	86 m x 86 m 1 Cell	3:1
6. Tuktoyaktuk	1981/82	94,300	53,100	7.0	2.1	0.5	1.3	102 m diameter 1 Cell	4:1
7. Arviat	1988	51,400	27,200	5.0	2.0	0.5	1.5	116 m x 116 m	2.5:1 to 3:1
8. Fort Smith	1991	2,400	N/A	3.0	N/A	1.1	0.6	29 m x 59 m	3:1

continued

TABLE 7-2   (continued)

Reservoir Location	Liner Description	Liner Cover	Subdrain System	Fenced	Cost of Reservoir	Unit Cost (per m³ water)	Significant Performance Problems
1. Pangnirtung	CSPE 1.3 mm (Installed 1976)	No cover	No	Yes	Approx. $200,000	$11.11	Heaving of liner due to air and groundwater. Liner torn due to ice action, lack of bedding, and high winds. High wind damages fence.
2. Arviat	CPE 0.8 mm	Sand on inside slopes only. 1.5 m at top to 0.0 m at bottom.	Yes	Yes	Approx. $400,000	$16.81	Scour of liner cover under reservoir fill pipe.
3. Pine Point	CPE 0.8 mm	1.2 m sand on inside slopes. Bottom: no cover.	Yes	Yes	$1,527,000	$3.29	Rain erodes liner cover.
4. Broughton Island	ELPO 1.3 mm	0.3 m of sand on inside slopes.	No	Yes (1981)	$322,900	$10.25	Heaving of liner due to groundwater. Liner torn due to ice action, lack of bedding, and high winds.
5. Fort Good Hope	CSPE 0.9 mm (Reinforced)	0.45 m of sand and gravel on inside slopes. Bottom: No cover.	Yes	Yes	$509,300	$19.29	No problems reported.
6. Tuktoyaktuk	CPE 0.8 mm	1.0 m sand on inside slopes. 0.3 m sand on bottom.	Yes	Yes	$2,900,000	$30.75	High winds blowing sand cover.
7. Arviat	CPE 0.8 mm	1.0 m sand on inside slopes only.	Yes	Yes	$1,278,000	$25.00	No problems reported.
8. Fort Smith	HDPE 2.0 mm	None.	No	Yes	$262,000	$109.17	No problems reported.

tiveness of a rock reservoir against other options. Rock is relatively thaw-stable, but not necessarily completely so. Fissures may be found sealed or wedged apart by excess ice and thus susceptible to thaw.

To date, no significant problems have been encountered in operating rock reservoirs in the Northwest Territories. Maintenance costs have been low.

Data on rock reservoirs are presented in Table 7-3.

7.5 Tank Reservoirs

7.5.1 Overview. Tankage is used in both raw water and treated water applications. As noted earlier, raw water storage requirements may run to nearly a full year's supply. In some cases treated water is stored, for the entire supply is based on supply reliability, quality and site considerations. The two applications tend to differ considerably in tank size and scale.

In raw water facilities, long-term storage may affect water quality and this needs to be considered. With treated water, the storage time usually is short and has by itself no significant effect on quality. Note, though, that change in temperature alters a water's tendency to corrode metals. Among other effects, warming water lowers the saturation concentration of oxygen.

Tanks of wood, steel and concrete have all been used in northern locations. Structural design generally follows standard southern practice, so is not discussed here. Northern considerations include foundation design, thermal analysis, insulation, and the need to prevent ice damage.

Ice buildup, either inside a tank, or on the exterior (due to leakage), can pose serious structural and safety risks. Elevated tanks have collapsed due to the weight of ice on supporting towers. A heavy layer of ice within a tank, rising and falling on a fluctuating water surface, can damage anything in its way. Tanks have ruptured or collapsed when interior ice perched at some higher level has fallen, at a time when water level in the tank happened to be at some lower level. A solid layer of ice falling within a tank can exert a considerable "piston" effect when it strikes the water surface. Positive protection against ice buildup is required.

Wood stave tanks are relatively easy to transport and erect, but tend to leak, especially in dry climates. Loose-fit liners have been installed inside leaky tanks as shown in Figure 7-2 (e.g., wood stave tank at Fort McPherson, NWT), but with mixed success. The liners are easily punctured and, while easily repaired, small defects are nearly invisible and extremely difficult to locate. Leakage problems tend to rule out wood tanks from most applications.

Concrete can be used for reservoir tankage where good quality concrete can be obtained at competitive cost, and where the foundation requirements for large concrete structures can be met. An advantage of concrete is its suitability to below-grade construction, which reduces exposure to severe winter weather. Concrete is quite durable if it is correctly designed, mixed, placed and cured using sound, well-graded aggregates. However, the opposite is true if any of these are lacking.

In many arctic communities, the choice reduces to welded or bolted steel tanks, wholly above ground. The remainder of this section deals mainly with steel tankage. The discussion can be applied fairly readily to concrete tankage, as appropriate.

7.5.2 Site and Geotechnical Requirements. The principal considerations are possible advantages of location, and geotechnical criteria.

Location. In raw water applications, the earlier discussion with respect to earth impoundments generally applies. With treated water, the main considerations are the layout of the distribution system and the community in general, and possible use of local topography to reduce pumping costs. Note that gravity pressure in the distribution system continues during power outages, a significant advantage in some locations, especially in preventing possible cross connections from contaminating drinking water.

Geotechnical. A steel tank is a large, heavily loaded, thin-walled structure which requires a stable and fairly rigid foundation. Significant differential movement, for instance due to thaw subsidence or to frost heave, may cause problems ranging from leaks (in bolted tanks especially) to rupture. Also, groundwater and rainwater need to be kept away from the underside of the steel tank, to reduce corrosion.

Tank foundations resemble building foundations, but there is a large difference in loading. A two-story wood frame residence has a footprint load equivalent to less than 0.6 m of water. Tanks have been placed on gravel pads topped with sand, asphalt or concrete, on timber sleepers, and on pile foundations. In permafrost locations, gravel pads have been cooled using ventilation ducts (capped in summer), and cryo-anchors. The design and subsequent moni-

TABLE 7-3   SUMMARY OF ROCK RESERVOIRS – NORTHWEST TERRITORIES

Reservoir Location	Year Built	Max. Reservoir Capacity (m³)	Usable Volume Under Ice (m³)	Max. Water Depth (m)	Design Ice Thickness (m)	Dead Storage Depth (m)	Freeboard (m)	Water Surface Dimensions (Reservoir Full)	Inside Slopes
1. Igloolik	1979	61,200	19,500	10.0	1.5	2.0	1.0	165 m x 330 m (Lake Surface) 35 m x 74 m (Blasted Hole)	Vertical
2. Coral Harbour	1980	26,700	17,500	9.0	2.0	0.8	2.0	40 m x 75 m 1 Cell	Vertical
3. Chesterfield Inlet	1991	32,175	21,038	10.0	2.0	1.5	3.0	110 m x 28 m	Vertical
4. Igloolik	1994	76,265	54,762	10.0	1.0	2.0	N/A	24 022 m²	Vertical

Reservoir Location	Liner Description	Liner Cover	Subdrain System	Fenced	Cost of Reservoir	Unit Cost (per m³ water)	Significant Performance Problems
1. Igloolik	No liner	Not applicable	N/A	Yes	$713,100	$11.65	No problems reported.
2. Coral Harbour	No liner	Not applicable	N/A	Yes	$583,100	$21.84	No problems reported.
3. Chesterfield Inlet	No liner	Not applicable	N/A	Yes	$1,041,044	$32.36	No problems reported.
4. Igloolik	No liner	Not applicable	N/A	Yes	$1,060,000	$13.90	No problems reported.

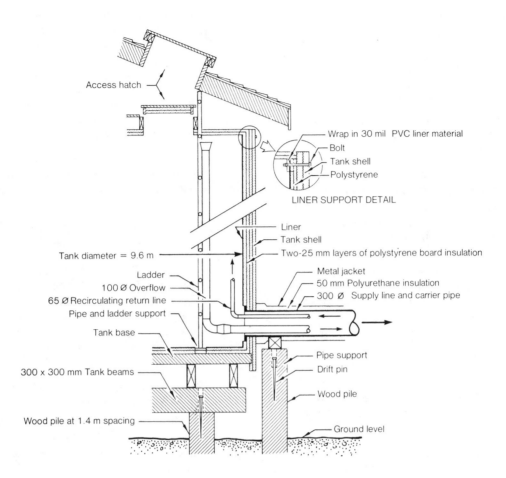

FIGURE 7-2   WOOD TANK INSULATION AND LINER ON INSIDE

toring of permafrost protective foundations is described in Section 3.

The geotechnical investigation preparatory to design will need to establish the thaw and freeze stability of foundation soils, availability and characteristics of granular fill and bedding sand, and the construction standards likely achievable in such matters as compaction of fill, bedding and placing insulation, and bedding culverts (ventilation ducts). If piles are to be considered, then the material, installation method, load transfer method and design load capacities also need to be determined. Finally, any specific equipment needs not covered by available resources need to be identified.

7.5.3 Thermal Design.   Methods and data used in heat loss calculations are described in Section 4. Note that usual daily fluctuations of water level in a treated water reservoir introduces some cold air. Also, the vent on a steel tank needs to be kept from

being blocked by frost, as vacuum can collapse the tank. Venting to a heated indoor location avoids this problem.

A treated water storage tank's operating temperature will likely be determined by the distribution system requirements. If the temperature is not already constrained, then:

- operating close to 0˚C will reduce total heat loss, which on a monthly basis is directly proportional to the difference between operating temperature and mean monthly air temperature;

- operating over 4˚C avoids stagnation of the coldest water at the surface through density stratification, and consequent ice-forming tendency;

- operating at relatively warm temperatures (i.e., over 10˚C) increases humidity and vent frosting.

An elevated tank is exposed the most to extremes of temperature, and to wind on all surfaces. Burial or partial burial of a reservoir is an advantage, where it can be achieved. Soil cover provides isolation from surface wind, a degree of insulation, and considerable thermal inertia. Steel and concrete provide essentially no insulation by themselves.

The thermal envelope of a tank should be as complete as possible, fully covering such features as hatches. Ladder supports are an example of (nearly) unavoidable penetration. Placing all pipes running to and from the tank in a common, insulated carrier or utilidor reduces surface area and heat loss. In a buried or partially buried design, insulation will be carried well underground and possibly entirely under the tank. In the case of a tank placed on a pad or sleeper foundation, an insulation layer is included in the foundation design. Bolted steel tanks can be obtained with double shell insulated roofs.

Insulation thickness is determined according to:

- best life cycle economics, balancing the increased capital cost of thicker insulation against the long-term benefits of energy saving; and

- operational safety considerations, cool-down and freezing times should the heating system fail during unpleasant weather.

Reliability of heat supply is important in cold-regions water system design. As a guide, temporary outage poses less risk to tanks than it does to pipes, owing to the much more favorable ratio of mass to surface area.

Separate heating of the head space above the water surface may need to be considered.

The momentum of water flow into the tank, for instance, recirculation flow returned to storage, can be used to maintain a mixed condition and avoid thermal stratification.

Heating of water systems is an ideal use for heat shed from diesel generators. Jacket water tends to be at around 65°C, slightly cool for exchange into building heating systems of typical design. A water reservoir provides a very large heat sink at a relatively cool, stable temperature.

7.5.4 Design Details.

Tanks. Design, prefabrication and erection of welded steel tanks has evolved into a specialty, parallel in some ways to the pre-engineered metal building industry. Structural design and detailing generally are best done by experienced manufacturers, working from layout drawings and performance specifications prepared by the purchaser.

Some of the design data necessary or useful to design and prefabrication are listed below.

- Service use. Climatic loads and exposure. Will the tank ever be allowed to cool down to winter ambient temperature, requiring the use of low-temperature steel? Design loads of insulation and cladding, and of any other items to be added by others, such as piping.

- Foundation type and rigidity. Underside moisture exposure.

- Volume requirements, and overall dimension limits. Note that tank dimensions are developed around standard widths of steel sheet or plate. In remote locations, erection procedures and equipment limitations need to be considered carefully.

- General arrangement, showing orientation of nozzles (connections), hatches, ladders and all other features to be included. Attachments to be provided, for example, for interior pipes, or exterior light fixtures. Thickness of insulation or cladding, which is needed to detail ladder support, nozzles, hatch necks, and other items that pass through the main thermal envelope.

- Exterior and interior prime coating or finish.

- Seismic considerations.

Steel Tanks. Most steel tanks, whether welded or bolted, are commonly manufactured in shell heights which are multiples of 2.4 metres. Standard capacities are selected, with diameters varied to obtain desired nominal volumes which conform to manufacturers' modular sizes. For field erection and loading considerations, tanks are normally specified less than 10 m high.

Bolted tanks are being used more frequently in Alaska for several reasons, including:

- tanks are supplied with a factory-applied coating system (provides a controlled coating application environment and avoids field surface preparation);

- construction can be accomplished by locally trained workforce;

- components can be packaged into conveniently sized units for transport;

- certified welders and weld testing is not required;

- adjustments in orientation can be made in appurtenances;

- installation can be performed under variable climatic conditions.

The design and construction of a flat, rigid foundation system is one of the most critical factors in successful installation of bolted tanks. Particular attention is required in bolt installation, assembly, torque and proper placement of gaskets and seals. Hydrostatic testing prior to installation of the insulation system should be conducted for all tanks. Control of overflows and surface drainage away from the foundation system should be considered.

Selection and design of bolted tanks must consider the following:

- foundation system;

- tank dimensions – location and spacing of piping and appurtenances;

- seismic zone and loads;

- roof pitch (usually 2:12) and projection (foam thickness 75 mm);

- access hole locations;

- pipes and fittings connections;

- overflow;

- ladder cage;

- location of roof entrance hatch;

- location and type of roof vent hatch;

- insulation system/thickness (75 to 125 mm);

- insulation of appurtenances and connections;

- siding and fastening system details;

- utilidor connection area and details; and

- shipping and crating considerations.

Factory-applied coating systems are specific to the tank manufacturer. Galvanized coatings have been sucessful, as well as factory-applied (thermal) epoxy systems. An evaluation of the type of system used should be considered during the design and purchase of the tank.

Tanks constructed in areas where the mean annual temperature is below 0°C should be located adjacent to the water treatment plant (WTP) so that tank water can be maintained above freezing through heat add systems. Tanks smaller than 38 m³ can usually be located inside the WTP more economically than outside. Outdoor tanks must be insulated and connected to the WTP with a utilidor which contains supply/return piping and a heat add loop. In most community projects in rural Alaska, this has been accomplished by circulating potable tank water through a dedicated heat exchanger that is in parallel and/or series with other WTP hydronic components. Circulation pumps that move water through the heat exchanger are controlled by an aquastat installed in a temperature monitoring well through the tank side wall. Colder tank water is passed through the heat exchanger, then returned to the far side of the tank through a long heat add pipeline mounted inside the tank. This helps promote circulation within the tank.

In locations where ambient temperatures and piped distribution systems combine to allow only minimal ice build up, uninsulated storage tanks are sited as they would be in temperate locations. Efforts should be made to locate these tanks at elevations which optimize gravity flow in suppling adequate distribution pressures. Piping should be such that flow from the treatment plant is through the tank. Tanks which "float" on line tend to have icing problems during low demand periods, and may have stagnation problems, difficulties in maintaining chlorine residuals, and reduced disinfection contact times.

Tank Appurtenances. Include an exterior ladder and cage, roof hatch, ground-level hatch, vent, overflow ports for instrumentation, and low-level drain for cleaning. The ground-level hatch provides access for scaffolding and other equipment used in liner application and in future replacement of the liner.

Provision sometimes is made for emergency drainage of the tank, in the event of extended outage of the heating system. There may be a need to provide for controlled thawing, and/or manual removal of ice. In most cases, however, it will be more cost effective and practical to focus on reliability of heating. A complete back-up system may be worthwhile.

A trickle overflow to a cold location will freeze, and must be avoided. An overflow to outdoors may need to be heat traced.

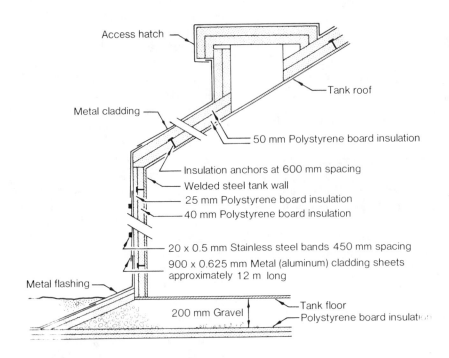

FIGURE 7-3   STEEL TANK WITH BOARD INSULATION AND METAL CLADDING

Connections to steel tanks and connection pipes need to be accessible, best located at the tank wall, not below a tank set on a gravel pad. Differential movement of piping and tank must be considered.

Insulation. Insulation materials commonly used include soil, polystyrene board, polyurethane board or sprayed-on, and spun glass batt. The choice is affected by many factors including exposure to moisture, attachment options, shipping and construction cost, among others.

Closed-cell polystyrene or polyurethane are used in potentially moist locations. High-density, closed-cell polystyrene boards with compressive strengths of up to 690 kPa are used beneath tanks resting on the ground.

Polyurethane is often used to insulate large steel tanks. It can be supplied as boardstock, or field sprayed. Field application greatly reduces shipping volume but requires proper surface preparation, trained and skilled workers, and environmental conditions that may in turn require temporary enclosures and heating. The surface must be free of rust, oil, grease and loose particles, coated with a compatible primer, and dry. The foaming process itself is sensitive to temperature and wind. Also, wind can lead to overspray and drift which damages surrounding structures. The following minimum conditions must be met throughout a field spraying operation:

- dry weather, i.e. no rain, mist, fog or condensation,

- wind (and gust) velocity less than 15 km/h,

- tank surface is dry, and its temperature not less than 15°C, with no direct heating permitted,

- air temperature not less than 10°C.

Note worker protection requirements for field spraying, which depend on the formulation used.

Boardstock insulations can be glued or strapped (typically with 40 mm by 0.5 mm stainless steel banding 450 mm on center) to the outside of the tank. Boards should generally be less than 75 mm thick to allow installation on curved surfaces. They should be installed in layers to the desired thickness. Joints should be staggered to prevent easy passage of outside air to the tank surface. Some details are shown in Figures 7-3, 7-4 and 7-5.

FIGURE 7-4   STEEL TANK WITH 75 mm
SPRAYED-ON POLYURETHANE
INSULATION – BARROW, ALASKA

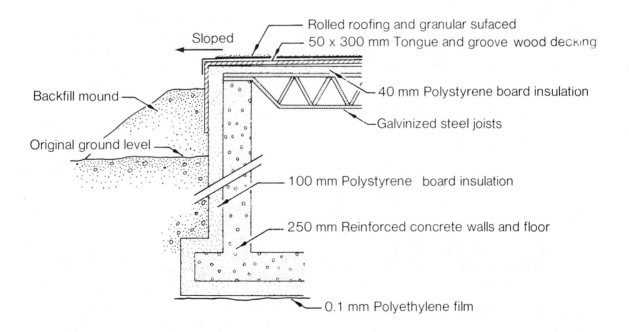

FIGURE 7-5   INSULATED BURIED CONCRETE TANK

Cladding. Insulation applied over above-ground surfaces needs to be protected against accidental mechanical damage, vandalism, birds and animals, and weathering. Compatible low-temperature elastomers sprayed over sprayed-on urethane provides protection against weathering, and helps prevent loss of insulating gas from cell voids. Metal cladding of aluminum or steel provides a good barrier against weather and reasonable defense against mechanical damage. Note that profiled sheets show deformities much less than flat ones, and that pebbled and flat-colored surfaces also hide defects better than smooth, shiny ones.

Liner Coating. Steel tanks used in potable water service need protection against corrosion. Waters from hard-rock drainage basins tend to be particularly aggressive.

Coatings technology is a complex and rapidly evolving specialty. Some general considerations are listed below:

- The liner should be FDA, AWWA or provincial agency approved for potable water service.

- In-service temperature range should include normal and possible extremes.

- Welding needs to be completed before the liner work starts. Note pipe hangers, insulation attachment, etc.

- Surface preparation requirements are likely to be stringent. Field sandblasting a tank interior to "near white metal" surface in a remote location is likely to be very costly compared to in-plant work, usually involving transport of the sand as well as equipment and personnel to the site. In bolted tanks, the liner can be factory-applied.

- Environmental requirements for application and curing – temperature, humidity and ventilation – vary between products, and need to be addressed carefully. In remote northern locations they tend to be both difficult and expensive to fulfill. Poor conditions are a primary cause of early failure. Direct heating of the tank wall results in uneven conditions and should not be permitted. Also note worker protection requirements.

- Quality control testing by specialists is strongly advised, with detailed reporting.

- Terms of warranty, and the soundness of the guarantor should be considered.

- The liner should be inspected every few years. Arrangements are needed to continue essential water service during inspection, repair, and eventual renewal of the liner.

Instrumentation. Needed instrumentation includes water level indication, high- and low-level alarms, and thermal probes at various levels. Air temperature above the water surface is likely to be of interest. External lighting may improve security and an internal light may assist inspection. Thermal monitoring of permafrost foundation soils may be part of the overall project.

7.5.5 Construction Aspects Affecting Design. The overall scheduling of projects in remote settings needs to be planned carefully, especially where transport is confined to a narrow window somewhere near the end of the summer construction season. Scheduling considerations, and limitations of equipment on-site both may affect the design in various ways.

The manufacture of a tank and its erection are often separate contracts, widely separated in both location and time. In the total project, the managing agency may in fact let and coordinate a number of entirely separate contracts, such as tank foundation, tank prefabrication, tank shipping, interim storage, tank erection, tank interior surface preparation and lining, tank exterior insulation and cladding, connections and controls, etc. Alert project management can reduce repetitive costs. Scaffolding, for example, will be needed for two or three of the activities listed above; separate mobilization each time will likely be expensive.

All components need to be packaged properly for shipping, and protected during storage on-site. Bent plates, for instance, can lead to assembly problems, and to leakage in bolted tanks.

Good quality control is needed in erection. Joint assembly is important in bolted tanks. Erection of large welded tanks involves construction technique, proper equipment, environmental conditions suitable for welding, and certified welders. Radiographic inspection of welded tanks is recommended. Vacuum testing will demonstrate absence of leaks.

Aside from difficulties in obtaining concrete, and in assuring quality control in mixing and placing, the long duration of a major cast-in-place concrete project is a drawback where the construction season is short.

7.5.6 Operational and Maintenance Considerations Affecting Design. The discussion below complements aspects which have been dealt with earlier.

Alaskan practice is to inspect tanks visually each year for:

- integrity of foundation,
- integrity of structure,
- integrity of internal coating, and
- leaks.

A tank reservoir storing raw water may need periodic cleaning. The method needs to be planned and provided for. This is especially true for raw water tanks.

If freezing occurs, it is likely that ice layers will be formed over a period of hours or days. With a fluctuating water level, layers may be anchored at various levels, separated by water or air gaps. The method and sequence of thawing needs to be planned and set out in the Operator's Manual.

Information on water tanks in Alaska is provided in Table 7-4. Similar information on water tanks in the NWT is provided in Table 7-5.

7.5.7 Cost Information. Capital cost data for Canadian projects are presented in Table 7-5. Figure 7-6 shows the costs of purchasing and erecting bolted steel water tanks as a function of tank size (Alaska Area Native Health Service, 1994; 1995a; 1995b). Figure 7-7 presents the costs ($/L) of welded steel water tanks as a function of tank size.

Example 7.1   Earthern Reservoir, Winter Water Source

Problem. Investigate the geometry of an earth-diked membrane-lined reservoir to serve as winter water supply for the community of Foxe Basin, NWT. The population is 1000, growth rate 2.2%/year, trucked delivery. Foxe Basin is a high arctic location, MAAT -14.4˚C, continuous deep permafrost. The reservoir site is a large, flat area, very poorly drained. Soils are saturated to ground surface, generally fine sand, sand and gravel with some finer material, reasonably compact, little excess ice. There is a large borrow area nearby and enough equipment on hand to move 2,200 m^3/day. However, the borrow area is frozen and unusable except dur-

ing an eight-week long summer construction season.

Other Data Given.

- Design horizon 20 years.
- Design water use rate (trucked delivery) 90 L/(p•d).
- There is a large shallow lake nearby, a suitable source for refilling. People in the community say it thaws mid-July most years.
- Mean monthly air temperature (MMATs), January-December respectively: -31.0, -32,1, -29.5, -20.9, -9.1, 0.0, 5.4, 4.6, -0.6, -10.5, -21.5, -27.4.
- The liner will be HDPE covered with 500 mm sand and 200 mm rip-rap on the sides, no cover on the floor.
- For dike stability, exterior side slopes will be 1:2.5.
- For liner cover stability, interior side slopes will be 1:3.5.
- Freeboard is to be 2 m, which allows for up to 1 m of settlement.
- Crest width is to be 7 m, for maintenance access. This includes the liner cover zone.
- Anticipated leakage rate is not known. No allowance is made in the preliminary calculations below.

Analysis by Static Method.

- Subdrainage is not practical at this site. Therefore, the reservoir base must remain above ground to avoid liner uplift, no excavation.
- Assume a free-draining base preparation and liner protection layer of clean gravel, 0.5 m thick.
- Analysis of the MMATs indicates that the average mean daily air temperature (MDAT) is above freezing on average, from mid-June to early mid-September. With due allowance for colder years, the reservoir refill window may be as little as six weeks. With due allowance for early fall or late spring, the reservoir needs to be sized for at least 10.5 months consumption, 11 months is safer: 335 days.
- Design year population is $(1,000) (1.022)^{20}$ = 1,545

TABLE 7-4   SUMMARY OF TANK RESERVOIRS – ALASKA

Location	Year Built	Tank Size (ML)	Type of Use	Tank Construction	Ground Condition	Foundation	Insulation	Notes
Savoonga	1976	3.79	Continuous use	Welded steel	Saturated, frozen organics, ice lenses, boulders	Drill/drive, steel pipe piles	89 mm sprayed urethane, clastomeric coasting	Some interior epoxy paint failure. May require repair.
Shishmaref	1976	1.14	Annual fill	Steel	Fine sand.	Sand pad	89 mm sprayed urethane, butyl coating	Epoxy coated, interior inspected 1980, no problems.
Unalakleet	1977	3.79	Continuous use	Steel	Gravel	Gravel pad	89 mm sprayed urethane, butyl coating	Epoxy paint failure, to be repainted.
Twin Hills	1977	0.212	Continuous use	Steel	Unfrozen tundra mat underlain by thawed silts with some gravel and sand	Gravel pad	89 mm sprayed urethane, butyl coating	No problems. Interior epoxy coating.
Stebbins	1978	1.92	Annual fill	Steel	Sand and gravel	Sand and gravel	89 mm sprayed urethane, butyl coating	No problems. Interior epoxy coating.
Grayling	1977	0.212	Continuous use	Welded steel	Unfrozen silts and gravels	Gravel pad	89 mm sprayed urethane, butyl coating	No problems. Interior epoxy coating.

continued

TABLE 7-4   (continued)

Location	Year Built	Tank Size (ML)	Type of Use	Tank Construction	Ground Condition	Foundation	Insulation	Notes
Kaktovik	1978	2.27	Annual fill	Welded steel	Ice rich and gravel, frozen	Wood piling 1.2 x 1.8 m spacing, 4.9 – 5.5 m deep	140 mm sprayed urethane, siding, and butyl roof	Interior piling failure, slurry not installed. Corrected by installation of interior post and pad supports.
Nenana	1979	0.568	Fill on demand (2 times/year)	Welded steel	Gravel	Select fill gravel pad with 50 mm insulation	89 mm polyurethane, siding, and butyl roof	No problems. Two coats of epoxy paint.
Eek	1979	0.238	Annual fill	Welded steel	Frozen silts and organics	150 mm dia. steel piling 5.5 m deep, 1.2 x 1.8 m spacing	89 mm polyurethane, siding, and butyl roof	No problems. Two interior coats of paint.
Nuiqsuik	1980	8.59	Annual fill	Welded steel	Frozen silts and organics	Select fill, gravel pad, insulation, mud sills and wood deck	140 mm preformed board stock, siding	No problems. Five interior coats of vinyl.
Russian Mission	1980	0.223	Fill on demand	Welded steel	Silts	Select fill, gravel pad	140 mm polyurethane spray, siding, butyl roof	No problems. Two interior coats of epoxy.
Emmonak	1993	1.20	Fill on demand	Bolted steel	Permafrost (frozen silt)	Select fill, gravel pad	50 mm polyurethane, 0.58 mm metal sheath	No problems. Two interior coats of epoxy.

continued

TABLE 7-4   (continued)

Location	Year Built	Tank Size (ML)	Type of Use	Tank Construction	Ground Condition	Foundation	Insulation	Notes
Clarks Point	1993	0.049	Continuous	Bolted steel	Gravel	Concrete pad	None	
Kokhanok	1995	0.189	Continuous	Bolted steel	Bedrock	Rock	100 mm polyurethane	
Chalkysik	1995	0.302	Fill on demand	Bolted steel	Silt	Pad with thermosyphons	100 mm polyurethane	
White Mountain	1990	0.378	Continuous	Bolted steel	Sand/gravel	Gravel pad	100 mm polyurethane	
Arctic Village	1990	0.378	Fill on demand	Bolted steel	Sand	Gravel pad	100 mm polyurethane	
Gulkana	1992	0.378	Continuous	Bolted steel	Sand	Gravel pad	100 mm polyurethane	
Scammon Bay	1995	0.378	Continuous	Bolted steel	Silt	Gravel fill	100 mm polyurethane	
Nulato	1995	0.473	Continuous	Bolted steel	Silt	On-grade	100 mm polyurethane	
Manokotak	1995	0.567	Continuous	Bolted steel	Bedrock	Pad	100 mm polyurethane	
Chevak	1995	0.567	Fill on demand	Bolted steel	Silt	Pad with thermosyphons	100 mm polyurethane	
Kaltag	1995	0.801	Continuous	Bolted steel	Silt	Pad with thermosyphons	100 mm polyurethane	
Tooksook Bay	1995	0.801	Continuous	Bolted steel	Silt	Gravel pad	100 mm polyurethane	
Eek	1995	0.801	Fill on demand	Bolted steel	Silt	Thermopiles	100 mm polyurethane	
Marshall	1989	0.801	Continuous	Bolted steel	Silt	Gravel fill	100 mm polyurethane	
Akiachak	1995	1.134	Continuous	Bolted steel	Silt	Pad with thermosyphons	100 mm polyurethane	
Noorvik	1993	1.202	Continuous	Bolted steel	Silt	Gravel pad	100 mm polyurethane	
Selawik	1994	1.202	Continuous	Bolted steel	Silt	Steel piles	100 mm polyurethane	
Togiak	1990	1.890	Continuous	Bolted steel	Sand	Gravel pad	100 mm polyurethane	
Kipnuk	1993	1.890	Fill on demand	Bolted steel	Silt	Thermopiles	100 mm polyurethane	
Kivalina	1994	2.079	Fill on demand	Bolted steel	Sand	Pad with thermosyphons	100 mm polyurethane	

TABLE 7-5   SUMMARY OF TANK RESERVOIRS – NORTHERN CANADA

Location	Year Built	Tank Size (ML)	Type of Use	Tank Construction	Foundation	Insulation	Cost ($CAN)
Yukon							
Watson Lake	1974	11.5	Continuous	Concrete two cells	Compacted native material	50 mm styrofoam 450 mm earth cover	151,000
Haines Junction	1975	4.80	Continuous	Elevated wood stave	H-piles	62 mm sprayed-on urethane foam	306,500
Whitehorse – Porter Creek/ Crestview	1976	5.92	Continuous	Concrete	Compacted native material	64 mm sprayed-on urethane, 150 mm earth cover	500,000
Pelly Crossing	1979	0.0045	Water/fire-truck fill				4,500
Dawson City	1979	8.20	Continuous	Bolted steel with baked enamel	Compacted native material	50 mm rigid foam, 600 mm earth cover	157,000
Faro	1980	29.5	Continuous	Concrete	Compacted native material	50 mm rigid foam, 600 mm earth cover	722,800
Carcross	1980	0.0045	Truck fill	Fibreglass		Contained inside heated building	13,000
Whitehorse – Hillcrest Zone II	1980	59.2	Continuous	Concrete two cells	Compacted native material	50 mm rigid foam 600 mm earth cover	808,300

continued

TABLE 7-5   *(continued)*

Location	Year Built	Tank Size (ML)	Type of Use	Tank Construction	Foundation	Insulation	Cost ($CAN)
Northwest Territories							
Pine Point	1972	4.54	Continuous	Burried concrete			207,600
Fort McPhernson	1975	0.45	Continuous	Wood stave	Piles	50 mm polystyrene	166,000
Resolute Bay	1976	0.45	Continuous	Bolted steel	Insulated gravel pad	50 mm polystyrene	190,000
Inuvik	1976	2.25	Continuous	Welded steel	Insulated vented pad	2 mm polystyrene	750,000
Grise Fiord	1978	1.41	Annual fill	Welded steel	Insulated gravel pad		350,570
Norman Wells	1978	0.91	Continuous	Welded steel	Gravel pad and cryo-anchors		301,931
Aklavik	1979	0.23	Continuous	Welded steel	Insulated gravel pad	50 mm polyurethane	
Yellowknife	1978	1.25	Continuous	Concrete	Bedrock	50 mm polystyrene	800,000
Cambridge Bay	1980	0.18	Continuous	Welded steel			138,019
Grise Fiord	1987	3.92	Annual fill	Welded steel	Insulated gravel pad	64 mm rigid	1,186,016
Rankin Inlet	1991	3.36	Continuous	Welded steel	Insulated gravel pad	50 mm polyurethane	1,442,810
Yellowknife	1991	6.00	Continuous	Concrete	Bedrock	38 mm polystyrene	3,633,655
Cape Corset	1992	0.54	Continuous	Welded steel	Insulated gravel pad	65 mm rigid	741,600

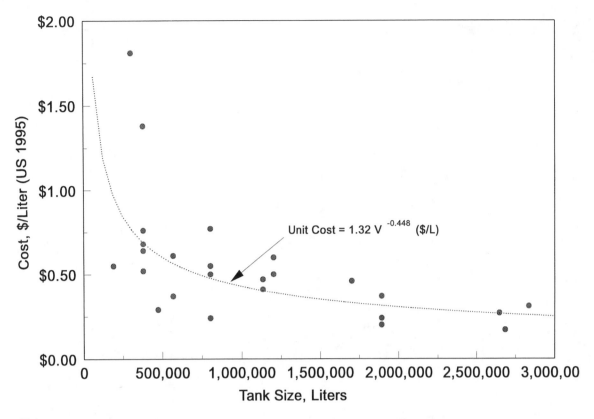

Unit Cost = 1.32 V $^{-0.448}$ ($/L)

Note:
1. Costs include 100 mm side insulation and 150 mm roof insulation with prefabricated metal exterior.
2. Foundation varies from gravel pad to pile supported elevated wood insulated platform.
3. Tanks are located in western and northern Alaska communities.

FIGURE 7-6   BOLTED STEEL WATER TANK COSTS

- Design year water use is: Daily basis: 1,545 pop.•90 L/(p•d) = 0.139 ML/d; Winter supply period: 0.139 ML/d•335 d = 46.6 ML

- Analysis of the MMATs (by conversion to an equivalent sine wave about the MMAT) (R) indicates that the air freezing index is 5568°C days per year.

- Ice thickness z (m) is calculated, assuming that the surface freezing index (SFI) is equal to the air freezing index (conservative, but only slightly so):

$$z = e•(SFI)^{0.5} \tag{7-1}$$

where,

e = coefficient, 8.1 to 11.0 (say 10) •10^{-5} m/(°C sec)$^{0.5}$

SFI = 5,568°C days • 86,400 sec/day

z = $10.0•10^{-5}$ $(5,568•86,400)^{0.5}$ = 2.19, say 2.2 m (as water, not ice)

- Dead storage below intake level is assumed to be 0.5 m, a reasonable figure based on experience.

- Total water depth is assumed, initially, to be 8.0 m. Investigation of a range of depths to optimize geometry is strongly advised.

- The water volume is the base section of a square pyramid, inverted, side slope S, is:

$$V = \int Adh = \int (2 \times Sh)^2 dh$$

$$= (4/3)S^2 (h_2^3 - h_2^3) \tag{7-2}$$

Since S = 3.5, V = 16.3 $(h_2^3 - h_1^3)$ (7-3)

Setting

h_d = top of dead storage, from pyramid point,

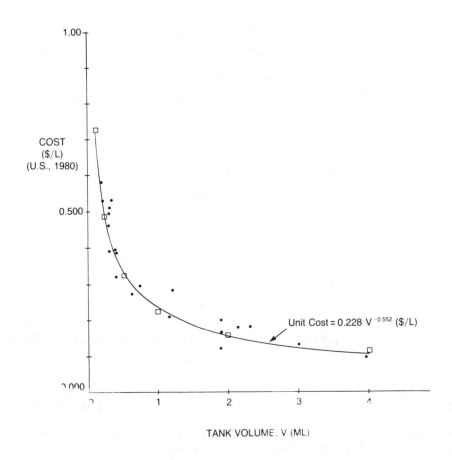

COST
($/L)
(U.S., 1980)

0.500

Unit Cost = 0.228 V$^{-0.552}$ ($/L)

0.000

TANK VOLUME, V (ML)

FIGURE 7-7   WELDED STEEL WATER TANK COSTS

h_w = underside of ice, from pyramid point, then,

$$h_w - h_d = 8.0 - 0.5 - 2.2 = 5.3 \text{ m} \qquad (7\text{-}4)$$

and

$$16.33 (h_w{}^3 - h_d{}^3) = 46,600 \text{ m}^3 \qquad (7\text{-}5)$$

Solving 7-4 and 7-5 together yields

h_d = 10.7 and h_w = 16.0 m

The reservoir floor is at h_d - 0.5 m = 10.2 m

Lengths along each side are 2•slope 3.5•h

At reservoir base 2•3.5•10.2 m = 71.1 m

At top of water (thawed) 2•3.5•18.2 m
       = 127.1 m

At inner crest 2•3.5•20.2 m = 141.1 m

At outer crest 141.1 m + 2•(crest 7 m) = 155.1 m

The total dike height including 0.5 m base preparation = 8.0 m water + 2.0 m fresh sand + 0.5 m base = 10.5 m.

The side length at the base is

155.1 m + 2•(slope 2.5) (10.5 m) = 207.6 m.

By static calculation,

total storage = 16.3 (18.2^3 - 10.2^3)

         = 80,667 m^3

ice (as water) = 16.33 (18.2^3 - 16.0^3)

         = 31,411 m^3

live storage   = 16.3 (16.0^3 - 10.7^3)

         = 46,601 m^3

dead storage = 16.3 (10.7^2 - 10.2^2)

         = 2,655 m^3

diked volume = 16.3 (20.2^3 - 10.2^3)

= 116,668 m³

The total dike volume including liner bedding and cover can be estimated readily, as the base of a truncated pyramid:

height = length/(2•slope) = 207.6/(2•2.5)

= 41.5 m

Truncated 10.5 m above ground,

41.5 m - 10.5 m = 31.0 m

V pyramid base = (4/3) S^2 ($h_2^3 - h_1^3$)

= (4/3)•2.5^2•[(41.5 m)³ - (31.0 m)³]

= 347,734 m³

Less diked volume	116,668 m³
Total earthwork	231,066 m³

Construction time at 2,200 m³/d, 105 days. Rough earthwork alone probably will take two seasons with existing equipment, and liner work a full third season. First filling probably not until the fourth season.

Note that in the trial design considered, total earth volume is twice the net storage yield. A range of reservoir depths should be tested to determine the optimum configuration that minimizes the total of earthwork cost and liner cost.

Check of Trial Design by Dynamic Drawdown

Dynamic drawdown is a time-step calculation. It takes into account the gradual reduction of the ice-growth plane, as the ice-water interface is drawn further down the sloped inner faces of the reservoir.

In the calculation below:

1. The reservoir is filled in summer. The earliest date for completion and cessation of filling is August 20, day 233 (of 365).

2. The example calculation is based on an average winter. In designing, extreme winters should be considered.

3. The mean daily air temperature falls below 0°C, on average, on September 12, day 256.

4. The time step chosen is 14 days.

5. Water consumption is 139.1 m³/d, 1947.4 m³ per time step. In the example, no allowance has been made for leakage. The computer printout shows the volume of water, or water and ice, remaining in the reservoir at the end of each time step.

6. The mean daily air temperature (MDAT) is calculated by a sine-curve approximation derived from the published mean monthly air temperatures listed earlier. The printout shows the calculated MDAT at the end of each time step.

7. Cumulative degree-days at the end of each time step are found by integrating the temperature sine function, from day 253 to each time period end. In the example, the sine-curve over each successive time step is approximated by a straight line.

8. The ice depth at the end of each time step is found from the degree-days column using the Stefan equation. Ice depths and volumes are calculated as water equivalent, not at ice density.

9. As the reservoir is drawn down, ice previously formed is continually beached and left behind on drying shorelines. It is assumed that this proceeds continuously by shearing along a close succession of vertical planes and there is no tilting of ice blocks.

10. As time passes, the level of the freezing plane is lowered within the reservoir, both by withdrawal of water and by loss of water volume to formation of ice. The area of the freezing plane diminishes significantly, due to the relatively low inner slope common in earth-diked reservoirs. The area of the freezing plane in a particular time step, which is used to calculate the volume of new ice formed in that time step, is based on its estimated level at the midpoint of the time step. Midpoint level is considered to be:

    - level at the end of the preceding time step,

    - less half the thickness of new ice growth in the time step,

    - less half the further drop in water level due to withdrawal (consumption) of water in the time step.

11. The volume of new ice formed found by multiplying ice thickness growth in the time step by the water surface area corresponding to the midpoint level found above. Total ice volume formed is accumulated.

12. Water volume left at the end of the time step is found by subtracting accumulated ice volume

from the current remaining total volume of water and ice.

13. Water volume left at the end of the time step is used to calculate water level at the end of the time step, from reservoir geometry.

14. Change in the concentration of dissolved minerals in a time step is considered to be approximately equal to

$$V_{wi}/(V_{wi} - 0.5 \cdot \text{consumption} - V_{gn})$$

where,

V_{wi} = initial volume of water, beginning of time step, consumption is the volume consumed in the time step,

V_{gn} = volume of new ice formed in the time step.

15. Concentration at the end of a time step is found by multiplying concentration at the end of the last time step by the new change.

The results of the dynamic drawdown method show that proper evaluation of water use and liquid water loss to ice allows more accurate evaluation of storage capacity.

7.6   References

Alaska Area Native Health Service. 1994. *Water Storage Tanks (WST) Design Considerations.* Technical Memorandum, Office of Environmental Health and Engineering, August 1.

Alaska Area Native Health Service. 1995a. *Costs Summary of Water Storage Tanks in Rural Alaska.* Internal Memorandum, Office of Environmental Health and Engineering, October 20.

Alaska Area Native Health Service. 1995b. *Water Storage Tank Foundations – Seismic Considerations* Office of Environmental Health and Engineering, February 27.

Cameron, J.J. 1976. Waste impounding embankments in permafrost regions: the sewage lagoon embankment, Inuvik, NWT. In: *Some Problems of Solid and Liquid Waste Disposal in the Northern Environment*, Environmental Protection Service, Rep. No. EPS-4-NW-760-2, Edmonton, Alberta.

Crist, B. 1993. Department of Municipal and Community Affairs, Government of the Northwest Territories, Inuvik, NWT. Personal communication, February 5, 1993.

Folkes, D.J. 1981. Control of contaminant migration by the use of liners. Presented at 5th Canadian Geotechnical Colloquium, 34th Canadian Geotechnical Conference, Fredericton, New Brunswick.

Kirby, P.C. 1979. *The Selection of Membrane Materials for Water Distribution and Allied Applications.* Water Research Center, Technical Report TR113, Medmenham, U.K.

Smith, D.W. and S.R. Justice. 1975. *Effects of Reservoir Clearing on Water Quality in the Arctic and Subarctic.* Institute of Water Resources, Rep. No. IWR-58, University of Alaska, Fairbanks.

Smith, D.W. and S.R. Justice. 1976. *Clearing Alaskan Water Supply Impoundments: Management, Laboratory Study, and Literature Review.* Institute of Water Resources, Rep. No. IWR-67, University of Alaska, Fairbanks, Alaska.

Smith, L.B., W.G. Notenboom, M. Campbell, S. Cheema and T. Smyth. 1989. Pangnirtung Water Reservoir: Geotechnical Aspects. *Canadian Geotechnical Journal*, 26(3): 335-347.

Stanley, S.J. and D.W. Smith. 1991. Reduction of ice thickness on northern water reservoirs. *Journal of Cold Regions Engineering*, 5(3): 106-124.

Thornton, D.E. and P. Blackall. 1976. *Field Evaluation of Plastic Film Liners for Petroleum Storage Areas in the Mackenzie Delta.* Environmental Conservation Directorate, Environmental Protection Service, Rep. No. EPS 3-EC-76-13, Ottawa, Ontario.

7.7   Bibliography

Alter, A.J. and J.B. Cohen. 1969. Cold region water storage practice. *Public Works Magazine*, 100(10): 109-111.

American Water Works Association. *American Water Works Association Standards.* Denver, Colorado.

Brown, W.G. and G.H. Johnston. 1970. Dykes in permafrost – predicting thaw and settlement. *Canadian Geotechnical Journal*, 7(4): 365.

Cohen, J.B. and B.E. Benson. 1968. Arctic water sorage. *Journal of the American Water Works Association*, 60(3): 291-297.

Department of Public Works. 1979. *Northern Municipal Services, Water and Sanitation Components Unit Cost for 1978-79 Projects*. Government of the Northwest Territories, Yellowknife, NWT.

Eaton, E.R. 1964. Floating plastic reservoir covers in the Arctic. *Solar Energy*, 8(4): 116.

EBA Engineering Consultants Ltd. 1977. *A Study of Spray-on Liners for Petroleum Product Storage Areas in the North*. Environmental Impact Control Directorate, Environmental Protection Service, Rep. No. EPS 4-EC—77-2, Ottawa, Ontario.

Foster, R.R. 1975. Arctic water supply. *Water and Pollution Control*, 3(3): 24.

Foster, R.R., T.J. Parent and R.A. Sorokowski. 1977. The Eskimo Point water supply program. In: *Proceedings, Utilities Delivery in Arctic Regions,* Environmental Protection Service, Rep. No. EPS 3-WP-77-1, Ottawa, Ontario, 553-575.

Johnson, G.D. (editor) 1981. *Permafrost – Engineering Design and Construction*. John Wiley & Sons, Toronto, Canada, 40 p.

Kay, W.B. 1977. *Construction of Liners for Reservoirs, Tanks and Pollution Control Facilities*. John Wiley & Sons, New York, N.Y.

Layton, J.A. and M.D. Gibb. 1978. Diving inspection of an arctic reservoir. In: *Proceedings, Conference on Applied Techniques for Cold Environments, Cold Regions Specialty Conference,* American Society of Civil Engineers, New York, 221-232.

Lumsden, T.W. and K. Siu. 1984. Pangnirtung water supply. In: *Proceedings, Third International Specialty Conference on Cold Regions Engineering,* Canadian Society of Civil Engineering, Montreal, Quebec, 307-322.

Nehlsen, W.R. and J.J. Traffalis. 1954. *Persistence of Chlorine Residuals in Stored Ice Water*. U.S. Naval Civil Engineering Research and Engineering Laboratory, Technical Note N-206, Port Hueneme, California.

Rice, E.F. and O.W. Simoni. 1966. The Hess Creek dam. In: *Proceedings, Permafrost International Conference, Lafayette, Indiana*. National Academy of Science, Washington, D.C.

Ryan, W.L. 1983. Design considerations for large storage tanks in permafrost areas. In: *Proceedings of the Fourth International Conference on Permafrost.*

Smith, L.B., A. Shevkenek, R. Millburn. 1984. Design and performance of earthworks water reservoirs in the Northwest Territories. In: *Proceedings Third International Specialty Conference on Cold Regions Engineering,* Canadian Society for Civil Engineering, Montreal, Quebec, 323-342.

Toman, G.J. 1967. Elevated water storage tank freeze-ups: joint discussion. *Journal of the American Water Works Association,*59(2): 166-168.

Wormald, L.W. 1972. Water storage tank failure due to freezing and pressurization. *Journal of the American Water Works Association*, 64(3): 173-175.

SECTION 8

WATER DISTRIBUTION

3rd Edition Steering Committee Coordinators

Vern Christensen

James A. Crum

3rd Edition Principal Author

Robert H. Boon

Special Contributions

Michael Mauser

Section 8   Table of Contents

Section 8   List of Figures

8 WATER DISTRIBUTION

8.1 Introduction

In this section the various methods of distributing water in cold regions from source to user are presented.

Water can be distributed by self-haul, vehicle haul, or piped systems. The system used in a given community depends on a number of factors including:

- government policy regarding health standards, levels of service, subsidy programs for both capital and operating costs, and community development strategies;

- geographic and physical site conditions such as community layout, building density and types, and the availability of water and other physical constraints;

- local factors such as population, lifestyle of residents, and economic base of the community; and

- the ability and willingness of the community to operate and maintain the utility systems.

The funding of both capital and operational costs for water distribution systems in northern communities is often different from that in southern communities. In both the United States and Canada, special programs exist to provide utility services in northern communities. The responsibility of the local government with regard to funding of both capital and operational costs is quite different from that considered normal elsewhere. For this reason, an economic analysis to select a utility system for a particular community must consider the capability of the local government to provide capital funds as well as its participation in ongoing operation and maintenance of the completed system. In the Northwest Territories, general terms of reference have been prepared for use in carrying out the economic analysis (Government of the Northwest Territories, 1986). This analysis is geared specifically to the funding and operating responsibilities in the Northwest Territories program and the availability of both capital and operating subsidies (Government of the Northwest Territories, 1989). In Alaska, capital funding is provided by the federal or state government, leaving the local government responsible for operating and maintaining the completed system. It is very important that the local people participate in the decision making process concerning the type of system most appropriate for the community, since they pay the cost of operation and maintenance (Ryan, 1977).

8.2 Prospective Water Consumption as a Design Consideration

With any type of water distribution system, water use is a key factor in system selection and design. Water conservation fixtures (for example, low water use showers and toilets) can reduce consumption and should be a part of building design for most northern communities. See Section 14 for detailed information on water conservation methods. The type of plumbing in the buildings and the water distribution system will have a significant impact on the quantity of water used. Experience in Alaska and the Northwest Territories has yielded the following figures for water use:

- houses with no indoor plumbing – 20 L/(p•d) or less;

- trucked water and sewer to homes with conventional plumbing – approximately 90 L/(p•d) (Table 5.2); and

- piped water and sewer systems – 225 to over 500 L/(p•d), if water is bled to the sewers.

Total community water use depends on the residential population water use, the commercial and industrial water use, bleeding required for freeze protection (which is being phased out in the NWT) and system losses due to leakage.

System designs incorporating fire protection capability are described in Section 15. Community population and development information is available from the appropriate government departments: in the Northwest Territories, the Department of Municipal and Community Affairs; in the Yukon, the Department of Community and Transportation; and in Alaska, the Department of Community and Regional Affairs.

8.3 System Options

The primary options for components of water systems in cold regions are illustrated in Figure 2-5. The options described in this section reflects current practice in cold regions but should not stifle innovation.

8.4   Self-Haul Systems

This method is practiced mainly in small communities of 100 people or fewer where little or no mechanization exists. The people obtain their water from a nearby lake or river and haul it to their house. Factors other than community size that also contribute to the practice of hauling water may include personal preference and cheap and convenient availability of water sources.

As the community grows, the increased population density and poor sanitation practices often lead to sewage contamination of the drinking water source. One solution has been to provide central watering points where safe water can be obtained. These watering points are usually only installed when the community does not have the necessary infrastructure, such as roads or water holding tanks in buildings, to accommodate a piped or trucked water and sewage system. Watering points have also been installed in communities that have neither the technical nor the economic resources to operate a more complex and expensive system.

Piped distribution systems for operation in the summer have been installed in some communities. With this system, uninsulated small-diameter pipelines are laid on top of the ground or buried just below the ground surface to distribute water to several faucets or hydrants in the community from which water can be obtained. This allows residents to obtain water closer to their home.

For several reasons water delivery to homes and buildings is preferred over self-haul:

- It is a more positive means of supplying housing units regularly with clean, safe water.

- It encourages increased water usage to facilitate better hygiene.

- Unsupervised watering points tend to be vandalized, and hence become unsanitary and nonoperational.

The higher capital and operating costs associated with trucked and piped systems must also be considered, however, economics cannot be allowed to override health concerns.

8.4.1   Watering Point Design. Watering points (Figure 8-1) may be provided in conjunction with trucked or piped water distribution, or they may be the only source of water for self-haul. Watering points are often installed in older parts of a community where houses do not have internal plumbing.

They are constructed as an interim solution to provide safe drinking water until water distribution to buildings is initiated.

Design objectives of the self-haul watering points are to:

- maximize cleanliness and sanitation in getting water to the container;

- maximize convenience relative to traditional water sources;

- minimize maintenance;

- provide durable design to minimize vandalism; and

- lower costs.

The failure rate of the self-haul watering point can be high, mainly because of vandalism and freeze-ups (the result of lack of supervision and management) and poor design. Consequently, water points are often abandoned and not repaired. Many watering points in Alaska are, however, still in operation after many years.

8.4.2   Types of Self-Haul Water Point Systems.   The types of self-haul systems are characterized by their location and accessibility. (See Section 12 for detailed information about central facilities.)

Exterior Watering Points.   Exterior watering points are usually located on the exterior wall of a small heated building. The building is typically a small water treatment plant, central facility, or remote watering point on a circulating water line. The building may house a well, water tank, piping, pumps and controls, or treatment equipment, depending on the type and extent of the facility. Older type watering points used nonelectric mechanical spring-loaded valves located inside the building and were operated by pulling a rope or chain extended from the valve through the wall. More recent designs have incorporated solenoid valves activated by a push button or coin box releasing water for as long as the button is held or for the preselected time set in the coin box. When the button is released or the coin box predetermined limit is reached, the solenoid valve closes and the remaining water in the line drains through the spout. Heat tape can be provided to prevent ice buildup in the discharge tubing which has been a problem for older designs. A typical watering point plumbing and heat tape detail is shown in Figure 8-1. A suitable splash pad of either gravel, slotted boards, or grating should be

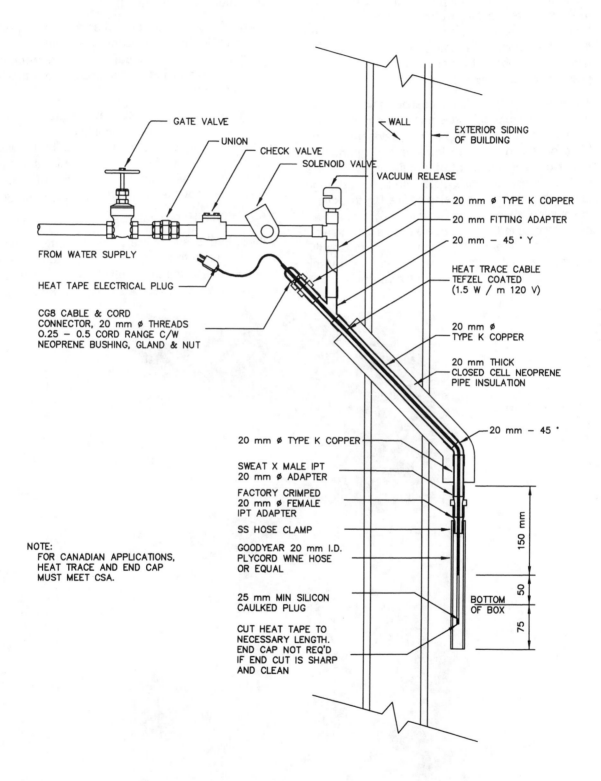

GATE VALVE

UNION

CHECK VALVE

SOLENOID VALVE

VACUUM RELEASE

WALL

EXTERIOR SIDING
OF BUILDING

FROM WATER SUPPLY

HEAT TAPE ELECTRICAL PLUG

CG8 CABLE & CORD
CONNECTOR, 20 mm ∅ THREADS
0.25 – 0.5 CORD RANGE C/W
NEOPRENE BUSHING, GLAND & NUT

20 mm ∅ TYPE K COPPER

20 mm FITTING ADAPTER

20 mm – 45 ° Y

HEAT TRACE CABLE
TEFZEL COATED
(1.5 W / m 120 V)

20 mm ∅
TYPE K COPPER

20 mm THICK
CLOSED CELL NEOPRENE
PIPE INSULATION

20 mm – 45 °

20 mm ∅ TYPE K COPPER

SWEAT X MALE IPT
20 mm ∅ ADAPTER

FACTORY CRIMPED
20 mm ∅ FEMALE
IPT ADAPTER

SS HOSE CLAMP

GOODYEAR 20 mm I.D.
PLYCORD WINE HOSE
OR EQUAL

25 mm MIN SILICON
CAULKED PLUG

CUT HEAT TAPE TO
NECESSARY LENGTH.
END CAP NOT REQ'D
IF END CUT IS SHARP
AND CLEAN

150 mm

50 mm

75 mm

BOTTOM
OF BOX

NOTE:
FOR CANADIAN APPLICATIONS,
HEAT TRACE AND END CAP
MUST MEET CSA.

FIGURE 8-1   WATERPOINT HEAT TAPE INSTALLATION

provided to allow dispersal of spilled water. Also, the area around the watering point should be level to assist users in getting water. In winter, ice buildup from spillage will require periodic removal to keep the watering point accessible. The watering point hose should be kept far enough off the ground to prevent contamination (Figure 8-2).

The exterior of the building must be as vandal-proof as possible, preferably constructed of logs, concrete, or heavy-duty siding. The entrance door should be kept locked to prevent damage and tampering. Water can be supplied to the building by either a well or a pipeline from a distant source. A storage tank in the building could also be filled by truck from the outside. Two outside lights (one light for redundancy) would indicate when the tank is full and alert the truck operator to stop filling. Failure to stop filling when the lights come on would result in the tank overflowing to the outside through a vent pipe. Simi-

lar on-line systems in conjunction with water distribution mains are also used.

Interior Watering Points. Interior watering points are similar to the exterior ones. They must be supervised to prevent vandalism or freeze-up, designed for spillage and for restricted access to other parts of the building.

Containers. An enclosed water container is recommended to prevent contamination. Containers over 20 L in capacity generally are too heavy when full. In practice, anything that can hold water, often an open water pail, is used. Disinfection of containers and water by the users should be encouraged.

8.5 Vehicle-Haul Systems

Vehicle-haul systems are those systems where a vehicle is used to transport water to storage tanks in individual buildings. Vehicle systems include tracked or wheeled vehicles equipped with a water

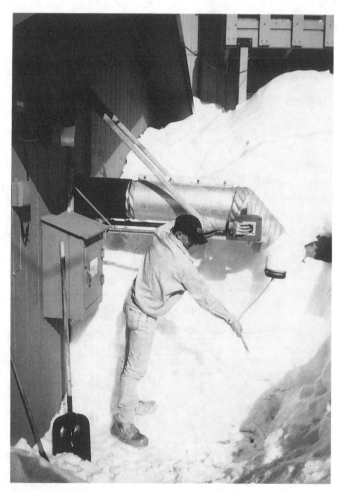

FIGURE 8-2   EXTERIOR WATERING POINT

tank, or a water tank mounted on a trailer towed behind a vehicle.

The trucked water delivery system is used in all or portions of most communities in northern Canada. The trucked system is relatively expensive to operate and maintain, but the capital investment is lower than that of a piped system. Operation and maintenance (O&M) costs for a vehicle haul system vary with individual community conditions. Water use, haul distance, frequency of service to buildings, size of domestic storage tank and vehicle tank size all influence capital and O&M costs. Additional information on trucked systems, and methods to estimate costs, are provided in Appendix C and in Cameron (1981).

In Alaska a smaller scale haul system using a 400 L wheeled insulated tank trailer pulled by an ATV vehicle is being used with some success. It allows individual families or a village employee to haul water from a watering point along the village boardwalks for storage in individual tanks in each home. This system is designed to provide a minimum amount of water to homes in villages that cannot afford the cost of delivery of larger quantities of water.

In a nonpiped community, the water loading point for the vehicle is usually a prefabricated building on the shore of a lake or river, or beside a well or a reservoir. The costs of trucked water delivery can be reduced by constructing a water supply pipeline from the water source to the community and a truck fill point within the community.

8.5.1 Water Loading Point Design. Objectives for the design of the water loading point are to:

- minimize potential for contamination;

- minimize spillage which would produce puddles and ice buildup around the water point;

- minimize maintenance requirements, snow removal needs and costs;

- create efficient and safe entry and exit routes taking into consideration prevailing winds and snowdrifting; and

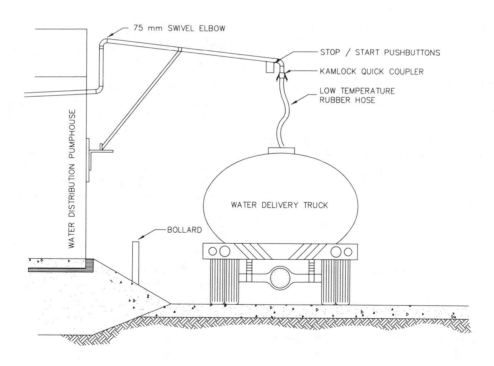

FIGURE 8-3   OVERHEAD TRUCK FILL POINT

FIGURE 8-4   WATER TRUCK FILL POINT

- provide efficient water truck fill rates.

Three types of systems exist: overhead pipe loading, hose nozzle fill, and interior building fill point. The overhead type is the most common.

Overhead Pipe Filling Systems.   Figures 8-3 and 8-4 show typical overhead systems. All water supplied should be metered.   Separate meters/key switches are sometimes required for each water delivery contractor. The portion of the loading arm located outside the building must drain after each filling to prevent freezing. From the high point, water drains back to the heated building or into the truck. Ideally, a swivel-type elbow would allow the loading arm to swing away if hit by a vehicle, however, this is difficult to insulate to maintain thermal integrity.

The system is operated by either an exterior key start switch, pushbutton on the overhead arm or pushbuttons in a lockable cabinet. Such arrangements limit unauthorized water use and wastage, and eliminate the need for truck drivers to enter the building.

A problem with this type of fill design is that invariably, the pump gets turned off when the truck tank is overflowing. This leads to ice buildup on the ground which makes parking for the truck difficult and often results in the siding of the building being damaged by vehicles.

Guard rails should be used to protect the building, and the water fill pipe should be located so that the downspout is far enough away from the building and high enough off the ground to compensate for seasonal snow and ice buildup. In some communities, more sophisticated controls allow the truck operator to preset the volume of water to be supplied or the time the pump is to be on, to increase efficiency and reduce spillage.

To prevent splashing during filling and to compensate for the seasonal variation in height, a flexible low temperature rubber hose should be connected to the bottom of the downspout (Figure 8-3). Start/ stop buttons on the overhead fill arm allow the operator to stop the pump when the tank is full without overflowing.

Hose Nozzle Filling System.   To prevent spillage and ice buildup, a fill point using a gas tank type fill hose nozzle has been used. A small door in the loading point building provides access to a hose which is pulled out through roller guides (Figure 8-5).

A platform is provided so the driver can take out the hose nozzle, put it in the top of the truck, and then turn on the key start. The nozzle is designed to turn off under back pressure. The hose must be long enough to reach the truck but not long enough that it could touch the ground. This type of truck fill is quite slow because of the relatively small hose and the need for the driver to handle it before and after filling.

While this design does not totally eliminate water spillage because of misuse, it reduces the amount of spillage compared to the overhead fill pipe. There is some spillage when the driver pulls the hose out of the tank and puts it back into the building.

Interior Fill Points. These are similar to the overhead type, except that the fill point is inside a building, usually the water truck garage.

Truck Filling Points. Design should provide a short turnaround time and economical truck operations. The recommended minimum loading rate is 16.5 L/s. While this rate does not provide full fire protection, as fire trucks can discharge water at three times this rate (Cameron, 1981), it is a reasonable rate since driving time generally forms the majority of truck turnaround time.

The truck filling points should be well-lit because in the winter it is dark during working hours.

The truck should be able to drive through the loading point, rather than having to back up. Backing up requires more time and often results in accidental damage to the building or support structure. To keep snow from accumulating in front of the loading point, the road should be built up higher than the surrounding ground and in line with the prevailing winds.

Where piped systems are available in a community, the water truck can obtain water from the fire hall or any other suitable building that has piped water service and appropriate connections for filling a water truck. However, a separate water loading point should be provided where truck filling is frequent. The water truck and the fire truck can both be stored in the same building with the filling point inside it. This is appropriate only in small communities where service is provided by the government rather than by contractors.

8.5.2 Vehicle Sizing and Design. Water trucks and sewage pumpout trucks are designed so that the holding tanks and filling or suction mechanism can be housed on a standard vehicle chassis.

Cost information is provided in Appendix B.

Tracked vehicle use in northern Canada and Alaska has been eliminated. It was found that in normal snow and road conditions a wheeled vehicle can do

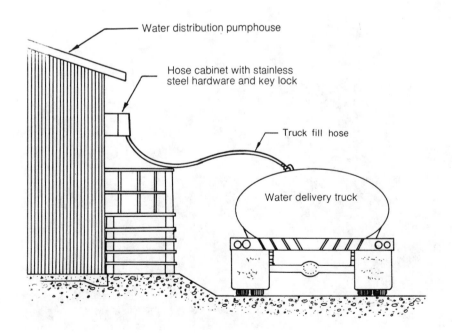

FIGURE 8-5   WATER TRUCK FILL POINT WITH NOZZLE FILL

the job adequately. In areas where poor roads or heavy snow and drifting conditions are prevalent, wheeled vehicles with four-wheel drive and large floatation tires are used successfully (Figure 8-6).

Use of tracked vehicles is discouraged because of:

- high initial cost;

- high annual maintenance costs and difficulty in obtaining parts;

- short life expectancy;

- lower payload; and

- slower speed and therefore longer travel time from fill point to buildings, and between buildings.

Tracked vehicles are only recommended for those communities:

- without roads, where only a tracked vehicle can get to the water source or the buildings;

- without snow removal; or

- where the road conditions prohibit the use of heavy wheeled vehicles.

Truck and Tank Sizes. A larger tank results in fewer trips to the fill point. However, large tanks re-

quire large vehicles. The condition of the roads in the community and to the fill point, road access to buildings and other local conditions determine the maximum size of tanks and vehicles. In communities of fewer than 250 persons, truck tank units of 2,225 litres capacity may be adequate. A minimum tank capacity of 4,450 litres is usually recommended, since the cost difference between the two is small and the larger truck provides more flexibility. Where the community infrastructure is still rudimentary (no mechanics, etc.), the 2,225 litre tank assembly can be mounted on a trailer chassis. These units are the same as those mounted on a truck chassis, except they are towed by a loader or other suitable vehicle (Figure 8-7). Another versatile arrangement for such small communities is the use of a flat-bed truck with a removable water delivery unit.

The number of vehicles required depends primarily upon total water consumption by the community, frequency of service to individual buildings, distance to the water source or fill point, vehicle characteristics and efficiency of operation. Equations to estimate the time and vehicle requirements, and the costs associated are presented in Appendix B.

Trucked systems are influenced by water consumption. In housing with only rudimentary household

FIGURE 8-6   TRUCK WITH FLOTATION TIRES

FIGURE 8-7   2,225 LITRE WATER TRAILER

plumbing (a 200 to 450 L water holding tank, a sink with a direct pipe to the outside and a honey bucket) water consumption is usually about 10 to 20 L/(p•d).

Housing with full plumbing (pressurized water systems with water holding tanks and sewage pumpout tanks) water use is higher. For example, private, five-person homes with full conventional plumbing on trucked systems in Yellowknife, NWT, have average water consumption of 87 L/(p•d). Water conservation measures can reduce this demand.

Since trucked system costs are directly proportional to the total volume of water delivered, trucked service becomes more expensive to operate as the residential and nonresidential water demand increases with higher populations or higher standards of housing and plumbing, or both. The installation of a piped supply line from the source to the community will reduce the vehicle travel time and may be more economical. A water distribution system to some or all of the buildings may also be more economical. An evaluation of any given system can be made using the rationale given in Section 2 (Planning) and in Government of the Northwest Territories (1989).

Vehicle Design. Water delivery vehicles and sewage pumpout vehicles have been continually modified and improved to meet the needs of northern conditions.

Water truck tanks and all working systems on them must be protected from freezing (Figure 8-8). Vehicle specifications are included in Appendix B.

For water vehicles using the pressure/vacuum system, the action is the reverse of that of a sewage vehicle, i.e., the tank is pressurized to force the liquid out. Such a system has a high flow rate and may have sufficient water flow and pressure to meet the Insurance Underwriter's requirements for fire trucks as well as for water delivery vehicles. Advantages of this system include elimination of the problems of pump maintenance and the possibility of freezing.

The building water and sewage connections and the hoses should be different in type or size to prevent inadvertent cross connection between water and sewer services.

8.5.3 Ice Haul. Ice as a water source and community ice haul as a means of water distribution should be considered only where no other normal water source is available.

In some communities, ice may be used as a personal water source during winter because of individual choice and preference. Some people prefer the taste of water obtained from ice to that of chlorinated water which is delivered. The taste of water from small lakes is affected by organics and the products of decomposition after prolonged storage un-

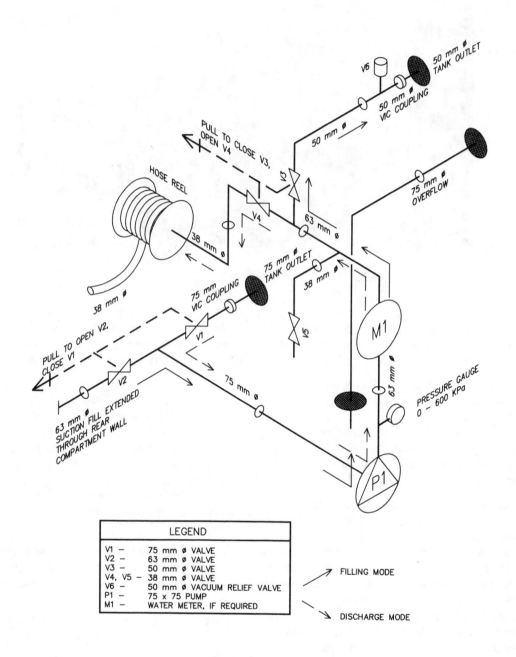

FIGURE 8-8   WATER TRUCK PIPING CONFIGURATION

der ice cover. Ice haul is therefore usually done by the individual; it is rarely supported and organized by the community.

8.6 Piped Systems

Whether piped systems are more economical than trucked systems must be determined for each individual community, for areas of a community and for individual buildings. Because of differing physical characteristics such as community layout, population density, and water use for certain buildings in a community, certain areas or buildings may still be more economically served by trucked systems.

The key requirement of a piped water and wastewater system is to prevent freezing. The total heat loss between the water source and the wastewater discharge point must be less than the total heat available before a change of state (i.e., liquid to solid).

This requirement can be summarized as follows (Yee and Smith, 1981; 1982):

$$Q_T = Q_{rw} + Q_{as} + Q_{au} - Q_{ldt} - Q_{ldu} - Q_{lc} - Q_{lwwt} - Q_L$$

where,

Q_T = heat loss or gain (J)

Q_{rw} = heat of raw water(s)

Q_{as} = heat addition at source and during distribution

Q_{au} = heat of input by users

Q_{ldt} = heat loss during treatment and distribution

Q_{ldu} = heat loss during use

Q_{lc} = heat loss during collection

Q_{lwwt} = heat loss during treatment and disposal

Q_L = heat for change of state.

Management of the heat balance of the system involves control of all heat inputs and losses. Usually components related to the acquisition, treatment, and distribution of water are the most critical to the successful operation of the system. The distribution component is that part of the system where failures are more frequent and the most costly to repair.

The items outlined below describe basic operation and maintenance mechanics that should be considered in the design of cold-region piping systems. Backup mechanisms are discussed under a separate heading.

Northern piping systems should be designed to:

- minimize energy input requirements for operation;

- be simple to operate, maintain and understand;

- be protected against mechanical damage, vandalism and severe climatic conditions;

- have a primary freeze protection mechanism with at least one backup, and be designed to have minimal damage if frozen;

- be easy to thaw when frozen, at any time of year;

- be easy to reactivate after thawing;

- be drainable in sufficient time to prevent/reduce freeze damage;

- provide easy isolation of sections and service lines at any time of the year;

- minimize on-site labor requirements; and

- allow maximum use of the short construction season.

8.6.1 Above- or Below-Ground System. The choice of an above- or below-ground piping system depends on the particular site conditions and requirements. Generally, below-ground systems are preferred where feasible and practical. The criteria for selecting above- or below-ground systems are presented in Section 2.

Piped systems, described later in this section, do not differ in operation whether the system is above or below ground. Only the appurtenances of the system vary.

Above-ground systems have been used where ground conditions and possible thawing of the ice-rich permafrost make the use of a buried system impractical or uneconomical. With the advent of more efficient insulating material and research into buried utilities in permafrost, more buried systems are being installed. In some cases such as temporary camps, above-ground utilities may be desirable. See Section 11 for a detailed discussion of utilidors.

In areas with no permafrost, the pipes could be buried below the depth of maximum seasonal frost penetration. (Methods for calculating frost depth are given in Section 4.) However, burial below the frost line may be impractical due to very deep frost penetration (i.e., in dry gravel) or expensive due to deep excavation (i.e., in rock). An analysis should be carried out to determine whether an insulated pipe buried just below the ground surface would be more appropriate, economical and easier to maintain than an uninsulated pipe buried below the frost line in the ground.

8.6.2 Hydraulic and Thermal Analysis. Hydraulic design of a water distribution system is readily accomplished using desktop computers and commercially available water system modeling programs. These include such programs as Cybernet™, Waterworks™ and others. Current programs have the added benefit of working in an AutoCad™ environment. Designers must satisfy themselves that any program used is suitable for the application (Jeppson, 1976; Hull, 1980; James and Robinson, 1980).

Thermal analysis of a water distribution system is more difficult than hydraulic analysis since there are no commercially available programs at the time of this writing to handle large distribution systems. The designer can utilize spreadsheets for such an analysis incorporating thermal calculations. Assistance may also be obtained from proprietary computer programs (Dupont of Canada Ltd, nd; Suncord Engineering, nd).

Hydraulic and thermal analysis are interdependent. The results of the hydraulic model provide the necessary input values for the thermal analysis which may, in turn, identify desirable changes in the hydraulic design.

8.6.3 Types of Piped Systems.

Single-Pipe Recirculation. The single-pipe recirculation system, whether above or below ground, is recommended as the best piping system for cold-region conditions. This system consists of one or more uninterrupted loops originating at a recirculating facility and returning to that facility without any branch loops (Figure 8-9).

Recirculation eliminates dead ends and the possibility of stagnant water which would freeze. To minimize energy losses, the recirculating system should be planned for the minimal length of piping required.

Advantage of System. A single-pipe system also allows positive simple control of water distribution.

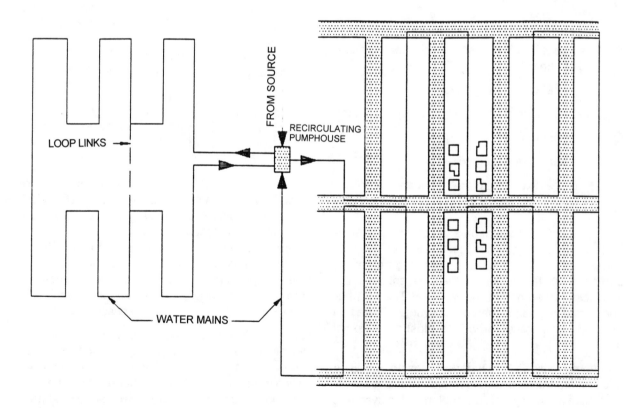

FIGURE 8-9   *LAYOUT AND LOCATION OF MAINS FOR SINGLE-PIPE RECIRCULATION*

Flow and temperature indicators on the return lines at the central facility are all that are required.

Under constant pumping, the heating requirement of make-up water is controlled by the supply and return temperatures. Normally water is pumped out at temperatures between 4°C and 7°C and returns above 1°C. The economic length of each loop is a function of ground temperature, insulation provided and heat input required. This length will vary depending on the site-specific parameters. In Greenland, water temperatures are held at 1°C, and down to 0.1°C on return, with continuous electric heat tracing controlled by very sensitive temperature sensing devices. However, this leaves a very narrow margin of safety for repairs. Such a narrow margin of safety is not recommended.

Disadvantage of System. A disadvantage of a recirculating system is the loss of service to all buildings on a loop in case of a shutdown. In practice, this usually is no problem, because temporary short loop links can be installed to complete the loop while the loop is repaired or periodically extended to meet the growth of the community. As the next extension is constructed, these links can be isolated and drained. In the case of a break or freeze-up, these abandoned links can be opened to reroute the flow of water and possibly isolate the break. Alternatively, flow in dead ends of uncompleted loops can be maintained, on a short-term basis, by bleeding at appropriate points. Because of the characteristics of recirculating looped systems, the engineer must work closely with the planner and community to make them aware of the utility system requirements.

Under fire conditions, the single-pipe recirculation system is usually designed to supply water from both the supply and return line. For this reason, the supply line and the return line are usually of the same size. Pipe size depends on whether fire protection is provided at all (at least for small communities) and on the level of fire protection.

Placement Relative to Community. The recirculating facility could be located at a water treatment plant or in a separate pumping facility. Figure 8-9 shows an ideal town layout for this system. The pretempering/recirculating facility or treatment facility, or both, is preferably located centrally. This allows the community to be divided into a number of single-looped sections going out in all directions from the central facility. By planning community growth, maximum efficiency can be derived with this layout. The most difficult and expensive layout would be a long strung-out community with the facility at one end, since this would result in increased pumping requirements and duplication of lines.

Placement Relative to Users. Back-of-lot mains (Figure 8-9) are preferred with respect to thermal considerations. The mains are subject to colder ambient temperatures and deeper frost penetration under snow-cleared roadways compared to undistributed areas. Few northern communities have paved roads. If the mains are placed in the gravel roads, the access holes are subject to physical damage during snow clearance.

Placing the mains at the rear lot line avoids these problems and permits service lines of equal length on both sides of the main. With mains in the road allowance, usually to one side, plus an 8 m distance required between the front of the house and the road right of way, average service line length would be 16 m (two 8 m pipes) on one side and 32 m (two 16 m pipes) on the other. With the mains at the rear lot line, where houses can be placed as close as 3 m from the lot line, an average service line length would be 6m (two 3 m pipes) in either direction. There is a significant saving to the lot owner in this layout of servicing which encourages development of piped services.

A further advantage of mains located along the rear lot line is that miniservice vaults for housing the appurtenances for the water line (valves and hydrants, freeze protection controls, etc.), can be elevated above grade. This results in easier access, especially during the winter.

Despite the advantages of back lot placement, most water and sewer mains are placed in the street due to accessibility and property easement concerns for maintenance operations.

Conventional, Water-Wasting Systems. In this type of system, the water line network is laid out conventionally. To ensure flow at dead ends, loops, and service lines, water is bled off and wasted to sewers at a number of points (Yee and Smith, 1981; 1982; Smith and Yee, 1983). Bleeding prevents freezing by wasting cooled water to the sewers and continually replacing it with warmer water from the supply system. Bleeding also helps to prevent freezing in the sewer systems by providing a continuous flow in the sewer (Figure 8-10).

Water-wasting systems must not be used where:

- the water quantity is limited;
- water requires expensive treatment; or

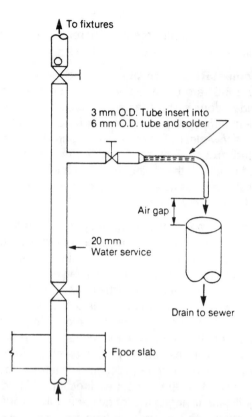

(a) Diagram of a basic water bleeder for household use

(b) A representative water bleeder connection

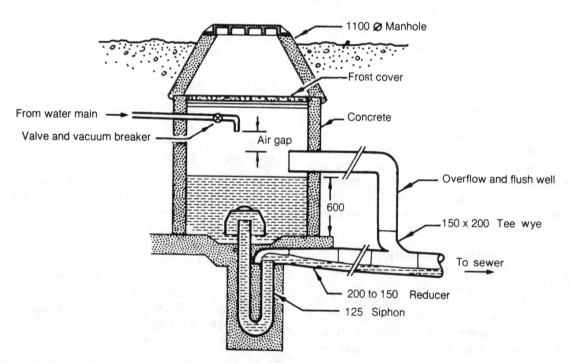

(c) Flushing access hole at end of system

FIGURE 8-10   WATER BLEEDERS

- heating of water is necessary.

Possible use would be where a relatively warm inexhaustible water source exists.

Disadvantages of this system include the inevitably high water consumption and high energy input. Consumption can exceed 4,500 L/(p•d) (see Section 5). Care must also be taken during design and installation of bleeders to prevent possible cross connections.

Water-wasting systems increase the sewage volumes, making it more difficult and expensive to treat the sewage. For example, the City of Whitehorse, Yukon Territory, has such a water-wasting system with its incumbent treatment problem of high volumes of dilute sewage. Whitehorse has the advantage of being able to tap relatively warm groundwater, which reduces operational costs.

Dawson City, also in the Yukon Territory, recently had new water and sewer systems installed. The old water system was a bleeder-type system, as is the new system. The decision not to change the type of system was based on the fact that the residents of Dawson were accustomed to this type of operation and would be unlikely to alter their habits. Design efforts were therefore directed toward controlling the bleeding rather than eliminating it. This system was designed using pipe size and length to regulate bleeder flow rather than resorting to valves which can be subject to tampering.

Several methods for reducing the quantity of water bled to the sewer have been used or studied successfully (Smith and Yee, 1983; Cotterill, 1983). These include a timer to switch the bleeder flow on and off, an orifice plate to control flow rate, and a storage tank. The storage tank pumpback system

is commercially available (Powergain Manufacturing Co., nd).

Single Pipe – No Recirculation. High-volume users, such as large apartment blocks or fish-processing plants, are strategically located at the ends of main lines to ensure a continuous flow in the line (Figure 8-11). Thus, a return loop is unnecessary. The use of this method requires close co-ordination with community planning. The noncirculating, single-pipe concept can also be used for lines with continuous heat tracing, usually only for short or special purpose lines. Examples of noncirculating, single-pipe systems can be found in Greenland where not only apartment blocks accommodating up to 1,000 people, but also fish-processing plants are located at the end of main lines. The potential for seasonal flow, low flows (e.g., schools on weekends) or occupancy of these buildings must be considered to ensure adequate flows year-round. In Canada and Alaska, small community sizes and the predominant use of single-family housing make it difficult to implement this type of system.

Dual-Pipe Pressure Differential System. This type of system consists of a large-diameter water supply line and a small-diameter return line with a pressure differential between the lines.

This system permits lines to be laid out in the normal manner. The water supply main is connected to the return line at the end of the line to ensure circulation. Service lines tap into the main, and water is returned via the return line of the service connections to ensure circulation through the service connections by the pressure differential between the two mains.

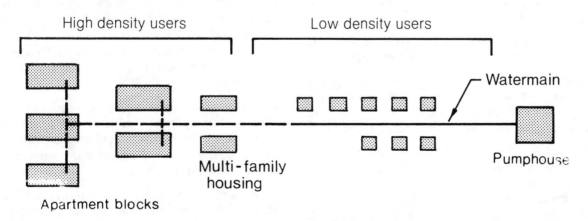

FIGURE 8-11   SINGLE-PIPE SYSTEM WITHOUT RECIRCULATION

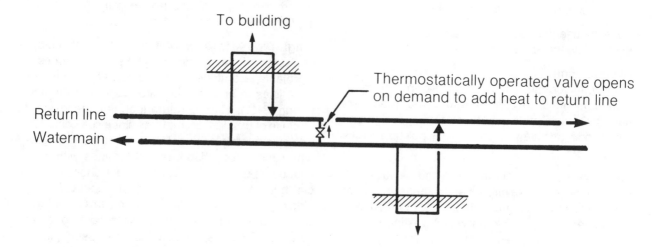

FIGURE 8-12   DUAL-PIPE SYSTEM

Heat can be supplied to the distribution system by heating the supply water or by using separate heating lines.

Control mechanisms for this type of system tend to be difficult, because varying consumption in different locations results in fluctuating pressure and flow rates in the supply and return mains. No-flow conditions are possible, which could allow freezing in portions of the systems. Thermostatically controlled solenoid valves may be required on "short circuit" branches between two lines at regular intervals to overcome this problem (Figure 8-12). Even with these or similar control devices, circulation of water cannot be guaranteed at all times.

Other disadvantages of this system include:

- the dual lines increase initial capital cost, and heat loss is high due to the greater pipe surface area; and

- the greater length and number of lines also increases the potential for line breaks.

However, more flexibility is allowed in isolating line breaks, keeping everything serviced and extending the system for short distances. Few, if any, new dual main systems are being installed today.

A variation of this two-pipe system was installed in Yellowknife in 1948 and 1949. Orifices (1.5 mm diameter) were used in each building and at the connection at the end of mains to the return mains. A pressure differential across the orifices ensures flow in all mains and service lines. This system is still in place in portions of the City. Disadvantages are as noted above and:

- wear of orifices allows higher flows and decreases pressure differential at the system extremes,

- homeowners have been known to remove their orifice or even install a pump on their service line, drastically affecting pressure differential in the mains.

8.6.4   Other Methods of Water Distribution. Other methods either still in the development stages or less frequently used are as follows:

Small-Diameter Water Distribution Lines can be a cost effective method of delivery for both small and large communities.

Small-diameter lines are pipes of diameters ranging from 50 to 150 mm. If full fire flow and hydrants are desired, the lines must be at least 150 mm in diameter. The method of circulation and freeze protection could be any one of the ways previously described.

Where small-diameter mains are used, more emphasis must be placed on alternative fire protection such as the building structure, materials and layout, as well as the use of sprinklers and hose cabinets in lieu of exterior hydrants.

Small-diameter systems save energy since less surface area means less heat loss. Also, greater flow rates in the pipes make the system more reliable. Small-diameter pipes can be placed near or directly

FIGURE 8-13   SMALL-DIAMETER WATER MAINS

under houses, practically eliminating service line problems.

Small-diameter piping systems are used extensively in Greenland (Figure 8-13) and in Alaska; 100 mm pipes have been used at a few locations by the U.S. Public Health Service.

Small-diameter lines have relatively short freeze times and are susceptible to flow interruptions.

Extended Main Circulating System.   With the extended main system the mains are extended or are located close to the building to be serviced. This eliminates a long service line, which is the most vulnerable part of most piped water systems. A short single-insulated service line with a short heat tape can then be used in place of circulation pumps or pitorifices normally used on single main recirculating systems. Operating costs are reduced since the higher flow rates in the mains required for pitorifice operation are not needed. The disadvantages of this type of system are increased capital costs and potential problems with access.

Thawing an abandoned or otherwise neglected service line is significantly easier because the length of the service line is much shorter than a typical service line loop length.

Several extended main circulating systems have been constructed in Alaska and are in operation in the Arctic and the subarctic, but more recent systems do not use this concept.

Summer Line Systems are useful in communities where freeze-protected winter water lines are not economical compared to trucked service. Local topography and conditions, as well as economics and convenience, may make uninsulated summer lines desirable.

Summer lines may be pressurized from the piped system or treatment plant, or may take advantage of sources of water that allow flow by gravity to the community. This can provide convenience to residents using individual water haul systems or eliminate the need for trucked water for up to four months by using inexpensive above-ground lines for water distribution. These lines provide unrestricted water for summertime uses such as gardening and lawn watering, however, the impact on sewage hauling costs must also be considered. In Yellowknife, sewage pumpout volumes increase by some 25% when summer water lines are in service.

Each fall, summer lines should be drained or blown out to reduce the potential for freeze damage. The lines must be properly disinfected at the beginning of each use season.

Some all-weather circulating distribution systems supplying several watering points have been constructed in Alaska. They allow individual haul supply points to be located closer to the individual homes.

8.6.5 Materials. Information on pipe materials can be found in Appendix A. Several articles have also presented information on the application of various designs for cold climate conditions (O'Brien and Whyman, 1977; Whyman, 1980; Cheema and Boon, 1985).

8.7 Service Lines

Service lines are kept from freezing through the use of both insulation and bleeding (either to waste or with pumpback, electrical heat tracing, or circulation in dual service lines. Circulation was first achieved though the use of dual mains maintained at different pressures and later in single-main systems through the use of pitorifices or small circulating pumps. Regardless of the freezing protection method, service lines must be able to withstand freezing and be easily accessed for thawing procedures.

8.7.1 Service Line Considerations. Copper service lines were historically used because they can be thawed electrically and are still used in many communities. Copper lines are still used in Fairbanks, Alaska because soil conditions there allow them to be inexpensively insulated in place with urethane spray foam. The use of 25 mm high density polyethylene (HDPE) for service lines has become quite common in recent years because the larger line size offers less flow resistance, requires a longer time to freeze and has more resistance to freeze damage than copper. HDPE service lines are commonly installed in an insulated conduit with a self-regulating heat-trace cable to allow thawing of the lines in the event of freezing and to provide a backup when flow in the lines stops. Figure 8-14 and 8-15 show typical services used by the Government of the NWT. Additional information can be obtained in Wilson and Cheema (1987).

Service lines can be thawed internally so it is advisable to ensure easy access to the lines and straight runs. It is also advisable to locate the tee to the house supply at the highest point so that any air which may accumulate is vented when water is used. Air traps can stop circulation in a line resulting in freezing and damage to the circulation pump.

The flow required in the service line depends on the heat loss rate and the permitted temperature drop in the service line. Usually supply and return lines for homes are the same size. Where demand is great and a large diameter supply is required, the return line can be sized just large enough to accommodate the required freeze protection flow.

Where possible, service lines should be buried and located in areas where snow will remain undisturbed. Above-ground service runs and service entry boxes should be avoided whenever possible. Close attention should be paid to above ground lines since they are particularly susceptible to freezing due to the much greater temperature differentials and due to the potential for loss of insulation or air infiltration. Directly insulating the service lines and using a thermostatically controlled heat trace is recommended in these areas.

8.7.2 Pitorifice Circulation. Pitorifices are upstream- and downstream-facing extensions on the corporation stops that produce a small differential pressure in the service lines when water is flowing past them in the main. Much of the pressure difference is due to impact pressure on the upstream facing pitorifice which functions like a pitot tube and some is due to the pressure drag effect on the downstream facing pitorifice which functions like an eductor. These effects are enhanced by their partial blocking of the main resulting in a local increase in flow rate and a small pressure drop in the main. Although spacing, size, and insertion depths should theoretically be as great as possible, practical concerns take precedence. Spacing is usually 150 mm, the size is usually the service line size, and insertion should be 50 mm or more (Figure 8-16).

The head produced by a pitorifice pair can be estimated by:

$$\Delta H = K_{po} \frac{V_m^2}{2g}$$

where,

K_{po} = a dimensionless constant which can range from 1.5 to 3.0, depending on the relative sizes of the pitorifice and the insertion depth of the pitorifice

V_m = velocity of the liquid

g = acceleration due to gravity.

The flow rate induced by this head depends on the service line lengths, diameter, and fittings. If the flow required is close to or within the turbulence range, do not use pitorifices.

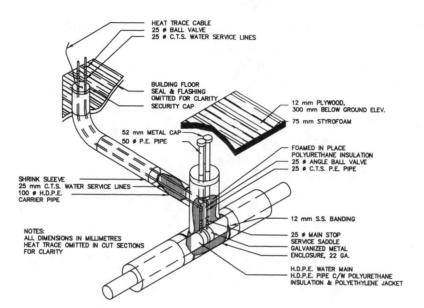

HEAT TRACE CABLE
25 ⌀ BALL VALVE
25 ⌀ C.T.S. WATER SERVICE LINES

BUILDING FLOOR
SEAL & FLASHING
OMITTED FOR CLARITY
SECURITY CAP

12 mm PLYWOOD,
300 mm BELOW GROUND ELEV.

75 mm STYROFOAM

52 mm METAL CAP
50 ⌀ P.E. PIPE

FOAMED IN PLACE
POLYURETHANE INSULATION
25 ⌀ ANGLE BALL VALVE
25 ⌀ C.T.S. P.E. PIPE

SHRINK SLEEVE
25 mm C.T.S. WATER SERVICE LINES
100 ⌀ H.D.P.E.
CARRIER PIPE

12 mm S.S. BANDING

25 ⌀ MAIN STOP
SERVICE SADDLE
GALVANIZED METAL
ENCLOSURE, 22 GA.

NOTES:
ALL DIMENSIONS IN MILLIMETRES
HEAT TRACE OMITTED IN CUT SECTIONS
FOR CLARITY

H.D.P.E. WATER MAIN
H.D.P.E. PIPE C/W POLYURETHANE
INSULATION & POLYETHYLENE JACKET

TYPICAL DOUBLE WATER SERVICE LINE (RECIRCULATING)

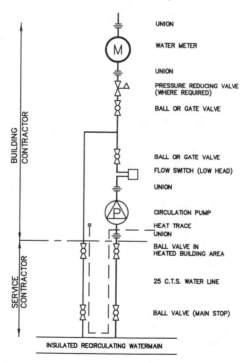

UNION

WATER METER

UNION

PRESSURE REDUCING VALVE
(WHERE REQUIRED)

BALL OR GATE VALVE

BALL OR GATE VALVE
FLOW SWITCH (LOW HEAD)

UNION

CIRCULATION PUMP
HEAT TRACE
UNION

BALL VALVE IN
HEATED BUILDING AREA

25 C.T.S. WATER LINE

BALL VALVE (MAIN STOP)

BUILDING CONTRACTOR

SERVICE CONTRACTOR

INSULATED RECIRCULATING WATERMAIN

WATER SERVICE RECIRCULATION SCHEMATIC

FIGURE 8-14   BELOW-GROUND SERVICE LINE

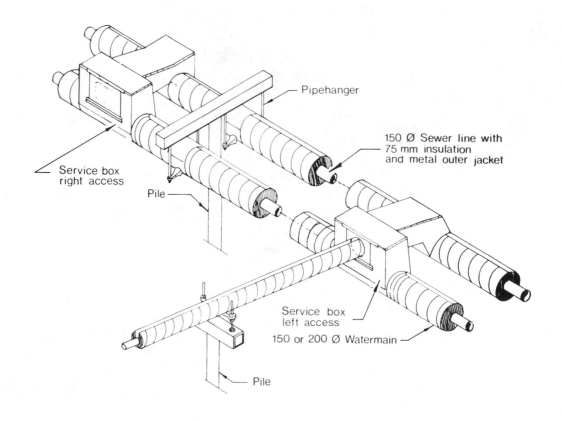

Pipehanger

150 Ø Sewer line with
75 mm insulation
and metal outer jacket

Service box
right access

Pile

Service box
left access

150 or 200 Ø Watermain

Pile

(a)   General view

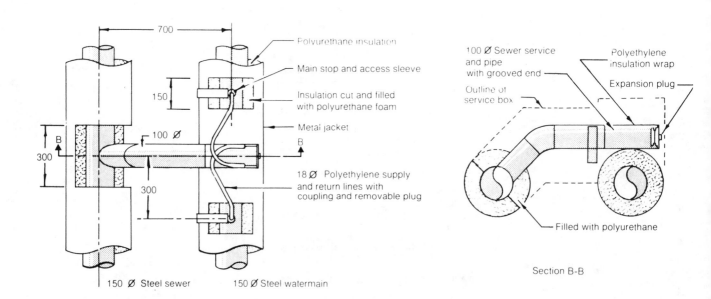

700

Polyurethane insulation

Main stop and access sleeve

150

Insulation cut and filled
with polyurethane foam

B

100 Ø

Metal jacket

B

300

300

18 Ø Polyethylene supply
and return lines with
coupling and removable plug

150 Ø Steel sewer

150 Ø Steel watermain

100 Ø Sewer service
and pipe
with grooved end

Polyethylene
insulation wrap

Expansion plug

Outline of
service box

Filled with polyurethane

Section B-B

(b)   Detail of connection

FIGURE 8-15   ABOVE-GROUND SERVICE LINE

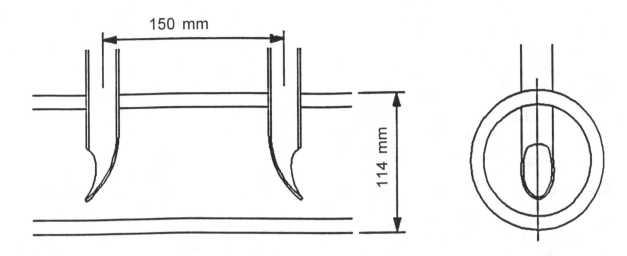

FIGURE 8-16   SCHEMATIC OF PITORIFICES IN A MAIN

Typically, pitorifices are used on service runs that are mostly buried and seldom longer than 23 m one way. Permitted temperature drops of about 1°C result in required flows in the main of 0.3 to 0.6 m/s. The use of larger diameter service lines, shorter runs, and better insulation allows lower flow rates in the mains.

Pitorifices were first used in 1953 (Westfall and Wallace, 1953) and are widely used in Alaska. They offer low capital and operating costs and are particularly well suited for relatively small-diameter looped main systems in places where electricity is expensive and power outages are frequent. If the power fails, circulation can be maintained using standby power at the pumphouse. Ensuring a minimum flow rate at all locations becomes more difficult and expensive where mains are networked and extended. Some Alaskan communities have started installing individual water service line circulation pumps in these areas.

For more information on pitorifice systems see Mauser (1994a; 1994b) and Johnson (1979).

8.7.3 Individual Circulation Pumps. Small circulation pumps commonly used for hot-water recirculation can be used to ensure flow in service lines. Grundfos (Clovis, California), Armstrong (North Tonawanda, New York), ITT Bell & Gossett (Morton Grove, Illinois), Taco Pumps (Cranston, Rhode Island), Laing Thermotech (Chula Vista, California), Hartell (Ivyland, Pennsylvania), March Manufacturing Inc. (Glenview, Illinois) and Little Giant (Oklahoma City, Oklahoma) all manufacture small pumps.

Some pumps use a wet rotor and others use a magnetic drive. Both designs allow full containment of water-lubricated seals and wet rotor design allows a high rate of waste heat transfer to the water being pumped. Hartell supplies pumps with highly efficient brushless DC motors and Ivan Labs (Jupiter, Florida) sells brushless DC drivers for March's model 809 magnetic drive pump. These pumps can use less than 10 watts, including losses from an AC/DC power supply.

The pumps should be constructed of corrosion resistant materials suitable for potable water use. While cast iron body pumps are cheaper, they are intended for closed-loop heating systems. Most of the pumps are intended for use with water at ambient temperature or higher. Year-round operation is recommended because homeowners frequently turn the pump off before the ground around the buried service is thawed, forget to restart pumps and, if a wet rotor pump is not being operated with cold water flowing through it, condensation may lead to corrosion of unencased windings.

Small pumps and small AC motors are very inefficient but since so little flow is required in a service line to prevent freezing, even the smallest pumps available usually provide far more than the required flow.

Proper orientation of these pumps is important in ensuring against air being trapped which leads to pump failure. Venting air from the pump and turning the rotor manually is sometimes required, so provision should be made for easy monitoring and access to the pump for these functions as well as for periodic replacement.

Individual circulation pumps permit easier distribution system design and longer service runs. Also, service lines are less likely to freeze because of higher service-line flow rates. The total system capital and operating costs are generally higher than those for a pitorifice system, with a greater burden borne directly by the homeowner in operating, maintaining and replacing the pumps.

8.7.4 Materials. In the NWT, the normal water service bundle includes two 25 mm polyethylene pipes with an electric heat-trace line strapped to them every metre with aluminum tape. This package is contained in a preinsulated high-density polyethylene pipe with appropriate jacket to protect the insulation. If the service bundle is exposed or used as an above-ground service, a metal outer jacket is recommended. Details of NWT above- and below-ground service box take-offs are shown in Figures 8-14 and 8-15.

8.7.5 Dual Servicing. When adjacent housing units have a common owner, as would be found in public housing, common services can be used as shown in Figure 8-17. The water service lines are so arranged that the supply line goes into one building through the circulating pump, out the return line to the next building, and so on. After exiting from the last building, the service line then returns to the water main. In other words, a small, water service loop connects the buildings using the same service bundle. This type of service increases the reliability of the service line by providing multiple circulating pumps and greater flow through the line. Capital costs are also reduced by as much as one-third.

8.8 Appurtenances

8.8.1 Hydrants.

Above-Ground Hydrants. Above-ground hydrant housings must be designed to fit the particular above-ground or utilidor design. They are generally of the siamese building wall-hydrant type. Figure 8-18 shows a typical NWT above-ground hydrant with its insulated housing. The distance from

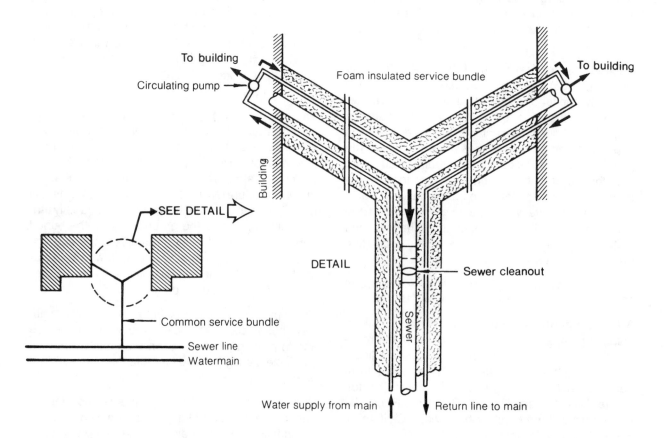

FIGURE 8-17   DUAL SERVICE

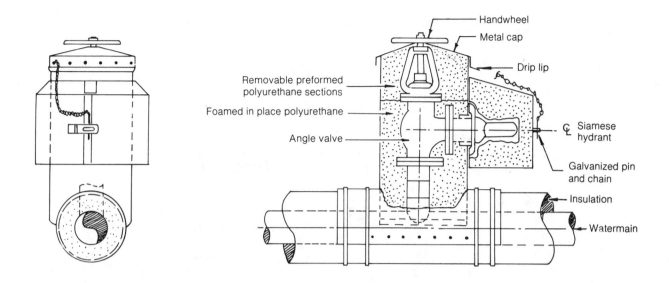

FIGURE 8-18   ABOVE-GROUND HYDRANT

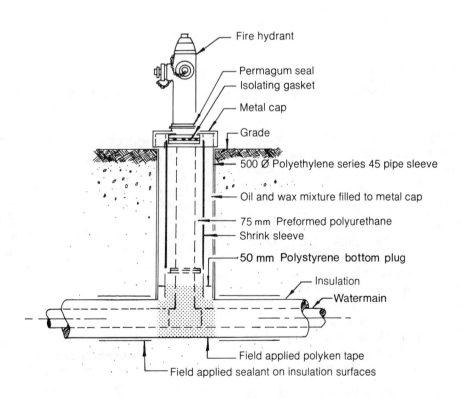

FIGURE 8-19   BELOW-GROUND HYDRANT

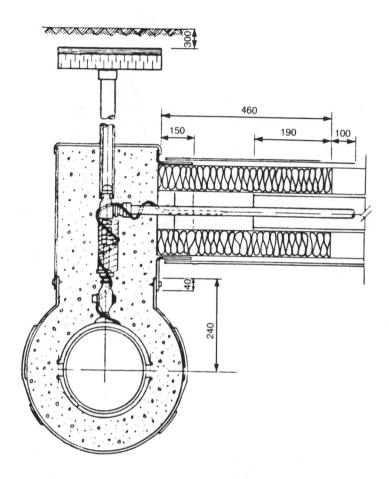

FIGURE 8-20   UNDERGROUND SERVICE VALVE AND BOX

the main is kept as short as possible so that heat conduction from the water in the main keeps the water in the hydrant from freezing. The hydrant housing is painted to easily identify it.

Below-Ground Hydrants.   The hydrant is normally on-line to minimize the possibility of freezing. A frost-isolating gasket is placed between the bottom of the hydrant barrel and the tee into the main. A typical below-ground hydrant installation is shown in Figure 8-19. The hydrant barrel is insulated with 75 mm preformed polyurethane and placed inside a 500 mm diameter polyethylene Series 45 pipe sleeve. The cavity between the sleeve and the insulated barrel is filled with an oil and wax mixture to prevent damage to the hydrant due to frost heave.

Isolating valves are normally installed on either side of the tee to allow for hydrant replacement or repair.

Water cannot be left in the hydrant barrel, since it would freeze. If the soil around the barrel does not freeze, the water can be allowed to drain out of a drain hole. Where the ground is frozen, as it would be in permafrost areas, the water in the barrel must be manually pumped out and replaced with an anti-freeze solution. An appropriate mixture of propylene glycol and water of a grade acceptable for potable water systems is pumped into the empty hydrant to prevent the hydrant from freezing. The potential of cross connections and contamination must be avoided.

Vault-Mounted Hydrants.   In recent years in the NWT most hydrants on buried water systems have been mounted in access vaults, either steel vaults as shown on Figure 8-22 or concrete vaults such as in Yellowknife. This arrangement, although more expensive than buried hydrants, allows easy repair when hydrants are damaged by traffic without having to excavate frozen ground.

8.8.2 Valves.   Above- and below-ground valves are basically the same, with the exception of the

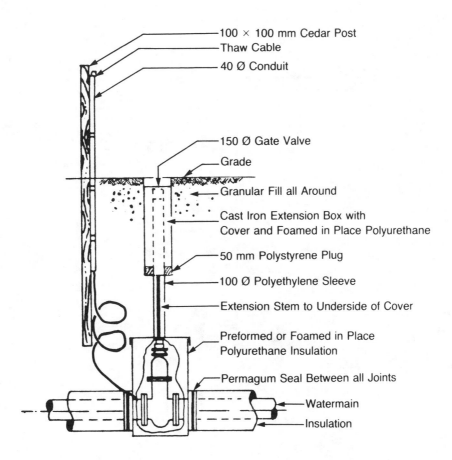

FIGURE 8-21   UNDERGROUND WATER MAIN VALVE AND BOX

valve box and operating stem details. Two details for underground installation are shown in Figure 8-20 and Figure 8-21.

Foamed-in-place polyurethane insulation is used with a Series 45 polyethylene sleeve over the operating stem, similar to the way hydrants are insulated. Where installations are buried or where the valve is completely insulated, valves are generally nonrising stem-gate valves. In other situations, such as in access holes, any appropriate valve can be used.

8.8.3 Metering. Standard meters are used to monitor the distribution system. These include magnetic flow meters, in-line gear meters, orifice pressure differential recording graph meters and paddle wheel flow sensors with remote meters.

Both the supply and return circulation mains must be metered. The difference in meter readings pro-

vides the amount of daily water consumption and provides a check for system leaks during low demand periods. Under fire conditions, a reverse flow meter or bypass is required on the return line where the water is supplied through both supply and return mains. A meter should also be installed on the supply line to the circulating loop to check the daily consumption.

Most northern communities are relatively small. This means the fire flows are as much as 10 to 15 times greater than the average flow. In this case, an orifice plate flow meter, which is satisfactory for fire flows, gives unsatisfactory results under normal conditions. If it gives a satisfactory results under normal conditions then the orifice plate is such that it restricts flows during a fire situation unless bypasses are installed. Both magnetic flow meters and gear-driven flow meters are very expensive for large-diameter pipes. With the large-diameter sup-

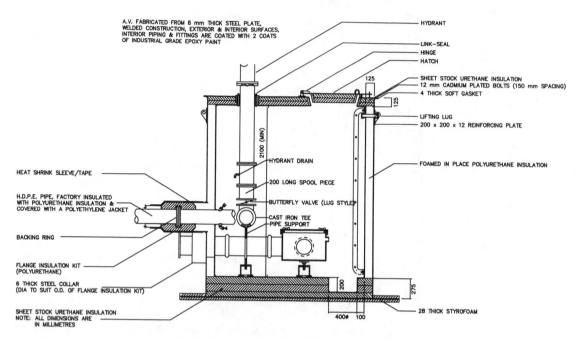

A.V. FABRICATED FROM 6 mm THICK STEEL PLATE,
WELDED CONSTRUCTION, EXTERIOR & INTERIOR SURFACES,
INTERIOR PIPING & FITTINGS ARE COATED WITH 2 COATS
OF INDUSTRIAL GRADE EPOXY PAINT

HYDRANT

LINK-SEAL

HINGE

HATCH

SHEET STOCK URETHANE INSULATION

12 mm CADMIUM PLATED BOLTS (150 mm SPACING)

4 THICK SOFT GASKET

LIFTING LUG

200 x 200 x 12 REINFORCING PLATE

FOAMED IN PLACE POLYURETHANE INSULATION

HEAT SHRINK SLEEVE/TAPE

H.D.P.E. PIPE, FACTORY INSULATED
WITH POLYURETHANE INSULATION &
COVERED WITH A POLYETHYLENE JACKET

BACKING RING

FLANGE INSULATION KIT
(POLYURETHANE)

6 THICK STEEL COLLAR
(DIA TO SUIT O.D. OF FLANGE INSULATION KIT)

SHEET STOCK URETHANE INSULATION
NOTE: ALL DIMENSIONS ARE
IN MILLIMETRES

HYDRANT DRAIN

200 LONG SPOOL PIECE

BUTTERFLY VALVE (LUG STYLE)

CAST IRON TEE
PIPE SUPPORT

2100 (MIN)

125

125

200

275

400ø 100

28 THICK STYROFOAM

STEEL PREFABRICATED ACCESS VAULT TYPICAL SECTION
NOT TO SCALE

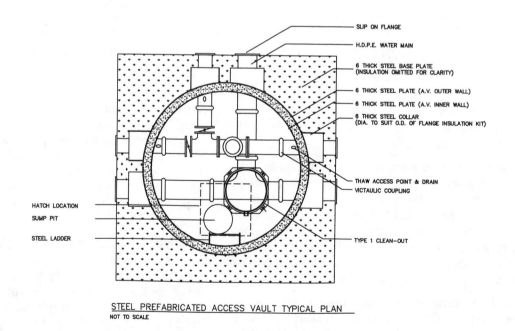

SLIP ON FLANGE

H.D.P.E. WATER MAIN

6 THICK STEEL BASE PLATE
(INSULATION OMITTED FOR CLARITY)

6 THICK STEEL PLATE (A.V. OUTER WALL)

6 THICK STEEL PLATE (A.V. INNER WALL)

6 THICK STEEL COLLAR
(DIA. TO SUIT O.D. OF FLANGE INSULATION KIT)

THAW ACCESS POINT & DRAIN

VICTAULIC COUPLING

HATCH LOCATION

SUMP PIT

STEEL LADDER

TYPE 1 CLEAN-OUT

STEEL PREFABRICATED ACCESS VAULT TYPICAL PLAN
NOT TO SCALE

FIGURE 8-22   MINISERVICE-CENTRE ACCESS VAULT

ply pipe, flow can be diverted through a small pipe and then either through a magnetic or a gear driven flow meter, and then back to the large-size supply pipe. During fires the meters are completely by-passed and flow is directed through the large sup-ply lines.

8.8.4 Access Vaults. Access vaults for both piped water and sewer systems have been con-structed of various materials. Common types are concrete, corrugated metal, and welded steel. All of these have various means of insulation.

Standard southern access holes are often too small and leak excessively; they are and always have been subject to damage. One alternative that evolved in the Northwest Territories is known as the miniservice-centre access vault.

The access vault shown in Figure 8-22 is a compre-hensive one, showing both water and sewer lines as well as a hydrant, but the same basic design can be used for any situation. The advantage of this vault is that maintenance of the distribution mains can be carried out in a climate controlled environ-ment. This includes maintenance of water line valves, hydrants and sewer cleanouts.

In some recent projects in the NWT, high-density polyethylene has been used in vault design. These vaults are limited to a maximum diameter, unlike steel or concrete, and are not insulated, necessitat-ing use of pipe insulation in the access vault.

Some of the features of such an access vault are as follows:

For hydrants:

- Standard hydrants can be used and operated in the normal manner.

- One isolating valve plus valve boxes and op-erating stem extension are eliminated, reduc-ing installation costs.

- Replacement is easier; no digging is required.

- They are relatively fool-proof compared to antifreeze-filled hydrants if heat is maintained in the vault.

For main valves:

- Easy access for operation, replacement or elimination of valve boxes.

For the access-vault structure:

- The access vaults are well-insulated for en-ergy conservation.

- They are large; minimum size is 1.2 m by 1.2 m depending on interior fittings.

- All exposed surfaces are metal or concrete to reduce vandalism and maintenance.

For the power point:

- A thermostatically controlled electric heater may be used to keep the temperature in the vault above freezing. However, this has proven to be excessively expensive.

For the sump pump:

- The sump pump may, if necessary, be located in a prefabricated steel sump pit set in the concrete base.

For the sewer cleanout:

- Closed sewer cleanouts are provided in the access vault. Covers have rubber gaskets to prevent leakage.

For water mains:

- Ready access is provided by 50 mm angled entries complete with ball valves.

8.8.5 Alarms and Safeguards. Alarms and safeguards play a vital part in the successful opera-tion of any cold-region utility system. These appur-tenances in a pumphouse are important consider-ations because of their contribution to the reliability of a cold-region water system.

Alarms indicate when an essential operation has failed. For example, a no-flow alarm warns of stopped (pump) circulation in a water main. Some of the more common alarm conditions are:

- no flow of water,

- low water level,

- low water temperature,

- high water level,

- low electrical voltage,

- chlorine gas leak detectors,

- low water pressure,

- power failure,

- combustion detection,

- heating failures, and

- fixed (high) temperature heat detection.

Alarms can be set to trigger by properly locating and limit-setting the following sensing devices:

- Flow switch that turns on a no-flow alarm when the flow of water against its paddle suspended in a pipe has stopped.

- Float switch that turns on or off a low-water-level alarm by the change in water level in a tank. This switching device may be used for either low- or high-water-level alarms.

- Temperature probes measure water temperature in a water line or tank and internally activate a switch at a preset temperature. The switch activates an alarm indicating low water temperature and initiates heat input.

- Low-voltage cutout devices open circuits to stop motors when the voltage drops more than 10 percent below the rated voltage for more than 10 seconds. The opening of the circuit can initiate a low-voltage alarm.

- Pressure switches turn on pumps or low-water-pressure alarms, or both, when the water pressure sensed by an internal element drops below a preset limit.

- Thermostat devices for regulating air temperature by activating or deactivating a heat source can also actuate low-air-temperature alarms to warn the operator of loss of heat in a building, utilidor, or well-head cover or activate high-air-temperature alarms to indicate a hot boiler room.

- Photoelectric-type smoke detectors that respond directly to visible smoke. They are unlike an ionization device which is more sensitive to particles of combustion at an invisible level of concentration common to the pumphouse environment. This device can activate a fire alarm powered by a DC source (battery).

- Thermal detector devices activate fire alarms only after the temperature of the internal thermal element exceeds the rated operating limit (normally limits can be preset from 80°C to 105°C). This fixed-temperature device is unlike the rate-of-rise detectors which are often too sensitive to temperature increases in cold-climate applications. With rate-of-rise detectors, for example, opening a garage door to a pump house storage area with below-freezing temperatures outside and then closing it may subsequently create too rapid a temperature rise, thus setting off the fire alarm.

These warning devices can assist water system operation, if used properly, to warn the operator of an impending freeze-up, a low-water condition, etc. The operator must be trained to respond to failure according to meaningful procedures. Many of the warning devices can operate a common audible alarm and activate their respective light on a panel to indicate the source of the problem.

Alarm panels should have test buttons to test the various alarm functions weekly and especially to test the standby DC battery power for a fire alarm system. Time delay relays such as with the low-voltage cutout device can prevent nuisance alarms during frequent minor electrical overload conditions.

The alarm panels can be consolidated into a central control panel containing alarm controls and lights, and switches for motors with "run" lights. Low-voltage cutout panels may also be included. Such centralized control panels however, increase the formidable task of training an operator how each alarm is tied into a specific warning device. While seemingly convenient to a highly trained operator, it can cause a training and maintenance problem with respect to obtaining action from an operator inexperienced in electrical controls in response to an activated alarm. Alarm silence buttons should be provided to give the operator time to calmly review the labelled relays and wires, and review the control schematics for a resolution of the problem.

The sequence of alarms can be arranged to have first an alarm light and then, as the situation becomes more crucial to resolve, have an alarm horn activated. There is no standard design for an alarm system applied to a pumphouse. Typical, accepted practices for sensing and measuring a function in the process of heating water, maintaining a water level, or maintaining suitable air temperatures in a pumphouse are being applied. Only alarms and controls that are necessary to resolve an emergency problem should be provided in a pumphouse.

Alarm systems for critical items should include a means to communicate such alarms to an operator who is not at the pumphouse. This may include such systems as autodialers which, when an alarm occurs, call a series of telephone numbers until a response is obtained. Such systems normally repeat the call if no action is taken to acknowledge the alarm on-site.

For remote sites where no telephone is available, radio-activated alarm systems can be utilized. These radios normally activate pocket pagers which the maintenance staff can keep on their person and at their homes during nonworking hours.

Safeguards that should be provided with all pump houses include standby power and standby heat. A permanent standby diesel generator can provide power to a pumphouse while portable generators may be plugged into remote electrical outlets to power an isolated pump. A standby power source is especially necessary if the local power source is unreliable. A permanent source of standby heat that does not require electricity to operate can be installed in pump houses or well houses, or in both. Oil, gas, or propane heaters must be adequate to keep all enclosed utility work areas above freezing.

The design must plan for the worst case when it comes to emergencies in cold climates. If the problem cannot be prevented, the impact on the entire system must be reduced by isolating the facilities or making them easier to deal with during the emergency.

8.9   Freeze Protection and Thawing

A critical part of design of a water distribution system is the prevention of freezing and, realizing that freezing may occur, providing a means to recover a frozen main with minimal damage as quickly as possible to minimize the interruption in service.

Freeze prevention can only be achieved by the addition of heat and limiting the loss of heat. Water must be kept flowing at a rate that ensures heat lost to environment and consumption is less than the heat added at the source facility. Alternatively, heat must be added to nonflowing pipes along its entire length at a rate above heat loss to the environment.

8.9.1   Backup Systems.

All systems should be designed on the principle that freezing is likely to occur during the life of the system. Therefore recovery procedures must be incorporated into the design. The primary backup freeze protection systems are:

- heat-trace systems;

- thaw-wire electrical resistance systems; and

- steam or hot-water thawing.

These systems are discussed in Appendix D.

The final alternative for protection of a distribution system is the ability to drain or pump out the pipes before the water in the pipes freezes. Designing a cold-region water distribution system to minimize freeze damage is a major challenge.

8.9.2   Freeze Damage Prevention.

The reduction of damage to water distribution pipes, when freezing does occur, has been discussed by McFadden (1977) and Cotterill (1983). The use of a diaphragm at the last point to freeze in each section of pipe was recommended as a good technique to reduce damage. The location of the proper point can be selected and somewhat controlled by careful analysis of the system and careful placement of additional insulation.

8.10   Disinfection

Key to the safe operation of any water delivery system is adequate and continuous disinfection of all water storage facilities, waterlines, transportation vehicles and equipment. Standards for disinfection are published by the American Water Works Association (AWWA, 1986) and should be adhered to.

8.11   References

AWWA. 1986. *Maintaining Distribution-System Water Quality.* American Water Works Association, Denver, Colorado.

Cameron, J. 1981. *Guidelines for the Preparation and Administration of Municipal Water and Sanitation Trucked Service Contracts.* Department of the Northwest Territories, Local Government, Yellowknife. 225 p.

Cheema, S. and R. Boon. 1985. *Water Distribution and Sewage Disposal Systems in the Northwest Territories.* Department of Public Works, Government of the NWT, Yellowknife. 88p.

Cotterill, R. 1983. Freeze protection of water service lines. In: *Proceedings, Cold Regions Environmental Engineering Conference,* University of Alaska, Fairbanks, 90-111.

Dupont of Canada, Ltd. nd. *PIPETEMP, Program for Thermal Analysis of Piping.* Engineering Division, Mississauga, Ontario.

Government of the Northwest Territories, 1989. *Water and Sanitation Subsidy Program Handbook.* Municipal and Community Affairs, Yellowknife.

Government of the Northwest Territories. 1986. *General Terms of References for a Community Water and Sanitation Services Study.* Department of Municipal and Community Affairs,

Community Works and Capital Planning, Yellowknife.

Hull, J.A. 1980. Thermodynamic analysis of the water distribution system in Inuvik, NWT. In: *Proceedings, Utilities Delivery in Northern Regions*. Environmental Protection Service Rep. No. EPS 3-WP-80-5, Ottawa, Ontario, 332-346.

James, W. and M.A. Robinson. 1980. Distribution pipe networks for arctic settlements. In: *Proceedings, Utilities Delivery in Northern Regions*. Environmental Protection Service Rep. No. EPS 3-WP-80-5, Ottawa, Ontario, 347-363.

Jeppson, R.W. 1976. *Analysis of Flow in Pipe Networks*. Ann Arbor Science, Ann Arbor, Michigan, 164 p.

Johnson, G.V. 1979. Pitorifice service loop calibration testing. In: *Proceedings, Applied Techniques for Cold Environments*. American Society of Civil Engineers, New York, 1053-1062.

Mauser, M. 1994a. Forty years in frozen ground – The Fairbanks water distribution system. In: *Proceedings, Seventh International Cold Regions Engineering Specialty Conference*. Canadian Society for Civil Engineering, Montreal, Quebec, 499-509.

Mauser, M. 1994b. *Arctic Water Distribution System Design Improvements*. Final Report for ASTF Project 19-1-0085. Alaska Science and Technology Foundation, Anchorage, Alaska.

McFadden, T. 1977. Freeze damage prevention in utility distribution lines. In: *Proceedings, Utilities Delivery in Arctic Regions*. Environmental Protection Service Report No. EPS 3-WP-77-1, Environment Canada, Ottawa, Ontario, 221-231.

O'Brien, E. and A. Whyman. 1977. Insulated and heat-traced polyethylene piping systems – A unique approach for remote cold regions. In: *Proceedings, Utilities Delivery in Arctic Regions*. Environmental Protection Service Report No. EPS 3-WP-77-1 Environmental Canada, Ottawa, 309-345.

Powergain Manufacturing Co. nd. "Aqua-Flo". Mi-Sask Industries Ltd., Saskatoon, Saskatchewan.

Ryan, W. 1977. Design guidelines for piping systems. In: *Utilities Delivery in Arctic Regions*. Environmental Protection Service Rep. No. EPS 3-WP-77-1, Ottawa, Ontario, 243-255.

Smith, D.W. and A. Yee 1983. *Evaluation of Water Bleeder Controls*. Environmental Protection Service Rep. No. EPS 4-WP-83-1, Ottawa, Ontario, 60 p.

Suncord Engineering HYDROTHERM. nd. Edmonton, Alberta.

Westfall, H.C. and J.R. Wallace. 1953. *Design Analysis, Water Distribution System, Fairbanks, Alaska*. R.W. Beck and Associates, Seattle, Washington.

Whyman, A.D. 1980. Pre-insulated high-density polyethylene piping – The evolution of a northern achievement. In: *Proceedings, Utilities Delivery in Northern Regions*. Environmental Protection Service Rep. No. EPS 3-WP-80-5, Ottawa, Ontario, 187-220.

Wilson, C.E. and S. Cheema. 1987. Water and sewer service connections in permafrost areas of the Northwest Territories. In: *Proceedings of the CSCE Centennial Conference*, Montreal, 2(1): 115.

Yee, A. and D.W. Smith. 1981. Evaluation of alternative water bleeder controls. In: *Proceedings, The Northern Community: A Search for a Quality Environment*. American Society of Civil Engineers, New York, 555-569.

Yee, A. and D.W. Smith. 1982. A study of water bleeder control alternatives. In: *Proceedings, Utilities Delivery in Cold Regions*. Environmental Protection Service Rep. No. EPS-82-6, Ottawa, Ontario, 152-179.

8.12 Bibliography

Alter, A.J. 1969. *Water Supply in Cold Regions*. Cold Regions Research and Engineering Laboratory, Monograph III-c5a, Hanover, New Hampshire, 85p.

Anonymous, 1987. College utilities corporation. *Water*, 28(2):38-41.

Armstrong, B.C., D.W. Smith and J.J. Cameron. 1981. Water requirements and conservation alternatives for northern communities. In: *Design of Water and Wastewater Services for Cold Climate Communities*. Pergamon Press, Oxford, 65-93.

Bainbridge, S. 1987. Operation, modification, and management of a municipal arctic water and wastewater system. In: *Proceedings of the Second International Conference, Cold Regions Environmental Engineering*. Department of Civil Engineering, University of Alberta, Edmonton, Alberta.

Bohlander, T.W. 1963. Electrical method for thawing frozen pipes. *Journal of the American WaterWorks Association*, 55(5): 602-608.

Bond, R. 1986. *Arctic Water Service Line*. U.S. Public Health Service Standard Design Notes, Anchorage, Alaska.

Buck, C. 1976. Variable speed pumping has many advantages. *Johnson Drillers Journal*, May-June.

Cameron, J.J. 1977. Buried utilities in permafrost regions. In: *Proceedings, Symposium on Utilities Delivery in Arctic Regions*. Environmental Protection Service and Department of Civil Engineering, University of Alberta, Report No. EPS 3-WP-77-1.

Cameron, J.J., Christensen, V., and Gamble, D.J. 1977. Water and sanitation in the Northwest Territories. *The Northern Engineer*, 9(4): 4-12.

Canada Department of Indian Affairs. 1970. *Handbook of Water Utilities, Sewers and Heating Networks Designed for Settlements in Permafrost Regions*. Translated from Russian by V.Poppe, Ottawa, Ontario. 107 p.

Capito, G., and Gajewski, B. 1991. Lessons learned from the Emmonak water and sewer project. *The Northern Engineer*, 23(1): 30-35.

Cheema, S. 1986. Buried water and sewer service connections in permafrost areas. *The Northern Engineer*, 18(2/3): 18-21.

Collins, J.T. and Jacobson, E. 1984. Alternate proposal for circulating water distribution systems in rural Alaska. In: *Proceedings, Third International Specialty Conference on Cold Regions Engineering*. Canadian Society of Civil Engineering, Edmonton, Alberta, April 4-6, 381-394.

Corwin, B.J. and R.E. Kniefel. 1983. Piping systems in permafrost areas. In: *Proceedings of the First Conference on Cold Regions Environmental Engineering*.Environmental Quality Engineering and Civil Engineering, University of Alaska, Fairbanks and Civil Engineering Department, University of Alberta, Edmonton.

Coterill, R. 1983. Freeze protection of water service lines. In: *Proceedings of the First Conference on Cold Regions Environmental Engineering*, Fairbanks, Alaska, May 18-20, University of Alaska, Fairbanks.

Curry, R. 1980. Thawing of frozen water services. In: *Proceedings, Second Symposium on Utilities Delivery in Northern Regions*, March 19-21, 1979, Edmonton, Alberta. Northern Technology Unit, Water Pollution Control Directorate, Environmental Protection Service, Environment Canada, Report No. EPS 3-WP-80-5, 310-313.

Dawson, R.N. and K.J. Cronin. 1977. Trends in Canadian water and sewer systems serving northern communities. In: *Proceedings of the Symposium on Utilities Delivery in Arctic Regions*. Environmental Protection Service and Department of Civil Engineering, University of Alberta, Report No. EPS 3-WP-77-1.

Dawson, R.N. and J.W. Slupsky. 1968. *Pipeline Research – Water and Sewer Lines in Permafrost Regions*. Canadian Division of Public Health Engineering, Manuscript Report No. NR-68-8, Edmonton, Alberta, 78 p.

Esch, D.C. 1990. Temperature and thaw depth monitoring of pavement structure. *The Northern Engineer*, 22(1): 4-11.

Farouke, O. 1982. *Thermal Properties of Soils*. U.S. Army Corps of Engineers, Cold Regions Research and Engineering Laboratory (CRREL) Monograph 81-1.

Gamble, D. and C. Janssen. 1974. Evaluating alternative levels of water and sanitation service for communities in the N.W. Territories. *Canadian Journal of Civil Engineering*, (1): 162-168.

Gamble, D.J. and Lukomshyj, P. 1975. Utilidors in the Canadian north. *Canadian Journal of Civil Engineering*, 2(2):162-168.

Gamble, D.J., 1977.Unlocking the utilidor. In: *Proceedings Symposium on Utilities Delivery in Arctic Regions*. Environmental Protection Service and Department of Civil Engineering, University of Alberta, Report No. EPS 3-WP-77-1.

Gerlek, S. 1982. *Small Circulating Pumps and Power Factor Controllers*. Alaska Area Native Health Service, Anchorage, Alaska, internal memorandum, January 4.

Gerlek, S. 1982. *Design Recommendations*. Alaska Native Area Health Service, Anchorage, Alaska, internal memorandum, February 18.

Gordon, R. 1973. *Batch Disinfection of Treated Wastewater With Chlorine at Less than 1°C*. U.S. Environmental Protection Agency, Rep. No. 660/2-73-0, Washington, D.C.

Government of the Northwest Territories. nd. *Equipment Specification No. 601, Municipal Water Delivery Truck Hydraulic Drive*. Department of Public Works, Yellowknife, NWT.

Grainge, J.W. 1969. Arctic heated pipe water and waste water systems. *Water Research*, 3: 47-71.

Haigh, C.J. 1986. *Insulation Study*. Unpublished report for City of Fairbanks Engineering Department, April.

Harris, B. 1987. *Pitorifice Tests*. Unpublished notes on test conducted for PHS.

Irwin, W.W. 1980. New approaches to water and sewer services in permafrost area – Norman Wells, NWT. In: *Proceedings, Second Symposium on Utilities Delivery in Northern Regions*. Environmental Protection Service, Environment Canada, Report No. EPS 3-WP-80-5, 507-542.

James, F.W. 1976. Buried pipe systems in Canada's arctic. *The Northern Engineer*, 8(1): 4-11.

James, F.W. 1980. Critical evaluation of insulated shallow buried pipe systems in the Northwest Territories. In: *Proceedings, Utilities Delivery in Northern Regions*. Environmental Protection Service Rep. No. EPS 3-WP-80-5, Ottawa, Ontario, 150-187.

James, F. 1977. Report on new Frobisher Bay utilidor phase I. In: *Proceedings, Symposium on Utilities Delivery in Arctic Regions*. Environmental Protection Service, and Department of Civil Engineering, University of Alberta, Report No. EPS 3-WP-77-1.

James, F.W. 1981. State of the art review – Water distribution and sewage collection in northern North America. In: *Design of Water and Wastewater Services for Cold Climate Communities*. Pergamon Press, Oxford, 55-64.

James, F.W. 1980. State of the art review – water distribution and sewage collection in northern North American. In: *Design of Water and Wastewater Services For Cold Climate Communities*. Proceedings of a Post-Conference Seminar held on 28th and 29th June 1980, in Edmonton, Canada in conjunction with the 10th IAWPR conference held in Toronto, Canada. Editors, D.W. Smith and S.E. Hrudey, Pergamon Press, 56-64.

James, W. 1980. Water distribution pipe networks for arctic settlements. In: *Proceedings Second Symposium on The Utilities Delivery in Northern Regions*. Environmental Protection Service, Environment Canada, Report No. EPS 3-WP-80-5, 347-363.

James, W. and A.R. Vieirn-Ribeior. 1979. *Arctic Hydrology Project, Baffin Island Field Program 1971 and 1972*. McMaster University, Hamilton, Ontario.

James, W. and M.A. Robinson. 1979. A computer program for designing pipe networks for arctic settlements. *The Northern Engineer*, 11(2): 4-11.

Johnson, G.V. 1978. Pitorifice service loop calibration testing. In: *Proceedings of the Conference on Applied Techniques for Cold Environments, Volume II*. American Society of Civil Engineers, New York.

Johnston, G.H. (editor) 1981. *Permafrost: Engineering Design and Construction*. John Wiley and Sons, Toronto, Ontario, 540 p.

Kent, D. and Hwang, C.T. 1980. Use of a geothermal model in northern municipal projects. In: *Proceedings Second Symposium on Utilities Delivery in Northern Regions*. Environmental Protection Service, Environment Canada, Report No. EPS 3-WP-80-5, 347-363.

Kill, D. 1974. Cost comparisons and practical applications of air lift pumpings. *Johnson Drillers Journal*, September-October.

Leman, L.D., 1980, Water and sewer utilities for Barrow, Alaska. In: *Proceedings, Second Symposium on Utilities Delivery in Northern Regions*, March 19-21, 1979, Edmonton, Alberta. Northern Technology Unit, Water Pollution Control Directorate, Environmental Protection Service, Environment Canada, Report No. EPS 3-WP-80-5, 484-505.

Martin, R.W., and J.F. Sahlfeld. 1984. Sewer/water connections for Barrow utilities system. In: *Proceedings, Third International Specialty Conference on Cold Regions Engineering*, April 4-6, Edmonton, Alberta. Canadian Society of Civil Engineering, 395-400.

Mauser, M. 1982. Operation and maintenance considerations for the design of arctic water systems. In: *Proceedings, Utilities Delivery in Cold Regions Symposium*. Environmental Protection Service, Environment Canada. Report No. EPS 3-WP-82-May.

McFadden, T. 1977. Freeze damage prevention in utility distribution lines. In: *Proceedings, Utilities Delivery in Arctic Regions*. Environmental Protection Service Report No. EPS 3-WP-77-1, Environment Canada, Ottawa, Ontario, 221-231.

McFadden, T. 1988. Thermal performance degradation of wet insulations In cold regions. *Journal of Cold Regions Engineering*, 2(1): 25-34.

Page, W.B. 1953. *Report on Tests Conducted to Design Pitorifices and to Measure Heat Losses from House Service Pipes*. Washington State College, Pullman, Washington, Arctic Health Research Center, Public Health Service, Department of Health, Education and Welfare.

Page, W.B. , L. Hubbs, and E.M. Lamphere. 1957. Report on the Operation of a Recirculating Water Distribution System at Fairbanks, Alaska. Environmental Sanitation Section, Arctic Health Research Center, U.S. Public Health Service, Anchorage, Alaska.

Poss, R.J. 1960. Distribution System Problems. *Journal of the American Water Works Association*, 52: 2.

Prentice, J. R. and G.A. Srouji, G.A. 1980. Waterworks systems NWT. In: *Proceedings, Second Symposium on Utilities Delivery in Northern Regions*, March 19-21, 1979, Edmonton, Alberta. Northern Technology Unit, Water Pollution Control Directorate, Environmental Protection Service, Environment Canada, Report No. EPS 3-WP-80-5, 409-425.

Roen Design Associates, Inc. and Bell-Walker Engineers, Inc. 1982. *Water Systems Master Plan*. City of Fairbanks, Alaska, Municipal Utilities System.

Rosendahl, G.P. 1980. Alternative strategies used in Greenland. In: *Design of Water and Wastewater Services for Cold Climate Communities*, Proceedings of a Post-Conference Seminar held on the 28th and 29th June 1980, in Edmonton, Canada in conjunction with the 10th IAWPR conference held in Toronto, Canada, Editors D.W. Smith and S.E. Hrudey, Pergamon Press, 17-24.

Ryan, W.L. 1973. Design and construction of practical sanitation facilities for small Alaskan communities. In: *Second International Permafrost Conference Proceedings*. National Academy of Science, Washington, D.C., 721-730.

Ryan, W.L. and K.C. Lauster. 1966. Design and operation of Unalakleet, Alaska water system. *Journal of the American Water Works Association*, 58(8): 1045-1051.

Santori, E. 1976. Winter maintenance of water mains and services. *Journal of the American Water Works Association*, 68(1): 19-21.

Shillington, E. and B. Miller. 1983, Fort Chipewyan water supply, treatment and distribution systems. In: *Proceedings of the First Conference on Cold Regions Environmental Engineering*. University of Alaska, Fairbanks and Civil Engineering Department, University of Alberta, Edmonton, Alberta, Canada.

Shillington, E.I. and G.D. MacKinnon. 1987. Barrow utility system, direct bury system – The alternative. In: *Proceedings of the First Conference on Cold Regions Environmental Engineering*, Department of Civil Engineering, University of Alberta, Edmonton, Alberta.

Smith, D.W. (editor) 1977. *Proceedings, Utilities Delivery in Arctic Regions*. Environmental Protection Service, Rep. No. EPS 3-WP-77-1, Ottawa, Ontario, 596 p.

Smith, D.W. (editor) 1980. *Proceedings, Utilities Delivery in Northern Regions*. Environmental Protection Service Rep. No. EPS 3-WP-80-5, Ottawa, Ontario, 542 p.

Smith, D.W. (editor) 1982. *Proceedings, Utilities Delivery in Cold Regions*. Environmental Protection Service. Rep. No. EPS 3-WP-82-6, Ottawa, Ontario, 419 p.

Smith, D.W. (editor) 1984. *Proceedings, Third International Specialty Conference on Cold Regions Engineering.* Canadian Society for Civil Engineering, Montreal, Quebec, 1201 p.

Spehalski, J.R. 1988. *Circulating Water System and Pitorifice Study*. Alaska Area Native Health Service, Anchorage, Alaska.

Srouji, G.A. 1978. Thermal analysis of water distribution systems. In: *Volume II Proceedings of the Conference on Applied Techniques for Cold Environments.* American Society of Civil Engineers, New York.

Thomas, J.E. 1988. The Barrow direct burrow utilities system design. In: *Permafrost Fifth International Conference Proceeding*s. Tapir Publishers, Trondheim, Norway.

U.S. Department of Health, Education and Welfare, Division of Indian Health. 1972. *Design Criteria for Indian Health Sanitation Facilities.* Anchorage, Alaska.

U.S. Environmental Protection Agency. 1962. *Manual of Individual Water Supply Systems.* Publication No. 24, Washington, D.C.

Vause, K.H., L.A. Esvelt and E.S. Jacobson. 1987. Development of innovative water and sewer facilities for a village, Emmonak, Alaska. In: *Cold Regions Environmental Engineering, Proceedings of a Second International Conference.* Edmonton, Alberta, 23-24 March 1987.

Wilson, C.E. and S. Cheema. 1989. Water and sewer service connections in permafrost areas of the NWT. *Canadian Journal of Civil Engineering*, 16: 188-196.

Zirjacks, W.L. and C.T. Hwang. 1983. Underground utilidors at Barrow, Alaska: A two year history. In: *Permafrost Fourth International Conference Proceedings.* National Academy Press, Washington, D.C.

SECTION 9

WASTEWATER COLLECTION

3rd Edition Steering Committee Coordinators

Vern Christensen

James A. Crum

Special Contributions

Gary Eddy

Tom Heintzman

Daniel H. Schubert

Jim Vogel

Section 9   Table of Contents

Section 9 List of Figures

Section 9   List of Tables

9 WASTEWATER COLLECTION

Wastewater collection systems remove wastewater from the vicinity of the users and transport it to the sewage treatment facility or disposal point. Arctic municipal waste collection systems are one of the most difficult facilities to design, construct, and operate. The sources of wastewater, types of collection systems, and design parameters are presented in this section. The main types of collection systems are individual bucket haul, bucket haul with collection vehicle, vehicle haul with house storage, and transport pipes. Table 9-1 summarizes the characteristics of wastewater collection systems.

9.1 Sources of Wastewater

Domestic and commercial water use, infiltration, and inflow are the main sources of wastewater. In northern areas nearly 100 percent of the delivered water ends up as domestic wastewater because there is very little lawn and garden watering or car washing. This also applies generally to the source of commercial wastes. An exception to this generalization would be an operation such as a cold-storage fish-processing operation or a cannery where a significant amount of the water may be used for cleaning floors and equipment and then discharged directly into the ocean. High sewage flows may be experienced where buried gravity lines and access holes are placed in areas with high groundwater levels or where water wasting is practiced. These conditions need to be evaluated and should be corrected before sewers and sewage treatment plants are designed. Inflow and infiltration studies should be performed, so that the problems can either be identified for correction or a plant designed to treat the more dilute and higher volume waste.

9.1.1 Domestic Waste.
The rates of water use given in Sections 5 and 10 show that sewage flows can vary considerably. They are lower in volume and higher in strength for camps and facilities with plumbing fixtures that constrain water use. Where water is wasted during the winter to keep service lines from freezing, the volume of wastewater can be extremely high and of very low strength. There is no substitute for actually monitoring sewage flows, or at least water use rates, before sewer or sewage treatment design is undertaken. A good evaluation of the present base flows followed by proper allowances for future growth and peak flows provides a good basis for the collection system design.

9.1.2 Commercial Waste.
Commercial waste output is often seasonal in cold regions. Canneries and cold-storage wastes occur during the summer fishing season only. Reindeer slaughtering and packing plant wastes have high strength and flow, but they usually last for only one or two months during the fall. The amount and strength of commercial wastes can be estimated using information from temperate areas for similar facilities. The kind and size of the solids in these wastes may require screening or comminuting prior to discharge to the sewer system.

9.1.3 Wastes from Public Institutions.
Nearly all communities will have schools and hospitals or health clinics. Estimates of wastewater flows must be obtained for schools by considering the number of students, the number of teachers in living quarters, and the type of facilities provided, e.g., swimming pools, flush valve toilets or regular gravity flush toilets, urinals, shower facilities, and boarding student dormitories. With hospitals or clinics, the number of beds and again, the type of facilities provided should be determined. Laundromats are another source of wastewater. These wastes are not much different from similar facilities in more temperate climates. See Section 5 for estimating expected water use at public facilities. Water conservation practices and low-water-use plumbing fixtures can result in increased sewage strengths.

9.1.4 Wastes from Water Treatment Processes.
Discharge of chemical wastes can affect wastewater treatment processes. The designer should review the justification for discharging wastes such as salt brine into the sewer system. If possible, removal and storage of water treatment plant wastes on-site may be a better solution rather than later treatment at a wastewater treatment plant.

If they cannot be eliminated from the sewer, waste characteristics and volume from water treatment plants must be determined or estimated. The highest flows, generally from filter-back washing, can be very sporadic and may require a holding tank for flow equalization before discharge to the collection system. Filter-back washing flows are generally low in BOD. Some water treatment chemicals can be of high or low pH and very corrosive. Water treatment plant wastes can cause interior corrosion problems such as wastes from a salt water distillation unit. Electrolysis can usually be reduced by use of du-

TABLE 9-1   CHARACTERISTICS OF WASTEWATER COLLECTION SYSTEMS

Type	Soil Conditions	Topography	Economics	Other
Gravity	Non-frost-susceptible or slightly frost-susceptible with gravel backfill.	Gently sloping to prevent deep cuts or lift stations	Initial construction costs high, but operational costs low unless above ground or lift stations used.	Lowest maintenance; high health and convenience; larger diameter pipes required; flushing of low-use lines may be required.
Vacuum	Most useful for poor soils or bedrock conditions; can be used with any soil type.	Level or gently sloping	Initial construction costs moderately high; operational costs moderately high.	Very low water use requirement; high health and convenience; special toilets available; can use existing low-use fixtures; requires vacuum station every 200 to 500 services; uses smaller diameter pipe; no exfiltration.
Pressure	Most useful for frost-susceptible or bedrock soils; can be used with any soil type.	Level to hilly topography	Initial construction costs moderate; operational costs moderately high.	High health and convenience; can use existing fixtures; no central facility necessary; individuals responsible for pumps; uses small pipes.
ATV Haul	Used in poor soil areas with boardwalks.	Level, gently sloping	Initial construction costs low; operation costs depend on water use – generally very high for comparable service.	Provides only a moderate health and convenience improvement due to small quantities of water used; operational costs are high depending on use rates.
Truck Haul	Most useful for frost-susceptible or bedrock soils; can be used with any soil type.	Level to hilly topography	Initial construction costs low; operation costs are high.	Moderate health and convenience improvement; operational costs usually must be subsidized.
Individual Haul	Used with any soil condition; boardwalks are necessary in some swampy conditions.	Level to hilly topography	Initial construction and operational costs very low.	Low health and convenience improvements due to low water use and differing individual disposal practices.

rable coatings, use of nonconductive fittings and piping systems. Where plastic or fiberglass reinforced plastic pipe is used, corrosion and electrolysis are usually not a problem.

9.1.5 Type of Pipe. Polyvinyl chloride (PVC) pipe, asbestos cement pipe (AC), and wood stave (WS) pipe was used in the early years of sewer construction in cold regions. These materials were periodically damaged in use, cleaning processes, and thawing after being frozen. Recent systems are typically being built with HDPE pipe which can be frozen full of water with a reduced probability of damage. Care is required during thawing of HDPE pipe to prevent melting and collapse of the pipe.

9.2 Undiluted Human Waste (Honey Bucket Sewage Haul Systems)

The honey bucket haul system consists of hand-hauled containers used to transport human excretement from the homes to a form of haul system that moves waste from the vicinity of the homes to a remote disposal or treatment site.

Where the collection system relies on the individual users to bring their wastes to a disposal point, the important considerations are the types of containers in which the waste is transported, and the facilities at the point of treatment. This topic is further discussed in Sections 14 and 16.

Community-haul sewage collection provides the collection of honey-bucket wastes and its transportation by a community or contractor-operated vehicle to a treatment disposal facility. Collection service is supported by an established service charge and is scheduled on either a routine cycle, or an on-call basis.

The procedure for disposal used at the individual dwelling or building can vary from simply emptying a honey bucket into a tank on a truck (hot pick-up), or into a holding tank adjacent to the building from which the wastes are later pumped and hauled to the disposal site.

9.2.1 All Terrain Vehicle (ATV) Haul Systems. During the 1980s a transportation system was developed for remote communities in Alaska which uses small all terrain vehicles (ATV) to pull a special trailer for sewage transport from the community. The trailer transports polyethylene vats stationed throughout the community into which the adjacent residents have deposited honey-bucket wastes (see Figure 9-1).

Transport is generally on wood boardwalks which are capable of supporting the loads of the loaded trailer and ATV. In the summer months the community operator places a cover on the vat prior to transport to prevent spillage. During the winter the operator can haul the frozen sewage in the vat without a cover. The frozen material is loosened by striking the side of the plastic container with a mallet before rotating the vat for dumping. This is possible because polyethylene plastic has the property of not developing a strong bond with the frozen waste.

During the winter months the trailer wheels can be replaced by skis and a snowmachine can be used for pulling.

The systems are used mostly to transport honey-bucket wastes. Greywater should not be disposed of individually near the homes but it often occurs.

Over 30 of these types of systems have been built in rural Alaskan communities and are operated by the individual communities with varying success. Where the operator and community government strongly support providing the service and keeping the local community environment clean, the systems operate well. Where these goals of the local government are not achieved, the operation is sporadic.

The operation of the local systems is supported by a local government subsidy or by a nominal service charge, typically $25 per month (US) per home.

9.2.2 Honey Bucket Haul Systems with Truck Transport. This type of system is used in communities with gravel roads adequate to support truck wheel loads and where snow removal equipment is available to clear truck access routes near the homes.

The waste is individually picked up from each house by the operator and deposited into the truck (Figure 9-2). In some cases the operator enters the house and removes the bucket from the restroom and in others it is placed on the front porch by the homeowner. If the home is not open, no service is provided that week.

This type of system is presently used in a few communities in Alaska and Canada, but is being replaced by more suitable systems which allow a flush-type toilet to be provided in the home.

9.3 Tanker Vehicle and Building Holding Tank Systems

9.3.1 House Plumbing Requirements. Most modern flush-and-hold vehicle-haul systems con-

FIGURE 9-1   ATV HAUL TRAILER AND CONTAINER – NAPAKIAK, ALASKA

FIGURE 9-2   TRUCK FOR COLLECTING HONEY BUCKET WASTES –
              BARROW, ALASKA

FIGURE 9-3   VACUUM PUMPING TRUCK

sist of a sewage holding tank located on or underneath the floor of the house into which the household wastes from sinks, lavatories, and toilets drain by gravity. The tanks are then drawn into the collection vehicle by vacuum (Figure 9-3).

The efficiency, operational costs, and feasibility of this type of collection system depends partly on the size of the holding tanks, haul vehicles, and road conditions in the community. If larger haul vehicles are used, tanks should hold around 1,000 L, at least 400 L larger than the water storage tank provided (Cameron, 1981). In Canada, household tanks with a capacity as large as 4,500 L have been used. It is is common practice to make the sewage holding tank up to twice as large as the water storage tank. The size of the collection vehicle and reliability of the service should be considered in sizing wastewater storage tanks. It is also important to provide the structural support in the house required to carry this additional load. The tank must be constructed with a large access hole with removable cover, so that the tank can be cleaned and flushed out at least once yearly. It must be protected from freezing by either being kept within the heated portion of the building or well-insulated and heated by an electric heating element or circulating hot water to prevent

any ice formation. In some instances, holding tanks have been buried in the ground beside or under the houses (Figure 9-4), but in ice-rich permafrost areas this may result in stability problems. Heat may even have to be added to tanks buried in the active layer. The tanks must be placed so that they can be emptied from the outside of the house.

The pumpout connection at the building must be sloped to have the sewage drain back into the tank after pumping so that it does not drain outside the house or remain in the line and freeze. The tanks must be vented into the house plumbing venting system or to the outside to allow for air escape or air supply as the tank is filled and emptied.

9.3.2 ATV Haul Systems – (Mekoryuk Prototype).
This type of haul system was developed to a prototype stage in the late 1980s in order to provide a system with an aesthetically acceptable flush toilet in the home but also meeting the requirement for transport of the waste by a haul system along light-duty boardwalks.

The system consists of a 350 L wastewater holding tank located either in or adjacent to the home. Waste is deposited through a low-water-use toilet (500 mL per flush). The holding tank requires emptying approximately once per month for toilet wastes only. If other fixture wastes are deposited into the system, the frequency of emptying is higher.

The interior-type tank is located in the heated space of the house and consists of a structural plywood container holding a reinforced rubber bladder. The tank is emptied by pressurizing the tank with a low-pressure blower which forces the sewage up the evacuation pipe and out through the wall and into the hauling tank. Components of this system are being patented by a Canadian manufacturing firm.

The outside tank installation is a super-insulated skid-mounted tank adjacent to the house that is connected to the house by an insulated arctic pipe. The size can be 350 L or larger. The outside tank is emptied by a similar low-pressure blower process forcing the sewage up the evacuation pipe and into the haul tank. The outside tank is fitted with electric heat tracing to prevent freezing during extremely cold weather.

The haul tank consists of a stainless steel insulated tank sized to hold slightly more than the 350 L holding tank. The tank is mounted on a two wheel trailer fitted either with wheels or skis. The trailer is pulled with an ATV or a snowmachine (Figure 9-5).

The operator of the haul tank carries a blower which is used to pressurize the holding tank to about 20 kPa. The contents of the tank discharge through a 65 mm flexible arctic hose into the haul tank. End connections are quick-connect type fittings. The haul tank is then transported along a road or boardwalk to a disposal facility (usually a lagoon). The sewage is dumped by gravity from the tank through a short hose.

A similar system is available with a haul trailer for water. Water is hauled from a watering point with an ATV vehicle and discharged with air pressure through a quick-connection hose into a holding tank in the home.

9.3.3 Pumper Truck Haul Systems. The type of vehicle used to carry the pumpout tank depends

FIGURE 9-4   INDIVIDUAL HOUSE
HOLDING TANK

FIGURE 9-5   ATV SEWAGE HAUL VEHICLE – MEKORYUK, ALASKA

on the conditions at the site. Where no roads exist, tracked vehicles may be necessary. However, tracked vehicles have much higher capital, operating, and maintenance costs compared to rubber-tired vehicles and should be avoided. Instead of tracked vehicles, four-wheel drive trucks with floatation tires can be used where snowdrifting and poor roads are prevalent.

The tanks on the vehicles to be used with a pumpout system may have to be insulated or surrounded by heated enclosures. These enclosure units on the back of the vehicle include the tank, pump, piping, and connection hose. The size of tank needed depends primarily on the distance to the disposal site and the sizes of the tanks in the houses. Larger tanks reduce operating costs, particularly where the disposal site is distant. The size and weight of the vehicle and thus of the tank, are limited to the layout of the community and the condition of the roads. The truck must maneuver close enough to the houses to pump out the tanks in a reasonably short time.

The number of trucks needed in a given community depends mainly on the number of buildings to be serviced and the volume of wastewater to be col-

lected (Cadario and Heinke, 1972). (Sample truck specifications are contained in Appendix B.)

9.3.4 Sewage Transfer from House Tank to Collection Vehicle.
Two methods of emptying the house tanks are being used: one method uses a vacuum tank, the other a sewage pump.

Vacuum Tank Method.   The tank on the vehicle is held under vacuum created by a small compressor. The contents of the house tank are withdrawn into the truck tank under vacuum. At the disposal point a three-way valve is turned to allow the pump to pressurize the tank, forcing the wastewater out (see schematic in Figure 9-6). The advantage of this method is that a compressor is used instead of a liquid pump resulting in less piping containing sewage, which could freeze.

Sewage Pump Method.   This method is similar except that a self-priming sewage pump is used instead of a compressor and the vehicle tank does not have to be pressure rated. Discharge from the tank is generally by gravity at the disposal site.

Characteristics of Suction Hose.   The suction hose must be long enough to extend from the holding tank to the most convenient parking place for

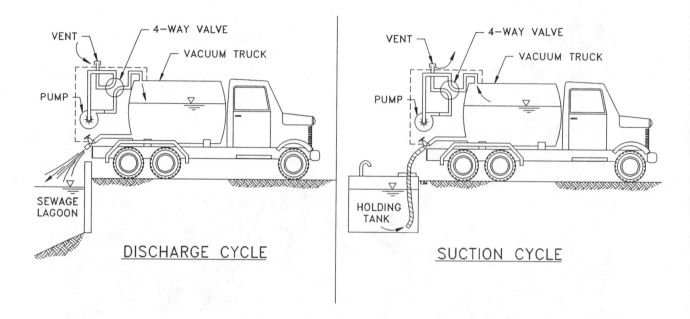

FIGURE 9-6   VACUUM SEWAGE TRUCK PUMPOUT PROCEDURE

the truck. It should be at least 75 mm in diameter and kept on a reel so that it can be rolled up inside the heated compartment on the truck. It must be noncollapsible under vacuum use and should be rugged enough that it can withstand being dragged along the ground between truck and houses. And it must be constructed of materials that are flexible and not brittle under extremely low temperatures.

The hose connections should be of the quick-disconnect type and should be of a size different from that of the water delivery hose to eliminate the possibility of cross-connection. Some health codes require the water and sewer connections at the house be separated by 1.5 metres.

9.3.5 Haul System Disposal Site. Where the disposal point is at a lagoon, a ramp, sheet-pile discharge face, splash pad, or similar protector should be provided to prevent erosion yet still allow the vehicle to deposit the wastes well within the lagoon. The disposal point should be adjacent to a drive-through area to shorten the unloading time as much as possible.

9.3.6 Haul System Maintenance Considerations. Vehicle-haul systems generally have higher operation and maintenance (O&M) and more re-

placement costs than piped systems because the vehicles have a shorter useful life.

A heated storage building should be provided for the vehicles while they are not being used or while they are being repaired. The building in which the vehicles are stored or emptied should also be equipped with water for flushing and cleaning the tanks. Extreme care must be taken to prevent any water system cross-connections, and proper disposal facilities for the wash water should be planned.

9.4 Piped Collection Systems

Conventional gravity sewers can be used in cold regions either as single insulated pipes above ground or below ground, or uninsulated as part of a utilidor. Normally, gravity sewers are the most cost-effective and have the added advantage of freezing up gradually and in layers because they rarely flow full. Therefore catastrophic failure is less likely to occur.

Gravity sewers are generally preferred because they use proven methods of construction and maintenance and have relatively low O&M costs. Conventional, readily available plumbing fixtures are compatible with gravity sewers and are relatively inexpensive. The primary disadvantage of gravity mains is the need to install pipes on a stable foundation for sewage to flow by gravity at design grades. Site conditions are not always favorable for the construction of gravity sewers, because of very flat or very hilly terrain, rock, ice-rich permafrost, or other conditions. In such cases, alternative sewerage may be feasible; e.g., both pressure sewers and vacuum sewers have been successfully installed and operated in cold regions.

By contrast, pressure and vacuum sewers both employ small-diameter pipe (32 to 150 mm in diameter). However, O&M costs of pressure and vacuum sewers are higher because they depend on an energy input to move the sewage. Water conservation is easier with these types of piping systems. However, plumbing fixtures are expensive, require maintenance, and may be hard to obtain.

9.4.1 Piped System Design Considerations.
Early in the design process several decisions are needed regarding :

- the layout and placement of the sewers;

- the temperatures of sewage, soil, and air;

- the exclusion of storm water;

- the use of alternative non-conventional sewerage; and

- provisions for preventing frozen sewers and for thawing sewers when freeze-up does occur.

Lines should be placed deep enough into the ground to prevent damage from surface loadings or else measures should be taken to prohibit surface loads. Each case should be evaluated according to the pipe manufacturer's recommendations. Substantial loads are imposed on pipes during frost penetration (Smith, 1976).

When collection lines cannot be buried because of soil conditions (as discussed in Sections 2, 3, 4 and 11), they must be installed on or above the surface of the ground or in utilidors. In most locations, the topography and building layout dictates that above-ground lines be on pilings (or a gravel berm) to hold the grades necessary for sewage flow by gravity. Above-ground lines are undesirable because they pose transportation hindrances (see Figure 9-7) and are subject to high heat losses, blocked drainage, and vandalism. They also create a cluttered look within the community. The lines of pressure or

FIGURE 9-7   *ABOVE-GROUND PILE-SUPPORTED WATER AND GRAVITY SEWER MAINS – BETHEL, ALASKA*

FIGURE 9-8   WATER AND VACUUM SEWER MAINS – EMMONAK, ALASKA

vacuum systems can be placed on the ground surface (see Figure 9-8), which minimizes the problems associated with above-ground lines. Sewage collection lines may be placed in utilidors with other utilities if local circumstances warrant.

Assuming the same design, heat losses of above-ground facilities are nearly three times higher than for the same line placed below ground because of the greater temperature differential between the inside and outside of the pipe.

Sewage temperatures are also an important design consideration. In camps or communities where the individual buildings have water heaters, sewage temperatures usually range from 10°C to 15°C. Where water heaters are not used and any hot water must be obtained by heating it on a stove, sewage temperatures range from 4°C to 10°C. With a gravity system more of this heat is lost between the user and the point of treatment than with a pressure or vacuum system. The sewage in a gravity system is longer in transit, and there is air circulation above the sewage in the partially filled gravity sewer pipes. Warm water may need to be dumped at the ends of low-flow laterals in a gravity system to eliminate freezing problems. Also, assuming the same amount of insulation, a greater percentage of the heat is lost

where the lines are above ground. Heat losses from sewage usually vary less from summer to winter with buried lines than with above-ground lines. (This is discussed in detail in Section 4).

Storm water should not be included in sanitary sewers. It is generally colder and therefore lowers the wastewater temperature, it overloads treatment facilities hydraulically, and it usually contains sand and grit which deposit in the collection lines.

9.4.2 Conventional Gravity Collection Lines. If lines can be buried and the topography is suitable, a gravity sewage collection system is the preferred type of sewer system.

Minimum grades are necessary to ensure adequate velocities in the pipelines. The minimum velocity is normally 0.6 m/s in wastewater collection mains.

If a small amount of settling or heaving of the soil is likely, use a minimum slope of 1.0 percent for all collection lines and bed the line with non-frost-susceptible sand or gravel at least 100 mm below the pipe (Figure 9-9). Smaller 100 mm service lines should be placed with a 2.0 percent minimum slope.

Collection Mains.   In many small communities, the recommended minimum size of a sewage collection line has been 150 mm, instead of the normal 200 mm. The logic supporting this recommendation is that the smaller diameter is more suitable for lower flows. However, maintenance considerations (i.e., method of cleaning) should be carefully evaluated prior to using the smaller pipe.

In addition to maintaining minimum slope requirements, provisions should be made to readjust the slope of a line if it traverses an area where movement is likely. With above-ground lines on piling this has been done by using adjustment blocks or threaded supports between the piling and the pipeline. They can be periodically adjusted to reestablish the proper slope. If the lines are in a utilidor, they can be suspended from the utilidor roof with adjustable turnbuckles or placed on adjustable yokes or supports.

Access Holes. Access holes require protection from frost heaving due to uplift forces on the sides or under the structure. Side uplift problems can be reduced by using a plastic film to break the freezing bond to the access hole (Figure 9-10) and securely connecting all access hole sections together with weld clips or bolts. They should also be insulated to retain heat in the sewage and prevent thawing of permafrost soils. Insulation should be a minimum of 75 mm of styrofoam or urethane. Insulation placed horizontally around the collar of the access hole reduces the heat loss from the surrounding ground and minimizes frost penetration. An insulated frost cover to reduce sewage heat losses should be provided inside. Access holes must have a firm foundation to prevent settlement. This could mean pilings under the access hole or, at least overexcavation and then backfill with gravel. In permafrost the access hole should be placed over insulation to reduce heat loss downward or adequate non-frost-susceptible soils provided to stabilize the foundation.

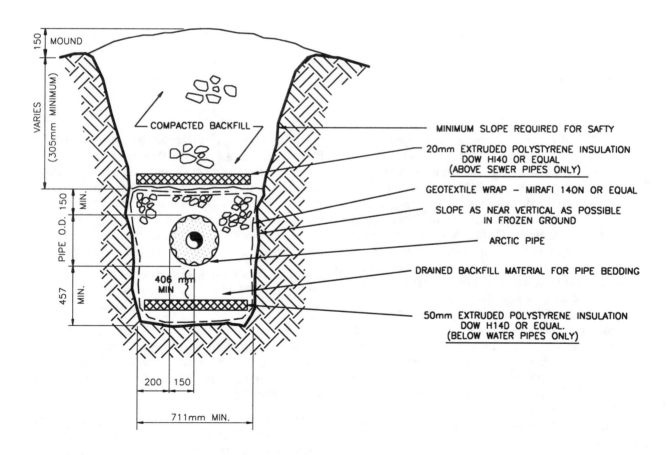

FIGURE 9-9   TYPICAL SECTION – GRAVITY SEWER MAINS

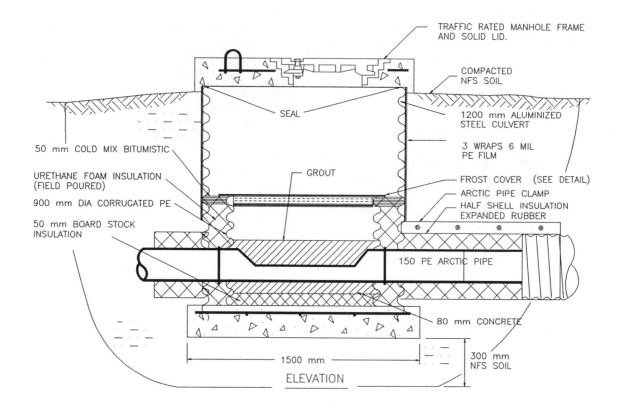

FIGURE 9-10   TYPICAL ARCTIC SEWER ACCESS HOLE

Conventional precast concrete access holes are very expensive to ship and install in remote northern areas. Bolted corrugated aluminum access holes can be used in areas with light traffic loads. They can be nested during shipment and air-freighted by small planes. Access holes for conventional 200 mm dia. sewer systems should be placed every 90 m or at any changes in grade or alignment. Also, solid access hole covers should be used to reduce air flow and prevent surface water entering during break-up or summer storms. Additional vaults should be provided at locations where periodic access to valves, cleanouts, or control structures is required. These structures are similar to insulated access holes with suitable access covers. They often are used to access both the water and sewer lines when the lines are close together as shown on Figure 9-11(a), (b), and (c). Often, smaller access holes (1,070 mm in diameter) are used as access vaults, cleanouts, and access ports.

Service Lines.   Figure 9-12(a) shows a typical service line connection to the side of a building for a cold-region gravity sewage collection system. Fig-ure 9-12(b) shows an under building connection. The line should slope at least 2.0 percent to the collection main. The example shown is a through-the-wall connection and is preferable to one going through the floor. The through-the-wall connection allows for more movement of the house without damage to the sewer service line and all house plumbing is easier to keep above the insulated house floor.

Maintenance Considerations.   Gravity sewers are liable to blockage by oversized objects, cloth, or plastic items which lodge in the sewer.

Frequently, grease builds up in sewers in cold regions. The problem develops to a greater extent downstream of where large amounts of grease are produced and in areas where sewers are subjected to a high cooling rate.

Gravity sewer lines should be thoroughly flushed every summer, ideally when the fire hydrants are being cleaned and flushed. Freezing often occurs in gravity sewers serving small communities, especially in areas of low flow or at dead ends. Thawing is possible using commercial hydraulic sewer clean-

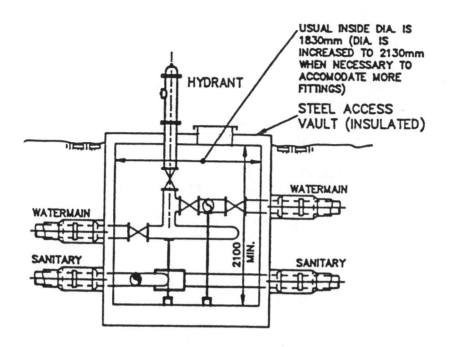

FIGURE 9-11(a)   Combined Access Structure

FIGURE 9-11(b)   Construction of an Access Vault – Rankin Inlet, NWT

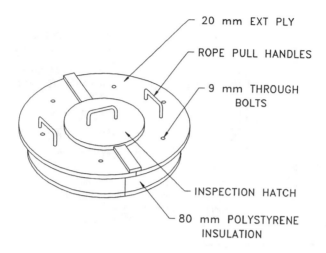

20 mm EXT PLY

ROPE PULL HANDLES

9 mm THROUGH
BOLTS

INSPECTION HATCH

80 mm POLYSTYRENE
INSULATION

FIGURE 9-11(c)   Access Vault Cover

ers. When these are not available, the preferred method for thawing sewage collection lines is to push a plastic line of small diameter through the frozen line, while circulating warm water. Metal lines can be steamed. A complete discussion about the thawing of lines is contained in Appendix D.

For long upper reaches of sewer systems with minimal service line flows, it may be necessary to install a self-flushing siphon on one of the laterals. They can be constructed in a building or in a service structure and set to dump a given quantity of warm water at specified intervals. The slug of warm water traveling down the sewer lines thaws any glaciering which may have built up because of slow trickles of water from normal use. The flushing also scours solids which may have been deposited because of low flows. Self-flushing siphons should not be used unless absolutely necessary because they waste water and create an additional O&M cost. If siphons are built, extreme caution must be exercised to prevent cross-connections with water lines. It is better to lay out the sewer system so that at least one large user is near the end of each lateral to eliminate the need for artificial flushing.

9.5   Pressure Sewage Systems

Soil conditions and community layout may make gravity sewerage impractical. Pressure or vacuum systems may then be considered as alternative sewerage methods (Ryan and Rogness, 1977). Pressure sewage collection systems usually have sec-

tions that operate under gravity. A small pump unit for each service in or near the building served provides the motive force for the pressure system. In the septic tank effluent pump (STEP) system, pretreatment is provided to the sewage with septic tanks at each dwelling before pumping. This allows for smaller mains (i.e., 100 mm to 150 mm) due to removal of most solids. Other pressure systems have pumps to lift the raw sewage directly into the system. Grinder pumps are often used for this purpose.

Advantages.   A significant advantage of the pressure sewer is that grades do not have to be maintained, and elevation differences throughout a community do not affect operation. Pipelines or utilidors do not have to be on piling or above the ground. They can be at or below the surface, and small movements due to frost heave or thawing do not affect operation. Construction costs for pressure systems are usually lower than either a gravity or vacuum system. Smaller connection lines can also be used with a pressure system and can be sized to handle any number of connections. Another advantage is that with the pump unit in or near each building, the O&M costs for the majority of the sewage collection system is paid for by the building owners, because they pay for the electricity and maintenance of their pump. This significantly lowers operational costs for the community and if a unit fails, only one user suffers.

Disadvantages.   The major disadvantage to the pressure sewer in cold regions is the requirement

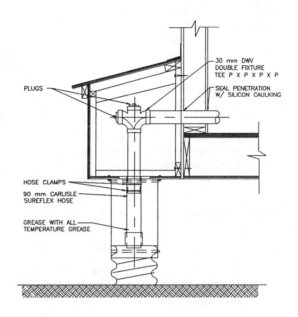

PLUGS

30 mm DWV
DOUBLE FIXTURE
TEE P X P X P X P

SEAL PENETRATION
W/ SILICON CAULKING

HOSE CLAMPS

90 mm CARLISLE
SUREFLEX HOSE

GREASE WITH ALL
TEMPERATURE GREASE

FIGURE 9-12(a)   Side of Building Sewer
Service Connection

FIGURE 9-12(b)   Under Building Water and Sewer Service Connections
— Rankin Inlet, NWT

to heat the utilidor or sewer. This is necessary because it is normally filled with water that can remain stationary for long periods and freeze. This requires that the service lines and main lines be heat-traced with heat tapes or heating loops. If any component of this heating system fails, the system is subject to a potential catastrophic failure. Pipelines that are full of liquid need to be heat-traced in extremely cold areas. Another disadvantage is the higher O&M costs for a pressure system than for a gravity system because of the pump units at each building. There are also potential freezing problems with the septic tank and discharge pump. A small-scale municipal pressure sewer was developed for Galena, Alaska but was later abandoned due to heat tape problems and high operational costs.

9.5.1 STEP Systems. A variation of the pressure system is called the septic tank effluent pump (STEP) system in which a septic tank is used to remove settleable solids from the sewage, and a submergible pump moves the liquid effluent into a combination pressure/gravity sewer. The septic tank sludge is pumped out regularly, i.e., annually. This type of system eliminates the need for an absorption field at each house or it can be used in areas where soil absorption fields have failed. Ultimate disposal can be at a community drainfield or lagoon. This type of system is operational in Wasilla, Alegnigik, Kokhanok, and Mcgrath, Alaska.

Advantages. Some of the advantages of this type of operation are: sewer service can be more economically provided to homes placed on small lots with poor soils; the treatment facility is less complicated and results in a lower operational cost for the community; less solids settle in the collection lines; and problems with the pressure pump plugging up are reduced.

Disadvantages. The disadvantages include: the freezing of septic tanks and pumps in extremely cold periods; the requirement of operating the heat tapes periodically on the pipelines; and the need for periodic pumping of septic tanks (and disposal of septic sludge). For these reasons this type of system is not used in extremely cold areas.

9.5.2 Individual Lift Pump Systems. This type of facility provides a raw sewage lift station at each user to transport sewage into the main line which could either be a pressure or gravity system.

Size, Pressure, and Layout Parameters. Pressure sewers must be sized to maintain a minimum of 1.0 m/s scour velocity. The minimum size for a collection line is 32 mm, which is the minimum size for one pump unit. If more houses are added than originally planned, velocities in the line increase. The effect of this velocity increase is an increase in head loss which, within limits is not a problem. Pressures should be held below 275 kPa in the layout. If the correct size is not available, the pipelines should be slightly undersized (higher velocity), rather than oversized to insure cleansing velocities. When more than one building is connected to one pump unit the hydraulic loading should be checked. Where possible, the collection lines should be constructed to drain to low points for emergency draining. In low-flow conditions (such as during the night in winter) when heat losses are extreme, warm water may occasionally be needed in the system or the heat tape system will need to be operated.

Valves. Air or any gas accumulation in pressure sewers increases both flow resistance and head loss, which pumps must overcome. Sources of air in pressure lines include improperly purged lines, malfunctioning pumps, release of gases from solution, and gas produced by biological activity. Automatic gas-release valves that permit release of accumulated gases without loss of liquid are available for sewer applications. These valves require regular maintenance or inspection and flushing to minimize clogging by sewage solids and greases. Gas can be released manually by opening a valve on a vertical riser at high points in the pressure mains. To be effective manual gas release must be regularly scheduled. Easy access is required for both manual and automatic types of gas-release devices.

Valves are also needed to isolate each pump and each service line to allow for repairs. Similarly, mainline valves and cleanouts are required. About 200 m is the recommended spacing for mainline valves and cleanouts. Cleanouts must be placed at the end of lines.

Pump Grinder Unit. A grinder pump reduces large chunks of solids to small pieces so the wastewater can pass through the pump impeller. The unit can shred all sorts of items flushed down the toilets including sewage solids, rags, and plastic items. One grinder pump unit can serve one or more buildings. Where it serves several buildings, the sewage drains into a common sump by gravity, and duplex pumps are provided. Each unit should have redundant controls and pumps. When the primary unit becomes inoperable, the redundant unit takes over. At the same time, it sets off a warning device (audible and visual) to alert the operator to the equipment breakdown. Standby power should be available in case of a power outage. The pumping unit sump should be well insulated and installed on a stable foundation, if they are placed outside or in the ground. As with access holes they must be protected from frost-jacking. Heat tapes or alternate heating devices are required on all piping exposed to anticipated freezing for extreme cold operation.

With a 750 W grinder pump unit and a water use rate of 190 L/(p•d), the pump would operate only about three times per person per day, with each operating period lasting about one minute. Thus, a five-person family would use about 0.5 kWh a day. At $0.25/kWh, this would only amount to about $3.75 per month per family for electricity. However, operating a heat-tape system required to keep a system thawed can take many times this amount of electrical power. Heating costs are usually much greater than the pumping costs. Standby power should be available in case of a power outage.

9.6 Vacuum Sewer Collection Systems

Vacuum sewers (Rogness and Ryan, 1977; Averill and Heinke, 1974) are a practical alternative to gravity or pressure sewers.

Advantages. Many of the advantages and disadvantages of pressure sewers also apply to vacuum sewers. Freedom from maintaining grades and the use of relatively small pipe sizes are among the greatest advantages. Also, special vacuum toilets are available which allow water savings of up to 90 percent compared to conventional flush toilets.

Disadvantages. Disadvantages include limits on vacuum lift, requirements for minimum vacuum pressure to operate household fixtures, and home owners' difficulty in maintaining the vacuum fixtures.

A vacuum system consists of a centrally located vacuum collection station, collection piping, and vacuum/atmosphere interface valves at toilets and sinks/drains. It is important to check with the individual manufacturers for design information and to adapt this to cold-regions applications.

The collection station contains a sewage collection tank, vacuum pumps, and sewage discharge pumps. A separate vacuum reservoir/moisture removal tank may be needed for most systems. Collection piping is constructed in "lift pocket" profiles between the collection station and the vacuum interface valves which are located near or inside residences. Fluid at rest reforms in "slugs" at lift pockets (profile changes) constructed at intervals along the mains.

Sewage enters the vacuum system when a vacuum interface valve opens. The valve remains open for a preset time allowing the stored liquid and a volume of air to enter the mains. Air pressure accelerates the sewage which becomes a foam-like material. Transport takes place only when a vacuum interface valve is open.

9.6.1 Types of Systems. Two basic configurations of vacuum sewers are currently being designed and installed:

- conventional gravity fixtures with exterior vacuum valve sump; and

- vacuum toilets and vacuum sumps with greywater valves.

Conventional Gravity Fixtures with Exterior Vacuum Valve Sump. This system collects all the sewage from conventional gravity fixtures in a sump located outside of the home. The sump contains the vacuum interface valve which allows sewage to enter the vacuum system. The sump is typically located in the public right of way and is maintained by the utility authority. A single valve sump may serve several homes. A photograph of the sump and interface valve is shown in Figure 9-12.

This is the most common type of vacuum system in the continental U.S. with some installations having over 3,000 interface valves and 12 vacuum stations. This type of system has not been built in the cold regions. The maintenance of exterior valve sumps and the required addition of frigid ambient air at the valve during operation make this option prohibitive in cold regions.

Vacuum Toilets and Vacuum Sumps with Greywater Valves. This type of system differs from the conventional system in the location of the interface valve and type of fixtures. A single vacuum line enters the home and is split into two lines. One provides vacuum service to a vacuum toilet, and the second to a vacuum greywater valve located at the end of gravity drain piping or an alternate provides a small sump located immediately upstream of the valve. The small pressure head in the gravity piping or sump activates the in-line valve. The toilet is activated by the user.

This type of system has been constructed in Noorvik, Alaska. The greywater valve is configured to use part of the gravity drainage pipe as the drainage sump. The municipal collection system consists of HDPE water and sewer pipes located in a common utilidor with heat provided by a circulating water distribution system with a glycol backup heating loop.

Another larger variation of this type of system has been constructed in Emmonak, Alaska. A small liquid storage sump is provided in each house immediately upstream of the greywater valve. The municipal collection system consists of a separate arctic carrier pipe used to house the HDPE vacuum collection piping and glycol heating lines. Electric thaw cables are provided throughout the system as a backup heating system. Several industrial camps in Alaska use small-scale variations of this type of vacuum sewer system.

9.6.2 Design Process. There are many parameters and usable guides in the design of a vacuum collection system. The following are some of the important considerations for the system components.

Vacuum Station. Location of the vacuum collection station is critical. It should be located centrally and at one of the lowest elevations of the present and future service area. Only about 600 TORR (mm Hg) vacuum pressure is available to lift and transport the sewage.

The collection tank should have a total volume of 1.5 times the operating volume, or 1.5 times the volume accumulated over 30 minutes of minimum flow. The total volume of the collection tank should not be less than 1,500 L. Vacuum reservoir/moisture removal tanks provide additional vacuum capacity and assist in preventing moisture from reaching the vacuum pumps. The recommended volume is 1500 L.

Vacuum pumps are sized to handle sewage flow adjusted to a 2:1 air/liquid ratio, plus a factor for gases generated from the sewage. Sewage discharge pumps are sized 20 percent greater than peak flow. The net positive suction head required should be scrutinized in the discharge pump selection since it is low due to the vacuum pressure on the pump inlet. Small-diameter pressure equalization lines are recommended from the outlet side of the discharge pumps to the vacuum tank. Special seals are required for the pumps to withdraw sewage from the collection tank under vacuum.

Collection Piping. Vacuum sewer design cannot be accomplished on the principles of simple fluid mechanics because sewage mixes with air, and is transported as a foam-like material. Before starting design, manufacturers of vacuum sewer equipment should be contacted for design parameters, tables,

limitations, and other aids. A brief description of the design process follows to help prospective designers in understanding the process.

Sewage is transported in the pipeline by air pressure entering through vacuum interface valves. Piping can be above or below grade and usually requires backup heating. Piping is laid on shallow slopes based on pipe size and profile changes. Small-diameter piping is used to transfer sewage from the vacuum interface valves to the collection station. The vacuum station is usually operated at 800 TORR (mm Hg) vacuum pressure. The vacuum interface valves require about 200 TORR (mm Hg) of vacuum to operate. Therefore, a total of around 600 TORR (mm Hg) vacuum pressure is available to transport the sewage. The total of elevation and velocity losses in the pipeline cannot exceed 600 TORR (mm Hg).

Vacuum sewers are usually laid in "sawtooth" profile in flat land or for uphill transport. Experience has shown that maintaining minimum grades of 0.2% improves system efficiency. Downhill transport follows the ground profile and sewage may flow by gravity. Inside plumbing is usually PVC or ABS Schedule 40 or 80 pipe 38 mm, 51 mm, 76 mm, or 102 mm in diameter and outside piping is usually HDPE. Fittings should provide smooth transitions and long radius bends.

Shutoff valves (ballcentric type) should be located at each branch sewer and at intervals not greater than 460 m along main lines. Profile changes are accomplished by using two 45 degree elbows joined by a section of pipe. Profile changes of transport pockets are usually limited to 0.2 to 0.3 m but could be as much as 0.6 m when close to the vacuum station. Sewer service branches are not recommended to join the main line within 1.8 m of a profile change. Where profile changes are less than 0.3 m head loss due to lift is computed at 50 percent of the lift. Friction losses are calculated from charts presented by manufacturers.

Regardless of friction losses, it is customary to limit the length of 75 mm pipe to not more than 240 m, and 100 mm pipe to not more than 610 m. Spacing between the "sawtooth" profile changes for 75 mm pipe should not exceed 120 m; 150 m spacing is used for 100 mm and 150 mm pipe.

The length of lines is governed by the total of static lifts and velocity head loss. Some systems have been built with more than 3,000 m of vacuum main line between the vacuum tank and the most distant interface valve.

Vacuum valves for sump operations require 130 TORR to 180 TORR (mm Hg) vacuum to properly operate. The vacuum sump must be installed in a heated environment and vented near the ceiling within the heated area of the house. Venting through the roof will allow cold air to enter the system and increase potential for freeze-up. Note that this practice may not follow some plumbing codes that are developed for nonvacuum systems. Some sump interface valves have a manual push button activator to empty the sump when desired (Figure 9-13).

There are two types of vacuum water closets (VWC): automatic and manual. The automatic has a timed vacuum valve operation and is operated with a push button. The manual VWC valve is foot operated and the valve will stay open as long as the foot lever is pushed down. Both VWC require 280 TORR (mm Hg) vacuum to properly operate. Each VWC should not be operated while sitting on it. The manual VWC is about half the cost of an automatic VWC.

Vacuum System Maintenance Considerations. A standby generator rated at 100 percent of power requirements for the vacuum station is required. Vacuum and pressure sewer systems are inoperable when the community power system fails. All pumps should be duplicated for backup.

9.7 Municipal Lift Stations

Nonresidential sewage lift stations are used mainly with gravity collection systems but could be used with pressure sewage collection systems, and even vacuum systems (to pump the waste from the collection tank to the treatment facility). Table 9-2 presents the advantages and disadvantages of each type of lift station.

9.7.1 Cold-Region Lift Station Adaptations. The following design features should be provided in sewage lift stations in cold regions.

- The outside of the station should be insulated with at least 100 mm of polyurethane or polystyrene with an outer covering to protect the insulation from moisture.

FIGURE 9-13   VACUUM INTERFACE VALVE

TABLE 9-2   CHARACTERISTICS OF SEWAGE LIFT STATIONS

Type	Advantages	Disadvantages
Submersible	Low initial cost; low maintenance requirements; station can easily be expanded; available in a wide range of capacities; little of structure is above ground so heat loss reduced; requires less appurtenances such as heaters, sump pumps, etc.	Difficult to make field repair of pump; requires specialized lifting equipment to remove pump.
Dry Well	Pumping equipment located away from wet well; good reliability; high efficiency; desirable for larger installations; easily maintained.	High initial costs; requires larger construction site; higher cost per litre capacity.
Wet Well	Low intial cost; high efficiency; wide range of capacity available.	Requires explosion-proof electrical motors and connectors; difficult to maintain.
Suction Lift	Good reliability; available for wide range of capacity.	Suction lift limtied to 4.5 m; decrease in priming efficiency as pump ages.
Pneumatic Ejector	For low capacity (150 L/minute); low head – for short distances; generally nonclogging; 380 L/minute usual maximum capacity.	Somewhat more complex than other types of stations; high maintenance; low efficiency.

• Insulation should be placed underneath the station to prevent settling due to the thaw of frozen ground. Visqueen (plastic) or some other bond breaker should always be used to reduce frost-jacking on the station sides within the active frost layer.

• If thawing and settling under the station is anticipated, adequate pile foundations penetrating into the permafrost are recommended (see Section 3).

• All stations must be attached to concrete slabs to provide sufficient weight to overcome the buoyancy of the station itself if it were completely submerged in water.

• Pressure-coupling type (flexible) connections are recommended at the inlet and outlet of the stations to allow differential movement without breaking the lines.

• All critical components, such as pumps and compressors, should be duplicated in each station. The controls should allow the operator to specify cycling of pump and compressor. The identical standby pump and compressor should be installed to take over if one or the other does not start.

• Alarms are an absolute necessity in any lift station.

The alarm (both visual and audible) warns the operator of malfunctioning. The alarms can also be set for low temperature, high sewage level, and high water level in the station (to warn of a sump pump malfunction). Major stations should have power failure alarms and standby power. The alarms can be tied back into a central alarm panel in the pumphouse, operator's house or work station, or to the fire or police department's phone.

• Standby electrical power should be provided for each major lift station.

• Inlet screens (easy to access and clean) must be provided to remove items that would clog pumps or check valves.

• Each lift station should be checked daily by the operator. The inlet screen must be checked daily and cleaned as necessary.

- Lift stations, whether submersible type or not, should be housed in a structure that can be heated and contains the electrical controls and alarms.

- All entrance access holes must extend above the ground surface sufficiently to be above any flooding or snowdrifts. Also, access hole entrances or building entrances must be kept locked.

- All pumps or motors should be supplied with running-time meters to accommodate maintenance schedules and the estimation of sewage volume pumped.

- Corrosion protection should be provided for the metal shell in each station. Sacrificial anode and impressed current corrosion protection systems may not work well when the ground surrounding the anode or lift station is frozen.

- The minimum size of lift stations and sizing criteria should generally follow traditional design practice unless heating design conflicts develop due to sewage sitting in lines or wet wells for extended periods results in the risk of freezing.

9.8   Force Mains

Force mains are pressure lines into which the sewage pumps discharge. They should be designed both to have scour velocities of 0.75 to 1.0 m/s during pumping and to retain adequate heat during pump-off cycles to prevent freezing. If this is not possible they may drain back between pumping cycles or be heat traced.

An option to keep the sewage moving is to time the pumping cycle so that the sewage remains in the force main for a limited, precalculated period.

The force mains should be pressure tested, and they should meet all the criteria of pressure water transmission pipelines. If they empty into a access hole, the entrance should be a least 0.7 m above the exit line. Force mains should be constructed of HDPE and provided with thaw ports at 100 M intervals or should be be electrically heat traced. The force main size should be determined considering capital and pumping costs and should be sized to handle ultimate flows. Minimum size should not be less than 100 mm diameter.

9.9   Building Plumbing

Service lines and their connections to buildings were discussed earlier in Section 9.4.2. The special plumbing needed with vacuum sewers was covered in the section dealing with vacuum collection systems.

The use of floor-mounted, rear-flushing toilets keeps all waste plumbing lines in the walls of the building and 100 mm above the floors. Wall-mounted, rear-flushing units are not recommended unless the wall is reinforced, because the wall has to support the entire weight of the toilet and the user. The floor-mounted, rear-flushing units must be carefully installed; over-torquing the flange bolts will crack the base. Tubs should be mounted on risers so that the entire trap and drain is 100 mm above the floor. This allows the tub to drain into the toilet discharge line in the wall. All fixtures should be placed on inside walls, which are warm on both sides. If possible, the sink should be placed on the opposite side of the bathroom plumbing wall to reduce the length of drain lines. All fixtures and lines should be installed, so that they can be drained or otherwise protected from freezing. Drainable traps should be used, and the user should be aware that antifreeze solution should be added to the toilet when there is danger of freezing. Bleed-off valves should be provided to drain at all low points in the house water supply plumbing. Home owners must be trained to maintain the facilities after they have been installed. They need to know how to order parts, how to repair leaky faucets and toilets, and what not to flush down toilets.

House vents frequently frost over in cold weather. They should be constructed with material of low conductivity, and they should increase in size as they go into the unheated attic, from 32 mm to 75 mm and from 75 mm to 100 mm and insulated.

9.10   Typical Construction Costs

Construction costs in remote arctic communities is highly variable and dependant on equipment needs, shipping costs, construction conditions, and foundation requirements. For example, shipping a large excavator and gravel supply to install a small amount of buried gravity sewer could result in high unit costs. Due to these variables in estimating, it is recommended that each project be individually evaluated and proper cost extensions be made to bracket anticipated costs. As an alternate, past experience at a particular site can provide estimates for future work.

Buried pipelines cost (per pipe) between $100 and $300 (US) per metre to install depending on the excavation conditions. Costs for an above-ground pipeline depend on the foundation required. If they can be laid directly onto the ground surface, they cost $50 to $400 (US) per metre. If they must be placed on pilings however, the costs could run as high as $500 per metre (US).

Small-scale utilidors, depending on the utilities included and the roads to be crossed, cost between $600 and $1,700 (US) per metre to construct (see Section 11).

9.11 References

Averill, D.W. and G.W. Heinke. 1974. *Vacuum Sewer System*. Indian and Northern Affairs Pub. No. OS-1546-EE-A, Ottawa, Ontario. (Also available from Deptartment of Civil Engineering, University of Toronto, Ontario.)

Cadario, P.M. and G.W. Heinke. 1972. *Draft Manual for Trucking Operations for Municipal Services in Communities of the N.W.T.* Department of Civil Engineering, University of Toronto, Ontario.

Cameron, J. 1981. *Guidelines for the Preparation and Administration of Municipal Water and Sanitation Trucked Service Contracts.* Department of Local Government, Government of the Northwest Territories, Yellowknife.

Gamble, D.J. and C.T.L. Janssen. 1974. Evaluating alternative levels of water and sanitation services for communities in the Northwest Territories. *Canadian Journal of Civil Engineering*, 1(1): 116-128.

Rogness, D.R. and W.L. Ryan. 1977. Vacuum sewage collection in the Arctic: Norvik, Alaska - A case study. *Proceedings, Symposium on Utilities Delivery in Arctic Regions.* Environmental Protection Service, Rep. No. EPS 3-WP-77-1. Ottawa, Ontario, 505-522.

Ryan, W.L. and D.R. Rogness. 1977. Pressure sewage collection systems in the Arctic. In: *Proceedings, Symposium on Utilities Delivery in Arctic Regions.* Environmental Protection Service, Rep. No. EPS 3-WP-77-1, Ottawa, Ontario, 523-552.

Smith, W.H. 1976. *Evaluation of Frost Loadings on Underground Pipe.* Ductile Iron Pipe Research Assoc., Oak Brook, Illinois, 9 p.

9.12 Bibliography

Alter, A.J. 1968. *Sewerage and Sewage Disposal in Cold Regions.* U.S. Army Cold Regions Research and Engineering Laboratory Monograph III-C5b, Hanover, New Hampshire, 106 p.

American Concrete Pipe Association. 1980. *Concrete Pipe Handbook.* Vienna, VA. 435 p.

EPA Technology Transfer Seminar Program. 1977. *Alternatives for Small Wastewater Treatment Systems. Pressure Sewers/Vacuum Sewers.* U.S. Environmental Protection Agency EPA-625/4-77-011, Washington, D.C.

Flanigan, L.J. and C.A. Cadmik. 1979. Pressure sewer system design, *Water and SewageWorks*, R-25-834.

Gamble, D.J., and C.T.L. Janssen. 1974. Estimating the cost of garbage collection for settlements in northern regions, *Northern Engineer*, 6, 1.

Gamble, D.J. 1974. *Wabasca-Desmarais Water and Sanitation Feasibility Study.* Prepared for the Northern Development Group, Alberta Executive Council, Edmonton, Alberta.

Highway Task Force. 1971. *Handbook of Stell Drainage and Highway ConstructionProducts.* American Iron and Steel Institute, Washington. 348 pp.

Metcalf & Eddy, Inc. 1981. *Wastewater Engineering: Collection and Pumping ofW astewater.* McGraw-Hill Book Co., New York. 432 pp.

Smith, D.W. (technical editor) 1977. *Proceedings, Utilities Delivery in Arctic Regions*, Environmental Protection Service, Rept. No. EPS 3-WP-77-1, Ottawa, Ontario, 596 p.

Smith, D.W. (technical editor) 1980. *Proceedings, Utilities Delivery in Northern Regions*. Environ. Prot. Service Rep. No. EPS 3-WP-80-5, Ottawa, Ontario, 542 p.

Smith, D.W. (technical editor) 1982. *Proceedings, Utilities Delivery in Cold Regions.* Environ. Prot. Service. Rep. No. EPS 3-WP-82-6, Ottawa, Ontario, 419 p.

Uni-bell Plastic Pipe Assoc. 1982. *Handbook of PVC Pipe.* Dallas, TX.

SECTION 10

WASTEWATER TREATMENT

3rd Edition Steering Committee Coordinators

Daniel W. Smith

James A. Crum

Special Contributions

Dennis Prince

Ron Kent

Daniel W. Smith

Section 10   Table of Contents

Section 10   List of Figures

Section 10   List of Tables

10 WASTEWATER TREATMENT AND DISPOSAL

10.1 The Receiving Environment

Knowledge of the environment, which receives wastewater discharges, is essential to establish proper treatment and disposal methods. This is true anywhere but particularly in the fragile northern environment. However, our present knowledge of how arctic and subarctic river, lake, estuary, and land systems respond to wastewater discharges is limited to some basic information about the physical and chemical conditions, and the ecology of streams and rivers.

The environmental conditions within northern rivers and lakes vary considerably with the time of year and location. Winter flows in many rivers are negligible, whereas summer flows may be high and carry heavy sediment loads. Ice flow is high during spring break-up period, and generally low to zero during the remainder of the year. During periods of ice cover, the concentration of dissolved oxygen (DO) progressively decreases. Near the mouths of most major rivers in Alaska, natural winter DO levels are well below generally recommended minima (see Table 10-1) (Schallock, 1974). The organisms in cold waters have adapted to these conditions either through reduced metabolic rates or migration.

Gordon (1970) showed in laboratory experiments that bacteria indigenous to cold waters exerted an oxygen demand at low temperatures, and the addition of organic and inorganic nutrients increases the rate of oxygen utilization. Murray and Murphy (1972) also studied the biodegradation of organic substrates at low temperatures. They found that the food to microorganisms (F/M) ratio was more important than temperature in determining the rate of DO depletion.

In another study, the effect of the ice cover on the DO content of the Red Deer River in Alberta was measured (Bouthillier and Simpson, 1972). In this river, which receives municipal effluents, the initial five day, 20°C biochemical oxygen demand (BOD_5) below the outfall and the rate constants established by BOD bottle tests were found to bear little relation to oxygen uptake in the ice covered, shallow river. Total organic carbon (TOC) was closely related to oxygen depletion. It was concluded that biological oxidation under river ice is a complex reaction. The bottom of the Red Deer River is mainly gravel and rock, giving a large surface area for the growth of microorganisms. These characteristics of the river bottom, which result in a benthic oxygen demand, are considered important to the oxygen regime of the river.

TABLE 10-1   WINTER DO CONCENTRATIONS IN SELECTED STREAMS (Schallock, 1974)

Stream/Location	DO (mg/L)	Saturation (%)	Date of Sample
Yukon River:			
near Eagle, AK (1664 km upstream)	10.5	73	March 1972
near Alakanak, AK	1.9	13	
Tanana River:			
near Fairbanks	9.9	69	February 23, 1970
at Yukon River	5.7	40	
near Fairbanks	10.8	75	March 5, 1970
at Yukon River	6.5	46	
Colville River at Umiat	7.5	52	March 1969
Kuparuk	8.4	58	March 1969
	8.4	58	February 1971

Assimilative capacities of northern rivers in the winter are undoubtedly lower, perhaps much lower than rivers of similar physical characteristics located in temperate zones. Oxygen use rates and reaeration coefficients at temperatures near 0°C need further development. In the interim, the aquatic integrity must be protected by both knowledgeable utilization of the limited data and common sense.

10.2 Guidelines for Municipal Wastewater Discharges

In general, winter characteristics of receiving waters are critical for the establishment of effluent standards and treatment requirements in cold regions. As elsewhere, a combination of basic effluent standards and specific stream standards must be taken into account to arrive at a specific solution. Effluent quality criteria used in a number of jurisdictions are shown in Table 10-2.

Guidelines for municipal wastewater discharges were established for the Northwest Territories (Northwest Territories Water Board, 1992) in 1992 and for the Yukon Territory (Yukon Territory Water Board, 1983) in 1983. They attempt to apply present knowledge about effects of waste discharges to the northern environment.

The State of Alaska and the U.S. Environmental Protection Agency require mandatory secondary treatment as a minimum treatment level. In 1979, however, the federal law was amended to allow for primary discharges into marine water under certain conditions. Large communities are required to submit very comprehensive applications under the 301(h) Waiver of Secondary Treatment Regulations. Native Alaska communities, because of their small size and isolation and because their wastewater is normally entirely domestic, are not required to submit a formal 301(h) application, but rather an abbreviated application to the appropriate Environmental Protection Agency office and Alaska Department of Environmental Conservation. Approval for a primary discharge depends on the volume and type of discharge and sensitivity of the receiving waters. In Canada, the appropriate territorial water board, provincial government agency, federal government agency or, in Alaska, the Department of Environmental Conservation, should be contacted to establish the required quality for discharge to available receiving waters.

10.3 Wastewater Characteristics

In general, the total quantity of wastewater tends to be close to the amount of potable water used by the community. Storm water should be excluded from the collection systems. Infiltration is usually not a factor in the Arctic since most collection systems are insulated and tightly sealed to ensure thermal integrity. Communities such as Anchorage, Alaska, and others in the subarctic practice water wasting and do encounter infiltration, but design considerations do not differ from conventional practice with high flows and low organic loadings. The wastewaters at most other cold-region installations tend to be lower in volume and higher in strength than at comparable facilities elsewhere. The wastewater from most cold-region facilities is essentially domestic in character, with the possible addition of laundry

TABLE 10-2   EFFLUENT QUALITY CRITERIA

Parameter	U.S. Monthly Average	U.S. Weekly Average	U.S. 1978 Adjusted	Canada Fed. Facility	Alberta*
Biochemical Oxygen Demand (BOD$_5$), mg/L	30	45	30	20	20 to 25
Suspended Solids (SS), mg/L	30	45	70	25	20 to 25
Fecal Coliforms, CFU/100 mL	200	400		400	–
pH	6 to 9			6 to 9	–

Note: In the Northwest Territories and the Yukon Territory effluent quality requirements are based on the type of discharge, location (i.e., river, lake) and the dilution available (Northwest Territories Water Board, 1992; Yukon Territory Water Board, 1983).
  * Values for mechanical plants, lagoons have a design standard specifying a configuration of 2 to 4S,1T,1L for daily flows over 250 m^3 and 0S,1T,1L for flows under 250 m^3 (Alberta Environment, 1988).

TABLE 10-3   TYPICAL QUANTITITES OF SEWAGE FLOW

Source		Quantity L/(p·d)
1. Communities and Permanent Military Bases		
a. 1,000 population with conventional piped water and sewage		
Thule Air Force Base, Greenland		303
College, AK		265
Fairbanks, AK		303
Ski resorts in Colorado and Montana		345
	Average	300
b. 1,000 population with conventional piped water and sewage		
Bethel, AK		265
DEW Line, Greenland		208
	Average	240
c. with truck-haul systems, conventional internal plumbing	Average	140
d. with truck-haul systems, low-flush toilets	Average	90
e. no household plumbing, water tanks and honey-bucket toilet	Average	1.5
f. same as (e) above but with central bathhouse and laundry	Average	15
2. Construction Camps		
North Slope, Ak (1971)		189
"Typical" Canadian		227
Alaska Pipeline (1976)		258
	Average	220
3. Remote Military with Limited Availability of Water		
McMurdo, Antarctica		151
Barrow, AK (DES Sta)		114
"Typical" Army Field Camp		129
	Average	130

wastes and extra amounts of garbage and grease from institutional kitchens.

10.3.1 Quantity. The resulting quantities of sewage flow depend on the type of installation and its permanence. Table 10-3 summarizes typical sewage flows for various cold-regions situations.

Separate facilities such as schools, laundries, restaurants, and hotels with conventional plumbing tend to have loadings similar to those in conventional temperate zone practice.

Projected data for the community should be used to establish a design value for per-person flow. The average values given in Table 10-3 may be used to

TABLE 10-4   CHARACTERISTICS OF BASIC WASTEWATER CATEGORIES

Parameter	Units	Undiluted (Heinke, 1973)	Moderately Diluted (Eggener & Tomlinson, 1978)	Conventional Diluted (Metcalf & Eddy Inc., 1979)	Greatly Diluted (Bethell, 1981)	Greywater (Hrudey & Raniga, 1981)
BOD_5	mg/L	–	460 280 to 700	220 110 to 400	55 40 to 60	–
COD	mg/L	110,400 80,800 to 134,800	1,000 700 to 1,300	500 250 to 1,000	–	(TOC) 210 40 to 900
Suspended solids (NFR)	mg/L	78,200 66,000 to 85,000	490 370 to 820	220 100 to 350	50 20 to 150	290 40 to 2,000
Total nitrogen	mg/L as N	8,100 7,300 to 9,500	–	40 20 to 90	(NH_3) 10 6 to 30	(NH_3/N) 1.4 8
Phosphorus	mg/L as P	1,200 1,100 to 1,400	–	8 4 to 15	3 2 to 6	9 4 to 20
Calculated flow*	L/(p·d)	1.2 1.1 to 1.4	170 110 to 290	360 200 to 730	1,500 1,300 to 2,000	310 50 to 2,300

All values rounded off from published data.
* Calculated based on 80 g BOD_5 per person per day and 90 g suspended solids (SS) per person per day (where applicable), modified activated sludge, and septic tanks. In some instances, lagoon treatment is followed by land disposal.

estimate quantities. These values are all for permanent residential personnel. If an accurate population count is not available, it should be assumed for design purposes that there are about 3.5 people per residence for military stations, and 4.5 people for remote civilian communities.

10.3.2 Quality. The physical, chemical, and biological characteristics of sewage in cold regions depend heavily on the type of installation and the sanitary facilities provided. Wastewaters in northern regions have been classified into four categories based on contaminant concentration (Smith and Heinke, 1981; Smith, 1982):

- undiluted (concentrated);
- moderately diluted;
- conventionally diluted; and
- greatly diluted.

In addition, greywater (domestic wastewater excluding toilet wastes) is commonly generated in conjunction with undiluted wastes. The basic characteristics for each classification of waste are presented in Table 10-4.

Undiluted Wastewater. Undiluted wastewaters are common in sparsely populated areas and primarily because bucket toilets (honey buckets) are used. Undiluted wastewater is usually disposed of in pits, dumps, or trenches. Occasionally they are macerated, incinerated, or composted.

Moderately Diluted Wastewater. This type of waste results from restricted water use: the principal sources are holding tanks, pipe-transported water/wastewater systems where water conservation is practiced, and vacuum systems. Workcamps and many small communities have such installations. Common methods of treatment include lagoons, extended aeration/activated sludge, physical/chemical processes, and septic tanks (where ground conditions permit).

Conventionally Diluted Wastewater. This category of wastewater is commonly referred to as municipal or domestic sewage. Methods used to treat conventionally diluted wastewater include lagoons, aerated lagoons, rotating biological contactors, modified activated sludge, and septic tanks. In some instances lagoon treatment is followed by land disposal.

Greatly Diluted Wastewater. Diluted wastewater occurs in cold regions primarily because of water bleeding and excessive infiltration. The practice of water bleeding is used for freeze protection of water distribution systems and sewage collection systems located in soils permanently frozen (permafrost) or seasonally frozen with deep frost penetration. Treatment processes used for dilute wastewater include both screening prior to discharge and lagoons. Rotating biological contactors and screening with ozonation prior to discharge are also effective.

Table 10-5 summarizes the information gained from seven studies on the relative contribution of various facilities to domestic sewage flow in communities and military field bases.

Concentration of Pollutants. The concentration of wastewater constituents varies with the amount of water and the type of facilities used. The actual per-person mass loading of organics and related substances should, however, be relatively constant for a given community. However, the mass loading is affected by lifestyle and diet. Table 10-6 gives estimated mass values for the major domestic wastewater sources. The values are based on a comparative analysis of a number of data sources.

The final item, "institutional garbage grinders", reflects the common practice at military stations and many construction camps of grinding most of the kitchen wastes for inclusion in the wastewater stream.

All the values in Table 10-6 are independent of the amount of water used for a particular activity. These values with data given in Tables 10-3 and 10-5 or other sources allow the determination of concentrations for a particular situation.

Temperature. The temperature of the raw wastewater entering the sewage treatment plant can be an important physical characteristic. The efficiency of most unit operations and processes can be strongly influenced by temperature. Temperature control is also necessary to prevent unwanted freezing either in the system or at the point of final discharge.

The energy level represented by moderate (10°C) to high incoming sewage temperatures should be considered a resource. The treatment system and its protective elements should be designed to take full advantage of this available energy. For example, the municipal treatment plant in Fairbanks, Alaska, extracts energy from the effluent with a heat pump, and this is used to heat the entire facility.

TABLE 10-5   DOMESTIC SEWAGE SOURCE (Percent of Average Daily Flow)

	Average (%)	Range
Communitites:		
Toilet, urinal	37	26 to 43
Shower, sinks	32	16 to 38
Kitchen	12	4 to 19
Laundry	19	7 to 30
	100	
Field Army Bases (population of 1,000-6,000):		
Toilets, showers, suites	60	
Kitchen	6	
Laundry	16	
Aircraft washrack	9	
Vehicle washrack	3	
Photographic wastes	5	
Hospital	1	
	100	

The temperature of raw wastewater is a function of the raw water temperature, the water and sewage system design and characteristics, the use, number and plumbing (hot-water) of buildings serviced, and the ambient temperatures. Some values for raw wastewater temperature are presented in Table 10-7.

10.3.3 Flow Variation. Institutional facilities in cold regions, such as military stations and construction camps, tend to have large flow variations, because a large portion of the population responds to the same schedule. The peak flow usually occurs around labor shift changes, when people are bathing and laundering.

Two such peaks occur at those installations operating on a continuous two-shift, 24-hour cycle. For institutional facilities the peak daily flow rate for design purposes should be three times the average daily rate.

Civilian communities and similar residential areas have less sharply defined flow variations. The time is dependent on transmission distance from the homes to the treatment system. The daily peak flow

TABLE 10-6   ESTIMATED SOURCES OF SEWAGE POLLUTANTS (g/(p·d))

Source	BOD$_5$	SS	Total N	Total P
Toilets	60.8	85.2	14.7	1.67
Bath/shower	5.33	5.12	0.31	0.04
Laundry	7.91	7.70	0.23	0.67
Kitchen	16.8	10.9	0.49	0.49
Subtotal	90.8	109.0	15.7	2.83
Institutional garbage grinders	59.0	58.8	1.31	0.95
Total	150.0	165.0	17.0	3.78

TABLE 10-7   RAW SEWAGE TEMPERATURES AT TREATMENT FACILITIES

| Location | Temperature (°C) | | | Notes |
	Winter	Summer	Average or Range	
Fairbanks, AK	0	2.8		Individual wells
Fairbanks, AK	11.8	10.9		Water main at 15°C
College, AK	18.9	18.3		Sewers in heated utilidor
Eielson, (AFB) AK	21.7	20.7		Sewers in heated utilidor
Juneau, AK	2.2	8.9		
Kenai, AK	8.0	10 to 14		
Homer, AK	3 to 5	9 to 10		
Dillingham, AK			3 to 4	April
Craig, AK			5	January
Kake, AK			4	December
Soldotna, AK	3	8 to 9		
Eagle River, AK			5	Initial operation, few services
Eagle River, AK			9	After four years
Inuvik, NWT		23		
Whitehorse, YT			3 to 15	Water main bleeders
Clinton, Creek, YT	17	22		
Emmonak, AK			28	Central facility, greywater only
Alaska			20 to 24	Construction camps (Alyeska)
Hay River, NWT			10 to 15	Airport facility

rate of these communities should be taken as two times the average daily rate. The U.S. Public Health Service uses a factor of 3.5 for designs in small communities (see Section 5).

Minimum flow rates are important to the design of grit chambers, monitoring devices, dosing equipment, etc. A minimum flow rate equal to 40 percent of the average rate should be used for design purposes.

10.4 Wastewater Treatment

The fundamental purpose of wastewater treatment is protection of human health, protection of the receiving environment and aesthetics. The guidelines, criteria, and standards set by the applicable regulatory agencies must be considered prior to undertaking any design.

10.4.1 Planning Considerations. The special factors for cold regions are not just responses to low temperatures but also include factors such as:

- remote locations;
- logistical problems during construction and operational resupply;
- permafrost and other unstable site conditions;
- lack of skilled manpower;
- very high energy costs;
- rapid turnover of operator personnel; and
- in some cases, unique environmental aspects.

Christensen (1982) provides general guidelines for the planning and preliminary engineering considerations for appropriate sewage treatment facilities and disposal systems for northern communities. Other references on this subject include Smith and Heinke, 1981; Smith and Christensen, 1982; Smith, 1982; Prince et al., 1995a; 1995b). The rapidly changing character of some northern communities through development complicates the planning of facilities for wastewater treatment.

Design Life of Facility. The design life has to be determined in the initial planning stages, since this factor determines the service life and treatment capacity requirements based on the present and future population. The design life for waste treatment systems is a function of the community or facility served.

Site Selection. Most of the basic site selection criteria, i.e., advantageous topography, suitable discharge point, are applicable to the cold regions. Distance requirements to avoid nuisance conditions are generally applicable in both temperate and cold regions. For example, lagoons should be no closer than 0.4 km from any residence. Dominant wind direction is also an important factor for facility siting because many treatment systems experience odor problems due to operational difficulties; even long distances may not protect the community from unpleasant odors if the facility is sited upwind. Of special concern is the presence of permafrost under the intended site and easy, continuous access to all critical points in the system.

The presence of permafrost strongly influences the design and economics of the type of system chosen. If convenient access cannot be provided and easily maintained, then wintertime maintenance will probably not be performed.

10.4.2 Effect of Cold Temperatures on Wastewater Treatment. Practically all operations processes used in wastewater treatment are affected by temperature through viscosity changes in the water or changes in chemical or biochemical reaction rates. An analysis during early design stages is necessary to predict the thermal status of major components in the system. If warm sewage is expected and the entire system is to be housed in a heated building, then conventional temperate zone practices can be used. If cold sewage is expected or significant temperature changes are predicted within the system, then adaptations are necessary in the design of unit operations and processes.

Mixing. Mixing is an important function in waste treatment. It is required for flocculation, for the continual suspension of solids in liquids, such as the mixed liquor in activated sludge, and for dissolving solids. All these mixing activities depend heavily on temperature because of changes in the viscosity of the liquid. Further information on mixing can be found in Section 6, since the general requirements are the same for water and wastewater.

Sedimentation. The settling of discrete particles in water is affected by the viscosity of the water. This is a factor in grit chambers and in primary settling tanks, when flocculant chemicals are not used. Detention times must be increased to compensate for lower settling velocities in colder water. The multipliers in Figure 6-1 can be used to adjust design detention times. The temperatures shown are for the fluid. Alternatives to adjusting design detention times include housing the tanks in a heated enclosure or preheating the incoming fluid.

The settling of higher concentration flocculant particles is not as greatly affected by temperature. At concentrations of 2,000 mg/L or less, temperature-induced viscosity effects are quite strong. At 6,000 mg/L and higher, the concentration of particles presents a greater influence than fluid temperature. The multipliers for the design of settling tanks and thickeners are presented in Section 6.

Density currents must also be considered in the design of settling tanks and thickeners. If the incoming fluid is a degree or two different in temperature from that of the tank contents, short-circuiting or excessive loss of solids or both may occur. Protective elements should be used, if possible, to maintain the temperature of the tank contents as close as possible to that of the incoming fluid. If soil conditions permit, settling tanks should be designed in the conventional manner as partially buried structures. In permafrost, particularly in ice-rich, fine-textured soils, above-ground tanks or tanks on special foundations are required. Temporary covers are recommended for buried tanks for winter operation in the Arctic and subarctic. Tanks above grade require sidewall insulation and covers, or enclosure in a protective structure.

Filtration. Filtration of wastewater is affected by temperature induced viscosity changes. The multiplier values from Figure 6-1 should be used to reduce filtration efficiency. Further details on filtration can be found in Section 6, since basic criteria are common for both water and wastewater.

Gas Transfer. The solubility of gases in water increases as the temperature decreases. Air, oxygen, and chlorine are the commonly used gases in wastewater treatment. The efficiency of aeration and other gas transfer operations follows conventional gas transfer theory:

TABLE 10-8   VARIATION IN SOLUBILITY OF SOME
CHEMICALS WITH TEMPERATURE

| | Solubility (g/L) | |
Chemical	20°C	0°C
Alum	873	723
Ferrous sulfate	120	60
Sodium hydroxide	527	287
Calcium hypochlorite	228	215

$$\frac{dc}{dt} = \alpha K_L a \left(\beta C_s - C_L \right)$$

where,

C = dissolved oxygen concentration, mg/L

subscripts: s = saturation

L = bulk liquid

$K_L a$ = overall O_2 transfer coefficient, 1/h

α = correction factor for $K_L a$

β = correction factor for C_s

However, the impact of temperature is on the C_s value and $K_L a$ values. Viscosity changes with temperature affect mixing and surface renewal which in turn affects the $K_L a$ value. These two factors tend to compensate for one another, so that the net practical effect is a little improvement in overall gas transfer efficiency with low temperature wastewaters. Section 6 provides additional details on aeration.

Adsorption.   Adsorption may be used with biological processes as well as in physical/chemical treatment with activated carbon. The rate of adsorption, as defined by the Gibbs equation, is inversely proportional to temperature, so adsorption should be more rapid at low temperatures. However, again the increased viscosity of the water at low temperatures decreases the contact surface, so there is no practical improvement in efficiency. Isotherms are modelled using the Freundlich equation.

Chemical Reactions.   Both metabolic and chemical reaction rates tend to be slower at low temperatures. These effects must also be considered in preparing chemical solutions for use in wastewater treatment. The solubility of most chemicals de-

creases as the water temperature decreases. Table 10-8 gives solubility values for typical treatment chemicals.

Disinfection.   The survival of pollution indicator microorganisms is longer in cold water than in warm water. Gordon (1972), Davenport et al. (1977), Putz et al. (1984) and Bell et al. (1977) have presented detailed information on the extended survival of these organisms in northern rivers. This information identifies the need for careful consideration of disinfection processes to ensure that water uses are not seriously affected.

The need for and the degree of wastewater disinfection to be achieved, and sometimes the agent to be used are specified by the appropriate regulatory agencies. Typical coliform contents of cold-region wastewaters are shown in Table 10-9. Chlorination is the most commonly used technique in the cold regions. Other alternatives have been investigated (International Environmental Consultants, 1978), but have so far not been used in the North. Testing has shown ozone to be very effective for the destruction of all types of microorganisms, unlike chlorine and UV (Given and Smith, 1979; Finch and Smith, 1987). Basic process criteria are similar to temperate zone practices with respect to contact time and the prevention of short-circuiting through properly baffled chambers. The selection of either gas or hypochlorite as the chlorine source and the related safety measures also follow standard practice, modified to account for logistical resupply problems for many cold-region locations.

Erosion-type chlorinators have been used at small, remote facilities in Alaska. It is sometimes difficult to maintain a specified dosage; however, they offer operational simplicity. They should be housed in a separate compartment constructed of noncorroding

TABLE 10-9   TYPICAL COLIFORM CONTENTS IN COLD-REGION
WASTEWATERS (Hrudey and Raniga, 1981)

Waste Type	Location	Quantity (L/(p•d))	Total Coliforms (number/100 mL)
Weak sewage	Dawson	5,915	1.0×10^5
Holding tank effluent	Aklavik	68	1.0×10^7
Honey bags	Aklavik	1.3	5×10^8
Lagoon effluent	Inuvik	545	1.0×10^5

materials, along with the tablet supply in a sealed container. In the past, extreme corrosion of metals and mechanical parts has occurred when exposed to the humid atmosphere of an enclosed treatment plant building.

Effluents from lagoons and other exposed treatment units are at or near 0°C under extreme winter conditions. Some thermal protection for the contact chamber is therefore essential.

Biological Processes.   Systems that have been successfully used in cold climates include: lagoons, both facultative and aerated; activated sludge; and attached growth systems. Each has special requirements for successful cold-region performance. Sludge treatment, both through biological and physical/chemical processes, is also affected by cold temperatures.

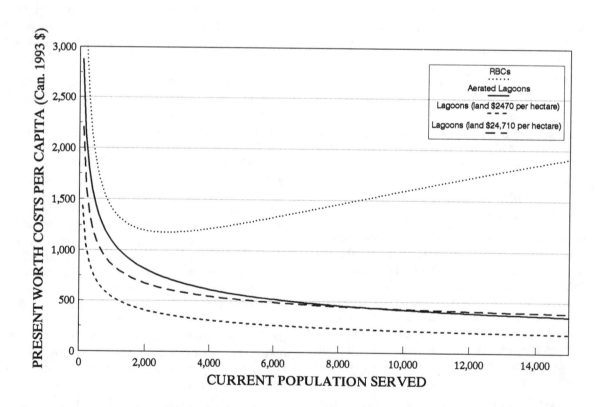

FIGURE 10-1   PRESENT WORTH COSTS PER CAPITA FOR TREATMENT
ALTERNATIVES

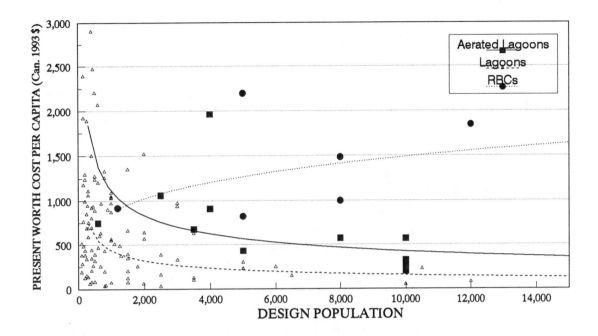

FIGURE 10-2   INDICATION OF CONFIDENCE FOR PRESENT WORTH COSTS PER
CAPITA

10.4.3   Treatment Processes in Cold Regions.

Lagoons.   Of the four types of lagoons (aerobic, facultative, anaerobic, and aerated) only the facultative, anaerobic and aerated lagoons are of real interest for most northern communities. Aerobic lagoons are not feasible for year-round use in northern areas. Aerated lagoons are used for special situations where operation and maintenance support plus the required power can be provided. They are not applicable to small, northern communities where low cost and simple operation and maintenance are of foremost concern.

A comprehensive critical evaluation of the operation and performance of lagoons in cold climates has been prepared Smith and Finch (1983) for Environment Canada. In 1988 a detailed guideline for lagoon design and operation was prepared (Heinke et al., 1988; 1991). A review of lagoon treatment in Alberta by Prince et al. (1993) for Alberta Environmental Protection provides information on effluent quality and the factors affecting effluent quality, costs, and public complaints (Prince et al., 1995a; 1995b).

Lagoons without aeration are often the treatment method of choice because of their simplicity and low cost of operation. Where remote communities are responsible for their own operation and maintenance without governmental assistance, these features are attractive (Schubert and Heintzman, 1994). In addition, the functional life of a lagoon can be expected to surpass that of a mechanical plant. The results of a cost survey of operating facilities in Alberta (Prince et al., 1993) are presented in Figures 10-1 and 10-2. The figures show that generally lagoons are the most cost effective treatment method. A report by UMA Engineering Ltd. (1993) concluded that in the NWT lagoons and lagoons with constructed wetlands are the most cost-effective treatment technologies that are capable of meeting the nontoxic effluent clause in the Canadian Fisheries Act.

Lagoon system designs are based either on organic loading or hydraulic loading. The design of intermittent discharge lagoons is usually governed by hydraulic loading because it is the volume of wastewater that must be stored that determines the size of the storage cells. With continuous discharge lagoons the common parameter for design is the organic loading. Facultative lagoons for northern climates are designed with low organic loading rates and long retention times. In Alberta, the general guideline of the regulatory agencies is to provide a 12-month retention time within the pond before discharge. Volumes of this size allow for total retention

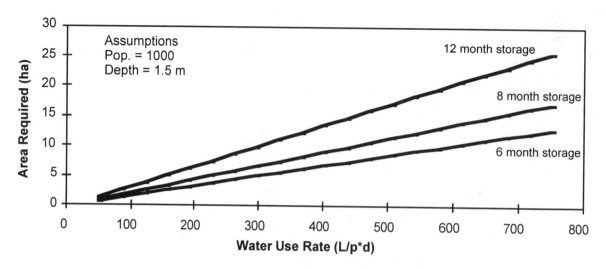

FIGURE 10-3   LAGOON AREA REQUIRED BASED ON HYDRAULIC LOADING

during the winter months when biological treatment is expected to be low, and treatment would essentially be only sedimentation (primary treatment). In Alaska, regulatory minimum retention times of at least six months and organic 18 to 40 kg $BOD_5/(ha\cdot d)$ are required. Figure 10-3 shows the area required by intermittent discharge lagoons designed based on hydraulic loading. Figure 10-4 indicates the areal requirements of continuous discharge lagoons designed based on the organic loading.

Discharge from winter retention lagoons can be accomplished by releasing the stored volume in the spring or fall, or by a controlled discharge over the course of summer and fall. The decision of when to discharge a containment lagoon is a matter of weighing the factors of receiving environment conditions and effluent quality characteristics. For a river the highest dilution capacity is available in the spring with the high flows due to runoff while the lowest dilution is in the fall (Figure 5-1). Lagoon effluent quality is poorest in the spring and best in the fall which is demonstrated by Figure 10-5. Generally one to two months of ice-free treatment will substantially improve BOD_5 in the effluent and still provide reasonably high flow in the rivers for dilution. Tables 10-10 to 10-13 show the dilution necessary for ef-

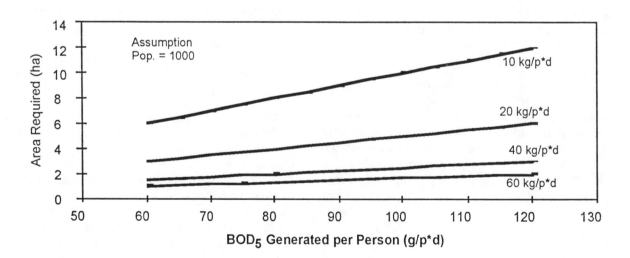

FIGURE 10-4   LAGOON AREA REQUIRED BASED ON ORGANIC LOADING

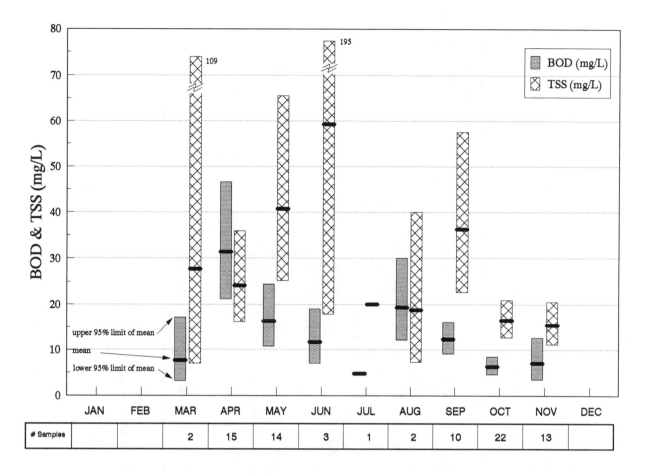

FIGURE 10-5   CHANGES IN LAGOON EFFLUENT QUALITY WITH MONTH OF DISCHARGES
                (Lagoons with 4S, 1T, 1L with 12 months storage, data collected in Alberta 1983-
                86)

fluents from lagoons, rotating biological contactors (RBCs), and activated sludge plants to meet surface water criteria as set out in the Canadian Water Quality Guidelines (CCREM, 1995). The average lagoon system with anaerobic cells requires roughly one to one dilution in the fall and one to six dilution in the spring while RBCs and activated sludge plants require roughly 1 to 1,000 dilution to meet receiving water criterion.

Data from Alberta indicates that effluent quality from control discharge lagoons is dependent on the season of discharge, storage time, and the presence of anaerobic cells. Table 10-14 gives an estimate of the effect of these significant factors. With consideration of these factors properly designed lagoons are capable of producing effluent quality superior to the effluent quality being produced at mechanical plants as shown in Figures 10-6 and 10-7. These

figures provide a summary of effluent quality from lagoons and mechanical plants in Alberta; the notation 4S,1T,1L stands for four short or anaerobic cells, one treatment cell, and one storage cell.

Oversummer storage lagoons have excellent disinfection capabilities due to natural disinfection processes and long retention times. Table 10-15 is a summary of microbial testing carried out by Alberta Environmental Protection. It shows that fall discharges are better than spring and anaerobic cells provide significant reductions of total and fecal coliform counts.

A grab sample of a discharge is sufficient to determine the effluent quality of an intermittent discharge lagoon. The results of daily samples taken from the discharge at the lagoon facility serving the Village of Legal, Alberta are presented in Figure 10-8. The figure shows that except for a few days at the be-

TABLE 10-10   DILUTION REQUIRED TO ACHIEVE RECEIVING WATER CRITERION – 0S, 1T, 1L LAGOONS

Variable	Effluent Concentration						Receiving Water Criterion (mg/L)	Reason for Criterion	Dilution Ratio Required			
	Fall			Spring					Fall		Spring	
	Mean (mg/L)	No. of Samples	95 Percentile (mg/L)	Mean (mg/L)	95 Percentile (mg/L)	No. of Samples			Mean	95 Percentile	Mean	95 Percentile
BOD	12.6	6	76.3	46.5	294.1	7	5.0	fish DO maintenance (Hickey 1989)	1.5	14	8.3	58
TSS	39.9	6	383.3	28.7	301.1	7	10	fish impact (CWQG)	3.0	37	1.9	29
Phosphorus	2.7	6	39.5	4.6	19.6	7	1.0	algal growth*	1.7	38	3.6	19
Ammonia	0.8	6	758.4	13.7	57.9	7	1.4	fish toxicity** (CWQG)	0.0	553	9.0	41
Nitrate/Nitrite	0.4	3	2,185.3	0.2	112,651.4	3	10 100	drinking water (CWQG) stock watering (CWQG)	0.0 0.0	218 21	0.0 0.0	11,264 1,126
TDS	1,627.3	6	5,540.2	1,164.8	4,934.5	6	500 3,000 500-3,500	drinking water (CWQG) stock watering (CWQG) irrigation**** (CWQG)	2.3 0.0 0.0	10 0.8 0.6	1.3 0.0 0.0	8.9 0.6 0.4
Total coliforms org./100 ml	***		***	***	***		1,000	irrigation (CWQG)				
Fecal coliforms org./100 ml	2.1E+02	13	8.6E+04	***	***		100 200	irrigation (CWQG) bathing/recreation (CWQG)	1.1 0.1	854 427		

* effluent limit when P removal required
** pH=8.0, 10°C
*** Coliform data is misleading, many samples exceeded the range of analysis of 8,000 org/100 ml.
**** Depends on crop and soil type, 3,500 mg/L used for dilution calculation.
Note: Dilution Ratio = (Effluent quality / Criterion) - 1

TABLE 10-11   DILUTION REQUIRED TO ACHIEVE RECEIVING WATER CRITERION – 4S, 1T, 1L LAGOONS

Variable	Effluent Concentration						Receiving Water Criterion (mg/L)	Reason for Criterion	Dilution Ratio Required			
	Fall			Spring					Fall		Spring	
	Mean (mg/L)	95 Percentile (mg/L)	No. of Samples	Mean (mg/L)	95 Percentile (mg/L)	No. of Samples			Mean	95 Percentile	Mean	95 Percentile
BOD	6.0	23.4	16	15.3	57.1	13	5.0	fish DO maintenance (Hickey 1989)	0.2	4	2.1	10
TSS	14.4	47.1	17	32.0	158.3	13	10	fish impact (CWQG)	0.4	4	2.2	15
Phosphorus	1.8	7.7	17	2.8	15.4	13	1.0	algal growth*	0.8	7	1.8	14
Ammonia	0.7	8.8	17	2.9	179.4	13	1.4	fish toxicity** (CWQG)	0.0	5	1.1	130
Nitrate/Nitrite	0.5	18.3	15	0.3	9.9	7	10 100	drinking water (CWQG) stock watering (CWQG)	0.0 0.0	1 0	0.0 0.0	0 0
TDS	893.6	3,482.5	17	838.6	4,952.1	13	500 3,000 500-3,500	drinking water (CWQG) stock watering (CWQG) irrigation**** (CWQG)	0.8 0.0 0.0	6 0.2 0.0	0.7 0.0 0.0	8.9 0.7 0.4
Total coliforms org./100 ml	5.7E+01	3.5E+03	7	1.6E+03***	3.7E+05***	15	1,000	irrigation (CWQG)	0.0	2.5	0.6	365.8
Fecal coliforms org./100 ml	1.7E+01	8.8E+02	16	6.7E+02***	3.9E+05***	15	100 200	irrigation (CWQG) bathing/recreation (CWQG)	0.0 0.0	8 3	5.7 2.3	3,895.8 1,947.4

*   effluent limit when P removal required
**   pH=8.0, 10°C
***   Coliform data is misleading, many samples exceeded the range of analysis of 8,000 org/100 ml.
****   Depends on crop and soil type, 3,500 mg/L used for dilution calculation.
Note: Dilution Ratio = (Effluent quality / Criterion) - 1

TABLE 10-12 DILUTION REQUIRED TO ACHIEVE RECEIVING WATER CRITERION – ACTIVATED SLUDGE PLANTS

Variable	Effluent Concentration						Receiving Water Criterion (mg/L)	Reason for Criterion	Dilution Ratio Required			
	Fall		No. of Samples	Spring		No. of Samples			Fall		Spring	
	Mean (mg/L)	95 Percentile (mg/L)		Mean (mg/L)	95 Percentile (mg/L)				Mean	95 Percentile	Mean	95 Percentile
BOD	18.5	38.5	9				5.0	fish DO maintenance (Hickey 1989)	2.7	7		
TSS	11.6	50.8	9				10	fish impact (CWQG)	0.2	4		
Phosphorus	2.8	18.3	9				1.0	algal growth*	1.8	17		
Ammonia	3.3	66.4	9				1.4	fish toxicity** (CWQG)	1.4	47		
Nitrate/Nitrite	4.7	52.8	9				10 100	drinking water (CWQG) stock watering (CWQG)	0.0 0.0	4 0		
TDS	622.4	1,450.2	9				500 3,000 500-3,500	drinking water (CWQG) stock watering (CWQG) irrigation**** (CWQG)	0.2 0.0 0.0	2 0.0 0.0		
Total coliforms org./100 ml	1.0E+06	estimated		***	***		1,000	irrigation (CWQG)	999.0			
Fecal coliforms org./100 ml	1.0E+05	estimated		***	***		100 200	irrigation (CWQG) bathing/recreation (CWQG)	999.0 499.0			

* effluent limit when P removal required
** pH=8.0, 10°C
*** Coliform data is misleading, many samples exceeded the range of analysis of 8,000 org/100 ml.
**** Depends on crop and soil type, 3,500 mg/L used for dilution calculation.
Note: Dilution Ratio = (Effluent quality / Criterion) - 1

TABLE 10-13   DILUTION REQUIRED TO ACHIEVE RECEIVING WATER CRITERION – ROTATING BIOLOGICAL CONTACTOR

Variable	Effluent Concentration Fall Mean (mg/L)	Fall 95 Percentile (mg/L)	Fall No. of Samples	Spring Mean (mg/L)	Spring 95 Percentile (mg/L)	Spring No. of Samples	Receiving Water Criterion (mg/L)	Reason for Criterion	Dilution Ratio Fall Mean	Fall 95 Percentile	Spring Mean	Spring 95 Percentile
BOD	17.5	36.2	10				5.0	fish DO maintenance (Hickey 1989)	2.5	6		
TSS	11.4	21.2	10				10	fish impact (CWQG)	0.1	1		
Phosphorus	3.7	8.0	10				1.0	algal growth*	2.7	7		
Ammonia	3.4	26.3	10				1.4	fish toxicity** (CWQG)	1.5	18		
Nitrate/Nitrite	5.7	20.7	10				10 / 100	drinking water (CWQG) / stock watering (CWQG)	0.0 / 0.0	1 / 0		
TDS	525.5	1,190.4	10				500 / 3,000 / 500-3,500	drinking water (CWQG) / stock watering (CWQG) / irrigation**** (CWQG)	0.1 / 0.0 / 0.0	1 / 0.0 / 0.0		
Total coliforms org./100 ml	1.0E+06 estimated				***		1,000	irrigation (CWQG)	999.0			
Fecal coliforms org./100 ml	1.0E+05 estimated				***		100 / 200	irrigation (CWQG) / bathing/recreation (CWQG)	999.0 / 499.0			

* effluent limit when P removal required
** pH=8.0, 10°C
*** Coliform data is misleading, many samples exceeded the range of analysis of 8,000 org/100 ml.
**** Depends on crop and soil type, 3,500 mg/L used for dilution calculation.
Note: Dilution Ratio = (Effluent quality / Criterion) - 1

TABLE 10-14   EFFECTS OF SIGNIFICANT FACTORS ON EFFLUENT QUALITY

Parameter	Season of Discharge (fall vs spring)	Storage Time (12 vs 6 months)	Anaerobic Cells (present vs absent)	Treatment Cell (present vs absent)
BOD	15 mg/L	–	9.0 mg/L	–
TSS	–*	–	–	–
Phosphorus	1.4 mg/L	–	–	–
Ammonia	4.6 mg/L	3.0 mg/L	–	–
TKN	4.0 mg/L	–	–	–

* "–" means the factor was insignificant at 95% level.

ginning and end of discharge, effluent quality remains fairly constant throughout the time of discharge and therefore a grab sample taken a few days after the discharge begins will be representative of virtually the entire discharge.

Multiple cells in lagoon facilities can improve effluent quality, however they also require that a thawed flow path be maintained between cells during the winter. This can be a very difficult problem, especially in the more northern locations. Also, construc-

tion costs are higher for multiple-cell lagoons than for single-cell lagoons of comparable volume. Figure 10-9 shows a typical two-cell arrangement. Figure 10-10 shows the use of a stop-log manhole for depth control. Not shown in the figure is a means for thawing both the line and the manhole, which should be provided.

Facultative lagoons in arctic and subarctic areas can generally expect to have a discharge, despite evaporation. While evaporation-type lagoons are not fea-

TABLE 10-15   SUMMARY OF TOTAL AND FECAL COLIFORM DATA FROM ALBERTA LAGOONS

Configuration		Fall Discharge		Spring Discharge	
		TC CFU/100 mL	FC CFU/100 mL	TC CFU/100 mL	FC CFU/100 mL
4S, 1T, 1L	Mean	57	17	1,627	666
	upper 95%	206	43	6,072	3,129
	# of samples	7	16	15	15
2S, 1T, 1L	Mean	232	74	5,355	1,643
	upper 95%	24,052	1,033	9,236	15,985
	# of samples	4	6	6	6
0S, 0T, 1L	Mean	3,276	214	1,202	253
	upper 95%	448,023	997	19,322	8,296
	# of samples	4	13	5	5
Lagoons Overall	Mean	245	56	2,021	681
	upper 95%	1,471	134	4,945	2,156
	# of samples	15	35	26	26

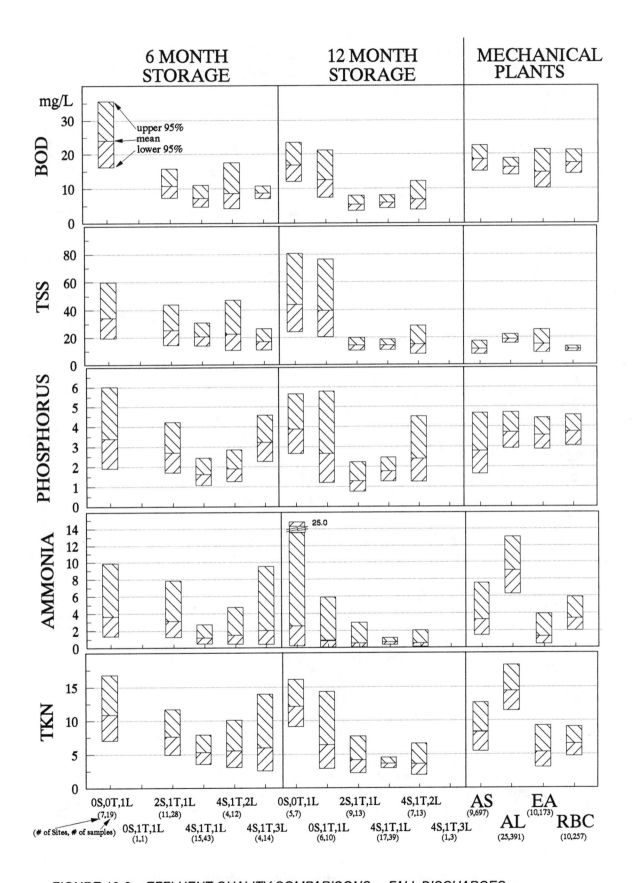

FIGURE 10-6   EFFLUENT QUALITY COMPARISONS – FALL DISCHARGES

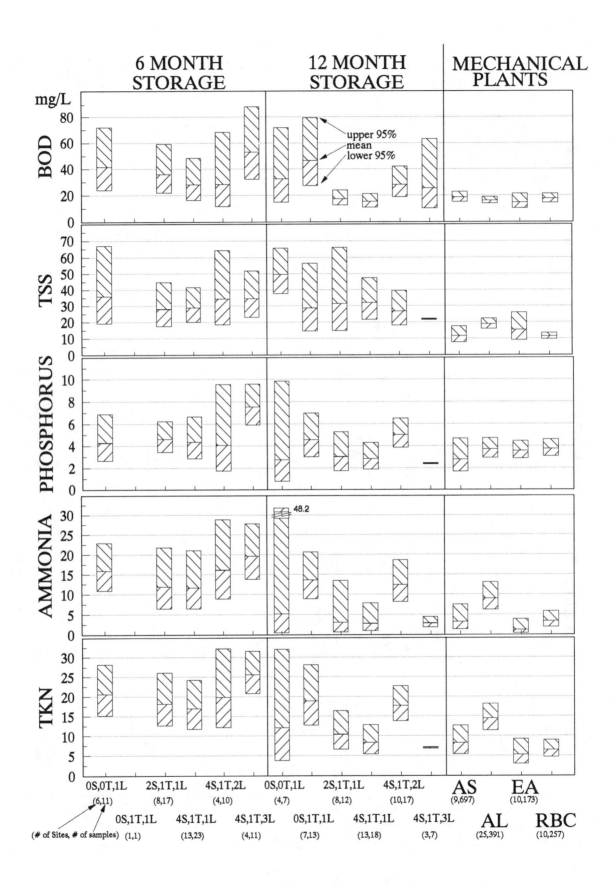

FIGURE 10-7   EFFLUENT QUALITY COMPARISONS – SPRING DISCHARGES

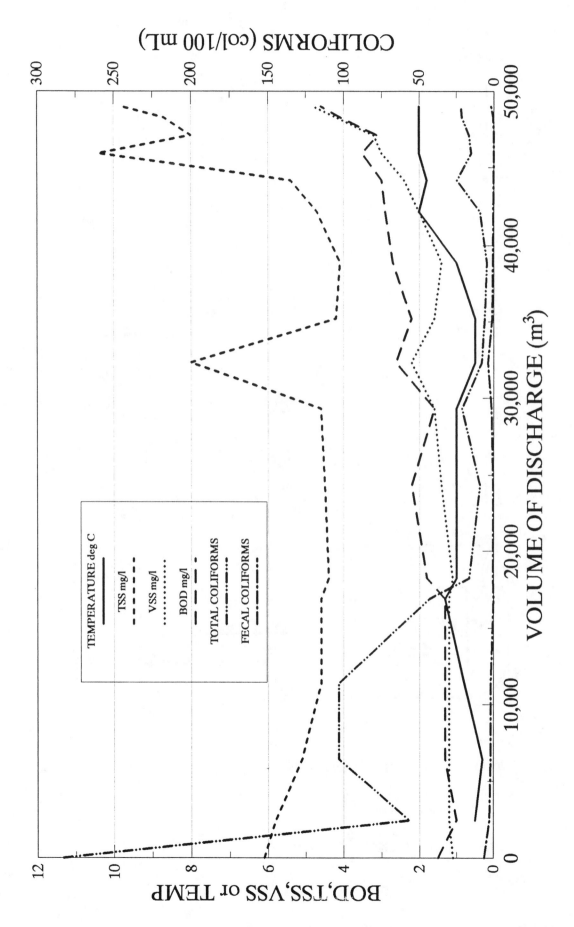

FIGURE 10-8   VILLAGE OF LEGAL LAGOON DISCHARGE

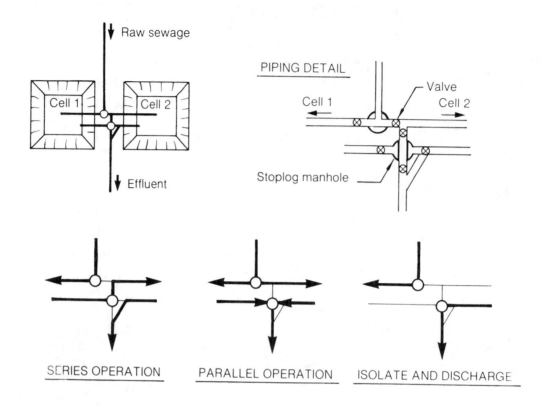

FIGURE 10-9   TWO-CELL FACULTATIVE LAGOON FLOW CONTROL

Calculation of basin volume for any rectangular basin
with sloped sides and round corners

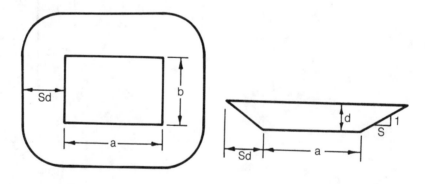

Volume = V = d [(a + Sd) (b + Sd) + .0472 S² d²]

Note: The last term (.0472 S² d²) can
          be dropped for preliminary estimates

FIGURE 10-11 DIMENSIONS OF LAGOONS

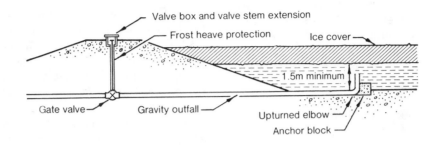

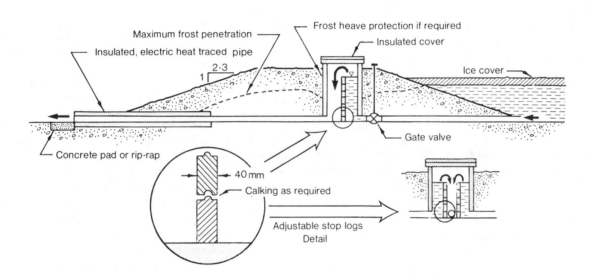

FIGURE 10-10 TYPICAL LAGOON DETAILS

sible, there may be areas where part of or even all of the flow from the lagoon can be absorbed in a percolation cell. In cases where the ground does not allow percolation due to permafrost or impervious soils, the effluent is sometimes discharged to the organic mat overlaying the impervious ground. The discharge spreads through the organic mat interstitially and drains gradually to lower areas. The organic mat is often very wet naturally, and the addition of wastewater is not readily apparent. This method has been successful where the hydraulic loading is low and where wastewater can be stored during periods when the organic mat is frozen. Little information is available to suggest what treatment

efficiencies can be expected with this type of land discharge.

The amount of ice cover expected on a facultative lagoon can be predicted by the methods described in Section 4. If sufficient data are not available for such calculations, a design depth of 1 m should be assumed for most subarctic locations. The design depth should be based on winter conditions and allow 300 mm of freeboard, plus the ice thickness, plus 1.5 m from the underside of the ice to the lagoon bottom. Maximum depth maintained by the stop-log manhole during the summer period would be 0.5 m.

Standard temperate zone construction techniques can be used, except that the effect of frost penetration and ice lens formation in embankments must be accounted for in the design. Often liners are required. Additional concerns are required where permafrost exists. Permafrost consisting of fine textured, ice-rich soils should be avoided, because thawing can result in failure or in the need for frequent repair and restoration of dikes and berms. However, in many locations there is no choice but to construct the lagoon in permafrost (see Figure 10-11 for sizing.)

If there are natural tundra ponds available, these can be used as a single-cell lagoon (Schubert and Heintzman, 1994). The use of such ponds is much more economical than construction of a lagoon with the normal fill or cut-fill method. Usually, a low dike has to be constructed around the pond to provide sufficient depth in the pond. Conversion of a natural pond into a lagoon will require approval by the appropriate regulatory agency. If no natural pond exists or cannot be used, a lagoon can be constructed by placing fill over the permafrost rather than attempting to partially excavate an area in frozen ground. Lagoons have been constructed on ice-rich soils containing large ice wedges. The visible ice wedges were excavated in the dike areas and back-filled with impervious soil to prevent leakage and dike slumping, which would occur when the ice wedges melt. The long-term thermal degradation effect however, of lagoons built on ice-rich ground needs to be considered.

Another method that has been used to construct lagoons in permafrost is a two stage operation. The first stage is to strip all of the insulating vegetation from the site to promote natural thawing. The second stage is the construction of the lagoon, after thawing has occurred to a suitable depth. Construction techniques for dikes and berms and the use of lining materials are the same as those described in Section 7.

Aerated Lagoons. Anaerobic lagoons have been used in the pretreatment of high-strength wastes and as a short retention pond preceding a facultative lagoon (Dawson and Grainge, 1969; Heinke, 1976; Magditch, 1985).

Supplemental aeration may be needed for such lagoons in very special cases. Examples might be where fish canning, animal slaughter, or other food-processing operations impose a brief but intense loading on the lagoon during the summer months. In these cases, small floating aerators or similar devices can be used during that period and removed for the balance of the year.

Aerated lagoons represent an intermediate alternative between facultative lagoons and activated sludge plants. Oxygen is supplied either with diffusers below the surface or with surface aerators. Lagoons are typically 2.4 to 6.1 m deep with detention times of 10 to 60 days. Advantages over short retention time facultative lagoons include reduced land area requirements, higher treatment efficiencies, reduced odors, and lower construction costs. Disadvantages include greater operation and maintenance requirements, greater energy requirements, the need for continuous and consistent operation to minimize odors and maintain effluent quality, and poorer effluent quality than 12-month detention time lagoons.

Aerated lagoons have been used successfully in cold-region applications. Federal facilities at Bethel, Eielson AFB, McGrath, and Ft. Yukon as well as more recent community facilities in Metlakatla, McGrath, and Nome have all provided effective wastewater treatment.

Details of design are similar to more temperate regions except for low temperature effects on rate constants and foundation and berm ice control issues. Designs can be based on complete mixing, in which sufficient energy is added by the aerators to keep the contents in suspension, or as partial aeration, in which only enough oxygen is added to maintain aerobic conditions. Design models are currently based on first-order BOD removal kinetics:

$$\frac{Ce}{Co} = \frac{1}{1 + \left(\dfrac{k\,\bar{t}_d}{n}\right)^n}$$

where,

k   =   first-order BOD reaction rate constant, days

$\bar{t}_d$ =   hydraulic detention time, days

n   =   number of equal-sized cells in series

Ce   =   BOD effluent concentration, mg/L

Co   =   BOD influent concentration, mg/L

Number of Cells vs. Surface Area

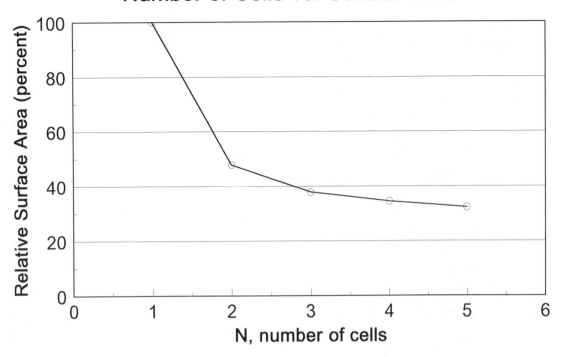

FIGURE 10-12 NUMBER OF AERATION CELLS VS RELATIVE SUR-
FACE AREA REQUIRED FOR EQUAL TREATMENT

In terms of the minimum hydraulic detention time,

$$\bar{t}_d = \frac{n}{K} \bullet \left[\left(\frac{Co}{Ce} \right)^{\frac{1}{n}} - 1 \right]$$

As the number of aeration cells increases, the total volume decreases. This is an advantage in cold climates as the total surface area is decreased, and heat loss is minimized. In addition, because of reduced loading for each subsequent cell, reduced aeration is required, resulting in a lower operating cost. As shown in Figure 10-12, minimal advantage is gained after three cells.

The critical factor in design is the selection of the reaction rate constant. Values of 0.1 to 0.15/day for temperatures of 1°C have been used. Design normally requires an iterative process of estimating the lagoon area, determining the winter temperatures, adjusting the reaction rate and determining a new surface area. Aeration requirements are normally based on summer temperature requirements, due to lower dissolved oxygen concentrations a higher temperatures.

Circular or square configurations have been used, although rectangular length to width ratios of 3:1 to 4:1 are suggested to allow for better hydraulic flow patterns. A final or polishing cell is desirable for removal of suspended solids.

Regulatory requirements in Alaska include: minimum detention times of 30 days, minimum dissolved oxygen concentrations of 2.0 mg/L, and a polishing cell with volume of 1/3 aerated cell volume.

Design of the aeration system requires special consideration with equipment housed in a heated structure to facilitate maintenance and servicing. Both surface and diffused aeration systems have been used.

Activated Sludge Variations. Systems in use in cold regions include conventional sludge and oxygen-activated sludge, contact stabilization, extended aeration, and oxidation ditches.

Many of these systems are enclosed in heated structures and receive warm sewage year round. Basic process design criteria for these situations do not differ from those of conventional temperate zone practice.

Special attention must be paid to system details, appurtenances, and process controls. The humidity inside such buildings is quite high because of exposed water surfaces. Condensation and icing can

TABLE 10-16   TEMPERATURE COEFFICIENTS FOR BIOLOGICAL TREATMENT

Process	θ Value	Temperature Range (0°C)
Oxidation pond	1.072 to 1.085	3 to 35
Facultative lagoon	1.06 to 1.18	4 to 30
Anaerobic lagoon	1.08 to 1.10	5 to 30
Aerated lagoon	1.026 to 1.058	2 to 30
Activated sludge	1.00 to 1.041	4 to 45
Extended aeration	1.037	10 to 30
Trickling filter (conventional)	1.037	10 to 30
Biofilter (plastic media)	1.018	–
Rotating biological contactor		
Direct filter recirculation	1.009	10 to 30
Final effluent recirculation	1.009	13+
Final effluent recirculation	1.032	5 to 13

occur on inner surfaces of exterior walls, doors, and windows. Control panels and similar elements should be located away from such exterior surfaces to avoid water or ice damage.

Ventilation of such structures is necessary, but the effect of the exhaust on the adjacent community must be considered. Ice fog, resulting in aesthetic and safety problems, can be created under extreme conditions. Dehumidification or heat recovery with induced condensation can control that problem.

Systems exposed to the weather or expected to receive low-temperature wastes may require even greater modification of the basic design criteria.

Conventional or Oxygen-Activated Sludge and Contact Stabilization.   These systems are probably enclosed, but they may nevertheless receive low-temperature wastewaters. Reaction rate coefficients are presented in Table 10-16 for use in the equation:

$$K_T = K_{20} \bullet \theta^{(T-20)}$$

Significant ice formation must be avoided in the aeration compartments of these systems. An ice cover inhibits atmospheric aeration and entraps mixed liquor solids. Both factors may reduce treatment efficiency. Tank design should provide minimum exposed surface area. Evaporation from liq-

uid surfaces and the cooling effect of the wind are major factors in heat loss during winter. An unheated protective shelter over the tanks reduces both and allows satisfactory performance. An alternative would be temporary tank covers and wind breaks during the winter period only.

Settling tanks and clarifiers associated with these systems should receive protection to reduce heat loss and inhibit formation of density currents. Protective elements in the northern temperate zone can be limited to those required for operator safety (prevent icing on walls, ladders, etc.) and to overflow weirs and scum removal points where freezing is most likely to occur.

Extended Aeration Systems.   These systems are available as prefabricated "packaged" units of up to 3,800 m³/d capacity. Smaller sizes are also available, completely installed in a prefabricated shelter, ready for installation. If warm sewage is received at an enclosed and heated treatment unit, basic process design criteria are comparable to temperate zone practice.

It is not necessary to provide a continuously heated building to maintain treatment efficiency; operator convenience and comfort are the only justifications for such energy inputs. If the incoming raw sewage is about 10°C or warmer, there is sufficient heat in the liquid to sustain the process. Protective elements

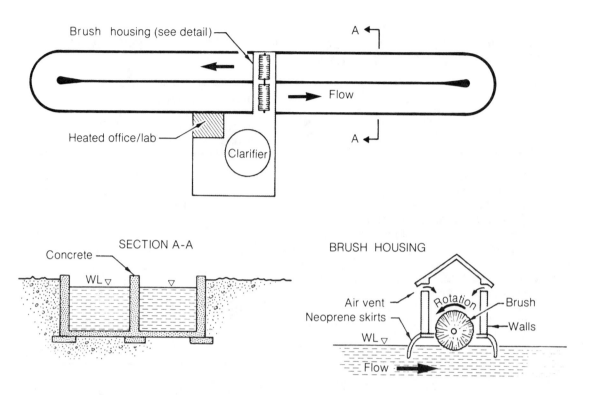

FIGURE 10-13 OXIDATION DITCH FOR THE SUBARCTIC

(i.e., tank covers, burial, wind breaks, unheated shelters) are useful in reducing heat losses. A standby heat source is recommended for emergency situations.

Extended aeration systems have been successfully operated with mixed liquid temperatures of 1° to 5°C, producing effluent BOD_5 and suspended solids of secondary quality. Mixed liquor concentrations tend to increase more rapidly under these conditions, therefore more frequent sludge-wasting is required. Design organic loadings (F/M ratio) of up to 0.08 (g BOD/g) MLSS•day and MLSS concentrations of 3,000 to 4,000 mg/L are recommended for low temperature operation. Pumps, motors, blowers, external piping, valves, and similar appurtenances require protective enclosures and heat (Given and Smith, 1977).

In the design of these units for military and construction camps, the potential for intermittent loading and wide fluctuations in population must be considered. Equalization tanks can be used to damp a strong daily variations in flow. Where regular population changes are expected, the design can provide two or more smaller units for parallel operation under

peak conditions. Only one unit would be operated under low flow conditions.

These small-scale systems are particularly sensitive to situations created by water system bleeding or infiltration; either can hydraulically overload the system. The potential for these conditions must be evaluated prior to design. Also the capability to operate the proposed type of system must be verified.

Oxidation Ditches have been successfully used in subarctic Alaska. Basic process design criteria are similar to those applicable elsewhere. Design for low-temperature operation should conform to extended-aeration criteria (Murphy and Ranganathan, 1974)

Modifications to temperate zone configurations are required to reduce heat loss. As shown in Figure 10-13, vertical side walls and a thin vertical centre island can be used to reduce surface area and heat loss in the ditch. Also shown in Figure 10-13 are the neoprene skirts for protective housing to prevent icing around the brushes. The clarifier units for such systems should be enclosed, but they need not be heated. Cautions regarding location of controls, etc. on exterior walls should be observed. All cautions

TABLE 10-17   *COST COMPARISON OF CONSTRUCTION CAMP WASTEWATER*
TREATMENT SYSTEMS*

Process	Amortized Annual Costs Equipment and Material ($1000 in 1976$)	Personnel (PH/d)**	Energy (kW/d)
Extended Aeration			
Capital Equipment	30.6	5 to 10	180
Parts and Chemicals	11.4 to 14.8		
(with 6 h equalization, pumps, blowers, aerobic digestor, chlorination, five year life at 15%)			
Physical Chemical			
Capital Equipment	31.3	24	400
Parts and Chemicals	19.4 to 71.9		
(with 12 h equalization, pumps, blowers, centrifuge, chemical feed, chlorination, five year life at 15%)			

* 95 m^3/d for Alaska pipeline construction camps
** PH = person-hours

and limitations stated for extended aeration apply here.

These systems are potentially applicable over the same range as that for extended aeration. They can be constructed on-site. Attempts at their prefabrication in the smaller sizes have been made in the past.

Rotating Biological Contactors (RBCs).   Of the types of attached-growth systems (trickling filters, rotating biological contactors (RBCs)) only the RBCs have been used in northern areas. Since attached-growth systems depend on a thin film of water for treatment, they are susceptible to freezing and must therefore be housed. The stability of RBCs under fluctuating organic and hydraulic loading, their low maintenance, low energy consumption and their simple operation make them suitable for application in communities where an effluent of higher quality than that produced in a short-retention lagoon is required. They have also found use in industrial camps (Given, 1984).

Configuration of these systems follows temperate-zone practice, with some form of preliminary clarification and final settling along with recirculation. Basic process design criteria are not unique to cold climates. Adjustments to the design rates for low-temperature wastewater should be made using the θ values in Table 10-16.

RBCs can also be designed to provide nitrification in addition to BOD removal and suspended solids removal (Chong, 1975; Forese and Heuchert, 1984; Forgie, 1984; Given, 1984).

Physical/Chemical Processes.   The biological systems described in the previous subsections are capable of secondary treatment at best. Physical/chemical systems can be designed to produce almost any level of effluent quality desired. In most applications they are used to remove organics, solids, and nutrients. The physical/chemical processes should follow preliminary biological treatment to convert soluble organics to biological solids. It is conventional practice in all climatic zones to install such units in heated buildings. Basic process design is not, therefore, unique to cold regions, except when low-temperature sewage is expected. Adjustments can then be made as described earlier. An alternative is to add heat to the raw sewage in an equalization tank inside a heated structure. This stabilizes and enhances performance of the system.

These units were selected for most camps during construction of the Alaska Oil Pipeline. Advantages claimed include:

- minimum space requirements;

- minimum site impact during construction and restoration;

- consistent high quality performance; and

- operation can be matched to varying flow or organic loading.

Disadvantages associated with this type of operation include very high costs due to intense labor requirements, chemicals, and energy.

Estimated chemical costs for a typical 95 m^3/d system ranged from $0.32 to $2.33 per 1,000 L. A comparison of annual costs, personnel, and energy is given in Table 10-17. The two systems compared are capable of producing generally equivalent effluents, since the physical/chemical units were only designed for organic (i.e., BOD) and suspended solids removal, not for removal of nutrients.

Based on experience to date, physical/chemical treatment should only be considered if very high effluent requirements exist in conjunction with site conditions that preclude other forms or combinations, of treatment.

Land Treatment Processes. Depending on the process and operational mode, these processes can have high levels of treatment. Not only removal of normal organics and suspended solids, but also removal of nutrients, metals, bacteria, and viruses is possible. References Black et al., 1984; Fahlman and Edwards, 1974; Sletten, 1977; Hartland-Rowe and Wright, 1974; Fetter et al., 1976; Smith, 1975; Pollen, 1983; Reed et al., 1977; Sletten, 1978 and U.S. Environmental Protection Agency, 1977, provide details.

Basic process design criteria are similar to those used in temperate regions. The land treatment component is usually preceded by some other form of treatment, typically a lagoon. Often additional storage may be required for inclement weather or other seasonal constraints. Treatment and storage functions can be combined in a multicell facultative lagoon.

The practical range of land treatment options is limited in permafrost regions. Shallow permafrost and extreme winter temperatures limit the horizontal and vertical movement of water in the soil profile. During the summer an overland flow process can oper-

ate under the conditions described. Basic process criteria is similar to temperate zone practice. During the operational period, high removal efficiencies for BOD, suspended solids, nitrogen, and metals are possible; the removal of phosphorus is moderate. However, an alternate discharge or storage is necessary for most of the year.

Slow-rate land treatment is feasible anywhere in cold regions where silviculture or agriculture is practiced. The wastewater is sprinkled or spread on the ground surface so that the vegetation becomes a component in the treatment process. The percolate can be recovered for reuse or allowed to move into the groundwater. This operation is limited to the period of unfrozen surface soil conditions with peak performance attained during the growth period of the vegetation. Alternative discharge or storage is required during the period when surface soils are frozen.

Rapid infiltration has the greatest promise for cold climates. It depends on relatively high rates of application on relatively coarse-textured, free-draining soils. These can be found throughout the cold regions in alluvial river valleys and coastal areas. Depending on the mode of operation, very high removal of BOD, suspended solids, bacteria, virus, and phosphorus can be achieved, and moderate nitrogen removal. Rapid infiltration can also be operated on a year-round basis in some locations, so that extensive storage is not needed. Treatment to primary effluent quality is needed to avoid rapid clogging of the soil surface. A facultative lagoon would more than suffice. Operation is similar to percolation ponds, except that numerous cells are alternately flooded and dried to improve treatment and to prevent clogging.

A thorough site investigation is required to ensure that the applied wastewater does enter the soil and move down and away from the application site. Experience in the Arctic and subarctic is limited, and optimum criteria are not available. Pilot tests are recommended for large-scale operations. Experience in northern New England suggests one hectare of basin surface for every 750 m^3/d of wastewater flow as a conservative estimate of annual applications.

In-Plant Sludge Treatment. In cold climates, large-scale conventional facilities, and facilities operating within a heated environment can be expected to produce sludge at rates similar to those in temperate zones. Typical values are summarized in Table 10-18.

TABLE 10-18   TYPICAL SLUDGE PRODUCTION RATES*

Process	Dry Solids (g/(p·d))	Solids Content of Wet Sludge (%)
Primary settling	54	6
Trickling filter (TF) secondary	18	4
Primary plus TF secondary	72	5
Pprimary plus high-rate activated sludge	82	5
Conventional activated sludge (AS) secondary	32	0.5 to 1
Primary plus conventional AS secondary	86	2 to 3
Extended aeration secondary	40	2
Lagoons	60	20**

* Average 24 h values, from temperate climate experience.
** High value due to long term consolidation of sludge on lagoon bottom. Data from partial mix aerated lagoons in Alaska.

Aerobic treatment systems designed for low-temperature operation tend to generate sludge at a rate similar to moderate temperature units, but because of lower metabolic activity, cannot oxidize it as fast. Therefore, a higher rate of accumulation occurs in the winter than that observed in temperate climates. As a result, winter sludge-wasting from exposed extended aeration plants is required about twice as frequently as in more temperate climates. Final sludge disposal is described in Section 10.5.5.

Temperature inhibitions on digestion increase the rate of sludge accumulation in all types of lagoons. Values of 0.25 to 0.40 m³/(1,000 p·d) have been reported for sludge accumulation in cold-region facultative lagoons. Sludge also accumulates at a faster rate in aerated lagoons. Available data indicate that the assumption of five percent of total lagoon volume for sludge is conservative, so that cleaning and sludge removal might be required every six to ten years.

The accumulation of undigested winter sludge imposes a high oxygen demand on a lagoon when the temperature of the liquid warms up in the spring. Supplemental surface aeration may be helpful for odor control during this period.

Schneiter (1982) found from a survey of 31 cold-region lagoons that 15 had total sludge accumulations ranging from less than 100 mm to more than 600 mm. Alaskan aerated lagoons accumulated sludge at a rate of about 25 to 50 mm per annum.

Procedures for sludge digestion follow temperate-zone practice. Additional heat or insulation must be provided to maintain anaerobic digestion in cold climates. Burial of digestor tanks is common in the subarctic to provide additional natural insulation. Aerobic digestion is less temperature sensitive but produces a less stable product.

Dewatering of sludge prior to disposal also follows conventional temperate-zone practice. One exception is the use of natural freeze-thaw cycles for dewatering. If sludge is flooded on exposed beds and allowed to freeze, the solid particles settle readily upon thawing. Approximately 50 percent of the total volume can be decanted from the beds as supernatant and returned to the treatment plant. The thickened sludge can either be left on the beds for complete drying or disposed of directly (Martel, 1989; 1994).

10.4.4 Operation and Maintenance. Repeated studies of treatment systems in cold regions have shown that performance does not achieve the design goals because of poor operation and maintenance (Given and Smith, 1977). Initial operator training is essential to successful system performance. Another critical element in the design process for wastewater treatment is the preparation of an operation and maintenance (O&M) manual. It is even more important in the Arctic and subarctic because of the general lack of and access to skilled personnel and the relatively rapid turnover of operators at both military and civilian treatment systems. In the military a one-year tour of duty is normal. This lack of continuity and experience results in a situation where the operator's only routine source of guidance and assistance is the O&M manual.

Contents of O&M Manuals. The basic requirements for process O&M may not differ from conventional temperate zone practice but the O&M manual applicable to cold regions should explain the requirements in greater detail. Special winter operational requirements must be clearly defined. These may include such tasks as draining lines, winterizing pumps, activating heat tapes, etc. Equipment for emergency or future use must be clearly identified. For example, a heat tape intended to thaw a pipe need not be continuously operated, and extra pumps and blowers may only be for standby.

Design for O&M. Attention to detail during the design of a system helps relieve or avoid subsequent operational or maintenance problems. The following details are those requiring attention:

- Alarm systems to indicate failure in the process (i.e., power failure, overflow) are essential, particularly for small units that receive only intermittent attention.

- The air circulation and heat distribution in an enclosed structure must be carefully planned. Often the temperature at floor level and in "blind" spots can be below freezing, resulting in the freezing of pipes and appurtenances.

- The depth and location of winter ice and snow cover must be considered to ensure easy access to all critical components.

- Pumps and other intermittently used facilities must either be protected or winterized each year.

- Coordination must be established with other critical elements in the community or camp to avoid unexpected or adverse effects on the system. Dumps or spills of toxic materials must be controlled. An improperly maintained or located grease trap could freeze or impose excessive organic loads on a system.

- Some remote systems may be subject to power failures and voltage variations. Provisions should be made to protect electrical

equipment from burning out because of low voltage.

- Location of the control panel must be away from the condensation laden atmosphere of the plant to avoid its malfunction due to icing or deterioration of the electrical circuitry.

10.4.5 Costs. Estimating costs of waste treatment systems in cold regions is difficult because of the limited data available and the extreme variability of logistical problems. Most cost data for temperate zones do not include the small-scale systems commonly required in the Arctic. The selection of appropriate indices and trend factors can also be a problem; extrapolation from temperate zone values is not recommended except for large communities.

Figures 10-1 and 10-2 summarize present worth value estimates for lagoons, RBC and aerated lagoons for Alberta. Smith and Finch (1983) reported that even for the most commonly used treatment process (lagoons) virtually nothing has been published on either construction costs or operating costs. They recommended that a cost compilation of northern wastewater treatment facilities be undertaken.

Table 10-17 provides a summary of cost information on wastewater treatment facilities built for the Alaska pipeline.

10.5 Wastewater Disposal

10.5.1 Disposal to Surface Waters. The type of receiving water bodies and the effluent standards are important factors in the type of outfall structure designed.

Receiving Waters.

Ponds and Lakes. Ponds are generally shallow; their depth is seldom more than two metres. Surface areas range from a few hundred square metres to a few thousand square metres, and retention times are often extremely long (one to twelve months). Freeze depths may vary from 0.5 to 1.5 m. Figure 10-14 shows ice thicknesses reported for different locations. Ice thickness can be estimated mathematically by assuming worst conditions: no snow cover and minimum heat input (Section 4).

The use of a pond to receive treated wastewater would convert it to a polishing lagoon. Posting to that effect and fencing are required. Local regulatory agencies should be consulted before this approach is adopted.

Lakes may have the ability to absorb a greater volume of wastewater. However, regional regulations

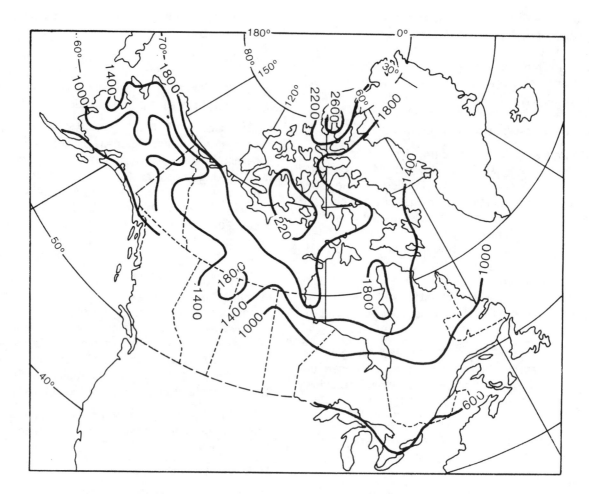

FIGURE 10-14 MAXIMUM ICE THICKNESS OBSERVATIONS, 1969-70 (Values in mm)

must be reviewed. Important considerations include surface area, depth, waste volume-natural inflow relationships, nutrients, flora and fauna, and benthic populations. Effluent quality, especially with regard to microorganisms, must be high if the lake is or may be used for fishing or recreation.

Streams and Rivers.   Northern streams and rivers are the most common receiving waters for wastewater. Flow, ice depth, and movement are the most important factors to be considered. DO conditions through the winter and downstream uses are of particular importance in selecting the type of outfall structure to be used. Mixing of the effluent into the river can have an effect on downstream water quality (Smith and Gerard, 1980; Putz et al., 1984).

Oceans.   Where possible, ocean discharge is desirable. The major advantage of disposal to the sea is dilution. Normally, the ability of oceans to absorb quality variations is very great. If initial dispersion

by outfall design and tidal movement is not obtained however, wastewater concentrates on the surface, because fresh water and sewage are less dense than salt water. The result is a visible slick (Heinke et al., 1990).

Outfalls.   Outfalls to surface water are the most common wastewater discharge technique. Design and operation of the discharge system requires understanding of the receiving water quality and quality of the discharge. It may be necessary to operate the discharge intermittently or seasonally. It is advisable to design the outfall pipe so that it does not pass through a water-air interface. This prevents damage during periods of ice cover.

Three basic types of outfalls exist: freefall from a pipe, submerged diffusers, and weir structures.

Freefall.   Freefall structures use an insulated pipe to transport treated wastewater from the treatment facility to the discharge point. Provisions must be made for freeze protection. Cold air penetration into the end of the pipe could create ice blockage of the line. The design of the outfall must consider ice movement due to tidal action, currents, near-by rivers, and wind action. Figure 10-15 shows the general design of an ocean outfall for the Arctic coast.

Submerged Pipe. The design of submerged outfalls has to deal with problems associated with ice scour, and freeze and ice damage to any portion of pipe crossing the air-water interface.

Weirs. During winter operation, the unfrozen effluent must move to the surface where cooling occurs. The effluent is then discharged to the surface of a stream or channel. Serious icing problems can result.

Intermittent Discharge. In the winter DO conditions in many streams preclude wastewater discharge. In other locations, the volume of effluent may be so small as to require slug discharge to prevent freezing.

Thermal Considerations. Depending on the location of the discharge system used, heat loss may be so great and the volume so small, that the effluent freezes before or right at the end of the discharge line. (The potential for freezing can be determined through procedures given in Section 4.) Such problems are most common when freefall systems are used. To eliminate this condition, treated effluent should be stored in a manner that prevents freezing, and discharged when sufficient volume has

accumulated so that complete expulsion from the discharge line is ensured. The discharge mechanism could be a siphon or pump; it must ensure that no trickle discharge occurs. Effluent ice mounds may develop below the discharge which may require mechanical removal.

Seasonal Discharge. Discharge of wastewater at times of maximum or substantial stream flow is highly desirable, because this dilution reduces the impact of even small amounts of oxygen-demanding materials to receiving water. The timing of this type of discharge varies with the location.

Flow Management Lagoon Discharge. Flow management lagoons are used to control discharge of organics or nutrients to streams or lakes to certain periods only. In Alaska, regulatory permits are issued for controlled discharge of lagoon effluents at specific times of the year.

Discharge of treatment facility effluent to a polishing lagoon may be necessary where the only available receiving water is sensitive to nutrient additions. Polishing lagoons provide additional BOD removal and phosphorus removal by chemical coagulation during ice-free periods (Finch and Smith, 1985). Conventional or pumped discharge can be used.

Total retention lagoons rely on percolation and evaporation to dispose of the accumulated water. Percolation can be provided by proper selection of the location of the lagoon. Gravel and sand gravel areas can be found or formed in some northern areas. The feasibility of percolation lagoons may de-

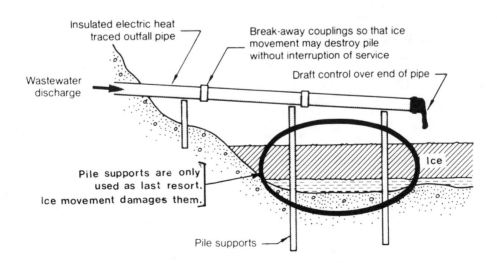

FIGURE 10-15 EXAMPLE OF AN OCEAN OUTFALL DESIGN

pend on the depth of the groundwater. The lagoon should be sized to hold wastewater accumulated during an entire winter (approximately 250 days). This allows for a particularly severe winter during which the percolation route is sealed off by freezing. Emergency discharge facilities should also be provided.

The water source of the community should be monitored for contamination, if groundwater is used. The percolate and groundwater flow patterns should be studied to ensure that no health problems will occur.

Net evaporation can be estimated only roughly from wet and dry bulb data and precipitation records. Normally, evaporation is estimated on the basis of evaporation pan tests and corresponding coefficients; however, very little data are available for the North. The lagoon design must discourage abnormal snow accumulation.

10.5.2 Subsurface Land Disposal. Several methods of wastewater disposal into soil are or have been practiced, including absorption field disposal of effluents from septic tanks or aerobic treatments, direct burial in pits, disposal in ice crypts or snow sumps, and dumping and covering in a landfill.

Septic systems can be used in some arctic and sub-arctic locations. The presence, depth and thickness of permafrost, the type of soil and its percolation rates, and the frost penetration depth influence the use of septic systems. General septic system design information is available from several sources, such as the *Manual of Septic Tank Practice*, and the *Private Sewage Disposal Systems Instructions to Installers* (Plumbing Inspection Branch, 1978) and is covered in more detail in Section 14 of this manual.

10.5.3 Land Disposal. Land application of wastewater requires maintenance of soil permeability. In some cold-climate areas this may be possible through special design; however, many areas are unsuitable. The presence of permafrost limits vertical movement of the wastewater, and winter freezeback of the active layer may prevent horizontal movement. It may be possible to design a suitable system in areas where the thawed zones extend to a water body that does not freeze completely.

Another approach warranting consideration is the use of a retention land disposal system, making use of land disposal during the warmer parts of the year and retention through the winter.

TABLE 10-19   RESULTS OF SPRAY APPLICATION OF LAGOON EFFLUENT AT EIELSON
AIR FORCE BASE (10 mm lysimeters, Sletten, 1977; Smith, 1975)

Test Cell	Application Rate (mm/wk)		TOC (mg/L)	BOLD (mg/L)	SS (mg/L)	FC (per 100 mL)
1974						
A	07.5	Mean	22	–	6	–
		Max.	47	–	10	–
		Min.	13	–	3	–
B	12.0	Mean	25	–	4	–
		Max.	34	–	8	–
		Min.	10	–	2	–
C	17.0	Mean	23	–	13	–
		Max.	36	–	28	–
		Min.	13	–	5	–
1975						
A		Mean	–	3	136	802
		Max.	–	8	511	2,400
		Min.	–	0	11	0
B		Mean	–	11	11	20,350
		Max.	–	50	29	102,000
		Min.	–	2	4	0
C		Mean	–	5	63	7,656
		Max.	–	18	428	36,000
		Min.	–	0	7	16

Complex installations may allow land disposal of effluent during the summer and fall, discharge during early winter, and ice mounding in late winter.

In general, land disposal is subject to the same constraints as land treatment; the processes are similar. The major difference is that land disposal systems are mainly concerned with getting the effluent into the ground and away from the site. The major concern is that the effluent moves in an acceptable direction and presents no hazards to public health.

Soil Conditions. Arctic areas underlain with permafrost are generally limited to noninfiltration land disposal techniques. Overland flow techniques may be suitable; however, no control studies have been conducted to confirm that. Essential considerations include winter storage requirements.

Suitable areas can be located for infiltration approaches in areas of discontinuous permafrost. Again, winter storage requirements are paramount. Design details can be found in the various land disposal manuals (Black et al., 1984; U.S. Environmental Protection Agency, 1977).

Loading Rates. With absorption fields, the allowable application rate is a function of the hydraulic capacity of the soil. Methods for estimating application rates are available from Black et al. (1984) and U.S. Environmental Protection Agency (1977).

The effects of the disposal of untreated wastewater in cleared areas were studied in the Northwest Territories (Fahlman, 1974). Only one application of wastewater was made to the test plots. The soils were identified as fine, glacio-lacustrine sands with no permafrost at the Fort Simpson site, and a 300 mm thick organic layer underlain by high ice content, clayey till at the Norman Wells Site. The loading rates varied from 13 to 203 mm, and results indicated no apparent adverse effects on the vegetation. There are indications that pioneer vegetation in disturbed areas with low nutrient availability may benefit from sewage irrigation. Suitable loading rates were not examined.

Slow-infiltration land application was studied at an interior Alaska location (Sletten, 1977). The soil, classified as a sandy silt, was able to accept 152 mm of secondary lagoon effluent per week during the summer test period. Performance data is presented in Table 10-19.

10.5.4 Wetlands Treatment and Discharge. Wetlands can provide effective treatment of domestic wastewater by further removal of organic material, suspended solids, nutrients and pathogens. Treatment to levels greater than secondary have been reported. Design criteria for hydraulic and organic loading is contained in the literature. Although naturally occurring treatment methods have predominantly been used, constructed wetlands may be an option.

Natural systems have included use of tundra ponds for wastewater treatment and disposal, natural depressions for retention of wastes, and tundra areas for land disposal of wastewater lagoon effluent. Due to the extended survival times of pathogens associated with cold climate conditions, access to disposal areas should be minimized (Schubert and Heintzman, 1974; Reed et al., 1988; Doku, 1993).

Area Affected. Discharge of settled wastewaters to a swamp at Hay River, NWT, was investigated in detail by Hartland-Roe and Wright (1974). Assuming no equilibrium, they estimated that the effect of the effluent could be detected in an area of 35 m^2 per each person-year of sewage discharge. If equilibrium conditions were assumed after five years of operation, it was estimated that 110 m^2 were ecologically influenced by each contributing person.

Effluents of higher quality would be expected to influence less area. In a study by Fetter et al. (1976), the polishing capability of a natural marsh of 156 ha with an average depth of 0.5 m was assessed. The wastewater flow was approximately 1,000 m^3 per day, which constituted about 20 percent of the flow into the marsh. The discharge was found to attain the reductions shown in Table 10-20.

Management. In general, swamp areas can provide effective polishing of effluent from treatment plants. If excess BOD and suspended solids are allowed to enter the area, normal aerobic-anaerobic relationships may be shifted severely to the anaerobic side. Such a condition could result in the release of excessive odors and decreased performance of the area.

Effects. The principal effect of swamp use is that the area must be assumed to be a part of the treatment system, and therefore access must be limited or discouraged. The area should be well posted. The signs should identify the area as part of the wastewater treatment system. Depending upon the quality of the effluent applied and the degree of disinfection, the area will function as a facultative lagoon.

The wastewater discharge increases the number of pathogenic bacteria present. It also adds heat to the system during the winter.

TABLE 10-20   EFFLUENT QUALITY IMPROVEMENT
THROUGH WETLANDS (Fetter et al., 1976)

Parameter	Percent Reduction
BOD	80.1
Coliform	86.2
Nitrate	51.3
COD	43.7
Turbidity	43.5
Suspended Solids	29.1
Total Phosphorus	13.4

Considerable icing may occur because of the shallow water. Ice buildup may retard break-up in the spring.

10.5.5 Undiluted Waste Disposal (Honey Bucket Lagoons).   Regulatory experience with the use of pits or bunkers for the disposal of undiluted wastes has halted the use of such facilities, and required construction of single-cell lagoon or disposal facilities, remote from the living environment of the community. Design criteria for facility sizing has varied from organic loading to volumetric loading requirements. Sizing criteria is based on an assumed volumetric generation rate of 1.5 L/(p•d) at 10 percent solids concentration for the selected design period. Several additional features require consideration in design, including:

- direction of the prevailing wind direction, and snow drifting tendency of the area with respect to the access ramp location needs to be considered;

- site selection should consider winter travel paths for subsistence activities, to allow fences to be marked for safety considerations;

- site location and future expansion potential in anticipation of facility upgrades to a higher level of service should be considered;

- dumping ramp and support structure configuration should consider vehicle turning and maneuvering;

- sizing should consider possible additional solids loading from bags and boxes, if part of the vehicle hauling system.

A typical single-cell disposal lagoon is shown in Figure 10-16 for Wales, Alaska. It consists of a single cell, sized for a design life of 20 years, an access ramp for haul vehicle transport and disposal, an emergency overflow, fencing, gates and warning signs.

10.5.6 Sludge Disposal.   All wastewater treatment systems require the disposal of sludge. This section presents only those techniques which make use of unique cold-climate conditions.

Sludge Drying/Freezing.   Sludge drying is a common practice at most wastewater treatment facilities. However, very few sludge drying facilities have been constructed in cold regions. Sludge dewatering by freezing has been investigated (Martel, 1989; 1994). It was found in one study that application of 100 to 150 mm of sludge during winter resulted in good freeze-assisted coagulation. In central Alaska and Canada, a total freeze depth of more than one metre can be expected.

Sludge Pits.   Sludge disposal pits are one of the most commonly used methods of disposal. Pits can be excavated in almost any material: snow, permafrost, various soils, and rock.

Several years ago, considerable work was done by the U.S. Army Cold Regions Research and Engineering Laboratory (CRREL) on the use of the ice crypts (Ostrom, 1962).

Snow pits were also studied by CRREL. In this case, sewage discharged to snow travels downward, forming a sump until it freezes. The depth of the sump depends on the rate of heat addition (volume and

FIGURE 10-16 WALES LAGOON

temperature of the sewage). The vertical pit develops until the percolating liquid freezes to form an impermeable bottom (Reed and Tobiasson, 1968).

The performance of pits in permafrost or periodically frozen soil was investigated by Heinke and Prasad (1973) in a two-year laboratory study. Their recommendation was that pits do not provide satisfactory treatment of human wastes (honey bags). The waste pit was found to serve as a holding tank at -5°C operation and little better when operated at +5°C for about three months each year. The sole significant change was a four-fold increase in the production of volatile acids. A significant increase in the heterotrophic bacteria population was observed each year in the latter part of the +5°C operation. No increase in psychrophilic bacteria was noticed and it was felt that pathogens were likely to remain viable in the pit for many years. It is considered better than disposal at an open dump.

Location of sludge pits is very important. They should be as far away as practical from the community, on a site that will not be needed for other purposes in the future, and which does not drain towards water supply sources. The pit should be sized on the basis of 0.5 m³ per person-year. For a community of 5,000 people, a pit of 30 m by 2.6 m by 1.2 m deep

would be required. A freeboard of 0.6 m should be allowed, the contents to be covered by 0.6 m of soil when the pit is full and a new one dug. Where contamination of ground and surface water may occur, the pit should be lined with a geomembrane to prevent seepage through the soil. Truck access for easy dumping is required.

Lagoon Sludge Disposal. Disposal of sludge accumulated in lagoons has seldom been considered. Removal of sludge from lagoons is infrequent ranging from once every 6 to 12 years. Normal practice is to move the sludge to a separate holding area, preferably near the landfill, and to cover it with 0.5 to 1.0 m of soil.

Landfill. Although landfill dumping of undiluted sludge and treatment plant sludge is practiced, it is not a recommended procedure without proper evaluation of the effects. Additional leachate formation due to the added liquid is a major concern. Principal considerations are proper isolation, frequencu of cover, soil conditions and permeability, groundwater depth and direction of movement, and access prevention for people and animals.

10.6 References

Alberta Environment. 1988. *Standards an Guidelines for Municipal Water Supply, Wastewater, and Storm Drainage Facilities.* Edmonton, Alberta.

Bell, J.B., W. Macrae and J.F.J. Zaal. 1977. *The Bacterial Content of the North Saskatchewan River in the Vicinity of Edmonton, Alberta.* Env. Can. Report EPS-5-NW-75-2, Edmonton, Alberta.

Bethell, G. 1981. *Start-up Characteristics and Performance Evaluation of an Anaerobic Sewage Lagoon, North of 60, Whitehorse, YT.* Environmental Protection Service, Environment Canada, Pacific and Yukon Region, Report 81-9, Whitehorse, Yukon.

Black, S.A., D.N. Graveland, W. Nicholaichuk, D.W. Smith, R.S. Tobin and M.D. Webber. 1984. *Manual for Land Application of Treated Municipal Wastewater and Sludge.* Env. Prot. Service No. EPS 6-EP-84-1, Ottawa, Ontario, 216 p.

Bouthillier, P.H. and K. Simpson. 1972. Oxygen depletion in ice covered rivers. *Journal of the Sanitary Engineering Division, ASCE*, 98(SA2): 341.

CCREM. 1995. Canadian Water Quality Guidelines. Canadian Council of Resource and Environment Ministers. Environment Canada, Ottawa, Ontario, approx. 1100 p.

Chong, T.M.Y. 1975. *Evaluation of Package Rotating Biological Contactor Treatment Plants for Northern Communities.* M.A. Sc. Thesis, Department of Civil Engineering, University of Toronto, Ontario.

Christensen, V. 1982. *General Guidelines for the Planning and Preliminary Engineering of Appropriate Sewage Treatment/Disposal Systems for Small Northern Communities.* Government of the Northwest Territories, Department. of Local Government, 12 p.

Davenport, C.V., E.B. Sparrow and R.C. Gordon. 1977. Fecal indicator bacteria persistence under natural conditions in an ice covered river. *Journal of Applied and Environmental Microbiology*, 32: 527-536.

Dawson, R.N. and J.W. Grainge 1969. Proposed design criteria for wastewater lagoons in arctic and subarctic Regions, *Journal of the Water Pollution Control Federation*, 41: 237-246.

Doku, I.A. 1993. *The Potential for Use of Wetlands for Wastewater Treatment in the Northwest Territories.* Masters Thesis, Department of Civil Engineering, University of Toronto, Ontario. 251 p.

Eggener, C.L. and B.G. Tomlinson. 1978. Temporary wastewater treatment in remote locations, *Journal of the Water Pollution Control Federation,* 50(12): 2643-2656.

Fahlman, R. and R. Edwards. 1974. Effects of land sewage disposal on subarctic vegetation. In: *Arctic Waste Disposal.* Report No. 74-10, Environmental-Social Program, Task Force on Northern Oil Development, Government of Canada, Ottawa, Ontario.

Fetter, C.W., Jr., W.E. Sloey and F.L. Spangler. 1976. *Use of a Natural Marsh for Wastewater Polishing.* Special Report of the Department of Geology, University of Wisconsin, Oshkosh.

Finch, G.R. and D.W. Smith. 1985. Batch coagulation of a lagoon for fecal coliform reductions. *Water Research*, 20: 105-112.

Finch, G.R. and D.W. Smith. 1987. Ozone disinfection of secondary effluent containing antibiotic-resistant *Escherichia coli. Canadian Journal of Civil Engineering*, 14(2): 234-238.

Forese, R.J. and K.R. Heuchert. 1984. RBC Wastewater treatment in arctic resource camps. In: *Proceedings, 3rd International Conference on Cold Regions*, Canadian Society for Civil Engineering, Montreal, Quebec, 659-674.

Forgie, D.J.L. 1984. Rotating biological contactor treatment efficiency at low temperatures. In: *Proceedings, 3rd International Conference on Cold Regions*, Canadian Society for Civil Engineering, Montreal, Quebec, 693-710.

Given, P.W. 1984. RBC Treatment of dilute wastewater. In: *Proceedings, 3rd International Conference on Cold Regions*, Canadian Society for Civil Engineering, Montreal, Quebec, 659-674.

Given, P.W. and D.W. Smith. 1977. *Critical Evaluation of Extended Aeration Systems in Arctic and Subarctic Regions.* Environmental Protection Service Rep. No. EPS 3-WP-77-10, Ottawa, Ontario, 74 p.

Given, P.W. and D.W. Smith. 1979. Disinfection of dilute, low temperature wastewater using ozone. *Ozone: Science and Engineering*, 1: 91-106.

Gordon, R.C. 1970. Depletion of oxygen by microorganisms in Alaskan rivers at low temperatures. In: *Water Pollution Control in Cold Climates Symposium,* R.S. Murphy (ed.),U.S. Environmental Protection Agency, Washington, D.C.

Gordon, R.C. 1972. Winter Survival of Fecal Indicators Bacteria in a Subarctic Alaskan River. U.S. Environmental Protection Agency Report No. EPA-R2-72-013, Corvallis, Oregon.

Hartland-Rowe, R.C.B. and P.B. Wright. 1974. *Swamplands for Sewage Effluents, Final Report*. Environmental-Social Committee, Northern Pipelines, Department of Indian Affairs and Northern Development, Ottawa, Ontario.

Heinke, G.W. 1973. Disposal of human wastes in northern areas. In: *Some Problems of Solid and Liquid Waste Disposal in the Northern Environment*. Environment Canada Report EPS 4-NW-76-2, 87-140.

Heinke, G.W. 1976. Disposal of human wastes in northern areas. In: *Some Problems of Solid and Liquid Waste Disposal in the Northern Environment*. Environment Canada Report EPS-4-NW-76-2, Ottawa, Ontario, 87-140.

Heinke, G.W. and D. Prasad. 1977. Anaerobic treatment of human wastes in northern communities. *Canadian Journal of Civil Engineering* , 7(1): 156-164.

Heinke, G.W., D.W. Smith, and G.R. Finch. 1988. *Guidelines for the Planning, Design, Operation, and Maintenance of Wastewater Lagoon Systems in the Northwest Territories. Volume I: Planning and Design, Volume II: Operation and Maintenance Manual*. Department of Municipal and Community Affairs, Government of the Northwest Territories, Yellowknife, 70 and 48 p.

Heinke, G.W., D.W. Smith and G.R. Finch. 1991. Guidelines for the planning and design of wastewater lagoon systems in cold climates. *Canadian Journal of Civil Engineering*, 18: 556-567.

Heinke, G.W., Smith, D.W. and R. Gerard. 1990. *Guidelines for Disposal of Wastewater in Coastal Communities of the Northwest Teritories*. Department of the Municipal and Community Affairs, Government of the Northwest Territories, Yellowknife, 52 p.

Hrudey, S.E. and S. Raniga 1981. Greywater management for isolated northern communities. In: *Proceedings, The Northern Community: A Search for a Quality Environment*, American Society of Civil Engineers, New York, 471-481.

International Environmental Consultants Ltd. (IEC). 1978. *Irradiation as an Alternative for Disinfection of Domestic Waste in the Canadian Arctic*. Report for Central Mortgage and Housing Corp., Ottawa, Ontario

Magditch, A. 1985. *Performance Evaluation of the Inuvik, N.W.T. Lagoon*. M.A.Sc. Thesis, University of Toronto, Department of Civil Engineering, Ontario.

Martel, C.J. 1989. Development and design of sludge freezing beds. *Journal of Environmental Engineering*, American Society of Civil Engineers, 115(4): 799-808.

Martel, C.J. 1994. Operation and performance of sludge freezing bed at Ft. McCoy, Wisconsin. In: *Proceedings, 7th International Cold Regions Engineering Specialty Conference,* Edmonton, Alberta, Canadian Society for Civil Engineering, Montreal, Quebec, 607-616.

Metcalf and Eddy, Inc. 1979. *Wastewater Engineering: Treatment, Disposal, Reuse*. McGraw-Hill Book Co., New York, 920 p.

Murphy, R., and K. Ranganathan. 1974. Bio-processes of the oxidation ditch in a sub-arctic climate. In: *Proceedings, International Symposium on Wastewater Treatment in Cold Climates*. Environmental Protection Service Rep. No. EPS 3-WP-74-3, Ottawa, Ontario, 332-357.

Murray, A.P. and R.S. Murphy. 1972. *The Biodegradation of Organic Substrates Under Arctic and Subarctic Conditions*. Institute of Water Resources Report No. IWR 20, Univ. of Alaska, Fairbanks.

Northwest Territories Water Board. 1992. *Guidelines for the Discharge of Treated Municipal Waste-*

water in the Northwest Territories. Yellowknife, NWT, 30 p.

Ostrom, T.R. et al. 1962. Investigation of a sewage sump on the Greenland icecap. *Journal of the Water Pollution Control Federation*, 34(1): 56.

Pollen, M.R. 1983. Arctic tundra as a wastewater receiving environment. In: *Proceedings, Cold Regions Environmental Engineering Conference*, University of Alaska, Fairbanks, 574.

Plumbing Inspection Branch. 1978. *Private Sewage Disposal Systems Instruction to Installers*. Alberta Labour, General Safety Services Division, Edmonton.

Prince, D.S., D.W. Smith and S.J. Stanley. 1993. *Evaluation of Lagoon Treatment in Alberta*. Report for Standards and Approvals Division, Alberta Environmental Protection, Edmonton, Alberta.

Prince, D.S., D.W. Smith and S.J. Stanley. 1995a. Performance of lagoons experiencing seasonal ice cover. *Water Environment Research*, 67(3): 318-326.

Prince, D.S., D.W. Smith and S.J. Stanley. 1995b. Intermittent discharge lagoons for use in cold regions. *Journal of Cold Regions Engineering*, American Society of Civil Engineers, 9(4): 184-194.

Putz, G., D.W. Smith and R. Gerard. 1984. Microorganism survival in an ice-covered river. *Canadian Journal of Civil Engineering*, 11: 177-186.

Reed, S.C., E.J. Middlebrooks and R.W. Crites. 1988. *Natural Systems for Waste Management and Treatment*. McGraw-Hill.

Reed, S. and W. Tobiasson. 1968. Wastewater Disposal and Microbial Activity at Ice-Cap Facilities, *Journal of the Water Pollution Control Federation*, 40: 2013.

Reed, S.C. et al. 1977. Land treatment of wastewater for Alaska. In: *Proceedings, 2nd International Symposium on Cold Regions Engineering*, University of Alaska, Fairbanks, 316-318.

Schallock, E. 1974. *Low Winter Dissolved Oxygen in Some Alaskan Rivers*. U.S. Environmental Protection Agency, Report No. EPA-660/3-74-008. Corvallis, Oregon.

Schneiter, R.W. 1982. *Cold Regions Wastewater Lagoon Sludge: Accumulation, Charac-terization and Digestion*. PhD Thesis, Utah State University, Logan. 405 p.

Schubert, D.H. and T.H. Heintzman. 1994. Use of tundra ponds for the treatment of wastewater. In: *7th International Conference on Wastewater Treatment*, Atlanta, Georgia, December.

Sletten, R.S. 1977. Feasibility study of land treatment at a subarctic Alaskan location. In: *Land as a Waste Management Alternative*, R.C. Loehr (ed.), Ann Arbor Science Publishers, Michigan.

Sletten, R.S. 1978. Land application of wastewater in permafrost areas. In: *3rd International Permafrost Conference*, National Academy of Science, Washington, D.C.

Smith, D.W. 1975. *Land Disposal of Secondary Lagoon Effluents, Pilot Project*. Institute of Water Resources Report No. 59, University of Alaska, Fairbanks.

Smith, D.W. 1982. Wastewater treatment for cold regions. *Technical Council Journal*, American Society of Civil Engineers, 108(TCL) 138-149.

Smith, D.W. and R. Gerard. 1980. Mixing and microorganism survival in the Slave River, N.W.T. In: *Proceedings, The Northern Community: A Search for a Quality Environment*, American Society of Civil Engineers, New York, 449-470.

Smith, D.W. and V. Christensen 1982. Innovative approaches to sewage collection, treatment and disposal: practices in northern Canada. *Canadian Journal of Civil Engineering*, 9(4): 653-662.

Smith, D.W. and G.R. Finch. 1983. *A Critical Evaluation of the Operation and Performance of Lagoons in Cold Climates*. Report for Environmental Protection Service, Environment Canada, Ottawa, Ontario. Also Environmental Engineering Technical Rep. No. 84-2, Department. of Civil Engineering, University of Alberta, Edmonton, 231 p.

Smith, D.W. and G.W. Heinke 1981. Cold climate environmental engineering. An overview. *Water Science and Technology*, 13: 3-16.

UMA Engineering Ltd. 1993. *Municipal Wastewater Treatment Technologies Capable of Achieving Compliance with the Fisheries Act in the Northwest Territories*. Report for Environmental Protection, Environment Canada, Yellowknife, NWT

U.S. Environmental Protection Agency. 1977. *Design Manual of Land Disposal.* Technology Transfer, Washington, D.C.

Yukon Territory Water Board. 1983. *Guidelines for Municipal Wastewater Discharges in the Yukon Territory.* Whitehorse, Yukon. 20 p.

10.7   Bibliography

Alter, A.J. 1969. *Sewage and Sewage Disposal in Cold Regions.* U.S. Army, Cold Regions Research and Engineering Laboratory Monograph III-C5b, Hanover, New Hampshire, 106 p.

Alter, A.J. 1973. The polar palace. *The Northern Engineer,* 5(2): 4-10.

Black, S.A. et al. 1984. *Manual for Land Application of Treated Municipal Wastewater and Sludge.* Environmental Protection Service Rep. No. EPS 6-EP-84-1, Ottawa, Ontario, 216 p.

Buens, G.E. 1970. Evaluation of aerated lagoons in cold climate. In: *Proceedings 22nd Western Canada Water and Sewage Conference,* 213-40.

Chen, R.L. and W.H. Patrick, Jr. 1980. *Nitrogen Transformations in a Simulated Overland Flow Wastewater Treatment System.* Special Report 80-16, U.S. Army Cold Regions Research and Engineering Laboratory, Hanover, N.H., 34 p.

Christensen, V. 1983. Status of water and sanitation facilities in the Northwest Territories. In: *Proceedings Utilities Delivery in Cold Regions,* Environmental Protection Service Rep. No.

Christianson, C. 1977. Cold climate aerated lagoons. In: *Proceedings 2nd International Symposium on Cold Regions Engineering,* Department of Civil Engineering, University of Alaska, Fairbanks,318-351.

Clark, S., H. Coutts, and C. Christianson. 1971. *Design Considerations for Extended Aeration in Alaska.* U.S. Environmental Protection Agency Rep. No. 16100 ExH 11/71, 213-236.

Clark, S.E., H.J. Coutts, and C. Christianson. 1970. *Biological Waste Treated in the Far North.* U.S. Federal Water Quality Admin. Rep. No. 1610. College, Alaska, 36 p.

Clark, S.E. et al. 1970. Alaska sewage lagoons. In: Proceedings 2nd International Symposium on Waste Treatment Lagoons. U.S. Federal Water Quality Administration, Washington, D.C., 221-230.

Coutts, H. 1972. *Arctic Evaluation of a Small Physical-Chemical Sewage Treatment Plant.* U.S. Environmental Protection Agency, Arctic Environmental Research Laboratory, Fairbanks, Alaska.

Dietrick, L. 1983. Wastewater disposal by stack injection. In: *Proceedings Cold Regions Environmental Engineering Conference,* University of Alaska, Fairbanks, 664-678.

Dubuc, Y., C. Roy, and R. Labonté. 1983. Treatment of wastewater by peatlands in the James Bay area, province of Québec. In: *Proceedings Cold Regions Environmental Engineering Conference,* University of Alaska, Fairbanks, 558-573.

Gaskin, D.A., A.J. Palazzo, S.D. Rindge, R.E. Bates and L.E. Stanley. 1979. *Utilization of Sewage Sludge for Terrain Stabilization in Cold Regions, Part II.* U.S. Army, Cold Regions Research and Engineering Laboratory Special Rep. 79-28, Hanover, N.H., 36 p.

Gordon, R. and C. Davenport. 1973. *Batch Disinfection of Treated Wastewater with Chlorine at Less Than 1°C.* U.S. Environmental Protection Agency Rep. No. EPA 660/2-73-005, Washington, D.C.

Hanaeus, J. 1984. Swedish field experience with chemical precipitation in stabilization ponds. In: *Proceedings 3rd International Conference on Cold Regions Engineering,* Canadian Society for Civil Engineering, Montreal, Quebec, 547-564.

Hoffman, R.W., M.H. Fawcett, D.E. Dorratcague. 1983. Wastewater disposal in Cook Inlet, a comparison to west coast receiving environments. In: *Proceedings Cold Regions Environmental Engineering Conference,* University of Alaska, Fairbanks, 599-609.

Jenkins, T.F., et al. 1981. *Seven-Year Performance of CRREL Slow-Rate Land Treatment Prototypes.* U.S. Army, Cold Regions Research and Engineering Lab. Special Report, Hanover, N.H., 25 p.

Johnson, R.A. 1980. Domestic sludge - a resource for Alaska. *Northern Engineer*, 12(1): 15-18.

Johnson, R.A., F.J. Wooding and J.W. Winslade. 1983. Agricultural utilization of domestic sludge in the subarctic. In: *Proceedings Cold Region Environmental Engineering Conference*, University of Alaska, Fairbanks, 649-663.

Jones, G.V. 1983. Dewatering and disposal of municipal sewage sludge, Fairbanks, Alaska. In: *Proceedings Cold Region Environmental Engineering Conference*, University of Alaska, Fairbanks, 638-648.

Komex Consultants Ltd. 1985. *Design and Construction of Liners for Municipal Wastewater Stabilization Ponds*. Alberta Environment, Edmonton.

Lehman, P.J. and D.D. Henry. 1982. Performance of an RBC facility in a cold climate. In: *National Conference on Environmental Engineering*, American Society of Civil Engineers, New York.

Lin, K.C. and G.W. Heinke. 1977. Plant data analysis of temperature significance in the activated sludge process. *Journal of the Water Pollution Control Federation*, 49(2): 286-295.

Miyamoto, H.K. and G.W. Heinke. 1979. Performance evaluation of an arctic sewage lagoon. *Canadian Journal of Civil Engineering*, 6(2): 324-329.

Middlebrooks, E.J., C.D. Perman and I.S. Dunn. 1978. *Wastewater Stabilization Lagoon Linings*. U.S. Army, Cold Regions Research and Engineering Laboratory Special Report, Hanover, N.H., 70 p.

Morrison, S.M., et al. 1973. *Lime Disinfection of Sewage Bacteria at Low Temperature*. U.S. Environmental Protection Agency Rep. No. EPA 660/2-73-017, Washington, D.C., 90 p.

Pollen, M.R. 1983. Arctic tundra as a sastewater discharge receiving environment. In: *Proceedings Cold Regions Environmental Engineering Conference*, University of Alaska, Fairbanks, 574-598.

Prasad, D. and G.W. Heinke. 1981. Disposal and treatment of concentrated human wastes. In: *Design of Water and Wastewater Services for Cold Climate Communities*. Pergamon Press, England, 125-136.

Putz, G., D.W. Smith and R. Gerard. 1983. Mixing zone investigations of continuous wastewater discharges in large northern rivers. In: *Proceedings Cold Regions Environmental Engineering Conference*, University of Alaska, Fairbanks, 610-637.

Reed, S.C. 1983. Nitrogen removal in wastewater treatment lagoons. In: *Proceedings Cold Regions Environmental Engineering Conference*, University of Alaska, Fairbanks, 440-458.

Reed, S.C. and R.S. Murphy. 1969. Low temperature activated sludge settling. *Journal Sanitary Engineering Division*, American Society of Civil Engineers, 95(SA4): 747-767.

Reid, B.H. 1977. Some technical aspects of the Alaska village demonstration project. In: Proceedings Utilities Delivery in Arctic Regions, Environmental Protection Service Rep. No.EPS 3-WP-77-1, Ottawa, Ontario, 391-438.

Rush, R.J. and A.R. Stickney. 1979. *Natural Freeze-Thaw Sewage Sludge Conditioning and Dewatering*. Environmental Protection Service Rep. No. 4-WP-79-1, Ottawa, Ontario.

Schubert, D.H. and T. Heintzman. 1994. *7th International Symposium of Individual and Small Community Sewage Systems*. Atlanta, Georgia, 13 p.

Smith, D.W. and P.W. Given. 1977. Evaluation of northern extended aeration sewage treatment plants. In: *Proceedings, 2nd International Symposium on Cold Regions Engineering*, University of Alaska, Fairbanks, 291-316.

Smith, D.W. and P.W. Given. 1980. Treatment alternatives for dilute low temperature wastewater. In: *Design of Water and Wastewater Service for Cold Climate Communities,* Pergamon Press, Ltd., England, 165-179.

Tilsworth, T., et al. 1972. Freeze conditioning of waste activated sludge. In: Proceedings 27th Industrial Waste Conference, Purdue University, West Lafayette, Indiana, 486-491.

SECTION 11

UTILIDORS

3rd Edition Steering Committee Coordinators

Gary W. Heinke

William L. Ryan

iii

Section 11   Table of Contents

Section 10   List of Figures

11 UTILIDORS

11.1 Introduction

A utilidor is a structure whose function is to contain the utility piping and wiring of a community or camp. Utilidors enclose utility piping such as water and sewer pipes as well as central heating, fuel oil, natural gas, electrical and telephone conduits. Single pipes or separate pipes on a common foundation are not utilidors. Utilidors may be located above or below ground, or at ground level. Factors to be considered when choosing between above- or below-ground utilities are detailed in Section 2. Above-ground utilities were often necessary in the past because the types of insulation in common use required dry conditions and accessibility. For the same reason, buried accessible (or walk-through) utilidors were used. The development and availability of near-hydrophobic, rigid foam insulations, such as polyurethane and extruded polystyrene, have now made buried systems feasible in many situations. Utilidors have been used in stable and unstable seasonal frost and permafrost soils as well as in snow (Tobiasson, 1971). They may be large enough to provide access for maintenance purposes or for use as an enclosed walkway, or they may be compact with no air spaces. Utilidors have been constructed in many locations (Tobiasson, 1971). Perhaps the best-known examples of above-ground utilidors are in Inuvik, NWT (Gamble and Lukomskyj, 1975; Leitch and Heinke, 1970; Cooper, 1968). They have also been extensively used below ground in high-density areas and in permafrost areas in Russia (Slipchenko, 1970; Porkhaev, 1965). Examples of various utilidors that have been constructed in cold regions are shown in Figure 11-1. The size, shape, and materials used depends on the number and types of pipes, local conditions and requirements, and economics.

Utilidors were commonly used in cold regions to house and support water and sewer piping, and other utilities, particularly where central heating pipes were used and for above-ground systems in general. Also of significance is the physical and thermal protection these utilidors offer the utility piping. As with other piping in cold regions it must be insulated to reduce heat loss, but often the insulation is located on the exterior casing of the utilidor rather than on the individual pipes. Other freeze-protection measures, such as recirculation to maintain flow in water pipes, may also be necessary. Most utilidors have some mutually beneficial heat transfer between the enclosed pipes. If central heating pipes are included, heat loss from these can be sufficient to replace utilidor heat losses and prevent freezing. But with this arrangement, temperature control within the utilidor is difficult and can lead to inefficiency and undesirably high temperatures in the water pipes. Freeze protection alone is not sufficient justification for using utilidors since individually insulated pipes could be used instead. Only where utilidors or utility tunnels are required for passage or transportation can overall economic benefit be derived from locating utility conduits within them.

Enclosed walkways are sometimes constructed between buildings of construction camps or military facilities or in conditions of high traffic and severe snowdrifting. Placing utility lines within these enclosed walkways should be considered. In some cases surface utilidors have been designed for use as sidewalks.

The U.S. Navy Civil Engineering Laboratory investigated above-ground utilidors that contained pipes for the distribution of freezable liquids (Hoffman, 1971; Smith, 1970); one of the pipes of the utilidor carried a heated liquid to act as heat source for the utilidor. A section of utilidor was built and low-temperature tests were conducted. A thermal performance analysis and a computer program were prepared for designing a utilidor for a specific location. Conclusions reached indicated that single-line, pre-assembled, insulated, and electrically heat-traced piping systems appear to be most cost-effective for one to six distribution lines. However, the cost advantage is so slight for four- to six-line systems that aesthetics or availability of an inexpensive source of heat may favour the use of a utilidor system for a specific facility.

In North America, utilidors are most commonly constructed above ground in permafrost areas. Although below-ground utilidors are preferred, above-ground construction may be necessary where the permafrost is ice-rich, where expensive excavation is necessary, or where equipment for the installation and maintenance of underground utilities is not available.

Utilidors have disadvantages. Their suitability for a particular location should be carefully assessed. Individually insulated pipes in a common trench or supported by a common pile (see Figure 9-18) may be a more appropriate and economical alternative (Hoffman, 1971; Smith, 1970). As a building, utilidors

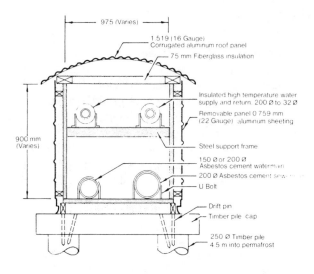

(a) *Utilidor with central heating lines – Inuvik, NWT (Gamble and Lukomskyj, 1975)*

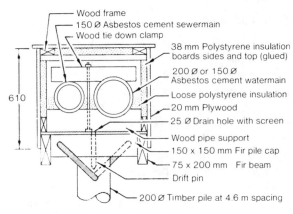

(b) *Plywood box utilidor – Inuvik, NWT (late 1960s through 1970s) (Gamble and Lukomskyj, 1975)*

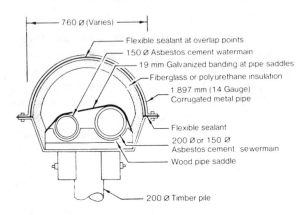

(c) *CMP utilidor – Inuvik, NWT (Gamble and Lukomskyj, 1975)*

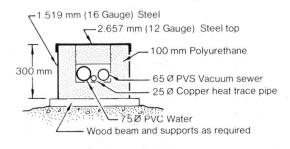

(d) *Utilidor with vacuum sewer – Noorvik, Alaska (Rogness and Ryan, 1977)*

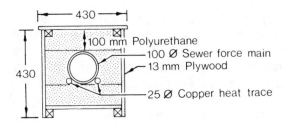

(e) *Single pipe with heat tracing – Noorvik, Alaska (Rogness and Ryan, 1977)*

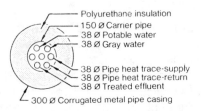

(f) *Utilidor with many small pipes – Wainwright, Alaska (Reid, 1977)*

FIGURE 11-1   *VARIOUS UTILIDORS INSTALLED IN COLD REGIONS*

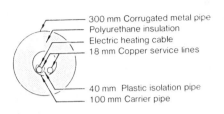

(g) Service connection (James, 1980)

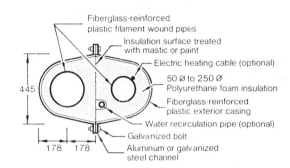

(h) Prefabricated utilidor (Gamble, 1977)

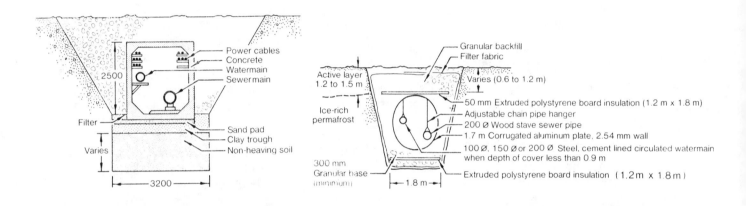

(i) Buried accessible utilidor – Mirryl, USSR
(Porkhaev, 1965)

(j) Buried accessible utilidor – Nome, Alaska
(Leman et al., 1978)

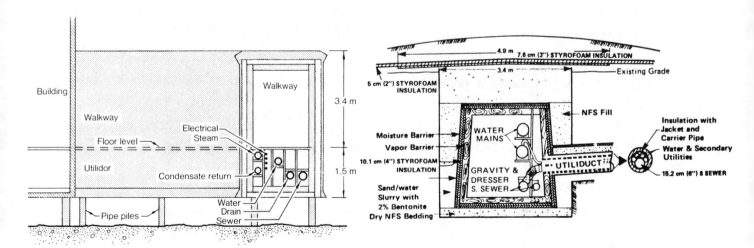

(k) Walkway utilidor – Cape Lisburne, Alaska
(Hoffman, 1971)

(l) Buried accessible utilidor – Barrow, Alaska
(Cerutti et al., 1982; Zirjacks and Hwang,
1983)

FIGURE 11-1   (continued)

may be required to meet certain codes for fire safety, ventilation, access and egress, and lighting.

11.2 Design Considerations

Inclusion of conduits for electric power lines, natural gas pipes, and telephone cables, along with water and sewer lines in utilidors can be cost-effective. This practice is generally recommended wherever practical. The addition of central heating lines however, may require expensive changes in utilidor design and operation.

Some of the problems encountered when other utilities besides water and sewer pipes are included in the utilidor are:

- jurisdictional problems in installation and maintenance;

- coordination between various utility agencies during planning and construction;

- differences in routing or other conflicting design criteria for various utilities or utilidor uses;

- difficulty or restrictions in expansion or changes to utilities;

- piping and utilidor design, particularly at intersections and bends;

- compatibility of materials; and

- increased size and perhaps the requirement for special materials; these increase the costs of utilidor construction and above-ground road crossings.

11.2.1 Design Objectives. Once the use of utilidors has been decided, the design objectives must be reviewed. The primary objective is to protect the utility distribution lines and to realize advantages that are not available to single pipes or wires.

Design objectives of utilidors include:

- protection of individual pipes and wires;

- taking advantage of heat losses or heat tracing to optimize operation and life cycle costs;

- reducing overall heat loss to protect the permafrost environment;

- aesthetic acceptability; and

- the need for access for modification or repair after initial construction.

11.2.2 Utilidors with Central Heating Lines. The benefits of central heating lines in utilidors include:

- the elimination of less efficient individual building furnaces;

- the elimination of a significant cause of fires;

- central fuel storage; and

- the possible use of waste heat (Section 17).

These pipes often require access, particularly for steam and condensate return lines. If they are buried, utilidors are necessary for access.

When only central heating lines are considered, utilidors or individually insulated pipes can be used. When central heating and water and sewer lines are installed in a utilidor, the central heating lines may give off enough heat to prevent freezing temperatures within the utilidor. This may eliminate or reduce the need to loop and recirculate the water mains, although circulation is always desirable in cold regions. Because the exterior of the utilidor is insulated, conventional, uninsulated water and sewer piping and appurtenances are normally used. For economical operation the central heating lines are usually insulated. Occasionally some insulation may be removed from around the heating pipes to provide enough heat during the coldest ambient temperatures. The utilidor in Figure 11-1(a) has 250 mm of insulation removed every 10 m. Also, insulation around water pipes prevents problems from occurring when the ambient air warms up (Cooper, 1968). Given that the heat source is constant and operating for all or most of the year (Hull, 1980), the interior temperature of the utilidor is often high, causing undesirably high water temperatures where the water in the pipes is stagnant or nearly so. Users have found it necessary to bleed large volumes of water to obtain cool water (Leitch and Heinke, 1970; Reed, 1977).

The heating of a large air space in the utilidor to prevent water from freezing is less efficient than the direct heating of water and its recirculation. This inefficiency is more significant for above-ground and large utilidors. In some situations, designing and operating the heating and the water and sewer systems independently is more efficient.

In large, open utilidors, thermal stratification can cause freezing in the lower pipes. Vertical air flow barriers may be necessary to prevent air currents along inclined utilidors (Leitch and Heinke, 1970). An example of thermal shielding of pipes and poor temperature distribution within a utilidor is shown in Figure 11-2 (Reed, 1977).

Utilidors for central heat distribution pipes and other utilities are expensive: the larger the size, the higher the material and construction costs, particularly for above-ground systems. Central heating is only economical in developments of relatively high population density, with nonheating utilidors servicing adjacent lower density areas. With adequate planning these utilidors can form the backbone of a utility delivery system serving a large community.

Central heating systems must provide continuous service. Therefore they should be looped so that sections can be isolated for maintenance or repairs. Long-range planning is necessary not only to incorporate future expansion but also to allow efficient staging without the need for temporary or unused lines to complete loops.

11.2.3 Utilidors with Pipe Heat Tracing.
Utilidors with pipe heat tracing are designed specifically to supply the amount of heat necessary to prevent freezing of the pipes in the utilidor. Low-temperature glycol solutions or other nonfreezing fluids are commonly used in small-diameter bare pipes (Section 4.4.3) to heat utilidors or thaw out frozen pipes.

11.2.4 Above-Ground Utilidors.
Below-ground utilities are usually preferred, but local conditions, operating requirements, and economics can make above-ground utilities and utilidors more feasible (Hull, 1980).

Above-ground utilidors should be as compact and as close to the ground as practical to reduce the obstruction to traffic. Low utilidors also reduce the elevation of buildings necessary for gravity sewer drainage. Utilidors with central heating lines are often large and require expensive wood or corrugated-metal arch structures at road crossings (Figure 11-3). Vacuum or pressure sewer systems may be feasible in undulating terrain, since the utilidor can either follow the ground surface or can be buried at a constant depth. These systems use pipes of smaller diameter than gravity systems. And where small-diameter water and sewer pipes are used, the size of the utilidor can be reduced accordingly.

Above-ground utilidors are located within the rights of way behind building lots where both physical and legal access are provided. Buildings should be located near the utilidor to reduce the length of ser-

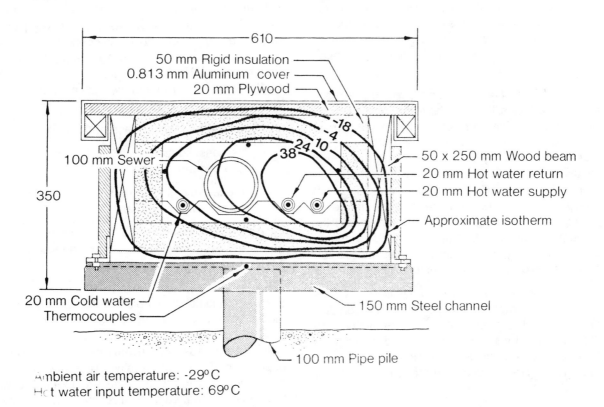

Ambient air temperature: -29° C
Hot water input temperature: 69° C

FIGURE 11-2   TEMPERATURE VARIATION IN A UTILIDOR WITH CENTRAL HOT-WATER DISTRIBUTION – FAIRBANKS, ALASKA (Reed, 1977)

FIGURE 11-3   ROAD CROSSINGS FOR A UTILIDOR WITH CENTRAL HEATING LINES – INUVIK, NWT

vice lines. A minimum of 3 m between building and utilidor is often specified for fire protection. In designing the street layout, the looping of water mains for recirculation must be considered together with the expense of road crossings and vaults, and the high unit cost characteristic of above-ground utilidors. An example of good planning is illustrated in Figure 2-7.

Where utilidors can be routed through the crawl space in buildings, the pipes have thermal and physical protection. This eliminates the service connection utilidettes which are the most freeze-susceptible portion of utility systems in cold regions, and reduces the length of utilidor and right of way required. However, with this layout the piping is not easily accessed, thereby increasing the danger of fires which could jeopardize the community utility system. Fire and smoke can travel along utilidors to other buildings. Fire-resistant materials and fire cutoff walls have been used to minimize this danger.

In some locations people walk on utilidors, but this is often discouraged by design and legislation, because the utilidors require increased maintenance then. In other locations, they have been designed as sidewalks, but this is often impractical because of the routing requirements for the utilidors and their elevation.

11.2.5 Below-Ground Utilidors. Although buried utilidors have been constructed in unstable soils in many large communities in northern Russia, they

have had limited application in the cold regions of North America and in the Antarctica (Tobiasson, 1971). Few buried utilidors have been constructed without the inclusion of central heating lines and perhaps other conduits besides water and sewer pipes. Notable exceptions to all of these generalities is the corrugated metal pipe utilidor for water and sewer lines in Nome, Alaska (Figures 11-1(j) and (k) and 11-4) and the system in Barrow, Alaska (Figure 11-5). The primary advantages of buried utilidors are:

- access for maintenance and repairs;

- the consolidation of utilities in a single structure and trench; and

- freeze-protection and ventilation to reduce the thawing of permafrost.

In some cases, individual pipes or utilidors without an air space (examples shown in Figures 11-1(h), 14-12, and 14-13) may be appropriate and more economical.

Below-ground utilidors are subject to ground movements from frost heaving or thaw settlement and groundwater infiltration. This has caused failures in ice-rich permafrost. To prevent progressive thawing of permafrost, open utilidors with natural ventilation have been used in USSR. Vents are opened during the winter to maintain subfreezing temperatures within the utilidor and to refreeze the foundation soils that were thawed the previous summer. Other foundation designs that can be used in permafrost ar-

FIGURE 11-4   BURIED UTILIDOR UNDER CONSTRUCTION – NOME, ALASKA

FIGURE 11-5   BURIED UTILIDOR UNDER CONSTRUCTION – BARROW, ALASKA

eas include the provision of supports, refrigeration, insulation, and improved foundation soils within the expected thaw zone (Section 3.4.1). Groundwater, meltwater, and water from pipe breaks must be considered in both unstable and stable soils. In the spring, flooding can occur in buried utilidors that have an air space, therefore drainage of the utilidor must be provided. Water-tight utilidors are desirable and usually required. Cutoff walls in the trench may be necessary in permafrost areas to prevent detrimental groundwater flow along the trench.

11.2.6 Camp Utilidors. Some work camps lend themselves well to the use of small box-type utilidors (Figures 11-1(b) through (e)). Utilidors in temporary camps can be installed above ground or at grade, and the top cover may double as a sidewalk between buildings. Utilidors may be routed through or under buildings to reduce the length of the utilidor housing required and to minimize heat loss. Thoughtful routing of surface utilidors also reduces the number of road crossings and minimizes the unseemly aspects. Designers must keep in mind, however, that routing utilidors through or under buildings can be disastrous in the event of fire. Not only can utility services be disrupted, but fire may also be conducted by some types of construction.

11.3 Components and Materials

Utilidors that are well designed and constructed with high-quality materials have a longer useful life, cost less to operate, and provide more reliable service than those of inferior design and construction. Utilidors that have performed best were constructed with a metal exterior casing, used closed-cell foam insulation, were structurally sound, and had a solid foundation (Gamble and Lukomskyj, 1975).

The basic components of a utilidor are:

- foundation;
- frame;
- exterior casing;
- insulation; and
- piping.

11.3.1 Foundation. Foundation considerations and design are different for utilidors that are below ground, at ground level, or elevated. Each requires site investigations and designs that accommodate, reduce, or eliminate the effects of frost heaving, settlement, and surface and subsurface drainage. A competent cold-regions geotechnical engineer should be consulted.

Above-Ground Utilidors. Above-ground utilidors must be supported to provide grades for gravity sewers and for draining both pipes and utilidors. Pipes must be adequately anchored to resist hydraulic and thermal expansion stresses. Where ground movements are within acceptable limits, utilidors can be installed directly on the ground, or on beams, earth mounds, sleepers, or posts.

In unstable soils, utilidors are commonly supported on piles adequately embedded in the permafrost (Section 3.4.2). The piles are dry-augered, if possible, because thawing the permafrost with steam or hot water increases the freeze-back period and frost heaving. Because of the light weight of utilidors, the vertical loads on the piles are relatively small, therefore frost heaving is usually the most significant design consideration for embedding piles. Lateral forces may be significant on some permafrost slopes, therefore lateral thermal expansion and hydraulic stresses must be considered at bends. Various types of piles have been used to support above-ground piping. The selection depends on the availability of local materials, the length of pile required, and the cost. Piles used in Inuvik, NWT, are usually rough timber poles embedded 4 to 6 m deep in the ground. At Norman Wells, NWT, frost heaving is severe, and steel pipe 100 mm in diameter is driven 12 m down and grouted to the fractured shale bedrock. Piles may cost from $200 to over $750 (Cdn.) each and may account for 10 to 20 percent of the total cost of above-ground utilidors. Small utilidors are usually placed on a single pile, but large utilidors may require double piling for stability. The utilidor structure must be adequately anchored to the pile foundation. Pile caps are often used to allow for poor alignment.

Below-Ground Utilidors. Buried utilidors in stable soils have conventional design considerations which include surface loads. Frost heave must be considered for shallow-buried utilidors. In ice-rich permafrost soils, thawing must be prevented or considered in the design (see Section 4.3.1).

11.3.2 Frame. The supporting frame must keep the above-ground utilidors rigid in the spans between one pile or other support structure and the next. The dead loads, live loads (including people), and stability of the utilidor must be considered in the structural design. Steel has been used in some instances, but wooden beams are more common. Solid utilidors may use the pipe, insulation, and shell for beam strength (Figure 11-1(g) and (h)). An economic analysis can determine the best spacing of piles to increase the beam strength of the utilidor frame. Utilidors supported by wooden beams in Inuvik, NWT, usually span 4.5 m. Steel beams or pipe can be used to span up to 7.5 m, which is generally the practical limit without special designs.

Buried utilidors must also have some beam strength, which is usually incorporated into the exterior casing to hold pipe grades and to span areas of poor bedding soil or poor foundation soil.

11.3.3 Exterior Casing. The primary function of the exterior casing is usually to hold the insulation in place and to protect the pipes from weather and physical damage. Where the exterior casing is an integral part of the structural strength and rigidity, it is usually made of metal or fiberglass with polyurethane insulation bonded to it.

The outer shell of above-ground utilidors should be designed for easy removal to provide access to the piping. This is necessary at appurtenances and desirable along the complete utilidor. The joints in the sections must be designed to seal against the infiltration of rain, snow, or air. Drain holes located in the bottom should be provided at intervals to allow drainage in case of a pipe break.

Materials that have been used for the exterior casing include corrugated and sheet metal, plywood or wooden beams, and fiberglass-reinforced plastic. Concrete has also been used for surface or buried utilidors. Although wooden box utilidors incur the lowest capital cost, they have a short life expectancy and the highest maintenance costs because they require painting and are susceptible to physical damage. Yet wood is easy to work with, and the utilidor, vaults, and service connections can be easily constructed on-site, even in cold weather. Wooden utilidors are sometimes covered with a thin metal sheet or asphalt paper. Metal utilidors may be difficult to fabricate and install, particularly at bends, junctions, and appurtenances, but they are rugged, have a longer life expectancy, and require less maintenance than the wooden exterior casing of utilidors.

Below-ground utilidors require a rigid exterior casing that is either watertight or designed to control seepage water, and provides the structural strength necessary to withstand soil loads and surface traffic.

11.3.4 Insulation. Insulation is probably the most critical component in all cold-region piping systems, including utilidors. Water can enter the utilidor at joints or from breaks in the pipes, therefore the insulation cannot be kept dry. Insulations that absorb

moisture lose their insulating value, particularly when the moisture freezes. Therefore moisture-absorbing materials such as asbestos fibre, rock wool, glass fibre, wood, sawdust, peat moss should not be used in utilidors. Asbestos fibre must not be used anywhere in the system. Expanded and extruded polystyrene and polyurethane are the most common insulations used. Ground or bead polystyrene and foamed in place polyurethane can be used to fill the voids in utilidors. The ground or bead polystyrene allows easy access to pipes for repair and removal. Fire-resistant insulations are preferred, since they reduce the spread of fire along utilidors. Thermal characteristics of various insulations are given in Section 4.

11.3.5 Piping. Leaks from pipes and joints in utilidors may cause water and icing damage to the utilidor, and there is also a danger of contamination and cross-connections. Most public health codes do not allow water and sewer pipes to be in proximity. Waivers or the adoption of new codes are sometimes necessary before utilidors can be used. In the Northwest Territories and Alaska, water and sewer pipes are allowed in a common trench or utilidor, but only if pressure-rated pipes and joints are used and all pipes are tested for zero leakage. Open sewers are not allowed, and sewer cleanouts must be capped.

Most types of pipe materials have been used in utilidors. Each has advantages and disadvantages. Rigid pipes with welded or equivalent joints may not require as much support or hydraulic thrust blocking at bends, hydrants, and intersections, but such design must allow for thermal expansion.

11.3.6 Cross-Sectional Design. The cross-sectional design and piping arrangements depend upon not only the access required, but also the type, size, and number of pipes to be included in the utilidor. Other considerations may include the routing of piping and the standard sizes of building materials. The utilidor size is often dictated by the maximum dimensions of pipe appurtenances, repair clamps and joints, and the spacing of pipes necessary at deflections and intersections. Pipes with smooth joints may be placed closer together, and a smaller cross section may be used. Installation, repairs, and maintenance requirements must also be considered.

11.3.7 Prefabricated Utilidors. Most utilidors are prefabricated to some extent, but all require some field construction. The primary advantage of prefabrication is the reduction in assembly time, which in turn reduces labor costs and facilitates installation within the short construction season. But prefabricated utilidors generally have high material costs. Prefabricated utilidors commonly combine the functions of the basic utilidor components. The pipes, rigid insulation, and exterior casing provide beam strength, rigidity, and thermal and physical protection.

One type of prefabricated utilidor consists of a carrier pipe, polyurethane foam insulation, and a bonded exterior casing of sheet or corrugated metal, or plastic with or without glass-fiber reinforcement, depending on the strength requirements (Figures 11-1(f) and (g)). This system is commonly used for small-diameter pipes and service connections to buildings. A completely prefabricated, glass-fibre-reinforced plastic, two-pipe utilidor system is illustrated in Figure 11-1(h). It is longitudinally segmented and has staggered joints to allow removal of individual pipes.

11.4 Appurtenances

Appurtenances such as hydrants, cleanouts, and bends are prefabricated modules that are inserted into the system where required (Gilpin and Faulkner, 1977).

Some appurtenances for piping systems and utilidors, however, must be specially designed or adapted. This applies particularly for above-ground utilidors, where special hydrants, valves, cleanouts, and bends may be necessary.

11.4.1 Above-Ground Hydrants.
In-line hydrants located directly on a tee in the water main are recommended. They may be installed without additional freeze protection, if the barrel is short enough that heat from the water main can prevent the valve from freezing. Leakage through the valve must be prevented since the freezing of water may damage the hydrant and make thawing necessary before use. Building-type fire hydrant outlets and butterfly valves have been used because of their small size and light weight. One design is illustrated in Figure 11-6.

The hydrant enclosure must be rugged, well insulated, and appropriately marked and painted. Access to the hydrants must be quick and easy, but it must also discourage vandalism (James, 1980).

11.4.2 Sewer Access.
For above-ground sewers and for water and sewer pipes within a utilidor, the sewer access cleanouts must be sealed to prevent cross-contamination. The flanged tees for pipes larger than 200 mm in diameter usually provide an adequate opening to insert cleaning or thawing equipment. Standard fittings for smaller pipes do not provide adequate access in both directions. Special fittings with larger slot openings that can be sealed have been used (Figure 11-7).

11.4.3 Vaults.
Vaults are the above-ground enclosures containing hydrants, valves, thrust blocks, intersections, bends, and other piping system appurtenances, including recirculation pumps, heaters, and controls. Access to these appurtenances is provided through the vaults, which may be either slight enlargements of the utilidor or small buildings. Usually they are individually designed and fabri-

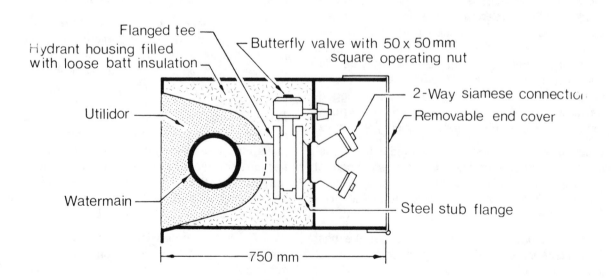

FIGURE 11-6   ABOVE-GROUND UTILIDOR HYDRANT

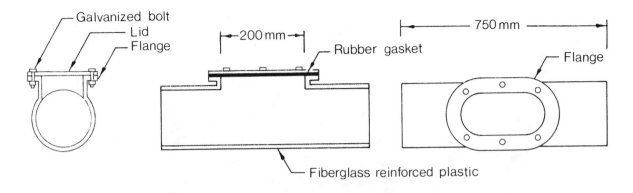

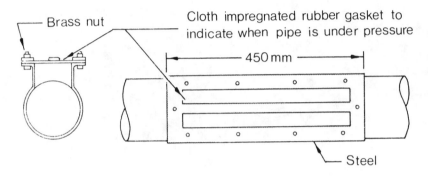

FIGURE 11-7   SEWER CLEANOUTS

cated. The use of thrust blocking and vaults at bends must be examined for each situation even where small-diameter pipes or rigid pipes are used. The vaults contain the expensive piping system appurtenances which can be a significant portion of the total utilidor cost; vaulted appurtenances accounted for 30percent of the cost for the two-pipe wooden box utilidor shown in Figure 11-1(b).

11.4.4   Utilidor Crossings.   Pedestrian and vehicle crossings must be provided where above-ground utilidors are used. The cost of these crossings depends on the size of the utilidor and its height above the ground. Large or high utilidors require bridge-type structures for road overpasses (Figure 11-3), for which the unit cost was approximately $35,000 (Cdn.) in Inuvik, NWT, in 1977. Smaller utilidors can be protected by less expensive corrugated metal pipe culverts.

The utilidor and road layout should minimize the required number of crossings. The plans for the locations of utilidor overpasses must take the roadway and drainage system into consideration. Steep approaches to the crossing can impair a driver's visibility. Long approaches can disrupt the surface

drainage system, and the overpasses tend to become drainage paths that accumulate garbage.

Utilidors and piping can also be elevated above roadways or buried at road crossings. These alternatives may require expensive lift stations, which can impede the complete drainage of the pipes and utilidor. Underpasses can be excavated; however, this is difficult and expensive in areas of ice-rich permafrost. Roadway excavation, permafrost protection, and utilidor reinforcement for an underpass in Inuvik, NWT, cost $70000 (Cdn.) in 1974.

Pedestrian crossings can be a part of the roadway crossings, but separate wooden stairways may be required at certain locations (Figure 11-8).

11.5   Thermal Considerations

The placement of pipes within a utilidor provides possible thermal benefits in that the proximity of water mains to warm sewer pipes and heating pipes, if they are included, reduces the risk of freezing. To reduce heat loss generally, the surface area of the utilidor has to be minimized. The size and shape of a utilidor may also be dictated by considerations other than heat loss. Most utilidors are no more thermally efficient than individual pipes with the same

FIGURE 11-8   PEDESTRIAN STAIRWAY OVER UTILIDOR

amount of annular insulation, and utilidors containing large air spaces are even less efficient. The effects of utilidor shape and size on heat loss are illustrated in Figures 4-12 and 4-13. Information required to estimate heat loss and freeze-up time for pipes and utilidors is presented in Section 4.

All exposed utilidor surfaces should be insulated. Thermal breaks or penetrations should be isolated from the pipes, and insulated flanges and extra insulation at pipe anchors used. Additional insulation should be provided at appurtenances and at vaults with surface areas larger than those of the utilidor. Freeze protection provided for utilidor piping can be similar to that used for single pipes; for example, heat tracing can be used to maintain a minimum temperature. Temperature control within the utilidor is difficult when heating or domestic hot-water lines are included.

While soil cover and snow provide some natural insulation for buried utilidors, shallow-buried and above-ground piping and utilidors are subject to extreme air temperatures. Therefore, they must be designed for the worst case in expected temperatures and wind conditions. The design must take into account:

- short freeze-up time;

- high maximum rate of heat loss;

- high annual heat loss; and

- expansion and contraction caused by changes in air temperature.

A warning about the last point: the most critical expansion and contraction problems occur when the system is started up, or when maintenance or emergencies require that the system be shut down and drained. Two conditions require calculations: (a) the maximum movement due to temperature changes and (b) the maximum stress, if movement is restrained. With metal pipe, it is generally not practical to restrain thermal movements. Maximum movement can be provided by the use of compression or sleeve-type couplings, "snaking" the pipe, expansion joints, or expansion loops. Most in-line expansion joints do not perform adequately under freezing conditions; however, a free-flexing bellows joint operates even with residual water frozen inside (Gilpin and Faulkner, 1977).

11.6 Maintenance

Repairs or replacement of piping and subsequent reclosure of the utilidor must be considered in the cross-sectional design and the materials selected. Extra materials and components used within the system must be available on-site. Standardization of components, materials, and design greatly facilitates maintenance.

Below-ground utilidors should be of the walkway-type for easy access. Alternatively, no air space should be provided in shallow-buried utilidors where access is gained by excavation. Above-ground utilidors need not have walkways. Access to pipes and appurtenances in them can be facilitated by removable panels.

Exposed utilidors need extra maintenance because of vandalism, accidents, and weathering. This must be considered in the design and the materials selected.

Repairs and service connections must be carried out only by trained personnel.

11.7 Costs

Capital costs, maintenance and heating requirements, and service life are important factors in utilidor design and in the selection of materials (Gamble and Lukomskyj, 1975; Leitch and Heinke, 1970). Capital costs can be reduced by lowering standards; however, this may be offset by higher operating and replacement costs. Above-ground utilidors must be particularly rugged to withstand vandalism and the rigorous climatic conditions. Numerous "low-cost" utilidors have not survived their intended life span.

The capital costs for a utilidor depend on the number, size, and function of the enclosed pipes, the degree of prefabrication, the foundation requirements, and local conditions. The cost breakdown for a two-pipe, wooden-box utilidor on piles in Inuvik, NWT (Figure 11-1(b)) is:

Engineering - design	7%	
- field supervision	10%	
total	17%	
Materials (including transportation)	33%	
Construction	50%	

The cost for the small utilidor in Figure 11-1(d) was approximately $230/m (Cdn.) in 1977. The wooden-box utilidor containing water and sewer piping, constructed on piles in a subdivision of Inuvik, NWT in 1976, cost $600/m (Cdn.) for straight portions. However, average costs range from $670/m to $805/m when the costs for vaults and road crossings are included. Large utilidors with central heating lines are more expensive. Costs for the utilidor shown in Figure 11-1(a) were over $1,200/m (Cdn.) in 1976.

Estimated annual maintenance costs in 1974 for utilidors in Inuvik, NWT, ranged from $34 to $91 (Cdn.) for each service connection. Recent cost estimates (1995) for small above-ground utilidors in Alaska have ranged from $700 per metre in Noorvik to $800 per metre in St. Michael and Savoonga.

11.8 References

Cerutti, J.L., W.L. Zirjacks, C.T. Hwang, and D.E. Bruggers. 1982. Underground utilities in Barrow, Alaska. In: *Proceedings, Utilities Delivery in Cold Regions.* Environmental Protection Service Rep. No. EPS 3-WP-82-6, Ottawa, Ontario, 358-382.

Cooper, P.F. 1968. *Engineering Notes on Two Utilidors.* Northern Science Research Group, Department of Indian Affairs and Northern Development, Ottawa, Ontario.

Gamble, J.J. 1977. Unlocking the utilidor: Northern utilities design and cost analysis. In: Proceedings, Utilities Delivery in Arctic Regions. Environmental Protection Service Rep. No. EPS 3-WP-77-1, Ottawa, Ontario, 99-130.

Gamble, D.J. and P. Lukomskyj 1975. Utilidors in the Canadian North. *Canadian Journal of Civil Engineering*, 22: 162-168.

Gilpin, R.R. and M.G. Faulkner 1977. Expansion joints for low-temperature above-ground water piping system. In: *Proceedings, Utilities Delivery in Arctic Regions.* Environmental Protection Service Rep. No. EPS 3-WP-77-1, Ottawa, Ontario, 346-363.

Hoffman, C.R. 1971. *Above-Ground Utilidor Piping System for Cold-Weather Regions.* U.S. Naval Civil Engineering Laboratory Technical Report R734, Port Hueneme, California.

Hull, J.A. 1980. Thermodynamic analysis of the water distribution system in Inuvik, NWT. In: *Proceedings, Utilities Delivery in Northern Regions.* Environmental Protection Service Rep. No. EPS 3-WP-80-5, Ottawa, Ontario, 332-346.

James, F.W. 1980. Critical evaluation of insulated shallow buried pipe systems in the Northwest Territories. In: *Proceedings, Utilities Delivery in Northern Regions*. Environmental Protection Service Rep. No. EPS 3-WP-80-5, Ottawa, Ontario, 150-186.

Leitch, A.F. and G.W. Heinke. 1970. *Comparison of Utilidors in Inuvik, NWT*. Department of Civil Engineering, University of Toronto, Toronto, Ontario.

Leman, L.D., A.L. Storbo, J.A. Crum, and G.L. Eddy. 1978. Underground utilidors in Nome, Alaska. In: *Applied Techniques for Cold Environments*. American Society of Civil Engineers, New York, 501-512.

Porkhaev, G.V. 1965. *Underground Utility Lines*. National Research Council of Canada, Technical Translation TT-1221, Ottawa, Ontario.

Reed, S.C. 1977. Field performance of a subarctic utilidor. In: *Proceedings, Utilities Delivery in Arctic Regions*. Environmental Protection Service Rep. No. EPS 3-WP-77-1, Ottawa, Ontario, 448-568.

Reid, B. 1977. Some Technical aspects of the Alaska village demonstration project. In: *Proceedings, Utilities Delivery in Arctic Regions*. Environmental Protection Service Rep. No. EPS 3-WP-77-1, Ottawa, Ontario, 391-438.

Rogness, D.R. and W. Ryan. 1977. Vacuum sewage collection in the Arctic, Noorvik, Alaska: A case study. In: *Proceedings, Utilities Delivery in Arctic Regions*. Environmental Protection Service Rep. No. EPS 3-WP-77-1, Ottawa, Ontario, 505-552.

Slipchenko, W. 1970. *Handbook of Water Utilities, Sewers, and Heating Networks Designed for Settlements in Permafrost Regions*. Northern Science Research Group, Department of Indian Affairs and Northern Development, Ottawa, Ontario, NSRG70-1.

Smith, C.K. 1970. *Thermal Analysis of Above-Ground Liquid Distribution Systems (Utilidors) for Polar Applications*. U.S. Naval Civil Engineering Laboratory, Technical Note N-1090, Port Hueneme, California.

Tobiasson, W. 1971. *Utility Tunnel Experience in Cold Regions*. American Public Works Association, Special Report No. 41, 125-138.

Zirjacks, W.L. and C.T. Hwang. 1983. Underground utilidors at Barrow, Alaska: A two-year history. In: *Proceedings Permafrost, Fourth International Conference*. National Academy Press, Washington, D.C., 1513-1517.

SECTION 12

CENTRAL FACILITIES

3rd Edition Steering Committee Coordinator

James A. Crum

Special Contributions

Robin Dalton

John Warren

Section 12   Table of Contents

Section 12   List of Figures

Section 12   List of Tables

12   CENTRAL FACILITIES

12.1   Definition

Many cold-climate communities are located where it is difficult to construct and operate water and sewer systems that provide service to individual buildings. Ice-rich permafrost, rock, lack of roads for truck haul, high capital and operating and maintenance (O&M) costs, and low-density housing are among the factors limiting the use of water supply and wastewater systems serving individual homes. In these situations, central facilities (often called "washeterias") have been used to provide several useful services. These include a source for treated water, showers, washing machines, clothes dryers, saunas, and rest rooms.

Over 65 central facilities have been built in rural Alaska. However, very few have been built in the Canadian Arctic because they do not meet the minimum water and sewer service levels, and government O&M subsidies are available to support higher levels of service. Also, favorable environmental conditions are available in the Northwest Territories (i.e., roads) so that the subsidy policies have resulted in truck hauled water and sewer service to individual homes where clothes washing, bathing, and showers are available. Almost all of the Alaskan facilities are located in areas where piped utilities are not presently available; the central facilities are considered interim facilities to be followed later by piped facilities.

This section outlines the design considerations for providing a central facility in a community where piped water and sewer services to individual houses are not presently economically or physically feasible. The following discussion details the level of service to be provided, sizing of utilities and services, energy conservation, fire protection, construction techniques, and costs.

12.2   Planning for a Central Facility

Central facilities can provide many services. Defining the types of service to be provided is one of the important considerations in the planning stages of a project. Decisions made during planning should consider:

- available resources including water supply, land, waste disposal site;

- operation, management, and maintenance capability of the community;

- local desires expressed by the community's willingness to participate in the planning, design, construction and operation of the completed facility;

- social customs, needs, and services that may be addressed by the central facility; and

- assistance available to the community including financial, administrative management, training, etc.

Other useful information to consider when planning a central facility can be found in Section 2.

Realistic consideration of these factors with the meaningful participation and involvement of the community in the decision-making process is essential to a successful project.

Consultation with people in the community, perhaps with the use of a planning questionnaire (Arctic Environmental Engineers, 1978) is essential to providing the best appropriate service. The consultation should provide answers to such questions as:

- How many hours per day and per week should the facility be open?

- How often will people use the services provided per day and per week?

- How many people are willing to pay for the services and how will the payment be made?

Given the community responses to these questions, some design planning estimates can be made.

The minimum level of service is a watering point at which people obtain safe water to drink. A full service facility provides:

- a watering point to obtain drinking water;

- a place to wash and dry clothes;

- bathing and toilet facilities; and

- safe treatment and disposal of wastes from the central facility (including toilets and washing machines, showers, and wash basins).

Saunas have been installed in some facilities to supplement the use of showers. In some parts of the North, especially among native people, saunas are a preferred method of bathing and a requested feature.

Central facilities may be constructed as the first stage of a more complete water and sewer system with service provided to individual buildings. Such central facilities should be designed so that its basic water supply and wastewater treatment systems may be expanded to provide increased service. Typical expanded services might be a truck-haul system or, in some cases, a water and sewer system piped to some individual buildings or to all community buildings.

Whenever possible, arrangements should be made for the central facility to serve buildings such as schools, health clinics, and community centers with piped building service. Since the revenue produced by serving these institutional buildings can be substantial and can offset the income required to operate the central facility, it can therefore reduce the individual user charges. The use of waste heat from local power plants can reduce the heating fuel required and increase system efficiency.

12.3 Design Space Considerations

The amount of floor space required for a central facility depends on the services to be provided and on the needed mechanical equipment area. Also, it will depend on how much water storage is to be provided within the central facility. Central facilities in Alaska have varied from 80 m^2 to 325 m^2 of total floor space. The increased area is often needed when water storage tanks or more sophisticated water treatment equipment are located inside the building. Table 12-1 provides the suggested number of service units for various populations. These numbers should be reviewed taking into consideration local needs and requirements.

Lack of sufficient space was a serious problem in some early central facilities. The early emphasis on saving space resulted in problems for both operators and users. Equipment and piping were so confined that vital repair and maintenance activities could not be undertaken without moving piping and equipment which often resulted in maintenance not being performed. Space considerations must now consider fire codes and access requirements for persons with disabilities.

12.4 Construction Techniques

Central facilities have been built in a modular format and shipped complete for erection on a constructed foundation. Other facilities have been built from components, or have been "stick built" on site. Each of these techniques has costs and benefits in construction times, quality, and local employment opportunities. When modular construction was used, there was a tendency to compact the equipment and service areas in the modules to facilitate shipping and reduce costs. This crowded space discourages use of the facility. See Section 2 for more details on construction techniques.

12.5 Water and Wastewater Service

A central facility must provide water that is chemically and bacteriologically safe, that is more convenient than other sources, and that tastes and appears better than alternative sources. The facility must also provide a sanitary means of wastewater disposal for wastewater generated within the central facility and for wastewater transported to the central facility.

12.5.1 Water Supply and Treatment. Providing a reliable water supply along with necessary treatment is essential to encourage use of the facility. The highest quality of raw water should be selected to be delivered to the central facility for final treatment so treatment costs and complexity are low-

TABLE 12-1   RECOMMENDED NUMBER OF CENTRAL FACILITY SERVICE UNITS

| Population | Washers | Dryers | Bathrooms | |
			ADA*	Standard
150	4	3	1	1
< 500	6	4	1	2
> 500	8	6	1	3

* American Disabilities Act handicap accessible

ered. Generally, water supply and treatment requirements for central facilities are not particularly unique compared with other water treatment processes needed in cold regions. Details on water supply and treatment systems are discussed in Sections 5, 6 and 7.

12.5.2 Water Storage. Storage capacity for treated water will depend on the availability and reliability of the water source, source flow capacity, and fire flow requirements. Storage may be adequate for a design flow of less than one day to over nine months when it must be drawn from an intermittent supply. The amount of storage should be sufficient to ensure a minimum level of service for the duration of any anticipated power outage or failure (such as a supply pump breakdown). Storage capacity amounting to about one day's total design flow has been used frequently for central facilities where water sources are reasonably available and where water treatment requirements are not unusually complex. Water conservation can ensure several days' reserve to provide minimum services such as drinking water supply and showers. Details for sizing and designing water storage systems can be found in Section 7.

12.5.3 Design Water Flows. Selecting the type of service (showers, laundry, saunas, sewage dump stations, washrooms, etc.) must be done on a case by case basis. Water use data collected on several washeterias indicated that practical treated water demands range between 9 to 31 L/(p•d) with an average demand of 18 L/(p•d) (Warren, 1993).

12.5.4 Wastewater Treatment and Disposal. Detailed wastewater treatment and disposal alternatives are discussed in Section 10. These alternatives, with modification, are appropriate for central facilities. For example, special consideration must be given to treatment of dilute laundry wastes, variations in effluent temperatures, foaming, shock loads from honey bucket wastes, and peak hydraulic flows during specific operating periods.

A major portion of the wastewater flow in the typical central facility comes from the washing machines. Laundry wastewater resembles domestic wastewater in many ways but it does not contain all the essential nutrients to sustain the organisms necessary for effective biological treatment. This is not a problem where domestic wastewater from honey buckets and toilets is added to the waste stream.

Sudsing detergents can cause excessive foaming in wastewater flows, so the use of low-sudsing de-tergents may be needed to control this problem. The high heat content of central facility wastes due to laundry (hot), and showers (warm) provides ideal flushing and warming to central sewer systems when it is located at the upper ends of arctic sewers. Forms of dosing mechanisms have been built for washer wastes to create surge flows.

Central facilities have been designed with honey bucket waste dump stations, but experience shows that using them is difficult. Many people find it difficult to carry their honey buckets to the central facility for disposal and prefer instead, to dump wastes at a disposal point nearer to their homes. A second problem is that honey buckets often contain material other than human waste (i.e., cans and trash), thereby requiring a separation area prior to the honey bucket sewer discharge or pump station. The disposal of these types of waste has the same considerations as that for a community (Section 10).

12.6 Equipment and Mechanical Aspects

When facilities are constructed in remote arctic areas, factors such as the isolation, an unforgiving climate, the low skills of the operators, and unreliable electrical power service may require that the design have less complex systems and backup systems to provide an acceptable level of reliability. However, some degree of complexity is necessary to address the difficult problems of water supply and waste disposal in arctic climates.

As a practical matter, a minimum of three washers and three dryers (Figure 12-1) should be provided and a minimum of two bathrooms, regardless of community size. This allows service to continue uninterrupted when one or more of these machines are out of service and allows for peak usage periods.

12.6.1 Heating System. An important design consideration for central facilities is choosing a heating system. The basic questions to be answered are:

- What types of heating are required (building heating, clothes drying, sauna, water heating, etc.)?

- What type of system is most appropriate – hot water, hot air, electric, or steam?

- What type of fuel system is most desirable, economical, and available?

- Should there be a single source of heat to meet the heating needs or should there be

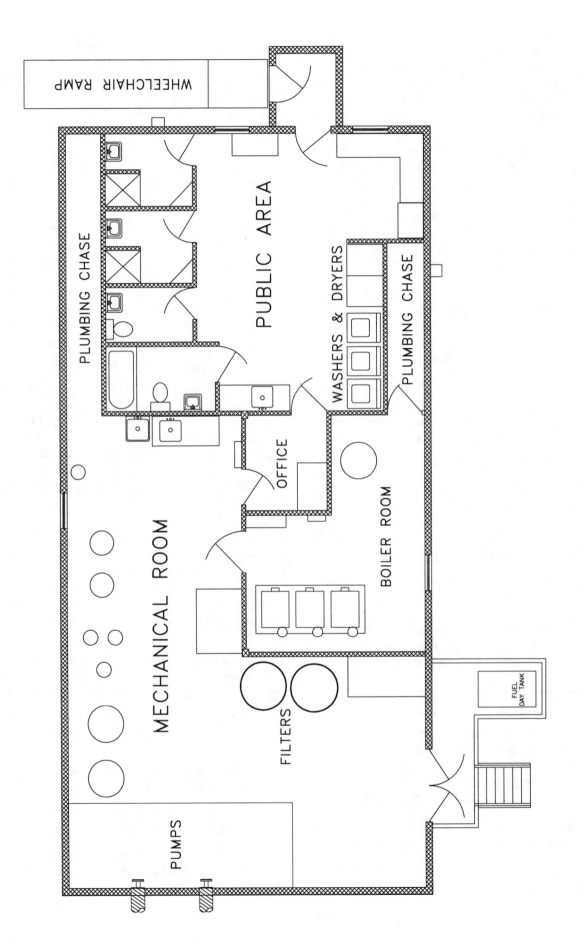

FIGURE 12-1 WASHETERIA FLOOR PLAN

separate sources of heat designed specifically for the point of need?

- How should heat recovery be included in the facility?

Answers to these questions require analysis of both the heating needs and the relative difficulty of maintaining the mechanical plant in a central facility.

Choosing the Type of Heating System. Circulating hot-water heating systems are the most popular in remote areas, probably because they are understood better than other systems. The primary disadvantage of a hot-water heating system is its susceptibility to damage during facility freeze-ups. Propylene glycol can be mixed with water in the system. This fluid can prevent hydronic water system freeze-ups, but requires more attention in handling and maintenance, and is slightly less efficient (10 to 20 percent) in heat exchange properties. Corrosion control inhibitors are often used, but they tend to break down at high temperatures, and maintaining the proper concentration requires testing by the operators.

Hot-water systems cannot provide the higher quality heat needed for saunas and can only marginally meet the higher temperatures required for clothes dryers. One way to achieve the higher quality heat for all services in a central facility and to avoid damage due to freezing is to use special organic fluids instead of water or glycol. These systems operate on the same principle as hot water but at higher temperatures. The fluids can be fairly expensive, and the plumbing system must be more elaborate. A sloppy installation of the plumbing system can make repair difficult, and faulty joints can leak hazardous chemicals into the central facility (Puchtler et al., 1976).

Steam has a relatively high capacity to carry heat; hence, a steam-heating system can readily meet all the needs in a central facility. The main disadvantages of a steam system are that people in remote areas are generally unfamiliar with the higher temperatures and mechanics of steam, and the hot pressurized vapor is more hazardous than hot air or hot water. This makes operation of the steam system more difficult and can result in higher maintenance costs and downtime compared with hot-air or hot-water systems. Pipe damage due to freezing is generally limited to low points or restrictions in the piping systems.

Hot air does not have the heat carrying capacity of water or steam, but it can be used effectively for dryers if separate heat sources are used and the furnace is close to the place where the heat is needed. Hot-air furnaces can also be used for building heat, although preheating make-up air and controlling building pressure becomes a problem. Also, ducting consumes more space than the plumbing for hot-water systems.

Regardless of the type of heating system selected, a nonelectric pot burner stove should be supplied for standby space heating in case an extended power failure or primary heating system failure occurs.

Central Versus Multiple Heat Sources. A central heat source, i.e., one duplex boiler system serving all heating needs for a central facility offers simplified maintenance requirements and reduced fire hazards compared with multiple heat sources, e.g., separate heating units for building heat, hot-water dryers, and other services. However, balancing such a system can be complex. Standby capability for a central heat source can be partially achieved by providing complete spare burners. Control systems for distributing heat from a single source can often be more complex than individual control for each of several heat sources.

Multiple heat sources meeting the specialized heating needs in the central facility is theoretically more efficient than a central source, since they produce different grades of heat, each designed for maximum efficiency. If multiple sources are selected, compatible equipment with interchangeable parts should be used. This reduces the need for a large inventory of spare parts and simplifies maintenance.

12.6.2 Incineration. Incinerators have been installed in central facilities for the primary purpose of disposing of sludges from waste treatment plants, honey-bucket wastes, and other solid wastes in a sanitary manner. Incineration theoretically offers an ideal solution to the problem of organic waste disposal, essentially eliminating the adverse environmental and health effects from these wastes. However, incinerators installed in central facilities are much too complex and costly to operate, rarely last more than a year, and are not recommended.

12.6.3 Washers. Commercial-type washers should be used. They range in capacity (weight of dry clothes that can be placed into them) from 6.8 kg up to about 16 kg. An average washer cycle is about 40 minutes, including loading and unloading time. The smaller units are significantly cheaper, but larger ones have proven particularly useful since they can handle bulky items such as sleeping bags, blankets, and parkas.

Horizontal axis washers tend to vibrate during use (Puchler et al., 1976). Therefore, they must be properly evaluated and incorporate a vibration dampening support system to prevent vibration damage to the building and foundation. Vertical-axis washers do not vibrate as much, but they are normally available only in the smaller sizes. In addition, top-loading, vertical-axis washers use about 40 percent more water for an average wash load than the front-loading, horizontal-axis machines (Cameron and Armstrong, 1979).

12.6.4 Dryers. Many types of dryers are available: hot water, electric, steam, hot air, and hot liquid. The choice should be based on operational costs and past experience.

Electrically operated dryers are the easiest to maintain but they are also the most expensive to operate. According to the Alaska Department of Environmental Conservation (ADEC) (Dowl Engineers, 1975), heat derived from electricity generated from fuel oil can cost over ten times as much as heat derived from bulk fuel oil. ADEC (Arctic Environmental Engineers, 1978) calculated that for an 8.2 kg capacity dryer requiring 5.9 kW and consuming electric power for 45 minutes/ load at $0.20/kWh (a very low electrical cost in remote areas), the electrical cost for drying alone would be $0.89/load. Comparable heating costs based on the use of oil were calculated to be about $0.11/load.

Hot-water operated dryers are a favored choice in central facilities because they can be connected directly to the hot-water furnace, and operators in remote communities are more familiar with hot-water heating systems. However, to provide sufficient heat to the dryers, the hot-water system must be operated at its upper limit for temperature and pressure and be balanced between the dryer units. This increases the chance of vapor locks in the system because of "flashing" (liquid to vapor).

Steam-operated dryers would appear to be a good choice because of the excellent heat carrying capacity of steam. However, few people in the remote areas are familiar with the principles and operating characteristics of steam systems. In addition, the relatively high operating temperatures and pressures require more maintenance, and the heat exchanger coils tend to require frequent cleaning to maintain efficiency.

Hot organic fluids such as "Dowtherm" or "Therminol" can also supply dryer heat (Reid, 1973; 1977). These systems can be efficient, although they operate at relatively high temperatures (177°C). One advantage, besides efficiency, is that freezing does not damage this type of heating system. But numerous operational, maintenance, and safety problems can occur, particularly if initial construction is substandard (Cameron and Armstrong, 1979).

A hot-air furnace with appropriate duct work can provide dryer heat. Such a system requires one less heat exchanger than steam-operated dryers, but an extra furnace is required (Arctic Environmental Engineers, 1978). It cannot be damaged by freezing, but ducting must be well insulated to reduce heat losses. (A comparison of advantages and disadvantages between separate and multi-purpose furnaces is given in Section 12.6.1). Dryers heated by hot-air furnaces are commercially available, but most systems are converted steam dryers connected to hot-air furnaces.

An extractor, which draws water centrifugally from the laundry, reduces dryer time, and saves energy. They tend to vibrate so it is essential to provide a solid base to which they can be securely anchored.

Dryers should be sized at least 1.5 to 2 times larger than the washers because people tend to put more than one washer load in a dryer. The appropriate dryer size can be determined after the washers have been selected.

Drying cycle times vary with the type of system used. Manufacturers' literature can provide this information. For hot-water systems, the cycle is about 45 minutes. Lower humidity during the winter leads to faster drying times. An additional 10 minutes to load and unload clothes can be used when estimating total cycle times.

Many acceptable brands of washers and dryers are on the market. After the type and size of equipment have been selected, the choice of model should be based on ease of repair and availability of spare parts.

12.6.5 Showers. Shower usage in public central facilities depends greatly on the cultural preference in the community. Some communities prefer steam baths to showers. For example, in the communities of the Lower Kuskokwim River in Alaska where steam baths are a traditional method of keeping clean, the rate of shower use is low. In interior Alaska, which does not have the steam bath tradition, use rates vary from eight to twelve showers/

person/month (rates are in communities without home showers).

Shower water consumption should be efficient to the extent practicable for several reasons:

- cost of heating water, and
- cost of water treatment.

Where the water source is limited, the efficient use of water is critical and extraordinary measures are necessary to minimize the waste of water.

Timers, pressure regulators, mixing valves, and low-flow shower heads are the most useful devices for conserving water (Section 14). Reid (1977) reports adequate and satisfying showers, using only 23 L/shower with these devices. Conventional shower heads use about 25 L/minute, whereas low-flow shower heads use only 5 to 12 L/minute.

12.6.6 Saunas. The sauna is a form of traditional steam bath accepted widely in the culture of the indigenous northern peoples. The operation of a sauna in the community's central facility, where it is a local custom, can provide a desired service and can generate substantial revenues. However, public saunas require a high-quality heat, reliable maintenance to prevent building deterioration, and consistent application of sanitation practices, so they must be managed properly. With a regulated heat source, plenty of insulation, adequate vapor barriers and venting of moisture, and consistent cleaning of the facility, a sauna can provide a desired service and produce sufficient revenue to help support the total central facility operation. Experience has shown that the saunas require adjacent showers for users to cool off as well as to rinse.

12.6.7 Restrooms. Most central facilities that provide more than just a watering point have restrooms for men and women, each with a wash basin and toilet. Some utilize unisex restrooms if space is limited.

Wash basins are required with toilet facilities. They should have automatic closing valves to minimize water wastage.

Where water is readily available at low cost, flush-tank toilets are appropriate. They are simple to operate and require no additional power source. Older flush toilets that use an average of 20 L/flush are wasteful and unnecessary since flush toilets that use only three to six L/flush are readily available (Cameron and Armstrong, 1979). The operation and maintenance costs are the same, but because of

the reduced volume, the costs of procuring, treating, and pumping the water, and of disposing of the wastewater are less.

Where water is limited, other types of toilets should be considered. Air pressure discharge types or vacuum toilets use only about 1.5 L/flush (Cameron and Armstrong, 1979). Recirculating chemical toilets use only about 0.07 to 0.2 L of water per flush (Cameron and Armstrong, 1979). Units are available which flush mechanically or electrically. Reid (1973) concluded that recirculating toilets installed in central facilities have been of poor design for the application. Refer to Section 14 for details on available toilet units. Substantial water savings can be realized with these systems, but the cost of maintenance may offset the cost of water savings.

Where water is scarce, greywater recycling has been tried for use in flush toilets. This substantially increases both the cost and the mechanical complexity of the facility. Greywater recycling has also led to foaming and odor problems and has generally proven unsatisfactory.

12.7 Heat Recovery

Central facilities offer a unique opportunity to capitalize on waste heat recovery. Many of the building mechanical components either use or produce significant quantities of heat and the relative closeness of components within each facility facilitates heat recovery. Heat recovery from dryer exhaust air via a heat exchanger provides recoverable waste heat used to supplement heating needs within the facility. Heat recovery often provides substantial low grade (low temperature) heat which is useful either for building heat or preheating purposes.

Several central facilities in Alaska are connected to the community power plant heat recovery system. This allows the facilities to utilize available waste heat to preheat water, preheat dryer air, or heat a large water storage tank. Care should be used in selecting and adjusting multiple heat sources to ensure the high grade heat sources do not add heat to the low grade sources. Heat recovery is discussed in Section 17.

12.8 Fire Protection

Municipal fire departments are often inappropriate for protecting central facilities in remote communities. Unless a fire is controlled within minutes after it has started, little can be done to save the building. Therefore, the design emphasis must be on fire resistant construction, early detection, and rapid automatic smothering of the fire. Fire protection equipment for central facilities must include smoke detectors, chemical suppression systems for critical fire hazard areas, and hand operated fire extinguishers. Although not required by code for these types of facilities, sprinkler systems connected to the water storage tank or a small pressure system are recommended. It is often difficult to meet design code requirements for pressure and flow on these sprinklers. However, even minimum sprinkler systems could save the facility.

12.9 Requirements Often Overlooked

One of the more frequently overlooked items for central facilities in remote areas is storage space. Transportation of bulky items such as chemicals and general supplies to remote communities is often limited to once a year. Hence, storage areas need to be sized to accommodate this quantity of supplies. Use of fire resistant cabinets should be considered for chemical storage. Another often overlooked requirement is space to work on pumps and motors and to perform other general repair and maintenance functions. Typically, a workbench with shelves is provided so tools can be stored. A utility sink and desk should be considered in the mechanical area for the operators to conduct tests, repair equipment and keep records. Laundromat areas should be provided with benches and adequate table space for folding clothes.

Vital components must be easily accessible. Piping must be arranged so that it does not interfere with basic maintenance and repair functions. Critical piping joints should not be located in walls; where this is unavoidable, removable panels should be provided. Cramped space in the user's portion of the facility discourages use. Easier access to washers, dryers, and other mechanical equipment can be achieved by including utility closets or removable wall panels in the design of a central facility. Chemicals that release corrosive vapors should be stored in areas designed for corrosive environments.

12.10 Costs

In 1991, the base cost for a full-service central facility including a well supply, moderate water treatment, and adjacent sewage lagoon located in a rural Alaskan community with a population of 400 and road access was $800,000 (US). Problems in water supply, wastewater disposal, and unstable soil conditions and remote shipping can easily add 20 to 50 percent to this figure. The most dramatic cost increase, however, is in water storage in those areas where obtaining water throughout the winter is difficult or expensive. A large storage tank (1.0 to 6.0 ML capacity) may be required to provide adequate storage during winter months.

Operating and maintaining central facilities in the North is costly. Operational costs include fuel, electricity, labor, chemicals, spare parts, and janitorial supplies. Operating revenues are generated from user fees assessed by the community for sauna, laundry, showers and other available services. A central facility does have the advantage of being a "cash and carry" business; the service (wash and dry clothes, shower, etc.) can be obtained only after paying first.

Connecting local schools and other institutions to a central facility and charging for piped water and sewer service can take advantage of any economy of scale to help generate the revenue needed to operate a central facility. Contracts between the community and these users of large amounts of water in the community results in a cost-effective method of providing these services.

Table 12-2 is a comparison of operating costs in central facilities in three Alaskan communities. Operational costs range from $25,000 to $70,000/year (US) depending on the size of community and costs of service. Experience has shown that the facilities generally operate slightly below a break-even point with a slight subsidy from municipal government being common.

12.11 References

Arctic Environmental Engineers. 1978. *Conceptual Design for Tanana, Alaska Facility*. Alaska Department of Environmental Conservation, Juneau, Alaska.

Cameron, J.J. and B. Armstrong. 1979. Water and energy: conservation alternatives for the North. In: *Symposium on Utilities Delivery in Northern Regions*. Environmental Protection Service, Rep. No. 3-WP-80-5 Ottawa, Ontario, 41-88.

Dowl Engineers. 1975. *Design Narrative for Pitkas Point Village Safe Water Facility*. Alaska Department of Environmental Conservation, Juneau, Alaska.

Puchtler, B. et al. 1976. *Water-Related Utilities for Small Communities in Rural Alaska*. U.S. Environmental Protection Agency, Rep. No. EPA-600/3-76-104, Corvallis, Oregon.

TABLE 12-2   EXAMPLES OF CENTRAL FACILITY OPERATING COSTS

	Brevig Mission	White Mountain	Koyuk
Population	138	180	277
Average monthly water usage (gallons)	54,000	90,000	50,000
Yearly O&M Costs			
Electricity	$4,261	$8,647	$5 174
	(20 290 KWH x $0.21/KWH)	(39 305 KWH x $0.22/KWH)	(23 518 KWH x $0.22/KWH)
Fuel	$11,609	$6,261	$7,560
	(5 277 gal x $2.20/gal)	(4 930 gal x $1.27/gal)	(5 400 gal x $1.40/gal)
Labor & benefits	$12,110	$21,304	$20,684
	(21 hr/wk x 52 wk x $11.09/hr)	(34 hr/wk x 52 wk x $12.05/hr)	(32 hr/wk x 52 wk x $12.43/hr)
Parts, supplies, chemicals	$3,383	$1,532	$1.994
Miscellaneous	$150	$6,684	$215
Total yearly cost	$31,513	$44,426	$35,627
Yearly Revenues			
Coin-ops (washers, dryers, showers)	$13,212	$14,971	$32,065
School, commercial, other	$4,261	$12,812	$733
Total funds collected	$17,473	$27,783	$32,796
Balance	($14,040) Deficit	($16,645) Deficit	($2,829) Deficit
Water prod. cost/gal	$0.04	$0.015	$0.033

Notes:
1. White Mountain has a circulating water line providing water service to homes from the central facility which results in the higher monthly water use indicated.
2. Cost per KWH in bush Alaskan villages has typically been subsidized by the state. KWH costs for Brevig Mission and White Mountain reflect subsidized costs based on 1992 rates.
3. Typically villages purchase bulk fuel for a break in price. Brevig Mission purchased fuel from the local store resulting in the higher cost per gallon indicated.
4. Equipment amortization is not included in the O&M costs shown. O&M costs are based on actual costs for the most part.
5. Water production cost/gallon reflects costs related to pumping, treating and storing treated water for use.

Reid, B.H. 1973. *Alaska Village Demonstration Projects: First Generation of Integrated Utilities for Remote Communities*. U.S. Environmental Protection Agency. Arctic Environmental Research Laboratory, Working Paper No. 22, College, Alaska.

Reid, B.H. 1977. Some technical aspects of the Alaska village demonstration projects. In: *Utilities Delivery in Arctic Regions*. Environmental Protection Service Rep. No. EPS 3-WP-77-1, Ottawa, Ontario, 391-438.

Warren, J.A. 1993. *Dryer Selection and Design for Alaskan Village Washeterias*. Technical Report, Alaska Area Native Health Service.

12.12 Bibliography

Alaska Area Native Health Service. 1985. *Washeteria Design*. Douglas Marx, Unpublished memoradum, Anchorage, Alaska.

Bailey, J.R. et al. 1969. *A Study of Flow Reduction and Treatment of Waste Water from Households*. Water Pollution Control Research Series 11050 FKE, Deptartment of Health, Education and Welfare, Washington, D.C.

Brown, C.K., et al. 1975. *Conceptual Design of an Environmental Service Module*. Report No.75-01 for Defense and Civil Institute of Environmental Medicine. Ontario Research Foundation, Mississauga, Ontario.

Given P.W. and H.G. Chambers. 1976. Workcamp sewage disposal, washcar - incinerator complex, Fort Simpson, NWT. In: *Some Problems of Solid and Liquid Waste Disposal in the Northern Environment*. Environmental Protection Service Rep. No. EPS-4-NW-76-2, Northwest Region, Edmonton, Alberta.

Mecklinger. 1977. *Servicing of Arctic Work Camps*. Department of Civil Engineering, University of Toronto, Toronto, Canada.

Nehlsen, W.R. 1962. *A Development Program for Polar Camp Sanitation*. Armed Forces Technical Information Agency, U.S. Naval Civil Engineering Laboratory Technical Note 476, Port Hueneme, California.

Sargent, J.W. and J.W. Scribner. 1976. *Village Safe Water Project in Alaska - Case Studies*. Alaska Department of Environmental Conservation, Juneau, Alaska.

U.S. Environmental Protection Agency. 1973. *Alaska Village Demonstration Projects*. Report to the Congress, Washington, D.C.

SECTION 13

REMOTE CAMPS

3rd Edition Steering Committee Coordinator

William L. Ryan

3rd Edition Principal Author

James O'Neill

Special Contribution

William O. Mace

Section 13   Table of Contents

Section 13 List Of Figures

Section 13   List of Tables

13   REMOTE CAMPS

13.1   Introduction

With the search for oil and gas a large number of industrial camps began to appear in Alaska in the early 1960s. The discovery of oil and gas at Prudhoe Bay along Alaska's North Slope in 1968 initiated a flurry of exploration and development activities. In the summer of 1969, a comprehensive survey (Alaskan Department of Health and Welfare, 1969) of 35 arctic camps and 30 inactive sites on the North Slope by Alaskan government officials and researchers reported crowded living conditions, although generally good dining facilities. Twenty-three camps used electric heat and all had some form of pressurized water system. Sewage was almost exclusively dumped on the ground or into shallow pits on the permafrost and garbage was dumped indiscriminately on the ground or into bodies of water. The same lack of planned disposal was also common in many arctic villages.

Concern by government, labor unions, and environmental groups quickly led to greatly improved conditions in remote camps and minimized detrimental effects on the environment. In new camps, housing is comfortable, dining facilities are excellent, recreational facilities are provided, water is safe to drink, package waste disposal systems are available for solid and liquid wastes, and incineration of combustible waste is practiced before disposal to controlled land sites (Kennedy, 1981). Scrutiny of remote commercial facilities and systems developed therein has also led to improved conditions in arctic villages.

13.2   Camp Classification

The terminology used to refer to the specific nature of a camp might vary from industry to industry, but whatever nomenclature is adopted, a camp by size, purpose, and operation conforms to one of the following classes:

- self-propelled;

- temporary;

- seasonally occupied; or

- permanent.

13.2.1   Self-Propelled Camps.   These camps provide housing and support facilities for a small number of people and their equipment. The camps are highly mobile because they can easily be transported on land, by sea, or by air, and may be completely self-contained regarding water supply, wastewater treatment, garbage disposal, and power generation. The camp can be operated during any season of the year. An example of a self-propelled camp would be the "cat trains" utilized by seismic crews. These are self-contained facilities for a relatively small crew. They may move a few kilometres each day or may stay at one location for weeks at a time.

Water is generally supplied from a built-in reservoir providing limited storage. In so far as these camps are highly mobile, site selections should be preplanned so that the camp is placed near available water supply sources: ponds, lakes, streams, clean snow or rivers as needed to maintain water supplies. Treatment of water is generally limited to filtration and disinfection.

Waste can be disposed of by direct incineration, either propane or electric; wastewater is collected in holding tanks or chemical toilets and discharged periodically at a suitable waste disposal site. Incinerating, electric toilets are often utilized in these facilities. These units are high-maintenance items and even though burnable liners are used, require frequent cleaning.

Facilities for water treatment, wastewater treatment, and disposal supplied for a mobile camp must be compact, lightweight, modular, and adapted for installation into a trailer, van, sled, tent, or barge.

Fire protection is generally limited to a supply of fire extinguishers, a smoke alarm system for dormitory areas if the facility consists of service units, and the use of fire-retardant materials. Windows in dormitory units should be large enough to allow for easy escape in the event of fire.

13.3.2   Temporary Camps.   Temporary camps are generally located at a site for the duration of a project and then upon project completion, removed or moved to another site. The size of a camp depends on the personnel requirements of the project; it can vary in size to accommodate 50 to 2,000 people (Figures 13-1, 13-2 and 13-3). Facilities provided depend on the duration of the camp. They usually include separate areas for recreation, commissaries, medical services, housing, food preparation, storage, and dining (Figures 13-4, 13-5).

In establishing temporary camps, on-site fabrication should be kept to a minimum to hasten camp installation. Temporary camps have tended to be modu-

FIGURE 13-1   ROAD/PIPELINE CONSTRUCTION CAMP

FIGURE 13-2   DRILLING RIG CAMP

FIGURE 13-3   126-PERSON CAMP – ALSANDS BASE CAMP, FORT MCMURRAY, ALBERTA

FIGURE 13-4   CAMP DINING AREA OF PREFABRICATED BUILDING

FIGURE 13-5   RECREATION AREA IN PREFABRICATED BUILDING

lar units that are factory-assembled and come to the site complete and ready for occupancy. Once at the site, these units can be installed on leveled gravel pads and connected to an external water supply, sewage collection piping, electrical power, and fuel supply as needed. Water supply and wastewater collection should be provided by above-ground pipe or utilidor systems which allow easy assembly, access, and disassembly to facilitate easy relocation even during periods not ideally suited for moving.

Treatment systems should also be of a modular design to allow for easy installation, connection to service piping and removal for relocation to the next camp location. These systems too can be factory-assembled as transportable units to reduce field setup time.

Alyeska, for example, has temporary camps that are moved onto pump station sites to handle population surges for seasonal projects. They typically rely on the existing water source but bring their own water treatment and wastewater treatment capacity.

13.2.3 Seasonally Abandoned Camps. Seasonally abandoned camps are those that, due to their purpose, are only required seasonally; for example,

logging and often mining camps. The season of operation dictates the requirements for the life support systems. If a camp is to operate only during the summer months, protection of the piping of water supply and sewage collection from freezing need not be extensive. The equipment has to be drainable to prevent ice damage if the camp is abandoned during winter. Traps and low points in piping that may be hard to drain completely can be protected with recreational vehicle (RV) type antifreeze. Use of plastic piping designed for freeze/thaw conditions can reduce seasonal camp start-up expenses. Water supply services must be carefully planned particularly for winter operation. Wastewater treatment and solid waste disposal requirements must be determined according to the camp location, size, and operational period.

13.2.4 Permanent Camps. Permanent camps are developed to serve a long-term function. Examples of this type of facility are highway maintenance camps, pipeline compressor stations, and military and mining operation camps.

Facilities provided for these camps can usually be constructed without concern for ease of disassembly. Utility service piping for water distribution and sewage collection can be buried if soil conditions permit. System designs can use buried reservoirs, surge tanks, and on-site construction of water and wastewater treatment systems. Facilities may be constructed, if soil conditions permit, of concrete floor slabs and designed utility buildings.

Treatment systems should be selected and designed to provide levels of treatment acceptable for the long-term nature of the camp. Water treatment may consist simply of filtration and disinfection, if the water source is of high quality. If the source is of inadequate quality or if the quality fluctuates with seasons, complete physical/chemical treatment and disinfection may be required. Reverse osmosis is today, a suitable, established technology for treating brackish or salt water at remote sites.

13.3 Planning Considerations

The planning of a camp is a multidisciplinary process requiring retrieval of many types of descriptive on- and off-site data including:

- determination of the advantages and disadvantages of various site locations;

- purpose of the camp;

- population;

- duration of stay and personnel fluctuation;

- accessibility of work area;

- proximity of existing transportation corridors;

- soil stability or foundation conditions for heated structures;

- flood potential;

- avalanche danger;

- water availability and quality;

- potential impact on local fish and wildlife habitat; and

- procurement of easements, liens, licenses or other legal permissions to proceed.

In the basic design of camp utilities, energy consumption for heat trace, freeze protection, etc. is typically not an issue. What is needed and desired is trouble-free utility service. The purpose of the camps is construction or resource extraction, not to be a model minimunicipal utility department. Access for ease of maintenance activities (thawing, rodding, etc.) is a must for all utility lines particularly if they are buried. The basic power generation package that comes with the camps is on-line 24 hours a day and the heat-trace load is insignificant in the long run.

In Section 2 (Planning Considerations) many of the specifics related to a well-planned facility are discussed.

13.3.1 Camp Population. Determination of the camp population and stability is probably the most important aspect affecting the design and performance of water and wastewater utilities. The water demand of camp personnel is determined from the average per person consumption rates, fire flow requirements, unusual demand from camp construction activities, and the application of a peaking factor to ensure sufficient supply during high-demand periods.

The amount of wastewater generated is in proportion to the average per person water consumption at camps because of the relatively closed-loop network of a camp (Rayner, 1980). Wastewater collection systems should not receive storm water drainage.

Underestimation of camp population can result in overdemand on the water systems, which can cause failure of the system to operate as designed (Eggener and Tomlinson, 1978). Overestimation of population results in oversizing of equipment, which could place systems in an operating condition which might be difficult to maintain.

Fluctuations in camp populations may be caused from rotational employment requirements such as completion of various trade works, scheduling of different phases of a project or rest periods. Seasonal holidays may cause periods of excessively low population. These situations could define periods when special precautions must be taken to ensure that systems do not freeze. Often creative and innovative procedures need to be developed and implemented to maintain biological populations in wastewater treatment plants during such low flow periods. Population fluctuations of 90% over a week period are not uncommon.

13.3.2 Facilities. Facilities provided in camps are related to the purpose, duration, and size of the camp. Highly mobile camps are usually small and minimally staffed, hence they are generally limited to a basic life support requirement. Large permanent camps might have facilities to meet the needs of all residents from community-sized water and wastewater treatment facilities to swimming pools, tennis courts, and shopping stores (Kennedy, 1981; Sherwood, 1963b).

The purpose of the camp dictates the ancillary facilities provided. The primary reason for constructing a camp is to provide a support base for work activities through the supply of housing and meals. Recreation, laundry, medical services, shops, warehousing, and office areas are provided with increasing camp size and duration. Agreements by construction trades councils and labor relations associations have established criteria for camp facilities in some parts of Canada. These agreements specify minimum requirements for lodging, recreation, commissaries, etc. (see Section 13.3.4).

13.3.3 Location Selection. Selecting a camp location not only requires a multidisciplinary approach but also a trading of related concerns of various areas of interest. Site location as selected by an engineer would provide for good foundation, ease for utility servicing, accessibility, reasonable proximity to a high-quality water supply, and accessibility by transportation routes. Others concerned with the location of a camp may place more importance on its proximity to the work site, necessity for limited movement, or accessibility, regardless of other disadvantages. The final site must be selected by consensus.

13.3.4 Regulations. Little regulation existed originally regarding facilities, construction, location, and movement of camps. Increasingly government, industry, and the public demanded regulations to guide camp construction and operation and to preserve the environment better.

In the Northwest Territories and Yukon, licenses must be obtained for water use, land use, and exploration. These must be obtained from the Department of Indian and Northern Development. They relate to legislated requirements under the Territorial Lands Act and Northern Inland Waters Act. Additionally, the Government of the Northwest Territories and the Yukon Territorial Government have several public health ordinances that pertain to lifestyles at camps.

Similarly in Alaska, lease permits and land development permits must be obtained from the appropriate state departments or the Bureau of Land Management and the Environmental Protection Agency. Water and wastewater systems must be approved by the Alaska Department of Environmental Conservation (ADEC) and operated in accord with all existing state and federal regulations. Systems utilizing surface water or groundwater influenced by surface water must be under the control of a qualified operator as defined by ADEC.

Labor unions together with various employer groups have developed camp rules, regulations, and standards that apply to unionized construction camps. These agreed-upon conditions outline inspection and approved requirements; grievance procedures; temporary and final camp structures; living accommodation, bedding, furniture, storage areas; ablution facilities; recreational facilities; commissary; catering specifications; conduct and procedural rules; and the overall review procedure. Included in these agreements is the requirement for a camp layout to receive review and approval before installation by the Trades Council with issue of a certificate of acceptance for the duration of the particular project (Alberta and NWT Building and Construction Trade Council and Alberta Construction Labour Relations Association, 1980).

The Province of Alberta, the only government to do so, provides a section in its Provincial Building Code on relocatable industrial accommodation which specifies general structural stability, heights and areas, fire safety, health requirements, general safety and identification requirements (Given and Chambers, 1975).

13.4 Water Supply and Treatment

Determination of water supply needs for camps follows the same considerations as those for communities: source identification and evaluation of water quantity, quality, and potential for treatment and distribution. These considerations are universal, although minor modifications to suit the specific needs of a camp may be required.

13.4.1 Estimating Demand. The anticipated demand for water for showers, laundry, food preparation, and waste flushing is commensurate with the anticipated camp population. Accurate population prediction is important for proper facility sizing. Table 13-1 presents a summary of water demand rates for various industrial and military camps. The demand varies considerably, especially the exploration base camp where the water lines are bled to prevent them from freezing.

The lifestyle in most camps is different from that commonly found in cities and villages. Shift work, often 12 hours a day, places severe demands upon life-support systems. A peaking factor should be applied to the average flow value for proper sizing of pumping equipment to handle an excessive demand.

Various peak-to-mean daily flow ratios have been calculated and reported in the literature; the ratios

TABLE 13-1   *WATER DEMAND VALUES FOR VARIOUS CAMPS*

Camp Type	Population	Water Demand	
		Range*	Average*
Drilling camp		83 to 227	132
Base camp (Trink, 1981)		121 to 348	200
Exploration base (Murphy et al., 1977)	40 to 100 w/o bleeding		250
	40 to 100 with bleeding		445
Alaska pipeline construction (Eggener and Tomlinson, 1978)	200 to 1,300		265
Alaska pipeline construction (Murphy et al., 1977)	200 to 400		257
Alaska drilling camp (Alaskan Dept. of Health & Welfare, 1969)			212
Correctional camp (Grainge et al., 1973)	44		
Hydro generation construction camp (Belanger and Bodineau, 1977)	4,000 summer		340**
	2,000 winter		
Artificial island (Heuchert, 1974)			108**
U.S. military camps (Lufkin and Tobiasson, 1969)			
Main base	3,000 to 6,000	442 to 514	514
Ice research camp	25		79
Other camp with snow melt for water supply	96 to 227		121
Other camp with steam to melt snow for water supply	85 to 200		189
Alaska drilling rig camps (North Slope) (Tilsworth and Damron, 1973)			313
Value most frequently quoted	44	227 to 681	149**

* flow rate (L/(p•d))
** wastewater flow rate (L/(p•d))

vary from 1.4 to 1.77 (Lufkin and Tobiasson, 1969; Murphy et al., 1977; Given, 1978). These values do not represent a drastic change from those found for the households in small communities.

In addition to life support, water requirements specific to the work camp activity, for example, equipment washdown, pressure testing, and fire protection must be included in the estimate of total camp water supply.

An evaluation of water usage of various facilities at an Alaskan drilling camp and base camp is shown in Tables 13-2 and 13-3. The percentage of water

TABLE 13-2   ALASKA DRILLING CAMP WATER USAGE – MARCH 1
to APRIL 25, 1971 (Tilsworth and Damron, 1973)

| Water Use | Volume of Water | |
	L/(p•d)	% of Total
Kitchen	26.1	19.7
Toilet and urinals	27.4	20.7
Showers	18.1	13.7
Wash basins	24.4	18.5
Laundry and utility	22.6	17.2
Camp administration	7.0	5.2
Unaccounted use	6.7	5.0
Total	132.3	100.0
Average camp population = 36.2		

consumed for kitchen use was small compared to that used for toilet flushing, showers, and laundry. The excessively high value of water use for the base camp trailer no. 3 reported in Table 13-3 may have been due to the convenient location of the trailer, and that contributed to greater use.

13.4.2 Source Location and Development. Identification of a suitable water supply and its development can be undertaken once the total demand

TABLE 13-3   ALASKA BASE CAMP WATER USAGE – OCTOBER
1971 to APRIL 1972 (Tilsworth and Damron, 1973)

| Water Use | Volume of Water | |
	L/(p•d)	% of Total
Kitchen	48.9	23.9
Wash trailer no. 1	15.9	7.8
Wash trailer no. 2	18.0	8.8
Wash trailer no. 3	62.2	30.4
Cold, cook's quarters	8.3	4.1
Hot, cook's quarters	2.2	1.1
Camp administration	18.6	9.1
Contractors	15.0	7.3
Sewage plant	15.3	7.5
Total	204.4	100.0
Average camp population = 43		

to support all activities of a camp is known. A suitable supply is one that can ensure continual supply throughout the operating period, whether seasonal or year round, regardless of fluctuations in camp population. If a camp population is reduced and lower flow conditions create potential freezing problems, methods for maintaining a continued supply should be incorporated into the design.

Temporary and seasonally abandoned camps need to identify a water source that can meet the needs of the camp for the duration of stay and through the season of operation. Site selection of temporary camps may be in locations where water quality degrades through part of the season. Storing a sufficient supply may be more economical than installing treatment equipment for improving the water quality.

Self-propelled camps need to identify and evaluate water sources along the route to be followed. If water of adequate quality and quantity is not available along the projected route, route changes or hauling of water from other sites and the expense of same must be considered.

More detailed information on water source location and development techniques can be found in Section 5.

13.4.3 Water Supply Treatment. Throughout most of the Arctic and subarctic the water quality varies with location, season, and type of water source. Frequently encountered waters may have high concentrations of iron, manganese, color, odor, turbidity, hardness, or alkalinity (Murphy et al., 1977). The degree and type of treatment required for a camp depends on the source, volume of supply required, and type of camp. In so far as dot (thaw) lakes are often utilized as water sources in the Arctic, special attention needs to be paid to seasonal changes in water quality. These lakes often average a depth of 2.1 to 2.4 m. The lakes freeze to a depth of 2.1 m and as the ice tends to be pure water, the salts continue to concentrate beneath the ice during the winter season. A lake source that tests less than 150 mg/L total dissolved solids in July or August may test greater than 3,000 mg/L total dissolved solids in February or March.

The microbiological characteristics of water sources are of concern. Routine disinfection ensures control of bacteriological contaminants, although it does not ensure freedom from contamination by parasitic organisms. Some microorganisms are not easily destroyed with chlorine disinfection. Some surface

waters in areas of Alaska and northern Canada may contain cysts or protozoans or eggs of tapeworms (28 to 56 mm in size), or both. The eggs can be transmitted into humans and cause hydatid disease (Murphy et al., 1977). The source of these parasitic eggs is from the feces of dogs, rodents, and foxes. It has been shown that these eggs remain infective after exposure to extremely cold temperatures and after ten minutes of exposure to either iodine or chlorine at concentrations of 100 mg/L. Currently the only means of removing these eggs from water is by a filtration process capable of removing particles 25 μm in size (Murphy et al., 1977).

The protozoan cysts and oocysts can cause gastrointestinal infections in humans. A number of animal carriers are possible. Microfiltration is the best practical means of eliminating these from a drinking water source.

Commonly used water treatment processes can vary from filtration through disposable cartridges or multimedia sand and gravel, complete physical/chemical treatment, to reverse osmosis. Ion exchange for treating and softening small flows has been used. Disinfection should be practiced with chlorine, ozone or other appropriate agents.

In designing water treatment equipment, the operating temperatures in which the equipment is to perform and the temperature of the water to be treated must be considered. Higher head losses are suffered in filters, and settling rates are reduced due to higher viscosities for cold water. Ion exchange efficiencies are reduced at low water temperatures, necessitating extra exchange capacity or more frequent regeneration of the exchange media. A more detailed review of these factors and reduced solubilities of water treatment chemicals is provided in Section 6.

The cost of water treatment varies widely with the degree and type of treatment provided. Operator time varies with the system provided and the variability of the raw-water quality. Table 13-4 provides a summary of costs for a water treatment system with a 189,000 L/d capacity, sized to suit the needs of a 400-person camp. The costs include treatment equipment, chemicals, and spare parts. Capital costs are amortized over five years at 15 percent. All equipment and chemicals are F.O.B. Anchorage, Alaska.

13.4.4 Water Supply Distribution. The important factors to be evaluated in determining the optimum means of distributing potable water relate to camp type, size, and location of the supply source.

TABLE 13-4   *COST COMPARISON OF CAMP WATER TREATMENT SYSTEMS (Trink, 1981)*

Required Process	Relative Annual Cost*	Person Hours Required (h/d)
Filtration (paper cartridge) chlorination	1.0	0.75 to 1.5
Filtration (pressure sand) chlorination	1.6	0.75 to 1.5
CFSF** (alum, polymer) chlorination	3.9	2 to 4
CFSF (acid, alum, polymer) chlorination	4.4	2 to 4
CFSF (carbon, alum, polymer) chlorination	4.0	2 to 4
CFSF (carbon, alum, polymer) chlorination	5.5	2 to 4
CFSF (alum, polymer), ion exchange softening, chlorination	7.9	3 to 5

Plant capacity 189,000 L/d
* reference cost $2,600 US, 1976
** coagulation, flocculation, sedimentation, filtration

A mobile exploration camp would probably rely in summer on short-term storage and small surface sources or in winter on melting snow. Hence, distribution is limited to a small pressure system to maintain personal conveniences operating from a small storage tank. Temporary camps aim to be easily relocated, hence utilities are normally installed above grade or just below ground level for easy installation and removal. The pipe joints are accessible and of a quick-connect type. Materials selection should be according to the duration of the camp installation. Materials suitable for installations are similar to those described in Section 11.5. Installations that will be relocated a number of times require recoverable systems for economy. Permanent camps more closely relate to the situation for servicing northern communities. Buried services can be used if site conditions permit.

13.5 Wastewater Treatment and Disposal

To protect the health of camp personnel, minimize the impact of a camp on the surrounding environment and make camp life more comfortable, wastewater collection, treatment, and disposal must be provided. Current legislation or guidelines in both the United States and Canada require camps to provide secondary sewage treatment. Proper selection of treatment techniques and identification of flows and sewage strengths are important to prevent process overloading and subsequent operational problems.

TABLE 13-5   *AVERAGE INFLUENT WASTEWATER CHARACTERISTICS FOR CAMPS**

Influent Parameter	Concentration (mg/L)		
	Low	Average	High
BOD_5	275	456	704
COD	696	1,078	1,321
Total nonfiltrable residue (NFR)	317	491	817

* based on 17 months of weekly 24-hour composite data for 20 camps

TABLE 13-6   WASTEWATER CHARACTERISTICS OF VARIOUS CAMPS

Camp/Type	Flow (L/(p·d))	BOD$_5$ (mg/L)	NFR (mg/L)	Temp. (°C)	Organic Loading and Other Parameters
Artificial islands	108	1,900	1,100		
Drill site (Kennedy, 1981)	108				
Post de la Baleine (Trink, 1981)	250 (Sept to Oct avg)		99	16	
	450 (Oct to Dec avg)	230	104	12	103 g BOD/(p·d)
Advanced military bases (Quasim et al., 1971)	246	400	400		
Various Alaska pipeline construction camps (Murphy et al., 1977)		490 to 931	316 to 891		
Drill rig camp (averaged data) Alaska, North Slope (Tilsworth and Damron, 1973)	147	610	376	30	90 g BOD/(p·d)
	1,230 (COD)	2,570 (TR)			pH 7.8
		915 (TVR)			
		232 (VNFR)			
Base camp (average)	204	590	670		120 g BOD/(p·d)
Alaska pipeline construction camps (Eggener and Tomlinson, 1978)	265	456	491	20 to 24	100 to 150 mg/L grease
		1,078 (COD)			
Hydro Quebec work camp (Belanger and Bodineau, 1977)	636	155	75	10 (winter)	
	277	262	135	24 (summer)	

BOD$_5$ – five-day biochemical oxygen demand
COD   – chemical oxygen demand
TR   – total residue
TVR   – total volatile residue
VNFR – volatile nonfiltrable residue
NRF   – nonfiltrable residue

13.5.1 Wastewater Characteristics.   Characteristics of wastewater generated by construction camp activities (Table 13-5) indicate that a considerable range exists and that wastewater is much stronger than expected when compared with normal domestic sewage (see Section 10).

The increase in strengths for camp wastes relates directly to the closed-loop collection system which provides little opportunity for dilution. Also, considerable loading occurs because high-quality meals are supplied and camp operations are at high stan-

dards. Flows within the treatment plant, returning to the headworks (supernate from centrifuges, etc.) often contribute to the apparent high strengths of sewage seen by the treatment processes.

A summary of various work camp waste flows and strengths is provided in Table 13-6. Considerable variation in concentration and flow are exhibited. Approximately 50 percent of the total influent BOD$_5$ is soluble (i.e., it passes through a 0.45 mm filter). Domestic sewage generally is quoted to have a soluble portion of approximately 22 percent. Oil

TABLE 13-7   *WASTE CHARACTERISTICS OF INDIVIDUAL WASTE STREAMS (Eggener and Tomlinson, 1978)*

Source	Fixtures	Waste Characteristics			
		L/(p·d)	BOD_5 (mg/L)	TNFR (mg/L)	Other
Dormitory	Showers, commodes, urinals, lavatories, personal laundry facilities, janitor's sink	222	200 to 300	200 to 300	high phosphate, high temperature
Kitchen	Coppers, preparation sinks, prewash table, garbage grinder, dishwasher, coffee urns, janitor's sump	17	2,000 to 3,000	2,800 to 3,800	high oil and grease, high temperature
Offices and recreation hall	Commodes, urinals, lavatories	8	350 to 450	400 to 500	
Utility building	Water treatment backwash, sinks, floor drains, cleanup sumps	6	200 to 300	300 to 400	
Miscellaneous	Bedding laundry, chemical	12	500 to 1,500	500 to 1,000	
Total		265			

and grease concentrations have been measured between 100 and 150 mg/L even with grease traps in the kitchen (Eggener and Tomlinson, 1978; Murphy et al., 1977; Trink, 1981).

An exploration base camp in northern Quebec with an average population of 44 people throughout a winter period used two 205 L drums as a grease trap from the kitchen sewer line and collected an average of 900 g/d of grease. This represented 65 percent of the total mass of grease generated at the camp. The remaining quantity was discharged in the wastewater with an average oil and grease concentration of 23 mg/L (Trink, 1981).

A breakdown of various camp waste stream contributions has been prepared from research conducted during oil pipeline construction in Alaska (Table 13-7). The high water usage in the dormitory areas was reported to increase during periods of colder weather when workers tended to have longer showers to warm themselves.

Variations in wastewater flow occur in domestic locations. The peaks in flow are clearly tied to camp shift changes and specific events. Data collected from five camps working on two 12-hour shifts are

presented in Figure 13-6. Data for some northern Quebec sources indicate that peak-to-mean-to-minimum ratios vary from 2.2:1.0:0.2 (Trink, 1981); others report peak-to-average flows varying from 1.4 and 2.5 up to 3.0 times (Eggener and Tomlinson, 1978; Murphy et al., 1977).

13.5.2 Treatment Systems.

Design Considerations.   Selection of a wastewater treatment system suitable for a camp must be based on an evaluation of the camp type, nature of the workforce, operation performance, process simplicity, equipment reliability, and variations in wastewater flow and strength. A more detailed review of waste treatment systems suitable for application to cold regions is provided in Section 10. In considering a system, factors specific to camp conditions, for example, flow equalization, high oil and grease concentrations, and higher soluble BOD_5 fraction must also be considered.

Camp size and type also place some restrictions on suitable treatment systems. Highly mobile camps require wastewater treatment facilities which are portable, quickly started up, easily drained, and relatively light in weight. If the camp itself is small, the

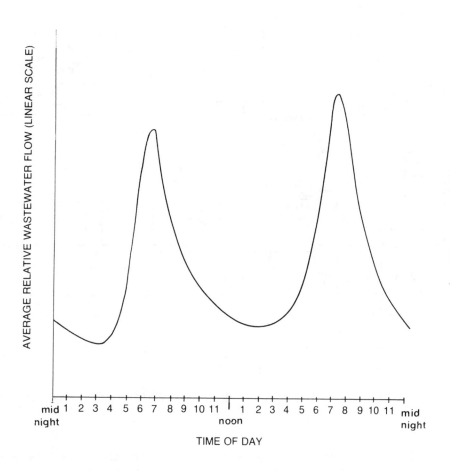

Based on eighteen 24-h observations at five camps working two
12-h shifts. Camp population ranged from 200 to 1800 occupants.

FIGURE 13-6   DIURNAL FLOW VARIATION (Eggener and Tomlinson, 1978)

use of holding tanks for periodic disposal at suitable sites should be considered. If there are only a few staff members, leach pits or nonfreezing recirculating toilets may offer an advantage.

Suitable treatment systems for permanent, temporary, and seasonal camps can be selected with more flexibility. A permanent location results in a wider choice of systems including biological units, which require a start-up period. But ultimately, camp location, population, specific environmental requirements, system operation, and economics are the determining factors in the selection of the treatment process.

Operation and Maintenance Experience.   Treatment systems that have been used include lagoons,

oxidation ditches, extended-aeration activated-sludge packages, ABF towers, rotating biological contactors, and physical/chemical package plants (Eggener and Tomlinson, 1978; Murphy et al, 1977; Heuchert, 1974) in conjunction with flow management reservoirs. Vaporization of the waste stream has been practiced at various pipeline pump stations.

Facultative lagoons have been used. At the Quebec Hydro LG2 camp site, lagoon removal efficiencies were below anticipated levels because of insufficient detention time in the facultative cell, underestimation of the oxygen requirement for summer loadings, and inadequate design of the shape of the sedimentation basin (Belanger and Bodineau,

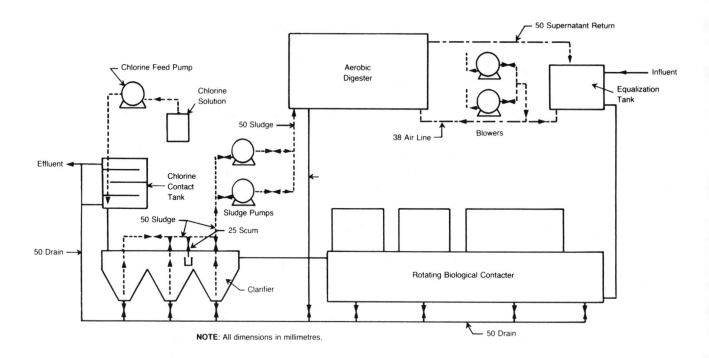

FIGURE 13-7   RBC CAMP WASTE TREATMENT PLANT FLOW SCHEMATIC (Heuchert, 1974)

1977). Properly designed facultative systems have generally been the more successful systems, because they require limited operator attention; and if adequately sized, the wastewater can be left until acceptable treatment has been achieved. Problems in siting and materials for construction may be limiting factors.

Oxidation ditches were used at some of the early Alaskan camps. Little performance data exist, although visual inspections of the systems indicate better performance may have been obtained at lower loadings. The camps that used these systems had populations of 30 to 50 people (Murphy et al., 1977).

Extended-aeration package treatment plants have been used quite extensively throughout the Arctic. Reports about their performance show that these systems are capable of effective removal of organics and suspended solids, if the systems are properly designed and operated. A typical installation would include comminution or screening, flow equalization, aeration, clarification, chlorine contact, sludge treatment, and disposal.

ABF towers followed by aeration chambers, clarifiers and filtration have been used quite successfully at large oil production camps on the North Slope of Alaska. The inclusion of a filtration step in the process makes it easier for these systems to operate in full compliance with permit limits and makes them more forgiving of the drastic population changes that are often inherent in remote camp operation.

Rotating biological contactors (RBCs) have been used at many sites in the American and Canadian Arctic to treat work camp wastewater. Laboratory analysis indicates effective removal of organic and suspended solids. Due to oversizing of facilities to process anticipated peak loads, camp RBC systems often are underloaded and denitrification in the clarifiers results in far larger volumes of sludge rising to the surface than would be found in a conventional setting. Modifications to standard designs for scum removal should be incorporated or retrofitted into RBC units for camp use. A typical installation is shown in Figure 13-7.

Vaporization of liquid wastes is also possible and can be cost effective when a high-grade source of waste heat is available. A form of vaporization termed "stack injection" is utilized at the northern pump stations of the Alyeska (Alaska) pipeline.

The mainline oil pumps are turbines driven by fixed jet engines which produce a high-volume, high-temperature exhaust stream. Domestic sewage is collected and processed through a "rotostrainer" to remove gross sewage solids. The solids are disposed of in a landfill. The liquid is then stored in an aerated sump until it is pumped/metered through an injection nozzle into the exhaust stream. The nominal flow rate is 0.25 to 0.38 L/s through the nozzle. The exhaust gas temperatures vary from 400°C to 480°C winter to summer.

Operational experience has shown that an air pressure assist through the injector nozzle is necessary to achieve atomization of the droplets such that complete vaporization of the fluid and die-off of bacteria are achieved. Parameters of operation have been developed through research and development after incidents of "yellow rain" were reported during initial operation. (See Section 10 for more information on wastewater treatment systems.)

Physical/chemical (P/C) package treatment systems were used in many of the Alaska pipeline construction camps. A study of 36 wastewater treatment plants, of which 23 were package P/C units, indicates that where these plants were preceded by aeration/equalization tanks or other biological pretreatment units, the systems effectively treat construction camp wastewaters. Costs are high and skilled operators are required to ensure the plants consistently meet secondary treatment standards. They fell short in handling excessive solids. Success in achieving high overall treatment with the P/C plants was attributed to soluble-to-cellular organic conversion in the aerated equalization tanks (Eggener and Tomlinson, 1978; Tilsworth and Damron, 1973).

Capital cost for the treatment systems including buildings, equipment and installations ranged from $5.33/L (U.S., 1977) for the 94,600 L/d units to $1.89/L (U.S., 1977) for the 946,000 L/d combination biological-P/C treatment units. Operation and maintenance costs ranged from $6.60 to $16.40/1,000 L of treated wastewater at design capacity. Labor costs are reported to account for 64 to 81 percent of these costs. Table 13-8 presents a summary of operation and maintenance costs for three types of P/C units of differing capacity (Eggener and Tomlinson, 1978).

A comparison of relative cost and personnel requirements for an extended aeration and P/C package plant (946,000 L/d) to treat wastes from a 380-person camp provides the breakdown shown in Table 13-9 (Murphy et al., 1977). All equipment from flow equalization to chlorination is included as well as sludge transfer, storage, and dewatering equipment. Equipment amortization is calculated at 15 percent over five years. Equipment was F.O.B. Anchorage, Alaska. Items excluded are sludge and effluent disposal, collection system, and lift stations.

RBC systems operate in Alaska where they serve 50- to 100-person camps. Two full-time operators

TABLE 13-8   OPERATION AND MAINTENANCE COSTS OF PHYSICAL/
CHEMICAL WASTEWATER TREATMENT PLANTS (Alberta
and NWT Building and Construction Trade Council and
Alberta Construction Labour Relations Association, 1980)

Unit Type	P/C (A)	P/C (B)	Bio-P/C
Capacity (L/d)	284,000	94,600	946,000
Unit Cost* ($/1,000 L)			
Labor	4.44	13.21	6.58
Chemicals	0.90	1.27	1.95
Miscellaneous	1.27	1.90	1.80
Total	6.61	16.38	10.33

* $US, 1977

TABLE 13-9   COST COMPARISON OF CAMP WASTEWATER TREATMENT SYSTEMS
(Murphy et al., 1977)

Process	Annual Cost ($US, 1976)	Person Hours Required	Electrical Requirement (kW/d)
Extended aeration	42,000 to 45,400	5 to 10	180
Physical/chemical	50,700 to 103,200	24	400

look after the entire service complex of such camps: water, sewage, incineration, generators, etc. They spend approximately four hours a day on maintenance, not including testing. Table 13-10 provides an indication of electrical requirements of RBC systems of various size (Smith and Straughn, 1971).

Rotating shifts and complete change of operating personnel on a scheduled basis create special problems in maintaining biological processes in remote camps. Often a crew will be on site for a one- to six-week period (depending on the camp and the crew schedules) to be replaced by a completely different crew, including both operation and supervisory staff for the next one- to six-week period. Even with excellent verbal and written communication between crews, this often results in major process changes by the incoming crew and often, subsequent process upsets. Attempts to alleviate these problems have included overlapping supervisory shifts, stag-gering crew rotations so some operators are on hand from the preceding shift and other innovative scheduling arrangements.

13.6   Solid Wastes

Solid wastes generated at camps can be classified into three general categories: domestic, industrial and hazardous wastes.

Domestic wastes consist of materials generated from kitchens, laundry facilities, living areas, and recreation facilities. These wastes are often similar to those generated in a community.

Industrial wastes are those materials of no immediately useful value generated from the camp effort and activity; for example, used equipment parts, pipe end cuttings, and building materials.

Hazardous wastes are defined by statute and regulations. Provisions for segregation, storage, and re-

TABLE 13-10   ELECTRICAL REQUIREMENTS OF
RBC SYSTEMS

Camp Population	Installed Power Requirement (kW)	
20 to 30	8	
40 to 60	9.4	2.0 m disc
70 to 100	11	
110 to 160	13	
170 to 270	17	3.2 m disc
280 to 400	19	

Maximum power consumed would more practically be 50 percent of the reported value due to 100 percent standby of equipment.

moval of wastes need to be made. Presently at Prudhoe Bay, such wastes are identified, segregated, stored in accord with regulations, packed and shipped to federally approved disposal sites in the lower forty-eight. Reduction of hazardous waste generated and concentration is an important part of each hazardous waste program.

13.6.1 Generation Rates.

Domestic-type solid waste generation rates quoted in literature applicable to most southern locations vary from 0.86 to 2.0 kg/(p•d), with 1.68 kg/(p•d) being the most typical. These values include wastes from residential and commercial activities. Domestic waste generated by camps has been reported to vary from 2.7 to 4.5 kg/(p•d) (Grainge et al., 1973). At Hydro Quebec camps in the James Bay area of northern Quebec, waste generation varied from 1.64 to 2.56 kg/(p•d) with the average being 1.0 kg/(p•d). The larger camps had a lower generation rate due to more efficient kitchen operations (Kovacs, 1981).

The higher generation rates for domestic-type wastes at work camps depend on the camp location and camp type. The rate is affected by the extra packaging required, a higher rate of wastage and larger per-person consumption of materials.

The industrial solid waste component of a camp is difficult to estimate and probably best assessed on a camp by camp basis. If the camp location provides access to suitable transportation, items of value should be considered for salvage and removed from the site. Where on-site disposal is to be practiced, segregation of industrial wastes from domestic waste should be considered. This would reduce the area needed for the disposal of domestic waste.

13.6.2 Collection and Disposal.

The method for collection of waste in part relates to the ultimate disposal techniques and camp size, although often a truck is used. Waste containers should have these characteristics: adequately sized, leak- and spill-proof, easily cleaned, inaccessible to insects, and animal resistant. Storage should minimize wildlife attraction and fire hazards. Transportation vehicles should be designed to prevent spillage. Provisions must be made to repair and maintain the storage containers. The nature of the work at remote sites often results in overloading of the containers, rough handling and subsequent structural failures. Tears and vents in the containers attract and allow ingress of foxes, ravens, etc.

Disposal options available vary; they can be haul-out, landfill, or incineration. Site conditions, genera-

tion rates, and environmental concerns are factors in determining the selected option or combination of options.

Costs reported by Hydro Quebec for disposal in a trench or modified landfill in northern Quebec was about $45/tonne (Cdn., 1980) and for an area landfill method about $100/tonne (Cdn., 1980) (Kovacs, 1981).

Incineration was favored in most areas where landfilling is not acceptable or where destruction of sewage sludge is a requirement. Ash disposal must however, still be considered. Incinerators can vary from single-chamber, unfired burners to controlled air units with auxiliary fuel supplies. Pathological, oil-fired incinerators have been popular in small permanent and semipermanent camps in the Arctic. Their use, however is labor intensive in that they must be manually charged and the load must be raked a number of times during the burn to break up clinkers and ensure complete combustion. Uncontrolled burning is not considered an acceptable practice and should not be carried out. Uncontrolled burning results in incomplete combustion of burnable material and the residue and odor from combustion can attract wildlife.

Incineration of sewage sludge was practiced in camps along the route of the Alaska pipeline to meet environmental regulations. Table 13-11 presents data about sludge generated from two types of physical/chemical treatment systems at 28 of these camps (Eggener and Tomlinson, 1978).

Co-incineration of solid waste and sewage plant sludge has been successfully practiced at Arco's Kuparuk facility. The sludge is dewatered using polymers and a filter press and mixed with solid waste to charge the incinerator.

While kitchen waste is comminuted in many camps and sent to the wastewater treatment facilities, others such as Alyeska segregate waste solids in the kitchens and do not allow them to be ground up and put into the waste stream for removal at the sewage treatment facility. Ideally grease is also excluded from the waste stream by the use of grease traps and scheduled cleanings. Grease may be sent directly to the incinerator for disposal after removal from the grease trap. These practices allow for improved operation of the wastewater facilities whether they are conventional plants, package plants, septic tank/drainfields or a screening mechanism prior to stack injection for vaporization.

TABLE 13-11   SEWAGE SLUDGE INCINERATION PERFORMANCE

Unit A	
Thickener underflow	average 6% solids
Incinerator temperature	760 to 870°C
Fuel required (diesel or waste oil)	3.34 L/kg sludge solids
Sludge generation average	0.32 kg/(p•d)
Ash formed	0.068 kg/kg of sludge solids
Unit B	
Thickened solids from centrifuge	average 8 to 10% solids
Operating temperatures*	870 to 980°C
For exclusive sludge burning – fuel consumed	2.92 L/kg sludge solids
Sludge generation averaged	0.32 kg/(p•d)

*Incinerated in a muffle chamber of combined refuse-sludge incineration.

Results of a study on solid waste generation, distribution, combustible and noncombustible portion, weight, volume, density and percentage of total weight at Dome Murphy Air Force Station, Alaska is presented in Table 13-12 (Smith and Straughn, 1971).

13.7 Layout and Construction

A camp layout should be selected not only to optimize orientation, personnel and traffic circulation, utility distribution, and disaster control but also to minimize the snowdrifting potential. Orientation should be perpendicular to the prevailing winds to prevent snowdrifting at the ends of the buildings. Drifting occurs between buildings where access is maintained through passageways (Sherwood, 1964).

Generally, a layout could be one of the following orientations (Sherwood, 1963b; 1964):

- circular or central core;
- in-line;
- company street (Figure 13-8);
- a V-plan (Figure 13-9); and
- parallel (Figure 13-10).

The central core plan with life-support systems placed in the center would result in a compact camp arrangement with a minimum of utility distribution lines and interconnecting tunnels. The orientation does not allow for easy expansion and accessibility, and it invites snowdrifting problems.

The in-line plan would have buildings placed end to end, creating a relatively drift-free arrangement with easy accessibility to the buildings. Walkways would be used to provide through-passageways. This layout would not be practical for large camps. Also, the disturbance and noise from through-traffic along walkways should be considered.

The parallel plan would have the buildings placed side by side with service cores connected by passageways for camp traffic. The camp would be oriented to keep the ends of all buildings easily accessible and free from drifting snow.

The company street plan where facilities are placed on either side of a street running through the center of the camp provides for easy expansion by extending the street or paralleling the layout with another street. This arrangement also results in a compact camp with buildings which can be easily connected by walkways, and with limited snow clearing along the streets. Outside access can be maintained to all buildings.

The V-plan is a layout evaluated by the U.S. Naval Civil Engineering Laboratory because of the interest in the resulting snowdrift pattern. Wind tunnel

TABLE 13-12   *RESULTS OF REFUSE QUANTITY SAMPLING AT MURPHY DOME AFS –*
1968 (Smith and Straughn, 1971)

Sources	Production (kg/d)		
	Range	Average Mass (kg)	Percent of Total (kg)
Living units			
Airmens' quarters	8.6 to 73	41	
NCO quarters	45 to 69	27	
BOQ	13 to 34	21	
		89	19
Recreational units			
Recreation building	0.45 to 59	25	
Bowling alley	0 to 11	4.5	
Hobby shop	0 to 12	2.3	
		32	7
Mess hall	160 to 280	220	47
Operations units			
Operations building	12 to 73	40	
Towers	0 to 12	2.3	
		45	9
Civil engineering units			
C.E. offices and shops	7.3 to 35	14	
Vehicle storage building	0 to 0.91	0.45	
Garage	5.4 to 160	40	
		55	12
Administration and supply units			
Supply warehouse	0.9 to 89	26	
Administration and dispensary	0 to 7.3	2.7	
		29	6
Distribution			
Combustible rubbish	85 to 320	200	48
Noncombustible rubbish	34 to 79	58	12
Garbage	140 to 220	190	40
Total mass (kg)	260 to 590	470	
Approximate volume (m^3)	3.5 to 8.3	5.6	
Approximate density (kg/m^3)	69 to 110	84	

model studies indicate that the interior of the V would be kept clear of snow for traffic circulation and accessibility to all buildings. A dispersed layout is achieved with this plan.

Area requirements for camps varies with available land, population, etc. An example of space requirements is a 500-person camp in Empress, Alberta that covered 1.6 ha and was designed with connecting central corridors between residential wings and the kitchen and recreation facilities (Rayner, 1980).

13.7.1 Facilities.   In Canada, all aspects of camp facilities (Figures 13-11 and 13-12) are detailed in the camp rules which are effective for the provinces of British Columbia, Alberta, and the Yukon and Northwest Territories. These regulations represent an agreement between the Building and Construction Trades Councils and the Construction and Labor Relations Associations. The provisions and equipment to meet these regulations has been reported to cost about $5,000/person (Cdn., 1981) and about $25/d (Cdn., 1981) for ongoing maintenance

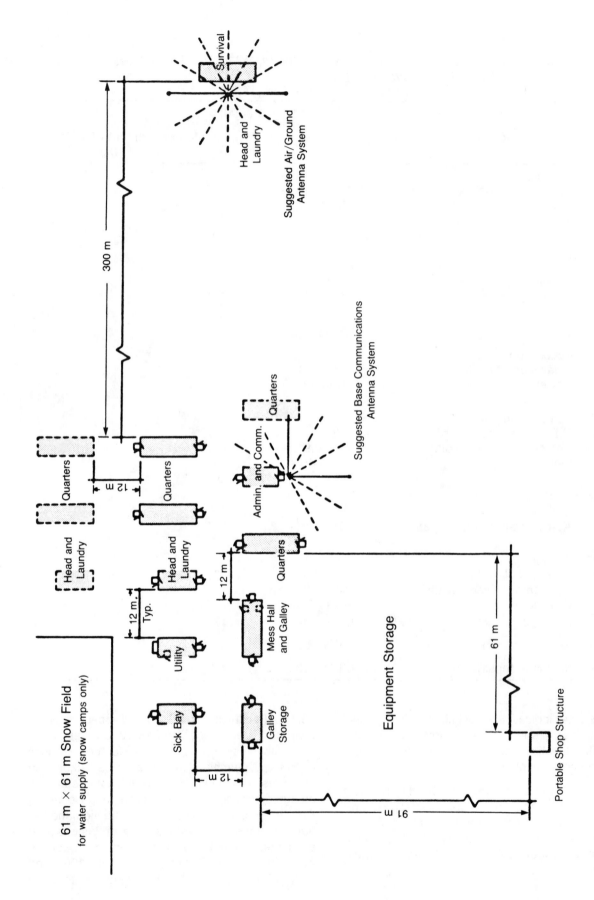

FIGURE 13-8   COMPANY STREET PLAN CAMP LAYOUT (Sherwood, 1963b)

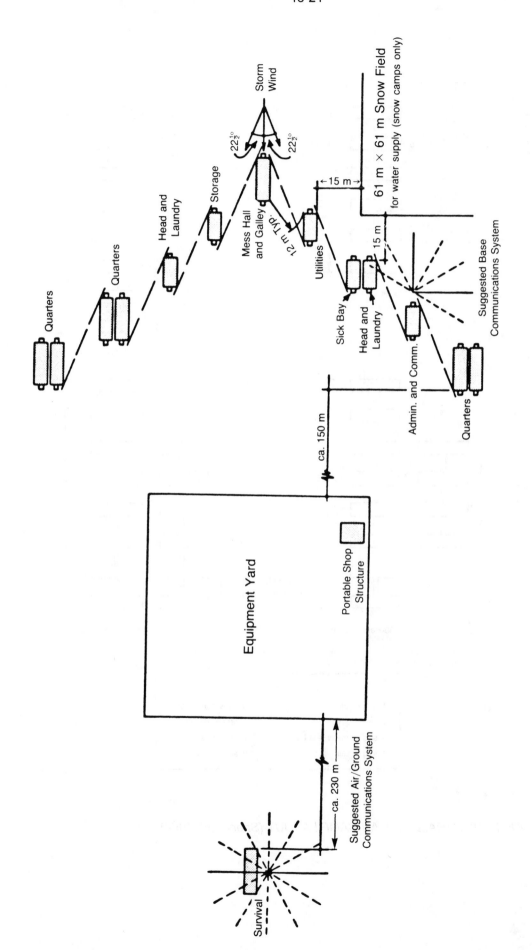

FIGURE 13-9   V-PLAN CAMP LAYOUT (Sherwood, 1963b)

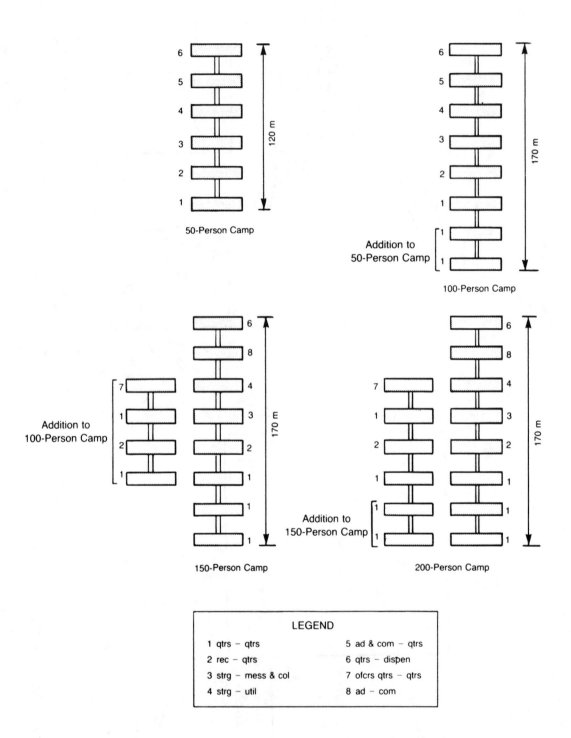

FIGURE 13-10 PARALLEL PLAN CAMP LAYOUT (Sherwood, 1964)

(Kennedy, 1981; Alberta and NWT Building and Construction Trade Council and Alberta Construction Labour Relations Association, 1980).

13.7.2 Construction. The supply of prefabricated units for location at a selected campsite can be designed to meet specific requirements; i.e., upgraded insulation, sizing for aircraft transport, etc. Suppliers manufacture these units on wheeled, skid-mounted, or bracketed carriages to meet the specific needs of a client. Wheeled units consist of a trailer assembly mounted on a wheeled frame with a protruding front towing assembly. Skid-mounted trailers have a welded steel frame for sliding the unit into place. These units are transported on a flatbed trailer or similar means. They provide the advantage that they can be pushed together for a more compact final assembly. Bracketed trailers are similar to skid-mounted units.

A heliportable shelter system for small exploration camps has been developed by the James Bay Energy Corporation. The units are fully transportable by helicopter of the Bell 205 size. The units weigh approximately 1,400 kg each and can be used as sleeping quarters, offices, toilet and shower storage areas, kitchen, dining rooms, housing for a pumping station, a small generating station, or work areas. Additional elements are corridor entrance, drop-in floor panel, flexible roof, ceiling and joint panels. The design eliminates mechanical and rigid joints as well as mechanical fasteners. Fire protection is provided not only by the autonomous nature of each unit, outside metallic surface on all four walls but also the installation of a smoke detector in each unit and a fire hose in each bathroom unit. Figures 13-13 and 13-14 provide an illustration of this system assembly and layout details (Sherwood, 1963a).

13.8 Energy and Supply Requirements

Operation of a camp in remote locations requires complete planning for all life-support facilities. As well as the previously discussed utilities (water supply and disposal of wastewater and solid wastes), power and supply requirements must be considered. Highly mobile camps require completely portable power-generating facilities.

Electrical power is required in a camp for water production and distribution, forced-air heating systems, kitchen facilities including food preparation and storage, wastewater treatment, as well as for the industrial demands of workshops and warehouses.

There is no rule of thumb to assist in feasibility planning; the needs of each camp must be calculated separately. Factors that have contributing effects on the energy demand are: the means to supply heat for accommodation areas, either electrical or fuel fires; the extent of water treatment required; the method of utility distribution; heat recovery, etc. Table 13-13 provides utility consumption rates of several military installations in Greenland where the average freezing index is 4,345°C·d. The facilities provided consisted of prefabricated wooden T-5 buildings suitable for arctic installations. Utility distribution was minimal and waste disposal consisted of discharge to isolated areas without treatment.

13.9 References

Alaskan Department of Health and Welfare, Federal Water Pollution Control Administration, Arctic Health Research Center. 1969. *Report of Survey – The Influence of Oil Exploration and Development on Environmental Health and Quality on the Alaskan North Slope.* Fairbanks, Alaska.

Alberta and NWT (District of MacKenzie) Building and Construction Trade Council and Alberta Construction Labour Relations Association. 1980. *Camp Rules and Regulations, 1980-1988.* Edmonton, Alberta.

Belanger, F. and R. Bodineau. 1977. Performance evaluation of the aerated lagoon at LG2 campsite. In: *Third National Hydrotechnical Symposium.* Canadian Society of Civil Engineering, Quebec City, May.

Eggener, C.L. and B.G. Tomlinson. 1978. Temporary wastewater treatment in remote locations. *Journal of the Water Pollution Control Federation*, 50(12): 2643-2656.

Given, P.W. and H.G. Chambers. 1975. *Workcamp Sewage Disposal Washcar-Incinerator Complex, Ft. Simpson, NWT.* Environmental Protection Service, Environment Canada, Edmonton, Alberta.

Given., P.W. 1978. *Report on Meetings with the Department of Environmental Conservation, State of Alaska, Fairbanks, Alaska, May 15-16, 1978, Regarding Waste Management for Northern Work Camps.* Northern Technology Unit, Environmental Protection Service, Edmonton, Alberta.

Grainge, J.W., R. Edwards, K.R. Huechert, and J.W. Shaw. 1973. *Management of Waste from Arctic and Subarctic Work Camps.* Report No. 73-19, Environmental-Social Committee

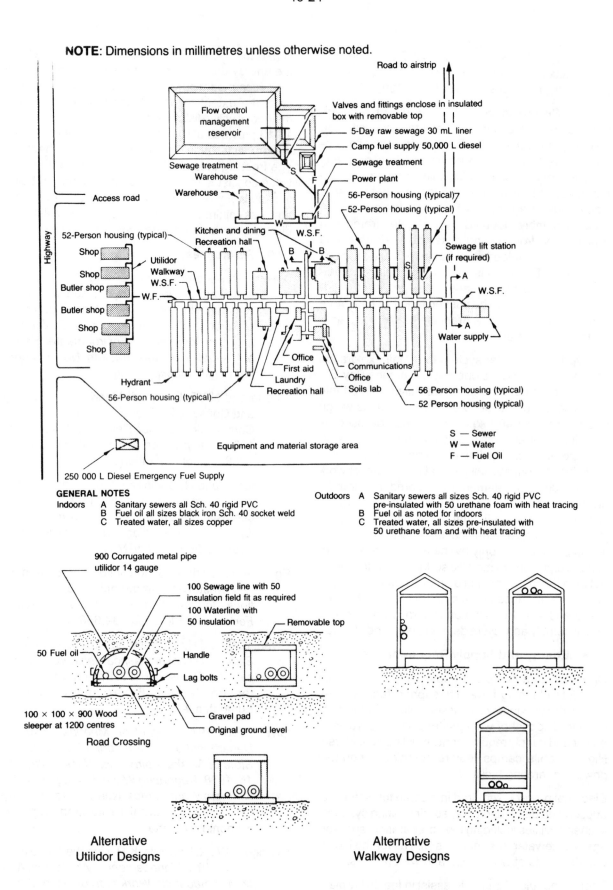

FIGURE 13-11 LAYOUT AND UTILITIES FOR A PIPELINE CONSTRUCTION CAMP

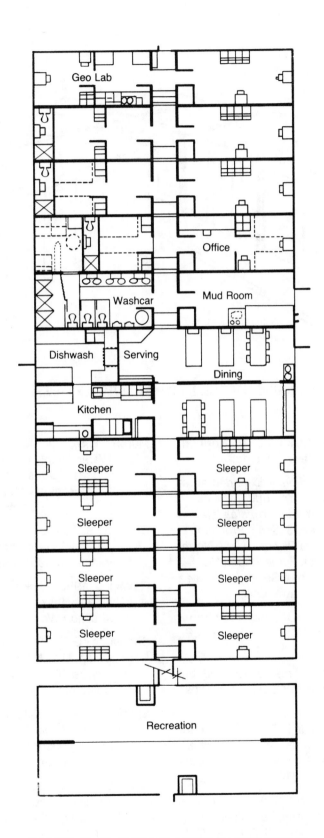

FIGURE 13-12 TYPICAL HERC CAMP LAYOUT

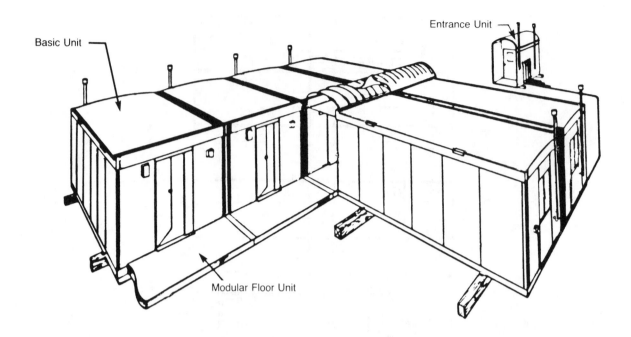

FIGURE 13-13 TYPICAL HELIPORTABLE CAMP ASSEMBLY (Kovacs, 1981)

Northern Pipelines, Task Force on Northern Oil Development, Ottawa, Ontario.

Heuchert, K.R. 1974. *Evaluation of Extended Aeration Package Sewage Treatment Plants on the Imperial Oil Limited Artificial Islands IMMERK and ADGO-F28. Some Problems of Solid and Liquid Waste Disposal in the Northern Environment*. Task Force Report No. 74-10 for the Environmental Social Committee Northern Pipelines, Ottawa, Ontario.

Heuchert, K. 1986. *Personal Communication*. PETWA Canada Ltd. 4120 - 23rd Street N.E., P.O. Box 481, Calgary, Alberta, T2P 2J1.

Kennedy, J. 1981. Camplife: it may be isolated, but these men lack nothing. *Construction West*, 4(2): 56.

Kovacs, J.M. 1981. Modular heliportable shelter system for small camps. In: *The Northern Community: A Search for a Quality Environment*. ASCE, New York, N.Y.

Lufkin, L. and W. Tobiasson. 1969. *The 50-Man Camp at Tuto, Greenland*. Technical Report 214, Corps of Engineers, U.S. Army, Cold Regions Research and Engineering Laboratory, Hanover, New Hampshire.

Murphy, R.S., G.V. Jones, and S.F. Tarlton. 1977. *Water Supply and Wastewater Treatment at Alaskan Construction Camps*. U.S. Army, Cold Regions Research and Engineering Laboratory, Hanover, N.H.

Quasim, S.R., N.L. Drobny and B.W. Valentine. 1971. Waste management systems for advanced military bases. *Sewage Works*, 118.

Rayner, J. 1980. Packing up and moving 1,600 km no easy task for a 500-man workcamp. *Construction West*, 60-61.

Sherwood, G.E. 1963a. *A 25-Man Pioneer Polar Camp*. U.S. Naval Civil Engineering Laboratory, Technical Report R267, Port Hueneme, California.

Sherwood, G.E. 1963b. *Specifications for a Temporary Polar Camp*. U.S. Naval Civil Engineering Laboratory, Technical Report N-540, Port Hueneme, California.

Sherwood, G.E. 1964. *A Temporary Polar Camp*. Technical Report R288, U.S. Naval Civil Engineering Laboratory, Port Hueneme, California.

Smith, D.W. and R.O. Straughn. 1971. *Refuse Incineration at Murphy Dome Air Force Station,*

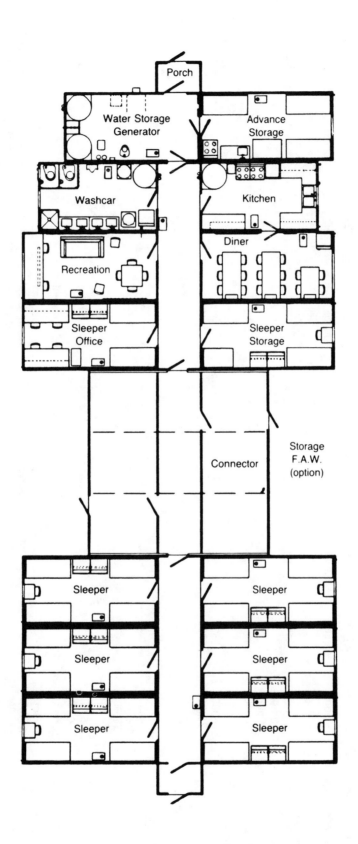

FIGURE 13-14 TYPICAL HELI CAMP LAYOUT
(Kovacs, 1981)

TABLE 13-13   UTILITY CONSUMPTION RATES FOR FACILITIES IN GREENLAND (from Lufkin and Tobiasson, 1969)

	Thule Air Base	Camp Tuto	Under Ice Camp	P-Mountain Radar Station	Artillery Site Near Thule	Loran Station at Cape Atholl	Heavy Swing	Camp Fistclench	Camp Century
Population	3,000 to 6,000	100 to 1,000	25	223	108	19	25	96 to 227	85 to 200
Total floor area, m²	256,400	19,230	531	10,150	9,316	1,500	129	1,162	3,789
Type of heating system	steam	diesel†	hot water*	steam	diesel†	forced hot air*	diesel† pot burners	diesel	electric
Diesel fuel (L/h) Avg	8,643	331	20	unknown	unknown	39	25	99	276
Max	10,390								
Min	7,508								
Avg L/(m²·d)	0.810	0.415	0.896			0.623	4.68	2.04	1.75
Electric power (kW) Avg	6,080	360	56	200	unknown	88	12		480
Max	7,630	590				103	15	<100	
Min	5,070	200				76			
Avg W/m²	23.7	18.8	10.8	19.7		58.7	92.6	<86	127
Steam (mJ/h) Avg	104,000	none	none	unknown	none	none	none	none	10.6
Max	156,000								
Min	69,000								
Water (L/(p·d)) Avg	515	unknown	79.5	unknown	unknown	397	37.9	121	189
Max	621					60.6 in winter			
Min	443								
Remarks	Not including BMEWS area		Small experimental camp in an ice tunnel near Tuto	Water trucked in from Thule		Inadequate water supply during winter	Includes prime mover fuel	594 L of fuel/day to melt snow for water	Steam used to melt ice for water

*   Heat reclaimed from diesel generators.
†   Electrically operated forced hot-air heaters

Arctic Health Research Center. Public Health Service, U.S. Department of Health, Education and Welfare, Fairbanks, Alaska.

Tilsworth, T. and F.J. Damron. 1973. Wastewater treatment at a North Slope industrial camp. In: *Fourth Joint Chemical Engineering Conference,* Vancouver, B.C.

Trink, D.T. 1981. *Exploration Camp Wastewater Characterization and Treatment Plant Assessment.* Environmental Protection Service, Environment Canada and Hydro Quebec, Report No. EPS 4-WP-81-1, Ottawa, Ontario.

13.10 Bibliography

Alter, A.J. 1970. Water and waste systems for North Slope, Alaska, 1 & 2, TAPE, *The American Plumbing Engineer*, April and May.

Anonymous. 1980. Arctic challenge. *Construction West*, July, 72-73.

Associated Engineering Services Ltd. 1973. *Report on Water Supply and Waste Treatment Requirements for Canadian Arctic Gas Pipelines Ltd. Construction Camps.* Northern Engineering Services Ltd., Calgary, Alberta.

Associated Engineering Services Ltd. *Waste Management for Northern Work Camps.* Prepared for Environmental Protection Service, Edmonton, Alberta.

Clark and Groff Engineers. 1962. *Part III. Final Report, Sanitary Waste Disposal for Navy Camps in Polar Regions.* U.S. Naval Civil Engineering Laboratory, Port Hueneme, California.

Drobny, N.L. 1967. *Polar Sanitation – Synthetic, Nonfreezing Waste – Carriage Media.* U.S. Naval Civil Engineering Laboratory, Technical Note N-916, Port Hueneme, California, August.

Eggener, C.L. and B.G. Tomlinson. 1978. Temporary wastewater treatment in remote locations. *Journal of the Water Pollution Control Federation*, 50(12): 2643-2656.

Given, P.W. 1978. *Report on Meetings with the Department of Environmental Conservation, State of Alaska, Fairbanks, Alaska, May 15-16, 1978, Regarding Waste Management for Northern Work Camps.* Northern Technology Unit, Environmental Protection Service, Edmonton, Canada.

Hoffman, C.R. 1965. *Requirements for Liquid Distribution Systems in Polar Camps.* U.S. Naval Civil Engineering Laboratory, Technical Note N-724, Port Hueneme, California.

Hoffman, C.R. and G.E. Sherwood. 1966. *Polar Camp Improvements – Water System Using a Hot-Water Snow Melter.* U.S. Naval Civil Engineering Laboratory, Technical Report R441, Port Hueneme, California.

Nehlan, W.R. 1952. *Review of Polar Camp Sanitation Problems and Approach to Development of Satisfactory Equipment for a Polar Region 100-Man Camp.* U.S. Naval Civil Engineering Research and Evaluation Laboratory, Technical Note N-032 U.S., Port Hueneme, California.

Sherwood, G.E. 1963. *Specifications for a 25-Man Pioneer Polar Camp.* U.S. Naval Civil Engineering Laboratory, Technical Note N-500, Port Hueneme, California.

Sherwood, G.E. 1967. *Review of Portable Structures for Polar Regions.* U.S. Naval Civil Engineering Laboratory, Technical Note N-913, Port Hueneme, California.

Tilsworth, T. and F.J. Damron. 1974. *Industrial Wastewater Treatment in Arctic Alaska.* American Chemical Engineers Symposium Series No. 144, 70: 227-236.

Valentine, B.W. 1972. *Self-Contained Sanitation Systems for 2 to 15 Man Polar Facilities.* U.S. Naval Engineering Laboratory, Technical Report R-0759, Naval Facilities Engineering Command, Port Hueneme, California.

Vining, C.V. and D.L. Hardy. 1978. Wastewater Treatment at Remote Sites in Alaska. IN: *Proceedings of the Conference on Applied Techniques for Cold Environments*, American Society of Civil Engineers, New York, N.Y.

SECTION 14

INDIVIDUAL ON-SITE SYSTEMS

3rd Edition Steering Committee Coordinator

James A. Crum

3rd Edition Principal Authors

Daniel H. Schubert

James A. Crum

Section 14 Table of Contents

Section 14 List Of Figures

Section 14 List of Tables

14 INDIVIDUAL ON-SITE SYSTEMS

14.1 Introduction

In some locations, individual on-site water supply and wastewater disposal systems may be a feasible alternative to community piped service. This section discusses systems for buildings, ranging from completely independent on-site water supply and waste disposal to facilities that depend on vehicle haul for water delivery and waste removal.

14.2 Water Supply

The basic objectives for on-site water supply systems are essentially the same as those discussed in Section 5 for community systems. Protection of health is the critical factor; this requires a reliable water source, adequate water quality, and systems for safe delivery, storage, and distribution (Gamble and Jansson, 1974).

14.2.1 Water Sources, Quality, and Treatment.
The traditional community water sources discussed in Section 5 also have potential for on-site use if economics favor their development. The money that can be committed for servicing a single building may limit the feasible depth of wells or construction of

surface impoundments or infiltration galleries. Because of their small size and relatively low demand, on-site systems can often take advantage of water sources that would not be adequate for a community supply. These include springs, rainwater collection, and shallow wells in marginal aquifers. On-site systems also have the advantage of more effective water conservation measures than those possible in conventional community systems. This is because at on-site locations occupants are more directly responsible for their own water supply and therefore recognize the benefits of conservation.

Typical design values for individual daily water usage are given in Table 14-1. They range from less than 10 litres/person/day (L/(p•d)) for self-hauled water to 150 L/(p•d) for a dwelling with a well and a conventional septic tank disposal system. Typical designs assume six persons per dwelling in the Alaskan Arctic and five persons per dwelling in the Northwest Territories. Systems dependent on vehicle haul are not constrained by availability of water at the source but rather by the size of the in-house water and waste tanks and the frequency of water delivery and waste pickup. Many of these communities

TABLE 14-1   TYPICAL DESIGN VALUES FOR HOUSEHOLD WATER/WASTEWATER

	Amount of Water Use (L/(p•d))	
	Average	Range
Self haul from watering point	<10	5 to 25
Washeteria	20	18 to 35
Truck delivery		
a. nonpressure water tank, bucket toilet, and community facilities for laundry/shower	15	5 to 25
b. nonpressure water tank, bucket toilets, no community facilities	25	10 to 50
c. nonpressure water tank, wastewater holding tank, no community facilities	40	20 to 70
d. pressure water system, low water toilet, waste holding tank, and community facilities	30	15 to 50
e. pressure water system, low water toilet, waste holding tank, no community facilities	90	40 to 130
Individual well with septic tank/drainfield	150	80 to 250

have central facilities for laundry and bathing , thus on-site water is needed only for drinking, hygiene, cooking, and possibly waste disposal (with little or no water used for toilets), for which 10 to 15 L/(p•d) should be adequate. Galena, Alaska, for example, has community facilities for laundry and showers with individual water and waste tanks in each of the 50 buildings. The average monthly water consumption per home is 1,440 L/month. Satisfying this water demand would only require two to three stops a month by the water-haul vehicle to keep the 700 L in-house water tank supplied. Costs of service may be a limiting factor for community-based haul systems in situations were the end user pays the full costs of delivery.

Water Wells and Pumps. Wells penetrating to aquifers in unfrozen materials or to bedrock are probably the most reliable water source. Procedures for well drilling and other installation details can be found in many references. In nonpermafrost locations, the well casing and any structure above permafrost may require frost-heave protection in the active layer, and the service lines to the house must be thermally protected to prevent their freezing. The temperature of groundwater sources in the Arctic and subarctic can be expected to be at about 0.5 to 1.5˚C year round which makes for a narrow margin of safety for freezing. Wells penetrating the permafrost have to maintain the integrity of the surrounding permafrost, while at the same time, prevent the water in the casing from freezing. The complexities of freeze protection

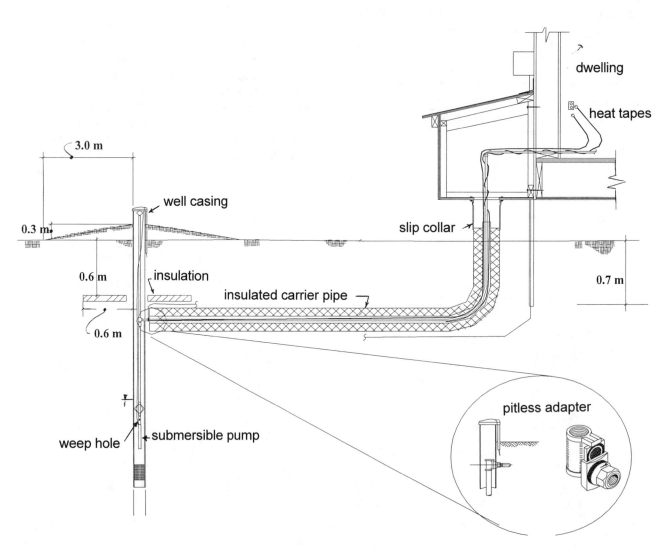

FIGURE 14-1   INDIVIDUAL WATER SYSTEM

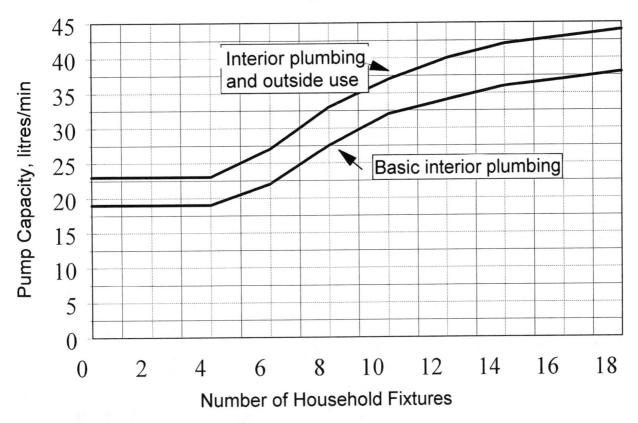

FIGURE 14-2   RECOMMENDED PUMP CAPACITY

are such that the use of subpermafrost wells for individual buildings requires special attention, artesian wells in particular. Recovery of water from a deep well requires electrical power and the use of a pump of appropriate capacity. Submersible pumps are most commonly used for this service.

In much of the Arctic, the permafrost is 100 m or more thick, and the active layer freezes down to the top of the permafrost. Shallow wells are obviously not effective. Shallow wells can be functional in permanently thawed alluvial soils adjacent to existing water bodies or in former stream beds. Design features of drilled wells are illustrated in Figure 14-1. Especially important is a small-diameter "weep" hole in the riser pipe at a location below the frost line and above the check valve. The weep hole prevents the water from freezing in the riser by allowing it to drain back between pump uses. Driven wells using a slotted well point are also effective for shallow water-bearing formations. Provision for air bleed is needed in buildings that use sealed bladder pressure tanks, to allow air flow in and out of the system to allow drain-back.

The needed capacity of the pump depends on the number of fixtures to be served. Figure 14-2 can be used to estimate recommended pump capacity for

a household based on the number of fixtures it contains. For example, a house with two sinks, a toilet, and a shower, i.e., four fixtures, should have a pump capacity of at least 20 L/min. The lower curve is for interior household uses only. If in addition, exterior uses such as agriculture or some fire protection or both are desired, then the upper curve should be used. These curves are based on the use of conventional fixtures with no water conservation measures employed; they represent the maximum requirement for cold-region applications.

The capacity of a well or spring is often less than the values indicated in Figure 14-2. In those cases, the well pump should be compatible with well capacity and deliver the water to in-house gravity storage. The gravity storage must be large enough to permit in-house pressure pumping and distribution at the rate indicated in Figure 14-2 for conventional services.

Large-capacity pumps are not required for in-house distribution from water storage tanks that are on a vehicle delivery system. The lower curve on Figure 14-2 provides adequate supply for in-house use. Figure 14-3 shows a typical pump system used for buildings in Alaska. The pump unit for a conventional house complete with plumbing fixtures would

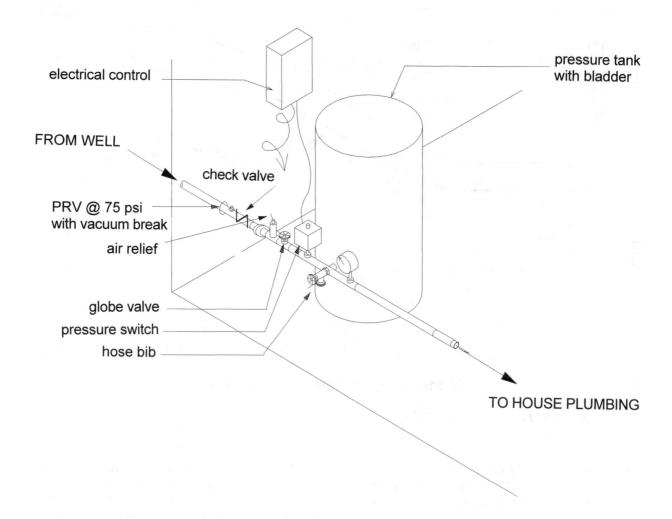

electrical control

FROM WELL

check valve

PRV @ 75 psi
with vacuum break

air relief

globe valve

pressure switch

hose bib

pressure tank
with bladder

TO HOUSE PLUMBING

FIGURE 14-3   TYPICAL PUMP PRESSURE SYSTEM WITH WELL LINE DRAINBACK

typically combine a 190 watt pump with a pressure tank.

Springs.   Natural springs may be available in the subarctic and may be developed into adequate on-site water supplies. The spring box structure is only provided to physically protect the spring and the quality of the water. It should be constructed of concrete or some other durable, relatively watertight material. The spring water should be admitted freely into the structure, but surface drainage must be excluded. There should be no openings in the structure for animal access. The top cover or access hole cover should be secured. Before constructing the spring box, the site should be excavated sufficiently to locate the true spring openings and to ensure a firm foundation. An insulated, screened overflow pipe should be provided. A service pipe which allows either gravity or pumped flow to the building should

be installed in the lower portion of the structure. Valves are needed to allow draining before maintenance. The housing is not usually sized to provide a significant amount of storage. If storage is needed, it is usually provided in the house or in a tank elsewhere. Thermal elements, such as insulation on the sides and cover as well as insulation and heat tape on the overflow and supply pipe are essential if winter operation is expected. The use of a perforated pipe in the water-bearing strata is suggested in aquifers with marginal productivity. Simpler construction is possible for aquifers with higher productivity; for example, the spring housing needs only to be an open-bottom container with impermeable walls. A variation of the construction can be used as a small-scale infiltration gallery next to rivers and streams (U.S. Public Health Service, 1980).

The open nature of some springs may limit the natural purification. A spring should not be used as a domestic supply without first determining suitable quality by taking a series of bacteriological tests. Periodic sampling should continue for the life of the spring. An increase in turbidity during break-up or after storms may indicate that surface runoff is reaching the spring.

Cisterns and Holding Tanks. The collection and storage of rainwater or snow meltwater is possible as a seasonal or a supplemental water source. It is a limited source since the amount of precipitation in the Arctic and subarctic is small. In central Alaska, for example, the rainfall in the warm months of the year averages about 160 mm. A minimal water usage of 10 L/(p·d) would require a catchment surface of 23 m²/p to supply the annual needs. Assuming no losses due to evaporation or other factors, the above equation can be used to estimate the minimum catchment area required:

$$A = \frac{q \cdot N \cdot d}{P}$$

where,

A = minimum catchment area required, m²

q = daily water usage, L/(p·d)

N = number of people

d = number of days during which water service is required, d/a

P = annual rainfall at the site (mm/a).

The catchment area from which rainwater is collected is usually the roof of the dwelling. The cistern is most feasible for schools and similar isolated structures with a relatively large roof area and limited summer water needs.

The quality of the water may be affected by the roof surface material, the collection gutters, and if used, the filtering devices. The first flow from the roof carries the maximum concentration of wind-blown particles. Manual or automatic devices to provide for the bypassing of this water should be used.

It is prudent to use filtering devices to remove suspended matter which may cause taste, odor, or color problems. Self-contained, prefabricated pressure filters are readily available in various capacities for home or similar on-site application. Precautions must be taken for filters for the removal of microbial contamination due to concerns over regrowth and breakthrough. Therefore, the holding tank or cistern should be disinfected periodically and the cartridge filters changed frequently (Bouthillier, 1950).

Melted Water. Snowmelting as an on-site water source in winter has been reported (Coutts, 1976). Snow collection can be labor-intensive. Water yield from melted snow or ice and the energy requirements for the melting depend on the density and temperature of the original material (Table 14-2). About 3.0 L of diesel fuel is required for each cubic metre of snow (at 0˚C), if the snow melt is 100 percent efficient. One cubic metre of snow (0˚C) produces about 300 L of water. At a water use rate of 10 L/(p·d), about 38 L of diesel fuel per person per annum would be required. To meet the energy requirements, the use of a heat source that would otherwise be wasted should be encouraged; for example, exhaust gases or chimneys provide a good source of heat which is typically wasted. Snow- or ice-melting should only be considered as a last resort or as a standby emergency system.

TABLE 14-2   ENERGY REQUIREMENTS AND WATER YIELD FROM ICE AND SNOW

Form	Density Yield* (g/m³)	Water Required** (L/m³)	Energy Required** (MJ/m³)	Volume Diesel (L)
Snow, new, loose*	85	85	29.8	0.8
Snow, on ground*	300	300	10.52	2.7
Snow, drifted and compacted	500	500	175.4	4.5
Ice at 0˚C	900	900	315.7	8.0
Ice at -40˚C	900	900	391.1	10.0

*   Assume 0˚C and temperature elevated to 4˚C.
** Assume no evaporation loss and 100% efficient heat exchange.

Surface Water. Freshwater ponds, lakes, and streams can be used for all or part of the year if available. If the water body does not completely freeze, an infiltration gallery similar to that used for springs may be used. Winter use of pond water drawn from under the ice may be limited because of the deteriorated water quality. Sometimes filling storage tanks during the summer months for winter use may be the best alternative.

Water from surface water sources should be filtered and disinfected prior to use. Occasionally chemical treatment for removal of suspended solids or color-causing materials may be required.

14.2.2 Water Quality. The quality of these on-site water sources is generally the same as that described in Section 5. Surface sources can be turbid and periodically contaminated by animals, birds or upstream human users. This is why filtration and disinfection should always be practiced. Groundwater sources often contain significant quantities of iron, manganese, and other minerals as well as naturally occurring organic compounds that make treatment difficult. Both surface water and groundwater sources tend to be low in temperature, which decreases the efficiency of chemical treatment and filtration. Readily available conventional equipment that was designed for use in temperate zones has been used including: water softeners, pressure filters often combined with potassium permanganate chemical feed for iron removal, disinfection equipment, and reverse osmosis units for demineralization. Households on vehicle delivery systems would not normally require individual treatment units since it is more economical to use a centralized community water treatment system of the types described in Section 5. Concern has been expressed over health effects of point of use home treatment units that contain activated carbon; they may serve as a media for microorganism growth unless properly maintained and serviced. Some health authorities do not allow point of use filters for domestic service.

14.2.3 Storage Tanks, Plumbing, and Piping. Storage tanks for water and wastewater are required for buildings on vehicle-haul systems. As indicated previously, gravity storage with pressure pumps for water tanks may also be necessary where the on-site water source has a low flow rate. In these cases the storage is sized to meet the daily need; and the low, but continuous, flow refills the tank during non-use periods.

Tanks can be buried where site conditions allow. Ice-rich, fine-grained permafrost soils generally re-quire tank construction above ground for stability. Tanks located above ground require thermal protection. The preferred locations for both water and sewage tanks are within the heated envelope of the dwelling, not only to take maximum advantage of the thermal protection but also to provide easy access for service and maintenance. Installation within the house imposes certain criteria to ensure maximum efficiency and cost effectiveness. The most critical criterion is the use of water conservation measures whenever possible.

Providing a vehicle delivery system to meet the commonly assumed water use rates (about 1,000 L/d for a family of four) for conventional facilities in warm climates is not feasible. At the usage rates shown in Table 14-1, the storage tank capacity should range from about 1,000 to 3,600 L to supply water for up to one week for four persons. Minimum single-family tank sizes can be estimated as follows, assuming twice weekly service:

Minimum water tank size = 675 x no. of bedrooms

Minimum sewage tank size = 1.5 x water tank size

In determining the frequency of service, the total cost of the vehicle delivery needs to be minimized. This total cost includes both the cost to provide service and the cost of the tanks. Delivering water every day will minimize the size and the cost of the building water tank but result in high delivery costs. Conversely, delivering water once per month will reduce the number of visits, but the customers would have to have larger, more expensive water tanks.

Determining a frequency of service to minimize the total cost is difficult because the cost varies with the water use, the cost of the water tanks, the cost of vehicle operation, relative building location, and other factors. As water use varies with each building, setting a fixed frequency of delivery is difficult. For single-family homes, a smaller range in water use as well as lower quantity of use occurs. By assuming a water consumption of 450 L per day for a household of five persons, it is possible to develop a relative cost chart (Cameron, 1996). Figure 14-4 shows one example of such an analysis. In this case, the lowest total cost occurs when water is delivered twice per week. The frequency of service to minimize the total cost of vehicle service to other buildings will vary with tank size and water use rate. Some general guidelines to determine the optimum schedule include the following:

- Service more than twice per week is only economical when water use in the building is sig-

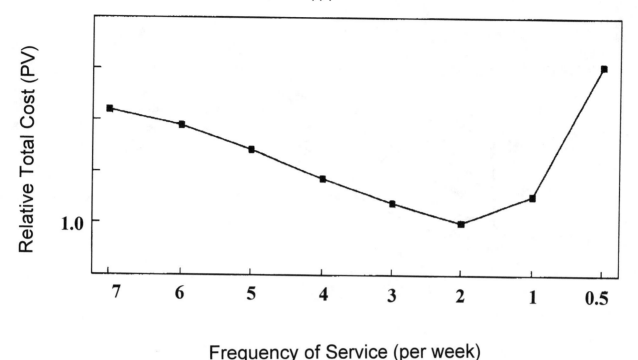

FIGURE 14-4   RELATIVE TOTAL COST FOR VEHICLE HAUL VS. FREQUENCY
OF SERVICE

nificantly higher than that of a single-family dwelling.

- The lowest service cost occurs when a full truck load of water can be delivered for each service. The service cost does not decline any further once the frequency is long enough such that a full truckload of water is dropped off with each servicing. For example, it costs the same to deliver five truckloads of water during one day as one truckload on each of five separate days.

- Specifying the frequency of service as opposed to providing whatever frequency of service is requested will result in a minimum overall system cost.

To reduce pumping requirements, the in-house water tank should be located as high as possible, and the wastewater tank as low as possible. Both tanks should be made of corrosion-resistant materials. The water tank should not impart tastes or odors to the contents. Plastic materials, particularly fiberglass or high density polyethylene, should be suitable for use with potable water, and meet established national standards.

The technical specifications for water and wastewater storage tanks at Galena, Alaska, include the following provisions: 5.0 mm nominal thickness fi-

berglass, one-piece unit, contact mold construction method, 13 mm radius of curvature at intersections of tank surfaces; tensile strength 36.6 MPa, flexural strength 40.0 MPa, impact strength 61.4 MPa; water tank fabrication with an isophthalic acid resin during lamination to eliminate taste and odors, and a paraffin surfacing wax for a curing agent on all interior surfaces; the wastewater tank will have a resin-rich interior surface for corrosion resistance. The location of these storage tanks and the related plumbing facilities of this Galena installation are shown in Figure 14-5.

The water tank fill line runs to a quick-connect fitting on the house exterior for connection to the hose from the water delivery vehicle. The pumpout line for the wastewater tank runs to a similar but different size fitting.

As shown in Figure 14-6, the wastewater tank is supported by the basic house flooring and decked over with plywood. Alternatively, the tank may be an insulated steel or plastic tank located outside the house. A low-water-use toilet (1.0 L/flush) is connected directly to the tank. Since the village may have public showers and public laundry, the fixtures in most buildings are limited to sinks and the low-water-use toilet.

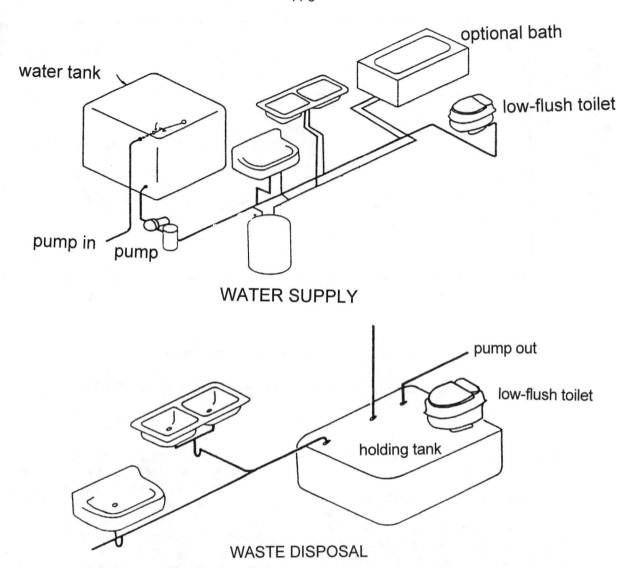

FIGURE 14-5   WATER SUPPLY AND WASTEWATER PLUMBING – VEHICLE-HAUL
SYSTEM

The plumbing details shown in Figure 14-5 are typical for cold-region installations. Copper lines, polyethylene, or polybutylene are commonly used for water service, and plastic with solvent-welded or compression joints, for wastewater.

All such plumbing should be located along interior walls, if at all possible, for maximum thermal protection. The water lines should be provided with drains at the low spots so that the system can be emptied during extended periods of cold or power loss or both, in the building. In addition to the normal sink traps, the wastewater lines should have clean-outs at critical locations in the system. Exterior walls must be penetrated so that the integrity of the wall insulation and vapor barrier remains sound. These exterior connections must be dry and empty when not in use. Exterior connections to on-site wells or springs must be insulated and heat traced. Figure 14-7 shows methods of making exterior connections of sewer pipes to on-site disposal systems or to community sewers.

The quick-connect connections and the hoses used for water and wastewater should be of different sizes to avoid the risk of cross-contamination. Both the water tank and the sewage tank must be vented to permit air movement during filling and pump-out. Household wastewater tanks have a typical capacity of 2,000 L, but some as large as 4,500 L have been used in Canada. It is common practice to install a wastewater tank from one and one-half up to twice the size of the water storage tank; although generally a tank volume of about 1,000 L or at least

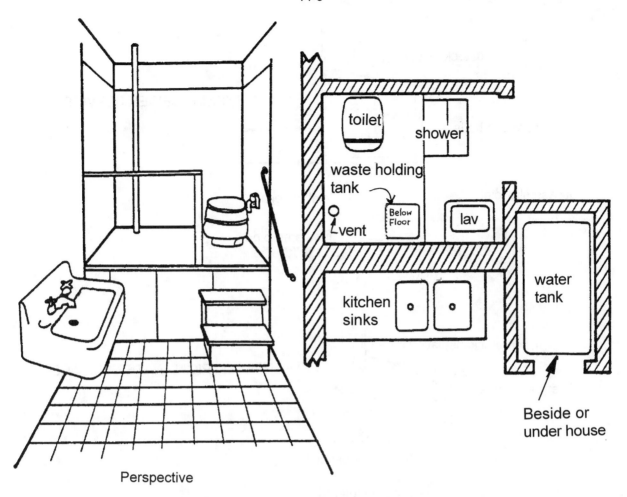

Perspective

FIGURE 14-6   BASIC HOUSEHOLD FACILITIES FOR VEHICLE-HAUL SYSTEM

400 L larger than the water tank should give adequate service.

14.3 Water Conservation Alternatives

Water conservation measures can include simple flow-control orifices in sinks and shower piping to limit flow. Automatic spring-loaded supply valves on sinks, and low water use toilets may be used. The toilet provides the major opportunity for water conservation since the conventional units without any modifications consume about 40 percent of the household water. Personal bathing in conventional showers or tubs uses about 30 percent of the water supplied. The balance is used in laundry (15 percent), kitchen (13 percent) and miscellaneous (2 percent) (Cameron and Armstrong, 1979; Armstrong et al., 1981).

14.3.1 Toilets. Toilets use more water than any other single fixture within the home. Toilets, which typically use 13 L/flush or less, can be easily modified by the homeowner to reduce water consumption during flushing. Modifications range from simple homemade devices such as weights or plastic bottle inserts, to inexpensive manufactured dams or dual-flush attachments. A more expensive modification, applicable for piped systems, replaces the reservoir tank with a small pressure tank.

Also a number of toilets designed to use as little as 1.5 litres of water are available. These recirculating toilets use the smallest amount of fresh water. They require an initial charge of water and chemicals or other additives. A number of toilet alternatives that do not require any water at all are also available. It is important to note that not all toilets are applicable or appropriate to every situation. For example, some mechanical-seal toilets must be located directly over a receiving tank, and a 1.5 L toilet should only discharge into a tank less than 25 m away through a sewer line with at least a three percent slope.

14.3.2 Bathing. Depending on the habits of the user, showering usually requires less water than tub bathing, particularly if an inexpensive flow-restrict-

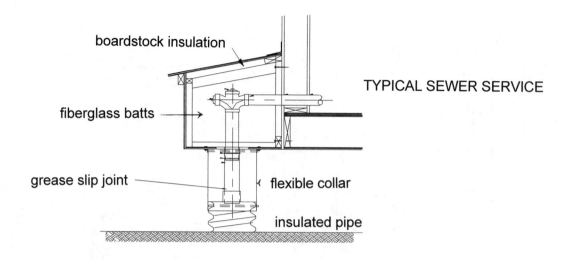

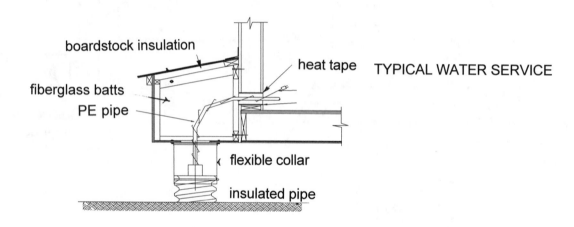

FIGURE 14-7 ARCTIC HOUSE UTILITY BOX INSTALLATION

ing insert or specially designed low-flow showerhead has been installed. Many low-flow showerheads give a satisfactory or even superior shower while saving a considerable amount of water along with the energy required for heating the water. Other specialty shower units or systems use very little water. Several add-on shower devices are available that can save water, and some increase convenience, comfort, and safety.

The use of hand-held showers can reduce water consumption by about 50 percent compared to the conventional fixed shower heads. A very significant saving in household energy use can be achieved by coupling a hand-held shower to an instant in-line water heater. The energy saving arises from the elimination of the conventional hot-water tank for heating and storing of water. The units are thermostatically controlled and provide instantaneous heating of water flowing through the unit. Both elec-

trical and gas-fired units are commercially available. The gas-fired units claim at least a 25 percent fuel savings compared to a conventional hot-water tank system. Single heating units can be located in the house at each point of hot-water use or a larger heater unit can be located at a central point.

14.3.3 Laundry. Numerous different brands of top-loading, automatic washers are available; some of them use considerably less water than others. The more efficient tumble action of the front-load washer makes it the lowest hot- and total-water user of the automatic washers. They are, however, more expensive, and consumer acceptance has been poor.

14.3.4 Kitchen. In the kitchen, dishwashing uses the most water. Washing dishes by hand can be done with very little water but may entail some inconvenience and extra effort. If an automatic dishwasher is used and always loaded to capacity for each full cycle of operation, water use is comparable to dishwashing by hand in a filled sink and rinsing under a free-flowing stream of water. In-sink food waste disposal units are a modern convenience that, if judiciously used, does not significantly increase domestic water use. Other kitchen operations such as drinking and cooking use small, relatively fixed volumes of water. Reductions in the wasting of water can be achieved by adopting habits such as keeping a container of cold water in the refrigerator. Various faucet attachments reduce the amount of water flow and hence water wastage. Water flow reduction of faucets also has the added benefit of energy savings since approximately 50 to 75 percent of the flow is heated water. Where water is corrosive, all faucets should be flushed before taking water for consumption, until the system is modified to correct the water characteristics (GCDWQ, 1995).

14.3.5 Economics. There are practical and technical limitations to the economic selection of water conservation alternatives for an individual building. All capital and operating and maintenance (O&M) costs associated with an alternative need to be discounted to obtain its present worth. Since the costs depend upon the unit costs for water, sewerage, and energy, the number of uses or volume used, and the O&M costs, each new and retrofit situation is different. The marginal unit costs (net of any subsidies) should be used to arrive at costs, but these are often difficult or impractical to obtain. Nevertheless, the following measures of water conservation are recommended.

For piped systems, toilets need not use over 6.0 L/flush. Low-flow showers and flow-control aerators are almost universally economical. Piped systems with preheating, excessive water pressures, or high treatment costs, and locations with very high electricity costs may find other devices economical.

For vehicle systems with marginal rates of $0.01 to $0.02/L, more restrictive alternatives are economical for households. Mechanical flush toilets should be used wherever possible. Where the sewage holding tank cannot be located directly below the toilet, recirculating toilets are usually the most economical despite the cost of chemicals. Toilets using over 6L/flush should not be installed. Low-flow showers, hand-held showers, flow-control aerators, mixing faucets, and a method to reduce hot-water pipe flushing, such as insulation, circulation, or a return line, are economical. Front-load laundry machines

are economical for new installations and use minimal amounts of water.

Where utility costs are very high or where the water supply is limited, more severe steps are necessary. Even nongravity piped sewer systems do not control excessive water use that is inherent with vehicle systems and central facilities. In addition to the recommendations pertaining to vehicle systems, devices such as spray and self-closing faucets, specialty shower systems, and timers on showers can be used. Water conservation is usually more economical than greywater reuse. Greywater reuse should only be considered for central facilities and where other considerations such as zero pollution are paramount. Reuse must be approached with caution due to the complex treatment systems needed, the controls that are necessary, and the risk of disease transmission during a malfunction.

14.4 On-Site Wastewater Management

Community sewage collection systems (pipes, vehicle haul) are not always possible or economical. It is often necessary to provide for on-site wastewater treatment and disposal. Community treatment systems are designed to meet regulatory requirements for both the protection of public health and the environment at the discharge point. On-site treatment systems must meet the same objectives.

14.4.1 Direct Disposal. The alternatives for direct, on-site disposal without some type of preliminary treatment are bucket and chemical toilet dump pits, pit privies, and vaulted privies. The use of bucket or chemical toilets is frequently chosen because of convenience and privacy. In nonpermafrost and many intermittent permafrost areas, a suitable location can be found for the excavation of a disposal pit. Pit location criteria should conform with those used for septic tank leach fields. One such criterion is the separation distance of at least 35 m from the water supply point and other water bodies. Groundwater depth should also be determined to ensure adequate separation between the bottom of the pit and the highest annual groundwater table. In addition, ground freezing patterns, direction and rate of groundwater movement, and soil types should be examined to ensure that the leachate from the pit cannot migrate to a water source during adverse conditions. Another important consideration is surface and interflow water movement during spring break-up. The pit should be located to minimize seepage inflow. If necessary and when soil conditions permit, a diversion ditch around the pit can be excavated. The pit should be about one metre in

TABLE 14-3   TYPICAL DOMESTIC WASTEWATER CHARACTERISTICS

Constituent	g/(p•d)	mg/L at	
		90 L/(p•d)	200 L/(p•d)
BOD$_5$	81.7	910	410
Suspended solids	90.8	1010	450
Total nitrogen	12.3	140	60
Total phosphorus	3.6	40	18
Fecal coliform* (no./100 mL)	3.0×10^9	2.2×10^6	1.5×10^6

* based on 2.0×10^7 fecal coliform/g of feces and 150 g feces/capita/d.

diameter and as deep as soil conditions and ground-water table location permit. Odor reduction is possible with periodic additions of layers of ashes or lime. A pit 1 m in diameter and about 2.5 m in depth used for human waste only may last a four-member family two to three years. Replacement pits should be 2 to 4 m apart to ensure soil stability and reduce infiltration from the old pit.

The pit privy is one of the earliest methods of human waste disposal. The main features of concern in cold regions are the criteria for locating the pit and the elevation of the entrance to account for snow accumulation. The privy building should be relocatable and the pit chamber should be vented to the outside.

Where pumpout equipment is available, vault toilets may be suitable. Vault toilets can be used where land use, groundwater, soil conditions, or proximity to surface water bodies restrict the use of conventional pits. They are not common for single-family dwellings but are extensively used for on-site sanitation at parks, playgrounds, and similar recreational facilities. The use of four vertical walls and a separate liner is usually cheaper and easier than attempting to construct a complete watertight structure on site. Both reinforced Hypalon and the more rigid cross-linked polyethylene have been successfully used as liners. If the thinner and more flexible Hypalon is used, a thin layer of concrete should be placed on the bottom for protection during the cleaning operations. The vault should have a maximum capacity of about 1,900 L, unless the design analysis shows that a larger capacity is warranted. A large-diameter access hole is essential in the vault cover for access during cleaning. The vault should be pumped out at the end of the recreational season

or more frequently if necessary. Some cleaning by hand tools may be necessary, since all sorts of undesirable objects tend to find their way into the vault and cannot be removed with the typical pumper vehicle.

14.4.2 On-Site Treatment. Individual facility wastewater treatment can be achieved by three different methods:

- composting;
- septic tanks (anaerobic); and
- aerobic units.

These processes are capable of treating different volumes and strengths of wastewater. Wastewater strength often varies with water use, as does the concentration of contaminants in wastewater. An estimate of the mass of waste contaminants generated by various sources in the home is detailed in Section 10. The average daily amounts are given in Table 14-3. These amounts can be divided by the amount of wastewater generated in the home to calculate concentrations. Concentrations based on 90 and 200 L/(p•d) are also listed in Table 14-3.

Composting. Non-water-carried treatment using a composting process can be accomplished with one of several different types of units. Some require electricity as they are electrically heated, or have ventilation blowers. Selection from the various types of composters requires careful evaluation for sizing, hydraulic loading, operation and maintenance requirements and cost of operations. Several types of units are described in the references. The composted material must be removed for disposal.

Acceptance of residential compost toilets in rural communities is limited for a variety of reasons. These include the expense of operation, the need for frequent tending, need for periodic addition of bulking agents, and upset of the biological process due to freezing. In addition, composting processes have a low tolerance for excessive liquid wastes, which tend to cause process upset.

Occurrence of anaerobic conditions is the generally accepted definition of compost process upset or failure. Upset of the process occurs when anaerobic conditions predominate within the reacting mass of waste. Upsets result in extremely obnoxious odors, which are not merely displeasing but may cause some individuals to become physically ill, forcing them to temporarily leave the household. Correction of the upset conditions requires physical removal of the accumulated mass of waste material and re-starting of the composting process. Composting can be of value as a sanitary waste disposal means only if:

- aerobic conditions are maintained by proper airflow and liquid/waste rations;

- ambient temperature within the compost toilet enclosure is maintained at 10°C or higher;

- bulking agents are properly applied at appropriate levels (approx. 0.0076 m³/capita/day);

- liquid overloading is not allowed to occur; and

- homeowners and users are willing and able to correct upset conditions.

These are stringent and demanding sets of minimum operating criteria. Each must be satisfied for satisfactory composting to be achieved. Other considerations include exclusion of trash and other nonbiodegradable discards.

The composting process can be effective as a means of sanitary waste disposal provided that the noted restrictions of use are both accepted and satisfied. It may be preferable to honey buckets in the absence of "higher" alternatives when applied under well-defined and controllable circumstances. This implies a limited application role in rural communities.

The composting toilets have a potential advantage in that they accept kitchen garbage in addition to human wastes which lessens solid waste disposal problems. The simplest installation of large units requires a very large warm space that has to be near the disposal points (toilet, kitchen). These composting toilets function properly when maintained. They are best suited where homeowners have voluntarily committed themselves to install and maintain them. They are less well suited to rental or institutional housing where the occupants may not have the personal commitment to their successful operation.

Septic Tanks. The most common method for on-site treatment is the conventional septic tank. Its principal function is to remove settleable and floatable material from the wastewater, and provide anaerobic treatment of solid and liquid components of the waste. Normal design requires 24-hour retention at average flow rates. Longer detention times should be used for cold-region applications in order to ensure better solids removal. This will improve the life expectancy of the drainfields. Minimum sizes and details of tank design can be found in the references. Where soil conditions permit (no permafrost or excessive frost penetration), the tank can be installed in the ground outside the dwelling. In cold regions, installation of hydrophobic insulating material such as polystyrene on the outside of the tank is often recommended. In warm climates, the accumulated sludge undergoes some anaerobic digestion. The resulting volume reduction allows a pumpout frequency of once every three to five years. In cold regions, an annual or biannual pumpout is usually recommended. Several rational methods for estimating the frequency of pumpout for septic tanks are available in the literature (Mancl, 1984; U.S. EPA, 1980; CMHC, 1978).

Where poor soil conditions exist (permafrost or deep frost penetration), septic tanks have to be installed within a heated environment (heated area under house or heated shed).

In extremely cold conditions, discharge from the septic tank to the absorption field or disposal site should be intermittent so that the discharge retains enough heat to prevent freezing in the discharge pipe. This can be done by using a siphon or a float-controlled pumped discharge.

The efficiency of contaminant removal by the septic tank varies with the tank design, operating temperature, and characteristics of the wastewater. Table 14-4 provides general information on some contaminants.

Aerobic Treatment. A variety of on-site aerobic treatment systems are available varying in capacity from single-family to institutional sizes. Treatment concepts include variations of the activated sludge

TABLE 14-4   CONTAMINANT REMOVAL EFFICIENCY BY SEPTIC TANKS
AND AEROBIC UNITS

	Contaminant Removal (%)	
Contaminant	Septic Tanks	Aerobic Units
BOD_5	14 to 52	60 to 72
Suspended solids	31 to 72	59 to 71
Total nitrogen	0 to 58	0 to 60
Total phosphorus	0 to 21	0 to 38
Fecal coliforms (reduction log.units)	1	2

process, principally extended aeration and attached growth processes such as trickling filters, rotating biological contactors (RBCs), and upflow filters. The major advantage of these aerobic processes is the higher level of treatment provided, which in turn permits a more effective final discharge. In particular, seepage pits and leaching fields may consider a smaller infiltration area when used with these aerobic units, because the higher-quality effluent reduces the potential for clogging. Sometimes direct discharge of these aerobic effluents to surface streams may be permitted.

All these aerobic systems have mechanical elements, and routine maintenance is required. They also require disposal of excess sludge at least once a year. Aerobic units require more thermal protection than septic tanks for the biological processes to operate at the design rates. Most aerobic units require pretreatment to remove gross solids such as trash, grease, and garbage grindings; therefore they may use a trash trap or a septic tank as a pretreatment step. Aerobic units are sensitive to variations in the quantity of flow and its composition. Their design may need to take into account the impact of water conservation measures on the composition of household sewage since the volume of water is reduced, yet the mass of pollutants remains about the same.

Table 14-4 compares contaminant removal efficiency from septic tanks and aerobic units. The aerobic units can achieve a higher level of BOD, and fecal coliform removal than septic tanks. The other parameters are about the same for either type of effluent. But the level of fecal coliform removal is not high enough to eliminate concern for public health, and the level of BOD and suspended solids remaining may not allow a more effective use of disposal fields and pits. Under optimum conditions, aerobic units can produce an effluent comparable in quality to conventional secondary treatment. Under practical conditions, however, they are subject to variations in flow, temperature, wastewater composition, and the degree of attention to operation and maintenance. Where the effluents from aerobic units are to be disposed of in seepage pits, leach fields, or other appropriate in-ground disposal sites, further treatment of these effluents is not justified. However, installations that discharge effluent to surface water may require additional treatment such as sand filtration. In general, the loss of solids with the final effluent is the major factor that keeps these units from achieving their theoretical performance efficiency.

An individual aerobic unit can be small and light, making it more easily installed within a dwelling, or protective cold climate structure. Since the effluent is aerobic, it should cause little odor.

Greywater/Blackwater Systems. The separation of greywater (sinks, baths, and laundries) from blackwater (toilets) is often described as a water conservation technique. The degree of conservation actually achieved depends on the type of fixtures used and not on the concept itself. The separation of greywater and blackwater may offer some advantages for treatment or disposal, but often the two flow streams are recombined for final disposal. The ultimate possibility is to eliminate the blackwater flow altogether by using non-water-carriage units such as composting or waste-incinerating toilets, or simple bucket and bag toilets.

Bucket and honey bag toilets are still in use in many villages and isolated locations in the cold regions because economics prevents the adoption of more modern concepts. Criteria for in-house management and community collection and disposal of honey bags can be found in Section 9 of this manual.

14.4.3 On-Site Disposal. The means of on-site disposal can varies with the site conditions. Disposal can take place by:

- direct discharge;
- leaching fields, drainfields, or beds;
- seepage pits; and
- mounded (raised-bed or at-grade) systems.

Direct Discharge. Direct discharge of effluent from septic tanks or aerobic units may be permitted by some regulating authorities. Such effluent from a single dwelling in a remote location may have a small impact on the receiving environment. Protection of health is the major concern for both the occupants of the source dwelling and its neighbors. This requires that the impact of the disposal site on water and food supplies as well as the potential for disease vectors (insects, birds, larger animals, etc.) be taken into consideration. Direct contact with sewage effluent should not be allowed.

In-ground disposal by leaching fields, drainfields or beds is the most commonly used technique in cold regions.

Drainfields. Figure 14-8 shows the design concept for a typical household drainfield or leaching field. A plastic pipe that is perforated is most commonly used. The perforations typically are 12 mm in diameter, located on 50 mm centers. The soil around and under the disposal fields must remain unfrozen throughout the winter for effective perfor-

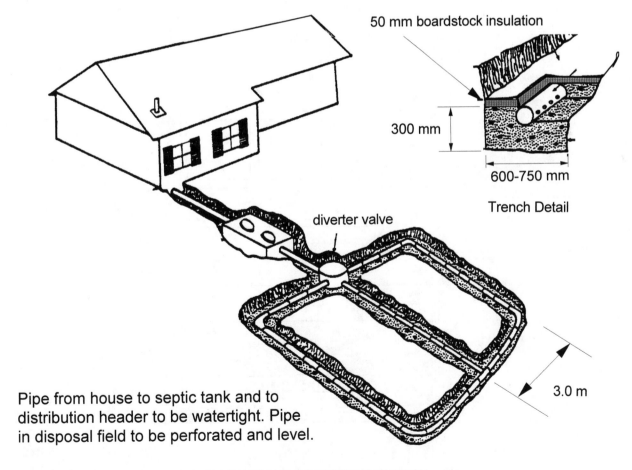

50 mm boardstock insulation

300 mm

600-750 mm

Trench Detail

diverter valve

3.0 m

Pipe from house to septic tank and to distribution header to be watertight. Pipe in disposal field to be perforated and level.

FIGURE 14-8   TYPICAL SEPTIC TANK AND DRAINFIELD SYSTEM

mance. Heat in the wastewater usually maintains unfrozen soil conditions in a continuously operated system. Problems can occur after long periods of nonuse in extremely cold climates.

Although the pipe and gravel bed should drain after each use, frost penetration into the in-situ soil can create an impermeable barrier. A natural or induced snow cover over the field area acts as an insulating blanket and maintains unfrozen conditions throughout the winter. In extreme climates with minimal snow cover, polystyrene insulation board over the trenches should be used. These disposal fields have been used throughout Alaska and Canada with marginal success. The failure rate in Alaskan communities is high. However, the typical failure is not due to freezing but often due to a faulty design, faulty construction, or carryover of solids from an improperly maintained septic tank. A typical trench system is shown in Figure 14-8, consisting of distribution piping on a continuous bed of sand or gravel.

Intermittent sand filters are commonly used in the United States where surface discharge of septic tank effluent is desired or necessary. They can either be completely buried and made inaccessible after construction, or constructed at or above ground for easier access during operation. Figure 14-9 illustrates a typical filter constructed with two compartments for alternating use with septic tank effluent Watson, 1978).

Septic tank effluent is applied at a rate of up to $0.2 \text{ m}^3/(\text{m}^2 \cdot \text{d})$. The sand used for filtering should have an effective size of 0.35 to 1.00 mm with a uniformity coefficient of less than 4.0.

The filter should be flooded at least twice a day to a depth of about 50 mm. Dual filters, each sized for total design flow, are recommended for effluent from septic tanks. A single filter sized for design flow is adequate for aerobic pretreatment. The hydraulic loading with aerobic pretreatment can be double the value used for septic effluent. The effluent charac-

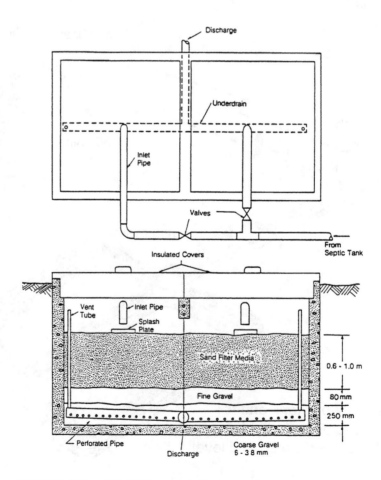

FIGURE 14-9   WASTEWATER SAND FILTER

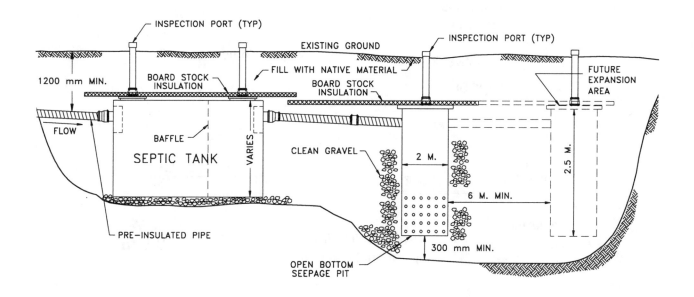

FIGURE 14-10 SEPTIC TANK WITH SEEPAGE PIT

teristics from these filters are quite good; BODs range from 8 to 23 mg/L and suspended solids are less than 10 mg/L.

An alternative system gaining acceptance in the United States is the use of manufactured "gravelless drainfield" systems. The system is provided in components consisting of a subsurface corregated pleat or arch unit approximately 1.0 m wide and 0.3 m high. Another type is made from 0.3 m perforated corrugated pipe sections, wrapped with geofabric material. Both types are made of HDPE and are lightweight, durable and resist crushing. They do not require gravel for drainfields, and are thus particularly useful in areas where gravel is hard to obtain.

Drainfields designed using these systems can either follow standards for a standard trench design or use bed design criteria. Manufacturers' recommendations on installation should be followed.

TABLE 14-5   *WASTEWATER APPLICATION RATES FOR CONVENTIONAL TRENCH AND BED DISPOSAL SYSTEMS*

Type	Soil Percolation Rate (mm/h)	Design Application Rate $(L/(m^2 \cdot d))$
Gravel, coarse sand	>1200	not suitable*
Coarse to medium sand	300 to 1200	50
Fine sand, loamy sand	100 to 300	36
Sandy loam, loam	50 to 100	25
Loam, porous silt loam	25 to 50	20

* Flow rate too high for adequate treatment. Replace existing material with 600 mm of sand or loamy sand.

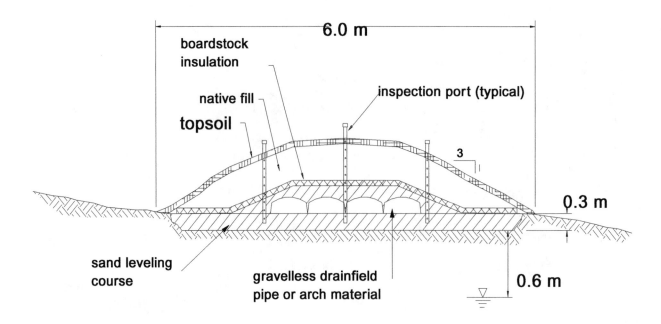

FIGURE 14-11 MOUND CONSTRUCTION

Seepage Pits.   A seepage pit is commonly used in much of rural Alaska. It requires deeper excavation than trenches or beds, but is more compact and requires less surface area and therefore is potentially more thermally efficient. Figure 14-10 shows typical construction details for a concrete seepage pit or crib.

Perforated, large-diameter concrete cylinders or concrete blocks installed with open joints are also used for pit construction. If more than one pit is needed, the space between them should be equal to about three times the diameter of the pit. Clogging occurs rapidly on the pit bottom, therefore the design is based on infiltration through the sidewalls of the pit. The area taken for design is at the interface of the natural soil and the gravel or rock backfill.

The required area is determined from the percolation test results in the natural soil. If the soil profile is layered and has different percolation values, a weighted average should be used for the design. If any layer has a percolation rate slower than 50 mm/h, it should not be included in the design calculation. Typically pits are 2 to 4 m in diameter and 3 to 6 m deep. Other dimensions are possible and depend on soil conditions, capabilities of the excavation equipment available, and structural stability of construction materials for the walls and cover.

The surface area required for trenches, beds, and pits is usually based on simple percolation tests of

TABLE 14-6   INFILTRATION RATES FOR DETERMINING BASE AREA OF MOUNDS

Soil Type	Percolation Rate (mm/h)	Design Application Rate (L/(m²•d))
Sand, sandy loam	>50	50
Loams, silt loams	33 to 50	30
Silt loams, silty clay loams	25 to 33	20
Clay loams, clay	13 to 25	10

the soil at the depth to be used in the system. Table 14-5 relates the wastewater application rates to the results of the percolation test and can be used for the design. For example, a percolation rate of 100 mm/h would allow a wastewater loading of 35 L/(m²•d) of infiltration area. A household with a 400 L/d design flow would require 11.4 m² of infiltration surface. This surface area is the sidewall of pit systems and the bottom area of trenches and bed systems. Some resgulatory agencies allow design credit for at least a portion of the sidewalls in a trench system. Gravel and coarse sands are excluded from Table 14-5 because water percolates too quickly through the material for adequate treatment to occur. In isolated remote locations, where water supply contamination is not possible, this criterion can be relaxed. Soils with high clay content usually have percolation rates below 25 mm/h and are not well suited for conventional pits, beds, or trenches.

Mounded or Raised-Bed Disposal Systems. Mounded or raised-bed disposal systems may be suitable for sites with high groundwater tables or shallow bedrock. The principle involves raising the application bed a sufficient distance so that the larger base area of the mound serves as the design infiltration surface. Sufficient distance between the bottom of the gravel layer and the water table or impervious layer must be maintained. Figure 14-11 shows typical design details. Many regulatory agencies have specific design requirements regarding size, slopes, and construction materials. Boardstock installation may be required over the top of the drainfield distribution pipe to prevent freezing in extreme climates. The design requires a two-step operation:

- percolation tests are run in the natural soils at the site and the base area of the mound determined from Table 14-6, and

- based on the type of soil used to construct the mound, the area of the application bed is determined.

Table 14-7 relates the most commonly used fill materials to the design rates and to the bed area. An alternative is to design the mound area according to Table 14-6, place the fill, allow it to consolidate for several months, run percolation tests in the consolidated fill, and use those values and those from Table 14-6 to design the application bed area. The design of the distribution bed should ensure even distribution of the wastewater over the entire bed area.

14.4.4 Residential Pumping Stations. In some situations, it may be necessary to lift or pump the sewage to a higher elevation or farther distance than is possible under gravity flow conditions. Small residential lift stations of various configurations, types of pumps, and associated equipment have been used. Commercially available package stations as well as custom-designed modules are available for use. Several components are common to both. These include the wet well, control and alarm panels, level sensors, valves, and pump unit. Effluent pumps can be installed in concrete, or in fiberglass, or plastic basins. The type of system, and site specific conditions determine facility type.

The control panel includes the start and run capacitors, disconnect switches and on/off controls. A method of level sensor to control pumping is also required, as are high level alarms. Since septic tank

TABLE 14-7   INFILTRATION RATES FOR MOUND FILL MATERIAL

Type of Fill	Characteristics (percent by mass)		Design Appplication Rate (L/(m²•d))
Medium sand	25%	(0.25 to 2.0 mm)	49
	30 to 35%	(0.05 to 0.25 mm)	
	5 to 10%	(0.002 to 0.05 mm)	
Sandy loam	5 to 15%	(clay content)	24
Sand/sandy loam mixture	88 to 93%	(sand)	
	7 to 12%	(fines)	49

effluent is corrosive, care must be taken in choosing components, including valves. Both flapper and ball valves, and means of disconnection to service and pump replacement are required.

Design considerations include minimizing heat loss from the wet well, protection of the equipment from freezing, and ensuring that the piping drains to minimize freezing.

14.4.5 Septage Disposal. A means of residential sludge disposal is needed for septic tank pumping, and for vehicle-haul facilities. In Alaska, state regulatory requirements include the need to assess and identify means of handling, treatment and disposal as part of the project. Many technological alternatives are available for cost-effective treatment and disposal. Several references are available that discuss reasonable alternatives and practical experience in hauled waste and septic tank sludge management (Tilsworth, 1984; Rezek, 1980). Several of the methods for small system use include lime stabilization followed by trench and cover, monofills, sludge drying beds, and freeze/thaw lagoons. Estimates of the quantity of sludge, the waste characteristics, and alternative evaluations are similar to those discussed in other chapters of this manual, and should be referred to for specific details.

Additional regulatory requirements related to sludge disposal are under development in Alaska, based on federal requirements.

14.5 References

Armstrong, B.C., D.w. Smith and J.J. Cameron. 1981. Water requirements and conservation alternatives for northern communities. In: *Design of Water and Wastewater Services for Cold Climate Communities*, D.W. Smith and S.E. Hrudey (eds.), International Association on Water Pollution Research, Pergamon Press, Oxford, England, 65-94.

Bouthiller, P. 1950. *Farm Water Systems and Sewerage*. Prairie Rural Housing Committee and Central Mortgage and Housing Corporation, Ottawa, Ontario, 74 p.

Cameron, J. 1996. *Water and Sewage Services Bylaws Handbook*. Department of Municipal and Community Affairs, Government of the Northwest Territories, Yellowknife, NWT.

Cameron, J.J. and B.C. Armstrong. 1979. *Water and Energy Conservation Alternatives for the North*. Deptartment of Local Government, Government of the Northwest Territories, Yellowknife.

CMHC. 1978. *Canada Mortgage and Housing Corporation Septic Tank Standards*, Metric Edition. Ottawa, Ontario, 25 p.

Coutts, H.J. 1976. Experiences with a snow melt water supply system. In: *Proceedings, Second International Symposium on Cold Regions Engineering*, Fairbanks, Alaska, 351-360.

GCDWQ. 1995. *Guidelines for Canadian Drinking Water Quality*. Health Canada, Ottawa, Ontario.

Gamble, D.J. and C.T.L. Janssen. 1974. Evaluating alternative levels of water and sanitation service for communities in the Northwest Territories. *Canadian Journal of Civil Engineering*, 1: 1.

Mancl, K. 1984. Estimating septic tank pumping frequency. *Journal of Environmental Engineering*, ASCE, 110(1): 283-285.

Rezek, J.W. 1980. *Septage Management*, U.S. EPA, ORD, Washington, D.C.

Tilsworth, T., et al. 1984. Cold climate septage management. In: *Third International Specialty Conference on Cold Regions Engineering*, Edmonton, Alberta, April 4-6, 851-9.

U.S. EPA. 1980. *Process Design Manual for On-Site Treatment and Disposal Systems*. U.S. Environmental Protection Agency. Rep. No. EPA-625/1-80-012, Washington, D.C., 391 p.

U.S. Public Health Service. 1950. *Individual Water Supply*. Publication No. 24, Government Printing Office, Washington, D.C., revised 1974. U.S. Environmental Protection Agency.

Watson, G.L. 1978. *Private Sewage Disposal Systems, Instructions to Installers*. Alberta Labour, General Safety Services Division, Plumbing Inspection Branch, Edmonton, Alberta, 58 p.

14.6 Bibliography

nd. *Recommended Standards for Individual Sewage Systems*. Health Education Services, Albany, NY.

1973. *Planning for an Individual Water System*. American Association for Vocational Instruction, May.

1974. *Manual of Grey Water Treatment Practice.* Ann Arbor Science.

1979. *Private Water Systems Handbook.* Iowa State University.

1981. *Guidelines for the Use of Sand Filters.* Washington State Department of Health, Olympia, Washington, August 2.

1990. *Cold Temperature Performance of On-site Sewage Disposal Systems in Yukon Territories.*

Clark, S.E., and D.P. Williams, D.P. 1983. Alaska on-site wastewater systems. In: *Proceedings of the First Conference on Cold Regions Environmental Engineering*, May 18-20, Fairbanks, Alaska.

Converse, J.C., and E.J. Tyler. 1990. *Wisconsin Mound Soil Absorption System Siting, Design and Construction Manual.* University of Wisconsin-Madison, January.

Kaplan, O.B. 1991. *Septic Systems Handbook.* Lewis Publishers.

Laak, R. 1987. On-site wastewater drainfields using light-weight in-drains in cold regions. In: *Proceedings of the Second International Conference on Environmental Engineering*, Edmonton, Alberta, March 23-24.

Lehr, J.H., T.E. Gass, W.A. Pettyjohn, and J. DeMarco. 1988. Domestic Water Treatment. National Water Well Association.

Olofssen, J. 1994. *Applicability of Composting Toilets to Rural Alaskan Villages: Phase I Guidance Document.* May 1, 1994 for US Public Health Service.

Oregon Department of Environmental Quality. 1975. *An Evaluation of Alternative On-site Sewage Treatment and Disposal.* April.

Perkins, R.J. 1989. *Onsite Wastewater Disposal.* Lewis Publishers.

Peterson, M. 1985. *Soil Percolation Tests in Alaska.* U.S. Public Health Service, Alaska Area Native Health Service, Anchorage, April.

Peterson, M. 1987. *Adsorption Area in Fine Grained Soils.* U.S. Public Health Service, Alaska Area Native Health Service, Anchorage, August.

Rignyk, R. Z. 1988. Peat Mound Drainfields for Residential Septic Treatment, *Northern Engineer*, Fall-Winter.

State of Alaska, Department of Environmental Conservation. 1982. *On-site Wastewater Disposal Study.* July.

Tilsworth, T. 1979. *The Characteristics and Ultimate Disposal of Waste Septic Tank Sludge.* Institute of Water Resources, University of Alaska.

U.S. Environmental Protection Agency. 1987. *It's Your Choice: A Guidebook for Local Officials on Small Community Wastewater Management Options.* Rep. No. EPA 430/9-87-006, Washington, D.C.

U.S. Public Health Service, Alaska Area Native Health Service. 1984. *Design Considerations for Septic Tanks in Remote Alaskan Communities.*

U.S. Public Health Service, Alaska Area Native Health Service, Anchorage 1989. *Domestic Septage Sludge Management and Disposal for Rural Alaska.* November.

Virarghavan, T. 1984. Temperature effects on on-site wastewater treatment and disposal systems. In: *Third International Specialty Conference on Cold Regions Engineering*, Edmonton, Alberta, April 4-6.

Winneberger, J.H.T. 1984. *Septic-Tank Systems: A Consultant's Toolkit.* Butterworth.

SECTION 15

FIRE PROTECTION

3rd Edition Steering Committee Coordinators

Gary W. Heinke

James A. Crum

3rd Edition Principal Author

Gary W. Heinke

Special Contributions

Cam Marianayagam

Daniel H. Schubert

John Warren

Section 15   Table of Contents

Section 15   List of Tables

15   FIRE PROTECTION

15.1   Review of Existing Situation in Northern Communities

The problems concerning fire protection in cold regions arise primarily from:

- the small size and isolation of communities;
- lack of finances (tax base) in most places;
- some social conditions resulting from the changing culture;
- the harsh climate;
- excessive response time; and
- inadequate water supply for firefighting.

It takes a certain minimum amount of equipment and trained personnel to provide fire protection for a community. The small population of northern communities therefore makes per capita costs of fire protection very high. Isolation makes sharing of equipment through mutual aid agreements impossible and sharing of personnel difficult or impossible. Fire prevention activities are therefore very important. Since very few, if any, serious fires occur in small communities during a year, keeping the need for fire protection in mind is difficult. This lack of interest contributes to a high turnover rate among volunteer firefighters, a lack of maintenance of equipment, and a general deterioration of fire prevention efforts.

In North America, fire protection is considered to be the responsibility of the local government. Therefore the money required for a public (or private) fire department must be raised locally. Because many northern communities have no tax base, they must depend on government assistance. In communities with a large industry, such as a mine, help is often found for the purchase of fire-fighting equipment. Usually these communities are sufficiently well organized to collect taxes so that even with a very small mill rate, such an industry adds substantially to the community's ability to raise money.

The effectiveness of a fire protection system depends on both the general public and the fire department. Although most fires can be prevented through reasonable care and by avoiding dangerous practices, this usually cannot be legislated and therefore depends on public awareness. In small traditional native villages the number of fires reported each year per capita is small. In larger northern towns with a larger transient population and a less traditional lifestyle, the number of fires per capita is large. In cities however, the number of fires per capita is slightly above average for similar southern communities. The physical climate has an effect both on the number of fires and the severity of fires. The long heating season coupled with dry conditions makes wood frame houses very susceptible to fire. A common cause of fire each year relates to heating devices. Throughout most of the North, furnaces may be turned on at any time of the year.

In many northern communities fire records are not as well organized and complete as in southern areas, except when fire deaths occur. This makes a comparison of fire loss statistics difficult. However, there has been considerable improvement in record keeping in the last decade. The present situation in each of the political regions varies, and depends mostly on the amount of government assistance received.

15.1.1   Alaska.   In Alaska, fire protection is primarily a local responsibility. A community must raise most of the money for fire protection itself. Grant funds are available though, through state revenue sharing. In a small community this does not amount to much and is generally used for things unrelated to fire protection.

A community with a well informed and active local government can obtain funding for fire halls, fire trucks, and other major purchases by applying for state and federal grants from various agencies. Sometimes specific appropriations can also be obtained through the state legislature. Larger communities can issue bonds or borrow money in addition to applying for grants. In any case, the money for operating and maintenance must be covered in the municipal budget.

With local initiative being the driving force behind fire prevention, no standardization exists. Large communities are typically well equipped, staffed, and trained.

Fire prevention is one of the main concerns of the state fire marshal's office. For projects involving public occupancy buildings, state regulations require the submission of plans to the state fire marshal for fire and safety code review. Here they thoroughly review building plans to ensure new construction meets the Uniform Fire, Uniform Building and Uniform Mechanical Codes. The state fire marshal's office also inspects public buildings. Large municipalities have their own fire prevention departments.

Some small communities also have well run fire departments, considering their limited resources. Most native communities have nominally organized fire departments. In small villages, fire prevention education and building inspections are provided from outside the village only at the request of the local government. Generally the service is free. A new program was introduced in 1980 by the Department of Public Safety which pays for a Village Public Safety Officer (VPSO). The VPSO is a local person, trained at the State's expense in law enforcement, emergency medical treatment, and fire prevention and suppression. These officers are employed on a full-time basis to promote public safety in the village in which they live.

Most villages have no building separation requirements, often no adequate roads, or even rights of way. Most housing consists of single-family units of wood frame construction. Interiors are lined with plywood rather than drywall for increased durability. In new federally financed construction, care is taken to provide space as well as heat protection around stoves and chimneys. In new public occupancy structures, ceiling-mounted fire extinguishers, located above fire burning appliances are frequently installed.

Water supply in many villages is self-haul from a central water point. Therefore, only fire extinguishers are available for fire suppression. When a fire occurs, everyone assists and uses whatever equipment is available. Quite a few larger villages have piped water supply systems, but generally these are not designed for full fire protection. There are very few truck-haul systems. Fire suppression systems must therefore depend on portable pumps and open water sources.

Larger towns (populations greater than 2,000) often have a well organized volunteer fire department with a paid fire chief. Such a community probably has a piped water system with hydrants or a truck-haul system for fire protection.

In cities, the fire departments have at least a full-time core staff with volunteers or a completely paid department. Usually mutual assistance agreements are drawn up with some of the smaller surrounding communities.

The fire departments in Alaska vary widely, making it difficult to generalize about their equipment. The main piece of equipment is a pumper truck with a water tank and hose bed (triple combination pumper or TCP). These trucks are similar to those that form the basis of most fire departments throughout North America. In addition, there may be minipumpers, high pressure pumpers, ladder trucks, hose tenders, and tank trucks. Communities with good equipment usually have adequate storage space for it. Alarm systems range from the church bell to telephone systems to public callbox systems.

Health corporations in Alaska are also active in fire protection. Health corporations emphasize fire prevention education and actively distribute smoke alarms with funding from state agencies.

The main feature of public fire protection in Alaska is its dependence on local initiative and resources. In general, communities of less than 1,000 may have almost nothing whereas larger communities can be quite well equipped and well organized.

15.1.2 Northwest Territories. In Canada, fire protection is also the responsibility of local government. Two major reports on fire protection in the Northwest Territories (NWT) have been prepared. The 1982 report (Heinke and Bowering, 1982) and the 1993 report (Heinke and Marianayagam, 1993) were used in the preparation of the monograph. In the NWT many local government responsibilities are handled by the Department of Municipal and Community Affairs (MACA) which is a branch of the Territorial Government. Since settlements and hamlets are not organized to raise taxes, they obtain most of their funding through MACA. Villages, towns, and cities are tax based and are primarily responsible for their own budgets. It is the policy of MACA to provide fire protection in all communities, usually in the form of equipment and money to operate a volunteer fire department. The fire prevention aspects of fire protection are, for the most part, handled by the NWT Fire Marshal's Office. Due to the small staff of the Fire Marshal's Office and the vast distances between communities, the effectiveness of fire prevention is rather limited, but substantial improvements have occurred in the last decade.

Fire department equipment provided by the government is related to the population of a community, the fire load in that community, and the availability of piped water. The type of equipment is similar to rural, North American standards, modified for cold region operation. In most communities the standard issue is a 2,950 L/min (625 gpm) front mount pumper with an on-board 4,540 litre (1,000 gallon) water tank. The truck package also includes all the necessary ancillary equipment such as hoses, ladders, and turnout gear.

In communitites with both a high fire load and piped water, a 4,750 L/min (1,050 gpm) midship pumper may be provided.

In very small communities a "duster" truck is provided. This is a standard 3/4-ton pickup truck mounted with a 135 kilogram (300 pound) dry chemical extinguisher. In these communities, the water truck may also be fitted with some extra hose to assist in firefighting.

Water supply in most of the smaller communities in the NWT is generally by tank truck. Some communities have partially piped systems, and the larger towns have piped systems with hydrants. The tank truck is kept full at night and responds to fires along with the fire truck. Fire department personnel consist of a volunteer fire chief and an organized group of volunteer firefighters. Training and experience varies widely; most volunteers are relatively inexperienced. Turnover is quite high in all but the large communities. Insufficient funds for the training of firefighters remains a problem in the NWT. Alarm systems based on public call boxes connected to a siren are still used in some communities. In larger communities fire departments are dispatched through offices that are operated 24 hours a day. In the smaller communities fire departments are dispatched through a special fire phone system. A fire report is called to the community fire number, which causes a dedicated telephone to ring in individual firefighters' homes. The person answering the telephone takes the information and activates the fire sirens. Other communities use a pager system. The maintenance of all alarm systems is difficult in cold regions.

The day to day operation and maintenance of the fire department has to be carried out at the local level. In small communities with very few serious fires in a year, alarm systems and fire trucks can break down due to neglected maintenance. Repairs are difficult because technical expertise is rarely available locally. For this reason much of the equipment has a relatively short life; one already shortened by operation in severe climatic conditions. Therefore some communities still have unacceptable fire protection systems.

New construction in the NWT must conform to the National Building Code (NRC, 1990a), the National Fire Code (NRC, 1990b) and the NWT Fire Prevention Act. Plans must be reviewed by the Fire Marshal's Office. That office is not staffed sufficiently for routine building inspection, but large public buildings are generally inspected before they are occupied. Most construction is of wood frame. Large separation distances (twice those required by the National Building Code) are required in almost all communities to compensate for the lack of well organized fire departments.

15.1.3 Yukon Territory. The fire protection policies in the Yukon Territory are very similar to those in the NWT. The Yukon Territory, however, has a better record of success. The main reasons are fewer and more accessible communities and a longer development history.

The Fire Marshal's Office administers the fire protection budget of the unorganized communities and the local improvement districts (LIDs). They are approximately equivalent to settlements and hamlets in the NWT. Equipment is fairly standard in each community. The smallest communities (fewer than 200 inhabitants) have only portable pumps and hoses. Most communities have a 2,270 L/min or 2,840 L/min triple combination pumper. All fire trucks are stationed in firehalls, some of which have meeting rooms and extra storage space.

Water supply in many Yukon communities is by piped distribution systems with fire hydrants. Although most water supplies would not meet Insurance Services Office (ISO) standards (IBC, 1977), the presence of fire hydrants does increase the effectiveness of the fire trucks. In most cases water trucks are also available for more outlying areas.

Alarm systems are very simple in the Yukon. They generally consist of a few call boxes, all directly connected to a siren. In only one very spread-out community is there any form of a zoned system. The systems have the usual electrical problems but because of the easier access to the communities, these can be readily fixed.

One of the main differences between the Yukon and NWT is that inspectors from the Fire Marshal's Office can keep in close contact with the communities. If a fire truck breaks down or if a fire chief is not doing his job properly, the inspector can drive to the community and rectify the situation. Fire trucks purchased by the government are driven to Whitehorse regularly for a major overhaul. This gives the fire trucks a much longer life.

Fire departments are organized like those in the NWT. The chief and firefighters are all volunteers and are paid only nominally for their efforts. Problems of high turnover and therefore limited experience of personnel exist. Training is usually left to each fire chief. Again, the ability of the inspector to be in communities helps in organizing training and practice drills.

Plan review procedures are similar to those in the NWT, but due to the smaller population and fewer communities, the job can be handled more thoroughly. The easier access to communities also facilitates ongoing inspection programs. There is also close cooperation between the mechanical, electrical, and boiler inspectors, so that an inspector, when travelling to a community, attends to specific problems that one of the other inspectors may have indicated.

15.1.4 Greenland. The population of Greenland is much more concentrated than that in northern Canada or Alaska. Most people live in larger towns with populations of 1,500 to 10,000. These towns are composed of many apartment buildings, row housing complexes, and single-family neighbourhoods. All towns have piped water supply and fire hydrants, although the amount of water available at a hydrant is often well below North American standards.

Fire protection in Greenland is quite different in many ways from that in North America in that it is a subsidized state program in most communities. Fire departments are generally well organized, well equipped, and well managed. They are also standardized to a very great extent. Fire departments are a municipal responsibility but with state government support.

Each town has a firehall with meeting rooms, workshops, and vehicle bays. Adequate fire equipment is available. The water supply is considered adequate by European firefighting practices, but would not be considered adequate in North America.

The job of fire chief is often a part-time paid position.

15.2 Requirements of Fire Protection Systems

The primary goal of a fire protection system is to save lives and reduce the property damage caused by fire. Property damages and loss of lives are called fire outputs; they are measured as losses per capita. Losses per capita can be further subdivided into two main components: number of fires per capita and average loss per fire. The losses can be measured in terms of direct dollar losses, direct plus indirect dollar losses, lives lost, insurance premiums minus

claims paid, or any other accounting method. For most purposes, to avoid confusion, direct dollar loss and number of deaths are measured (Switzer and Baird, 1980).

Fire outputs are reduced through the establishment of an integrated fire protection system. These are called fire protection inputs and are usually subdivided into: fire prevention and fire suppression. Fire prevention activities are those which reduce the probability of a fire occurring and also include the passive control of fires through building design. Fire suppression is the active control and extinguishing of fires already out of control. Both are an important part of fire protection systems but are usually considered separately and often provided by different agencies. Fire prevention inputs should reduce the number of fires per capita and also have an effect, through passive control, on the average losses per fire. Fire suppression inputs mainly affect the average losses per fire through active control.

The effectiveness of these fire protection inputs in reducing fire outputs is influenced by physical and social conditions within each community. Some conditions relevant to northern communities are climate, population, water supply, degree of isolation, type of construction, economic base, and lifestyle of the community. Therefore, one must consider the relevant community conditions and likely changes in these conditions when determining needed fire protection system components.

15.2.1 Legislation and Ordinances. Three levels of government must be considered: the federal, the regional (provincial, state, or territorial), and the local governments. Each has some input. In North America, fire protection is primarily a local government responsibility. Regional or federal laws require neither a fire department in towns nor building codes or zoning laws.

Fire Prevention. Public forms of fire prevention, such as public education through advertising, are generally provided by regional and federal governments, but they are not required to do so. During the 1980s the Offices of the Fire Marshals in NWT, Yukon and Alaska were very active in public education on fire prevention (Heinke and Bowering, 1982).

Private forms of fire prevention, which include building quality and life safety measures, require legislation at some level. Generally this is done at the local level through bylaws or ordinances requiring adherence to nationally recognized codes. This is done at the regional level, especially in the North where a large portion of the population is without a real local government.

Fire Suppression. Public fire suppression at one time was provided by private service clubs, but this is uncommon now. Generally the fire department is a division of the local government in accordance with a local ordinance or bylaw. The legislation is required to allow the local government to raise and spend money on a fire department. In the Canadian territories, public fire suppression funding is provided by the regional government where the local government cannot afford it.

Private forms of fire suppression, such as sprinkler systems, fire extinguishers, and smoke alarms, have to be legislated if they are to be part of a dependable fire protection system. Government subsidies for the purchase of such equipment can be an incentive. Where the government owns or constructs many of the buildings, such equipment is included as government policy, particularly for smoke alarms and fire extinguishers in all buildings. In the NWT, pilot projects in selected communities are now recommended to accurately determine the costs to install, inspect, operate and maintain sprinkler systems in new and existing homes (Heinke and Marianayagam, 1994). Automatic sprinklers drastically reduce the response time in the event of a fire.

Law Enforcement. Any legislation requires enforcement if it is to be effective. It is currently difficult in the North to enforce adherence to building codes because of a lack of supervisory staff. If more fire prevention and suppression strategies were legislated at the regional level, staff to do inspections would have to be increased. Some improvement in this situation has occurred during the 1980s.

15.2.2 Fire Prevention: Reduction in Numbers of Fires.

Public Education. Public education is the main form of fire prevention. It is sponsored by all levels of government and also by the insurance industry. Its main purpose is to make people aware of common hazards, encourage care, and also to teach some "first aid" fire suppression. This is very important in the North where the domestic environment has been changing rapidly.

Public education is available through advertising and industrial films on television and radio or through public meetings conducted by a qualified individual. Working with children in schools is also important for long-range results and sometimes even for more immediate results since children may influence their parents' attitude toward fire safety. It is important that a qualified fire professional provide the instruction in schools. Leaving it to teachers, no matter how dedicated, is not acceptable. It is the responsibility of the regional/territorial fire marshals to prepare material for public education. Much progress has been made in the North during the 1980s. Examples include: Learn Not to Burn Program; fire safety seminars; annual fire safety contests for the public; calendars with fire safety messages in local languages; posters and handout fire safety pamphlets; and regular teaching in all schools.

Other Prevention Activities. Other fire prevention activities include the supervision and approval of electrical and mechanical installations and the handling of combustibles and explosives. The purpose is to reduce the chances of mechanical or electrical fires caused by faulty installation. This again is very important in the North where specialization among contractors is limited. Qualified personnel are required to see that installations are according to standard practices and therefore safe. Because qualified people are not likely to be available at the local level to do this, the regional government must participate. Even then, regional personnel are seldom available in remote or isolated areas; therefore, care must be taken in both design and choice of contractors to ensure reasonably safe mechanical or electrical installations.

15.2.3 Reduction of Fire Losses – Passive Means. The passive reduction of fire losses is achieved through careful design and choice of materials and contents. It works on the premise that fires will start and therefore must be controlled. The purpose of building codes is to help designers keep risks to acceptable levels for occupants and neighbors of a building. Building codes are set up along with minimum standards to limit the spreading rates of heat and smoke sufficiently to allow occupants to escape. The codes also stipulate building separations required to control the spread of fire to other properties. Adherence to these nationally recognized codes is often required by legislation because safe construction is usually more expensive and might be short-circuited if not legislated.

In the North, fire prevention is more important than in the South because of the difficulty in providing good fire suppression. In some areas the building codes assume a reasonably quick response by the fire department. In those northern communities where this cannot be accomplished, stricter building requirements are necessary including, in some jurisdictions, greater building separation distances to reduce the risk of spreading fire.

15.2.4 Reduction of Fire Losses – Active Means.

Fire suppression is the reduction of fire losses through active means such as detecting, controlling, and extinguishing fires that have started, as well as rescuing people from burning buildings. Privately maintained fire suppression components include smoke detectors and alarms, sprinkler systems, fire extinguishers and, in some large industrial complexes, private fire departments. The major area of interest is in public fire suppression, the duty of the public fire department (ISO, 1974).

Fire Detection Devices. The main purpose of fire detection devices is to provide an early warning of a fire condition so that occupants can escape and the fire can be extinguished while still small. Ionization and photoelectric-type smoke detectors are two types of detection devices (Gratz and Hawkins, 1980; Smoyer, 1981). The ionization type is the most common in residential use. Smoke detectors can be either wired into the electrical system of the house or battery operated. For new construction, wired-in detectors are preferred, because batteries tend to be either "borrowed for other uses" or not replaced when run down.

In major buildings such as hospitals or schools, fire detection devices should be connected to the fire department alarm system or at least to a loud siren on the outside of the building. This enables a faster response by the fire department.

Sprinkler Systems. Sprinkler systems were generally only used in large buildings in the past, because installation cost and maintenance were considered too high for residential buildings. Recent developments in sprinkler technology have resulted in a quick-response-type residential sprinkler, specifically for one- and two-family homes (Kung et al., 1980). They have a response time from five to fifteen times faster than standard sprinklers. Criteria for these residential sprinkler systems NFPA 13D (for one- and two-family homes) and NFPA 13R (for multiple-family homes) are given in (National Fire Protection Association, nd). NFPA 13D does not require that a residential sprinkler be connected to a municipal water supply in communities where a piped supply does not exist. Because the primary purpose of a residential sprinkler system is life safety, a ten-minute water supply complete with a smoke alarm system is considered adequate. The storage tank volume required is about 1,000 litres, to provide ten minutes of flow for two sprinklers at 50 L/min each, together with an automatically operated pump. A combined storage tank system for household water supply and for fire protection is acceptable, provided that the tank volume is sufficient for both uses. A combined storage tank system must have a low water level shut-off on the domestic supply line to guarantee at least 1,000 litres for fire protection.

Where municipal piped distribution systems can serve as the water source, increasing the size of the service connection piping may be necessary. Increasing the size of the service connection piping will depend on system pressures and distances. Building sprinkler systems connected to municipal water systems require rubber-faced check valves to isolate the sprinkler system piping from the municipal system.

The use of automatic residential sprinkler systems in all new one- and two-family homes and retrofitting of existing homes is recommended by some studies because of the continuing high fire death rate in the Arctic. It should be noted that the mandatory use of residential sprinklers is not required in Canada or Alaska. In the NWT, the situation is sufficiently different to warrant the testing of residential sprinkler systems on a pilot project basis in selected communities (Heinke and Marianayagam, 1994).

Annually about 40 voluntary installations of automatic residential sprinkler systems in one-family homes have occurred in the NWT since 1988. The majority of the sprinkler systems are wet systems. This provides minimum delays in water application, and less dependence on additional mechanical/electrical systems. Maintenance problems so far have been minimal. Freeze protection is no more of a problem than for regular household plumbing.

Fire Extinguishers. A portable fire extinguisher can provide a quick knock-down of flames. The most common are the multipurpose, dry chemical types. They must be able to be serviced and refilled locally. The cartridge-operated type has found the greatest acceptance throughout cold regions. Ceiling-mounted chemical dump fire extinguishers, mounted over fire use appliances, are also becoming more popular. In the past, halon fire suppression systems were common, but with legislation in the United States banning production of these chemicals, other types of chemicals are now being used.

Usually each household is responsible for providing its own fire extinguisher. The local or regional government can help by providing them at subsidized prices. Refills and maintenance can be provided free by the local fire department. In very small communities, larger public fire extinguishers located throughout the community are often much more effective than a pump and hose layout system. Portable extinguishers must be used quickly and on fairly small fires. Communities depending on portable fire extinguishers should therefore have a reliable fire detection system in place.

15.3 Public Fire Suppression

Public fire suppression is generally handled by an organized local fire department. It is a labor intensive activity and depends on volunteers in all but the large communities, which may be able to afford some paid firefighters. Different sizes and types of communities have different resources and different firefighting requirements.

Where regional government input to local fire departments is expected, communities should be classified according to available resources and requirements.

15.3.1 Community Classification.

Type A – Small (fewer than 1,000 inhabitants); mostly native; subsistence economy. Growing slowly or not at all, because young adults leave for larger communities. Large proportion of children. Relatively traditional lifestyle. Typical of NWT settlements and hamlets, and communities in western and northern Alaska.

Type B – Small (fewer than 1,000 inhabitants); highway-type communities. More related to a wage economy. May have one or two main employers; i.e., mine, sawmill, highway services. Growth is related to the availability of services. Greater nonindigenous portion of the population. Less isolated than Type A. Typical of Yukon and south-central and eastern Alaskan communities.

Type C – Large (more than 1,000 inhabitants); village or town status. Local government with ability to pass bylaws and raise money through taxes. Wage-related economy. Fairly transient population from smaller surrounding communities. Often a regional government centre, transportation and distribution centre. Typical of a number of Alaskan communities and about ten communities in northern Canada.

Type D – Cities (population greater than 10,000); several of these are in Alaska. In northern Canada only Whitehorse and Yellowknife are in this category.

It has to be understood that these are fairly broad classifications with undefined boundaries. One should consider not only the present state of the community, but also its future. The main significance is that Type A and B communities have no tax base and therefore cannot afford to buy and operate equipment. Type C communities can raise taxes and so can at least operate a fire department. Type D communities are not really different from other southern communities of similar size and therefore have similar resources and requirements.

Some other special conditions which may take on more significance than the type of community are road conditions, type of water supply, and the existence of large buildings or special fire hazards. Another type of community is industrial camps. Because of the usual presence of trained people, the compactness and new construction of camps, their fire prevention and protection is normally less of a problem than for communities of equivalent size.

15.3.2 Alarm Systems. A small, centralized community needs nothing more than a centrally located call box directly connected to a siren. In slightly larger or more spread out communities there should be several call boxes (approximately one for every 50 people), all directly connected to a siren. Such systems are simple and relatively dependable. In larger communities (more than 500 habitants) however, a problem of determining the exact location of the fire with this type of alarm system may arise.

If telephone coverage is good, the community can rent alarm systems from the telephone company. It is basically a dedicated number that rings the phone in all volunteers' houses or work places. It can also be directly connected to the town siren. The main advantage is that the telephone company owns the system and contracts to maintain it. The capital cost is very small compared to a community call box system. The only operating cost is the rental fee to the telephone company.

If there is a telephone manned for 24 hours in the community (answering service, weather station, hospital desk, water pumphouse) it is often better to have that telephone's users serve as dispatcher. Radio pagers have been found to be more effective than a siren, provided a central dispatcher operates them.

The maintenance of any of these alert systems is difficult. Sirens not activated daily are filled with wind-driven snow, which can freeze the motor armature solid, so that when an alarm is required, the siren burns out instead of sounding. Time clocks installed to exercise sirens at a regular time each day, end up going off in the middle of the night because of power outages, or because of cycle changes causing clocks to run fast or slow. Fire phones have to be moved as firefighters change, which may require a trip to the community by the telephone company, often with much delay and considerable cost. Pagers and radio repairs can seldom be done in the community, so the potential for the system to be down for lengthy periods is real.

15.3.3 Firefighting Vehicles and Related Equipment.
The fire truck must be compatible with the road conditions, the required fire flow, and especially the water supply. It must also be fairly simple to operate and maintain.

In small Type A communities that have no fire hydrants and probably no water trucks, the largest building is usually a school of wood frame construction. An automatic sprinkler system provides the best possible fire protection. There are also, typically, only one or two fires a year in the community. It is not feasible in this type of community to protect the school or other large building adequately with fire trucks. Experience has shown that where very few fires occur, the fire truck is left to deteriorate or is used for other purposes. Instead, a large pickup truck or van with one or two good-sized portable pumps and a hose bed is more practical. A water tank truck, if available, carries the initial water supply. The truck also carries fire extinguishers and other small equipment. If no water truck is available, then enough large-diameter (75 or 100 mm) hoses and pumps are required to carry water from an open source.

A portable pump should be able to supply at least one 65 mm hose stream or two 38 mm hose streams, providing about 1,150 L/min with 345 kPa of pressure at the nozzle. The large-diameter hose is to transport the water to the fire scene and the small-diameter hose is for firefighting. In hilly areas, extra pumps may be required to deliver water at sufficient pressures.

Small communities should concentrate on built-in fire prevention (passive means) and building separations rather than on the ability to fight large fires. The major buildings should have fire extinguishers and a sprinkler system on a pressure tank or pump.

Some Type A communities, especially larger ones, have good roads and may even have a piped water supply with fire hydrants. Where no fire hydrants exist, at least a good water source, such as an ad-

equately sized storage tank near the centre of the community, should be available for water truck refill or for portable pump and hose relay. Since there may be a number of larger commercial buildings, a small fire truck is useful. Pump size should be at least 2,270 L/min, preferably 2,840 L/min. The water tank can be as large as 4,090 litres.

Minimum storage for an adequate truck relay fire protection system is about 60 m³. This is equivalent to an average flow of about 450 L/min for two hours plus 10%.

Type B communities are in many ways similar to the larger Type A communities. They generally have good roads and a more organized water supply. The presence of some industrial or commercial interests may make more money available for the purchase of equipment. Again, it must be recognized that not all fire flow requirements of a building can be met by one fire truck. Therefore, important buildings should have sprinkler systems and fire extinguishers.

The maintenance of equipment is generally better in Type B communities because of their lesser isolation. In the Yukon, which consists mostly of Type B communities, government fire trucks are driven to Whitehorse for regular overhauls.

Type C communities require at least one fire truck and more in larger communities. Most of these communities have a piped water supply with fire hydrants, at least in the core area. Outlying areas, which may include some commercial or even light-industrial buildings, are serviced by water truck. If more than one pumper is required, a large pump should be on one (3,800 to 4,650 L/min) and a small pump on the other pumper (2,840 L/min). A fire department water truck is also useful for fires in outlying areas. Private water contractors probably do not want to have to respond to all alarms, but it would be prudent to have an agreement with them to be on standby during alarms in case they are needed to relay water. If large warehouse or storage facilities are located outside the area serviced by fire hydrants, portable pumps should be available, preferably on site, with hoses stored nearby.

In Type C communities, maintenance is not as big of a problem as it is in Type A or B communities because of the greater resources, including locally available training facilities and the greater frequency of fires. The communities are still isolated by southern standards and replacement parts are slow to arrive and are very expensive. Often the fire department has a full-time fire chief or at least a full-time maintenance person to keep the equipment in good order. Most communities of this type are serviced by fire hydrants and one or two pumpers.

Type D communities are large enough to determine their own needs. They are basically not much different from southern communities of a similar size. Water supply is usually a problem in outlying areas and therefore northern cities often run a larger tanker operation than they would in the South. The spread-out nature of some cities may require additional fire stations and more fire trucks per capita than is usual in more contained cities. This makes the costs of running a northern fire department high, besides the additional expense of operating in a harsh climate.

15.3.4 Firehalls. All types of communities require single-purpose firehalls. The firehall space may be physically part of another complex but it should be kept separate and locked. Its main purpose is to provide storage space for firefighting equipment. It must be well organized with nothing missing. In small communities the firehall tends to be used as a public works garage or vice versa. This should be discouraged because maintenance, along with keeping equipment in order, is one of the major problems in these situations. The garage portion is used constantly and the firehall only intermittently. The firehall portion is soon overrun by material from the garage. Tools, boots, hoses, etc. may be borrowed, lost, or misplaced and not readily available in an emergency. One solution is to erect a locked partition between the two areas. When expansion is required, the partition is removed and one function is moved to a new facility.

Firehalls must provide space for such functions as washing, thawing, and drying the hose, a work area for refilling and maintaining fire extinguishers, and room for training and administration. One should be able to carry out all activities indoors. Also, an adequate water supply and floor drains for washing, thawing, and refilling fire trucks and tankers should be included in the design of a firehall.

Four types of firehalls are being built in the NWT having vehicle bays varying in size from 45 m² to 65 m², training areas varying in size from 45 m² to 60 m², and storage and office space of about 30 m² (usually on two levels). Firehalls in Alaska vary in both size and layout. Design of firehalls in Alaska are generally based on the characteristics of the population being served.

A good firehall can help develop a "clubhouse" atmosphere, an advantage to the morale and effectiveness of the volunteer fire brigade that should not be underestimated, especially where not many fires need be fought.

15.3.5 Fire Department Organization. Most fire departments are of the volunteer type, except in the cities. In Type C communities (towns) at least a paid full time or part-time fire chief should be hired. This should be someone with some formal firefighting training, along with the ability to handle the administration of a fire department. The fire chief should work closely with the regional fire marshal to make sure fire prevention activities such as plan review, building inspection, and public education are carried out.

In Type A and B communities a volunteer fire chief usually has only a little more training or experience than the rest of the volunteers. This is not a good situation, but to keep a fire chief in a full-time position is usually not feasible. One alternative to consider is a variation of the VPSO program in Alaska. A full-time paid position could be created, which includes a variety of public safety duties such as fire chief, emergency medical services, bylaw officer, search and rescue coordinator, and public safety education coordinator. Training in each area would be part of the job requirements. In the NWT some hamlets have a paid position combining bylaw officer and fire chief.

In small fire departments the fire chief is the most important element in the fire protection system. To have a paid fire chief, at least on a part-time basis, can prove a good investment for a community. Lesser payment to firefighters is also important.

Problems with high volunteer turnover will probably always exist in small communities. The best that can be done is to hold regular practice drills that have local relevance and also to try to develop a professional attitude through local and regional firefighters' associations.

In Type A communities, problems develop during hunting and fishing seasons when most of the men suddenly depart. This should be planned for and a core of volunteers who do not plan to leave can provide some backup.

Fire Department Training. Training is a problem in the North because of the isolation of communities and the high turnover of local staff. At least the fire chief and an assistant chief should receive tactical fire training at a professional training centre in a course relevant to their type of community. The fire chief should also receive training in administration, fire reporting and investigation, and building inspection. Volunteers are trained in the community by the fire chief. Fire instructors from nearby towns or regional fire officers can help the local chiefs in setting up drills and practices.

In Alaska, regional training centers are available to any organization that wants to provide training in fire suppression. They were built by the state but are owned and operated by the local government. In northern Canada no facilities specifically suited for training in firefighting exist. In the NWT, the inadequate training of firefighters remains one of the major obstacles to the success of fire protection.

Fire Department Financing. In a city, the greatest portion of the fire department's budget goes to firefighters' salaries. A well run volunteer fire department is therefore a great savings to the community. Firehalls and fire trucks are now available in most NWT and Yukon communities. Some Alaskan communities have been able to raise money for equipment privately. This indicates a strong local interest in fire protection, ensuring that equipment is maintained and volunteers are motivated. The only other alternative for unincorporated communities is to obtain grants from the regional government. In Alaska these grants have to be sought out and applied for by each community. In Canada it is territorial government policy to provide capital and operating grants to settlements and hamlets.

Larger communities can also obtain grant money, but they can usually raise money themselves through normal municipal channels.

15.4 Water Supply for Fire Protection

Traditionally, the public fire department uses the community water supply and distribution system as its water source. This is also true in the North; however, the amount of water available in many existing systems is very limited. Typically, only large-diameter piped distribution systems with large reservoirs and emergency pumping capacity can supply standard fire flows. Existing systems for organized water supply systems in the North consist of:

- self-haul where water is collected by individuals from water points or a community reservoir;
- truck delivery where water is delivered by truck to consumers; and

- piped systems where water is delivered by a network of water pipes.

Many communities may have a combination of two or all three alternatives. New systems, which use piped water delivery, may be designed similar to systems in the south. These systems are very expensive; as a result, only the larger communities can obtain them.

Mandatory use of sprinkler systems for all buildings (including single-family homes), new construction and retrofitted, may become legislated in some areas in the future. Sprinkler systems are the most effective way of providing immediate fire protection. In some areas sprinkler systems may result in a reduced need for regular fire protection.

Central reservoir water tanks, used with the self-haul systems, may be easily tapped as a water source for firefighting, if designed properly. Only a connection for a portable pump is required. Where the water source is intermittent or a well is of very low yield, care must be taken that an untrained fire department does not drain the reservoir before extinguishing the fire. Two solutions to this problem are to have either a larger reservoir or a separate fire reservoir. Communities of this type would not have a pumper fire truck but rather a truck carrying portable pumps, large-diameter hose to lay to the fire and regular hose for one or two attack lines.

In self-haul systems, water reservoirs are usually not very large. To require 60 m^3 of stored water for fire protection may require tripling the existing storage capacity; however, if the system is to be used for fire protection then there should be enough water available to supply the fire department for at least one hour at an average flow of 450 L/min or a total volume of 30 m^3. Another approach is to determine how much water can be spared for emergency fire protection and then arrange a system that only uses that much water. For many fires this may be a sufficient quantity.

Trucks used in truck delivery systems may be used to supply a pumper truck with water if more than one tank truck is available. Water flow with this type of system is limited; however, it is still useful to knock down flames in short bursts and extinguish a fire in its early stage. The full tank truck and the fire truck must arrive at the scene at the same time to ensure that the fire truck is not restricted by lack of water in the first important minutes. Water delivery trucks should respond to all fire alarms unless the fire department has a tanker truck. Water delivery trucks and tanker trucks should also be kept full when off duty and parked in heated garages.

In typical piped distribution systems, measures for providing water for fire suppression are usually included in the design. Such measures increase the cost of the distribution system by requiring a minimum pipe size of 150 mm diameter, fire hydrants every 100 m, and a larger storage capacity of about 435 m³ (3,625 L/min for two hours). If such measures are not included, the water system is less expensive to build, and does not change the day-to-day benefits of having piped water. Such systems designed without provision for fire protection are often called small-diameter systems because they are characterized by piping smaller than 150 mm in diameter in much of the distribution network.

Water flow for fire suppression may be distributed by several methods, including gravity storage and distribution. Gravity storage and distribution is the most reliable method, but requires larger storage and main line sizes. High-flow pressure pumps may be used on the distribution system when limits on the configuration reduce the options on main sizes or available system pressures.

A small-diameter system has benefits to fire suppression that are similar to those in a trucked delivery system. Standpipes provided throughout the community are capable of delivering water at rates of 450 L/min to 900 L/min. These are used to supply a fire truck in a manner similar to water trucks, but with the advantage of continuous supply. The reservoir should contain a two-hour emergency supply of water in addition to the normal storage requirements.

When considering the total cost of a piped water and sewer system, the cost difference between a small-diameter system and a regular large-diameter system amounts to about five to ten percent of the total cost. The only sections of the system that are less expensive are the storage reservoir, the distribution pumphouse, and the distribution network. This cost difference depends on the layout of the community, but in only a few cases would it be more than about ten percent of the total water and sewer capital and operating costs.

Fire protection should be included in the design of the initial water system. It should be designed to provide at least 3,600 L/min of water to hydrants located at regular intervals throughout the community.

Arctic water distribution systems, which circulate water for freeze protection, typically consist of small-diameter pipe. Small-diameter pipe is used to maximize water velocities in the pipe and to minimize pumping requirements for circulation. High water velocities are required in the main line to maintain adequate circulation in water service lines that use pit orifices to create circulation. Typical water main sizes for circulating systems are 100 mm and 150 mm diameters. High-pressure pumps may be required to provide adequate flows for fire suppression due to high head losses in small-diameter pipe. In many cases a combination thaw port/drain valve can be provided to act as a fire hydrant on circulation systems.

Cold region communities may have open water sources nearby which can be used by the fire department. A fire truck can draft water for a height of at least 3 m at its maximum rate. However, using the fire truck to draft the water leaves it too far away from the scene of fire; also it is often difficult to get a fire truck that close to an open water source without building and maintaining separate fire access lanes. Therefore, if open water sources are to make up a major portion of the water supply for fire suppression, large trailer mounted, portable pumps can be used to relay water to the fire scene. Such units should also include a hose tender capable of quickly laying large-diameter hose from the pump to the fire truck.

In high-risk areas or near large buildings situated close to water, a portable pump and hose in a small shed near the water may be built. In communities without roads or without fire trucks, a few such installations can make a quick attack possible.

The open water source must be maintained throughout the winter. A simple way of doing this is to freeze an oil drum, with the top and bottom removed, into place. The top can then be covered with an insulated wood cover. Snow and ice accumulation requires periodic removal of the cover. This method, of course, does not work in shallow lakes or rivers where the water may freeze to the bottom.

Another method involves introducing air bubbles at the bottom of a lake through a small air compressor and lines to keep water surfaces ice-free by circulating warmer water from the bottom of the lake to the top. This principle has been used in harbor facilities but may need some experimentation before implementation in a small community.

TABLE 15-1   FIRE STATISTICS

	NWT 1970s	NWT 1980s	Alaska 1970s	Alaska 1980s
Annual Fires (per 1,000 pop.)	5.0	3.0	4.7	5.9
Fire Death Rate (per year per 100,000 pop.)	15.6	10.9	9.2	4.4
Annual Dollar Loss (per capita per year)	42 (CND)	93* (CND)	64 (US)	74 (US)

* In real dollar terms, the increase from the 1970s to the 1980s is $16/capita because $42 in the 1970s inflates to $77 in the 1980s.

A comparison of fire statistics for the NWT and Alaska are provided in Table 15-1 for the 1970s and 1980s.

In the NWT, the number of fires per 1,000 population decreased from an average of 5.0 during the 1970s to 3.0 during the 1980s, which is now about the same as in southern Canada, an encouraging result. However, while the fire death rate per year per 100,000 population decreased from 15.6 to 10.9 it still remains four to five times higher than in southern Canada, in spite of the fact that many communities have experienced no deaths during a 10-year period.

In Alaska, the number of annual fires per 1,000 population actually increased from 4.7 to 5.9; however, the fire death rate per year per 100,000 population decreased significantly.

The annual dollar loss per capita per year has increased both in the NWT and Alaska. The increase may appear to be more pronounced in the NWT than Alaska. However, the increase in "real" dollars is not significant when inflation (residential and nonresidential construction) is factored in.

15.5 References

Gratz, D.B. and R.E. Hawkins. 1980. *Evaluation of Smoke Detectors in Homes*. U.S. Fire Administration, Washington, D.C.

Heinke, G.W. and E.J. Bowering. 1982. *Recommended Guidelines Community Fire Protection and Prevention North of 60*. Department of Local Government, Government of the N.W.T. and Government of the Yukon Territory, Yellowknife, N.W.T. 210 p.

Heinke, G.W. and C.C. Marianayagam. (Editors). 1994. *Fire Protection in the Northwest Territories*. Department of Municipal and Community Affairs, Government of the Northwest Territories, Yellowknife.

IBC. 1977. *Water Supply for Public Fire Protection*. Insurance Bureau of Canada, Public Fire Protection Survey Services, Toronto, Ontario.

Insurance Services Office. 1974. Grading Schedule for Municipal Fire Protection, New York, N.Y.

Kung, H.C. et al. 1980. *Sprinkler Performance in Residential Fire Tests*. U.S. Fire Administration Washington, D.C.

National Fire Protection Association. 1987. *Automatic Sprinkler Handbook*. 15th Edition, Quincy, Massachusetts.

NRC. 1990a. *National Building Code of Canada - 1990*. National Research Council Canada, Associate Committee on the National Building Code, Ottawa, Canada.

NRC. 1990b. *National Fire Code of Canada*. National Research Council Canada, Associate Committee on the National Fire Code, Ottawa, Canada.

Smoyer, N. 1981. *Background Information for a Research Program on Use of Smoke Detectors*

Smoyer, N. 1981. *Background Information for a Research Program on Use of Smoke Detectors in Rural Alaska Residences*. Alaska Council on Science and Technology, Juneau, Alaska.

Switzer, R.W.A. and D.M. Baird. 1980. *Study on Fire Prevention and Control Systems in Canada*. National Research Council Canada, Ottawa, Ontario.

15.7 Bibliography

NRC Joint Task Group. 1990. *Report on Mandatory Sprinklers in Houses to the Standing Committees on Fire Protection, Houses and Small Buildings, and Occupancy*. Ottawa, Ontario.

OEH&E. 1984. *Boiler Room Fire Code Requirements*. Office of Environment, Health, and Engineering, Anchorage, Alaska.

OEH&E. 1987. *Fire Detection/ Fire Suppression Equipment Guideline*. Office of Environment, Health, and Engineering, Anchorage, Alaska.

OEH&E. 1989. *Fuel Storage Requirements per the USEPA, Uniform Fire Code and State Regulations*. Office of Environment, Health, and Engineering, Anchorage, Alaska.

Viniello, J.A. 1985. Residential and quick response sprinklers - you need to know the difference. In: *NFPA Sprinkler Quarterly*, No. 52.

SECTION 16

SOLID AND HAZARDOUS WASTE MANAGEMENT

3rd Edition Steering Committee Coordinators

Gary W. Heinke

James A. Crum

3rd Edition Principal Author

Gary W. Heinke

Special Contributions

Sukhi Cheema

Section 16   Table of Contents

Section 16   List Of Figures

Section 16   List of Tables

16 SOLID AND HAZARDOUS WASTE MANAGEMENT

16.1 Introduction

The purpose of this section is to identify those aspects of solid waste management that may be unique or present different problems because of cold climate conditions. To provide continuity, some general aspects of solid waste management have been included. The generation, collection, and disposal of solid waste, and the cost of solid waste management in cold climates are covered. The emphasis is on domestic/municipal sources of wastes. See Section 13 for information on camps.

The objectives of solid waste management, north or south, are to have a positive attitude towards and awareness of proper waste reduction, handling, and disposal. This is followed by a system to collect and dispose of wastes in the most economical manner for a given situation, without creating hazards, nuisances, or aesthetic blights for people or the environment. To achieve these objectives, each component of the system, namely, awareness, positive attitude, storage at the source, collection, and treatment/disposal, must be properly carried out. All of these components have to be properly planned and managed.

The disposal of solid wastes in cold regions results in problems similar to those in warmer regions. In principle, the concept appears simple, but in practice it is difficult to realize.

The disposal of hazardous wastes in northern regions has often been neglected. However, recent investigations and strategies have resulted in progress for hazardous and solid waste management.

16.2 Existing Systems

The disposal of solid wastes has often been a neglected area of a community's municipal services. Concerns over public health and increasing environmental awareness have resulted in more stringent requirements. Many communities have reassessed their solid waste disposal systems, including reduction in waste production and improvements in collection and disposal. Old disposal sites that proved to be inadequate in location and capacity are now being replaced by better planned landfill sites. An update of the status of solid waste management in NWT communities (Heinke and Wong, 1990a) and guidelines (Heinke and Wong, 1990b) have been prepared. In Alaska, the Department of Environmental Conservation, Hazardous and Solid Waste Management Section has prepared planning guidelines for solid waste management in Alaskan communities (Alaska Department of Environmental Conservation, 1992). They have also prepared a popular written document (Drum, 1992) on the implementation of such systems. It is found that regulations for larger warm-climate landfills are often difficult to adapt to small remote arctic communities.

In Alaska, most of the 550,000 residents live in cities. Eighty-five percent of the municipal solid waste is disposed of in seven facilities that process over 20 tons per day. However, the remainder of Alaska's population is spread across the state in small rural or remote communities. It is estimated that there are about 290 smaller municipal waste disposal sites which serve individual communities. The state and federal governments have strict regulatory standards for construction and operation of these disposal sites. It is estimated that less than two dozen of the smaller sites meet current design and operational standards. Currently the municipal, state and federal levels of government are trying to match often rigid regulatory requirements with the economic realities of municipal government operation in rural Alaska.

16.2.1 Collection System. An important objective of any system of solid waste collection is to transport wastes from their point of generation to a disposal site in a safe and effective manner. This may be achieved in a number of different ways. Evaluation of existing systems indicates that many communities have developed a common method of solid waste collection which has proven to be adequate considering the severe climatic conditions, equipment limitations, and the type of wastes generated.

In many communities, empty oil drums are used to store refuse prior to collection. Oil drums are usually readily available in most northern communities and are large enough to hold several days' accumulation. Residents place domestic wastes in these oil drums at the roadside near their homes. In some communities, the oil drums are placed on truck-height stands, one or two stands per block, for easier handling by pickup crews (Figure 16-1). Oil drums are not be easily blown over by strong winds and can be used as safe containers for burning combustible wastes in communities where burning of wastes for the purpose of volume reduction is

FIGURE 16-1   GARBAGE CART AND STAND

permissible. Some communities have established ordinances (bylaws) to forbid burning of wastes in drums to eliminate the smoke and fly ash from blowing over the community.

In larger communities oil drums are emptied directly onto the collecting vehicle for transport to the dump site (Figure 16-2). In some smaller communities the drums are transported to the dump where they are emptied and then returned to the residences. The use of oil drums has disadvantages in that they are heavy, hard to handle, and are not covered. They are not allowed for use in some larger communities such as Yellowknife, Kotzebue, Bethel, and Nome.

Garbage is collected on a regular basis with the frequency varying amongst the different communities depending on the size of the community, refuse vol-

FIGURE 16-2   GARBAGE COLLECTION

FIGURE 16-3   AN EXAMPLE OF UNSATISFACTORY
STORAGE OF GARBAGE AT THE HOME

ume, storage at each building, and the prevailing weather conditions. A minimum level of service of once weekly is generally sufficient to maintain acceptable sanitary and aesthetic conditions within a community. In some communities, collection may occur two times per week or once every other week (Figure 16-3). The type of collection vehicle also depends on the size of the community, type of roads, and winter road maintenance. The collection vehicles may range from an open cart pulled by an all terrain vehicle (ATV) or truck to a full sized packer truck for larger communities (Figure 16-4). Pickup trucks are commonly used to collect solid wastes in smaller communities.

Some residents prefer to haul their own garbage to the disposal site despite the availability of regular service.

Human waste in plastic bags called "honey bags" could be considered solid waste. However, honey bags should not be collected along with the domestic solid wastes nor should they be disposed of with the domestic solid wastes. Residents place honey bags in separate containers, often oil drums cut in half or polyethylene storage vats that are placed by the roadside for collection. Government planners in the NWT and Alaska hope to phase out the honey-bag system by the year 2000, replacing it with holding tanks or piped systems. It is probable that not all communities will be converted to these systems

within this time period and therefore, provisions should be made to accommodate honey-bag disposal in future plans where applicable.

Once a year following snowmelt, each community should organize a spring cleanup to collect and dispose of loose refuse that has accumulated in the community during the winter months. Also, during this period metal wastes and large, bulky items such as old appliances or discarded snowmobiles should be disposed of by the collection crews.

16.2.2   Disposal Site.   The solid waste disposal site is often a neglected area of a community's municipal services, receiving only a fraction of the attention or funding required to properly construct and maintain it in a safe and efficient manner. In the past, solid waste disposal sites frequently developed close to the community for convenience and economic reasons, without consideration for the potential problems of smoke nuisance, water contamination, health risks, aircraft safety, and aesthetics. The effects of such neglect and increased waste quantities generated are evident today. Increasing pressures from concerned residents have forced some communities to close inadequate disposal sites and to establish new sites designed and located to avoid such problems. Many of the existing sites do not meet the current regulations regarding minimum distance from airports.

FIGURE 16-4   SOLID WASTE COLLECTION

Most communities dispose of solid wastes at an open dump/modified landfill. This involves piling waste at a designated site and occasionally burning the piled waste to reduce the volume. If local cover material is available, the piled waste may be covered to prevent the spread of wind-blown debris and scavenging by animal and insect vectors. This method of solid waste disposal is used because it requires little or no initial site preparation and minimal maintenance requirements. For some communities, severe climatic conditions and the lack of available cover material or equipment has rendered open dumping the only method currently feasible. Large bulky wastes should be segregated from the other wastes at the disposal site and can be a visual nuisance if they are not disposed of in an orderly manner.

The type of community refuse generated depends on the extent and type of activities in the community. For most small communities in arctic regions, solid waste is primarily domestic in character. Since combustible materials comprise the greatest propor-

tion of domestic waste, burning is frequently practiced at the disposal site to reduce volume.

Honey bags present a special disposal site problem. In some communities honey bags are treated as liquid waste, so they are disposed of at the lagoon site. In other communities, honey bags are considered a solid waste and are taken to the solid waste site for disposal. Any new construction in communities where the honey-bag system operates should include in its plans a separate area for honey-bag disposal in the vicinity of the solid waste site.

16.2.3   Concerns with Existing Systems. Existing solid waste disposal systems should take into account important factors regarding public health and the environment, and operation and maintenance.

Public Health, Physical and Environmental Concerns – Several planning and design concerns are recognized: location, area required, cover material, fencing, and oil and battery disposal.

Location – Poorly located solid waste disposal sites present numerous problems for operators and community residents alike.

- Too close to the community:
    - unpleasant odor or smoke,
    - aesthetically unpleasant.
- Too close to important water bodies:
    - community water supply affected,
    - fishing affected.
- Too far from the community:
    - cost of building and maintaining an access road is high,
    - travel time between disposal site and community too great, and
    - access road restricted by blowing snow or flood waters.

Proximity to Airports – An issue of particular concern is the separation distance between solid waste disposal sites and airports. The Canadian Manual of Airport Bird Hazard Control developed by Transport Canada recommends that garbage dumps containing food garbage should not be located within an 8 km radius of an airport, because of the potential danger of bird flocks to aircraft. At the present time, none of the NWT disposal sites conform to this guideline. These guidelines were used to develop a compromise policy to keep future solid waste disposal projects a distance of 3 km from airports.

In Alaska federal EPA and FAA regulations request that a separation distance of 1,500 metres be maintained between new refuse sites and small airports using piston-driven aircraft. A distance of 3,000 metres is requested for larger runways that turbo fan and jet aircraft use.

Area Required – Insufficient area has often been provided for many sites, requiring frequent relocation. Criteria used in the past to plan these sites and the disposal technique need to be reconsidered and updated wherever possible. The area selected must handle generated waste for a 20 year planning horizon.

Cover Material – Adequate cover material is often not readily available in northern communities. Under these circumstances, coverage is not practiced or only practiced infrequently when cover material can be hauled from surrounding sand bars or gravel pits. In the winter months, snow can often be used to cover the waste to reduce the occurrence of windblown debris. Adequate cover material for use throughout the facility life should be identified.

Fencing – Complete or partial fencing around the working area of the waste disposal site is often cited by operators and planners as a desirable design feature. Fencing should be implemented at all solid waste disposal sites where wind-blown debris and unauthorized access are problems. At present many Canadian and Alaskan sites are fenced. Snow accumulation and soil stability are important considerations in fencing design in Arctic areas.

Oil Disposal – Presently no clear guidelines or requirements exist regarding the disposal of domestic or commercial waste oil; consequently, their disposal is treated in a variety of ways by individual communities. Some communities prefer to dispose of waste oil along with combustible wastes using the oil as a fuel during burning; others spread the waste oil on roadways to control dust. Some new solid waste disposal sites are being designed with separate waste oil areas in which waste oil is deposited in specially designed holding cells with oil-resistant liners. Some communities use waste oil as a special heater fuel to heat municipal repair shops.

Operation and Maintenance Concerns – Acceptable methods of operation and maintenance are not practiced at the majority of solid waste disposal sites in cold regions. Various factors are responsible for this condition including: low priority given to it, inadequate funding, a lack of equipment, a lack of trained personnel, and severe climatic conditions. Regardless of the reasons, the main issues of concern are the potential threats to public health and the environment.

Public health is threatened when wastes are indiscriminately dumped and allowed to accumulate without sufficient coverage. Such conditions promote disease transmission by animal and insect vectors which frequent the sites in search of food. Ravens, seagulls, bears and dogs are often seen scavenging through the garbage at dump sites. Animals and insects are not the only concerns; residents themselves are known to frequent the sites in search of recyclable wastes. Not only do individuals risk their own health, but there is also the risk of liability to the site owner/operator for any accidents that may occur while people are on the grounds. On the other hand, scavenging is one way to recycle "wastes".

Air pollution is also a concern. Smoke from burning garbage and products of incomplete combustion can become trapped in the lower atmosphere during temperature inversions and can aggravate respiratory problems or cause eye and skin irritations. In most communities, the disposal sites are unattended and access to the sites is unrestricted. Consequently, unauthorized burning of waste by individuals has often created an uncontrolled burning problem.

Attempts to alleviate some of the concerns associated with the operation and maintenance of existing disposal sites have been implemented with moderate success. A very successful example is that at Norman Wells, NWT. Here the town foreman implemented a system of waste segregation and a more effective operation and maintenance system at the existing site (Whiteman, 1990). However, in other cases an adequate solution to a problem may only be accomplished by relocating or redesigning the solid waste disposal system. For example, relocating a site to a closed depression or an old quarry that is not part of the surface watershed could provide an effective disposal site without contamination of important water bodies.

Unauthorized access to a disposal site can be controlled by erecting fencing around the perimeter of the site. The fence serves a dual purpose. In addition to keeping unwanted scavengers off the site, it also prevents wind-blown debris from leaving the site and littering the surrounding area. Fences must be designed to be clogged with debris and still not be blown over by the wind. Signs should be posted at disposal sites to warn people of potential hazards and to facilitate easy and proper disposal of solid wastes, hazardous waste and recyclables. It may be practical to use the fence only to contain debris and not install gates to prevent access. Gates are often impractical with drifting snow and are soon removed or damaged.

16.2.4 Assessment of Current Systems.

Northwest Territories – A questionnaire survey of all NWT communities was conducted in early 1990 to help assess the current status of solid waste management. The conclusions are based on solid waste collection and disposal responses from 53 out of the 61 NWT communities surveyed in 1989 and 1990. A summary of the relevant results from the study follows:

1. Substantial progress has been made in solid waste collection and disposal practice during the past decade. Over one-third (23) of the communities obtained a new disposal site since 1985.

2. About one-third of the communities (22) have less than five years before expansion or relocation of their disposal sites must occur. Of these, nine are essentially out of capacity in one year.

3. The minimum design life for a new disposal site should be 20 years. There is evidence that some of the sites constructed since 1985 will have to be expanded or relocated soon, indicating a much shorter design life was actually used.

4. The modified landfill method (Section 16.4.1) is the minimum acceptable standard for NWT communities. Only 15 of the communities surveyed currently meet that standard. Lack of sufficient gravel or other cover material in 24 other communities is the main reason why they do not meet the standard. Five communities require substantial improvements.

5. All communities surveyed meet the minimum requirement of once per week collection for all residents. In most communities, collection occurs twice weekly.

6. Approximately 75 percent of the communities surveyed still have bagged sewage "honey bags" to some degree. The minimum acceptable NWT level of service for honey-bag collection is five times per week with no two consecutive no-service days. Only three communities do not meet this standard. In five other communities honey bags are not collected separately from other solid wastes, or residents must bring the honey bags to the disposal site.

In 1982, 58 percent of the solid waste disposal facilities had public health deficiencies while 32 percent had environmental deficiencies. Typically, public health deficiencies were related to a generally uncontrollable solid waste disposal site or a lack of bagged sewage segregation at the site, whereas environmental deficiencies involved poor management capability or an unorganized self-haul system of waste disposal.

A comparison of the 1982 and 1990 NWT data for the total of 44 communities indicated a significant reduction in both public health deficiencies and environmental deficiencies has occurred in NWT communities. Public health deficiencies dropped from 58 percent to 18 percent and environmental deficiencies dropped from 32 percent to 11 percent.

In Alaska, the seven larger landfills, each handling in excess of 20 tons per day, meet current state and federal codes. Of the 289 smaller disposal sites located in rural and remote communities, it is estimated that less than 12 meet these codes. This record is not good, but one must consider that the codes are not considered by most as achievable, and many of the smaller communities are striving to meet some form of management of their solid wastes.

16.3 Objectives of Solid Waste Management

16.3.1 Basic Objectives.

Public Health and Safety – The primary objective of any solid waste management system, regardless of climatic conditions, is to collect and dispose of wastes in the most economical manner and also to protect public health and provide reasonable protection to the environment. Potential health risks exist at all stages of the waste management system from storage to collection to disposal, although some have more deleterious effects than others.

Wastes left uncovered in storage barrels or at open dump sites provide a food source for disease vectors. Under the proper conditions, infestations of disease-carrying bacteria, insects, rodents and birds will breed and increase the risk of disease transmission to humans. It should be an objective of a waste management system to minimize the potential for vector breeding through proper disposal.

Communicable diseases transmitted from human fecal wastes remain a concern in communities that still rely on the honey bag system of human waste disposal for all or part of their disposal needs. Consequently, the health implications associated with broken bags or improper disposal methods either through direct or indirect contact, must be considered in the planning of those waste management systems where a honey bag system will continue to exist in the near future.

Burning of combustible wastes is frequently practiced in open drums by individuals and by operators at the disposal site. This has the advantage of significantly reducing the volume by 40 to 70 percent. Individual burn barrels help when collection is suspended for days as a result of inclement weather conditions. However, the smoke generated can be a nuisance to residents of the community. Smoke from open burning contains chemicals that are known to be irritants to the eyes and human respiratory system; some are considered to be carcinogenic. Any nuisance caused by smoke will be aggravated during periods of atmospheric inversion conditions.

Of particular concern to aviation safety is the potential danger to aircraft from bird flocks. Solid waste disposal sites may increase bird populations in the area and alter flight routes. Therefore, appropriate separation distances between airports and solid waste disposal sites must be considered along with the location of traditional nesting areas.

Environmental Protection – A solid waste management program must consider environmental impacts, as improper design and operation can lead to environmental damage. Since all waste disposal operations will have some impact on the local environment, it should be an objective of the management program to minimize this impact.

The degree to which a solid waste disposal site can adversely affect the environment is influenced by the characteristics and quantities of wastes as well as site location. In general, municipal solid waste disposal operations in small communities produce relatively local environmental impacts. Data collected on the characteristics of garbage in northern communities reveals that 65 percent of domestic solid waste is composed of nonhazardous paper and food wastes (Heinke and Wong, 1990a). In many communities, the greatest problem stems from the burning of garbage and the adverse effects of smoke and soot. In some communitites, surface and groundwater contamination may be an additional concern.

The collection and disposal of hazardous wastes with regular community wastes is normally not permitted, nevertheless small quantities of hazardous wastes such as paint and batteries are sometimes collected and disposed of with the community solid waste. Under certain conditions co-disposal of minor amounts of hazardous wastes in modified landfill sites will not result in significant environmental hazard (Quaye and Heinke, 1992).

To assess the environmental impact of an existing waste disposal site, proper field investigations need to be carried out. Abnormalities in growth or color of vegetation caused by vegetation stress, smoke from burning garbage, soot, dust, gas and odor are some of the more visible impacts that can be observed on a routine site evaluation. If the problem is thought to be severe, a water sampling program and/or a leachate collection and analysis program may be necessary. The outcome of these tests may indi-

FIGURE 16-5   SPRING CLEANUP – EMMONAK, ALASKA

cate a need for further detailed hydrogeological studies.

It may be necessary to provide site modifications or change operating procedures to control adverse environmental effects. Often regulations, based on worst-case conditions or uniform minimum standards, require modifications to existing landfills that are actually not necessary when evaluated by proper field testing.

Aesthetics – Maintaining or enhancing the aesthetics of a community is a major objective for an effective solid waste management system. Freezing temperatures, untrained personnel, frozen ground and a lack of cover material are no longer sufficient excuses for improper waste disposal. Clean workplaces, homes, and communities are important for the morale as well as the health of residents. This is most clearly reflected in some communities by the annual spring cleanup (Figure 16-5). During the winter, garbage that had not been collected because of adverse conditions is often mercifully buried or frozen in snow only to be exposed during spring melt. The accumulated wastes usually produce such public disfavor that annual spring cleanups are organized to rid the community of the wastes in a proper manner.

Open dumps, in addition to being an eyesore for the residents of a community are also a source of blowing paper, dust and odor. General negligence and a lack of proper containment facilities result in public disfavor and concern. This is a particular concern in smaller communities where the disposal site is often located adjacent to the main road leading to the community and presents a glaring eyesore for residents and visitors alike. Exposure to such negligence can breed carelessness and eliminates any incentive to keep the community clean. An effective solid waste management system should promote a responsible attitude toward solid waste disposal by example.

16.3.2   Disposal Objectives.

Reduce Air Pollution – There are generally two types of emissions that result from solid waste disposal sites: gaseous and particulate.

Gaseous emissions are produced primarily by the biodegradation of organic material. Over 90 percent of the gas volume produced is methane and carbon dioxide. Both gases are potentially harmful. Methane can diffuse through the relatively porous fill material and may accumulate to explosive concentrations. Carbon dioxide is soluble in water and reacts to form carbonic acid which can dissolve mineral matter, particularly carbonates in refuse, soil and rock.

In cold climates, levels of gaseous emission are not usually significant since the cold climate makes biological degradation of putrescible matter extremely slow. Hence, gaseous emissions from solid waste modified landfill sites in northern communities are not as important an issue as they are in more temperate regions.

Of greater concern are particulate emissions. Controlled open burning is permissible at some disposal sites of northern communities. In most northern communities, open barrel burning is permissible. The resulting smoke contains sizable particles of soot and pieces of charred paper can be an occasional nuisance.

Reduce Water Pollution – The potential for surface and groundwater pollution is one of the primary environmental concerns associated with the land disposal of solid wastes. Factors such as location, nature of wastes deposited, and substandard operational procedures are recognized as principal contributors to the problem of groundwater pollution.

The contaminated liquid effluent produced at waste disposal sites is referred to as leachate. The composition and quantity of leachate is affected by the type of waste deposited and the manner in which it is landfilled, the climatic and hydrogeological characteristics and the conditions of the landfill itself: age, chemical and biological activity, moisture, pH. Since these conditions vary, leachate characteristics can vary significantly from one site to another or within a site.

There are several mechanisms by which contaminants leach from a waste disposal site:

- downward movement of surface water into the solid waste;

- mixing of liquid and solid wastes; and

- contact of buried waste with a fixed or laterally moving water table.

As leachate percolates through the underlying strata, many of the organic, inorganic and biological constituents originally contained in it will be biologically altered or removed by the filtering and adsorption properties of the materials in the strata. The extent of this action depends on the characteristics of the soil, especially the clay content. It is neither correct nor safe to assume that leachate will be diluted by groundwater since very little mixing occurs in aquifers; flow is laminar.

In order to reduce the risk of groundwater pollution, it is necessary to understand the way the biological, chemical and physical characteristics of the soil and geologic materials interact with the wastes (Leskiw et al., 1985). Also, a knowledge of the amount and projected use of water resources in the area and the direction of groundwater movement is essential. Interaction should be avoided if possible.

Improve Aesthetics – The severe northern climate, combined with permafrost soil or rock, prevents many northern communities from operating a sanitary landfill. Although the engineered disposal site is increasingly popular, by far the most common form of solid waste disposal in northern communities is still the open, uncontrolled garbage dump. As long as the great majority of these sites remain accessible to users any time of day or night, without any form of supervision, it can be expected that disposal sites will be messy.

By locating the site out of view and downwind of the community, the aesthetics of the occupied community can be vastly improved. However in situations where economics or resources dictate site location, other mechanisms of improving aesthetics must be considered. Regardless, the basic objectives are to control odor, blowing paper and trash, and to achieve this in a clean and orderly manner.

Well-defined storage and disposal areas are essential at refuse disposal sites. Clear, concise signs to define storage areas, disposal areas and the types of wastes to be deposited in each, are beneficial for the users and aid in the orderly disposal of wastes.

When a disposal site has reached capacity, steps must be taken to ensure that the facility is properly closed and that the general aesthetics of the area are restored.

Reduce Unsupervized Scavenging and Increase Recycling – Scavenging at unsupervised community disposal sites not only endangers the health and safety of individuals but it may also lead to liability claims against the owner or operator of the disposal site. For these reasons it is in the best interest of all parties concerned to consider methods by which unsupervised scavenging can be reduced or even eliminated.

On the other hand, scavenging is often the only practiced method of recycling wastes. A separate area next to the solid waste deposit site can be set aside for discarded materials that are likely of value to others, and set hours of operation for scavenging/recovery, when supervision is possible. Waste disposal operation should consider fencing to restrict access, and supervision when addressing salvage operations.

Fencing – Fencing around solid waste sites is recommended. It should be installed where it is necessary and practical to control access to the site. Perimeter fencing has been used with success in some communities. In order for it to be an effective means of control, the fence must be well maintained and entrance gates must be installed to restrict unauthorized access after hours. To successfully restrict access for bears, either use an electric fence or bury the refuse.

Hours of Operation – The hours of operation should be established and the hours posted at the entrance to notify users of the site. If the site is supervised, the hours of operation may be limited by the operating budget. At all other times the site should be closed to restrict unauthorized access.

Supervision – Supervision during normal hours of operation could prevent unauthorized scavenging. Unfortunately, the waste volumes in many small communities do not make continuous supervision practical. When sufficient cover material is available, more frequent covering, burning, or compaction of wastes may be a more effective method of reducing scavenging than supervision.

16.4   Disposal Method

16.4.1   Disposal Alternatives.   The disposal objectives of an effective solid waste management program can be realized in a number of different ways. Some of the methods that are currently being used or have been used in the past for solid waste disposal in northern communities are:

- open dump/landfill;
- modified landfill;
- sanitary landfill;
- burning and landfill;
- incineration and landfill;
- milling and compaction;
- ocean disposal; and
- recycling and reuse.

The choice of a suitable method is site specific, depending on the size of the operation, equipment availability, economics, terrain conditions and public acceptance.

Open Dump/Landfill – The open dump/landfill is, without question, the most utilized method of waste disposal in northern communities.

The open dump/landfill, as the name implies, is not an open dump in the strict sense nor a sanitary landfill but a combination of the two methods with certain characteristics borrowed from each.

The true open dump is typically an unsupervised, uncontrolled operation without regular covering or compaction of wastes. Wastes of all types are dumped at a designated site without any attempt to segregate the domestic wastes from the human or bulky wastes. Any site operations which may be practiced typically include only the cleanup of the access road and any necessary work to keep the working area accessible. As a result, in spring and early summer, there are often unsightly, offensive accumulations of exposed garbage and sewage which are a hazard to public health. Generally, open dumps are a nuisance and are an unfavorable method of waste disposal.

The positive virtues of the open dump, particularly its ease of operation and low cost, are highly appealing to northern communities where such factors as severe weather, permafrost, and lack of equipment add extra burden to the already difficult task of effective waste disposal. As a consequence, the current system has more or less evolved to one in which wastes are progressively dumped according to some plan and then covered periodically with earth fill and compacted whenever possible. Separate areas are established at the site for bulky wastes and honey bag disposal. By introducing a limited form of control, the disposal site can accomplish the objectives of reducing air pollution, reducing water pollution, and improving aesthetics to a degree that is acceptable to the residents. Unfortunately, this method fails when the control breaks down; cover or compaction is not provided as often as needed or indiscriminate dumping takes place. Under these conditions the site degenerates to an open dump.

Modified Landfill – In some respects, the modified landfill operation shares characteristics of the open dump/landfill. In both cases refuse is periodically compacted by mechanical means and covered with a layer of earth or other suitable material. Although the schedule for compaction and cover is typically more frequent for a modified landfill operation, it must still deal with the problems of exposed refuse, attraction to animals and birds, surface and groundwater contamination, open burning and scavenging.

The distinguishing characteristic of the modified landfill is that it is well planned whereas the open dump/landfill receives very little planning if any at all. Every aspect of the modified landfill is engineered, from its conception to its closing. The site is carefully selected, disposal areas are identified, cover materials are stockpiled, and access roads are constructed. In some cases drainage facilities are constructed and fencing is installed.

Three methods of organizing a modified landfill are recognized: the area method, the trench method, and the depression method. These will be examined in detail in Section 16.6.1. In all cases the wastes are deposited in a planned sequence at specific locations and then compacted and covered with suitable cover material.

Modified landfills are appropriate for small populations where it is not feasible to have continuous supervision or dedicated equipment at the site to continually cover the small daily or weekly accumulations of waste.

If the modified landfill is well planned, designed, operated and maintained, it can provide a relatively inexpensive disposal method and a good level of service for sparsely populated communities.

Burning and Landfilling – Authorization for burning of combustible solid wastes within the community is under the discretion of the local government and state regulatory agencies in Alaska. At the disposal site, garbage is often burned providing it does not create a significant nuisance or hazard from smoke, odor and fire. Burning and landfilling is a beneficial method of waste disposal because it can reduce waste volumes by 40 to 70 percent, can reduce the amount of windblown material, and can render garbage somewhat less accessible and attractive to foraging birds, animals and people. However, not all wastes are combustible, and many wastes should not be burned, such as animal carcasses, rubber and plastic materials and used engine oils. These should be separated from the combustible wastes prior to burning.

The following processes for burning and landfilling have been recognized:

Open Burning – Open burning at the disposal site must be controlled to prevent significant hazards from smoke, odor and fire. This requires isolating a safe area within the site to contain the burning debris and burning only when wind conditions prevent smoke, odor, and ash from blowing in the direction of the community.

Trench Burning – Trench burning has limited potential in northern regions since it may only be prac-

ticed in communities where soil conditions and equipment availability permits the construction of trenches. This method of disposal reduces the nuisance of blowing debris and ashes since excavated material can serve as cover material.

Burning in Oil Drums – Oil drums used to store wastes prior to collection may serve a dual purpose by containing burning wastes. This practice can provide a level of volume reduction almost as great as open burning. However, if proper ventilation is lacking, reduced combustion and the subsequent dangers of smoke, sparks and fly ash can arise. The burning may be initiated by the resident or by the operator at the waste disposal site.

Controlled Trench Burning – Controlled trench burning is similar to trench burning except that ventilation is mechanically provided to enhance the combustion process. Air pollution is considerably reduced as a result and by placing a screen over the mouth of the trench, ash release is reduced.

Sanitary Landfill – The sanitary landfill is an operation in which solid wastes are deposited and compacted in a controlled area and then covered with a layer of soil at the end of each working day. The sanitary landfill is becoming the acceptable standard of landfill practice in North America for municipal sites serving populations greater than 5,000.

Operations involve dumping wastes on the working face of a designated site then spreading and compacting wastes into layers 0.3 to 0.6 m thick. At the end of each working day a final layer of soil 0.15 m thick is spread over the exposed wastes. When the design depth of the area is reached, a final layer of cover material at least 0.6 m thick is applied.

Daily covering requires a considerable volume of cover material. A volume of cover equivalent to 20 to 25 percent of the total waste volume is necessary to maintain the recommended schedule demanded by sanitary landfilling.

In the majority of northern communities this volume of cover is not available. In addition, sanitary landfilling requires that earth moving equipment be available on a daily basis. This is not always possible particularly in small communities where there may be only one bulldozer. Also, the volume of wastes generated is generally insufficient to warrant daily covering and in many cases, weather conditions do not allow it.

Incineration – Incineration reduces solid waste to a readily transportable and easy to dispose of inert residue by high temperature burning in an incinerator. An incinerator is composed of a furnace into which refuse is charged and ignited.

In the past, incineration seemed to be a promising alternative to landfill since heat from incinerators could be harnessed for building heating systems, domestic water heating, and electrical power generation. It is estimated that normal domestic garbage has an energy equivalence of 11.6 MJ/kg. However, the mechanical complexity of incinerators and the possibility of breakdown often precludes their use, particularly in remote communities where service and parts may be unavailable for prolonged periods.

There have been several instances where incinerators have been installed in small communities but soon abandoned. Incinerators are successful in industrial or resource development complexes. A large refuse incinerator has operated at Prudhoe Bay, Alaska for years. Other units are being operated in Tuktoyaktuk and Norman Wells by Esso Resources Ltd.

Some small communities in Alaska have successfully used a burn cage at the local landfill. The refuse is placed with a front end loader into this steel structure located over a disposal trench. The refuse is burned at the convenience of the operator. The ash and material that does not burn falls through the bottom grill into the disposal trench. This process significantly reduces the amount of refuse, but creates smoke and air pollution.

Shredding/Bailing – Wastes may be preprocessed by volume reduction using such equipment as shredders, hammer mills, and impact mills. The purpose of these methods is to make the handling and disposal of the wastes cheaper and hence more manageable prior to landfilling or incineration. This process was utilized in Anchorage, Alaska and later abandoned after the mill was damaged in a explosion. The waste is currently compacted for transport and landfilled.

A more practical method of volume reduction is the refuse bailer. A bailing facility was recently completed at Yellowknife, NWT, and Unalakleet, Alaska. These installations provide a promising method to reduce the high costs of landfilling. The Unalakleet installation allows the storage of the bailed refuse near the bailer so it can be hauled to the landfill site when better weather permits. Benefits of volume reduction have not generally been considered great enough to offset the high capital and operating costs associated with such facilities in northern communities. As future regulatory pressures dictate improved operations, these processes may become more acceptable.

Other Alternatives – Ocean disposal of municipal wastes has been practiced in some of the communities in the past, but this practice is no longer permitted in most northern countries.

Recycling is gaining popularity throughout North America in cities where the quantities of recoverable material and populations are large enough to complement the recycling programs and there is a demand for the recovered material, i.e., metals, glass, newsprint. In northern communities, recycling is generally practical only for direct reuse by the scavengers. The small volumes of recoverable material, lack of local markets, great distances between communities and high shipping costs make normal recycling programs uneconomical. In some communities, stockpiled bulky waste items such as used vehicles and scrap metal may periodically be shipped to southern recycling industries, but this is usually done at considerable expense.

Disposal Selection – Figure 16-6 outlines a simplified disposal selection scheme. If adequate cover material is available, a sanitary or modified landfill method of solid waste disposal should be selected, otherwise the open dump will prevail. In most northern communities, population size and waste production is too small to necessitate daily covering, thus the modified landfill method of solid waste disposal with periodic covering should be the method of choice. Three types of modified landfill methods are recognized and shown in Figure 16-6. It should be noted that if burning of wastes is permitted in the community, burning and landfilling or incineration can be incorporated with any of the disposal methods outlined.

16.5 Guidelines for Planning a Disposal Site

16.5.1 Community Conditions. The community population, characteristics of the solid waste, solid waste volume, solid waste collection, and design life of the disposal site will influence the methods and equipment used in landfilling operations.

Community Population – Future demands for waste disposal services and the economics of alternative methods of providing services depend to a large extent on the size of the population to be served, now and in the future.

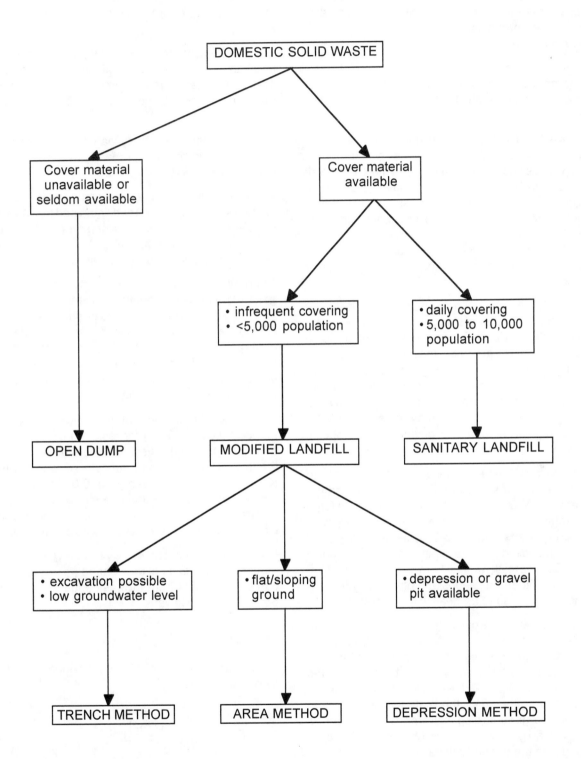

FIGURE 16-6   DISPOSAL SELECTION

Population data and forecasts for northern communities are normally provided by the planning studies done by an organization of the state, territorial, or federal government.

Characteristics of Solid Waste – The solid waste generated by northern communities can be characterized as either domestic, industrial or commercial waste.

Domestic Waste (including honey bag waste) – Domestic wastes are comprised of typical household wastes such as food, packaging materials, cardboard, and household articles. Bulky goods such as discarded stoves, empty oil drums, miscellaneous equipment, and vehicles are also considered domestic wastes. Undiluted human waste in plastic bags from bucket toilets, called honey bag wastes, continue to be generated in many of the smaller communities.

Industrial Waste – These wastes vary with the type of industry. Typical industries include oil exploration, mining, wood processing, and fishing. Some industrial wastes are not disposed of at community dumps with domestic wastes since they may contain potentially toxic and hazardous wastes requiring special handling and disposal methods. In such cases, individual industries may be responsible for operating and maintaining private waste disposal facilities.

Commercial Waste – These wastes are generated by local businesses, services, and government agencies. Wastes of commercial origin include large packaging materials, oil drums, building materials, paper, and arts and crafts wastes. Normally, they can be disposed of at the community waste disposal site. Commercial wastes from cleaning establishments, photography shops and others may be hazardous and need to be dealt with separately.

Bulky wastes can be of a domestic, industrial or commercial origin. Large metal items including discarded vehicles, snowmobiles, appliances, oil drums, machinery, and holding tanks should be hauled to the bulky waste disposal area. Although construction wastes are often bulky, this material can be deposited in the general disposal area where it can be recovered by scavengers or burned and covered along with the general refuse.

Until recently, little reliable information existed regarding solid waste composition in northern communities. Recent information for six NWT communities is shown in Table 16-1, based on waste surveys for Iqaluit, Pangnirtung and Broughtan Island

(Heinke and Wong, 1990c), and Inuvik, Fort Mcpherson and Arctic Red River (Quaye and Heinke, 1992).

Solid wastes generated in most small northern communities are primarily domestic in character. Table 16-1 shows that the major components of the refuse are food, cardboard and other paper products, and plastic.

The composition of honey bag wastes has been analyzed indicating solids range from six to nine percent (Heinke and Prasad, 1980).

A study conducted on several communities in central Alaska connected by road, indicated that a waste volume of around 2.3 kg per person per day could be expected (Tilsworth, 1982). No study to determine the composition of waste has been conducted.

Solid Waste Volume – Knowledge of the waste generation rates and quantities is necessary in determining the site capacity. Ideally, such parameters should be based on historical data of solid waste generation for the community under investigation. When such information is unavailable, waste volumes can be estimated on the basis of data available from similar communities.

Until recently the Government of the NWT recommended 0.010 m^3/(p·d) for residential solid waste, and 0.001 m^3/(p·d) for school waste based on earlier information. Based on two recent studies (Cameron, 1993) an average of 0.014 m^3/(p·d) for residential or municipal waste should be used for northern communities. Any industrial and bulky waste needs to be estimated separately and added.

If site-specific total community refuse volume information is not available, the volume (m^3 in any year) can be estimated from the residential volume and population as follows:

$$365 V P_1 (1+G)^n + 0.084 V P_1^2 (1+G)^{2n}$$

Total community refuse volume (m^3) during the planning horizon is:

$$\frac{365 V P_1}{\ln(1+G)} \left[(1+G)^{PH} - (1+G) \right]$$

$$+ \frac{0.084 V P_1^2}{2\ln(1+G)} \left[(1+G)^{2PH} - (1+G)^2 \right]$$

TABLE 16-1   DATA SUMMARY OF WASTE COMPOSITION
FOR SIX NWT COMMUNITIES

Component	Percent by Weight	
	Average	Range
Food	19.6	21.4 to15.9
Cardboard	10.9	14.4 to 8.6
Newsprint[1]	2.1	6.0 to 0.3
Other Paper Products	15.3	18.5 to10.2
Cans	4.8	6.7 to 2.5
Other Metal Products	5.5	7.4 to 3.9
Plastic, Rubber, Leather	12.2	14.3 to 8.8
Glass, Ceramics	4.1	6.5 to 1.7
Textiles	3.7	4.4 to 2.8
Wood[2]	11.3	20.0 to 4.5
Dirt	3.8	4.8 to 2.5
Diapers	6.7	11.6 to 3.5
Total	100.0	

Notes:
1   The higher number is for Inuvik and Iqaluit, which have a local
     paper and daily southern newspapers. The other communities
     are all less than 1%.
2   The percentage of wood varies depending on the season when
     construction work is going on.

where,

V = average residential refuse volume, $m^3/(p \cdot d)$

P_n = population in n^{TH} year

G = average community population growth rate,
decimal percent

PH = planning horizon, years.

Example

The volume of general refuse generated by a community with an initial population of 1,000 and a one percent growth rate during a 20-year planning horizon is as follows:

V = 0.014 $m^3/(p \cdot d)$

P_1 = 1,000 persons

G = 0.010 persons/year (one percent per year)

PH = 20 years

$$\frac{365(0.014)1000}{\ln(1+0.01)}\left[(1+0.01)^{20} - (1+0.01)\right]$$

$$+ \frac{0.084(0.014)1000^2}{2\ln(1+0.01)}\left[(1+0.01)^{2(20)} - (1+0.01)^2\right]$$

$= 514,000 \times [0.21] + 59,000 \times [0.47]$

$= 108,000 + 28,000$

$= 136,000\ m^3$

Note that this is an uncompacted volume. Through compaction and burning this may be reduced to one-third to one-quarter of the uncompacted volume.

16.5.2 Solid Waste Collection.

Collection Schedule – The minimum level of service for solid waste collection should be once weekly per residence. This level of service is sufficient to maintain acceptable sanitary and aesthetic conditions within a community without requiring special storage containers.

Bagged sewage should be collected five times per week with no two consecutive days without service. Bagged sewage should be collected and disposed of separately from other community solid waste.

Collection of bulky wastes does not require frequent service so collection scheduling could be monthly, seasonally, or on demand.

Collection Vehicle – The size and type of the collection vehicle will depend on the volume of refuse and available funding. Vehicles may range from half-ton pickups capable of hauling oil drums to the disposal site, to compactor vehicles. Increasing the capacity of the collection vehicle reduces the frequency of trips to the disposal site, and so reduces the truck mileage, labor time and fuel consumption. The most cost-effective choice in a garbage truck ranges from half ton to one ton in capacity; factors such as convenience and alternate uses should be considered in making the choice. Collection vehicles should be covered to reduce the problem of wind-blown debris.

Vehicles for bagged sewage disposal should be capable of containing the bags in such a manner so as not to expose the operators to potential health hazards resulting from broken bags during the collection process.

Crew Size – The number of crew members per collection vehicle depends on the volume of waste collected and the type of collection vehicle. The optimum crew size is usually a driver plus one or two helpers, but a single operator can be employed in small communities where waste volume is small. More helpers are needed in small communities because the collection vehicle is more basic and not automated.

Design Life – The community plan, if available, should be referred to during the design stage of the disposal site. The plan typically considers the community land use needs, the extent of the present land use, and the direction of community growth.

Unless otherwise specified, the disposal site should be designed for a 20-year planning period. Methods for estimating the annual total requirements for the 20 year design life are shown in the previous example.

Compaction and burning reduces garbage volume and influences the design life of the disposal site. Compaction can reduce waste volume by a 3:1 ra-

tio, and if burning is permitted a volume reduction of 4:1 can be expected.

16.5.3 Siting Criteria. The criteria for selecting an acceptable solid waste disposal site in a northern community are similar to those for other communities. Those points of special importance to northern communities include the following:

Proximity to Airport – Transport Canada has established guidelines for the development of landfills in the vicinity of airports (Soberman et al., 1990). The guidelines recommend a minimum separation of 8 km between airports and municipal solid waste sites that include food storage disposal. The 8 km separation distance is measured from the boundary of the waste disposal area to the center of the runway. The rationale for the separation distance is to reduce the potential for bird aircraft strikes resulting from birds feeding at the landfill site or by birds migrating to the site.

It has been recognized that the 8 km separation distance is excessive for NWT communities where the volume of air traffic is small and the bird density is low. At present, none of the community landfill sites comply with the 8 km guideline. The average distance for NWT communities in 1986 was about 2 km.

Based on a study prepared to investigate the problem of landfill proximity to airports in the Northwest Territories (Soberman et al., 1990) a tentative guideline for a minimum separation distance of 3 km has been established. This separation distance is site specific and will vary with the volume and type of aircraft using the airport facilities, the density and type of birds visiting the landfill site, and the migratory pattern of the birds.

The community waste disposal site should be generally located upwind of airports and in areas where birds attracted to the site will not pose a hazard to aircraft.

In Alaska a minimum separation standard for planning of 1.5 km for piston aircraft and 3 km for turbo fan and jet aircraft is generally accepted. There is a provision for waiver after site-specific studies are conducted on available alternatives and possible impacts.

Geology and Terrain

Geology – Knowledge of the geology of a site is required to predict the extent of the potential contaminant zone and the potential for excavation. This requires a geological study of the site to determine the stratigraphy, soil types, and soil grain sizes.

Terrain – The surrounding terrain is an important factor when evaluating the suitability of a disposal site. This is often accomplished through air photo examination and field examinations. By making use of natural barriers, major landforms, depressions, and drainage patterns, the earthwork required for site preparation, operation and maintenance can be minimized.

Availability of Cover Material – For practical and economic reasons, the location of the solid waste disposal site is influenced by the availability of cover material. A minimum cover material thickness is necessary to properly maintain a site. Cover material volumes are dependent upon the surface area to be covered and the thickness of the soil needed to achieve the final design thickness after compaction. Minimizing the area of exposed waste will reduce the cover material requirement.

Whenever possible, the use of on-site cover material should be maximized. When sufficient cover material is unavailable, borrow material must be brought to the site, thereby increasing operating costs.

Cover material can be any type of soil or other suitable material that compacts well, does not crack excessively when dry and is relatively free of organics and large items.

Geotechnical Factors – A hydrogeological study of the proposed area is necessary to establish site suitability with respect to the following site-selection criteria:

- water precipitation/infiltration rate;
- groundwater flow system;
- hydraulic gradient;
- hydraulic conductivity;
- groundwater chemistry;
- existing wells and uses; and
- anticipated loadings.

The extent or complexity of the hydrogeological investigation will depend on the size of the disposal site and the specific site conditions.

Other geotechnical aspects which may require consideration are foundation problems arising from thaw sensitive and heave-prone permafrost soils, soil erosion and possible flooding.

Landfill sites located on permafrost can affect the groundwater and ground ice balance by altering the thermal properties of the soil. A site used for landfilling operations is subject to heat generation as a result of vehicular traffic, removal of surface soil for cover material and burning of wastes. The heat generated causes the ground ice to melt and subsequently increases effective soil pressure of the region which can result in consolidation and possible slope failures.

Climatic Factors – Climatic factors of concern when locating a disposal site are limited to atmospheric conditions such as predominant wind direction, wind speed, and the likelihood of localized temperature inversions.

Short term air pollution caused by burning wastes can be reduced by locating sites downwind from the community.

Temperature inversions occur when a cold air mass is trapped in a topographic low point such as a valley by a warmer air mass above. When this situation arises, upward flow of air is prevented and airborne pollutants become trapped near ground level. Areas susceptible to temperature inversions should be avoided when locating disposal sites.

Accessibility/Land Use Pattern – The location of the disposal site must be accessible to vehicular traffic year round. Access roads subject to frequent flooding or snowdrifting can be a nuisance for users of the site and may even suspend operations if the conditions are severe enough.

The construction of an all-weather access road is a major expense in the design of a solid waste disposal site. Where possible, it is recommended that existing all-weather roads be utilized or extended to meet the requirements. Often a common road is used for sewage disposal and solid waste disposal, making maintenance and snow clearing costs more efficient and cost effective.

A major factor in siting criteria is land use patterns and land ownership. Many technically desirable refuse disposal sites are rejected by community residents who are opposed for "not in my back yard" (NIMBY) reasons. Careful planning and public involvement in site selection are required to resolve these political problems.

MODIFIED LANDFILL METHODS OF OPERATION
Area Method – Sloping Ground

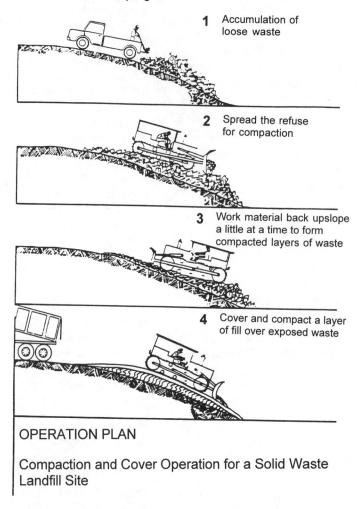

1 Accumulation of loose waste

2 Spread the refuse for compaction

3 Work material back upslope a little at a time to form compacted layers of waste

4 Cover and compact a layer of fill over exposed waste

OPERATION PLAN

Compaction and Cover Operation for a Solid Waste Landfill Site

FIGURE 16-7   AREA METHOD ON SLOPING GROUND

16.6   Guidelines for Design of the Disposal Site

16.6.1   Modified Landfill Disposal Option.

The basic landfilling methods of a modified landfill disposal site depend on the characteristics of the site such as topography, the amount of available cover material, hydrogeologic conditions and the amount and types of solid wastes to be handled. Also, the containment of leachate must be addressed in a manner similar to more southern locations. Any one or a combination of accepted methods listed below may be planned.

Area Method – The area method of a modified landfill is used on flat or sloping ground where soils are unsuitable for excavation or a high groundwater table exists.

For sloping ground, solid waste is dumped out of the collection vehicle and allowed to accumulate on the working face of the slope. Periodically the waste is spread and compacted along the working face using a bulldozer. When the design thickness of compacted waste is attained, a final layer of cover is spread and compacted over the exposed waste. The slope is then ready to receive the next layer of waste. See Figure 16-7 for the sequence of operations on sloping ground.

For flat ground, the refuse is end-dumped onto the slope of the working face of the designated area or garbage cell. Each cell is sized to receive an annual volume of compacted waste. Compaction and spreading of waste is achieved by means of a bulldozer. Periodically, a layer of cover material is spread over the exposed waste and compacted. This process of spreading, compaction and covering is re-

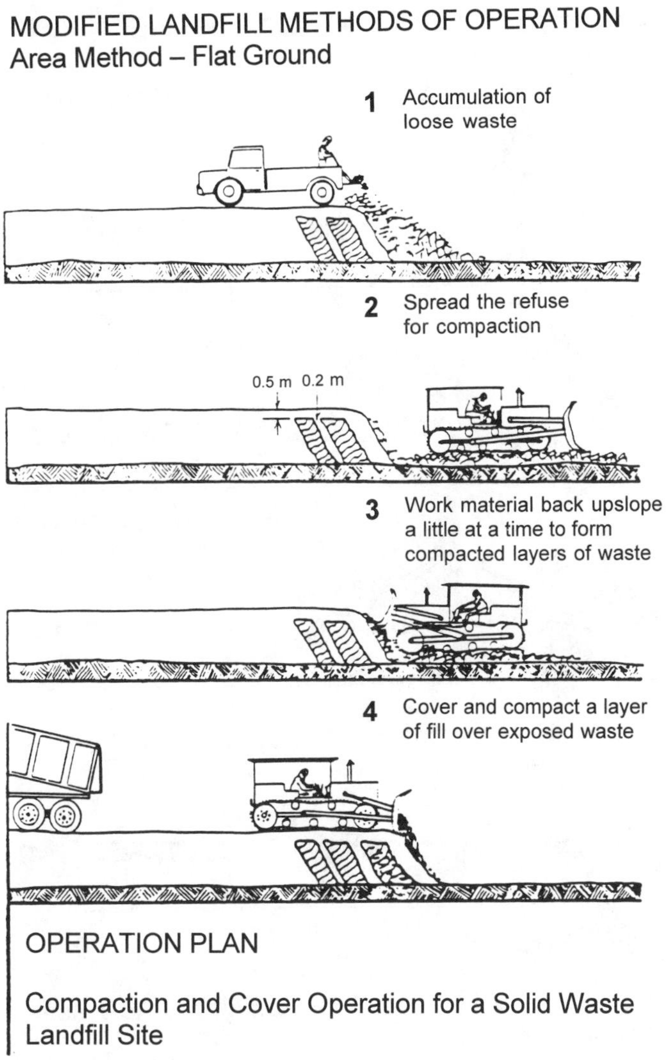

MODIFIED LANDFILL METHODS OF OPERATION
Area Method – Flat Ground

1 Accumulation of loose waste

2 Spread the refuse for compaction

0.5 m 0.2 m

3 Work material back upslope a little at a time to form compacted layers of waste

4 Cover and compact a layer of fill over exposed waste

OPERATION PLAN

Compaction and Cover Operation for a Solid Waste Landfill Site

FIGURE 16-8   AREA METHOD ON FLAT GROUND

peated as wastes accumulate along the working face until the design capacity of the cell is reached. At this point, the compacted waste should then be covered with a final layer of fill. Recommended thicknesses are 0.15 m to 0.2 m on the slope face and a minimum of 0.5 m on the top layer. Since the top layer will eventually serve as a driving surface for collection vehicles, it should also have a granular topping as needed. See Figure 16-8 for the sequence of operations on flat land.

Trench Method – The trench method is used where the soils can be excavated and groundwater levels are deep. In this method, a trench is excavated and wastes are deposited into the trench. The size of the excavated trench should be designed to contain the annual volume of compacted waste for the community. The wastes should be spread and compacted periodically and then covered with a thin layer of cover material. The cover material for the trench method is the excavated material which should be stockpiled nearby. Each cycle of spreading, compaction and covering constitutes a lift. After several lifts, when the trench is full, a final layer of cover is spread and compacted over the trench. Another trench should be excavated as planned and the process repeated for the new trench. The following factors affect the capacity of a trench: availability of land, quantity of solid waste generated, groundwater table, and depth to permafrost or rock.

See Figure 16-9 for the sequence of operations of the trench method.

Depression Method – The depression method is used where a depression is available and filling to

MODIFIED LANDFILL METHODS OF OPERATION
Trench Method

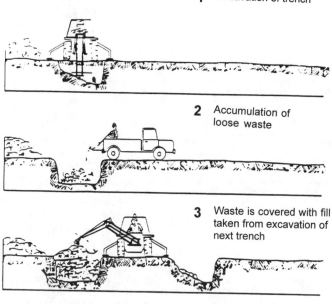

1 Excavation of trench

2 Accumulation of loose waste

3 Waste is covered with fill taken from excavation of next trench

4 Waste and fill are compacted

OPERATION PLAN

Compaction and Cover Operation for a Solid Waste Landfill Site

FIGURE 16-9   TRENCH METHOD

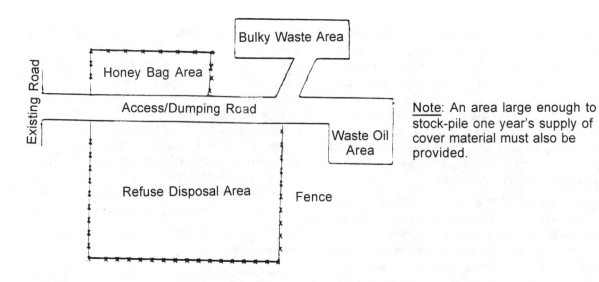

Existing Road

Bulky Waste Area

Honey Bag Area

Access/Dumping Road

Waste Oil Area

Refuse Disposal Area

Fence

Note: An area large enough to stock-pile one year's supply of cover material must also be provided.

FIGURE 16-10 TYPICAL SOLID WASTE DISPOSAL FACILITY LAYOUT

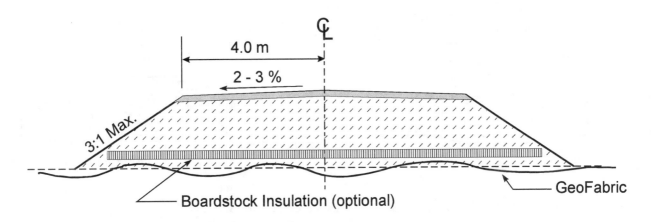

FIGURE 16-11 TYPICAL ACCESS/DUMPING ROAD CROSS SECTION

surrounding elevations or original ground contours is desirable. The depression may be a natural depression or a constructed feature as a result of mining or quarrying operations.

The procedure of landfilling in the depression method is similar to that of the trench method but because excavation is not necessary, cover material may not be available on-site and must be transported to the site from borrow locations. The cost of this method of modified landfill is more attractive than the trench method since excavation and equipment costs are saved. Providing proper drainage facilities for this method can be a problem.

16.6.2 Design of Site Facilities. The solid waste disposal facility must include separate areas for disposal. Construction wastes such as wood, insulation and other combustible refuse can be disposed of in the refuse disposal area while bulky wastes such as automobiles, old furnaces, holding tanks and other large noncombustible refuse should be disposed of in the bulky waste area. In communities where honey bags are used, another area separate from the refuse disposal area and the bulky

waste disposal area must be provided for their proper disposal. Also, if required, a special area should be provided for waste oil disposal. A typical solid waste disposal facility layout is illustrated in Figure 16-10.

Refuse Disposal Area – Methods for determining the volume of refuse generated and the type of modified landfill operation for the selected site are outlined in Sections 16.5.1 and 16.6.1 respectively.

For a community of 1,000 people the volume of general refuse generated in 20 years is calculated from the example in Section 16.5.1 to be 136,000 m^3. This may be reduced to as much as 1/3 to 1/4 of the uncompacted volume, but cover material will add to the depth depending on frequency and extent of material used. A minimum area of 150 m x 150 m with a depth of 2 m (uncovered) is required to satisfy the 20-year planning horizon.

The access or dumping road must be constructed to facilitate collection vehicles from the entrance of the site to the unloading area. A typical cross section is shown in Figure 16-11.

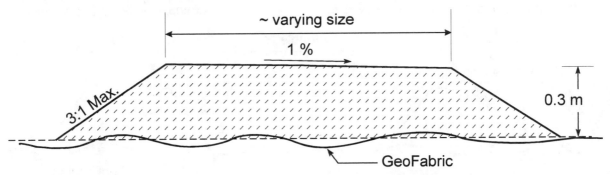

FIGURE 16-12 TYPICAL BULKY WASTE PAD CROSS SECTION

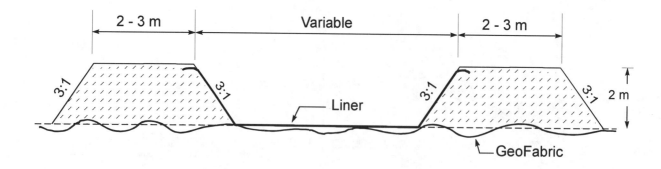

FIGURE 16-13 TYPICAL HONEY-BAG DISPOSAL CELL CROSS SECTION

Bulky Waste Area – The bulky waste disposal area consists of an elevated pad of granular material. Gravel or coarse sand can be used as fill since it provides good drainage. A typical bulky waste pad cross section is illustrated in Figure 16-12. A minimum area 20 to 50 m wide by 20 to 50 m long is recommended. The size depends on the needs of the community so bulky items can be stored for later back shipping, compacting, or ultimate disposal.

Honey-Bag Disposal Area – The volume required for honey-bag disposal can be determined on the basis of a production of 0.5 m³/person per year (Heinke and Prasad, 1980). The disposal area consists of a honey-bag disposal cell surrounded by embankments of acceptable granular material. A typical honey-bag cell cross section is illustrated in Figure 16-13. An alternative approach to honey bag disposal is to treat the waste as sewage and provide a facilitative sewage disposal lagoon for treatment. The choice between these approaches depends on the applicable health and environmental regulations.

Waste Oil Area – The waste oil area consists of a cell surrounded by an embankment and lined with an oil resistant liner. Different liners require different installation procedures. It is best to consult with the liner manufacturer for details. A typical waste oil cell cross section is illustrated in Figure 16-14. Maintenance of the oil disposal pit is required to pump the water from under the oil to waste or the floating oil will eventually overflow. Some facilities use drums to store this waste so it can be exported later for final disposal.

Battery Storage Area – Many small disposal sites have an above-ground storage box to hold waste lead/acid batteries. When adequate quantities are collected they can be exported and some costs recovered to offset shipping costs.

16.6.3   Design of Access Roads.   An all-weather road must be provided from the community to the disposal site. This road must be designed to safely accommodate vehicular traffic and the anticipated loads. The road should be aligned to minimize snowdrift accumulation and provide for adequate drainage.

A typical cross section for road design is illustrated in Figure 16-11.

16.6.4   Design of Site Drainage.   Off-site runoff should be diverted around the refuse disposal site. Surface water runoff from the site must be diverted away from water supply sources and recreational

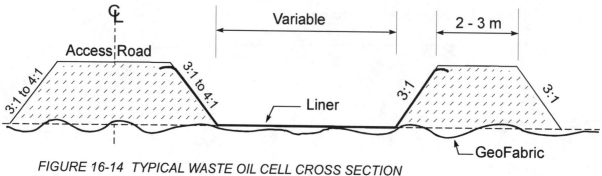

FIGURE 16-14 TYPICAL WASTE OIL CELL CROSS SECTION

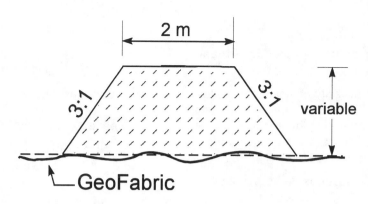

FIGURE 16-15 TYPICAL DRAINAGE CONTROL BERM CROSS
SECTION

water bodies. This is achieved by installing temporary or permanent drainage control berms as required. A typical drainage control berm cross section is illustrated in Figure 16-15.

Infiltration of surface water through the refuse and into the groundwater can be reduced by providing 0.6 to 1.2 m of final cover and performing periodic grading.

Planned maintenance of the drainage channels and periodic filling of surface depressions to prevent ponding is recommended.

16.6.5 Design of Site Fencing. Fencing is used to control or limit access to the landfill site and to control the spreading of blowing garbage. Fencing may be portable or permanent and may be woven or chain linked. Wooden fences are not recommended as they can be a potential fire hazard during on-site burning of wastes. Installation of snow fencing may be beneficial during the winter.

A lockable gate at the site entrance to limit access to specific hours is desirable but often leads to maintenance problems due to deep snowdrifting and refuse piled by the gate for the operator to handle the next morning. Regulated access should only be planned for larger communities.

Regular maintenance of the fence should be planned to ensure that it retains its effectiveness.

16.7 Regulatory Review

The planning and design of solid waste modified landfill sites must comply with the relevant federal,

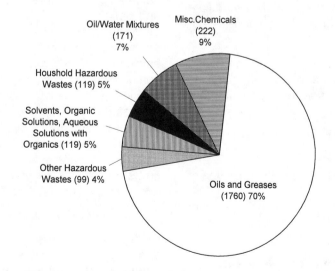

FIGURE 16-16 ESTIMATED QUANTITIES OF HAZARDOUS WASTES IN NORTH-
WEST TERRITORIES (tonne/year) (Heeney and Heinke, 1991)

state, or territorial guidelines, as well as with any community regulations.

The operational responsibility of small refuse sites is generally held by the local government. In order to have a successful operation there must be community acceptance of the regulatory requirements and proper local community operational procedures.

16.8 Overview of Hazardous Wastes in Northern Regions

16.8.1 Working Definition of Hazardous Wastes.
For the purpose of the monograph the following simplified definitions of hazardous waste materials are used. It divides hazardous wastes into seven categories, as follows:

1. **Flammable** – Liquids or solids that can ignite (e.g., waste fuel, turpentine, paint thinner).

2. **Corrosive** – Substances that wear away at many materials (e.g., vehicle battery acid, caustics, drain cleaners).

3. **Reactive** – Materials that can create an explosion or produce toxic vapors (e.g., explosives, ammunition).

4. **Poisonous** – Substances that are poisonous or lethal (e.g., waste oil, mercury, cyanide, arsenic, pesticides, antifreeze).

5. **Infectious** – Materials that can cause disease (e.g., medical wastes, bagged sewage, sewage sludge, animal carcasses).

6. **Environmental** – Materials that bioaccumulate and persist in nature (e.g., contaminants, DDT, PCBs).

7. **Radioactive** – Substances that emit ionizing radiation or objects contaminated with particles that emit ionizing radiation (e.g., uranium, ionization smoke alarm).

This definition has been adopted because it is simple and includes most pertinent waste categories.

In Canada, for regulatory purposes the definition of the Transportation of Dangerous Goods Act (TDGA) is used. In Alaska, the State of Alaska Department of Environmental Conversation maintains a listing of hazardous wastes.

16.8.2 Problems Associated with Northern Areas.
Until very recently the issue of hazardous wastes and their proper disposal, while recognized in principle, was neglected. Three recent studies have been carried out in the Northwest Territories:

- a hazardous waste management strategy for NWT communities (Bryant, 1991);

- a survey, analysis and guidelines for hazardous wastes in the NWT (Figure 16-16) (Heeney and Heinke, 1991); and

- co-disposal of hazardous and solid wastes in the NWT (Quaye and Heinke, 1992).

These three reports provide the majority of the material presented in Section 16.8.

Many people believe that northern regions with their vast land area and small population could not possibly have a hazardous waste problem. However, since the arrival of the first Europeans, nonperishable materials have been brought into the North, and due to the high cost of transportation, many of these items are still there.

Standard solutions used for the disposal of hazardous wastes in southern areas are, in most cases, of little use in northern areas. Extreme climate, small waste volumes, large distances between communities and lack of the infrastructure to deal with hazardous wastes, require the modification of southern solutions.

In industrialized regions, hazardous waste control systems have evolved as disposal problems were progressively revealed through the discovery of serious pollution incidents. With this in mind, it is unrealistic to aim for a transplant of technically and administratively sophisticated controls that are not standard even in more prosperous regions (Wilson and Balkau, 1990). It is more realistic to try to improve on waste management in other areas. After the infrastructure, resources and public awareness have grown more favorable, the progression to more complex control systems becomes easier. Interim solutions allow early measurements to be made of the waste stream and can provide immediate relief, but these solutions should be seen as a first step leading to more permanent measures. The disposal suggestions made in this section are intended as a first step, not as a complete, long-term solution (Heeney and Heinke, 1992).

16.8.3 Hazardous Waste Inventory.

Northwest Territories – An industry, business and community survey was conducted in 1990 throughout the NWT to determine the quantities and types of hazardous wastes generated in the NWT (Heeney and Heinke, 1992). Approximately 2,500 tonnes of hazardous wastes are generated each year by the industries, businesses and approxi-

TABLE 16-2   ESTIMATED QUANTITIES OF HAZARDOUS WASTES GENERATED IN
THE NORTHWEST TERRITORIES (Heeney and Heinke, 1992)

Waste Classification	Quantity Generated Per Year	Percent of Total
Organic sludges and still bottoms (no oil)	6,000 L	–
Solvents and organic solutions	58,000 L	2
Oils and greases	1,760,000 L	70
Oil and water mixtures	171,000 L	7
Organic and oily residues	60,000 L	3
Miscellaneous chemicals	222,000 kg	9
Paint and organic residuals	33,000 L	1
Aqueous solutions with organics	61,000 L	3
Cleanup residuals*	35,000 t	*
Household hazardous wastes	119,000 kg	5
Total	2,500,000 kg	100

Note: A density of 1 g/cm^3 was assumed for liquid wastes.
* Not included in final total and % calculations

mately 57,000 people (Table 16-2 and Fig. 16-17). The majority (70 percent) are waste oil and petroleum products. The survey excluded biomedical wastes, PCBs, spill wastes and wastes at abandoned mines and military sites because they are under direct federal jurisdiction. Human sewage and drilling muds were also not quantified.

Most of these wastes are produced by municipalities and small businesses (see Fig. 16-18) thus generation of wastes is distributed across the territory.

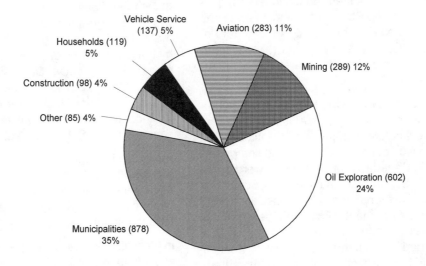

FIGURE 16-17 QUANTITIES OF HAZARDOUS WASTES GENERATED BY
PRODUCING CATEGORIES (Heeney and Heinke, 1991)

FIGURE 16-18 VILLAGE HAZARDOUS WASTE STORAGE

This makes centralized waste disposal facilities impossible or costly to run, considering that the NWT covers an area about 40 percent the size of continental USA.

PCBs — A number of initiatives in recent years have been completed by both territorial and federal governments to clean up PCB contamination and PCB materials in the NWT (Collins, nd). An estimate of 98.6 tons of PCBs currently in use (almost all in transformers) and 335.9 tons are stored at about 20 locations.

The Distant Early Warning (DEW) Line is a series of radar stations stretching from Alaska to Baffin Island at approximately 70 degrees north latitude. The construction was done in the 1950s and comprised 42 stations in Canada. All of the stations were operated by the U.S. Air Force. In the early 1960s, 21 of the Canadian stations were abandoned in accordance with an agreement (circa 1955) between the governments of Canada and the United States. This agreement had no requirements for environmental cleanup, and large quantities of hazardous materials were left at each site. After abandonment, these sites reverted to the crown under the custodianship of the federal Department of Indian Affairs and Northern Development (DIAND).

In 1985, the Government of Canada completed a program for the removal of hazardous materials from the sites. The cleanup involved the removal of surface contamination (PCBs and petroleum products), and the identification of areas where buried wastes could pose some environmental risk.

In the mid-1980s, it was decided that the DEW Line would be upgraded into a new radar network called the North Warning System (NWS) to be operated by the Government of Canada. The NWS would include 16 of the active DEW sites, and the remaining five active DEW sites would be abandoned. Prior to the conversion, the U.S. Air Force removed all PCB-containing equipment from the stations and sent it to the U.S. for disposal.

Plans are in progress for the cleanup of PCB contaminated soils at all 21 active sites. Consultants have been retained to assess the degree of contamination and to design the detailed cleanup plans for each station.

In Alaska, similar cleanup operations are in progress at abandoned DEW Line sites as well as at many active military installations.

16.8.4 Technologies for Waste Treatment and Disposal. At the present time there are no proper facilities for the treatment, long-term secure storage and disposal of hazardous wastes in the NWT or Alaska, except for temporary storage areas managed by local governments, state, or federal agencies and a few operating industries.

Small-scale and relatively low-cost technologies have been identified that may be applicable for northern regions and include the following:

Waste Minimization – In many cases the most economical and environmentally safe methods for the disposal of hazardous wastes begin with waste minimization. This reduction of waste volume can be accomplished either by not creating the waste or by extracting useful products or raw materials from it (Alexander and Interante, 1989). Some examples are solvent recovery and reuse, burning of waste oil for heat recovery, good housekeeping measures, and segregation of wastes to make recovery easier.

Co-disposal – Co-disposal is defined as the conscious deposition of difficult wastes (including hazardous wastes) with domestic garbage by mixing, so that normal reactions in a landfill site are not impaired and the quality of the leachate produced at the site is not significantly worse than the leachate produced at normal municipal landfill sites. The essence of co-disposal is that hazardous wastes are not the major waste input (Finnecy, 1988). A recent study (Quaye and Heinke, 1990) examined the potential for proper co-disposal for a number of hazardous wastes in the NWT. Currently several types of hazardous wastes are disposed of at many open dumps and modified landfill sites, but not according to proper procedures. The study found that co-disposal practiced at most landfills, with limited amounts of hazardous wastes, is not detrimental. However, currently co-disposal of hazardous wastes with domestic wastes is officially not permitted in Canada. The Government of the NWT decided in its strategy for hazardous waste management that co-disposal is considered an interim solution until the problem can be addressed, not a permanent one.

Solidification – Solidification is a technology that combines wastes with materials that tend to set, to form a solid product. In this way the hazardous constituents are trapped within a solid matrix (Batstone et al., 1989). One of the most applicable technologies is in-drum processing. In this process, solidifying binders such as Portland cement are added to wastes contained in a drum or other container. After mixing and setting, the solidified waste is normally disposed of in the drum (Wiles, 1986). An additional

benefit is that it makes use of empty oil drums, which now plague many arctic communities.

Incineration – In general, incineration of hazardous wastes is an expensive process requiring costly, technically advanced equipment and large volumes of wastes. These large incineration facilities should be constructed in central locations with wastes transported to them, which is difficult in northern areas. However, two types of incineration may be applicable: open pit incineration and mobile incineration.

Open pit incinerators provide an efficient method of destroying wastes such as oils and organic solvents, particularly when it is not financially feasible to use sophisticated incinerators (Batstone et al., 1989). This equipment is constructed in an open pit with a tiled floor. Perforated pipes under the floor supply forced under-fire air. Forced air is also directed over the top of the fire to permit particles to be returned to the combustion zone, thus reducing particulate emissions.

Mobile incinerators are useful when the cost of transporting wastes to central treatment facilities is excessive (Batstone et al., 1989). They also create less public opposition because people view the siting of the unit in their community as temporary, and also providing a solution to a problem that may have plagued them for years (Hunt, 1990).

Costs – There is very little cost information available at this time. The cost of handling, collecting and disposing of hazardous wastes by backhaul to applicable disposal sites can vary greatly with the type of waste, quantity, hazard rating and location. Costs for shipment to disposal facilities in southern Canada or the south 48 states is about $500 per 45 gallon drum or labpack. PCB waste shipping costs can be as high as $2,250 per drum.

Regulatory Issues – The application of regulations to solving solid waste problems is not always successful. In Alaska, requirements concerning solid waste facilities are administered by the state, based on regulations established with the U.S. Environmental Protection Agency.

These regulations are applicable to all states and U.S. territories. Minimum standards are contained under subtitle D of the federal Resource Conservation and Recovery Act. This legislation provides national standards for three general classes of municipal disposal sites based on daily quantities of waste processed. These regulations detail siting factors such as flood plain, wetland, airport safety, seismic impact zones, and wellhead/aquifer protection zones, as well as operating criteria relating to hazardous waste, cover material, explosive gas control, surface water quality, wildlife control, groundwater monitoring, record keeping, and financial assurance. In addition, post-closure monitoring and corrective action standards are developed.

Exemptions and modifications for Alaskan cold-region facilities are under discussion at the present time, and will be evolving as regulators try to address many of the complex issues facing solid waste management in the North.

16.9 Management Plans

In Alaska, a permit is required for the operation of a municipal solid waste landfill. An important aspect of such operation is the development of a management plan. Important aspects of a community management plan include the following:

Codes and Ordinances – Local codes and ordinances (bylaws) can be established which address the:

- statement of objectives of solid waste management;
- responsibility of the local government;
- methods of handling complaints and enforcement of rules;
- methods of disposal of hazardous wastes, batteries, or oil products; and
- general rules of community aesthetics.

Consolidation of Waste – A plan is documented to control the landfill use in a manner that allows waste to be consolidated into an area where it can be compacted and covered in an orderly manner. This allows the entire landfill area to be efficiently utilized and extends the life of the site.

Control of the Site – This part of the plan varies substantially with site size. It could consist of directive signs to help the public to utilize a small rural site properly or it could consist of a method to restrict the site to the actual operators. Waste would be deposited by the general public at an exterior disposal point.

Waste Minimization – This part of the plan presents the concept of reducing the waste stream volume to extend the site life. Methods to do this consist of recycling aluminum cans, storage of car batteries separate from the waste stream for later export, storage and control of salvageable materials,

and individual burning of waste in burn barrels prior to disposal.

Community Commitment – This presents the goals and principals of waste disposal in the community. Past solid waste disposal requirements did not consider these concepts or their relationship to regulations from state or provincial governments.

Visual Monitoring – This part of the plan presents how the landfill site is to appear along with the general appearance of the community. This allows residents of the community and adjacent property owners to have some assurance of the quality of life that can be expected from their neighbors and community residents. When conflicts exist, these requirements can be reviewed to determine if acceptable procedures are being maintained.

Closure Planning – This part of the plan presents the procedure or method that is to be followed in the shut-down and ultimate closure of the refuse site. It establishes the responsible parties and general procedures that will be followed to close the landfill. Future assurances and responsibilities to the general public may be established in this plan.

16.10 References

Alaska Department of Environmental Conservation (Hazardous and Solid Waste Management Section).1992. *Solid Waste Management Planning Guidelines for Alaska Communities.* January.

Alexander, H.D. and J.V. Interante. 1989. Waste minimizaation: Canadian study. In: *Proceedings, 21st Mid-Atlantic Industrial Waste Conference.* Technonic Publ. Co., Lancaster, Pennsylvania, 76-92.

Batstone, R., Smith, J.E. Jr., and D. Wilson. 1989. *The Safe Disposal of Hazardous Wastes: The Special Needs and Problems of Developing Countries.* World Bank, Washington, D.C.

Bryant, W.J. 1991. *A Hazardous Waste Management Strategy for NWT Communities.* Report for the Department of Municipal and Community Affairs, GNWT, Yellowknife.

Cameron, J.J. Personal Communication. 1993. *Solid Waste Management Costs for all NWT Hamlets (1988/89).* unpublished.

Collins, nd. *Summary Report on the Clean-up of PCBs in the NWT by the Federal Government.* Environment Canada, Yellowknife, NWT.

Drum, M. 1992. *Trash Management Guide*. Alaska Health Project, Alaska Department of Environmental Conservation.

Finnecy, E.E. 1988. The case for co-disposal of hazardous waste with municipal waste. In: *Hazardous Waste: Detection, Control, Treatment.* R. Abbou (ed.). Elsevier Science Publishers B.V., Amsterdam, The Netherlands.

Heeney, P.L. and G.W. Heinke. 1991. *Guidelines for the Collection, Treatment and Disposal of Hazardous and Bulky Wastes in the NWT.* Department of Municipal and Community Affairs, Government of the Northwest Territories, Yellowknife, NWT.

Heeney, P.L. and G.W. Heinke. 1992. Disposal of hazardous wastes in Canada's Northern Territories. *Canadian Journal of Civil Engineering*, 19: 866-810.

Heinke, G.W. and D. Prasad. 1980. Anaerobic treatment of human waste in northern communities. *Canadian Journal of Civil Engineering*, 7: 156-164.

Heinke, G.W. and J. Wong. 1990a. *An Update of the Status of Solid Waste Management in Communities of the NWT.* Department of Municipal and Community Affairs, Government of the Northwest Territories, Yellowknife, NWT.

Heinke, G.W. and J. Wong. 1990b. *Guidelines for the Planning, Design, Operation and Maintenance of Solid Waste Modified Landfill Sites in the NWT. Vol. I (Planning & Design), and Vol. II (Operation and Maintenance).* Department of Municipal and Community Affairs, Government of the Northwest Territories, Yellowknife, NWT.

Heinke, G.W. and J. Wong. 1990c. *Solid Waste Composition Study for Iqaluit, Pangnirtung and Broughton Island of the NWT.* March.

Hunt, S., 1990. The federal PCB destruction program: an approach to the study of mobile PCB incineration in Canada. In: *Proceedings, 12th Canadian Waste Management Conference*, St. Solens, NFLD. Environment Canada, Ottawa, Ontario, 49-57.

Leskiw, E.J., D.W. Smith, and D.C.C. Sego. 1985. The effect of frost penetration on moisture movement in solid waste landfills. In: *Proceedings 1985 Annual Conference Canadian Society for Civil Engineering*, Montreal, Quebec, 301-321.

Quaye, F.A. and G.W. Heinke. 1992. *Co-Disposal of Hazardous and Solid Wastes in the NWT.* Report for the Department of Municipal and Community Affairs, GNWT, Yellowknife.

Soberman, R.M., M. Lovicsek and G.W. Heinke. 1990. *Guidelines for the Separation of Solid Waste Disposal Sites and Airports in the Northwest Territories.* Department of Municipal and Community Affairs, Government of the Northwest Territories, Yellowknife, NWT.

Tilsworth, T. 1982. Solid waste management in remote Alaskan communities. In: *Utilities Delivery in Cold Regions*. Environmental Protection Service Rep. No. EPS 3-WP-82-6, Ottawa, Ontario, 329-344.

Whiteman, D. 1990. *Solid Waste Management at Norman Wells, NWT.* Department of Municipal and Community Affairs, Government of the Northwest Territories, Yellowknife, NWT.

Wiles, C.C. 1986. Treatment of hazardous wastes with solidification/stabilization. In: *Alternatives to Land Disposal of Hazardous Waste*. E.T. Oppelt, B.L. Blaney and W.F. Kenner (eds.). Air Pollution Control Association, New Orleans, Louisiana, 60-70.

Wilson, D.C. and F. Balkau. 1990. Adapting hazardous waste management to the needs of developing countries - an overview and guide to action. *Waste Management and Research*, 8: 87-97.

16.11 Bibliography

Alter, A.J. 1969. *Solid Waste Management in Cold Regions*. Scientific Research Data and Reports, Department of Health and Welfare, State of Alaska, College, Alaska, 90 p.

Cameron, J.J. 1981. *Guidelines for the Preparation and Administration of Municipal Water and Sanitation Trucked Service Contracts.* Water and Sanitation Section, Department of Local Government, Government of the Northwest Territories, 120 p.

Greenland Technical Organization. 1980. *Guidelines for Housing and Human Settlement Planning and Design Criteria in Cold Climates: Greenland.* Regional Office for Europe. World Health Organization, Copenhagen, Denmark, 167 p.

Grundwaldt, J.J., T. Tilsworth and S.E. Clark. 1975. Solid waste disposal in Alaska. In: *Proceedings, Environmental Standards for Northern Regions*. Institute of Water Resources Rep. No. IWR No. 62, University of Alaska, Fairbanks, 331-356.

Heinke, G.W. 1973. *Solid Waste Management in the Canadian North*. Department of Civil Engineering, University of Toronto, Ontario, 145 p.

Heinke, G.W. 1984. *Public Health Effects of Municipal Services in the Northwest Territories*. Department of Local Government, Government of the Northwest Territories, Yellowknife, 45 p.

Kelton, K. 1975. *Comprehensive Plan for Solid Waste Management*. Alaska Department of Environmental Conservation, Juneau.

Leskiw, G. 1981. *Elements of Engineering Design for a Sanitary Landfill*. Water Management Branch, Publ. No. 4, Alberta Environment, Edmonton, Alberta, 54 p.

Miller, R.E. and O.E. Dickason. 1982. Management of solid waste in cold regions: resource recovery potential. In: *Utilities Delivery in Cold Regions*. Environmental Protection Service Rep. No. EPS 3-WP-82-6, Ottawa, 345-357.

Straughn, R.0. 1972. The sanitary landfill in the subarctic. *Journal of the Arctic Institute*, 25: 1.

Watmore, T.G. 1975. Problems of waste disposal in the Arctic environment. *Industrial Wastes*, 21(4): 24.

Wiles, C.C. 1986. Treatment of hazardous wastes with solidification/stabilization. In: *Alternatives to Land Disposal of Hazardous Waste*. E.T. Oppelt, B.L. Blaney and W.F. Kenner (eds.). Air Pollution Control Association, New Orleans, Louisiana, 60-70.

Wilson, D.C. and F. Balkau. 1990. Adapting hazardous waste management to the needs of developing countries - an overview and guide to action. *Waste Management and Research*, 8: 87-97.

SECTION 17

ENERGY MANAGEMENT

3rd Edition Steering Committee Coordinator

William L. Ryan

Vern Christensen

3rd Edition Principal Author

Eugene Bjornstad

Section 17   Table of Contents

Section 17 List Of Figures

Section 17 List Of Tables

17   ENERGY MANAGEMENT

17.1   Introduction

Energy is required to maintain not only life but also a reasonable standard of living. When designing utilities in cold regions, local weather conditions, energy sources and distribution, and methods of conservation require careful consideration. These tasks of total energy management are covered in this section.

17.1.1   Standard of Living.
Concomitant with an improved standard of living is the improvement in the quality and safety of the housing environment. The provision of heat and electricity is often considered of equal, if not of greater, importance than the provision of safe water and sanitation services. In reality, all four utilities along with a properly designed and constructed building are essential constituents of a good, healthy living environment.

17.1.2   Importance of Energy Management.
The climate in northern regions is among the most severe on the planet. Much energy is consumed to attain an adequate standard of living. Energy is used for heating, electricity production, transportation, and communication.

Even though considerable coal, oil, and natural gas deposits have been confirmed in the North, energy costs remain high. This is due in part to the high costs of production and the need in some cases to transport some of the raw materials to southern processing plants. Not only is the energy use higher in the North but the cost per unit is higher as well. Therefore good energy management fosters the use of less energy, and savings in the cost of heating and electricity.

17.1.3   System Components.
Although many energy requirements exist for those living in the North, the discussion here is limited to building heating, electricity generation and use related to water and wastewater operation.

17.2   Climatic Considerations

The climate and the resulting impact on the environmental components are what make cold regions unique. The severe climatic conditions of winter make operation of utilities difficult because materials and equipment get damaged, and ground and water freeze. The charts and maps showing climatic factors for Canada and Alaska are included in Appendix E. Additional and more detailed information is available (Hartman and Johnson, 1978; Walker, 1979; ASHRAE, 1993; Argue, nd).

17.2.1   Temperature and Indices.
The heating, freezing, and thawing indices are calculated using temperature information. The method of computing °C·d indices is demonstrated in Section 3.

Temperature Information.   Temperature information is needed for any design or construction in the Arctic (Hartman and Johnson, 1978; Argue, nd). Appendix E provides the mean annual air temperature for Canada and Alaska and gives the mean January and July air temperatures. These are usually the highest and lowest temperature months in the arctic. Design temperatures for various communities throughout Canada and Alaska are also presented in Appendix E. The design temperatures indicate the lowest temperature that can be expected in a particular location and provide a measure of the needed capacity of heating systems.

Heating Index.   The heating index (HI) is the number of °C·d below 18°C. This is considered to be the heat that must be supplied at that location in an average year. For design purposes, the one-in-ten-year heating °C·d value should be used. This value can be estimated by using the values given in Appendix E.

Freezing and Thawing Indices.   See Section 3 for an explanation of the freezing and thawing indices. Appendix E presents freezing and thawing indices for Canada and Alaska.

Freeze-up and Break-up Dates.   These dates for Canada and Alaska are presented in Appendix E.

17.2.2   Solar Data.
The amount of light energy reaching a location varies with latitude, topography, cloud cover, and twilight (Hartman and Johnson, 1978; Walker, 1979; ASHRAE, 1993; Boyd, nd; McCullagh, 1978). These factors are involved in the important parameters discussed below:

- **Hours of Daylight.**   The hours of daylight are not the same as the hours of sunlight because of the effects of topography and cloud cover. Appendix E provides the hours of daylight for latitudes 45° to 75°N.

- **Sun Path Diagrams.**   These give the overhead position of the sun, both altitude and azimuth, for any date and time at several latitudes. Latitudes between those shown can

TABLE 17-1   PERCENT LIGHT DURING A YEAR

Latitude (°N)	Twilight (%)	Sunlight (%)	Sunlight and Twilight (%)
0	3.1	53.3	53.4
25	3.5	50.5	54.0
35	3.9	50.6	54.5
45	4.7	50.8	55.4
55	6.2	51.0	57.3
60	8.1	51.3	59.4
65	10.4	51.9	62.3
70	10.9	52.0	62.9
75	8.9	51. 9	60.8

be interpolated. Again sunlight at a given spot varies according to local topography. The values given in Appendix E are maximum values.

Amount of Twilight and Sunlight. The total time during which light reaches the earth in a year is greater at the Arctic Circle than at the equator. However, the light arrives primarily during the summer. Table 17-1 indicates the percentage of sunlight and twilight in a year for different latitudes.

The reason for the greater amount of light at the higher latitudes is that the sun crosses the horizon at a shallow angle and consequently takes longer to rise or set. This lengthens the day slightly and the twilight considerably. Also, refraction of the sun's rays by the atmosphere lengthens the period of day-light by making the sun visible even when it is below the horizon. The higher the latitudes the greater the percentage of sunlight received there during the summer (Table 17-2). Since the sun's rays strike the ground at an angle in the Arctic, the energy received on the ground surface is less than it would be if they struck directly from overhead as they do at the equator. Thus, radiation intensity is lower but, because of the longer daylight hours, the total amount of radiation received during the year is approximately equal to that at lower latitudes. Appendix E provides the estimated number of hours of bright sunlight per year and the net solar radiation for Canada and Alaska.

17.2.3   Wind. The basic driving force for wind is the unequal heating of the earth and atmosphere. Air flow patterns are dictated by the rotation of the

TABLE 17-2   LATITUDE VS. TOTAL
                         ANNUAL SUNLIGHT

| | Total Annual Sunlight | |
Latitude (°N)	Summer (%)	Winter (%)
55	63	37
60	66	34
65	71	29
70	79	21

earth and by the terrain. Inventories of average wind speeds are available (Noble and Van der Hoek, nd; Adelaav and Assoc., 1981). Appendix E gives general speed information about mean annual wind. The installation of an anemometer is recommended for locations where data is not available.

17.2.4 Other Information. Information about rainfall (Hartman and Johnson, 1978; Walker, 1979; U.S. Department of Commerce, nd) and snowfall may be found in Appendix E.

17.3 Energy Source Assessment

Primary energy is the energy content of the resource in its natural state. Secondary energy is the amount of energy as the fuel is delivered to the consumer. The difference between the two is lost primarily in processing, transmission, and delivery.

Renewable energy is generally associated with solar energy, such as wind, hydropower, or wood; and other phenomena, such as tides and geothermal heat.

Nonrenewable energy sources are those which are essentially consumed and not replaced in a reasonable length of time. These include oil, natural gas, and coal.

The nuclear energy source is a self-decaying element; nuclear energy, however, is not being used as a commercial power source in arctic regions.

Some of the materials classified as renewable may for practical purposes be nonrenewable. For example, peat (a form of biomass) has a long regeneration time in the Arctic. Similarly, wood cannot be produced in many cold regions at a rate considered renewable.

17.3.1 Sources and Uses. Energy is used for heating, electricity generation, and transportation. The source of energy for each use depends on a variety of economic and site conditions. For heating, oil, gas, coal, wood, geothermal heat, and even electricity are used. Electricity generation makes use of the same energy sources (except for wood) as well as moving water. Transportation is limited to petroleum products and gas.

All the major energy sources are produced in cold regions from raw materials found there. But distribution costs are often so high that it is more economical to transport the energy source from southern locales.

17.3.2 Energy Materials. Each of the major sources are reviewed below with regard to their properties and the limitations on their use.

Oil. The oil products of common use in cold regions are: fuels such as gasoline, diesel, heating, kerosene; jet fuels; lubrication oils; hydraulic oils; and greases.

Gasoline is used primarily for transportation in the Arctic and subarctic areas: in automobiles, all-terrain vehicles, snowmobiles, and outboard motors. In remote areas, gasoline is generally 10 to 15 percent more expensive than diesel or fuel oil. (Only in larger communities is gasoline available in bulk.) Gasoline does not experience the troublesome viscosity increase with lower temperatures as oils do (Figure 17-1). This advantage, however, is offset by the ease with which water can enter gasoline in storage tanks or in vehicles because of the condensation which is readily formed by the temperature changes in cold regions. Gasoline deicer or antifreeze (isopropyl alcohol) should be added to tanks regularly to absorb the moisture.

Several factors cause a decrease in fuel economy in cold regions during the winter, unless the following precautions are taken:

1. Preheating engines can reduce pollution and fuel usage. At a water temperature of 30°C, cylinders and rings wear at a rate of between 0.025 mm and 0.050 mm per 1,600 km, whereas at a water temperature of 77°C, wear is nil. An automobile using 16 L/100 km when the engine is warm, uses up to 26 L/100 km when it is cold. Even at the higher electrical costs in cold regions, preheating an engine is economical for the fuel savings alone. This applies even at temperatures above -18°C.

2. For an automobile travelling at 83 km/h at normal operating temperature, a 5.6°C drop in air temperature increases fuel consumption by 2 percent. This can be improved in several ways:

   a) A conventional belt-driven cooling fan runs all the time and uses at least 5.2 kW. Conventional fans can be replaced with electric ones which are operated on a thermostat. This keeps the engine operating near its optimum temperature and reduces fuel consumption by approximately 10 percent.

   b) Insulating the underside of the hood keeps the engine warmer. Putting cardboard, or some other material, over the radiator is not advisable because that can

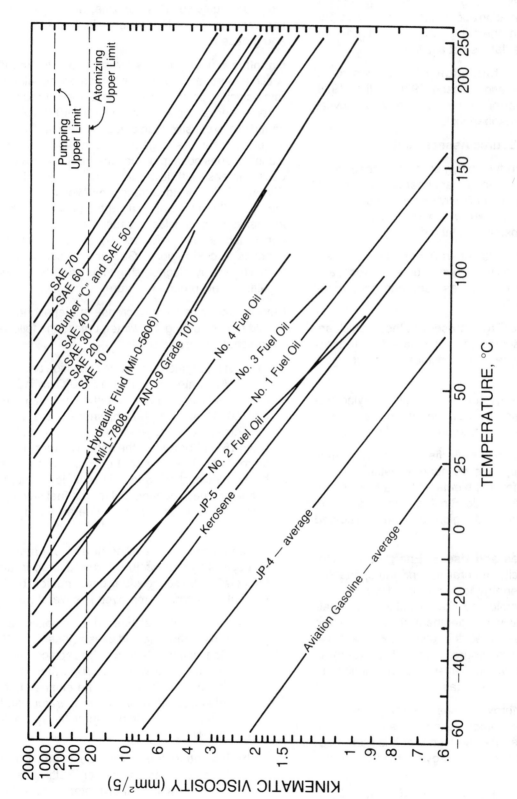

FIGURE 17-1   KINEMATIC VISCOSITY AS A FUNCTION OF TEMPERATURE FOR VARIOUS FLUIDS*

* The values presented in this figure are average values and suppliers can provide the actual viscosity values for their fuel. Also, at temperatures less than the fuel's cloud point, the relationships alone are no longer straight lines and should not be used.

c) Higher-temperature thermostats (such as 90°C) should be used to keep the engine temperature up for more efficient operation as well as for the comfort of the passengers.

d) Keep tires at the recommended pressure in the winter because the air in them contracts as it gets colder.

3. Snow or dirt road surfaces increase the rolling resistance of vehicles which increases fuel consumption.

4. Use lightweight oils and lubricants in transmissions and differentials. Instead of SAE 90 which is commonly used, use SAE 75 or SAE 75W90 multiweight. Synthetic multiweight motor and lube oils can significantly improve mileage over regular oils.

5. Air filters must be kept clean.

6. Battery warmer plates can be strapped to the bottom of the transmission to pre-warm the transmission and reduce drag.

Diesel fuel oil is available in several different grades. Canadian designations differ from the U.S. designations for the grades of diesel. Generally, only diesel fuels no. 1 and no. 2 should be used in diesel engines. Diesel fuel properties are presented in the ASTM fuel classification and selection chart (ASTM D-975). Heating fuels are presented in ASTM D-396. Engines operating at altitudes over 1,500 m require the next lighter class of fuel oil than would be used at sea level.

Another important property to watch for is the "cloud point" of the fuel. This is the temperature at which wax crystals begin to form in the oil. The cloud point of the selected fuel should be at least 5.5°C below the lowest expected fuel temperature. ASTM D2500 provides the procedure for determining the cloud point of a fuel.

Ignition quality, represented by the octane number, is a measure of the ability of fuel oil to ignite spontaneously in engine cylinders. Higher octane numbers (as with higher octane numbers for gasoline) minimize the tendency of the engine to knock. The average diesel engine requires fuel with an octane number of 45 or greater.

The pour point determines the temperature at which a fuel can be transferred, and is determined by the procedure in ASTM D-97. Appropriate upper limits for pumping are 1,000 mm^3/s and, for mechanical atomizing, 20 to 30 mm^3/s.

Fuel oil that cannot meet the viscosity requirements for the anticipated temperature has to be warmed before use. Suppliers also produce what is commonly called "arctic" diesel which is primarily no. 1 fuel oil with varying quantities of JP-4 (jet fuel) added to lower the viscosity at extremely cold temperatures.

Heating fuel oil properties are provided in Table 17-3. These are covered by ASTM Standard D-396.

Kerosene and jet fuels (JP-4 and JP-5) are slightly more highly refined than diesel fuel oils. However, most of the above discussion for fuel oils also pertains to them.

Motor oil is available in various weights with different properties, uses, and viscosity-temperature relationships. "Cold soaking" is defined as the difference in difficulty between starting an engine having been left in a cold environment for several weeks versus one that has been left for only a day or two. This difference is caused in part by the waxes which precipitate from most oils after a period of time at temperatures below -20°C. The wax that precipitates is slow to redissolve upon heating. This means that an engine which contains precipitated wax must be heated to a higher temperature before starting than does an engine that has only recently become cold.

Hydraulic oil commonly used in cold regions has properties shown in Figure 17-1. In the selection of a hydraulic oil one should also consider the pressure under which it will operate.

Approximate Comparison of Diesel Oil Grades and Designations:

U.S. Grade Designation	No. 1	No. 2	No. 3	No. 4	No. 5
Canadian Grade Designations	Type AA Type OO	Type A	Type B	Type C	Type D

TABLE 17-3   PROPERTIES AND USES OF HEATING FUEL AND DIESEL OILS

Grade	Gravity (API)	Density (kg/m^3)	Heating Value (MJ/L)
No. 1	38 to 45	833 to 800	38.2 to 37.1
No. 2	30 to 38	875 to 834	39.5 to 38.2
No. 4	20 to 28	933 to 887	41.3 to 39.9
No. 5 (light)	17 to 22	951 to 921	41.8 to 40.9
No. 5 (heavy)	14 to 18	968 to 945	42.4 to 41.6
No. 6	8 to 15	1010 to 965	43.5 to 42.2
Grade No. 1	Light distillate; used in burners of the vaporizing type; high volatility assures that evaporation of the fuel oil proceeds leaving a minimum of residue; used extensively under cold-region conditions.		
Grade No. 2	Heavier distillate than Grade No. 1; used in pressure-atomizing type burners which spray oil into combustion chambers; atomized Grade No. 2 is used in most domestic burners in temperature climates; also used in medium-capacity commercial-industrial burners.		
Grade No. 4	A light residual or a heavy distillate; used in burners designed to atomize oils of higher viscosity than domestic burners can handle; its permissible viscosity allows it to be pumped and atomized at relatively low storage temperatures.		
Grade No. 5 (light)	A residual fuel of intermediate viscosity; used in burners designed to accommodate fuel more viscous than Grade No. 4 without preheating; in cold climates, preheating for both burning and handling may be necessary.		
Grade No. 5 (heavy)	A residual fuel more viscous than Grade No. 5 (light); preheating necessary for burning, even in temperate climate.		
Grade No. 6 (bunker C)	High-viscosity oil; industrial use requires preheating in tank to permit pumping and additional preheating at burner to permit atomizing.		

Greases, brake fluids, etc. should be selected for the cold temperatures to be encountered. Arctic-type brake fluid (SAE 1707E) should be used.

Oil Storage.   The basic considerations for oil storage in cold regions are the same as those for water storage (see Section 7). The most common method for storing large quantities of petroleum products is in large, above-ground storage tanks of welded steel. These tanks should be constructed to the standards set by the American Petroleum Institute, Standard No. 650. The structural design of the tanks themselves is usually not significantly different from that in a more temperate climate, except that roofs must be designed to handle the design snow load, and suitable low-temperature steels should be used. All welds should be carefully inspected by using x-ray or ultrasonic techniques.

Tank appurtenances in arctic climates also vary from those in temperate climates. Water contamination is to be avoided because of the problems caused when water enters the fuel lines and filters of equipment and then freezes there. A water draw-off connection should be supplied on the tank. This draw-off valve must be a nonfreeze type. The valve should be located inside the tank, provided its temperature never falls below freezing or the valve must be heated to prevent its freezing and breaking.

Some tanks are fitted with vapor-saving breathers which tend to freeze up in cold weather unless they are equipped with a flexible frost- and ice-resistant diaphragm material. Breather screens tend to ice or frost over during periods of extreme cold, and it is probably best either to remove them during winter or to heat them to prevent ice formation.

Oil tanks are usually kept at lower temperatures than water storage tanks and thawing of foundation materials is not as serious a problem. If necessary, the oil temperature should be kept below 0°C so that heat lost from the oil to the soil does not thaw the permafrost. In fact, it is a good idea to paint tanks with a reflective paint, such as silver, to reduce the solar heat gain. Also, oil tanks, unlike water tanks, are usually not insulated.

Flexible "bladder" tanks have also been used to store petroleum products in the Arctic, especially for temporary storage. Large tanks can be flown to remote sites and filled immediately. There have been cases of damage by vehicles running into or over the tanks. Such tanks should therefore be fully diked so that in case of puncture the oil is contained.

Bolted-steel tanks have been used in several installations in the Arctic for fuel storage. It is important that the predrilled pieces be shipped to a remote site without being warped or bent during handling.

Welded-steel oil-storage tanks cost about $0.20/L (US, 1995) installed in remote villages or camps. Small tanks (less than 60 m³) are somewhat costlier, while large ones (over 3,500 m³) should cost less. These costs include diking with liner in locations having fill material readily available. Double wall tanks are typically three to four times more expensive.

Oil Transportation.   Pipelines are the most efficient method of transporting large quantities of petroleum products. New installations use welded joints. If lines must be moved from time to time, screwed union joints must be used. The design of expansion joints or loops depends on the type of pipe, length of straight runs, and the amount of movement that can be tolerated. Rubber hoses are often used where flexible lines are necessary, for example, between barges and shore facilities.

All pipes crossing streams or roads must be completely protected from damage by vehicles and ice. Also, above-ground or surface lines should be painted bright orange to reduce collisions between vehicles (mainly snow-machines) and the pipeline.

Pressure testing taps should be supplied at convenient points along all oil pipelines so occasional pressure tests can be accomplished. All seasonally used lines, such as those from barge to storage tanks, should be tested before each use. All pipelines of more than 1.0 km should be fitted with flow meters at each end so a complete fuel inventory can be maintained.

The approximate cost to install a 50 mm steel welded oil pipeline on the ground surface at a remote community or industry would be $200/m (US, 1996).

Petroleum products are also transported by tank truck, usually in and around communities or sites; by barges, usually over long distances on rivers or at sea; or by air to remote sites not served by barges or roads. Under emergency conditions, petroleum products are flown to any remote site.

All petroleum-product pipelines and storage facilities should be well grounded to dissipate any static electricity that builds up. Do not ground electrical distribution or generation equipment to fuel lines. Vehicles transporting fuel or being filled with fuel should be well grounded during any fuel transfer operations. Corrosion potential should be checked at all points considered for a ground. Cathodic protection in cold regions is nearly identical to that designed and supplied in a more temperate climate.

Fuel spill and containment practices are similar to those in warmer climates. All fuel storage facilities must be diked. The largest cause of fuel spills is human error and negligence of operating personnel. An important initial step is to have a thorough and well understood fuel spill contingency plan, so that response can be immediate to reduce environmental and property damage as much as possible. Absorbents should be on site for use in the containment and clean-up of fuel spills. Peat moss can be used for this purpose. Weather conditions, topography of the site, soil properties, and type and volume of oil spilled must be evaluated in predicting the spread of the oil spilled and the amount that would infiltrate the ground.

When spilled, oil first flows over the surface and then moves downward, saturating the organic layer of the soil and finally, penetrating to the subsurface water or the upper surface of the permafrost. Dikes should surround all tanks, and the diked area, including the area under the tanks, should be impermeable to prevent seepage. Lining that area may be necessary for it to be impermeable. Regulations on capacities required within the dikes vary, and the designer should check the applicable environmental protection regulations for the requirements. Typical requirements call for 150 percent of the capacity of the largest tank within the dike plus 25 percent of the capacity of the remaining tanks. Methods for lining are discussed in Section 7. Plastic liners are probably the most satisfactory solution, unless clay is available locally. Liners of polyethylene, polyurethane, polyvinylchloride (PVC), and chlorinated polyethylene have all been used successfully. Liners should be installed over the entire surface in new tank installations, before the tanks are constructed.

Natural and Processed Gas. Natural gas and processed gases are used for heating, food preparation, and electricity generation. It is a clean and low-maintenance fuel. It is usually the fuel of first choice, where available, because it does not require air pollution prevention equipment.

Propane gas is extracted from natural gas. It is used throughout the Arctic and subarctic, especially for cooking. It can be transported and distributed in bottles or tanks mounted on trucks or barges. It has a combustion rating almost twice as high as that of natural gas. The one major problem with propane in cold regions is its vaporization characteristics: propane remains a liquid below -43°C and can cause fires or explosions if this liquid is allowed to pass into piping and appliances. Propane tanks exposed to these temperatures must be enclosed in a heated structure or themselves be heated in some way. If located in a structure, adequate ventilation is necessary to prevent a buildup of explosive vapors. Also, at subzero temperatures, the expansion of the propane as it passes through the valve and regulator can cause a drop in temperature which can result in freezing of the regulator and valve. This can be prevented, however, by the addition of 100 to 200 mL of alcohol to each 20 L propane bottle before it is filled.

Transportation of Natural Gas. Natural gas can be transported overland for long distances by:

- pipeline in the gaseous form;
- pipeline in the liquid form; and
- rail, road, or aircraft in the liquid form.

Natural-gas piping may be buried, installed at ground level, or placed overhead, depending upon the site conditions and the overall concept and usage of the installation. The types of soil at the site, depth and structure of the permafrost, water content, surface drainage, etc., must be determined as part of an assessment of the site conditions.

Where soil types and the natural drainage patterns permit, direct burial of the gas distribution system is the most desirable installation from the viewpoint of economics, security, and area utilization. Where soils consist of frost-susceptible materials, and where surface water forms pools, the gas piping should not be direct-buried. If piping is installed in soil which is frost susceptible or water saturated, the water can freeze to the pipe, forming ice anchors. A section of piping secured between two ice anchors can fail (usually at the connecting points) because of axial stress from further temperature reduction.

Where direct-buried gas piping must cross roadways, the piping should be installed in a protective sleeve or culvert. The piping within the sleeve should be provided with supports or spiders which position the gas pipe concentrically with the sleeve. Where the ends of the sleeves are buried, the ends should be sealed tightly to the gas pipe with flexible boots and drawbans or clamping devices to prevent infiltration. Several types of these boots are available. Where the ends of the sleeves extend upward due to area grading or to road elevation, the ends may be left open. The sleeve or culvert should be pitched slightly to ensure proper drainage of moisture.

The design and installation of the gas transmission and distribution system must include provision for the expansion and contraction of the piping caused by changes in temperature of the soil or the gas, or both. Where the system supply is obtained from an above-ground transmission line, the gas temperature may range from -50°C to 27°C. The range of ground temperature, of course, is less. It also changes more slowly. Flexibility may be worked into the system layout, making use of L- and Z-shaped runs, with properly located anchors. On long straight runs, expansion loops must be provided. Legs of the loops should be encased in suitable resilient material to allow space for pipe movement when required. The loops should be prestressed when installed, and adjusted so that each loop is in the neutral position at minimum design temperature.

Slip-type expansion joints or couplings are not recommended for use in gas distribution systems in cold regions. At extremely low temperatures, the seals of the joints harden and lose the elasticity required for the sealing effect. The use of slip joints or couplings can be avoided by proper use of bends, loops, or offsets, thereby eliminating an unnecessary risk of system failure.

Drip pockets, for moisture or gas condensate accumulations, with blow-off pipes and valves are required at all low points in the system and at the base of risers. Natural gas in cold regions may absorb sufficient moisture to require dehydrating.

A pipeline installed just high enough above the ground to be exposed to the normal snow level is called a ground level system. It is the most simple and economical system, but it is vulnerable to damage. In general, the concept can only be used in areas requiring a limited number of service connections and where the piping can be routed to avoid planned traffic patterns.

Coal. Coal has been intermittently mined commercially in Alaska for over 80 years, yet the reserves are essentially untouched, primarily because of the high cost of extraction and transportation of the coal relative to oil or natural gas. The same holds true for the deposits in the Northwest Territories west of Norman Wells. Low sulfur (0 to 5 percent) and ash content (4 to 25 percent) characterizes these coals.

The means of storage and transport of coal is about the same in both arctic and temperate climates. However, unloading and storing coal in cold regions pose special problems. Coal is usually washed at the mining operation prior to loading for shipment. If the coal is shipped in rail cars, it must be thawed before the rail car can be emptied. This is usually accomplished in an enclosed shed over the tracks at the power plant. Steam probes may be necessary to thaw the coal more quickly. Once unloaded, If the coal is then stored outside, steam probes again or bulldozers, or both, are necessary before the coal can be dumped in the inside storage bins at the plant. Freezing can still be a problem if the temperature of the coal is very low when it is brought in. This condition is usually remedied with shakes attached to the chutes or bins.

Most boilers can be adapted to burn coal or oil or a combination of both at the same time. For remote northern villages, boilers or gasifiers that burn wood, coal, or oil could eliminate the need for standby sources of fuel, because one of the three would always be available.

Hydropower. Hydropower resources in the Arctic and subarctic areas of Alaska and Canada are considerably underdeveloped (Adelaav and Associates, 1981; Shira, 1978).

There are several hydropower-generating facilities in subarctic areas, but very few in the arctic areas of the USA and Canada. Ice problems at penstock entrances and the effect of large dams and reservoirs on permafrost must be overcome. Russia has several large hydroprojects in the permafrost areas of Siberia. Some information on hydroelectric projects in the arctic is presented in Section 3.5.

Hydropower can be subdivided into low-head ($\leq$18 m) or high-head (>18 m) hydropower. Low-head installations require higher flows to produce the same power as high-head installations. Also, low-head turbines are not as efficient as high-head turbines.

Without some storage the utilization of low-head hydropower in the Arctic is usually limited. Energy needs are greater in the winter, but the stream and river flow rates are sharply reduced at that time. Without water storage, a complete standby power source derived from oil must be provided, which greatly increases capital costs. The head available depends on the elevation of the dam or intake and the elevation of the turbine. The power (P) available in kWh can be estimated by:

$$P = \gamma H Q e$$

where

P = power, W

H = head, m

Q = flow, m^3/s

e = efficiency

γ = specific weight = 9.81 kN/m^3 at 4°C

Considering generator efficiency and pipe head losses, overall efficiency is typically 60 to 70 percent.

Wood. The renewability of wood depends on the rate of tree growth. In many cold regions only black spruce grows and does so very slowly because of the soil and climatic conditions. In much of the Canadian Arctic, especially on the alluvial plains of the MacKenzie River basin and along some rivers in Alaska, significant stands of white spruce can usually be considered renewable. Generally, wood should probably not be considered as a significant, renewable resource in most of the arctic areas of North America. Along the north coast, only drift wood is available. Many areas of the subarctic, however, contain renewable wood resources.

The average heat value of wood drops quickly with increasing moisture content as can be seen in Figure 17-2. For instance, at 50 percent moisture content, 13 percent of the heat value of the fuel is required just to vaporize the moisture.

The cost of wood in small arctic villages varies considerably, depending partly on the distance it must be hauled and the sophistication of the harvesting methods. Costs (US, 1980) range from $80 to $130 per cord (1.2 by 1.2 by 2.4 m) in three small villages

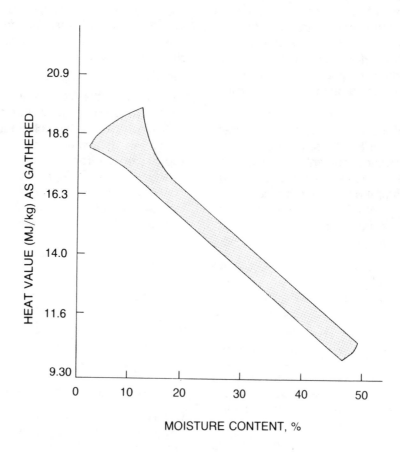

FIGURE 17-2   *HEAT VALUE VS MOISTURE CONTENT FOR WOOD*

studied in northwestern Alaska with the highest cost for wood harvested by the most labor-intensive method: hand-felling with a chainsaw and yarding with a snow-machine (Wind Systems Engineering, 1981).

In addition to direct burning and wood gasification, wood can be used to produce methanol by destructive distillation. Methanol can be substituted for gasoline in snow-machines, outboards, etc. The process is relatively simple and can be carried out in small plants. Whether white or black spruce would be satisfactory for methanol production is still to be investigated.

Geothermal.   Geothermal energy is abundant, but currently economical in only selected locations. The temperature increase with depth below the earth's surface (geothermal gradient) is about 0.03°C/m. In areas of volcanic activity or hot springs, the gradient can be many times higher than this value. The

most common method of tapping this resource is simply by using the hot water.

Geothermal energy is environmentally attractive because, unlike fossil fuels, few atmospheric pollutants are generated. There is, however, significant potential for thermal or chemical pollution. Most geothermal waters are corrosive and contain high brine concentrations. One possible solution to these geothermal pollution problems is to re-inject the water after the heat has been removed. The corrosiveness is a result of high temperatures coupled with high mineral contents. All piping and heat exchangers must be designed to handle this highly corrosive water.

Water heated geothermally to 150°C or higher is necessary for power production. Lower temperatures, however, could suffice for space heating. Space heating or greenhouse heating can be accomplished with water temperatures ranging between 10°C and 100°C. Many hot springs in the Arctic and subarctic produce water in this range. Usually with water temperatures of less than 40°C, fossil fuels would have to be used to boost the heat for domestic heating. Conversion efficiencies for space heating are relatively high: 70 to 90 percent. For electrical power generation, however, efficiencies drop to between 5 and 25 percent.

Many pollution problems can be alleviated by using a "down hole" heat exchanger. In this situation a heat exchanger is installed in a deep well. Water can then be pumped down from the surface, through the heat exchanger and back to the surface for use. The corrosive water is never brought to the surface.

The hot springs in the arctic areas of Alaska seem to be relatively noncorrosive and free of toxic substances. The water is usually low in salt content but high in silica.

Where geothermal resources are available, their use for space heating should be considered.

Wind. An average wind velocity of 19 km/h is considered necessary for a wind generator to be feasible. Actual anemometer readings at the specific site are essential for the proper design of a wind generator.

Small wind generators and water-pumping windmills have been used for over 150 years. These 1 to 2 kW units have been used to charge batteries for lights on farms and for individual houses in remote communities. They have been used in cold regions to run small well pumps and heat tapes. The larger electrical power generating units are relatively new, and many are now being tested throughout the world. Most of these units, and even the smaller, older units, have not received the rigorous testing needed to make modifications so that the units can operate relatively maintenance-free in arctic and subarctic areas. Problems must still be solved with respect to high winds, ice formation on the blades, governors that withstand ice and cold weather, synchronous inverters, and other electrical components, before the units can be relied upon in remote northern locations. Appendix E provides some general wind speed information for Canada and Alaska.

A wind generator consists basically of a tower of sufficient height to place the blades above the turbulent-flow winds caused by surface obstacles, blades attached to a shaft which is turned by the wind, and a generator to produce a direct current. The amount of energy that can be extracted from the wind is directly related to the power available in the wind stream. In arctic areas, the wind is usually stronger in the winter than in the summer, and the power in the wind is a function of the cube of the wind velocity. For example, winds in Kiana, Alaska, average about 13 km/h in the summer and 21 km/h in the winter (Wind Systems Engineering, 1981). Thus there is about four times more power available in the winter. The community power demand is about three times greater in the winter than in the summer.

Wind stream power can be calculated from (Cheremisinoff, 1981):

$$P = 1/2 \, krV^3A$$

where

P   =   power in windstream, kW

r   =   density of air, kg/m^3

    =   1.201 kg/m^3 at 101.325 kPa and 21°C

A   =   unit area of windstream, m^2

V   =   wind velocity, km/h

k   =   2.143 x 10^{-5} for the above units.

Extracting this power from the wind stream can be accomplished using horizontal- or vertical-axis wind machines. The most commonly used device is the horizontal-axis machine. The maximum power extracted by this type is:

$$P_{max} = 1/2 \, krV^3 A_t e$$

where

A_t =   projected cross-sectional area of turbine blades, m^2

e   =   efficiency of power extraction

    =   0.593 maximum theoretical

As air density decreases with increasing altitude the energy in the wind decreases correspondingly. For instance, at 150 m the energy in the same wind is only 86 percent of what it would be at sea level. The energy increases significantly (by the cube) with wind speed. For example, if the wind speed doubles, the energy available increases eight times. Using the average annual wind speed in the equation can produce erroneous results, because the average of the cube of many different wind speeds is greater than the cube of the average wind speed.

Wind speed can vary significantly with height (Figure 17-3). As a rule, towers for residential wind generators should be at least 20 m high and not located

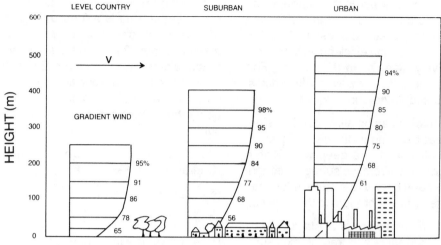

Percentage of wind speed over rural, suburban and urban terrain.

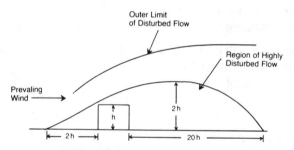

Effects of an individual building on air flow.

FIGURE 17-3   AIR FLOW AROUND OBJECTS

where trees, buildings, or other obstacles can affect the air flow. The approximate variation of wind speed with height is given by the formula:

$$V = V_o (H/H_o)^{1/a}$$

where

V_o = measured wind speed at a height of 10 m, km/h

V = estimated wind speed, km/h

H = new height, m

H_o = original height, m (usually 10 m, the standard anemometer height)

a = constant which varies with the surface roughness

= 7 over smooth terrain

= 2.5 over rough terrain, buildings, etc.

= 6 average value often used

Doubling the height over smooth terrain increases the wind speed by only about 10 percent, whereas doubling the height over rough terrain increases the wind speed by about 80 percent.

Theoretically the maximum efficiency is about 60 percent. The very best systems to date can recover little more than 40 percent of the energy available, and most small wind generators operate at 20 to 30 percent efficiency.

Wind power generators must be properly located. It is extremely important to avoid turbulent areas and find a location having a 19 km/h or greater annual average wind velocity. Units larger than about 3 kW require trained personnel to install and operate. Wind generators can add units incrementally as demand for power increases. They tend to be high in capital cost but low in operating costs.

Three main types of power generators are available for wind machines: direct current generators, induction generators, and synchronous alternators.

Most wind generators are constructed to cut out at wind speeds over 64 km/h. This must be taken into consideration when computing available energy. For more detail on the design and placement of wind generators see a study by Peters (1981).

The cost per kWh for electricity from wind generators depends heavily on the wind speed. Yearly operating costs are generally reported to be in the range of two to four percent of the capital costs (Adelaav and Associates, 1981; Peters, 1981). Routine maintenance is very important to the safety of personnel and continued operation of the machine (Peters, 1981). There have been many failures because of the lack of proper and timely maintenance.

Solar. There are over 200 more hours of sunlight at the Arctic Circle than there are at the equator; however, the sunlight does not have the intensity because of the angle at which the rays strike the Earth's surface. Also, the sunlight comes mostly during the summer months, and the electrical or heating needs are during the winter. However, heated water is needed for clothes washing and bathing year-round, and solar energy can be a sufficient source of heat.

Figures 17-4(a) and (b) indicate the wave length and relative energy levels for sunlight. A lesser proportion of the total solar energy is available at any given time in any location under cloud cover and in the shadow of surface obstructions such as trees and buildings. Figures are available from some weather records which provide cloud cover estimates for different communities throughout North America.

Incident sunlight can be converted directly to electricity by photovoltaic cells, because they have the photoelectric properties of a semi-conductor. The solar cells are connected in modules which are then connected in arrays to provide the desired amount of electricity. These solar generators have been used for charging batteries to operate, for example, remote microwave repeaters, navigational aids, and well pumps. Each cell can supply current at voltages up to 0.5 Vdc. At lower voltages the current supplied is nearly independent of voltage, but varies with light intensity. Thus, each cell 90 mm in diameter produces 1.5 A. The maximum power deliverable to an external load is typically 11 to 12 per-

cent of the total solar energy incident on the cell. To obtain higher voltages, cells must be connected in series. Higher currents are obtained by connecting cells in parallel.

Some means of storage is necessary for cloudy days and night-time. This storage is usually supplied through deep-cycle, lead-acid storage batteries. Nickel-cadmium batteries have somewhat better charge retention. An inverter is used to change DC to AC. Inverters are about 90 percent efficient. The storage batteries act as buffers between the solar array and the load, supplying power to the load during periods of little or no sun and accepting charge from the array during periods of high sunlight.

To obtain the maximum solar energy year-round, the solar panels should be tilted toward the sun at an angle equal to the latitude of the location. By setting the tilt angle of the panels equal to the latitude plus 10°, additional energy can be gained during the winter. It is not difficult to design the panels so that the angle of tilt can be changed from summer to winter.

Other Energy Sources. Numerous other energy sources could be considered under special circumstances in a given location. These include biomass, tidal power, and ocean thermal energy. Systems for exploiting these sources are currently not in use in cold regions.

17.3.3 Heat Value of Fuels. The value of fuels normally used in communities is related to the heat available. Table 17-4 provides information on the heat value of common energy sources.

17.3.4 Energy Costs. The cost of energy in rural areas is particularly expensive due to the small population base located in very isolated and remote regions. Rural utilities have limited opportunity to take advantage of economies of scale which would otherwise reduce unit costs. Electricity rates in rural areas are generally five to ten times those of more urban areas, and range from $0.15 (US) per kwh on the North Slope to over $1.00 (US) per kwh in the interior of Alaska. Because of the high costs, electric use is much less. Annual average use in rural Alaska was 335 kwh per month, compared with 820 kwh per month nationwide (US).

While there are several cooperative organizations that operate power supplies, many are small utilities, owned by municipal, native corporations or tribal entities. Other communities rely on power from schools, canneries or other businesses. In the 1970s and 1980s, the state of Alaska provided grants and

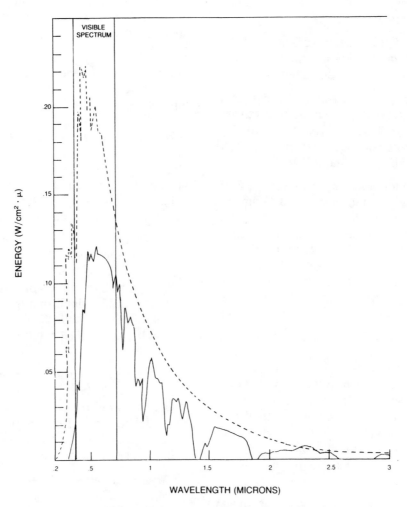

(a) Relationship between energy and wavelength

	Region	Wave-length (μm)	Cum.%	
3%	Ultraviolet C	0	.51	
	" B	.28	1.97	
	" A	.315	9.03	
44%	Visible Blue	.4	24.9	
	" Green Yellow	.51	38.4	
	" Red	.61	48.8	91% usable Solar Radiation
	" Infrared	.7	66.5	
53%	Infrared II	.92	76.7	
	" III	1.12	85.5	
	" IV	1.4	93.02	
	" V-VIII	1.9 00	100	

(b) Composition of Solar Energy

FIGURE 17-4   COMPOSITION OF SUNLIGHT

TABLE 17-4   HEAT EQUIVALENTS OF VARIOUS
             ENERGY FORMS

Energy Form	Heat Value (MJ)
Petroleum (per L) @ 15°C	
Crude oil	38.5
Liquefied petroleum gases	27.2
Motor gasoline	34.7
Aviation gasoline	33.6
Aviation turbo fuel (jet fuel)	35.9
Kerosene	37.7
Diesel and light fuel oil**	38.7
Heavy fuel oil and still gas	41.7
Petroleum coke	42.4
Propane and butane	25.6
Natural gas (per m^3)	37.3
Coal (per kg)	
Anthracite	29.6
Imported bituminous	30.0
Canadian bituminous	29.3
Sub-bituminous	19.8
Lignite	15.4
Coke (per kg)	26.2
Coke oven gas (per m^3)	18.6
Electricity (per kWh)	3.6
Woods* (per kg)	
Spruce	18.7
Willow	16.6
Driftwood	18.2
Cottonwood	17.1
Birch	25.6
Hemlock	18.7
Peat (dry)***	21.6

Notes:
*   As collected (approximately 14% moisture content).
**   A breakdown of the properties (including heat value)
     of the different grades of fuel oil is included under
     Section 17.3.1.
*** Approximately 85% moisture in field, must be dried.

TABLE 17-5   GENERALIZED ELECTRICITY COSTS IN ALASKA

Community Population Range	Electricity Costs ($/kWh) (US, 1995)	Communities in Alaska
>5,000	0.06 to 0.12	Fairbanks, Juneau, Anchorage
3,000 to 3,800	0.12 to 0.25	Nome, Bethel
700 to 2,500	0.24 to 0.60	Dillingham, Fort Yukon, Kotzebue
<500*	0.48 to 1.00	

* Very small communitites have even higher costs.

loans to assist in developing power throughout the State. The state legislature created the Alaska Power Authority to assist in developing the needed infrastructure. Several evaluations concluded that there was no other viable alternative to the use of diesel generators for energy generation in most locations.

In 1980, the State began a power cost assistance program for rural residents to balance the assistance provided to the more urban areas which were provided with larger power generation and intertie projects. In 1984, the program became known as the Power Cost Equalization (PCE) program. PCE pays part of the cost for residential electric use as well as electricity used for some municipal water and sewer systems. This subsidy provides a financial reimbursement to utilities on consumer bills up to 700 kwh per month used. The reimbursement is reviewed by the Alaska Public Utilities Commission based on a formula determined by state regulations and statutes. This subsidy varies, and has ranged from a low of $0.021 per kwh in Skagway to $0.347 per kwh in some interior communities. This results in an annual average reimbursement to consumers of $637 (US, 1995). The annual cost of this program has ranged from $15 to $20 million, with a total expenditure of about $240 million to date.

Electrical generation costs average about 30% for fuel, 25% for operation and maintenance, 10% for administration and management, and the remainder for miscellaneous expenses.

Table 17-5 shows the general cost for electricity in Alaska as it relates to population size. The actual costs and availability in each community or village can be obtained from Alaska Power Commission and the Northern Community Power Commission in the Northwest Territories and the Yukon.

In most Canadian territories, the cost of electricity is subsidized in one form or another for remote communities. The amount of the subsidy, in each case, is difficult to identify because many are indirect and absorbed by other government programs. Electrical rates charged in the NWT are presented in Table 17-6. In addition to the above, a Federal Power Support Subsidy provides that all diesel-serviced settlements receive the first 700 kWh/month at the Yellowknife rate (Table 17-6(b)). Present Yellowknife electrical rates are 7.3¢/kWh (Cdn, 1982) for a private residence, and 6.9¢/kWh (Cdn, 1982) for up to 50 kWh and 7.5¢/kWh (Cdn, 1982) for electricity consumption in excess of 50 kWh for commercial and government users.

Heating fuel and gasoline costs in the NWT depend heavily on transportation costs. The current cost of heating fuel, diesel, and gasoline in each community in the NWT is available from the NWT Government. The cost varied by region as shown in Table 17-6(c).

17.4   Heating

Heating is one of the most important mainstays of an acceptable standard of living in cold regions. Heating requirements and methods are covered in this subsection. Heating systems include fireplaces and stoves, gravity and forced warm air units, hydronic units, and electrical baseboard units.

17.4.1   Demand.   The heating requirements of a building are determined by the difference between inside and outside air temperature; wind and sun have a lesser, more temporary effect. The thermal efficiency of a building is determined by its design and construction which, in turn, affects its annual heating requirement. The retrofitting of existing build-

TABLE 17-6(a) POWER COSTS IN THE NORTHWEST TERRITORIES (1980)

Region	Consumer Cost	Production Cost	Subsidy
Domestic Use (¢/kWh; Cdn, 1980)			
Fort Smith	6.25	8.40	2.15
Inuvik	8.25	16.36	8.11
Baffin	11.09	20.43	9.34
Keewatin	14.08	20.83	6.75
Commercial Use (¢/kWh; Cdn, 1980)			
Fort Smith	9.38	10.11	0.73
Inuvik	10.49	14.83	4.34
Baffin	15.81	18.71	2.90
Keewatin	16.36	18.80	2.45

ings is an important step in saving energy and reducing heating costs.

Typical residential heating requirements, where the heating demand is based on the heating index, are:

	Heating (MJ/°C·d)	Hot Water (MJ/d)
Single-residence houses	32.4	43.2
Multiresidence buildings	21.6	43.2

The domestic hot-water requirements are nearly constant throughout the year. The design heating requirements can be obtained by computing the difference between the heating index base temperature and the winter design temperature. The daily peak load on a heating system is influenced mainly by the effects of "morning warm-up" after "night set-back" and by the heating needs during dropping ambient temperatures in the evening. The ratio of daily peak to average daily load can be about 1:1.5.

Typical commercial and institutional heating requirements are as follows:

Schools, offices (all sizes) $0.23 \text{ MJ}/(m^2 \cdot °C \cdot d)$

Shops and retail stores (including food stores) $0.22 \text{ MJ}/(m^2 \cdot °C \cdot d)$

For central or district heating facilities, the distribution lines tend to be more expensive than the heating plant. Thus the compactness of the buildings to be served can have a significant effect on the feasibility of using a central heating system. Dispersed residential areas must usually rely on individual heating.

TABLE 17-6(b) YELLOWKNIFE RATE SUBSIDY

Region	Acutal Consumer Cost (¢/kWh; Cdn, 1980)	Additional Subsidy (¢/kWh; Cdn, 1980)
Fort Smith	5.97	0.28
Inuvik	6.99	1.27
Baffin	8.49	2.60
Keewatin	10.28	3.80

TABLE 17-6(c) FUEL COSTS IN THE NORTHWEST TERRITORIES

Region	Consumer Cost	Production Cost	Subsidy
p-50 Heating Oil (¢L; Cdn, 1980)			
Fort Smith	0.42	0.55	0.13
Inuvik	0.38	0.46	0.08
Baffin	0.41	0.51	0.10
Keewatin	0.47	0.60	0.13
Gasoline (¢/L; Cdn, 1980)			
Fort Smith	0.47	0.63	0.16
Inuvik	0.43	0.54	0.11
Baffin	0.45	0.56	0.11
Keewatin	0.50	0.63	0.13

Cooking and hot-water output reduces the heating load by 0.074 GJ/m². A study (GNWT, 1981) to consider district heating in Canadian mining communities used figures of approximately 397 MJ/m² for new, 150 m² houses. It was stated that the domestic hot-water load is assumed to be constant throughout the year and equal to about ten percent of the annual heat energy consumption in the community including residential, commercial, and industrial needs (no heating index given).

17.4.2 Generation. Heat is generated by boiler plants which are characterized by being low in initial cost, but high in maintenance costs compared to electricity-generating plants. They can be for heat alone, or part of the heat can be used to generate electricity (cogeneration). Cogeneration plants are more efficient with respect to fuel and operating costs than individual electricity-generating plants. Both types of plant can be fired by natural gas, oil, coal, or almost any other fuel. The type of fuel used depends on its cost and availability in the vicinity of the community or camp. In the arctic areas of Canada and Alaska boiler plants are usually oil-fired. Many construction camps use natural gas, as do the communities of Barrow, Alaska, and Norman Wells, NWT.

17.4.3 Central or District Heating. Most heating facilities in arctic communities serve individual structures. The only community-wide central heating system is in Inuvik, NWT, and even there the heat distribution system serves only the central core area of town. Central heat has not proven feasible in the past, because the communities usually consist of widespread, single-family houses, the price of oil has not been high enough to offset the heat distribution costs, and nearly all the communities have fewer than 2,000 inhabitants (GNWT, 1981). District or central heating has been used in many temporary and permanent camps throughout the Arctic and Antarctic. The buildings are usually closely spaced and connected to the central plant by utilidors. The utilidors do not obstruct the personal and vehicular mobility as they would in a community.

A detailed energy balance is necessary before a central heating facility is designed. An example for a small community was presented for the community of Wainwright, Alaska (Reid, 1980).

A single heat source requires standby capability. Experience has shown that complete standby is required for any critical component of a remote facility where repairmen and parts are not available 24 hours a day. Another disadvantage of a single source is that the individual controls for several small heating sources serve different components of a facility. If multiple sources are selected, compatible equipment with interchangeable parts should be used.

17.4.4 Heat Pumps. When an indigenous source of low-temperature heat is available through heat pumps, it offers an independent alternative to electricity used for the heat pumps, which is generated from renewable resources. Examples of low-temperature sources are ocean water, groundwater and geothermal water (with temperatures of at least 5°C). Sources with higher temperatures are advantageous, because the coefficient of performance depends also on the temperature of the heat output; lower output temperatures are found along the coast of southeastern Alaska, southeast of Valdez, and the west coast of Canada. Suitable groundwater temperatures are found in southwestern Canada and at sporadic points in northern Canada and Alaska.

17.4.5 Boiler Efficiency. Boiler efficiency becomes more critical with rising fuel costs. Boiler efficiency is directly related to combustion efficiency, which is determined by the percentages of air, oxygen, and carbon dioxide in the stack gas. Boilers should operate at about 80 percent efficiency. The actual efficiency can be determined from the curves shown in Figure 17-5 for (a) natural gas or (b) oil-fired boilers. To use these curves, the CO_2 and O_2 in the stack gas must be measured and the net stack temperature (actual stack temperature coming directly out of the fire box minus the air temperature in the room in which the boiler is operating) must be measured. Figure 17-6 is a plot of the relationship between the CO_2 and O_2 percentages of the stack gas and the excess air, and the type of fuel for a properly operating boiler. As long as the stack is designed to overcome the resistance in the boiler and the breeching, and the firing rate of the forced draft burner does not exceed the stack capacity, the forced draft burner neither creates a positive pressure in the combustion chamber nor causes leakage of combustion products into the room. Control of the draft is important and should be accomplished by either an electric overdraft control system or a barometric damper. With natural gas as a fuel, the barometric damper should be of the double action type. A gas burner should neither burn with more than 20 percent excess air nor have carbon monoxide among the products of combustion. Oil burners should meet the same specifications and emit a maximum of no. 2 smoke as measured on the Ringelmann Scale.

Most oil-fired boilers operate with a supply and return oil line including a circulating pump, a filter, and a check valve. A boiler designed for a circulating supply system must not be used on a single-pipe (noncirculating) supply system. In cold regions the circulating supply line from the day tank helps to warm up the oil in it. Day tanks and oil storage tanks feeding boilers should be located and installed ac-

cording to local codes and the rules of fire inspection services. If any part of the tank is located above the level of the burner, an anti-siphon device must be used to prevent flow of oil in case of a break in the oil lines. Connections to buried tanks must be made with swing joints or flex-connects to prevent the pipes from breaking, should the tank settle. Where more than one boiler is tied to the same tank, they can have a common return line, but each boiler should have its own suction line. Always provide a tee and plug in the suction line at the highest point to aid in priming the pump and to expel air that may accumulate there. All oil tanks should be vented to the atmosphere. Fire isolation valves should also be provided on each line.

In discussing boiler efficiency, one needs to distinguish between seasonal and steady-state efficiency. The latter appears on Figure 17-5 and may be below 75% for older units and as high as 95% for modern condensing units. The former, which is an average over the entire heating season, may range from just above 60% for older forced draft burners to around 90% for a modern condensing unit. As shown on Fig. 17-6 A, the efficiency falls dramatically for fractin burner on times less than around 30%. Hence, it is important to size a heating unit such that it is normally firing at least 30% of the time during the main part of the heating season. To further reduce off-cycle losses, stack dampers can reduce stack losses when the burner is not firing. This technology is appropriate for gas and oil-fired systems. Even though the term "boiler" is used in the preceding discussion, these considerations also apply to furnaces. The former applies to heating units where the heat from the combustion gases is transferred to water and the latter to forced air systems.

It should be noted that efficiencies here are defined in terms of heat delivered divided by the heating value of the fuel. Within this definition, there are two subgroups depending on whether one assumes the water vapor condenses as the combustion gases exchange heat with the working fluid. If they do, the heat available in the fuel is called the "higher heating value." For oil and natural gas having heating values around 20,000 Btu/lbm (138K Btu/gal for fuel oil or 2,200 Btu/scf for gaseous propane), the difference between the higher and lower heating values is around 10%. Both gas- and oil-fired condensing units can be obtained which have steady-state efficiencies over 90%; they recover the latent heat of vaporization before the exhaust products leave the unit. Since propane or natural gas are very clean fuels, the condensing units should not present prob-

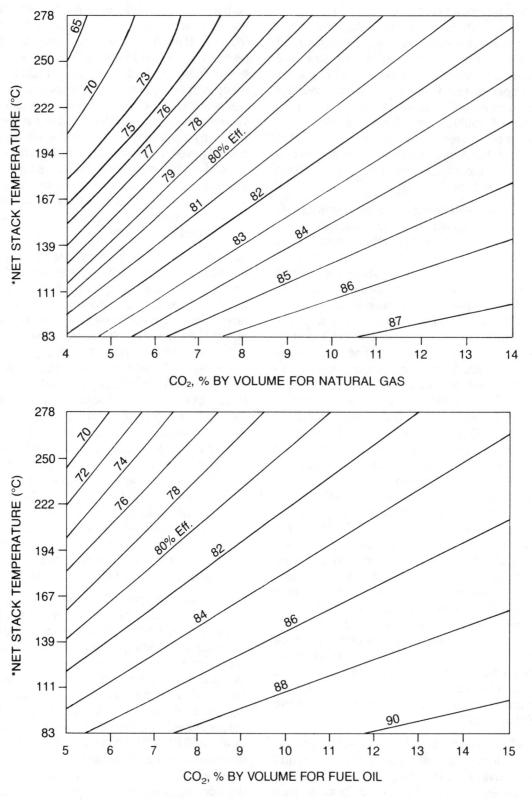

*Net = Actual Minus Ambient Temperature

FIGURE 17-5   COMBUSTION EFFICIENCY CURVES SUITABLE FOR BOILERS

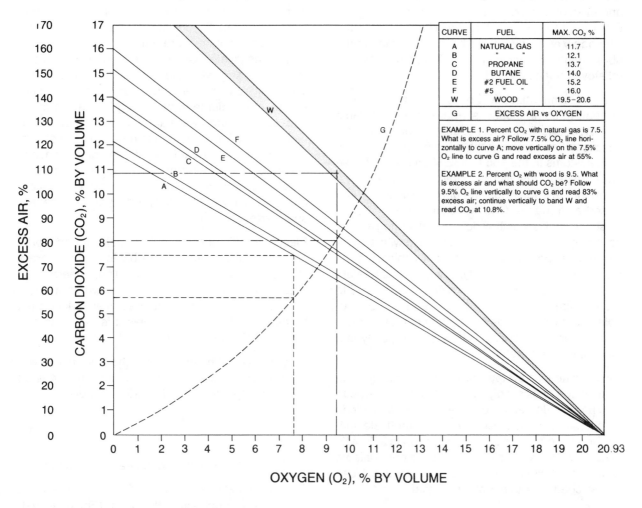

FIGURE 17-6   CO₂–O₂ RATIO CURVES FOR OILS, GASES AND WOOD (after Gordon-Piatt Energy Group, Inc., 1996)

lems in terms of corrosion in the exhaust ducting. For oil-fired systems, however, there may be difficulties with corrosion in the exhaust stack if the temperature falls below the dew point.

A system that is mounted outside of the building being heated, may have icing problems during periods of extreme cold. If the relatively warm exhaust comes in contact with the cold intake vents, condensation and subsequent icing can occur. Such problems can be addressed by ensuring that the exhaust is ducted away from the intake area or using the exhaust to preheat the intake air.

17.4.6 Heating for Individual Buildings. Most communities in cold regions have individual heating. In some cases several school buildings are connected to a central heating plant, but usually each

building has its own heating plant. Bottled propane is used for cooking in the cold regions of Alaska. Average use was reported to be three 45 kg bottles per year. Some problems are encountered with moisture freezing in the regulators at temperatures below -18°C. This can be eliminated, if alcohol is added to the cylinder (100 to 200 mL per 20 L) before it is filled, or if the regulator is located in a heated area. This use of bottled propane is supplemented by white-gas stoves or wood stoves.

Very few homes are actually heated by propane. Where wood is available, approximately 50 to 60 percent of the household heat can be supplied by wood. Many homes in the remote villages have homemade barrel stoves for burning wood. These stoves are made from used fuel drums ranging in size from 20 to 200 L. Many have baffles, dampers, and even ovens. Many newer homes are using airtight cast-iron wood heaters. They have primary and secondary air metering, fire-brick lining, and removable ash pans. They are more efficient than the barrel stoves but cost significantly more.

In most communities fuel oil is the primary source of heat. In some locations where coal is available nearby, coal is the primary source of heat. The most successful systems are the oil, gun-fired, hydronic heat systems in the 105 to 132 MJ/h range. These units are being installed in many of the new houses. Some of the earlier housing units included oil-fired forced-air furnaces. One reason these units are not so successful is that they do not work during electrical outages. However, as a cook stove with a cooking surface and oven or a sheetmetal cabinet around a sheetmetal pot, these pot burner, oil-fired heaters are very popular. They are inefficient compared to the gun-fired models, but they are more dependable and easier to repair. They do not require electricity to operate, because they use a gravity-fed, mechanical carburetor.

Hydronic heating systems are available that can use either wood or oil simply by switching fuel. They can also provide domestic hot water, if desired, and can function at reduced efficiency even without electricity. Wood hydronic units can produce up to 295 MJ/h of heat. In cases where the operator is not available to stoke the boiler, on weekends, for example, it can be switched over to run on oil for that period.

For larger installations, such as schools and hospitals, heat must usually be supplied continuously, and fuel cannot be fed intermittently. For wood an automatic continuous feed system is necessary. These systems feed chipped wood with a particle size 13 mm by 50 mm by 50 mm or smaller. The chips are blown into a storage silo from where they drop into the combustion chamber. Silo storage systems suffer a major problem in cold regions; when the wood is chipped, the ice particles in it melt and then refreeze in the silo, unless the silo is located in a heated area. This refreezing causes the silo hopper to plug. Instead of chips, there are systems which feed and burn logs about 200 mm by 500 mm in size. The supply conveyor can be sized to feed up to five cords of logs between stoking. This method has the advantage that it can be manually fed if the conveyor (or chipper) breaks down.

Heat exchange in boilers can be accomplished using fire tubes or water tubes. In fire-tube boilers, hot gases from combustion circulate in tubes surrounded by water. They are limited in size but very efficient. Water-tube boilers are more common and consist of tubes filled with water around which the hot gases circulate. Care must be taken not to run large quantities of cold water through a boiler which is not operating, because it will cause condensation in the combustion chamber. Older wood boilers control only the volume of fuel going in, leaving moisture content and heat values to fluctuate. Newer models of boilers monitor and regulate fuel volume, air flow rate, and heat flow rate. In some locations no additional safety precautions for wood-burning boilers prevail other than those applicable to oil-fired ones as well, except that wood units must have a spark arrestor in the flue.

Electric heat is possible but normally not economical in the Arctic, except possibly in areas where a cheap source of electricity is available nearby (hydropower, etc.) or where off-peak power that would otherwise be wasted can be used. If electricity must be generated using fossil fuels it is never as efficient as using the fossil fuel itself for heat.

Much heat can be gained through south-facing windows, even in the winter. The amount of heat gained can be computed by the following formula:

$$q = (SC)(SHGF) + U(T_o - T_i)$$

where

q =   total instantaneous heat gain, $kJ/(m^2 \cdot d)$

SC =   shading coefficient = 1.0 for a single sheet of double-strength glass. (For other values see ASHRAE (1993).)

$SHGF$ = solar heat gain factor ($kJ/m^2 \cdot d$).

These vary with latitude, orientation, and time (ASHRAE, 1993). The values presented are for the "average cloudless days" at that latitude and may have to be adjusted, if the location under consideration is overly clear or overly cloudy.

T_o =   outside temperature, °C

T_i =   inside temperature, °C

U =   coefficient of heat transmission, $kJ/(m^2 \cdot d \cdot °C)$.

This coefficient is available in ASHRAE (1993) for various windows.

Greenhouse windows should face due south and be inclined at an angle on the horizontal so that the sun during the spring and fall strikes the glass at nearly 90°. This allows for the most efficient collection of solar energy during the critical growing period of a month or so in the spring, and in the fall to extend the growing season.

17.4.7 Heat Distribution. With heating for individual buildings the fuel distributed is usually in the form of natural gas, fuel oil, or wood. This is the condition in most of the remote villages in Alaska and Canada. Wood is usually delivered by truck or snowmobile. Fuel oil is usually delivered by truck. Natural gas, where available, is distributed by pipe.

Natural-Gas Distribution. Several items should be considered regarding the distribution of natural gas. Gas pipe fractures usually occur at tension cracks of the soil where the pipe is placed under tensile and shear stresses at the same time. Most new installations use high molecular weight, high-density polyethylene pipe (PE 3306) which retains its flexibility and strength in low temperatures. There have been very few problems with this pipe. It is easy to install, and corrosion is not a problem. However, the pipe has been limited to a 690 kPa working pressure.

Ice can form very quickly in natural-gas lines, when the moisture content of the gas is high and the temperatures are below freezing. The moisture content of the gas should be kept below 96 g/L of gas. There are instruments available to determine the moisture content of natural gas. Usually devices remove moisture at the wellhead, but they can fail or be overloaded. Once ice starts to form, it is hard to control. The line can be slugged with antifreeze, which prevents more ice from being formed and melts ice that has formed.

Gas distribution systems must comply with national pipeline safety standards. Strict requirements hold for joints and valves. Ball valves with 90° operation full-open to full-close and with removable handles are recommended in some cases. Also, all lines near roadways must be well protected with barricade piling.

The Athabasca area in Alberta has had several years of successful experience with plowing-in some 1,450 km of 150 and 50 mm high-density polyethylene pipe for a rural gas distribution system. The pipe, in coils up to 180 m long, is pulled on a trailer drawn by a crawler tractor with a plow on the rear. The pipe is continually fed as the plow moves along and placed at a depth of 1.1 m. The joints of successive coils are butt-fused. About 4.8 km of pipe can be placed a day. It requires as many as four tractors to pull the

plow depending on the soil. Frost penetration is about 2 m in the area which consists primarily of marsh and muskeg. There is some permafrost in the muskeg and, in those areas, a trench is dug with a hydraulic backhoe and the line laid by hand. The plowing operation can continue until the ground is frozen 150 to 200 mm deep in the fall.

Hot-Air Distribution. In cold regions a heat distribution system using hot air is not as desirable as one using fluids. Air does not have the heat carrying capacity of water or steam, and the ducting requires considerably more space than the plumbing for hot-water systems. Hot-air furnaces are sometimes used for individual buildings, but controlling air pressure in the buildings can become a problem. Also, in central facilities or camps, air is effectively used for clothes dryers and saunas.

Hot-Water Distribution. Hot-water heat distribution systems are the most popular in remote arctic regions. One of the reasons is that they are simpler for the layman to understand and maintain. They are being used to distribute heat in buildings and through insulated pipes and utilidors throughout entire camps and communities. The major disadvantage to hot-water (or steam) distribution systems is the potential for freezing during power outages. They can be protected, however, using an antifreeze (glycols) water solution, but this reduces the efficiency of the system.

Design information on heat distribution systems is available from the American Society of Heating, Refrigeration and Air-Conditioning Engineers (ASHRAE, 1993).

It must also be remembered that ethylene glycol is toxic to humans and animals, therefore cross-connections with domestic water systems must be avoided. Propylene glycol should be used if there is any possibility of cross-contamination. All glycol solutions are corrosive to zinc, and they can leak through joints and pump seals that would not leak water at the same pressure. The inhibitors added to control corrosion tend to break down at high temperatures. Keeping the required concentration of the corrosion inhibitors becomes another chore for the operator. Special organic fluids instead of the glycols and water are available as antifreeze, but they are expensive, and plumbing systems must be more elaborate to handle higher temperatures. Faulty joints have been known to leak hazardous fumes into the building. Mechanical seal pumps should be used. Some boiler manufacturers void their warranties if glycol solutions are used. A 50 percent mixture of glycol and water requires an increase in flow rate of about 14 percent to achieve the same heat transfer as would be achieved with pure water.

The required system flow rate is:

$$Q = \frac{q}{p \bullet c \bullet \Delta T}$$

where

Q = flow rate, m^3/s

q = total heat transfer rate needed, kJ/s

r = density of fluid, kg/m^3

c = specific heat capacity of fluid, kJ/kg•˚C

ΔT = temperature difference, ˚C

Most heating systems operate between 80 and 99˚C.

The viscosity is also increased when glycol is used (Figure 17-7. This viscosity increase influences the operation of pumps and the friction losses determined for water. The size of the expansion tank must be increased as shown in Figure 17-7(k). Normally the expansion tank is 10 to 15 percent of the total volume contained in the system. The boiler for equipment heat exchange must be increased in size when using ethylene glycol because of its lower heat transfer characteristics. Figure 17-7(l) provides correction factors for sizing boiler components.

Glycols also have a higher boiling point than water but are completely soluble in water at any mixing ratio. Glycols, by themselves, have a pH of about 9.5 but when mixed with water on a 1:1 basis, the pH drops to about 9.3. At high temperatures (>93˚C) it can decompose into acidic compounds, lowering the pH and causing corrosion. Figure 17-8 gives the freezing points for various mixtures of water and antifreeze.

Propylene glycol has been approved by the U.S. Department of Agriculture for direct ingestion. The corrosion inhibitors added to glycols create a film over all metal surfaces which tends to reduce heat transfer efficiency. As shown in Figure 17-7, the viscosity of glycol solutions changes drastically with temperature. Glycols do not provide permanent protection, because they tend to break down with time, especially at the high temperatures of a boiler system. They must be monitored for corrosion potential and decay in glycol strength. Measurement of the density indicates the percentage of glycol remaining and, thus, the freeze protection. Measurement of pH indicates the corrosion potential. De-

pending on the operating temperatures, a two- to four-year life can be expected.

Steam for Heat Distribution. Steam has a high capacity for carrying heat, and is therefore used in heat distribution. The primary disadvantage of using a steam system is that people in remote areas are generally unfamiliar with the higher temperatures and pressures which must be used. Pressurized vapor is also more hazardous if an untrained maintenance staff is making repairs. This makes operation of a steam system more difficult and results in higher maintenance costs and longer downtime compared with hot-air or hot-water systems. With steam, however, damage due to shutdown and subsequent freezing is usually confined to traps and low points.

Heat Storage. Heat sinks can be used to take advantage of off-peak power to store heat for use at a later time. Heat sinks are an absolute necessity when solar energy is used. Sinks consisting of rock, concrete, water or salt solutions can be used. The effectiveness of a heat sink is based on its volumetric heat capacity (see Section 4).

17.4.8 Air Pollution Problems. The utilization of energy can diminish air quality both directly by combustion processes and indirectly by the way fuels are procured, transported and in the fabrication of the prime movers. Generally, for renewable energies such as wind, any air pollution would be associated with the latter. The focus here is on direct impacts.

Of the six pollutants specified as part of the National Ambient Air Quality Standards (carbon monoxide, lead, nitrogen dioxide, particulates, ozone, and sulfur oxides), carbon monoxide is of major concern in Alaska. Because Anchorage and Fairbanks have been in noncompliance for CO, beginning in the winter of 1992-93 each was required to initiate an oxygenated fuels program for motor vehicles. In urban areas, motor vehicles typically contribute 90% of the total CO emissions. After numerous health-related complaints from the users, Fairbanks was granted a waiver from using oxygenated fuels in December of 1992. If oxygenates are not used in the following winter, Fairbanks must devise an alternative strategy to allow it to come into compliance with the nine part per million (ppm) 8-hour NAAQS.

A major contributor toward high CO levels is the very stable air that can exist in northern communities in the winter. This stable air is associated with strong

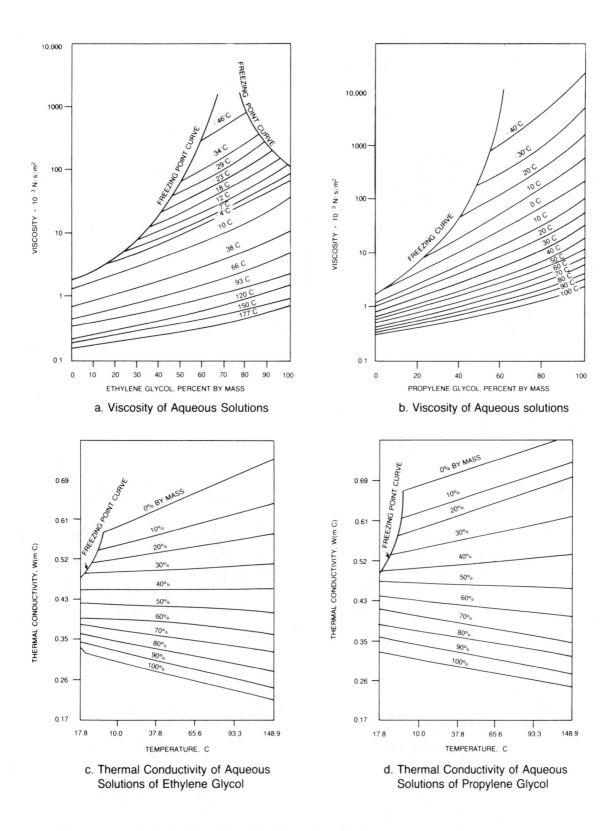

a. Viscosity of Aqueous Solutions

b. Viscosity of Aqueous solutions

c. Thermal Conductivity of Aqueous
Solutions of Ethylene Glycol

d. Thermal Conductivity of Aqueous
Solutions of Propylene Glycol

FIGURE 17-7   GLYCOL SOLUTION CHARACTERISTICS

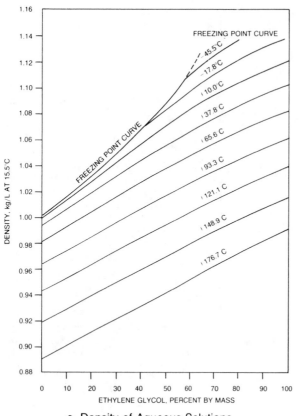

e. Density of Aqueous Solutions
of Ethylene Glycol

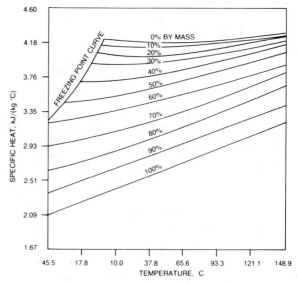

f. Density of Aqueous Solutions
of Propylene Glycol

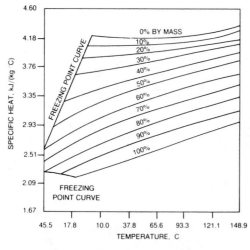

g. Specific Heat Capacity of Aqueous
Solutions of Ethylene Glycol

h. Specific Heat Capacity of Aqueous
Solutions of Propylene Glycol

FIGURE 17-7   Continued

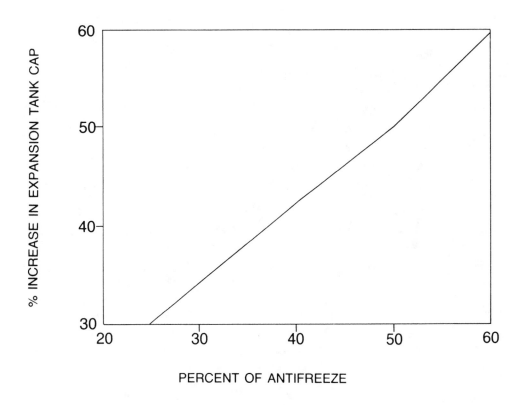

PERCENT OF ANTIFREEZE

k. Increase in Expansion Tank Size

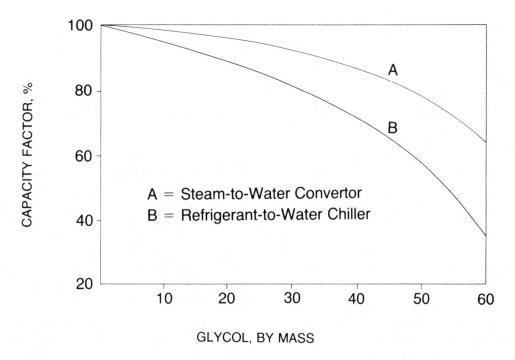

GLYCOL, BY MASS

l. Boiler Sizing

FIGURE 17-7   Continued

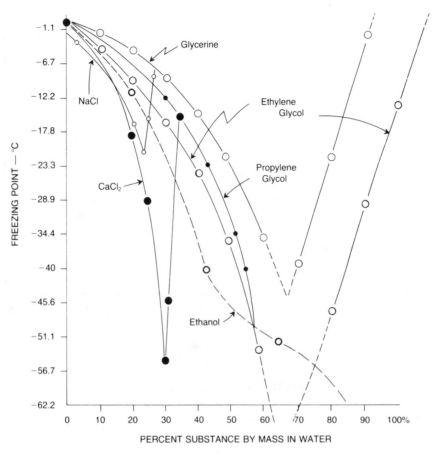

FreEZING POINT — °C

PERCENT SUBSTANCE BY MASS IN WATER

*Data from: Hndbk of Chem & Phys, & from ASHRAE GUIDE

FIGURE 17-8   FREEZING TEMPERATURES OF WATER SOLUTIONS

temperature inversions (Fig. 17-9). The inversions in Fairbanks are as large as those found anywhere in the world, at times exceeding 0.3°C/m in the lowest 30 m. This plus the high emissions from motor vehicles caused over 100 violations of the 8-hour standard in the mid-1970s. But, with improvements in motor vehicles with respect to emissions, the annual number of exceedences in the early 1990s was less than five.

The stable air coupled with the production of H_2O mainly in combustion reactions has led to severe ice fog conditions in some northern communities during the coldest parts of the winter. Ice fog consists of tiny (around 5 micron diameter) ice crystals formed when water vapor condenses into droplets and then freezes. This condensation starts to occur at temperatures less than -20°F and leads to se-

vere ice fog at temperatures below -40°F in Fairbanks. This is because the ability of air to hold water vapor which diminishes rapidly as the temperature decreases. For example, air can hold about 250 times as much water vapor at temperatures of 75°F than at -45°F. The combustion of one kilogram of gasoline produces about 1.4 kg of water vapor. Besides the other pollutants associated with ice fog, the diminished visibility can be a safety hazard to drivers and pedestrians.

Diesel engines are lower emitters of carbon monoxide and moisture than gasoline engines, but are high emitters of hydrocarbons. As more and more exhaust gas and pollutants are added to the layer between the ground and the inversion (Figure 17-9), the ice crystals and pollutants increase in density. Fairbanks, Alaska, has severe ice "smog" problems

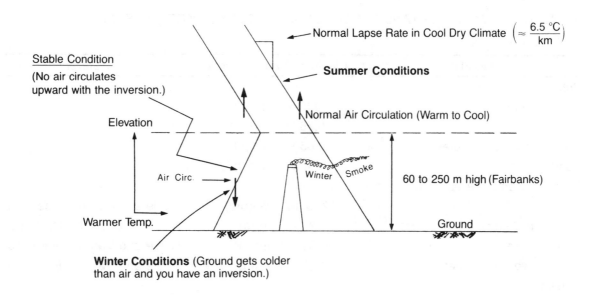

FIGURE 17-9   TEMPERATURE INVERSION

because the town is located in a valley so that the air remains trapped. About 100 m above town the air circulates enough to keep the air clean. Hydrocarbons are not a critical problem in Fairbanks however, carbon monoxide levels often reach three times the national standards. Lead values in Fairbanks have been over four times the EPA air quality standard, but this problem will probably be resolved with the phasing out of leaded gasoline. Smoke stacks leave a trail (Figure 17-9) under inversion conditions. The largest source of ice fog in Fairbanks is the cooling ponds of the generating plant and the resulting unfrozen stretches in the Chena River.

17.5   Electricity and Its Generation

Management techniques that shift power demands from peak to off-peak periods are desirable. To this end it is advantageous to mix baseload and peak-load plants. Baseload plants are characterized by high capital cost but low energy cost, whereas peak-load plants have the opposite characteristics and are used only intermittently.

17.5.1   Demand.   The energy consumption for lights and appliances for a typical residence in the Arctic is approximately 12 kWh/d during the summer and 24 kWh/d in the winter. If domestic hot water is produced electrically, the consumption of electricity during summer or winter increases by approximately 12 kWh/d. These figures can increase by 5 to 10 kWh/d per car for engine heaters in the winter. The ratio of the daily 15-minute peak to average daily load is 2 or 3 for a community and 10 for an individual residence. Studies in three villages near Kotzebue, Alaska, show a peak of twice the average daily load. Average daily loads for winter are about twice those for the summer. Commercial and institutional electrical requirements are more cyclical throughout the day than the residential requirements. Some typical requirements are listed in Table 17-7.

The big difference between remote arctic community utilities and the national averages for more temperate climates is the absence of substantial commercial and industrial loads, except for isolated instances. In 1995 the electricity use breakdown by percentage in Alaskan villages was:

Schools	21
Residences	47
Other public facilities (sanitation, etc.)	17
Commercial	15

17.5.2   Generation.   Generating plants are usually the most expensive part of an electrical system from an initial as well as an operating standpoint, especially in small cold-region communities. The generating plants typically depend on fuel oil for energy. They must be designed to provide four times

TABLE 17-7   MEASURED AND RECOMMENDED ELECTRICITY REQUIREMENTS

Location/User	Electrical Requirements	Reference
Schools and Offices		
>1,500 m^2	0.5 kWh/(m^2•d)	
<1,500 m^2	0.75 kWh/(m^2•d)	
Shops and Retail Stores	0.85 kWh/(m^2•d)	
Food Stores	1.4 kWh/(m^2•d)	
Northern Remote Areas (Canada)*		
Alberta	7 to 14 kWh/(p•d)	Noble and Van der Hoek, nd
British Columbia	21 kWh/(p•d)	Noble and Vsn der Hoek, nd
Yukon Territory	31 kWh/(p•d)	Noble and Van der Hoek, nd
Northwest Territories	29 kWh/(p•d)	Noble and Van der Hoek, nd
Alaska		
Atkasook	74 kWh/(d•house)	Alaska Energy Authority, 1989
Kiana	9 kWh/(d•house)	Alaska Energy Authority, 1989
Ambler	9 kWh/(d•house)	Alaska Energy Authority, 1989
Shungnak	9 kWh/(d•house)	Alaska Energy Authority, 1989
Greenland	7 kWh/(p•d)	
	~30.5 to 40 kWh/(d•house)	

*   Power costs subsidized by government.

(or more) the average daily demand, unless the distribution system can be tied into a network whereby peak power can be supplied by other sources.

The main concept behind having large central generating units instead of many small units throughout a community is tied to an economy of scale — primarily lower operation and maintenance costs. However, in remote areas one must balance this advantage against the need for backup generating capability, if the main unit fails. For example, in a situation where a 600 kW unit may be more economical, two 300 kW units or even three 200 kW units could supply backup in case one unit were to break down. The small units also allow the generating capacity to be more closely matched to the load. During the night one 200 kW unit can probably generate all the power needed; the additional units can come on- and off-stream as the demand during the day requires. Several smaller units in a central generating facility also allow for the maintenance of one or more units while the others carry the load. In any

one community and even in communities in the same general area, all generators should be from the same manufacturer to permit an interchange of parts.

A dependable source of electrical power is very important for the operation of nearly all other utilities, especially sanitation facilities. Currently, due to the difficulty in operating and maintaining diesel-powered generators, reliable sources of electricity are difficult to maintain in rural communities in cold regions. Often operators are poorly paid, and the available funds are rarely sufficient to support the required preventive maintenance. Yet, paid operators and preventive maintenance are extremely important because of the difficulty in obtaining parts and skilled mechanics in remote communities.

Electrical power generated with diesel-powered generators in remote communities can vary greatly in voltage and frequency. These fluctuations cause problems with controls and motors unless high- and low-voltage and frequency protectors are included.

This protection is necessary for any major installation and should be of such design that they can be easily adjusted and maintained.

Whenever electricity is generated, the utilization of the waste heat should be considered. About two-thirds of the energy of the fuel ends up as waste heat, much of which is recoverable. Waste heat recovery and utilization are discussed in detail later in this section.

Costs can be reduced (energy saved), and reliability of the electrical power in the remote communities of Alaska and Canada can be improved by (Wind Systems Engineering, 1981):

1. Regionalization of the electrical utilities whereby each region has electricians and mechanics that can go to a community needing assistance on short notice. The regional centers would also be able to maintain an inventory of spare parts and even spare generators that could be flown to a community or site on short notice.

2. Generator units should be standardized (i.e., the make and size) by region to reduce the necessary inventory of repair parts. Having three or more smaller units permits them to be run separately or together to meet the slack or peak demands. This requires automatic switching controls so the units can run synchronously.

3. The distribution system should be divided up into three or more circuits so that startup can be sequenced to reduce the peak demands and thus the required generating capacity. This would also allow repairs without shutting down the entire community. The three circuits should be divided according to demand (possibly one each for residences, school, and utilities).

4. Consideration should also be given to battery banks to store off-peak electricity.

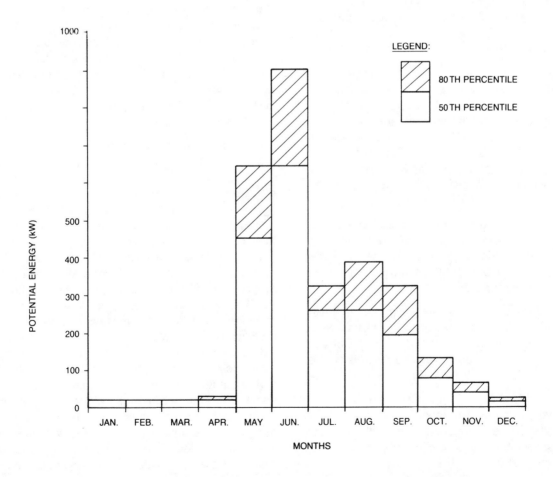

FIGURE 17-10 PROBABLE POTENTIAL ENERGY FOR A SMALL STREAM

TABLE 17-8   SMALL TURBINE EFFICIENCIES

Turbine Type	Head (m)	Output (kW)
Pelton	over 100	50 to 2,000
Turgo	15 to 200	50 to 2,000
Francis	15 to 200	500 to 2,000
Banki (crossflow)	2 to 170	50 to 1,000
Propellor (Kaplan)	2 to 15	500 to 2,000
Pumps*	–	up to 150

* Run backwards as turbines

Hydropower. Hydropower is probably the most maintenance-free form of electricity generation. Several studies have been undertaken to determine the hydropower potential for arctic regions (Adelaav and Associates, 1981; Ott Water Engineers and U.S. Army Corps of Engineers, 1981; Adelaav, 1981; Stockbridge, 1925).

Hydroelectric plants are independent of fuel logistics and use a renewable resource. A significant disadvantage of hydropower (especially low-head) in arctic regions is that the flow rate for streams and even rivers (Figure 17-10) is reduced to nearly zero in the winter. This is when the power needs are the greatest, and some storage of water or electricity would probably be necessary for these critical periods.

Hydroelectric power can be supplied to a community from two major sources: grid extensions from large hydroelectric projects, and development of units smaller than 5 MW (microhydro plants) to serve individual communities.

Despite the cost advantages of large-scale developments, small-scale or microhydro plants are more suited to the cold regions of Alaska and Canada. However, in the subarctic regions large-scale development and grid extensions are possible (Stockbridge, 1925).

Microhydro turbines vary according to type and size and exhibit different efficiencies, depending on head and flow conditions as shown in Table 17-8 (Benson and Rizzo, 1980).

Microhydro plants, because of their small size, lend themselves to prepackaged generation units, which significantly reduces the installation time and cost.

Care should be taken when developing a hydropower system (Bettine and Retherford, 1981). The specific components include the intake, the penstock, the turbine/generator, and the governor. Protection should be provided in the form of overheat and overspeed shutoffs in the turbine and pressure switches to bypass (or "dump") water before the penstock pipe is damaged.

Diesel Generators. The most common form of power for generating units found in cold-region communities or sites is diesel-fired power plants of various sizes. They are available in basically two configurations: two- and four-cycle. The two-cycle engines have a higher output-to-weight ratio but poorer fuel economy than the four-cycle engines. Most medium-sized units operate at 1800, 1200, 900, or 600 rpm and are coupled to a generator having the proper number of poles to generate 60 Hz electricity. The lower-speed units usually have low power-to-weight ratios but a longer life and greater reliability. Because of the weight of the units and their individual parts, however, they are hard to repair in remote locations. The 1200 or 1800 rpm units offer higher output-to-weight ratios, and repairs are often easier and cheaper because of the lighter engine components. Liquid-cooled units are more popular than air-cooled units. Waste heat recovery is easier with liquid-cooled units. To ensure the proper operation of a unit, adequate ventilation must be provided. Since outside temperatures can range from -57°C to 35°C in the Arctic, the ventilation system of the generators must be sufficient to handle these

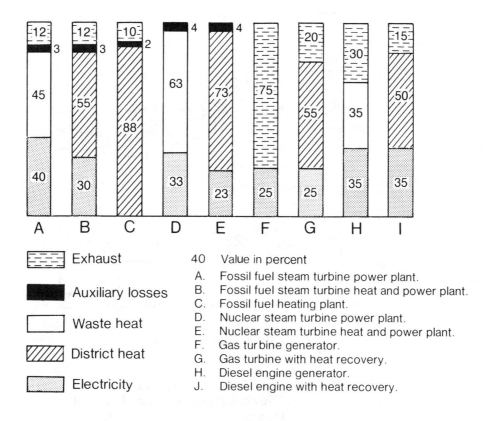

	Exhaust	40	Value in percent
	Auxiliary losses	A.	Fossil fuel steam turbine power plant.
		B.	Fossil fuel steam turbine heat and power plant.
	Waste heat	C.	Fossil fuel heating plant.
		D.	Nuclear steam turbine power plant.
	District heat	E.	Nuclear steam turbine heat and power plant.
		F.	Gas turbine generator.
	Electricity	G.	Gas turbine with heat recovery.
		H.	Diesel engine generator.
		J.	Diesel engine with heat recovery.

Note: The output called "District Heat" represents recoverable
thermal output in the temperature range of 80 to 120°C

FIGURE 17-11 OUTPUT CHARACTERISTICS OF VARIOUS COMMON
ELECTRIC GENERATION SYSTEMS

temperature extremes. All installations should be protected by dry chemical dump fire-suppression systems.

The shaft of the engine is connected to an alternator which turns at a constant rate generating 60 Hz electricity. The frequency output is proportional to the speed at which the generator turns, therefore speed must be carefully controlled. Standby capacity in remote areas should be complete. An indoor tank holding enough fuel for 24 h of operation at full load should be provided. If such an indoor day tank cannot be provided and temperatures are often below -18°C, heat tracing should be installed on pipelines in case it is needed. Most engines require a circulating fuel line which also helps provide some heat to the day tank, even if it is located outside. Heat tracing is also necessary where the outside temperature drops below the pour point of the fuel

being used. All water-cooled systems should be filled with a permanent-type antifreeze solution. Combustion air must come from the inside and, for diesel engines, amounts to 3 m³/kWh of output. For gas turbines it can be as high as 55 m³/kWh of output.

The output characteristics of various electrical generating systems are shown in Figure 17-11.

Figure 17-12 presents a typical heat balance for diesel engines. The most common cooling system from which waste heat can be feasibly recovered is the air plenum system. With this system, all engine radiators discharge heated air into a plenum. The plenum air is then either discharged to the outside or blended with outside air and recirculated. Motorized dampers are used to regulate the amounts of outside air and recirculated air. The dampers can be controlled by a thermostat to provide for heating the

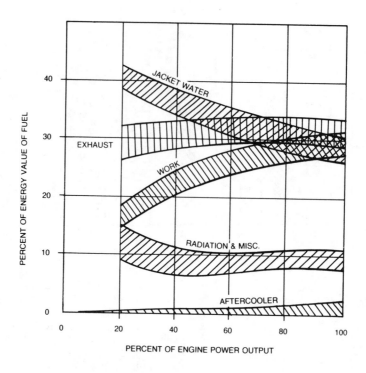

FIGURE 17-12 HEAT BALANCE FOR DIESEL
ENGINES (from Caterpillar Technical
Manual)

generator building, if desired. The radiator of each engine must be fitted with a gravity shutter to prevent backflow when the engine is off. Air-to-air heat exchangers can also be used to exchange heat between the outgoing exhaust air and the incoming makeup air.

The main advantages of diesel-driven generators compared to gas turbines, in the lower power ranges (less than 10 MW), are:

• lower fuel costs in most areas;

• lower maintenance costs; and

• lower fuel storage costs.

Diesel generators are available in sizes from 2 or 3 kW up to 36 MW. Units up to about 500 kW are classed as light plants and require maintenance more often than larger units. Light plants require overhauls every 10,000 to 20,000 hours, whereas larger units require maintenance every five to six years, because they usually operate at a lower rpm.

Fuel costs are directly related to efficiencies. The engine efficiencies vary from 25 to 37 percent at full load, and actual efficiencies seldom exceed 35 percent when generator excitation, mechanical coupling, and ancillary equipment are considered. The cost of fuel per kWh can be computed as follows:

$$\text{Fuel cost} = \frac{\$/L \cdot 0.093 L/kWh}{e}$$

where:

e = efficiency of a generator.

Some special considerations must be taken into account for extremely remote installations where generation units have to operate for longer periods of time without an operator (at communication sites, for example):

• Air-cooled diesel engines seem to be more reliable, especially where they may be exposed to freezing temperatures. An air-cooled unit is about 12 percent more efficient than a water-cooled unit of the same size, because power is used for pumping coolant. Also, the blower fan is about two-thirds the size of that for a water-cooled unit. It is estimated that 40 percent of all engine failures are related to the cooling systems.

- Generators have been installed where they must operate for up to 6 months without an operator being present. The engine oil must be continuously filtered in a bank of filters and replaced at a metered rate to provide the required oil changes. The used oil is then burned in the engine with the fuel, so that there are no waste or disposal problems. This has increased the life of the injectors in many cases.

- These small diesels run well at any point between 0 and 120 percent of their rated continuous load but extended running at less than 60 percent of their load leads to carbon-fouling, which shortens the life of the engine.

Specific problems encountered during part-load operation include: hydrocarbon buildup, rings sticking, plugged injectors, glazed pistons and cylinder walls, and unburned oil in the exhaust. Together, these result in a shortened engine life less than 1,000 hours in extreme cases. To prevent these problems, it is very important to size generators to prevent the problem from arising in the first place. Other solutions recommended by those involved with engine maintenance include: using dummy loads during periods of low load, periodically operating near full load to clean out deposits, and derating the fuel system.

Fuel storage facilities require conventional safety precautions. In addition, ease of access during winter and the special protection of exposed piping need to be considered to prevent damage from conditions such as snowdrifting and ice flows during break-up (Figure 17-13).

Although diesel-generators only account for 0.85% of the total U.S. utility capacity, the 277 MW units installed in Alaska represents 17% of its capacity. Because of the large heating load compared with the electrical load in many northern communities, it is prudent to consider operating diesel-generators in a cogeneration mode. There are a number of packaged systems, typically of a few hundred kW size, produced in the U.S. In Alaska, there are 30 to 40 operational heat recovery systems in the villages with almost all practicing heat recovery from the jacket water only. The typical system includes a shell and tube or plate heat exchanger to transfer heat from the jacket water to an external glycol-water loop that includes arctic pipe. This in turn transports the hot fluid to the end user which is usually a public facility such as a school. The entire system is custom designed for a given site (Cuzme, 1990).

Gas Turbines. Gas turbines can include both natural-gas and oil-fired turbines operating on the Brayton cycle. They have the advantages of being small, lightweight, and low in capital costs; but they have the disadvantage of high fuel consumption which, in turn, usually means that more fuel storage is needed. There are two types: the light-duty, jet-engine type, and the heavy-duty, industrial type. The light-duty units require much maintenance and are generally used only for peak power needs, while the heavy-duty types are used for continuous power generation; they have lower maintenance costs.

Full-load efficiencies range between 20 percent for the smaller units (2 to 3 MW) to 28 percent for the larger units (15 MW). Anything less than full load greatly reduces these efficiencies. They are lighter and easier to handle, and replacement units can be flown in, as was done in Bethel, Alaska, the day after the power plant burned to the ground. Gas turbines also have the advantage in cold regions that at lower temperatures, more than the rated power can be obtained. For example, many of the offshore oil platforms are powered by 820 kW International Harvester Saturn gas turbines. Tests on these units show that they put out 145 kW more at an ambient air temperature of -26°C than at 21°C. A disadvantage of the turbines is that if they take in foreign particles such as ice with the intake air, it can completely destroy them.

In a total energy system, gas turbines are used along with heat recovery equipment such as heat exchangers, heat recovery boilers, and heat pumps. With gas turbines, heat and electricity recovery can be achieved in simple-cycle or combined-cycle systems. In a simple-cycle system, a gas turbine can be connected to either a nonfired or supplementary fired heat recovery boiler. Depending on the boiler used, heat can be recovered either as hot water or steam in heat-to-electricity ratios of 1.4:1 to 3.0:1. Combined-cycle systems include gas turbines, heat recovery boilers, and steam turbines.

Steam Turbines. The fuel for a steam turbine can be anything that burns: wood, oil, coal, and peat. Steam turbines are used in the cold regions where a fuel is readily available. The fuel is used to fire a boiler which produces steam. The steam is then run through a turbine which is rotated by the force of the steam on its blades. Fairbanks, Alaska, has a coal-fired plant. The disadvantages of the coal-fired

(a)   Barge Bulk Fuel Fill Line with I-Beams for Ice Deflection – Ambler, Alaska

(b)   Diesel Fuel Lines and Tanks – Selawik, Alaska

FIGURE 17-13 FUEL STORAGE FACILITIES

plants are the thermal pollution and air pollutants which require more expensive cooling towers, ponds, and precipitators for abatement. The high capital costs for a coal-fired plant usually make them uneconomical at capacities below 150 MW. Also a lead time of up to four years is required to build a plant. The greatest asset of the coal-fired plant is the cheap source of fuel. Efficiencies are about 30 to 40 percent which however, can be greatly increased if central heating is combined with power production (cogeneration).

If heat is not recovered, it can create severe thermal pollution problems in its disposal. Several studies are being carried out which use this waste heat for agriculture by warming the ground or by heating greenhouses.

17.5.3 Distribution and Transmission.   Reliability is more important in cold climates than in warmer climates because of the dependence on electricity for pumps, fans, and controls for heating. A power outage of a few hours could cause millions of dollars in damage. Standby power is usually installed, but power reliability is still critical. Dual-feeders, looped circuits, parallel transformers, and dual busses can improve reliability. These should be designed into and installed in any cold-region distribution system.

Electrical transmission and distribution lines can be overhead or underground. As with pipelines, it is usually best to go underground, if soil conditions are favorable.

Buried lines are not subject to vandalism, but they can be damaged by soil movement, especially if buried in frost-susceptible soil. During the winter when the soil cracks with tension (see Section 3), buried electrical lines have been pulled apart where the line on each side of the crack was frozen tightly in the soil.

If buried, electrical lines (Hunt, 1981) have to be buried according to standards. In nonpermafrost areas, the lines are placed below the depth of the frost, especially in frost-susceptible soil. In permafrost or if the lines must be placed in the frost zone, they must be bedded in non-frost-susceptible gravel, so that the surrounding soil does not bond to the wires.

Overhead electrical distribution systems should be selected and designed for the basic economic factors that normally dictate standard voltage, conductor, transformer, and substation sizes. Increased safety factors and cold-region considerations are necessary for reliable electrical distribution. Extreme wind chill factors make routine repairs slow and difficult. Modular apparatus designs should be used where possible. This allows the components that are malfunctioning to be replaced and brought inside for repair. Cable terminal buildings, substations, and transformers should be designed to protect equipment from the elements. Controls may have to be located in a heated compartment to function properly.

The primary problem encountered with poles is frost-jacking. The reader is referred to Section 3 for analysis and remedies for frost jacking. The minimum depth to which poles should be set is 10 percent of the pole length plus 1.2 m. This depth may need to be increased to prevent frost jacking.

All poles in the Arctic must be imported. The wood used and its quality should be selected based on the permanency of the line. The section of the pole in the active layer and below, if permafrost is not present, should be treated with a preservative and bond breaker. In the subarctic, it is best to treat the entire pole. The methods of placement are discussed in Section 3. Crossarms, hardware, and insulators are designed similar to those in other areas, but cold-temperature effects on the materials must be considered. The entire overhead distribution system must be accurately designed for wind conditions at the site. Stranded wire with a polyethylene waterproof jacket is recommended for overhead lines (GNWT, 1981). The minimum size wire should be No. 4 AWG medium-hard, drawn copper or the equivalent in aluminum. Aerial cables of the bundled or self-supporting types are not recommended because of the exacting limitation of insulation flexibility in areas where extreme cold and high-velocity winds occur.

Conventional overhead-type transformers are satisfactory for use in cold regions. Switch gear and circuit breakers (conventional) should be installed in heated areas of generating stations or buildings. Unheated, outdoor shelters are not satisfactory.

Grounding is probably the most difficult electrical problem to overcome in arctic areas. The accepted standard ground resistivity value used in zero-phase-sequence calculations is 100 Ω•m. Tests by CRREL in permafrost around Fairbanks, Alaska, revealed values of 7,000 to 21,000 Ω•m (Figure 17-14).

Grounding of electrical systems is necessary for two reasons:

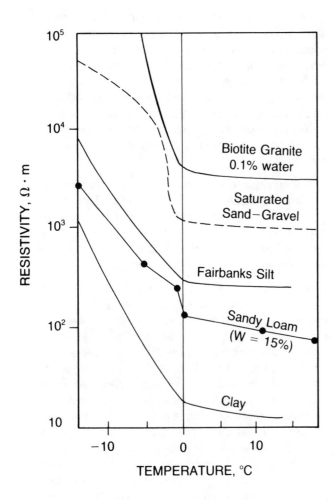

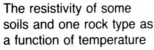

The resistivity of some soils and one rock type as a function of temperature

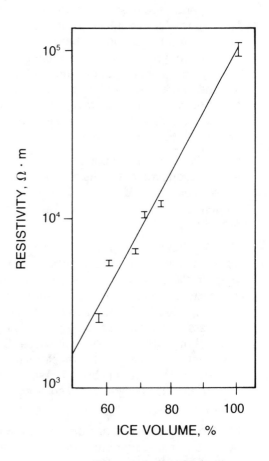

The relationship between the ice content in permafrost and resistivity.

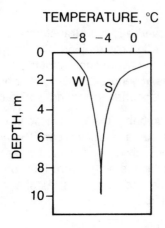

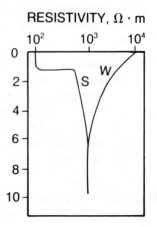

Temperature and resistivity profile during summer and winter, typical of permafrost conditions.

FIGURE 17-14 RESISTIVITY OF FROZEN SOILS

- Safety requirements call for the grounding of all exposed metal parts of switches, structures, transformer banks, fences, etc., so that a person or animal touching or near any of the equipment cannot receive a dangerous shock if a high-voltage conductor flashes to, or comes in contact with, the metal item; and

- Provide voltage stability of the system.

In cold regions, grounding practices assume a more critical nature because of the uncertainties associated with frozen soil resistivities and discontinuities. Locations in more temperate conditions which require the installation of a safe grounding design are normally dealt with generally accepted engineering principles which have been described extensively in many studies and standards.

Grounding designs should provide for the safety of personnel working in and around energized electrical facilities and also for the general public who could be accidently exposed to unexpected shock hazards associated with these facilities. Additionally, a safe grounding design would also permit fault electric currents to not exceed equipment and apparatus limits and therefore not interrupt service to customers.

A central premise upon which a safe grounding design is based is that no ground fault current will occur until or unless a ground return circuit exists. The existence of a ground return circuit depends upon a number of different, but interrelated criteria – metallic conductors, ground rods and electrical equipment locations.

A shock accident is possible when a person is in a position close to a ground return circuit and when a relatively high fault current exists, or when a high potential gradient exists near points close to this person. There must also be insufficient contact resistance which could limit shock current through the person's body.

In designing grounding systems, the current magnitudes passing through the human body should be limited to less than what has been recognized as the threshold for the fibrillation of the heart. Studies by various authors indicate this is in the range of 60-100 milliamperes (mA). In addition, consideration is also made for the duration of this current. Tests indicate that very brief exposures reduce the possibility of injury or death. Dalziel (1972) and others in studies assume that 99.5% of all persons can safely withstand shock energy based on the following formulas if the person weighs 50 kg and 70 kg respectively:

$$I_{B50} = 0.116/\sqrt{t_s}$$

$$I_{B70} = 0.157/\sqrt{t_s}$$

where

I_B = rms magnitude of current through the body

t_s = duration of the current exposure in seconds

Standard 80-1986 shows examples of nonfibrillating currents for these body weights and a three second shock current as 107 mA for 50 kg and 91 mA for 70 kg.

The resistance in a circuit completed at the right position and time by a person depends on a number of circumstances. There are two basic situations where a person could complete a circuit during the occurrence of an electrical fault. A fault saturates the earth with currents, and these result in high potential gradients. The first situation in which a person might find himself is walking or standing with his feet apart in the areas of these gradients. In this instance the shock current would flow through both legs. Another situation might be where the person is touching electrical apparatus with a hand and standing in the vicinity of the potential. Here the shock current would flow through the feet and hand.

Standard 80-1986 uses a value of 1,000 W to represent the resistance of the body in these cases. Using this value in design calculations to limit possible shock currents to less than the above mentioned nonfibrillating currents is the aim of safe grounding systems. Formulas using these concepts result in maximum allowable step and touch voltage criteria. The possible actual step and touch voltages should be less than these values.

As stated before soil resistivity has a significant effect on the ground return circuit. Resistivity is greatly affected by the moisture content of the soil and rises appreciably when less than 15%. On the other hand, studies indicate little effect if the content is over 22%. The critical factor in cold region is the effect of freezing. Resistance increases abruptly as the temperature passes from + to - temperatures. Studies by Sunde (1968) suggest using a modeling technique with the assumption of two layers for the soil environment. In the situation where the upper layer has more resistance than the lower layer (a case where the frozen layer overlays the unfrozen) the model predicts lower maximum allowable step and touch voltages than if the soil were considered a uniform single layer.

Actual practice in cold regions by engineers or designers is to be very conservative in the system design parameters. Methods to decrease the resistivity of the soil include the following.

Mats consisting of a nest of buried conductors are often used for grounding. Each individual piece of equipment is a substation or powerhouse that has its own connection to the grounding mat. The mat has to be kept in the thaw bulb under a building or beside a lake or river if possible. Connections to the mat should be of heavy copper (at least 4/0 stranded conductor). They should be protected against mechanical damage. To equalize the ground potentials around a substation, the various ground cables or busses in the yard and substation should be bonded together with heavy multiple connections and tied to the ground of the main station. This reduces the possibility of an appreciable voltage difference to ground occurring between the ends of control cables, communication wires, or other conductors which run from the switchyard to the substation. Heavy ground currents, such as those which may flow in a power transformer neutral during ground faults, should not be localized in ground connections of a small area, since the potential gradients in the earth around the ground connections may be dangerous. Ground mats composed of heavy bonded cables and covering a large area are the safest and most satisfactory means of reducing potential gradients in the earth's surface at large substations where heavy currents to ground can occur.

Another objective in substation grounding applies to grounded neutral systems and concerns the limitations of the rise of potential of the earth, in the immediate vicinity of the station ground mat, above absolute earth potential during faults involving ground. This rise can be attributed to the resistivity of the earth and the electrical resistance of the ground connection and may endanger communication lines entering the field of influence. A substation may be located on ground that is underlain with permafrost or solid rock of low conductivity. The ground mat should be embedded in topsoil or, in the case of permafrost, in the active layer. Even though individual ground rods may show normal resistance when tested by the usual probe methods, the overall ground resistance at the station may be relatively high when measured from a point remote from the substation. The station ground resistance enters into the zero-phase-sequence condition which means the resistance of the station ground can vary directly as the system voltage. The lower the system voltage, the lower the station ground resistance should be. For instance, a 500,000 kVA ground fault at 132 kV would correspond to about 2,200 A in the ground connection; whereas for a fault at 13.2 kV, there would be about 2,000 A in the ground connection. If the ground resistance has a 3.1 ohm resistance to absolute earth, the internal resistance drop would be 220 V (caused by the 132 kV fault), which would be negligible. However, the 2,200 V internal resistance drop due to the 13.2 kV fault current can cause serious trouble to communication lines entering the station if they are not isolated. Also, the 2,200 V drop in the component of voltage in the zero-phase sequence would cause a rise in voltage to the neutral on the order of 25 percent in one of the unfaulted phases of the three-phase system.

Any method of grounding should be booth economical and easy to maintain. But these two criteria are difficult to meet.

One method is to sink a deep well, place the grounding rod inside, and fill the surrounding free space with a salt solution. The salt solution must be continually replenished.

Engineers in Russia have used a ground mat laid out and constructed on lake ice during the winter and an additional mat was located in the active layer near the substation. The two mats were connected. When the ice melted, the one mat settled to the bottom of the lake. This installation provided a very good resistivity of 12 $\Omega \cdot$m but was expensive to construct and would not work unless the station was located near a lake that does not freeze to the bottom.

Another method that has provided an acceptable ground is to drill several wells around a proposed substation at least 150 m into the permafrost. An explosive charge is then set off at the base of the well to produce cracks and fissures in the permafrost. A copper-weld steel rod is inserted to the depth of the well, and the voids, cracks, and the well itself are filled with a sodium bentonite solution which is then allowed to set. Sodium bentonite absorbs water and swells to many times its original volume. After setting it is incompressible and impermeable and has an electrical resistivity of 50 $\Omega \cdot$m. A conventional substation ground mat should be constructed in the active layer and connected to the long ground rods.

TABLE 17-9   TOTAL COSTS FOR LARGE GRID-
CONNECTED INSTALLATIONS (Adelaav
and Associates, 1981)

Generating Capacity (MW)	Costs ($/kW)(US, 1981)
30 to 500	1200 to 1300
75 to 150	900 to 1200
300 to 350	750 to 900

Note: Costs exclude transmission costs and interest payments.

The other possibility is to design the electrical distribution system based on the premise that a satisfactory ground cannot be economically obtained. With fairly large installations on ice or permafrost where pole lines are to be used and the distribution voltage is 3-phase, 60 Hz, 2,400 to 4,160 V, the fourth wire is connected to the neutral at the wye-connected generators and carried as a non-current-carrying neutral throughout the primary system. It is bonded to secondary neutrals of single-phase transformer banks or to the secondary neutral of three-phase delta-wye-connected transformers.

Note: This non-current-carrying neutral should be connected to the primary side of transformer banks. All transformers are connected at cross phases of the primary system for 4,160 V. This method can also be used for smaller systems such as a 277 to 480 V primary system.

For ground of individual buildings, multiple ground rods can be used. Resistances to ground for various shapes of ground rods, etc. are shown in Table 17-9. They must be separated by a distance at least equal to or greater than the rod depths. Rod depths can have a significant effect on changing the resistance. Resistance decreases as depth increases in most soils. Rod diameter, however, has little effect on the resistance obtained. As with larger grounding installations, bare wire can be buried and run in the ground connecting rods. The rods should be placed in thawed areas such as under streams, rivers, or lakes that do not freeze to the bottom. Another possible location is the thaw bulb beneath the building itself.

Considerations for Preventing Transmission Failures.   Arctic environments, especially near the coast, present severe storm combinations that can cause failures of transmission lines and make nor-

mal service access impossible. (Aeolian vibration control is covered in Stockbridge (Fuller, 1981).) Dumbell torsional dampers and armour rods should be used at all primary cable conductors to prevent fatigue failure. The "galloping" conductor phenomenon is apt to occur anywhere. The conductor spacing must allow for elliptical loop dancing in single conductors to prevent short-circuit phase-to-phase faults. With the present state of the art, galloping cannot be eliminated. In severe galloping areas, conductors should be dead-ended at each pole; or preferably, the line should be rerouted or placed underground.

Wind and ice buildup can cause severe damage to towers, insulators, and other hardware. Damage can be minimized by the use of slack spans in relatively short runs of 90 m or less, or by using midspan spacers in larger runs of up to 300 m.

In most cases, arctic areas do not have a recorded history of meteorological data or even spot data in locations where transmission lines must be located. For this reason conservative designs must be used. Wind speeds can exceed 320 km/h and estimates of up to 480 km/h have been made for ridgetops in southeast Alaska. Anemometers rated at 320 km/h have been found twisted off or destroyed.

Ice can build up several millimetres thick on power lines which adds greatly to the load. The Alaska Electric Light and Power Co. of Juneau has measured 1.1 m of ice with 240 km/h winds at higher elevations in the area. This amounts to a wind loading alone of 2.68 kPa. Freezing rain with temperatures slightly below freezing are ideal conditions to produce icing. Rime ice (hoar frost) builds up under ice fog conditions, but this loading is usually not significant because of the low density of the ice and its quick removal by wind. Most reports assume that ice buildup is concentric with the conductor. However, there are several cases where the actual buildup was shaped like an airfoil because the ice was formed with the wind blowing. An airfoil shape over 175 mm long has been observed. These conditions can cause galloping and, if the resonance frequency is reached, the line and supporting structures can be destroyed.

In rural cold-region communities, efficiencies of scale can often work to advantage, if communities can be tied together using electrical transmission lines, with electricity supplied by larger generators instead of smaller units in each community. The main problem lies in the reliability and cost of the transmission lines. Also, electrical distribution and transmission is about 95 percent efficient. The losses occur in the form of heat in the transformers and lines.

The single-wire ground return transmission (SWGR) system uses a single overhead energized conductor and uses the earth as the return conductor (Flanders et al., 1981). The wire is supported on A-frame structures using long spans (180 m). The A-frame provides transverse stability from gravity alone and does not need to penetrate the surface. Longitudinal stability is derived from tension in the conductor itself with dead-end anchors located a maximum of every 10 spans. The A-frames can be built using local spruce, imported utility poles, pipe, or other metal components. A line has been in operation for approximately one year between Bethel and Napakiak, Alaska (13.6 km). It has provided single-phase power to Napakiak with very few maintenance problems. Three-phase equipment can be successfully operated using three-phase converters. These converters (roto-phase, etc.) should be thoroughly

checked before installation in a remote location as they seem to have maintenance and operation problems. The line operates at a voltage of 14.4 kV. The dead-end anchors were drilled into the permafrost to a depth of 2.4 m. This system cost about 1/3 of an equivalent conventional (3-phase, 4-wire) line of similar capacity. The cost breakdown is 20 percent for materials and 80 percent for installation. The effectiveness of the SWGR system is determined by the ability to establish adequate grounding at each end of the line. Permafrost thaw bulbs located under lakes and rivers can provide enough thawed soil area to reduce the earth electrode resistance to a satisfactory low value.

Costs. The percentage breakdown of the total cost of electrical service in a remote arctic community is shown on Figure 17-15. These statistics were compiled in Alaska by the Alaska Rural Electrical Cooperative Association.

17.6 Energy Conservation

Much can be accomplished to reduce energy costs by conservation: modifying existing buildings, utilities, and transportation modes to use less energy or considering new energy-saving approaches when designing new facilities (Argue, nd).

17.6.1 Determining Energy Losses. It is important to determine where energy losses occur so that existing buildings can be corrected or designs for new buildings changed. The locations of energy losses in buildings are shown in Table 17-10.

Infrared photography can be used to analyze existing buildings for areas of high heat loss. The resultant thermograms measure heat radiated from the

TABLE 17-10   BUILDING ENERGY LOSSES

	Energy Losses (%)	
	Single-Family Residence	Large Multistory Office Building
Walls	27	12
Doors and Windows	26	44
Roof	26	4
Floor	6	3
Air Exchange	25	37

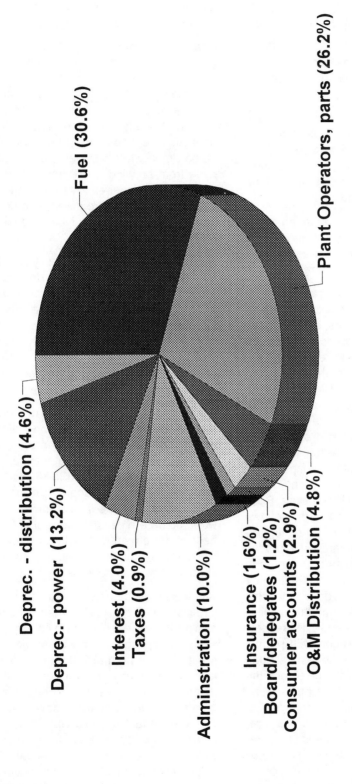

FIGURE 17-15 TOTAL COST OF ELECTRIC SERVICE BY ITEM (for calendar year 1994)

object. High heat-loss areas appear as white or red areas; low heat-loss areas appear as black, purple, or blue areas. In black and white photos, areas of high heat loss appear white or light and low heat-loss areas black or dark. These colors and shades are reversed for a thermogram taken from the warm area looking toward the cold area.

Thermographic analysis of rooftops at night can help identify areas of damp or missing insulation for subsequent repair.

Setting thermostats back reduces the energy use of a building. Thermostats are available that automatically set the temperature back at night or while no one is at home. Store owners should set thermostats at night back as far as they can without causing damage to the merchandise.

17.6.2 Correction of Losses. Once the areas of high heat loss have been determined, they should be repaired. A variety of government grants, loans, and tax credits are available for these repairs.

Heat losses through infiltration-exfiltration can be reduced by weatherstripping and caulking around windows, eaves, foundations, doors, etc. Dampers should be installed in all fireplaces, so that they can be shut tightly when not in use to reduce the escape of large quantities of heat. Fresh air is replenished through cracks, windows, and when a door is opened. Anytime the temperature in a building or room is different from that outside, pressure differences occur between the inside and outside because of the different air densities. When the inside temperature is higher than the outside temperature, a negative inside pressure is produced and colder air flows inward through cracks at lower levels. Positive inside pressure is produced and warm air flows outward at higher levels. Based on the configuration of the building and the temperature differential, ASHRAE (1993) provides formulas and curves to estimate the pressure differential and resulting air exchanges. Wind blowing on one side of a building also creates pressure differentials and thus infiltration and exfiltration. With these pressure differences, large amounts of air are exchanged with people entering and leaving the building. The general quality of construction determines, to a large extent, the amount of infiltration and exfiltration a building experiences. A certain amount of air exchange is necessary to keep humidity levels low enough to prevent condensation on windows and walls and to remove pollutants. This amount varies with personal preference, the use of the room, and the size of the room. Older conventional, residential buildings nor-

mally have air exchange rates of one to two exchanges per hour. New energy efficient houses usually have 0.2 to 0.5 exchanges per hour. A consensus is developing that 0.5 exchanges per hour is necessary for a healthy environment and for preventing damage to buildings.

Again, the number of air exchanges needed depends on the uses of the building, but it is better to design and construct using 0.5 as a criterion. Exhaust fans can be used in kitchens and bathrooms to increase the air exchange during times of cooking, etc. Usually a kitchen exhaust fan should be capable of exhausting 0.066 m³/s to prevent the carbon monoxide, nitrogen dioxide, and formaldehyde from building up. Many local building codes list minimum requirements for ventilation of occupied spaces; these requirements should be met. Table 17-11 gives some typical ventilation requirements for different uses.

Heat loss around doors is usually equal to or greater than the heat loss through the door itself. Heat loss around sliding doors is about three times that around an equivalent swinging door because of the construction of the frames. Sliding doors are also usually all glass, which adds to the heat loss. All doors should be well weatherstripped, including heated garage doors. Double-entry ways can reduce heat loss considerably (Fisher, 1973).

Doors and windows are areas of high energy loss even discounting the infiltration-exfiltration losses discussed above. Metal-covered, insulated doors, with no metal from inside to outside, have the lowest heat loss for all doors. Solid-wood doors are often used for exterior doors, but they have higher heat losses than the insulated metal doors. Solid-wood doors are usually impossible to insulate. The best metal-covered, insulated doors are usually wood-framed with insulation between the metal sheets. They also have better fire ratings.

Windows should be used sparingly in northern construction. They cost approximately 10 times as much as a normal wall would cost to construct for that area. Heat loss through a single-pane window is 15 times greater than a normal 100 mm insulated wall. Heat loss is nine times greater for double-pane windows (6 mm air space between panes) and six times greater for a triple-pane windows than for a wall for that area. In fact, it would take about eight panes of glass with air spaces to equal an insulated, 50 mm wall. The glass itself has essentially no insulating value. The insulation is almost entirely due to the enclosed air space. Practicality and energy conser-

TABLE 17-11   OUTDOOR AIR REQUIREMENTS[a]

Application	Smoking	Fresh Air Replenishment		
		Recommended (m³/s per person)[b]	Minimum[c] (m³/s per person[b])	m³/s per m² of Floor[b] Minimum[c]
Apartments		0.009	0.005	
Banking space	occasional	0.005	0.004	
Barber shops	considerable	0.007	0.005	
Beauty shops	occasional	0.005	0.004	
Brokers' boardrooms	very heavy	0.024	0.009	
Cocktail bars		0.019	0.012	
Corridors (supply or exhaust)				1.28×10^{-3}
Department stores	none	0.004	0.002	0.25×10^{-3}
Directors' rooms	extreme	0.024	0.014	
Drug stores[e]	considerable	0.005	0.004	
Factories[d,f]	none	0.005	0.004	0.50×10^{-3}
Variety stores	none	0.004	0.004	
Funeral parlors	none	0.005	0.004	
Garages[d]				5.08×10^{-3}
Hospitals				
Operating rooms	none			10.16×10^{-3}
Private rooms	none	0.014	0.012	1.68×10^{-3}
Wards	none	0.009	0.005	
Hotel rooms	heavy	0.014	0.012	1.68×10^{-3}
Kitchens				
Residence				10.16×10^{-3}
Restaurant				20.32×10^{-3}
Laboratories[e]	some	0.009	0.007	
Meeting rooms	very heavy	0.024	0.014	6.35×10^{-3}
Offices				
General	some	0.007	0.005	1.27×10^{-3}
Private	none	0.012	0.007	1.27×10^{-3}
Private	considerable	0.014	0.012	1.27×10^{-3}
Restaurants				
Cafeteria[e]	considerable	0.006	0.005	
Dining room	considerable	0.007	0.006	
School rooms[d]	none			
Shop, retail	none	0.005	0.004	
Theatre[d]	none	0.004	0.002	
Theatre	some	0.007	0.005	
Toilets (exhaust)[d]				10.16×10^{-3}

[a]   Taken from present-day practice
[b]   This is contaminant-free air
[c]   When minimum used, take the larger of the two
[d]   See local codes which may govern
[e]   May be governed by exhaust
[f]   May be governed by special sources of contamination or local codes
[g]   All outside air recommended to overcome explosion hazard of anaesthetics

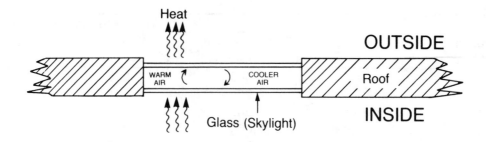

FIGURE 17-16 CONVECTION CURRENTS BETWEEN GLASS PANES

vation should play a role in determining the window area in building designs. Specifications for windows should state allowable air transmission values for windows.

Heat loss through windows can also be reduced considerably by the addition of insulated shutters which can be closed at night or anytime the window is not being used. Even the proper placement of curtains on the inside can reduce heat losses. A common, efficient shutter is a sandwiched panel with 50 mm of polyurethane in the center. These can be placed on the outside of the house on hinges or slides. They must be designed so they will not ice up and be impossible to close. Total heat losses for a building can be reduced by over 20 percent for an average building by the addition of insulated shutters and using the proper open-closed cycle. The shutters can be operated manually or by electric motors with timers or photosensors. In large buildings, automatic operation is desirable.

Skylights contribute to high energy losses. They act as a heat pump, pumping heat out of the house as shown in Figure 17-16.

Considerable energy savings can be realized by the proper design and installation of the heating and ventilation system for a building. Most heating systems, especially in larger buildings, need to pull in some outside air for combustion.

An air supply system for a northern building should always combine the make-up type and the recirculation type. It should be periodically checked and adjusted, so that excessive outside air is not being drawn in. Using 100 percent outside air could increase fuel costs 25 to 50 percent. Many heating systems have an electric, preheating coil in the outdoor air intake. If possible, such coil should be replaced with an air-to-air heat exchanger (up to 80

percent of heat may be recoverable) or, if that is not possible, the coil must not be allowed to overheat the intake air.

Internal sources of heat are lights (incandescent lights provide 90 percent of their energy in the form of heat and only 10 percent as light, and fluorescent lights provide about 25 percent as light), equipment, passive solar gains, and people (each person gives off about 300 J/s of heat during normal activity). Walker (1979) gives more information on heating load analyses and heat balance modeling for large buildings.

Moisture and condensation are critical means of heat loss, because latent heat is carried in the moisture which migrates with the air. Condensation occurs on windows, walls, or other objects when their surface temperature is lower than the dew point. The dew point is the temperature at which air becomes saturated when cooled under constant pressure with a constant vapor content (humidity). Living areas should be kept at 35 to 50 percent minimum humidity for maximum comfort. This value varies with individuals and their health.

Moisture within a wall or ceiling destroys materials and decreases the thermal resistance of the insulation. Vapor barriers must be installed on or as near as possible to the warm surface, so that the barrier temperature is always above the dew point. Then condensation cannot take place. As with heat, which flows from areas of higher temperature to areas of lower temperatures, so vapor or moisture travels from areas of high vapor pressure (warm areas) to areas of low vapor pressure (cold areas). The vapor barrier is added on the warm side of the insulation to prevent this flow of vapor into the insulation and wall where it would condense and ruin the insulation. No matter how carefully the vapor barrier is installed or how resistant it is to vapor penetration, some moisture does enter the insulation. This moisture must be allowed to pass through the wall and the outside, without being trapped in the wall where it would cause damage. The exterior of a wall must never be painted with an interior paint; exterior paints are designed to "breath" and allow this escaping moisture to pass.

Even though foil is nearly impermeable to moisture transmission, the foil backing on insulation cannot be relied upon to provide an adequate vapor barrier because of the many discontinuities such as the studs. At least one layer of 0.10 or 0.15 mm polyethylene film should be provided directly under the inside wall surface. Extreme care should be taken not to tear or puncture it when installing wiring and plumbing. Any necessary holes should be thoroughly sealed with a nonhardening caulking and wall-to-floor or ceiling joints should be lapped and folded.

Due to the many difficulties associated with the measurement and accuracies of values, vapor migration computations are less accurate than ordinary structural or mechanical computations. Thus vapor migration calculations are more of an index of a problem than a precise design technique.

Moisture in buildings can come from several sources and create many problems (Zarling, 1978). Most common building materials will absorb water, expanding as they do. They then usually shrink upon drying, often to less than their original dimensions. Action similar to frost heaving in soils can completely destroy a brick or block wall, concrete slabs, or roofing membranes if a source of water is close by. Water enters cracks in roofs, walls, and floors where it freezes and forces the cracks further open. These cracks can be caused by thermal expansion and contraction. The moisture in building is caused by several factors:

- moisture absorbed by materials during construction;
- precipitation on an unroofed building; and
- water vapor produced by the normal use of a building.

Water leakage through a roof can also create problems in a structure. Wind can force water under the shingles, even on a sloped roof. Ice dams on the eaves can also back water up under the shingles and into the wood base of the roof. Wood rot can be a major problem if the moisture content of wood components exceeds 35 percent.

17.6.3 Electrical Energy Conservation.

Controllers are available to regulate the intensity of fluorescent lights according to the light coming in through windows from the outside. They also extend the useful life of fluorescent lights. The payback on these units is estimated to be less than three years. A unit to operate four 40 W bulbs costs approximately $50 (US).

Electrical energy can be conserved by using fluorescent lights over incandescent lights. Incandescent lights convert only about 10 percent of the electrical input into light, whereas fluorescent lights convert about 25 percent. For example, one 40 W fluorescent lamp provides about 3,200 lm of light, but one 100 W incandescent light provides only 1,740 lm. The fluorescent lamp is rated for 18,000 hours of life; the incandescent bulb is only rated for 1,000 hours. The remainder of the energy from each is in the form of radiated or convected heat. An even more efficient light source is the halide (sodium halide) lamp. In Table 17-12 the efficiencies of the various lights are shown in lumens per watt used. The choice of light source must depend on other factors as well, such as the mounting height and location, the use of the light and the color rendering index (CRI). The CRI is measured on a scale of 0 to 100 with higher numbers being more like a reference source (not necessarily natural sunlight). The CRI could be considered a "quality of light" factor.

Most lighting is manufactured to operate at a minimum of -29°C. Temperatures lower than this have a direct influence on the life and efficiency of the various lights normally used outdoors (Gauss, 1974). The problems, which must be considered, include:

- the ionization of the fill gas;
- the starting voltage and time; and
- the freezing temperature of the equipment.

The best light for use in cold weather is actually an incandescent light because of its fast starting. Fluorescent lighting should not be used below -18°C because the mercury freezes in the tube. Mercury and metal halide lights are very good for light output, but starting time and voltage increase significantly below -29°C. For instance a 1,000 W mercury lamp takes 375 V to start at -29°C, and 775 V at -45°C. To produce these higher voltages, special ballasts and capacitors are required. High-pressure sodium with special capacitors are probably the most efficient light for use under arctic and subarctic conditions (Gauss, 1974).

TABLE 17-12   *EFFICIENCY AND QUALITY OF LIGHTING*

Source	Typical Efficiency of Lamps	
	Range (lm/W)	Color Rendering Index (CRI)
Low-pressure sodium (35W to 180W)	80 to 150	10
High-pressure sodium (70W to 1,000W)	64 to 130	20
Metal halide (175W to 1,000W)	67 to 116	50
Fluorescent (30W to 215W) 600 min to 2,400 min, U-shape, cool white	54 to 76	59 to 97
Mercury (100W to 1,000W) color-corrected	35 to 59	45
Incandescent, reflector, pair, indicator	17 to 23	95+

An electrical energy savings can also be realized by providing more switch circuits. This allows fewer lights to be on when only a couple are actually needed. In large commercial buildings even in the North, electrical lighting is probably the largest single source of internal heat. Ventilation, even air-conditioning equipment, may be needed to dissipate the heat produced by lights.

A few high-power lamps should be used instead of more low-power ones. For example, a single 100 W incandescent lamp produces more light than two 60 W incandescent lamps. The so-called long service lamps have longer life, but they are also less efficient in light production.

Power factor controllers can reduce the electrical use of motors and lighting. Motors larger than 1 kW and lighting larger than 15 W with an inductive reactance load component should have a power factor of not less than 85 percent. If power factors are less than 85 percent a power factor controller should be added to raise the power factor to at least 90 percent. For example: a 3.7 kW motor, 60 percent loaded, running 24 hours a day would normally consume 24,511 kWh/a. The addition of a power factor controller would reduce this consumption to 23,040 kWh/a or save about 6 percent. The controllers cost around $200 (US) for a single-phase motor and $300 (US) for a three-phase motor depending on the type of enclosure required.

17.6.4   Heat Recovery.   All excess heat should be considered for recovery and used at another time or in another application which can reduce the overall, year-round energy use.

Cooling ponds, where used, can serve beneficial uses also. The ones for the Elmendorf AFB heating plant in Anchorage, Alaska, serve as a hatchery for raising salmon fry. Cooling water can also be used for other beneficial uses such as heating the ground to enhance agriculture, heating greenhouses, and melting ice and snow on roads. Central heating plants tied into electrical generating plants are especially efficient as they provide an outlet for the waste heat from the generating plant.

All electricity generated using heat (steam or gas turbines, oil-fired generators) also produces waste heat as a byproduct. This heat is usually unavailable to the primary generation process and can amount to about twice as much energy as that converted to electricity by the process.

Shutcliffe (1982) presents and discusses five types of waste heat recovery units. They are economizers, heat pipes, heat pumps, regenerative units, and absorption chillers. A common use of heat exchangers is the preheat boiler feed water or intake air. Boilers fired by coal or oil produce sulfur oxides in the flue gases. As the heat recovery unit transfers heat from the exhaust to the incoming air or feed water, the temperature in the stack drops. This must be controlled, because the sulfur combines with water vapor and condenses as a hot, dilute acid that can cause severe corrosion problems in the stack or the heat exchanger, or both, if the exhaust temperature drops too low. There are several advan-

TABLE 17-13   ENERGY IN EXHAUST

| System | % of Fuel | |
	Energy in Exhaust	Temperature of Exhaust (°C)
Oil or gas turbine	65 to 84	430 to 540
Diesel engine	35	150 to 320
Home furnace	30 to 40	177

tages and disadvantages of the heat recovery units (Shutcliffe, 1982).

Stack gases with temperatures lower than 149°C do not have economically recoverable heat. Table 17-13 is a summary of the proportion of the input fuel energy that is lost in the exhaust and typical exhaust temperatures for three processes.

Air-to-Water Heat Recovery.   Heat is recovered from exhaust gas by using water or water/glycol heat exchangers. These are essentially water-tube boilers without a combustion chamber. These heat exchangers can also be used to recover up to 50 percent of the net heat input to an incinerator.

Water-to-Air or Water-to-Water Heat Recovery. Heat can be recovered from the cooling water of the engines driving electrical generators, or nearly any other internal-combustion engine. Approximately one-third of the energy of the initial fuel is dissipated through the jacket cooling system. These systems can be as basic as the conventional radiator on the engine through which the air in the building is circulated. To heat water in a tank, a heat exchanger can be installed in the cooling system lines through which a water, water-glycol, or other heat exchange fluid is circulated with the engine radiator remaining as a back-up for auxiliary heat dissipation in the event that heat is, for some reason, not withdrawn by the heat exchanger. These units are commercially available. Care must be taken to maintain the incoming and outgoing temperatures to the engine jacket so that the engine will not be damaged.

A small unit for individual automobile engines has been developed (Fisher, 1973). It continually collects and stores heat from the cooling system of the engine as the car is running. After the engine has stopped and shortly before restarting, a small pump (using less than 1 A/h from the car's battery) circulates the coolant through the storage device and the engine block (as an in-line tank heater does). No auxiliary heating is necessary for periods of time less than 24 h even at -20°C temperatures. The storage unit is a major component of the system.

Air-to-Air Heat Recovery.   Air-to-air heat recovery devices have been manufactured for large-flow industrial applications for several years. However, small units which would be applicable to an individual house are still being developed. The four basic types of air-to-air heat exchangers are: rotary, coil recovery loop, heat pipes, and plate type. Figure 17-17 shows schematics and presents the major features of each type. Rotary heat exchangers recover sensible heat and total enthalpy as they can transfer moisture. The other three types transfer sensible heat only. The heat of condensation can be recovered in all four types but drainage provisions must be included. This can be a large factor because about 2,300 kJ are produced per kilogram of condensation. Depending on the temperatures the condensate can create severe frosting and icing conditions in some situations. ASHRAE (1972) provides tables which can be used to predict the effect the moisture contents of the air flows has on frost formation in the exchanger. Frosting can be controlled by preheating the incoming air or, what is more commonly done, by diverting part of the air supply around the exchanger.

The efficiency of a heat exchanger can be determined by comparing the total heat in the supply air with the total heat of the exhaust air. This value is not a "true" saving of energy, because most exchangers require some energy to move the air, heat the incoming air, or pump heat exchange fluid.

Rotary or heat-wheel exchangers have been used for several years to remove heat from the exhaust gases of thermal power-generating plants and then use it to preheat the combustion air. They consist of

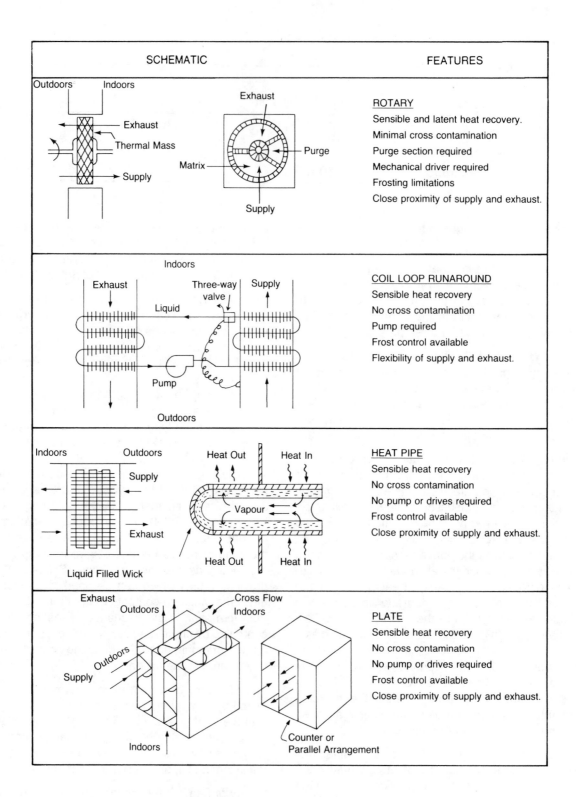

FIGURE 17-17 TYPES OF AIR-TO-AIR HEAT EXCHANGERS

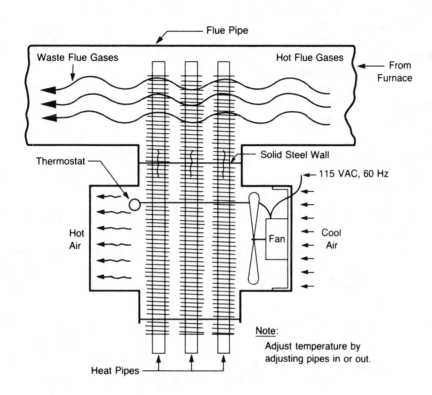

FIGURE 17-18 STACK ROBBER

a revolving cylinder filled with a porous material (usually aluminum or stainless steel foil strips) having a large surface area (Figure 17-17). The exhaust and supply sources parallel each other. As the wheel rotates (about 20 rpm) the medium is alternately exposed to the two flows. This enables the heat from the exhaust air to be stored momentarily until supply air can extract the heat. Moisture is transferred by using a hydroscopic material in the wheel. Frosting usually occurs on the discharge face of the exhaust side when the supply air is colder than -6°C. Filters are usually used to prevent dirt and lint from entering the exchanger. With a section of the wheel set aside for a purge channel, cross-contamination is usually less than one percent.

Coil recovery loops consist of finned tubing in coils that gather heat from the exhaust air by raising the liquid in the tube. This liquid is then pumped to a finned tube area in the supply air duct where it losses the heat (Figure 17-17). The return liquid from the supply air side should be warmer than -1°C to prevent frosting. This is usually accomplished by placing a three-way valve between the supply and exhaust finned tubing, which can recycle some of the liquid through the exhaust side until sufficient heat is stored in the liquid. Thermostatically controlled solenoid valves can be used to accomplish this. The exhaust side should be equipped with a condensation drain.

Heat pumps (Figure 17-17) contain a working fluid such as freon, carbon dioxide or ammonia. The tubes contain a small amount of liquid under a vacuum. It "boils" in the stack end, the vapor expands and travels to the cooler end where the vapor pressure is lower. A small-scale heat pump unit for installation on house chimney flues is available as shown in Figure 17-18. These units extract heat from the flue gasses and release it in the surrounding room. They can be used to heat a garage, or crawl space or to increase the efficiency of a home heating system 10 to 20 percent. They are relatively maintenance-free and should pay for themselves in less than a year.

Plate exchangers are made up of ducts where the exhaust air flow and supply air flow are separated by a thin wall through which the heat passes (Figure 17-17). The more the flows are separated and channelled beside one another, the greater the efficiency. There are three types of flow arrangements: parallel, cross, and counter. Counter-flow types provide the greatest temperature differential for maximum heat flow, but cross-flow types are generally easier to duct and thus are often more cost effective. Dust accumulations on the channel surfaces reduces the efficiency, therefore frequent cleaning may be necessary.

17.7 References

Adelaav, M. 1981. *Energy Demand and Supply in the NWT*. Prepared for the Government of the NWT and Department of Indian Affairs and Northern Development, Yellowknife.

Adelaav and Associates. 1981. *Community Specific Energy Supply in the Yukon and NWT*. Prepared for the Department of Indian Affairs and Northern Development, Ottawa, Ontario.

Alaska Energy Authority. 1989. *Power Consumption Information in Remote Alaska Villages*.

Argue, R. nd. *The Well-tempered House – Energy Efficient Building for Cold Climates*. Renewable Energy in Canada, 415 Parkside Drive, Toronto, Ontario, M6R 2Z7.

ASHRAE. 1993. *ASHRAE Handbook of Fundamentals*. American Society of Heating, Refrigeration and Air-Conditioning Engineers Inc.

Benson, C. and K. Rizzo. 1980. Air pollution in Alaska. In: *Weatherwise*, 33: 210-216.

Bettine, F.J. and R.W. Retherford. 1981. Single wire ground return transmission for rural Alaska. In: Proceedings, Northern Community: A Search for a Quality Environment. American Society of Civil Engineers, New York, 409-424.

Boyd, D.W. nd. *Converting Heating Degree-Days from below 65˚F to below 18˚C*. Building Research Note No. 98, National Research Council of Canada, Ottawa, Ontario.

Cheremisinoff, N.P. 1981. *Fundamentals of Wind Energy*. Ann Arbor Science Publishers, Ann Arbor, Mich.

Cuzme, J. 1990. *The Benefits of Waste Heat Recovery from Small Power Plants in Arctic Sanitation Facilities*. Office of Environmental Health and Engineering, Public Health Service, Anchorage, Alaska.

Dalziel, C.F., 1972. Electric shock hazard. *IEEE Spectrum,* February, 41-50.

Fisher, E. 1973. Engine preheating with saved heat: something for nothing. *Northern Engineer*, 5: 3.

Flanders, S.N., J.S. Buska, and S. Barrett. 1981. Window performance in extreme cold. In: *Proceedings, Northern Community: A Search for a Quality Environment*. American Society of Civil Engineers, New York, 396-408.

Fuller, W. 1981. What's in the air for tightly built houses? *Solar Age Magazine*, June.

Gauss, E. 1974. Cold soaking. *Northern Engineer*, 6(2): 21-23.

GNWT.1981. *DISTRICT Heating in Canadian Mining Communities*. Government of the Northwest Territories, Research Rept. No. 9, Yellowknife.

Gordon-Piatt Energy Group, Inc. 1996. Technical Information, Engineering Section (7E80.3), P.O. Box 650, Winfield, Kansas 67156-0650.

Hartman, C.W. and P.R. Johnson. 1978. *Environmental Atlas of Alaska*. University of Alaska, Fairbanks, Alaska, 95 p.

Hunt, A.J. 1981. *Energy Costs and Prices – NWT, 1980.* Trans-serv Limited, March.

McCullagh, J.C. (ed.) 1978. *The Solar Greenhouse Book*. Rodale Press, Emhaus, Pa.

Noble, N. and R. Van der Hoek. nd. Introduction to Small Wind Energy Systems in Alaska. State of Alaska, Department of Commerce and Economic Development, Division of Energy and Power Development.

Ott Water Engineers and U.S. Army Corps. of Engineers. 1981. *Regional Inventory and Reconnaissance Study for Small Hydropower Projects in N.W. Alaska.*

Peters, K. 1981. *Wind Power in Alaska*. University of Alaska, Anchorage.

Reid, B.H. 1980. *Alaska Village Demonstration Projects*. U.S. Environmental Protection Agency Rep. No. EPA-600/3-80-039, Corvallis, Oregon, 55 p.

Shira, D.L. 1978. Hydroelectric powerplant siting in glacial areas of Alaska. In: *Proceedings, Applied Techniques for Cold Environments*.

American Society of Civil Engineers, New York, 59-76.

Shutcliff, W.A. 1982. Air-to-air heat exchangers for houses. *Solar Age Magazine* 3.

Stockbridge, G.W. 1925. Overcoming vibration in transmission cables. *Electrical World*, 86: 26, (December 26).

Sunde, E.D. 1968. *Earth Conduction Effects in Transmission Systems.* New York: McMillan.

U.S. Department of Commerce. nd. Weather Bureau, Washington, D.C.

Walker, H.V. (ed.) 1979. *Energy Conservation Design Resource Handbook.* The Royal Architectural Institute of Canada, Ottawa, Canada.

Wind Systems Engineering. 1981. *Shungnak, Kiana and Ambler – Reconnaissance Study of Energy Alternatives.* Alaska Power Authority, Anchorage, Alaska.

Zarling, J.P. 1978. Air-to-air heat recovery devices for small buildings. In: *Proceedings, Northern Community: A Search for a Quality Environment,* American Society of Civil Engineers, New York, 381-395.

17.8 Bibliography

Alaska Energy Authority. 1992. *Rural Alaska Bulk Fuel Assessment Program – Summary Report and Recommendations.*

Alaska Energy and Engineering, Inc. 1995. *City of Selawik Waste Heat Recovery System – Design Analysis.*

Alaska Power Authority. 1988. *First Annual Statistical Report of the Power Cost Equalization Program.*

Alaska Power Authority. 1989. *Village Power System Survey.*

Alaska Rural Electric Cooperative Association, Inc. 1996. *Affordable Power in Rural Alaska.*

American Institute of Architects. 1974. *Energy Conservation in Building Design.* 1735 New York Avenue, N.W., Washington, D.C.

Bryson, R.A. and F.K. Hare. 1974. *World Survey of Climatology, Volume 11, Climates of North America.* New York: Elsevier, 420 p.

Canada Mortgage and Housing Corporation 1981. *The Conservation of Energy in Housing,* NHA 5149 81/02, Ottawa, Ontario

Climatic Information for Building Design in Canada. 1977. National Research Council of Canada. NRCC No. 1556, Ottawa, Ontario, 24 p.

Davis, A.J. and R.P. Schubert. 1977. *Alternative Natural Energy Sources in Building Design.* Passive Energy Systems, P.O. Box 499, Blacksbury, VA., 24060

DeHarpporte, D. 1983. *Northwest, North Central and Alaska Wind Atlas.* Van Nostrand Reinhold, New York. 122 p.

Dept. of Environmental Conservation (latest edition). *Village Sanitation in Alaska,* State of Alaska and U.S. Public Health Service, Juneau, Alaska.

Fletcher, R.J. and G.S. Young. 1976. *Climate of Arctic Canada in Maps.* Boreal Institute for Northern Studies, Occasional Publication No. 13, Edmonton.

Hare, F.K. and M.K. Thomas. 1979. *Climate Canada.* 2nd Ed., Toronto: Wiley, 230 p.

Hutchens, D. 1993. Legislature Funds, Interties & PCE, June Ruralite.

Hutcheon, N.B. and G.O.P. Handsgord. 1983. *Building Science for a Cold Climate.* Toronto: Wiley. 440 p.

Janz, B., D.G. Howell, and A. Serna. 1982. Wind Energy in the Northwest Territories. Yellowknife: Government of Northwest Territories. 108 p.

Johnson, R. 1988. *Cogeneration and Diesel Electric Power Production.* Institute of Northern Engineering, School of Enginering, University of Alaska, Fairbanks. Prepared for Alaska Dept. of Transportation and Public Facilities, Report no. AK-RD-90-09.

Johnson, R., C. Hok, and M. Bauer. 1987. *Economics and Reliability of Diesel Electric Generators.* Institute of Northern Engineering, School of Engineering, University of Alaska, Fairbanks. Report prepared for Alaska Department of Transportation and Public Facilities Rsearch Section.

Johnson, R., M. Anderson, E. Lilly, and C. Hok. 1988. Implementation of CALINE4. Institute of Northern Engineering, School of Enginering, University of Alaska, Fairbanks. Prepared for Alaska Department of Transportation and Public Facilities, Report no. AK-RD-89-01.

Locklin, D. and H. Hazard. 1980. *Technology for the Development of High Efficiency Oil-Fired Residential Heating Equipment.* Brookhaven National Laboratory, Rpt. No. 51325.

Mazria, E. 1979. *The Passive Solar Energy Book.* Rodale Press, Emmans, PA, 1979.

Thomas, M.K. 1953. *Climatological Atlas of Canada* National Research Council, Ottawa, Ontario, 256 p.

Tobiasson, W. and Greatorex. 1996. *Ground Snow Loads for the USA.* U.S. Army Cold Regions Research and Engineering Laboratory, Hanover, New Hampshire.

APPENDIX A

PIPING OPTIONS

3rd Edition Steering Committee Coordinator

James A. Crum

Special Contributions

Mark Stafford

Daniel Schubert

Appendix A   Table of Contents

Appendix A List Of Figures

Appendix A   List of Tables

A   PIPING OPTIONS

A.1   Introduction

In cold regions, pipe materials are subjected to extreme temperature fluctuations (as much as 70°C), corrosive natural conditions, poor foundation materials, (particularly in thermally sensitive soils), and rough handling during shipping and installation. Buried pipe systems have different requirements from above-ground systems. The application factors at each location should be assessed before selection of pipe materials. The selection of a piping system for any water or sewer utility requires evaluation of the following criteria:

- **Service Suitability:** physical, chemical, biological, electrical, and thermal characteristics of the internal and external environment, internal and external structural loadings; and

- **Economics:** initial capital costs, fabrication or manufacturing lead time, transportation costs, installation complexity (number of different installation skills involved), and operation and maintenance costs (including availability of replacement parts).

The variety of piping materials is extensive. The most common materials are:

- **Metals:** ductile and cast iron, brass, steel alloys, copper, and aluminum;

- **Plastics:** polyethylene (PE), polyvinyl chloride (PVC), polybutylene (PB), polypropylene (PP), acrylonitrile-butadiene-styrene (ABS), teflon, nylon, etc.;

- **Mineral-Based:** prestressed concrete, reinforced concrete, and vitrified clay;

- **Composites:** wound and wrapped fiberglass, and asbestos cement (AC).

Typical properties of these materials are shown in Tables A-1 and A-2. More information about these pipes, their system configurations and costs are covered in detail by many publications: manufacturers' literature, trade association handbooks, industry standards and various textbooks and professional journals.

A.2   Plastic Pipes

Plastic pipe is relatively new compared to pipes of other materials. Some types of plastics are more ductile at low temperatures. Plastic pipes do not corrode at temperatures encountered in water and sewer lines, but some types deteriorate when exposed to sunlight. Some have a large coefficient of expansion which definitely must be taken into account in design. Some types become weak at high temperatures so steam or water over 60°C must not be used to thaw plastic pipes. Plastic pipe is lightweight, about 18 kg per length of 100 mm nominal diameter. Because it is relatively smooth, it has head losses lower than most other pipe materials. The Hazen-Williams coefficient, C, is 150 for plastic pipes. The properties of different plastics are given in Tables A-1 and A-2.

A.2.1   Polyvinyl Chloride (PVC).   Polyvinyl chloride Type I has relatively good characteristics at temperatures used in water supplies. It has lower impact resistance but higher chemical resistance and pressure ratings than PVC Type II or polyethylene (PE) at a given temperature. Exposure to sunlight and cold conditions can make PVC pipe brittle over time. PVC piping systems can be installed by inexperienced labor in the field. Type II has higher impact resistance but a lower pressure rating for the same dimensions. Also, PVC Type II has a higher minimum temperature rating, i.e., -18°C.

PVC is unaffected by corrosive soils, most fluids, sea water, and oils at temperatures below 21°C. Pressure ratings and abrasive resistance drop off quickly at higher temperatures. The standard pipe dimension ratio (SDR) is the ratio of pipe diameter to wall thickness. Seven standard values are 15.5, 17, 21, 26, 32.5, 41, and 64.

Hydrostatic design stress (HDS) is the maximum tensile stress in the wall of the pipe due to the hydrostatic pressure inside. PVC pipe is rated according to the following equation:

$$\frac{2(HDS)}{P} = (SDR) - 1$$

where,

P   =   pressure rating of pipe, kPa

HDS   =   hydrostatic design stress, kPa

SDR   =   standard dimension ratio.

The most common PVC pipe materials are:

- Type I, Grade HDS = 13,800 kPa @ 23°C (No. PVC 1120),

TABLE A-1   TYPICAL PROPERTIES OF PIPE MATERIALS

Material	Density, kg/m³	Coefficient of Thermal Expansion, 10⁻⁶/°C	Thermal Conductivity, W/(m•°C)	Tensile Strength, MPa (ASTM D 638)	Modulus of Elasticity, GPa (ASTM D 790)
Plastics:					
Thermoplastics†	900 to 1,800	50 to 180	0.14 to -0.5	21 to 50	1.0 to 3.1
Thermosets†	1,300 to 2,000	22 to 31	0.19 to 0.26	62	0.90 to 19
Metals:					
Aluminum	2,700	23	220	76	69
Brass	8,500	19	120	500	100
Copper	8,750	17	390	220	120
Ferrous Metals**	7,200 to 78,500	12 to 16	3 to 52	160 to 1,400	90 to 270
Concrete	1,800 to 2,500	11	–	1.4 to 3.5	14 to 34

*   Values given are approximate and should be considered only as a guide.
** This group of materials comprises cast iron and a variety of steels
†   Thermoplastics can be reversibly softened by heat and hardened by cooling. Conversely, once manufactured, thermosets char upon heating; they undergo largely chemical rather than physical changes.

TABLE A-2   TYPICAL PROPERTIES OF SOME PLASTIC PIPE MATERIALS*

Type of Plastic	Density, kg/m^3 (ASTM D 792)	Coefficient of Thermal Expansion, 10^{-6}/°C (ASTM D 696)	Thermal Conductivity, W/(m·°C) (ASTM C 177)	Tensile Strength, MPa (ASTM D 638)	Compressive Strength, MPa (ASTM D 695)	Flexural Strength, MPa (ASTM D 790)	Modulus of Elasticity, GPa (ASTM D 638)
Polyvinyl chloride (PVC)	1,380	50	0.16	48.3	62.2	99.8	3.1
CPVC	1,540	79	0.14	50.3	106.9	99.8	2.5
Polyethylene PE (UHMW)	950	149	0.50	23.4	–	19.3	0.48
Polyethylene (PE)**	920 to 950	130 to 180	0.33 to 0.50	12.0 to 19.3	–	11.7 to 13.8	1.4 to 10
Acrylonitrile butadiene styrene (ABS)	1,040	101	0.20	37.9	53.1	68.9	2.1
Polypropylene (PP)	910	68	0.19	33.8	58.6	58.6	1.0

* These data are representative values; pipe materials differ in properties depending on formulation and manufacturing process.
** Low, medium and high density (Type I, Type II and Type III).

- Type I, Grade HDS = 13,800 kPa @ 23˚C (No. PVC 1220),

- Type II, Grade HDS = 6,900 kPa @ 23˚C (No. PVC 2110),

- Type IV, Grade HDS = 11,000 kPa @ 23˚C (No. PVC 4116).

PVC is also available in Schedules 40, 80, and 120, which roughly corresponds to iron pipe sizes. This sizing procedure results in small-diameter pipe having a much higher pressure rating than a larger pipe in the same schedule. Both of these general types (SDR or schedules) are designed for pressure applications. PVC pipe designed specifically for use in gravity sewers is thin walled. Experience has shown thin-walled pipe to be more breakable and should be used with caution in cold regions.

PVC pipe is covered by the following American Society for Testing and Materials (ASTM) specifications:

- PVC pipe and fittings (SDR classes) in D-2241;

- PVC pipe and fittings (Sch. 40, 80, 120) in D-1785;

- PVC sewer pipe and fittings in D-3034.

A.2.2 Acrylonitrile-Butadiene-Styrene (ABS).
Acrylonitrile-butadiene-styrene has also been used for sewer mains, service lines, and drainfield installations. This pipe is not as readily available as PVC in some sizes. In general, it has a higher impact strength than PVC but requires a thicker wall to be equivalent to PVC in pressure rating. Ratings and nomenclature for different types of ABS pipe are basically the same as those for PVC.

ASTM Standard specifications covering ABS pipe are as follows:

- ABS pipe and fittings (SDR classes) in D-2282;

- ABS pipe and fittings (Sch. 40, 80, 120) in D-1527;

- ABS sewer pipe and fittings in D-2751.

Recommended design information is also available from manufacturers and should be followed closely. Advantages, disadvantages, and recommended joints are basically the same as those for PVC pipe. ABS pipe is used quite extensively for building plumbing. It is more susceptible to damage from sunlight and the atmosphere (ozone) than PVC.

A.2.3 Polyethylene (PE).
Polyethylene is flexible and impact resistant even at low temperatures, particularly for materials of higher molecular weight. It is available in various molecular weights.

Pre-insulated PE pipe has become quite popular in cold regions. One significant advantage of high-density PE (HDPE) is that the water in the pipe can go through several freeze cycles without causing damage to the pipe.

It can be joined with compression-type fittings or clamps, or it can preferably be butt fused (essentially welded) together. Recent experience in electrofusion couplings has been good and they are suitable for tight areas such as joints made in an excavated trench. Machined bell connectors have been successfully used for PE sewer main joints.

By itself, it is not rigid enough under most conditions to hold proper grades in an above-ground gravity sewer line. It does gain rigidity when it is covered with urethane insulation and an outer covering of PE or metal, either wrapped or corrugated.

A.3 Metal Pipes

The greatest disadvantages of metal pipes are probably the lack of corrosion resistance and the weight of the pipe. Sewer pipes are usually epoxy or cement lined and also coated on the outside. The advantages of metal pipes include high strength and rigidity. There is danger of splitting metal pipes when their liquid content freezes solid. Bedding does not have to be as carefully prepared as with other types and they can tolerate some movement. Metal water lines can be thawed electrically.

Some of the characteristics of various metal pipes are presented below.

A.3.1 Cast Iron.

- A 6 m section of 100 mm pipe weighs about 227 kg.

- It cracks or breaks if mishandled (dropped, etc.).

- Hazen-Williams roughness factor: C ≈ 125.

- Corrosion resistance with concrete lining and tar coating is excellent.

The *Handbook of Cast Iron Pipe*, published by the Cast Iron Pipe Research Association contains detailed information on cast iron pipe.

A.3.2 Ductile Iron.

- It is a little lighter than cast iron.

- Hazen-Williams roughness factor: $C \approx 125$.
- It is resistant to breaking and bending damage.
- Corrosion resistance with concrete lining and tar coating is very good.

It is similar to cast iron pipe, but the metallurgy has been changed to make it more ductile and flexible. The pipes are covered by the following ASTM standard specifications:

- cast iron and ductile iron pressure pipe - in A-337;
- cast iron soil pipe and fittings - in A-74.

American National Standards Institute specifications for ductile iron pipe are A21.51 for the pipe and material, A21.6 for the pipe coating, A21.4 for the cement lining, and A21.11 for Tyton-type gaskets.

Although other types of joints are available, bell and spigot joints with rubber gaskets are generally the best choices for cast and ductile iron.

A.3.3 Steel.

- It is less corrosion resistant than either ductile or cast iron.
- It is somewhat lighter than either ductile or cast iron, and more flexible.
- It can be welded.
- Hazen-Williams roughness factor: $C \approx 125$.

It should be galvanized and lined with cement or epoxy linings to be used for water services.

A.3.4 Copper. According to observations in the field, buried water service lines of Type K drawn copper (19 to 25 mm in diameter) when full of water, can be frozen several times without bursting. This is probably due to the fact that only a single ice plug has formed. Excess pressure buildup as the volume of ice expands is relieved by transmission through the remaining liquid to the distribution system or house plumbing. However, frost penetration is uneven, as is the grade of such lines, and intolerable pressures are possible if growing ice plugs bracket an unfrozen section. Still, the ductility of copper and its tremendous internal working pressures in service line sizes (9.07 MPa at 65°C for 19-mm diameter Type K (9) prevents bursting in many cases, if the rate of freezing is slow.

A.4 Asbestos-Cement and Concrete Pipes

Asbestos-cement (AC) and concrete pipes are brittle and should not be used where any movement can occur. Damage can also occur during shipping. They are only available in short lengths and they weigh about 8.6 kg/m for 100 mm pipe. There are few uses for these types of pipes for small-diameter pipe lines in cold regions.

A.5 Wood Stave Pipes

An advantage of wood stave pipe in cold regions is that it can usually take freeze-thaw cycles with little damage. Also, the thick wood walls offer a degree of insulation.

It is relatively corrosion-free except for the spirally wound wire on the outside, which must be coated if it is exposed to corrosive conditions. It is relatively expensive and is only available in short lengths. Woodstave pipes are not used significantly in cold regions.

A.6 Pre-Insulated Pipe

Pre-insulated pipe is a three-component fluid transport system consisting of:

- an inner-core pipe;
- insulation; and
- an outer jacket.

A fourth component, a freeze protection or thawing system, is often included. Figure A-1 shows schematically some of the more frequently used pre-insulated pipe configurations.

Figures A-2 through A-11 describe in detail one kind of pre-insulated piping system: pre-insulated polyethylene for water distribution and transmission.

Pre-insulated pipe is usually shallow buried (3 to 5 outside diameters to the invert) but has been laid directly on the ground surface (especially for temporary facilities) or supported to grade above ground on piles and bracing. Up to 75 percent of the initial cost of a buried pre-insulated pipeline in the Arctic is for on-site installation including excavation, shoring, drainage, bedding, laying, joining, and backfill.

A.6.1 Inner-Core Pipe. Virtually every kind (and diameter) of pipe can and has been used in the Arctic as the inner-core fluid carrier in pre-insulated piping. Usually, more than one inner-core pipe is specified (e.g., for heat tracing or for electrical control cables), and fabrication costs for such configurations are higher.

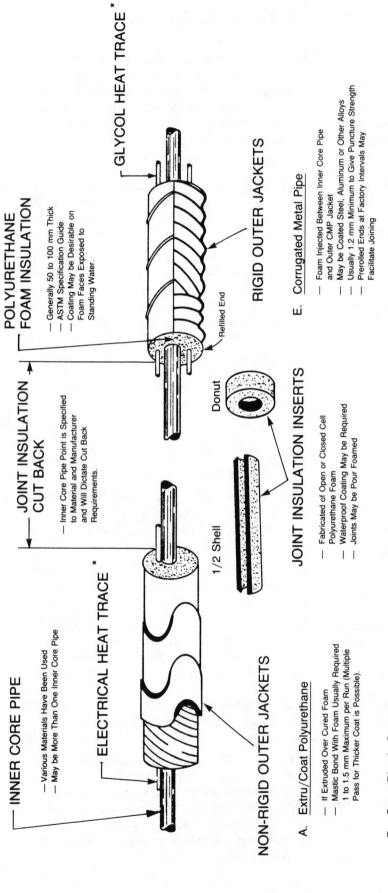

INNER CORE PIPE
— Various Materials Have Been Used
— May be More Than One Inner Core Pipe

ELECTRICAL HEAT TRACE *

NON-RIGID OUTER JACKETS

A. **Extru/Coat Polyurethane**
— If Extruded Over Cured Foam
— Mastic Bond With Foam Usually Required
— 1 to 1.5 mm Maximum per Run (Multiple Pass for Thicker Coat is Possible).

B. **Outer Plastic Carcass**
— Foam Injected Between Inner Core Pipe and Outer Carcass.
— 3.8 mm Polyethylene is Typical but Polyvinyl Chloride has Been Used Also.
— Inner Surface Preparation May be Required for Jacket Foam Bond.

C. **Polyurethane Elastomers**
— Applied Over Cured Foam
— May be Field Applied

D. **P. E. Tape Wrap**
— Usually Applied Hot for Shrink Fit

JOINT INSULATION CUT BACK
— Inner Core Pipe Point is Specified to Material and Manufacturer and Will Dictate Cut Back Requirements.

1/2 Shell

Donut

JOINT INSULATION INSERTS
— Fabricated of Open or Closed Cell Polyurethane Foam
— Waterproof Coating May be Required
— Joints May be Pour Foamed

OUTER JACKET JOINT COUPLINGS

Heat Shrink Sleeves
— Smooth Outer Jacket Required for Water Tight Joint

Metal Bands
— Moment Transfer Capabilities
— May be Difficult to Achieve Water Tightness

POLYURETHANE FOAM INSULATION
— Generally 50 to 100 mm Thick
— ASTM Specification Guide
— Coating May be Desirable on Foam Faces Exposed to Standing Water.

GLYCOL HEAT TRACE *

Refilled End

RIGID OUTER JACKETS

E. **Corrugated Metal Pipe**
— Foam Injected Between Inner Core Pipe and Outer CMP Jacket
— May be Coated Steel, Aluminum or Other Alloys
— Usually 1.2 mm Minimum to Give Puncture Strength
— Prerolled Ends at Factory Intervals May Facilitate Joining

F. **Spiral Wound Pipe**
— Can be Wound Over Cured Foam
— May be Coated Steel, Aluminum or Other Alloys
— Usually 1.2 mm Minimum to Give Puncture Strength

*(Optional, See Appendix G)

FIGURE A-1(a) PRE-INSULATED PIPE — COMMON SYSTEM CONFIGURATION

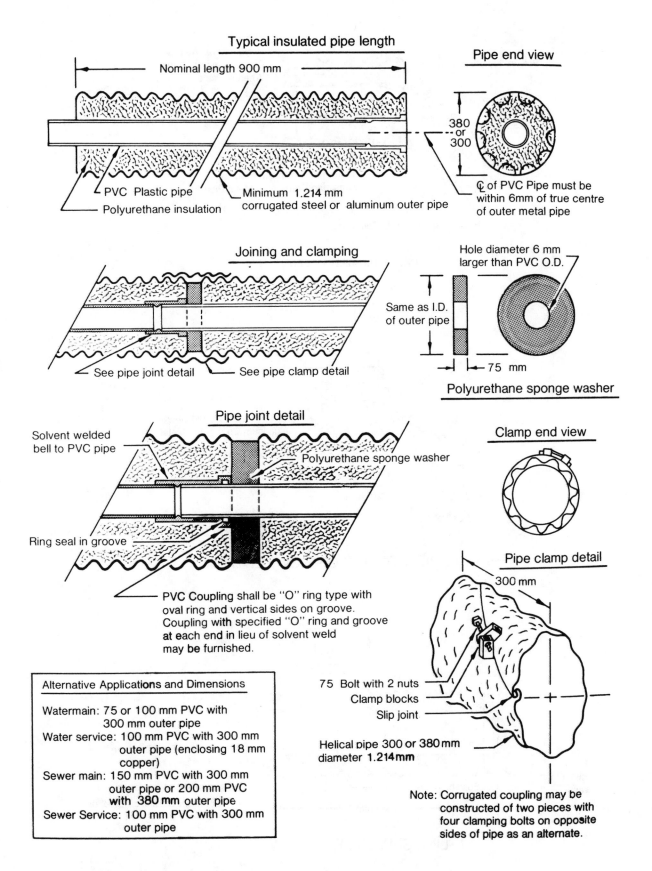

Typical insulated pipe length

Nominal length 900 mm

Pipe end view

380 or 300

PVC Plastic pipe

Polyurethane insulation

Minimum 1.214 mm corrugated steel or aluminum outer pipe

℄ of PVC Pipe must be within 6mm of true centre of outer metal pipe

Joining and clamping

See pipe joint detail

See pipe clamp detail

Hole diameter 6 mm larger than PVC O.D.

Same as I.D. of outer pipe

75 mm

Polyurethane sponge washer

Pipe joint detail

Solvent welded bell to PVC pipe

Polyurethane sponge washer

Ring seal in groove

PVC Coupling shall be "O" ring type with oval ring and vertical sides on groove. Coupling with specified "O" ring and groove at each end in lieu of solvent weld may be furnished.

Clamp end view

Pipe clamp detail

300 mm

Alternative Applications and Dimensions

Watermain: 75 or 100 mm PVC with 300 mm outer pipe
Water service: 100 mm PVC with 300 mm outer pipe (enclosing 18 mm copper)
Sewer main: 150 mm PVC with 300 mm outer pipe or 200 mm PVC with 380 mm outer pipe
Sewer Service: 100 mm PVC with 300 mm outer pipe

75 Bolt with 2 nuts
Clamp blocks
Slip joint

Helical pipe 300 or 380 mm diameter 1.214 mm

Note: Corrugated coupling may be constructed of two pieces with four clamping bolts on opposite sides of pipe as an alternate.

FIGURE A-1(b) PRE-INSULATED PIPE COMMONLY USED IN ALASKA

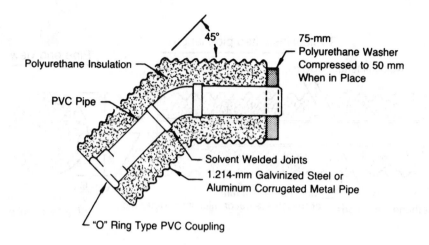

PREFABRICATED 45° BEND

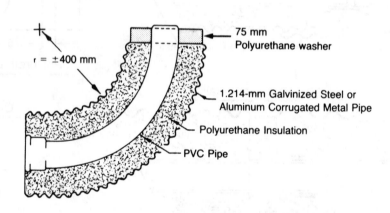

PREFABRICATED 90° LONG SWEEP BEND

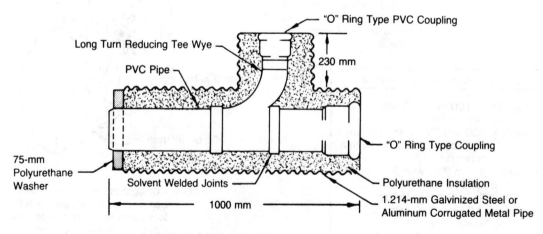

IN-LINE SEWER SERVICE WYE

FIGURE A-2   FITTINGS FOR INSULATED PIPES

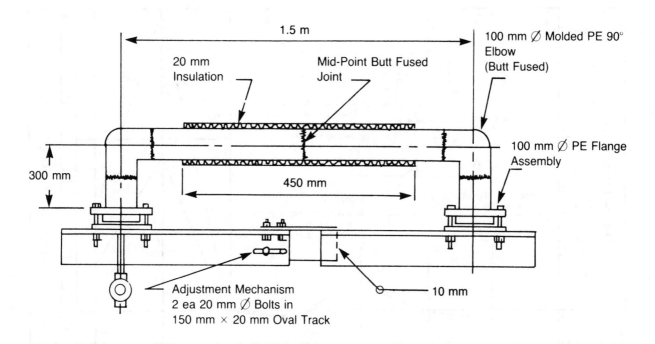

1.5 m

20 mm
Insulation

Mid-Point Butt Fused
Joint

100 mm ⌀ Molded PE 90°
Elbow
(Butt Fused)

100 mm ⌀ PE Flange
Assembly

300 mm

450 mm

Adjustment Mechanism
2 ea 20 mm ⌀ Bolts in
150 mm × 20 mm Oval Track

10 mm

TOP VIEW

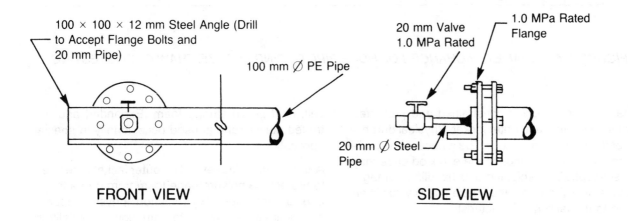

100 × 100 × 12 mm Steel Angle (Drill
to Accept Flange Bolts and
20 mm Pipe)

100 mm ⌀ PE Pipe

20 mm Valve
1.0 MPa Rated

1.0 MPa Rated
Flange

20 mm ⌀ Steel
Pipe

FRONT VIEW

SIDE VIEW

FIGURE A-3   ALASKAN INNER-CORE PIPE FREEZE/THAW TEST ASSEMBLY

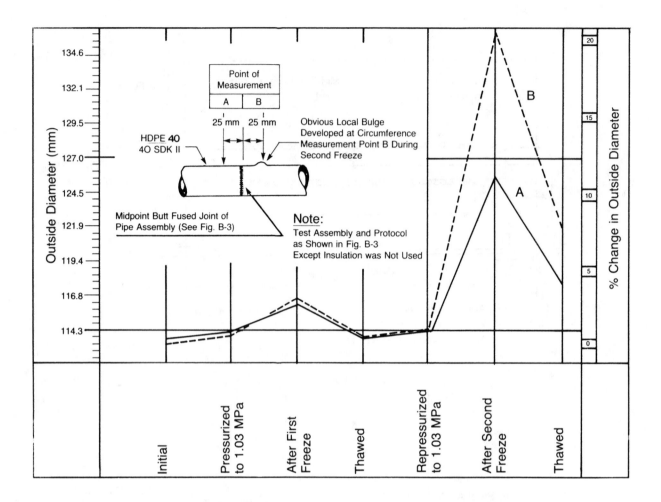

FIGURE A-4   DIAMETER CHANGES OF HDPE PIPE DURING FREEZE/THAW CYCLE TESTING

The inner-core pipe is constrained and protected by the insulation and outer jacket. Physical durability of the core piping may not be a necessary qualifying characteristic, though the exposed ends may be susceptible to shipping and handling damage. But the ability to withstand freeze-thaw cycles is an important qualifying characteristic.

A.6.2 Insulation.   In the past, insulation materials for piping systems used in the Arctic have included moss, sawdust, sand, mineral-wool asbestos), blown silica (foamed glass), and fiberglass batting. Most of these materials are not used today because of cost, fabrication practicality, field construction costs, and waterlogging problems. Currently, the thermal insulation in pre-insulated piping systems is almost exclusively low-density, closed-

cell, rigid plolyurethane foam. Expanded and extruded polystyrene is used occasionally in special applications.

A.6.3 Outer Jacket.   The outer jackets are used to protect the polyurethane foam from degradation due to ultraviolet (UV) radiation, physical abuse, chemical attack, and water intrusion, especially in conjunction with freeze/thaw cycles. Also, imperviousness to the gases found in the cells of some polyurethane foams prevents their diffusion and replacement with air and moisture which decreases thermal effectiveness. The outer jacket can also be instrumental in providing beam stiffness and shell rigidity for the pre-insulated pipe composite.

There are two main types of outer jackets used in pre-insulated pipe systems: flexible and rigid. Sev-

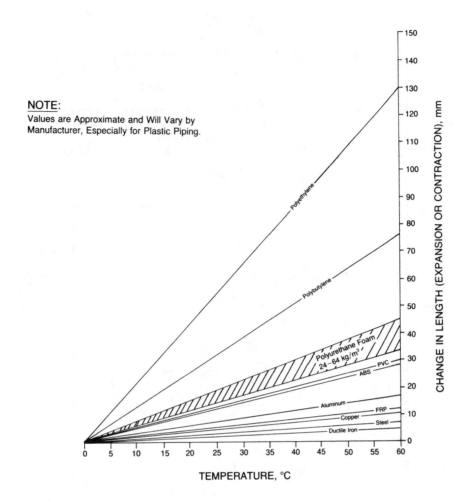

NOTE:
Values are Approximate and Will Vary by
Manufacturer, Especially for Plastic Piping.

FIGURE A-5   THERMAL EXPANSION AND CONTRACTION OF
VARIOUS PIPE MATERIALS

eral options are available within each category. All dictate certain requirements and constraints on jacket joint configuration, available beam and circumferential strength, necessary protection during shipping, handling weight, in-trench and out-of-trench joining capabilities, and bending radius.

Nonrigid Jackets.   Two major categories of materials comprise the most commonly used outer skins in flexible pre-insulated piping: polyethylene and polyurethane elastomers. Urethane elastomer coatings are made by numerous manufacturers, and many coatings meet the basic jacketing requirements although some are sensitive to puncture. This must be considered when packing for shipment and when selecting bedding and backfill. Field application of polyurethane elastomers over field-insulated

pipe has been tried with success in Alaska; however, it is field labor-intensive and relatively expensive.

Flexible PE jackets occupy three subgroups:

- In cases where pipe foaming is done in special fabricating forms, the PE is extruded/coated directly onto the finished insulation. One pass thicknesses of jacket material are typical; several passes are possible. Current practice dictates an intermediate layer of mastic to promote adhesion to the foam, ensuring composite behavior. Such pipe is vulnerable to shipping damage, and manufacturers recommend heat-shrink sleeves, and tape to patch outer jacket tears and perforations.

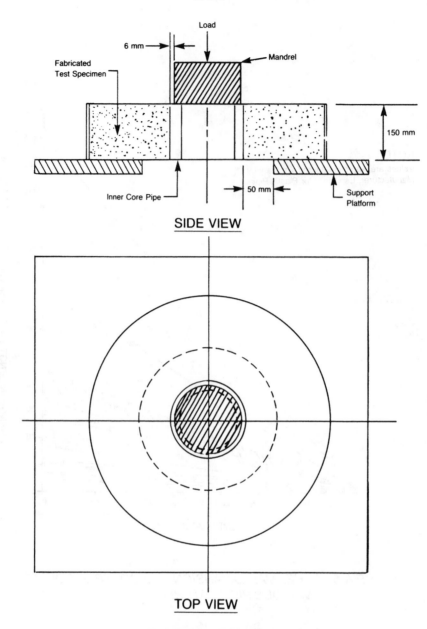

SIDE VIEW

TOP VIEW

TEST PROCEDURE

1. ONE TEST REQUIRED PER EACH 1200 m OF PIPE PRODUCED OR FRACTION THEREOF.
2. TEST TO BE CONDUCTED ON TWO SPECIMENS AT DIFFERENT TEMPERATURES:
   — ONE AT 21 ± 3°C
   — ONE IMMEDIATELY AFTER RETRIEVAL FROM A −54°C (MIN.) ENVIRONMENT. TEST SPECIMEN SHALL BE BROUGHT TO THIS LOWER TEMP. IN 4 HOURS OR LESS AND REMAIN THERE FOR AT LEAST 24 HOURS PRIOR TO TESTING.
3. WEIGHT SHALL BE APPLIED TO MANDREL NO SLOWER THAN 22 kg (MIN. UNTIL THE INNER CORE PE PIPE IS PUNCHED COMPLETELY THRU THE TEST SPECIMEN.
4. RECORD FINAL LOAD, VISUALLY INSPECT INNER CORE PE PIPE AND SAVE

RESULTS:

THESE SHALL INCLUDE THE ULTIMATE LOAD, PHOTOS, AND RECORD OF VISUAL INSPECTION, SUCCESSFUL TEST SPECIMENS SHALL EXHIBIT FOAM TO FOAM (OR FOAM TO PIPE FOAM SURFACE FILM) SEPARATION OR TEARING LEAVING NO INNER CORE PE PIPE SURFACES BARE.

FIGURE A-6   ALASKAN INNER-CORE PIPE/INSULATION FOAM BOND TEST APPARATUS

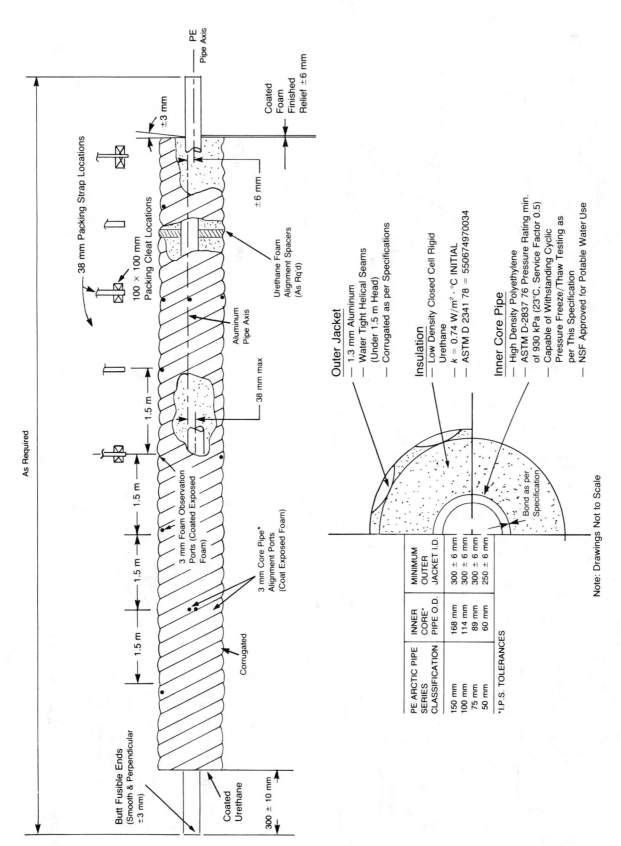

FIGURE A-7   INSULATED PIPE EXAMINATION

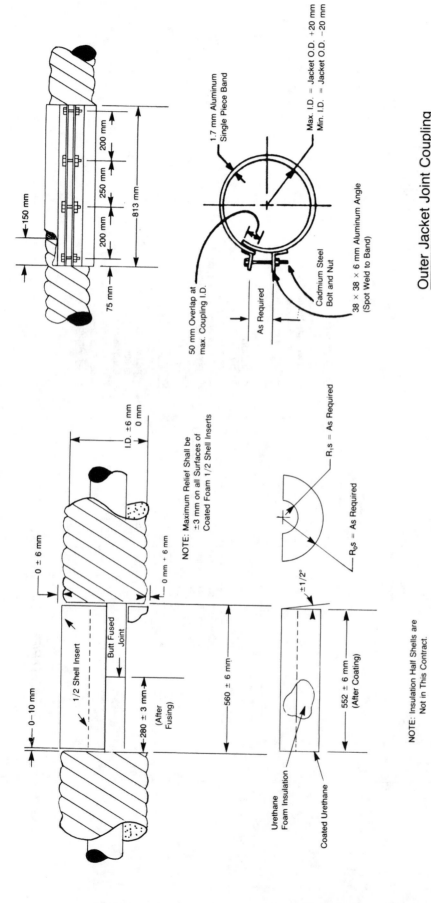

Outer Jacket Joint Coupling

1.7 mm Aluminum Single Piece Band

Max. I.D. = Jacket O.D. +20 mm
Min. I.D. = Jacket O.D. −20 mm

38 × 38 × 6 mm Aluminum Angle (Spot Weld to Band)

Cadmium Steel Bolt and Nut

As Required

50 mm Overlap at max. Coupling I.D.

NOTE: Drawings Not to Scale

I.D. ±6 mm
0 mm

NOTE: Maximum Relief Shall be ±3 mm on all Surfaces of Coated Foam 1/2 Shell Inserts

0 ± 6 mm

0 mm + 6 mm

R₁s = As Required

R₀s = As Required

±1/2°

1/2 Shell Insert

Butt Fused Joint

0–10 mm

280 ± 3 mm
(After Fusing)

560 ± 6 mm

552 ± 6 mm
(After Coating)

Urethane Foam Insulation

Coated Urethane

NOTE: Insulation Half Shells are Not in This Contract.

1/2 Shell Joint Insulation Inserts

150 mm

200 mm

250 mm

200 mm

813 mm

75 mm

Insulation

— Low Density Rigid Closed Cell Urethane
— k = 0.74 W/m² · °C INITIAL
— 241 kPa Compressive Strength min.
— ASTM D 2341 78 = 550675970034

Coating

— Material Characteristics as per this Specification

FIGURE A-8   INSULATED PIPE JOINTS

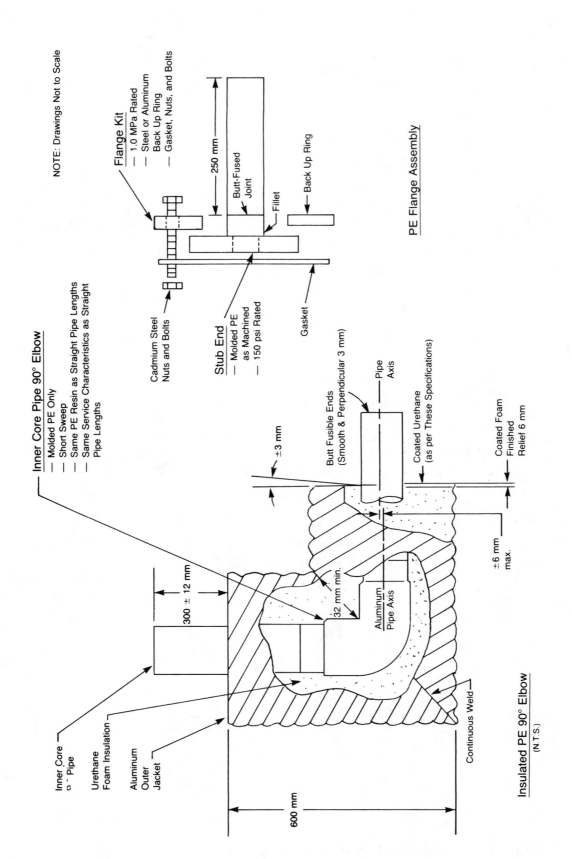

NOTE: Drawings Not to Scale

Flange Kit
— 1.0 MPa Rated
— Steel or Aluminum Back Up Ring
— Gasket, Nuts, and Bolts

250 mm

Butt-Fused Joint

Fillet

Back Up Ring

Cadmium Steel Nuts and Bolts

Stub End
— Molded PE as Machined
— 150 psi Rated

Gasket

PE Flange Assembly

Inner Core Pipe 90° Elbow
— Molded PE Only
— Short Sweep
— Same PE Resin as Straight Pipe Lengths
— Same Service Characteristics as Straight Pipe Lengths

Inner Core 6" Pipe

Urethane Foam Insulation

Aluminum Outer Jacket

300 ± 12 mm

32 mm min.

Aluminum Pipe Axis

Continuous Weld

600 mm

± 3 mm

Pipe Axis

Butt Fusible Ends (Smooth & Perpendicular 3 mm)

Coated Urethane (as per These Specifications)

Coated Foam Finished Relief 6 mm

± 6 mm max.

Insulated PE 90° Elbow
(N.T.S.)

FIGURE A-9   INSULATED PIPE 90° ELBOW

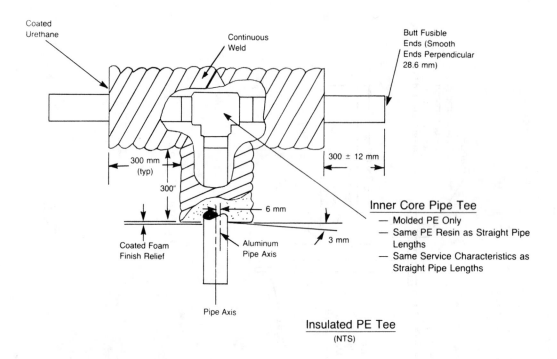

Coated
Urethane

Continuous
Weld

Butt Fusible
Ends (Smooth
Ends Perpendicular
28.6 mm)

300 mm
(typ)

300"

6 mm

Coated Foam
Finish Relief

Aluminum
Pipe Axis

3 mm

300 ± 12 mm

Inner Core Pipe Tee
— Molded PE Only
— Same PE Resin as Straight Pipe
  Lengths
— Same Service Characteristics as
  Straight Pipe Lengths

Pipe Axis

Insulated PE Tee
(NTS)

Inner Core Pipe 45° Elbow
— Same PE Resin as Straight Pipe Lengths
— Same Service Characteristics as Straight
  Pipe Lengths
— Molded or Fabricated
— If Fabricated:
  1. Use the Next Smallest SDR as the straight lengths
  2. Fabricated Elbow to be Wrapped in Fibreglass to a min. Thickness of 6 mm
  3. Fibreglass Shall Extend a min. of 1/2 Pipe O.D. Past Last Fabricated Butt Fused
     Joint

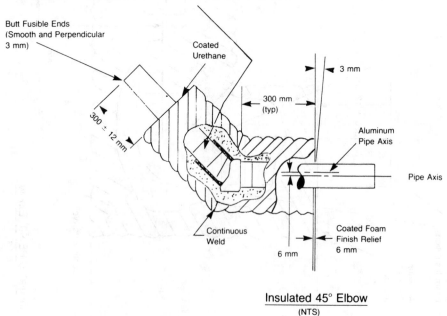

Butt Fusible Ends
(Smooth and Perpendicular
3 mm)

Coated
Urethane

3 mm

300 ± 12 mm

300 mm
(typ)

Aluminum
Pipe Axis

Pipe Axis

Continuous
Weld

Coated Foam
Finish Relief
6 mm

6 mm

Insulated 45° Elbow
(NTS)

FIGURE A-10   INSULATED PIPE 45° ELBOW AND TEE

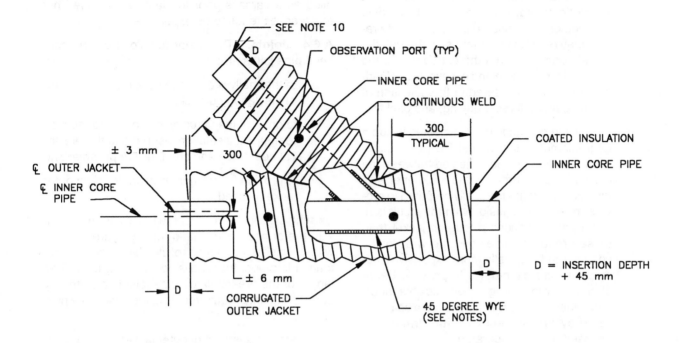

NOTES:

1. DRAWING NOT TO SCALE
2. SEE SPECIFICATIONS FOR ADDITIONAL REQUIREMENTS
3. WYE SHALL BE OF SAME RESIN AS INNER CORE PIPE
4. WYE MAY BE INJECTION MOLDED OR FABRICATED
5. INJECTION MOLDED WYES SHALL BE 1100 kPa RATED
6. FABRICATED WYES SHALL COMPLY WITH THE FOLLOWING:
    - USE SDR11 INNER CORE PIPE
    - ENCASE ENTIRE WYE IN FIBERGLASS (6 mm THICKNESS)
    - INTERIOR OF ALL INNER CORE PIPE SHALL BE SMOOTH
      WITH NO LIPS AT RUN/LEG AND BUTT FUSION JOINTS
8. OPENNING AT THE 45 DEGREE SHALL BE SMOOTH AND UNIFORM
    WITH NO LIP LONGER THAN 5 mm.
9. PIPE SPIGOT END SHALL BE BEVELED PER SPECIFICATION.
10. CLEANOUT LEG OF WYE TO BE 100 mm DIAMETER UNLESS OUTLINED WITHIN BID SCHEDULE.

FIGURE A-11   INSULATED PIPE 45° WYE

- Spiral PE tape wrap is also commercially available. Generally, the PE tape is applied hot in two counter-wound and overlapping layers with integral adhesive to ensure a shrink-tightened, waterproof bond. Bending this kind of pre-insulated pipe in an acute radius may result in water intrusion through the jacket seams. As with the extruded PE, the potential for puncture and excessive soil loadings must be considered in the selection of shipping procedures, bedding, and backfill.

- Finally, the core pipe may be nested inside a large-diameter, thin-walled PE pipe or carcass. The foam is then injected between the core pipe and carcass. The slick inner surface of the carcass is usually prepared to provide maximum adhesion with the insulation. The system provides a tougher jacket than those produced either by tape wrap or by coating (and is usually more expensive), but shipping crates must still provide beam strength for long lengths with flexible inner-core pipes. PVC outer jackets are also made but they are subject to brittle fracture if stressed at low temperatures.

Rigid Jackets. While usually more expensive than the flexible PE-jacket system, there are inherent advantages in a rigid outer pipe. The obvious ones are: less potential for shipping damage and beam strength for maintenance of grade. Rigid outer jackets allow pre-insulated pipe to be installed in poor soils (low bearing strength, thaw instability, etc.) and in areas with marginal bedding materials.

There are two basic types of rigid outer casings: metal pipe culverts (slid over the core pipe prior to foaming), and spirally wound metal or fiberglass jackets (which may be wrapped on the finished foamed product).

Corrugated aluminum culvert has been the most frequently used rigid outer jacket in Alaska. Its cost and relative immunity to corrosion have been influencing factors, although galvanized steel and other metal alloy corrugated pipe (CMP) and large-diameter tubing are now being used.

Spirally wound metal jackets applied after foaming lack corrugations and thus fall between the CMP and flexible outer jackets in circumferential crushing strength. The inside face of the metal can be knurled to form a mechanical bond with the foam. The typical thickness of these metal jackets is 1.2 mm. Thinner skins may be more susceptible to puncture and buckling-type failures.

Wound fiberglass jackets (which may also be supplied as a carcass prior to foaming) are used but this process tends to be expensive.

A.6.4 Joints. The major functions of pre-insulated pipe joints are:

- to thermally protect the core pipe in the area of insulation cutback;

- to prevent the intrusion of water into the pipe foam, joint insulation inserts, and the electrical heat trace and channel if applicable; and

- to provide for moment transfer if a rigid outer jacket is used.

Existing practice is to cover the inner-core pipe joint with preformed half-shells of rigid polyurethane or with a washer or "donut" of flexible polyurethane foam. Field-pour foaming is occasionally used. The length of the pre-insulated pipe joint is dictated by the working area requirements of the inner-core pipe joint.

There are two systems of outer jacket joints or couplings: rigid and flexible. The flexible category consists of PE heat-shrink sleeves and occasionally, heat-shrink PE tape.

Heat-shrink sleeves require smooth surfaces to achieve their unique mechanical bond. For water tightness, they come highly recommended for smooth jackets. They have difficulties conforming to the acute helical spirals of CMP pipe, however, and do not allow moment transfer.

Rigid jacket joints consist of metal clamp and band assemblies. Depending on the configuration, they are not always watertight. For this reason, if the application subjects the pre-insulated pipe to a wet or submerged environment, coating the exposed foam faces at the pipe ends and the insulation inserts may be required.

Some Alaskan designs are proposing a more rigid outer PE exterior pipe with an electrofusion joining process for both the inner and outer pipes. Insulation is foamed in-situ. This process may result in a stronger watertight system, but very little practical experience has been developed on this technique.

A.7 Bibliography

Barber, D. A. 1990. Unpublished report on *Arctic Pipe Deflection / Pile Spacing*. U.S. Public Health Service, Anchorage, Alaska.

Blaga, A. 1981. *Use of Plastics as Piping Materials.* Canadian Building Digest, National Research Council of Canada, Ottawa, Ontario.

Blaga, A. 1981. *Thermoplastic Pipe.* Canadian Building Digest, National Research Council of Canada, Ottawa, Ontario.

Copper Development Association, Inc. *Copper Tube Handbook.*

Dupont Testing Laboratories Dupont, Inc. 1979. *Properties of Rigid Urethane Foams,* Mississauga, Ontario.

Gerlek, S. 1981. *Pre-Insulated Piping Systems.* Unpublished report. U.S. Public Health Service, Anchorage, Alaska.

Kaplar, C.W. 1969. *Studies for Rigid Thermal Insulation (Draft Report).* U.S. Army Cold Regions Research and Engineering Laboratory, Hanover, New Hampshire.

Lambacher, H. 1973. Written Communication of test results on 60-inch 60 psi HDPE Sclair pipe. Dupont Canada, Pipe Division, Customer Technical Center, Pipe Testing Laboratory, Kingston, Canada.

Laurin, D.E. 1979. *Expansion/Friction Analysis.* Nova-Tech Engineering Inc., Edmonds, Washington.

Mace, W.O. 1982. Personal communication. U.S. Public Health Service, Anchorage, Alaska.

Smith, W.H. 1976. *Frost Loadings on Underground Pipe.* Water Technology and Distribution.

APPENDIX B

VEHICLE-HAUL SYSTEM TIME AND COSTING EQUATIONS

3rd Edition Steering Committee Coordinator

D.W. Smith

3rd Edition Principal Author

J.J. Cameron

Appendix B   Table of Contents

Appendix C   List Of Figures

APPENDIX B VEHICLE-HAUL SYSTEM TIME AND COSTING EQUATIONS

B.1 Vehicle-Haul System Model

This appendix presents a mathematical model to estimate the resources and cost for vehicle haul systems. The model is based on the procedures and equations developed by the Department of Local Government, Government of the NWT (1992), based on work by Gamble and Janssen (1974). Guidelines for trucked service contracts were prepared by Cameron (1981).

The model is applicable to vehicle water delivery, sewage pumpout, honey-bag collection and garbage collection systems. These vehicle-haul systems are basically the same and can be analyzed by using the same general model and equations. To illustrate the model, a brief description of the water delivery procedure follows.

The vehicle must fill up its tank at the water source, travel to the community, deliver water from building to building by filling the individual building tanks until the vehicle tank is empty, and then return to the water source. This process is repeated when the water tanks in the buildings have been used or on a fixed schedule.

The time required to perform each task can be represented by equations which quantify the characteristics of the system. For example, beginning at the source, the time to fill the vehicle tank can be expressed as the time for the pump to fill the truck tank (i.e., the tank size divided by the pumping rate) plus the time required to start the pump, and hook and unhook the hoses. The travel time between the water source and the community is the distance divided by the vehicle speed. The time to service a building is the time for the pump to fill the building tank (i.e., the tank size divided by the pumping rate) plus the time required to start the pump, hook and unhook the hoses and record the volume delivered.

Once the total annual time required to service the buildings is calculated, the resources (e.g., labor, vehicles, fuel, garages) and the cost to deliver the service can be determined using values for performance and cost.

The same steps are required for the sewage pumpout, honey-bag and garbage collection systems, although the system parameters such as tank size are different.

B.1.1 General Equations.

The following system of equations applies to a particular year under consideration, and for a particular set of community, building and vehicle characteristics. The costs are based on equivalent annual costs (i.e., all capital costs are annualized). The following circuit time is based on servicing a group of buildings that have the same characteristics (tank size, water use, frequency of service, etc.). The total time required is the sum of the time required to service each of the distinct building types in the community.

Circuit Time

$$CT = CST + CSTT + CBT$$

where:

CT = circuit time: time to service a defined group of buildings once (hours)

CST = circuit source time: time at the source or disposal point to turn around, hook/disconnect, pump (hours)

CSTT = circuit travel time: time to travel between the community and the source/disposal point (hours)

CBT = circuit building time: time to service the buildings (hours)

<u>Calculating Circuit Source Time (CST)</u>

$$CST = EL \cdot NB \cdot \left[\frac{C \cdot CSF}{VS \cdot VUF} \right] \cdot \left[\frac{VS \cdot VUF}{60 \cdot R} + \frac{TT}{60} \right]$$

where:

EL = efficiency of labor. The effective hours of work are less than the hours available due to travel, breaks, and maintenance. An EL of 1.5 is a conservative value that should be easily obtained on average.

NB = the number of buildings in the defined group to be serviced (with similar characteristics)

C = container size for the buildings, e.g., water tank, garbage barrel (litres, m^3)

CSF = container safety factor, for under-utilization of the container volume and safety factor (usually taken to be 0.85)

VS = vehicle size, e.g., 4,500 litre water truck, 5 m³ garbage truck

VUF = vehicle use factor for under-utilization of the vehicle container volume (usually 0.95 for water trucks and 0.80 for garbage trucks)

R = rate of filling or emptying the vehicle container at the source or disposal point (litres per minute)

TT = time for turn around at the source or disposal point, i.e., time required to hook up, disconnect and turn around, exclusive of time emptying or filling the vehicle (minutes)

Calculating Circuit Source Travel Time (CSTT)

$$CSTT = EL \cdot NB \cdot \left(\frac{C \cdot CSF}{VS \cdot VUF} \right) \cdot \frac{2D}{S}$$

where:

D = travel distance to and from source or treatment/disposal point (kilometres)

S = speed of vehicle to and from source or treatment/disposal point (kilometres)

Calculating Circuit Building Time (CBT)

$$CBT = EL \cdot NB \left[\frac{DB}{1000 \cdot SB} + \frac{C \cdot CSF}{60 \cdot RB} + \frac{NTB \cdot TTB}{60} \right]$$

where:

DB = average distance between building (building hookups) (metres)

SB = average speed that the vehicles travel between the buildings (kilometres/hour)

RB = rate of filling or emptying the vehicle container while servicing the buildings in the community (litres per minute)

TTB = average turnaround time while servicing each building, excluding the filling or emptying time (minutes)

NTB = number of vehicle trips required to fill or empty the building container

$$= 1, \text{ if } \left[\frac{C \cdot CSF}{VS \cdot VUF} \right] \leq 1, \text{ i.e., the building container is smaller than the vehicle contai}$$

$$= \left[\frac{C \cdot CSF}{VS \cdot VUF} \right] \text{ if } \left[\frac{C \cdot CSF}{VS \cdot VUF} \right] > 1$$

Combining these equations, the circuit time equation becomes:

CT = CST + CSTT + CBT

$$= EL \cdot NB \cdot \left[\left(\frac{C \cdot CSF}{VS \cdot VUF} \right) \cdot \left(\frac{2D}{S} + \frac{VS \cdot VUF}{60 \cdot R} + \frac{TT}{60} \right) \right.$$

$$\left. + \frac{DB}{1000 \cdot SB} + \frac{C \cdot CSF}{60 \cdot RB} + \frac{TTB \cdot NTB}{60} \right]$$

Total Hours. The total hours per year required to service the buildings in the group is:

THRS = CT · F

where:

THRS = total hours per year required to provide the service to the group of buildings

F = frequency of service to the buildings (times per year)

Frequency. The frequency of service can be specified (e.g., once per week, 52 times per year) or service can be provided when required. For the latter, the servicing cycle must be repeated as soon as the building container has been emptied or filled by the user and for a given consumption or generation rate and building container size, the servicing frequency is:

$$F = \frac{building\ consumption/production\ per\ year}{building\ container\ size}$$

$$= \frac{VPBY}{C \cdot CSF}$$

For a residential building group the frequency could be expressed as:

$$F = \frac{NOPB \cdot VPCD \cdot 365}{C \cdot CSF}$$

where:

F = the number of visits required per year to adequately service each building

NOPB = average number of persons per building (real or equivalent),

VPBY = the volume of water consumed or waste generated per building per year (water = litres/d; garbage = m^3/d)

VPCD = the volume of water consumed or waste generated per occupant per day (water = litres/(p•d); garbage = m^3/(p•d))

For garbage and honey bag collection, frequency is usually fixed by policy and is not a variable.

If the frequency of service that is provided to the buildings is specified, the time equation can be re-written in terms of the specified frequency of service and the volume of service provided as follows:

$$THRS = EL \cdot V \cdot$$

$$\left[\frac{2 \cdot D}{S \cdot VS \cdot VUF} + \frac{TT}{60 \cdot VS \cdot VUF} + \frac{1}{60 \cdot R} + \frac{1}{60 \cdot RB} \right]$$

$$+ EL \cdot F \cdot NB \cdot \left[\frac{DB}{100 \cdot SB} + \frac{TTB}{60} \right]$$

where:

V = the total volume of service (i.e., consumption or production) per year for all buildings in the group

For residential buildings the volume can be expressed as:

V = NB • NOPB • VPCD • 365 d/yr

The above equations provide the total hours required to service a group of buildings with similar characteristics (e.g., water use, container size, frequency of service). The total hours to service all buildings in a community is the sum of the hours calculated to service each separate group of buildings in the community. Computer models can be written to loop through each building type. A quick estimate of the total hours can be calculated by assuming there is only one type of building which has "average" characteristics.

Number of Vehicles. The number of vehicles required is equal to the total hours per year required to service all the buildings divided by the maximum hours per year that each vehicle is available.

$$NV = \frac{THRS}{HPVPY}$$

where:

NV = exact number of vehicles required

NPVPY = hours available per vehicle per year

The available hours should exclude days that service is not provided and an estimate of down time for maintenance. For example, if a normal work day is 7.5 hours, no service is provided on Sunday and half a day is required for maintenance, then each vehicle is available for 2,145 hours per year (i.e., 7.5 hr/d • 5.5 d/wk • 52 wk/yr).

The actual number of vehicles required is:

INV = the number of vehicles required, i.e., the integer greater than NV

Vehicle Costs. Knowing the number of working hours and the number of vehicles required permits the calculation of the total annual capital and operations and maintenance (O&M) cost for vehicles.

Vehicle Annual Capital Cost

VCC = INV • VRC

and

VACC = VCC • VCRF

where:

VCC = vehicle capital cost (dollars)

VACC = vehicle annual capital cost (dollars)

VRC = current vehicle replacement cost (dollars)

VCRF = vehicle capital recovery factor

and

$$VCRF = \frac{\dfrac{DR}{100}\left[1 + \dfrac{DR}{100}\right]^{VEL}}{\left[1 + \dfrac{DR}{100}\right]^{VEL} - 1}$$

where:

VEL = vehicle economic life in years (generally four to ten years)

DR = discount rate expressed as a percentage. This is the actual interest rate charged on a capital loan or the rate established by government to evaluate projects (typically 8% net of inflation).

Vehicle Annual O&M Cost

VAOMC = VSC + VOC

where:

VAOMC = vehicle annual O&M cost (dollars)

VSC = vehicle service cost, which includes painting, major repairs, overhaul (dollars)

VOC = vehicle operating cost, which includes fuel, oil, minor maintenance (dollars).

and

VSC = INV • VRC • VSF

where:

VSF = vehicle service factor. As a general guide use:

= 0.21 for tracked water/sewage vehicles

= 0.15 for wheeled water/sewage vehicles

= 0.30 for garbage and honey bag vehicles

and

VOC = VOCPH • THRS

where:

VOCPH = BHP • (FR • FUEL + MISC)

VOCPH = vehicle operating cost per hour (dollars/hour)

FR = fuel consumption rate (litres per kilowatt hour). As a general guide use:

= 0.12 for wheeled vehicles (gasoline)

= 0.19 for tracked vehicles (gasoline)

FUEL = fuel cost (dollars/litre)

MISC = miscellaneous operating cost factor. As a general guide use:

= 0.011 for wheeled vehicles

= 0.013 for tracked vehicles

BHP = brake horsepower of vehicle (kilowatts). As a general value BHP is approximately one-fifth of the engine horsepower.

Note: VSF, FR and MISC were determined according to Canadian Construction Association cost calculation methods (Canadian Construction Association, 1976).

Vehicle Total Annual Cost

VTAC = VACC + VAOMC

where:

VTAC = vehicle total annual cost

Labor Cost. Labor cost includes direct and indirect costs for drivers and helpers. Management and administration costs could also be included.

LCPA = LCPH • THRS

where:

LCPA = labor cost per year (dollars)

and

LCPH = (WD + WH • NH) • LBF

where:

LCPH = labor cost per hour (dollars/hour)

WD = hourly wage of driver (dollars/hour)

WH = hourly wage of helpers (dollars/hour)

NH = number of helpers

LBF = labor benefits factor. This factor converts the hourly wage rate into the actual payroll, i.e., worker's hourly wage plus employer's contributions to vacation, health, pension and other benefits plus miscellaneous items. LBF is usually about 1.25.

Parking Garages. Cost includes capital and O&M.

PGACC = PGCC • PGCRF

where:

PGACC = annual capital cost of parking garage (dollars per year)

PGCC = capital cost of parking garage (dollars)

where:

PGCC = PGBSF • VSR

PGBSF = base cost per square metre of floor space in the parking garage. This value is a function of the location, garage size and other factors (dollars/m²)

VSR = vehicle space requirement (m²); typically VSR equals approximately 50 to

80 m² per vehicle. Parking space is not usually required for garbage vehicles.

PGCRF = parking garage capital recovery factor, i.e.,

$$PGCRF = \frac{\dfrac{DR}{100}\left[1+\dfrac{DR}{100}\right]^{PGEL}}{\left[1+\dfrac{DR}{100}\right]^{PGEL}-1}$$

where:

PGEL = parking garage economic life in years (generally 15 to 25 years)

Parking Garage Annual O&M Cost

PGAOMC = PGCC • PGOMF

where:

PGAOMC = parking garage annual O&M cost

PGOMF = parking garage O&M factor. As a general value use 0.06.

Parking Garage Total Annual Cost

PGTAC = PGACC + PGAOMC

where:

PGTAC = parking garage total annual cost

Building Containers. The cost of building containers is usually borne by the customer but this cost should be included in an economic analysis where the total cost of service is desired.

ACCB = NB • CUC • C • NCI for trucked water, sewage pumpout and garbage service

= NB • CUC • C • NCI • F for trucked honey bag service

AACCB = ACCB • ACBCRF

where:

ACCB = associated capital cost to buildings (dollars)

AACCB = associated annual capital cost to buildings (dollars/year)

CUC = container unit capital costs. It is a function of container size and type of container, i.e., water, sewage pumpout, garbage, honey bag (dollars per m³)

ACBCRF = associated cost to buildings capital recovery factor

and

$$ACBCRF = \frac{\dfrac{DR}{100}\left[1+\dfrac{DR}{100}\right]^{ABCEL}}{\left[1+\dfrac{DR}{100}\right]^{ABCEL}-1}$$

= 1.0 for honey-bag collection only

where:

ABCEL = building container economic life in years (generally 15 to 25 years)

AAOMCB = ACCB • AOMFB

where:

AAOMCB = annual associated O&M cost for buildings (dollars/year)

AOMFB = annual O&M factor for building containers. As a general value, use 0.02, i.e., 2% per year.

Total Annual Associated Costs for Building Containers

ATACB = AACCB + AAOMCB

where:

ATACB = associated total annual cost for building containers (dollars)

Vehicle-Haul System Total Costs. The total cost is the capital and O&M costs for vehicles, garages, building containers and labor.

Total Capital Cost

TSCC = VCC + PGCC + ACCB

TSACC = VACC + PGACC + AACCB

where:

TSCC = trucked system capital costs (dollars)

TSACC = trucked system annual capital costs (dollars/year)

Total Annual Operations and Maintenance Cost

TSAOMC = VAOMC + LCPA + PGAOMC + AAOMCB

where:

TSAOMC = trucked system annual O&M costs (dollars/year)

<u>Total Annual Cost</u>

TSTAC = TSACC + TSAOMC

where:

TSTAC = trucked system total annual cost (dollars/year)

Note: By manipulating the equations the total cost can be broken down into the cost related to servicing the buildings and the cost related to the source/disposal.

Cost per Unit Service

<u>Water Delivery and Sewage Pumpout</u>

$$CPL = \frac{(TSTAC - ATACB)}{V}$$

where:

CPL = average cost per litre (dollars/litre)

<u>Garbage and Honey Bag</u>

$$CPP = \frac{(TSTAC - ATACB)}{52 \cdot NB \cdot F}$$

where:

CPP = average cost per pickup (dollars/pickup)

Note: This garbage and honey-bag collection cost could be expressed as a cost per volume; however, cost per pickup is a more significant statistic.

B.1.2 Application of Vehicle-Haul System Equations. The equations presented in this appendix can be used to estimate the cost of trucked water delivery, sewage pumpout, garbage and honey-bag collection services for individual or groups of buildings, or for whole communities. The model can also be used to optimize the design or sizing of components and equipment for a particular location. For example, various vehicle sizes, within the limitations of the road system and maneuverability requirements, can be assessed. Service frequency and building container size can be matched to identify the least-cost system. Technical improvements to increase the efficiency can also be assessed.

The equations and their applications are particularly suited to evaluation by computer or programmable calculator. By repeating the general equation, the cost for each of the building types in a community can be calculated and summed to obtain the total system cost. The costs over a planning horizon, say 20 years, can be calculated by incorporating future changes due to growth and increased water demand. The present value of the annual costs (capital and O&M) over the planning horizon can be calculated.

The most significant variable, with respect to the cost of vehicle-haul water and sewage service, is the quantity of water consumed and waste generated. The quantities necessary for sanitation, convenience and future demand must be carefully and realistically estimated. Many low-water-use fixtures and appliances will be economical and should be incorporated into building plumbing. The benefits of water conservation can be quickly identified using this model.

Portions of communities with high-water-use consumers or compact or high density areas, may be more economically serviced by piped systems. Conversely, housing that is spread out, or where other conditions make it expensive to pipe service, at least during the winter, may be more economically serviced by a vehicle-haul water and/or sewage system. The cost of vehicle-haul servicing is relatively insensitive to housing density as can be demonstrated with this model.

A breakdown of the total delivery costs of vehicle-haul water and sewage systems is presented in Table B-1. These values are for typical conditions and should only be used for preliminary estimating purposes.

B.2 Data Supplement for Vehicle-Haul Systems

Data on vehicle-haul systems in the Northwest Territories has been compiled by the Department of Municipal and Community Affairs, Government of the Northwest Territories. Analysis of the data shows that the values for a number of parameters can be assumed to be the same for the typical vehicles being used and for the typical worker performance and community conditions. Also, some parameters, such as the distance between buildings, are relatively insensitive to the total cost and reasonable values can be assumed. This data is summarized in Table B-2. It should be noted that unusual local conditions or equipment may result in some parameters differing somewhat from those presented in this table. The analyst should assess the community in question to ascertain the reasonableness of the parameters.

TABLE B-1   BREAKDOWN OF ANNUAL COST OF VEHICLE-HAUL WATER DELIVERY OR SEWAGE PUMPOUT*

Item		Portion of Total Costs
Vehicle capital	14%	(4 year life)
Vehicle O&M	25%	(repairs, fuel, etc.)
Garage capital	9%	(construction, 10 year life)
Garage O&M	4%	(heat, repairs, etc.)
Labor	48%	($13/hour for two workers)
Total	100%	

* Assumes a discount rate of 8 percent. The cost of building tanks is not excluded.

B.3 Simplified Equations

A number of the parameters in the detailed general equations are not very significant, are common to most community conditions, or are a function of the equipment. If the typical values presented in Table B-1 are assumed (including a vehicle size of 4,500 litres and efficiency based on helper), the truck equations become greatly simplified as follows:

Water Delivery

for a given building container size:

$$THRS = 0.0051 \, POP \cdot VPCD \cdot D +$$
$$0.080 \, POP \cdot VPCD +$$
$$55.61 \, POP \cdot VPCD/C$$

and for a given frequency of service:

$$THRS = 0.0051 \, POP \cdot VPCD \cdot D +$$
$$0.080 \, POP \cdot VPCD +$$
$$6.73 \, FW \cdot NB$$

Sewage Pumpout Collection

for a given building container size:

$$THRS = 0.0051 \, POP \cdot VPCD \cdot D +$$
$$0.056 \, POP \cdot VPCD +$$
$$55.61 \, POP \cdot VPCD/C$$

and for a given frequency of service:

$$THRS = 0.0051 \, POP \cdot VPCD \cdot D +$$
$$0.056 \, POP \cdot VPCD +$$
$$6.73 \, FW \cdot NB$$

where:

$THRS$ = total number of hours required to service the people or buildings within the defined group (hours)

POP = population (real or equivalent) within the defined group

$VPCD$ = volume of water consumed or waste generated per person per day (litres/(p·d))

D = distance to the source or disposal point (kilometres)

C = building container size (litres)

FW = frequency of service (times per week)

NB = number of buildings in the defined group

Using these simplified equations the labor and vehicle requirements and costs can be quickly assessed. Also, the effects of changes in the most significant factors can be quickly assessed.

B.4 References

Cameron, J.J. 1981. *Guidelines for the Preparation and Administration of Municipal Water and Sanitation Trucked Service Contracts.* Department of Local Government, Government of the Northwest Territories, Yellowknife.

TABLE B-2   TYPICAL VALUES FOR VEHICLE-HAUL SYSTEM PARAMETERS

Parameter Description	Name	Water Delivery	Sewage Pumpout	Honey Bag Collection	Garbage Collection
Efficiency of labor	EL	1.5	1.5	1.5	1.5
Building container size	C	varies	varies	0.03 m^3	0.02 m^3
Container utilization factor	CSF	0.85	0.85	0.85	0.85
Vehicle speed to/from source/disposal	S	50 kmph	50 kmph	50 kmph	50 kmph
Rate of filling/emptying vehicle container	R	450 lpm	450 lpm	$0.086 \text{ m}^3\text{pm}$	$0.26 \text{ m}^3\text{pm}$
Turnaround time at source/disposal	TT	4.0 min	4.0 min	1.0 min	1.0 min
Vehicle speed between buildings	SB	10 kmph	10 kmph	10 kmph	10 kmph
Distance between buildings	DB	30 m	30 m	30 m	30 m
Rate of filling/emptying building container	RB	180 lpm	340 lpm	$0.018 \text{ m}^3\text{pm}$	$0.085 \text{ m}^3\text{pm}$
Turnaround time while servicing buildings					
without helper	TTB	3.0 min	3.0 min	1.0 min	1.0 min
with helper	TTB	5.0 min	5.0 min	0.25 min	0.25 min
Service cycle frequency	F	156 per year	104 per year	260 per year	52 per year
Vehicle size (typical)	VS	4,500 litres	4,500 litres	3.8 m^3	1.4 m^3
Vehicle utilization factor	VUF	0.95	0.95	0.85	0.85
Vehicle cost	VC	$105,000	$80,000	$50,000	$28,000
Vehicle weight	VW	4,300 kg	4,300 kg	2,950 kg	2,950 kg
Vehicle economic life	VEL	6 to 10 years	6 to 10 years	6 to 10 years	6 to 10 years
Vehicle service factor	VSF	0.15	0.15	0.15	0.15
Vehicle break horsepower	BHP	55 kW	55 kW	45 kW	45 kW
Fuel consumption rate	FR	0.12	0.12	0.12	0.12
Miscellaneous operating cost factor	MISC	0.10	0.10	0.10	0.10
Number of helpers	NH	0 to 1	0 to 1	0 to 2	0 to 2
Parking garage economic life	PGEL	20 years	20 years	20 years	20 years
Building container economic life	ABCEL	20 years	20 years	N/A	5 years
Building container cost	BCUC	$2.00 per litre	$2.00 per litre	$0.15 per bag	$50.00 each

Canadian Construction Association. 1976. *Rental Rates on Construction Equipment.* Ottawa, Ontario.

Gamble, D.J. and C.T.L. Janssen. 1974. Evaluating alternative levels of water and sanitation service for communities in the Northwest Territories. *Canadian Journalof Civil Engineering* , 1(1): 116-128.

Government of the Northwest Territories, 1992. *General Terms of Reference for an Engineering Pre-Design Report on Community Water and Sanitation Systems.* Department of Local Government, Government of the NWT, Yellowknife.

APPENDIX C

SNOWDRIFTING AND SNOW LOADS

3rd Edition Steering Committee Coordinator

William L. Ryan

3rd Edition Principal Author

Wayne Tobiasson

Appendix C   Table of Contents

Appendix B List of Tables

APPENDIX C   SNOWDRIFTING AND SNOW LOADS

C.1   Introduction

Snow is an important climatic feature in cold regions. It is a difficult material to describe quantitatively since its properties vary considerably. Classical hexagonal snowflakes are broken apart by the wind and sintered together by vapor diffusion. Large grains grow larger and small grains grow smaller as intergranular bonds develop and a stronger, more stable material is formed. At one point in its metamorphisis snow can resemble noncohesive beach sand and at another, hard ice. Because it is encountered close to its melting point, temperature changes greatly affect its properties. The creep rate of snow increases dramatically as it is warmed. These changes affect snowdrifting and the behavior of snow on roofs.

C.2   Drifting Snow

Winds, during or after snowstorms, can cause snow to drift. While snow is drifting, travel and other outdoor activities may have to be curtailed. Snowdrifts develop on roads and runways, and in among buildings creating obstructions and hazards. Drifts on roofs create heavy loads and snow injestion can block air intakes and damage heating coils.

Detailed descriptions of snowdrifting processes are available (Mellor, 1965; Kind, 1981; Tabler et al., 1990; Tabler, 1991). The wind speed at which snowdrifts is a function of the size and shape of the snow particles, the amount of cohesion between them and the roughness of the surface. Loose, unbonded snow may drift at wind speeds as low as 3 m/s (measured 10 m above the surface) but densely packed, firmly bonded ("age-hardened") snow may remain in place with winds in excess of 30 m/s. Winds just strong enough to cause drifting cause slow movement of particles one over the other along the surface. This action is termed "creep." As the wind's velocity increases, particles are lifted up off the surface 10 to 20 mm. They fall back to the surface a short distance downwind. When they hit, they may cause other particles to be ejected and move in the same fashion. This action is termed "saltation." Higher wind speeds cause snow particles to be plucked from the surface and held in suspension for long periods. This action is termed "turbulent diffusion." Cold dry snow can be carried up to eye level by winds of about 20 m/s. Very high winds can create a snow-laden turbulent boundary layer over 100 m high. In blizzards, appreciable concentrations of snow can be transported great distances by turbulent diffusion. Winds of 15 m/s can transport about four times as much snow as winds of 10 m/s can. The amount of snow being transported diminishes with height above the surface. About 80% of snow that is blowing at 10 m/s is within 0.1 m of the surface. At 15 m/s about half the blowing snow is within the first 0.1 m. Above 20 m/s less than 20% of the blowing snow is within 0.1 m of the surface.

As snow drifts, it forms dunes, sastrugi, barchans, waves and ripples on the surface. Dunes are long rounded deposits with their long axis aligned with the wind direction. Sastrugi are sharp-edged ridges also aligned with the wind. Sastrugi 1 to 2 m long are common features on windswept snowfields. Barchans look like classic boomerangs with the tails pointing downwind. They may be a metre or two across.

Waves and ripples are transverse undulations. Waves are formed by high winds and have wavelengths up to 10 m and heights up to 0.2 m. Ripples are low features with wave lengths of 0.1 to 0.4 m. They form in winds not strong enough to create sastrugi. All these surface features slowly move downwind under the continued action of wind.

In the vicinity of large abrupt obstacles that severely disturb flow, pronounced snowdrifts and scour can occur. Drifts form in areas of aerodynamic shade (i.e., eddies).

C.2.1   Measures to Reduce Snowdrifting.

Past observations of snowdrifting at a site, photographs in particular (Fig. C-1), provide important indications of potential snowdrifting problems that will be created if the site is modified (e.g., by construction of new buildings, by rerouting of roads, or by storage of materials). Information on the direction, intensity and frequency of winter winds is essential to determine the relative importance of each wind direction. Through the application of snow transport calculations (Tabler, 1989), which estimate the volume of driftable snow based on the surrounding upwind terrain and vegetation, the relative importance of each wind direction at a specific site can be determined.

Snowdrifting around groups of structures can be difficult to predict but the following general guidelines (Departments of the Army and Air Force, 1971;

FIGURE C-1   SNOWDRIFTING AMONG THE BUILDINGS OF CAMP TUTO, GREENLAND. WINDS ARE FROM THE UPPER LEFT.

Williams and Waechter, 1991b) can minimize problems:

- Use trees, shrubs, snow fences, structures, or earth/snow berms to precipitate snow before it reaches the site proper. Where storms may occur from any direction, provide protection from all quadrants.

- Place major roads parallel to the wind and raise the surface of the road above the immediate surroundings by 0.5 to 1.5 m.

- Do not locate roads directly upwind or downwind of large obstructions. Where possible maintain at least a 30 m clearance upwind and 60 m downwind.

- Locate parking lots beside roads to act as buffer zones. Do not place parking lots among buildings. Expect additional snow accumulation around parked vehicles and provide ample room for snow storage on the downwind end of a lot away from roads. Alternatively, use landscaping or other suitable drift control measure to act as a buffer for the parking lots.

- Parking aprons should be placed beside, not upwind or downwind of hangers and garages.

- Orient surface structures with their longest dimension parallel to the wind. Doors placed on the downwind end of the structure will be subjected to suction forces during drift formation and will be rapidly blocked with drifted snow. Those on the upwind face can be difficult to seal, but will be clear of snowdrifts. Doors are best located along the sides, toward the upwind end.

- Orient large garage doors nearly parallel to the wind even if this results in a building orientation perpendicular to the wind. Adjust this orientation slightly to assure that the doors are not in the lee of the upwind corner of the building.

- Place structures in rows perpendicular to the wind with enough space between them to permit effective snow removal. [This is a controversial recommendation since obstructions (e.g., snow fences or buildings) cause less drifting when placed parallel to the wind. Also, rows perpendicular to the wind can create long snowdrifts downwind of the gap between the buildings (Irwin and Williams, 1985). The advantage of perpendicular placement is that one builidng is less likely to cause drifting on another.] If a second row is necessary, place those structures directly downwind of those in the first row.

- Locate priority buildings toward the downwind end of the facility where they are afforded protection by less important upwind structures.

- Provide snow dumping areas to eliminate large piles of snow and windrows within the facility or community. Piles and windrows act as obstructions and can increase future snow removal requirements unless the snowdrifts they create have been accounted for.

In permafrost areas, heated buildings are often elevated above the surface to prevent thawing of sup-porting soils. The space under such buildings must be kept clear of snow that would retard winter cooling of these soils. Such elevated buildings reduce snow deposition against the downwind building face and can be made more aerodynamic to further reduce drifting under and around them (Floyd, 1974; Kwok et al., 1993). In the high Arctic and in Antarctica where snowdrifting can be quite a problem, it may be appropriate to elevate all buildings, even those supported on solid rock or other thaw-stable material, just to minimize snowdrifting. However, where buildings will be constructed on-grade due to local conditions, or perhaps functional requirements (e.g., garages), the adverse effects of snowdrifting can be accounted for through building orientation and the location of doors away from areas of aerodynamic shade. New Australian stations in Antarctica (Fig. C-2) use this approach (Incoll, 1991).

Figure C-3 shows on-grade and elevated passageways at Cape Lisburne, Alaska. The on-grade passageway is drifted in but the gravel pad under and around the elevated passageway is clear. An on-grade building is shown in Figure C-4 and Figure C–5 shows an elevated building.

FIGURE C-2   NEW AUSTRALIAN BUILDINGS IN ANTARCTICA ARE ON-GRADE AND ARE ORIENTED TO REDUCE DRIFTING (Incoll, 1991)

FIGURE C-3   ELEVATED PASSAGEWAYS BETWEEN BUILDINGS TO REDUCE
SNOWDRIFTING – CAPE LISBURNE, ALASKA

FIGURE C-4   ON-GRADE BUILDING

FIGURE C-5   ELEVATED BUILDING

C.2.2 Control Devices. Snow fences and other control devices can be employed to collect snow, keep an area scoured clear of snow or deflect snow-laden air away from an area to be protected (Mellor, 1965; Tabler, 1989; Meroney and Meroney, 1989). Snow fences are used to protect roads, increase snow cover on agricultural land, and increase snow used as a water supply (Farmwald and Crum, 1986; Slaughter et al., 1975) as was done at Barrow and Shishmaref, Alaska (Fig. C-6). Most fences have a 0.15 to 0.40 m gap at their base (the gap is often 10% of the fence height). Most fences are of open construction with about half of their windward surface of solid material (i.e., they have a "density" of about 50%). Heights range from 1 to 4 m. They are most efficient when placed at right angles to the wind. However, when used as deflectors they are placed at an angle to the wind. Multiple rows of fences used to collect snow should be spaced about 30 fence heights apart. Downwind fences can be of a lower "density" than the first upwind fence. Wood, metal and, more recently, plastics are being used in the construction of snow fences. Experience has shown that plastic fencing requires greater care in the design and construction of connection details as plastic is more susceptible to wind damage (Williams and Waechter, 1991a). Trees and shrubs can also be effective snow collectors (Finney, 1937).

One form of blower fence resembles a table top tilted 15 to 40° with the high end upwind. The tilt constricts flow and accelerates it in the area immediately downwind of the blower. Another form of blower fence consists of a series of tilted slats arranged horizontally like a Venetian blind. Drifts can form across the area to be protected at either end of blower fences. On highways such drifts and the high winds downwind of blower fences can create hazards.

Scoops have been used at the edges of upper roofs to reduce snowdrifting on narrow lower roofs (Fig. C-7) and to keep walkways and entrances scoured clear of snow (Williams, 1989). However, scoops may not be beneficial if the long, low roof is wide since it would simply relocate the snowdrift further downwind on the lower roof.

FIGURE C-6   SNOW FENCES INSTALLED ON THE WINDWARD SIDES OF A
WATERSHED TO AUGMENT THE WATER SUPPLY – SHISMAREF,
ALASKA (Farmwald and Crum, 1986)

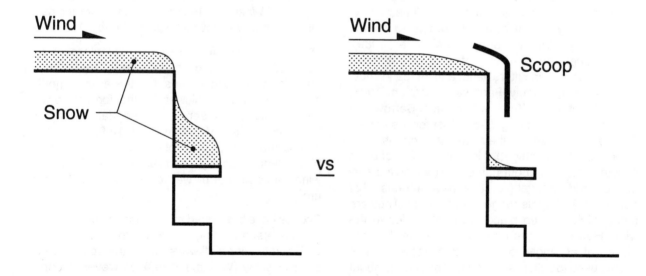

FIGURE C-7   SNOW SCOOPS USED TO REDUCE DRIFTING ON LOWER ROOFS

FIGURE C-8   SNOWDRIFTS MODELLED IN A WIND TUNNEL

Snowbanks created by the plowing of roads are, unfortunately, very effective snow fences that cause the depressed area between (i.e., the roadway) to rapidly collect drifting snow. In open country, roads should be elevated 0.5 to 1.5 m above grade wherever possible. The height elevated should be at least equal to the maximum depth of snow on the ground expected each winter. If side slopes are steeper than about 1:6 (9°) a wedge of snow will accumulate on either side until the surface slope is reduced to about 1:9 (6°).

C.2.3 Modeling. Flumes and wind tunnels are being used to model snowdrifting (Fig. C-8) (Anno, 1989; 1984; Kind, 1986; Iversen, 1979; Isyumov and Mikitiuk, 1992; Peterson and Cermak, 1989; O'Rourke and Weitman, 1992; Irwin and Williams, 1985). Snow is represented by sand or ground walnut shells in water flumes and by sawdust, bran, sand, sodium bicarbonate, activated clay or glass beads in wind tunnels. All of these physical models are an imperfect representation of nature but they are, nonetheless, very useful in helping to understand snowdrifting. They are particularly good at evaluating alternatives (e.g., positioning of entrances, air intakes, and exhausts). Several studies have been conducted of snow loads on roofs (Williams, 1989; Isyumov and Mikitiuk, 1992; Peterson and Cermak, 1989; O'Rourke and Weitman, 1992).

Physical modeling has been used for arctic town planning (Morrison, Hershfield, Theakston and Rowan Ltd., 1975) and to reduce snowdrifting problems in communities such as Baker Lake, NWT (Adam and Piotrowski, 1980) As the modeling of entire communities presents scaling problems when trying to replicate important snowdrift characteristics, modern-day designers augment modeling studies with field and local experience.

Analytical (numerical) modeling is beginning to be used to predict snowdrifting on roofs (Irwin et al., 1992; Irwin and Gamble, 1989). Currently it is often used in conjunction with physical modeling. Limited comparisons of the model simulation results to full scale have been undertaken (Irwin, 1994); however, additional comparisons are needed between full-scale observations, physical models and numerical models to advance the state of this art.

C.3 Snow Loads

In Canada the basic reference for snow loads is the latest edition of the National Building Code of Canada (Associate Committee on the National Building Code, 1990a). In the USA (including Alaska) the basic reference for snow loads is the latest version of ASCE Manual 7 *Minimum Design Loads for Buildings and Other Structures* (American Society of Civil Engineers, 1995).

Both documents use extreme-value statistical analysis of weather records of snow on the ground to estimate ground snow loads with a specific annual probability of being exceeded. The ground snow loads determined by this probabilistic approach are converted to roof loads by factors that account for the wind exposure of the site and the thermal, geometric and aerodynamic characteristics of roofs. These factors are based on research findings, case studies (Schriever et al., 1967; Lutes and Schriever, 1971; Taylor, 1979; 1980; O'Rourke et al., 1983) and experience. For the most part they have not been obtained by rigorous statistical analysis. Their strength is that they have been periodically updated and subjected to consensus review by the design professions of both countries. While the Canadian and USA methodologies are similar, they are not identical.

C.3.1 Meteorological Database.

The Canadian meteorological database is for the maximum annual depth, not load, of snow on the ground at 1,618 stations. A Fisher-Tippet Type 1 distribution was used to determine extreme value depths with a 3.3% annual probability of exceedance (i.e. a 30-year mean recurrence interval). Other information was used to estimate the unit weights of this snow for various regions. In Canada new snow has a unit weight of about 1 kN/m³. The unit weight of old snow ranges from 2 to 5 kN/m³. North of the tree line a unit weight of 2.94 kN/m³ is used to convert ground snow depths to ground snow loads. South of the tree line and east of the Continental Divide a unit weight of 2.01 kN/m³ is used. West of the Continental Divide the unit weight varies with location from 2.55 to 4.21 kN/m³.

These unit weights were estimated from measurements made no more frequently than four times a month. Some rainfall is included in them but since many intense rains percolated down through these snow packs between measurements, they do not adequately represent short-term load increases due to rain on snow. For this reason a separate rain load is included when determining roof snow loads. This added load varies from place to place and is based on one-day maximum rainfall measurements during the time of year when ground snow loads maximize at 2,100 locations across Canada. Values are for a 3.3% annual probability of being exceeded but they are constrained to be less than or equal to the snow load mentioned above.

These rain and snow loads that, in combination, are used to determine snow loads on roofs are tabulated for numerous places in Canada in the *Supplement to the National Building Code of Canada* (Associate Committee on the National Building Code, 1990). Normalized ground snow load maps are also available for most of Canada from the Atmospheric Environment Service (AES) in Downsview, Ontario. For a few sparsely populated areas the maps have not been published but ground snow load recommendations can be obtained for a nominal fee from AES. Call (416) 739-4365 for more information.

In the USA the meteorological database consists of 204 stations where both the depth and load of snow on the ground has been measured frequently for many years and another 9,800 stations where only the depth of snow has been measured. For each station where depths and loads have been measured, maximum seasonal values have been used with a log-normal distribution to determine extreme values for a 2% annual probability of exceedance (i.e., a 50-year mean recurrence interval). The relationship between these depths and loads for all 204 stations was used to develop the following equations (one for Alaska and one for the "Lower 48" states) that relate "50-year" depths and loads (Tobiasson and Greatorex, 1996):

Alaska $\qquad L = 1.75D^{1.39}$

Lower 48 $\qquad L = 1.97D^{1.36}$

where,

D = 50-year depth of snow on the ground, m

L = 50-year snow load on the ground, kPa.

These equations were then used to determine 50-year ground snow loads at the other 9,800 stations where only depths are measured. Ground snow loads for design purposes are presented on a map in ASCE 7-95 (American Society for Civil Engineers, 1995) except for Alaska where values are tabulated. The map contains many areas where extreme local variations prevent meaningful mapping. In such areas site-specific snow load case studies are needed to establish a 50-year ground snow load. The

CRREL report *Ground Snow Loads for the USA* (Tobiasson and Greatorex, 1996) presents the database and methodology for doing this.

The USA measurements of snow on the ground have been made frequently enough so that they contain more of the rain that falls on snowpacks than the Canadian measurements do. Thus a separate rain load is not added to the ground snow load. However, since not all rain on snow has been captured in the USA measurements, roofs with a slope of less than 1.19°(1:48) have 0.2 kPa rain-on-snow surcharge loads added just before establishing their final design snow loads. Roofs with greater slopes retain rain-on-snow loads for shorter periods and thus do not need the surcharge.

C.3.2 Snow Loads on Roofs. In the USA, ground loads are first converted to those on a flat roof by way of exposure, thermal and importance factors. These factors acknowledge the influence of the exposure of the site (e.g., urban and suburban areas), the exposure of the roof (e.g., tight in among conifers), and roof thermal characteristics (e.g., unheated building) on roof snow loads. The importance factor establishes higher roof loads for certain buildings (e.g., places of assembly) where greater than normal risks are incurred and lower loads for other structures (e.g., agricultural buildings) where risks are less than normal.

Experience has indicated that in areas of the USA with low ground snow loads, the flat roof snow load determined this way is too low. For this reason the resulting flat roof snow load cannot be less than minimum values which vary with the ground snow load. For sloped roofs the flat roof snow load is modified by a factor that varies with the slope, the thermal condition of the roof, and the slipperiness of its surface (Sack, 1988). If a roof is slippery it may qualify for snow load reduction but only if its surface is unobstructed and there is a place for snow on the roof to slide to. The methodology acknowledges that ice dams occur on certain types of roofs and they obstruct sliding snow. Other slope factors are included for roofs with curved surfaces. For multiple-folded plate, sawtooth, barrel vault and other roofs that result in valleys, no load reductions are allowed due to slope (in the USA). Such load reductions are allowed in Canada in some cases.

In Canada the step of establishing a flat roof snow load is not taken. Instead, the slope of the roof is included in the single step from a ground load to a roof load. Exposed roofs without obstructions qualify for load reductions and those north of the tree line qualify for even larger reductions. The thermal condition of the roof is not a consideration in the Canadian code. This is one reason why the snow load on most flat roofs is about 80% of the ground snow load in Canada and only 70% of the ground snow load in the USA. However, if that roof were over an unheated building in the USA the roof load would be about 85% of the ground load. The Canadian code also allows further load reductions for sloped roofs with unobstructed, slippery surfaces.

Partial and Unbalanced Loads. Both the USA and Canadian documents include partial loads and unbalanced loads. Partial loads result when wind scours snow away from portions without increasing loads on other nearby portions. Partial loads can produce stress increases, stress reversals and changes in deflection (that may influence drainage) for certain types of structural systems (e.g., cantilevered roof joists). In both documents it is assumed that one part of the roof contains the full load as discussed above and the rest of the roof contains only half the load.

Sunlight tends to decrease loads on roofs that face the sun and winds tend to decrease loads on windward roofs and increase loads on leeward roofs. Since it is not possible to define wind direction with assurance, winds from all directions are generally considered when establishing unbalanced loads.

In both countries the windward side of the roof is considered snow-free when unbalanced loads are considered. Loads on leeward portions vary with roof exposure (in the USA), and roof slope (in both countries). Further reductions are allowed in both countries if the roof has an unobstructed slippery surface. Unbalanced loading diagrams are available for hip, gable and curved roofs (in both countries). Other unbalanced loading diagrams are given for multiple-folded plate, sawtooth, barrel vault and other roofs that result in valleys where excess snow may collect by drifting, creeping or sliding. For such roofs no reduction in unbalanced load due to slope is allowed in either country.

Drift Loads. Most failures associated with snow loads on roofs are caused by drifted snow. Large drifts can develop on lower roofs (Fig. C-9); thus it is extremely important to consider drift loads when designing roofs (Taylor, 1984; O'Rourke et al., 1985; 1986). In areas that are exposed to winds over much of the winter (e.g., the high Arctic) it is often appropriate to modify the geometry of the building to eliminate areas of "aerodynamic shade" where large drifts will form.

FIGURE C-9   THE PEAK SNOW LOAD IN THIS DRIFT WAS 6.2 kPa, THE UPPER ROOF
SNOW LOAD WAS 0.8 kPa AND THE GROUND LOAD WAS 1.0 kPa.

In the USA and Canada the size of the design drifts on lower roofs is a function of the difference in height between upper and lower roofs and the magnitude of the ground snow load. In Canada the influence of the size of the upper roof on drift size is acknowledged but not quantified. In the USA the sizes of "upwind" and "downwind" drifts are influenced by the length of lower and upper roofs, respectively (American Society of Civil Engineers, 1995).

Criteria from both countries acknowledge that upper roofs can cause drifting on lower roofs located some distance away. Both also acknowledge that for roofs of unusual shape or configuration, wind-tunnel or water-flume tests together with full-scale field experience may be needed to help define drift loads.

Sliding Snow. In general, situations that permit snow to slide onto lower roofs should be avoided. Where this is not possible, the extra load of the sliding snow should be considered. Since snow may slide off roofs that do not qualify as being unobstructed and with a slippery surface, it is prudent to assume that any upper roof sloped to eaves is a potential source of sliding snow.

Property has been damaged (Fig. C-10) and people have been killed from snow and ice falling from roofs. Situations which compromise public safety should be avoided (Taylor, 1983). Devices that hold snow on roofs (i.e., snow guards) may be needed to protect certain areas (Fig. C-11). When determining snow loads, a roof with snow guards is considered to be obstructed no matter how slippery its surface is. Snow guards must be designed to resist substantial forces. They may also be needed to protect plumbing vents and other roof penetrations from damage by sliding snow. For roofs that slope to cold eaves, such problems can be minimized by configuring them as well-insulated and ventilated cold roofs with as much slope as possible and with overhangs at eaves just wide enough to prevent wetting of walls (Mackinlay, 1988; Tobiasson, 1994). Another approach is to use internally drained low-slope membrane roofs since they avoid the problems associated with ice damming and falling snow and ice. Protecting such membranes by placing them under

FIGURE C-10   ICE FALLING FROM A ROOF CRUSHED THIS VAN

FIGURE C-11   ALUMINUM ANGLE SNOW GUARDS USED TO HOLD SNOW ON A
SLIPPERY METAL ROOF

ballast, filter fabric, and extruded polystyrene insulation has been quite successful in cold regions (Tobiasson, 1994).

C.4   References

Adam, K.M. and R. Piotrowski. 1980. Solving snow-drifting problems at Baker Lake, NWT, using snow-modelling techniques. In: *Utilities Delivery in Northern Regions*, Environment Canada Economic and Technical Review Report EPS 3-WP-80-5, 394-408.

American Society of Civil Engineers. 1995. *Minimum Design Loads for Buildings and Other Structures*. ASCE Manual 7, New York, NY.

Anno, Y. 1984. Requirements for modelling of snow drift. *Cold Regions Science and Technology*, 8: 241-252.

Anno, Y. 1989. Snowdrift wind tunnels in Japan. In: *Proceedings, First International Conference on Snow Engineering*, CRREL Special Report 89-6, Hanover, NH, 191-198.

Associate Committee on the National Building Code. 1990a. *1990 National Building Code of Canada*. National Research Council of Canada, Ottawa, Ontario.

Associate Committee on the National Building Code. 1990b. *Supplement to the National Building Code of Canada 1990*. National Research Council of Canada, Ottawa, Ontario.

Departments of the Army and Air Force. 1971. *Arctic and Subarctic Construction: Buildings*. Army Technical Manual TM 5-852-9 (USAF Manual AFM 88-19, Chapter 9), Washington, DC.

Farmwald, J.A. and J.A. Crum. 1986. Developing a community water system for Shishmarif, Alaska. In: *Proceedings, Fourth International Conference on Cold Regions Engineering*, ASCE, New York, 597-608.

Finney, E.A. 1937. *Snow Control by Tree Planting*. Michigan State College Bulletin 75.

Floyd, P. 1974. The North Slope center: how it was built. *The Northern Engineer*, 6(3).

Incoll, P. 1991. The development of Australian Antarctic building types. In: *Proceedings First International Offshore and Polar Engineering Conference*, ISOPE, 434-442.

Irwin, P and S. Gamble. 1989. Predicting snow loading on the Toronto Skydome. In: *Proceedings, First International Conference on Snow Engineering*, CRREL Special Report 89-6, Hanover, NH, 118-127.

Irwin, P. 1994. *Hybrid Physical/Computer Modelling of Snow Drifting*. Special ASCE Publication on the Modeling of Windblown Snow and Sand, ASCE/ISSW Snow Science Workshop, Snowbird, Utah, October 13.

Irwin, P.A. and C.J. Williams. 1985. Snowdrift models. *The Northern Engineer*, 17(3): 4-11.

Irwin, P.A., C.J. Williams, S.L. Gamble and R. Retzlaff. 1992. Snow load prediction in the Andes mountains and downtown Toronto – FAE simulation capabilities. In: *Proceedings, Second International Conference on Snow Engineering*, CRREL Special Report 93-27, Hanover, NH, 135-146.

Isyumov, N. and M. Mikitiuk. 1992. Wind tunnel modelling of snow accumulations on large-area roofs. In: *Proceedings, Second International Conference on Snow Engineering*, CRREL Special Report 92-27, Hanover, NH, 181-194.

Iversen, J.D. 1979. Drifting snow similitude. *Journal of Hydraulics Division*, ASCE, 105(HY6): 737-753, Proc. Paper 14647.

Kind, R.J. 1981. Snow drifting. In: *Handbook of Snow Principles, Processes, Management and Use*. Pergamon Press, Toronto.

Kind, R.J. 1986. Snow drifting: a review of modeling methods. *Cold Regions Science and Technology*, 12: 217-228.

Kwok, K.C.S., D.J. Smedley and D.H. Kim. 1993. Snowdrift around Antarctic buildings – effects of corner geometry and wind incidence. *Journal of Offshore and Polar Engineering*, 3(1): 61-65.

Lutes, D.A. and W.R. Schriever. 1971. *Snow Accumulation in Canada: Case Histories II*. Research Council of Canada, NRCC Report 11915, Ottawa, Ontario.

Mackinlay, I. 1988. Architectural design in regions of snow and cold. In: *First International Conference on Snow Engineering*, CRREL Special Report 89-6, Hanover, NH, 441-455.

Mellor, M. 1965. *Blowing Snow.* Cold Regions Research and Engineering Laboratory (CRREL) Monograph III-A3c, Hanover, NH.

Meroney, B. and R. Meroney. 1989. Snow control with vortex and blower fences. In: *Proceedings, Fourth International Conference on Cold Regions Engineering*, CRREL Special Report 89-6, Hanover, NH, 286-296.

Morrison, Hershfield, Theakston and Rowan Ltd. 1975. *Wind and Snow Accumulation Patterns in the Area of Norman Wells, NWT.* Task Force on Northern Oil Development, Report No. 74-38, 28 p.

O'Rourke, M. and N. Weitman. 1992. Laboratory studies of snow drifts on multilevel roofs. In: *Proceedings, Second International Snow Engineering Conference*, CRREL Special Report 89-6, Hanover, NH, 195-206.

O'Rourke, M., P. Koche and R. Redfield. 1983. *Analysis of Roof Snow Load Studies: Uniform Loads.* Cold Regions Research and Engineering Laboratory, CRREL Report 83-1, Hanover, NH.

O'Rourke, M., R. Speck and U. Stiefel. 1985. Drift snow loads on multilevel roofs. *Journal of Structural Engineering*, ASCE, III(2): 290-306.

O'Rourke, M., W. Tobiasson and E. Wood. 1986. Proposed Code provisions for drifted snow loads. *Journal of Structural Engineering*, ASCE, 112(9): 2080-2092.

Peterson, R. and UJ. Cermak. 1989. Application of physical modelling for assessment of snow loading and drifting. In: *Proceedings, First International Conference on Snow Engineering*, CRREL Special Report 89-6, 276-285.

Sack, R.L. 1988. Snow loads on sloped roofs. *Journal of Structural Engineering*, ASCE, 114(3): 501-517.

Schriever, W.R., Y. Faucher and D.A. Lutes. 1967. *Snow Accumulation in Canada: Case Histories I.* National Research Council of Canada, NRCC Report 9287, Ottawa, Ontario.

Slaughter, C. et al. 1975. *Accumulating Snow to Augment the Fresh Water Supply at Barrow, Alaska.* U.S. Army Cold Regions Research and Engineering Laboratory, Special Report No. 217, Hanover, NH.

Tabler, R. 1989. Snow fence technology – state-of-the-art. In: *Proceedings., First International Conference on Snow Engineering*, CRREL Special Report 89-6, Hanover, NH, 297-306.

Tabler, R.D. 1991. Snow transport as a function of wind speed and height. In: *Proceedings, Sixth International Conference on Cold Regions Engineering*, ASCE, New York, 729-738.

Tabler, R.D., J.W. Pomeroy and B.W. Santana. 1990. Drifting snow. In: *Cold Regions Hydrology and Hydraulics*, ASCE, New York, 95-145.

Taylor, D.A. 1979. A survey of snow loads on roofs of arena-type buildings in Canada. *Canadian Journal of Civil Engineering*, 6(1): 85-96.

Taylor, D.A. 1980. Roof snow loads in Canada. *Canadian Journal of Civil Engineering*, 7(1): 1-18.

Taylor, D.A. 1983. *Sliding Snow on Sloping Roofs.* National Research Council of Canada, Canadian Building Digest 228, Ottawa, Ontario, 4 p.

Taylor, D.A. 1984. Snow loads on two-level flat roofs. In: *Proceedings, 41st Eastern Snow Conference*, Eastern Snow Conference, Washington, DC.

Tobiasson, W. 1994. General considerations for roofs. In: *Moisture Control in Buildings*, American Society for Testing and Materials, ASTM Manual 18, Philadelphia, PA.

Tobiasson, W. and A. Greatorex. 1996. *Ground Snow Loads for the USA.* U.S. Army Cold Regions Research and Engineering Laboratory, Hanover, NH.

Williams, C. 1989. Field Observations of Wind Deflection Fins to Control Snow Accumulation on Roofs. In: *Proceedings, First International Conference on Snow Engineering*, CRREL Special Report 89-6, Hanover, NH, 307-314.

Williams, C.J. and B.F. Waechter. 1991a. Arctic snowfences – a big solution for a big problem. In: *Proceedings of the 48th Annual Eastern Snow Conference, Guelph, Ontario*, June 5-7, 307-313.

Williams, C.J. and B.F. Waechter. 1991b. Planning for sun, wind and snow effects from urban highrises to Arctic communities. In: *Extended Abstracts, Third International Symposium on*

Cold Region Development (ISCORD '91),
Edmonton, Alberta, June 16-21, 191.

APPENDIX D

FREEZE PROTECTION, THAWING AND HEAT TRACING

3rd Edition Steering Committee Coordinator

James A. Crum

Special Contributions

Daniel Schubert

Appendix D   Table of Contents

Appendix D List Of Figures

Appendix D List of Tables

APPENDIX D   FREEZE PROTECTION, THAWING AND HEAT TRACING

D.1 Freeze Protection and Thawing

Regardless of design, the probability of freeze-up of many arctic utilities is high. While it is prudent and often mandatory to provide for thawing, the first design consideration should always be to ensure that the pipelines can be returned to service after a freeze-up. Many kilometres of pipe in the Arctic have been abandoned in place because of their almost complete destruction during a freeze. Freeze protection and thaw recovery are critical elements for all exposed utility components. Provision should be made for systems design with consideration of both a primary system and secondary or backup system (Peskator, 1980).

D.1.1 Freeze Protection. Freeze damage to containers of fluid, including pipes, tanks and fittings, occurs because of the expansion of water when it changes to ice. Passive measures for freeze protection to accommodate or prevent these volume changes include:

- metal or plastic disk-rupture parts;
- overinsulated sacrifice zones;
- expansion tanks;
- drain ports;
- drainback of forcemains and well pumps during periods when not pumping.

The use of polyethylene plastic or small-diameter, annealed copper pipe reduces breakage and cracking during a freeze-up because of their material properties. Other passive methods include insulation systems for tanks and piping.

The most common means of primary freeze protection is the circulation of water through the main and the addition of heat to the water through heat exchangers. Circulation through service lines is maintained by the pitorifice system or small circulation pumps. The secondary means of freeze protection is through the use of heat tracing. This is discussed in the following sections.

D.2 Heat Tracing

Heat tracing is any means by which heat is added continuously along the length of a pipeline in a controlled fashion. Heat tracing may be the primary means of flow maintenance, it may be a secondary or backup system, or it may provide thaw capabilities. Many schemes have been devised to maintain flow in pipelines carrying fluids subjected to ambient temperatures below the freeze point. Two of the most common include the use of heated fluid-based systems, and electric heat-tape installations. Manufacturers' representatives and catalogues should be consulted for the most recent information on products and services.

D.2.1 Thermal Requirements. The objective of any heat tracing system is either to thaw a frozen pipe or to achieve and maintain a specified fluid temperature within a pipe. The pipe may be pre-insulated, buried, above ground, or bare inside an insulated utiliduct or building. The necessary heat input to achieve a specific internal pipe temperature is a function of:

- the temperature of the external environment (heat sink); and
- the thermal geometry of the system.

Heat tracing is generally rated in watts per linear metre. Modeling the thermal aspects of heat-traced systems requires several tasks:

1. Specifying the internal working temperature.

   The fluid temperature inside the carrier pipe is job specific and depends on the design objectives. In the preliminary analysis it is important to choose pipe and insulation materials that are compatible at the working temperature required. Since heat loss is proportional to the temperature differential across the system, a further consideration should be to minimize fluid temperatures and proportionally lower operating costs.

2. Choosing the location and assessing the temperature of the ultimate heat sink (or "outside" temperature).

   A heat sink is characterized by a temperature lower than the contents of the carrier pipe which remains constant during the time period investigated; it is a thermal environment not significantly influenced by the imposition of the heat-traced system. For above-ground applications the heat sink is usually the surrounding air. For buried pipes and those in below-ground utilidors it is usually the surrounding frozen soil. However, a more refined analysis will allow for heat transfer through the ground and snow cover to the colder winter air above. Note that except for the case of deeply buried pipes, the temperature of natural environment heat sinks in the Arctic vary cyclically, as will the heat loss of pipes installed in them.

3. Determining the insulating significance of materials between the pipe contents and heat sink.

4. Deriving or applying the appropriate mathematics for the system considered.

The thermal characteristics of common materials for building, piping, insulation and frozen/unfrozen soil, as well as the mathematical expressions for heat loss through various geometric configurations can be found in Section 4.

Once an appropriate model has been determined (see Section 4), the steady-state heat loss equation for any system is ultimately a function of the internal temperature and that of the heat sink. By setting the heat loss at the perimeter of the system equal to the heat input of the heat trace (i.e., steady-state conditions), a chart similar to that shown in Figure D-1 can be generated.

In choosing the input power or energy, the variability of the heat-sink temperature must be considered. The heat-trace system has to be able to keep the pipe contents above freezing during the coldest ambient temperatures. In addition, for nonvariable output heat tracing, when the heat sink is at its warmest, the resulting internal temperature must not be detrimental to the pipe material, the insulation, or other system components. This is particularly critical for plastic pipe materials.

Some heat-tracing methods lose heat over their operating distance; circulating glycol loops lose heat by conduction and electrical heat tapes lose output by voltage drop. In a pipe with standing or slowly moving fluid contents, the result is a significant drop in the temperature profile along the length of the heat-tracing alignment. The temperature and potential freeze-up time of the coldest spot in the pipeline should be investigated.

Finally, if the heat trace is activated after the pipe contents have cooled for some reason, the steady-state temperatures predicted by the above analysis are not reached for some period of time; it may be a matter of hours or even days. This phenomenon (described in the section on thawing) is a function

of the heat-trace rating and the thermal inertia of the system (OEH&E, 1984; 1987).

D.2.2 Methods of Heat Tracing. There are as many different kinds of heat-tracing systems for pipelines as there are manufacturers. Their capital costs, shipping costs, installation costs and complexity, space requirements, and maintenance costs (including necessary skills) vary widely. All of these issues, including the most economically available energy form (fuel oil, waste heat, or electricity), dictate which heat-tracing system is the most feasible and reliable for a particular location and application (Corwin, 1984).

Glycol. In the Arctic, aqueous glycol solutions were first used successfully in heat-tracing systems for utilidors. As a replacement fluid in heating loops (i.e., instead of steam, hot water, or brine), glycol exhibits the added advantages of remaining liquid at freezing temperatures (in the event of system shutdown) and of having a higher boiling point at atmospheric pressure, which allows for greater rates of heat transfer given similar system configurations (Dow Chemical Company, 1990).

Glycol heat tracing has been used in conjunction with single insulated pipelines, and often where waste heat from electric generating plants is available.

The use of glycol or other antifreeze solutions protects the heat-trace piping and appurtenances during a freeze-up, allows start up during winter, provides a means of thawing pipes quickly with high temperatures, and allows for the economical manipulation of the heat-trace temperature (resulting in higher efficiency as the ambient or heat sink temperature changes).

Table D-1 lists some of the physical properties of seven kinds of undiluted glycol. Ethylene and propylene glycols are the most commonly used in the Arctic.

Glycols are used in a mixture with water, and are formulated to remain liquid at the minimum expected ambient temperatures. The resultant properties of these aqueous solutions have important ramifications in the design of glycol heat-tracing systems. Figure D-2 and Table D-2 show the freeze points, viscosity, and specific heat capacity of various glycol/water mixtures.

Most aqueous glycol solutions, however, do not freeze (crystallize); rather, they supercool (set up to amorphous glass like solids). Sharply defined freeze

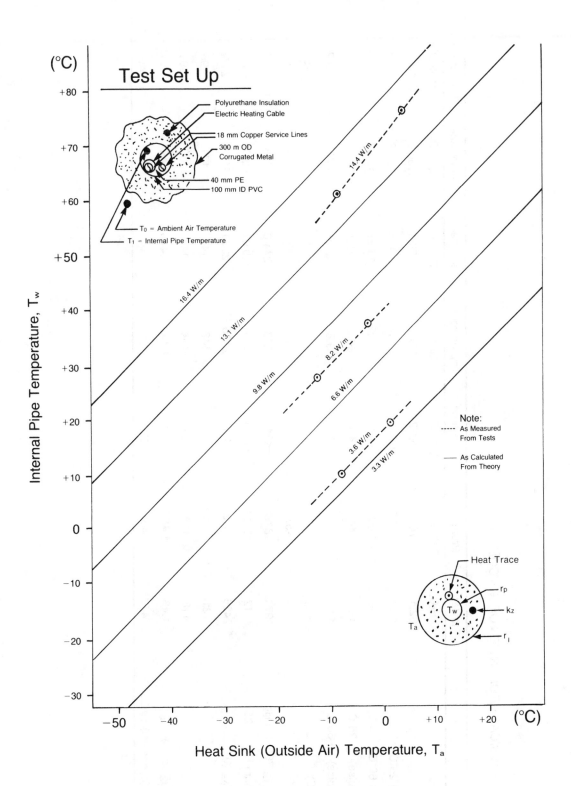

FIGURE D-1   STEADY-STATE INTERNAL TEMPERATURES OF AN ABOVE-
GROUND INSULATED PIPE WITH CONSTANT OUTPUT HEAT
TRACING

TABLE D-1   PHYSICAL PROPERTIES OF GLYCOLS

	Ethylene	Propylene	Dipropylene	Tripropylene	Diethylene	Triethylene	Tetraethylene
Formula	$C_2H_6O_2$	$C_3H_8O_2$	$C_6H_{14}O_3$	$C_9H_{20}O_4$	$C_4H_{10}O_3$	$C_6H_{14}O_4$	$C_8H_{18}O_5$
Molecular weight	62.1	76.1	134.2	192.3	106.1	150.2	194.2
Density (g/cc) @ 25°C	1.11	1.033	1.023	1.016	1.113	1.119	1.12
Density (g/cc) @ 60°C	1.085	1.007	0.996	0.992	1.088	1.092	1.092
Viscosity (centipoise) @ 25°C	16.5	44	74.1	56.2	28.2	37.3	44.6
Viscosity (centipoise) @ 60°C	4.68	8.5	11.1	9.8	6.99	8.77	10.2
Freezing point (°C)	-13				-8	-7	-5
Pour point (°C)	-57		-39	-41	-54	-58	-41
Boiling point (°C) @ 760 mm Hg	197.3	187.2	231	268	244.8	288	327
Vapor pressure (mm Hg) @ 25°C	0.12	≤0.22	<0.03	<0.01	<0.01	<0.01	<0.01
Specific heat (J/kg•K)	2428	2512	2428	2135	2303	2219	2177
Flash point (°C)	121	104	124	143	147	171	196
Fire point (°C)	118	104	127	154	143	166	191
Refractive index @ 25°C)	1.43	1.457	1.439	1.442	1.446	1.454	1.457
Surface tension (dynes/cm) @ 25°C	47	45	33	34	44	45	45

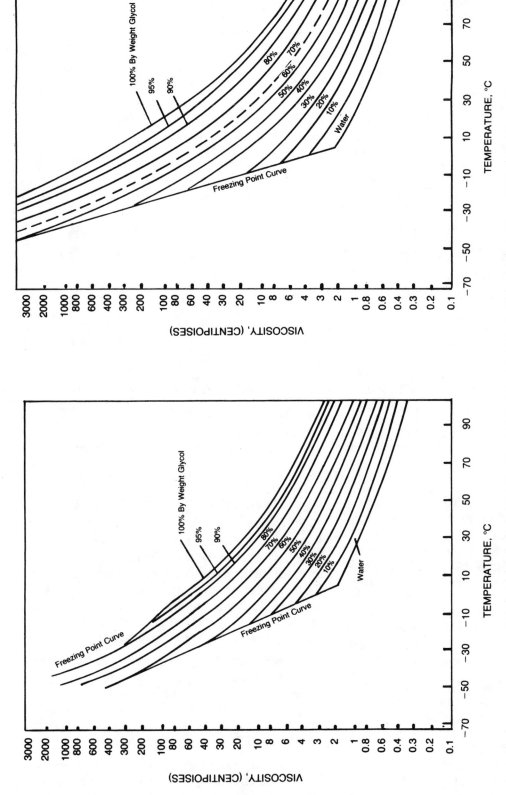

Note: Freeze Point is the First Occurrence of Solid Crystals; a Slushy Solution that Still Flows Exists at that Temperature.

a) Viscosities of Aqueous Ethylene Glycol Solutions

b) Viscosities of Aqueous Propylene Glycol Solutions

FIGURE D-2   PROPERTIES OF AQUEOUS GLYCOL SOLUTIONS – VISCOSITY

TABLE D-2   SPECIFIC HEAT (cal/(gm·°C))

Temp, °C	% by Volume								
	10	20	30	40	50	60	70	80	90
	Ethylene Glycol								
0	0.941	0.901	0.858	0.813	0.766	0.717	0.666	0.613	0.557
20	0.949	0.912	0.872	0.829	0.785	0.738	0.689	0.638	0.585
40	0.958	0.923	0.885	0.845	0.803	0.759	0.712	0.664	0.612
60	0.966	0.934	0.899	0.861	0.822	0.78	0.736	0.689	0.640
80	0.974	0.945	0.912	0.877	0.840	0.801	0.759	0.714	0.668
100	0.983	0.956	0.926	0.893	0.859	0.821	0.782	0.740	0.696
120	0.991	0.967	0.939	0.909	0.877	0.842	0.805	0.765	0.723
	Propylene Glycol								
0	0.966	0.939	0.907	0.869	0.826	0.777	0.722	0.661	0.592
20	0.974	0.949	0.920	0.885	0.844	0.798	0.746	0.687	0.621
40	0.810	0.960	0.933	0.900	0.863	0.819	0.770	0.714	0.651
60	0.989	0.97	0.946	0.916	0.881	0.840	0.793	0.740	0.680
80	0.997	0.981	0.959	0.932	0.899	0.861	0.817	0.767	0.709
100	1.005	0.991	0.972	0.948	0.918	0.882	0.841	0.793	0.738
120	1.012	1.002	0.985	0.964	0.936	0.903	0.865	0.819	0.767

points can only be obtained under controlled conditions and if seeding and agitation are properly employed. In service, glycol mixtures turn slushy as the temperature is depressed, becoming more viscous until they eventually fail to flow. This is particularly significant when sizing circulation pumps, which must operate under various temperature ranges. Figure D-3 compares the pump head for a given pipeline length and diameter required for 60-percent ethylene and propylene glycol solutions over a typical range of arctic service temperatures.

In the design of glycol heat tracing for use in potable water systems, it should be noted that ethylene glycol is toxic, and cross-contamination must be prevented. Double-wall heat exchangers and reduced pressure backflow preventors are required. Propylene glycol, which is considered nontoxic (see Table D-3), can be used where the risk of contamination is significant. The differences in viscosity and

specific heat capacity should be compared and considered.

In general, the mechanical configuration of glycol heat tracing takes the shape of many standard hydronic systems. It consists of a heat exchanger from the main heat reservoir, circulating pump(s), expansion tank, specialty valves and controls. The output heat exchanger is the heat-trace tubing which runs in a loop next to the pipe requiring heat.

The mechanical design and component selection recommendations for glycol systems are covered in detail by many technical and manufacturers' publications. Excellent reference material is available from the American Society of Heating Refrigeration and Air Conditioning Engineers (ASHRAE).

While the total system efficiency of a boiler-fired glycol heat-tracing loop is greater than that for electric resistance heat tapes, the maintenance requirements (including heated space for the boilers,

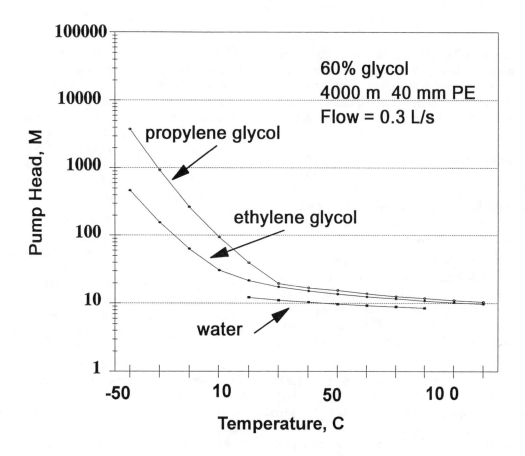

FIGURE D-3   PUMP HEAD VERSUS TEMPERATURE
FOR AN AQUEOUS GLYCOL SOLUTION

TABLE D-3   TOXICOLOGICAL CONSIDERATIONS OF VARIOUS GLYCOLS

Glycol	Single Oral Lethal Dosages	
	Rat LD50	Human LD100
Ethylene	5.5	1.6
Diethylene	14.8	1.1
Triethylene	22.0	–
Tetraethylene	31.8	–
Propylene	32.5	Nontoxic**
Dipropylene	14.8	–
Tripropylene	3.0*	–

Note: LD is a lethal dose ratio (e.g., an LD50 dose results in death to
      50 percent of a subject population.)
*   LD100
** Considered safe for use in foods and pharmaceuticals.

pumps, etc.) are often cited as their main disadvantage. In addition, because of the increased pump capacity requirements, installation costs, and loop leaking potential, glycol heat tracing along lengths greater than 600 to 900 m requires special design considerations.

Electric Heat Tapes. One of the earliest forms of electric heat tracing was the use of arc welding units to thaw metallic piping (Section D.3). Frequently, however, high temperatures developed, which damaged insulation and other system components and created fire and safety hazards.

Electric heat tracing using heavy, rigid cable in mineral insulation (MI) has a long history of use in the cold regions. These high-watt-density single-wire resistance-type heaters have seen decreasing application as the development and use of high-temperature-sensitive (+65°C maximum) plastic piping has increased.

More recently, efficient electric heat tracing in the 6 to 33 W/m range with various types of jacketing materials have appeared (approximately $12 to $15 per metre, USA 1995) for quantities to 600 m. Commonly they are constructed of parallel resistance bus wires that provide an integral closed circuit. Thus, they may be cut in the field to any desired length,

and power need only be applied to one end. Constant watt per metre (W/m) output heat tracing has resistance heating wires wrapped around the bus wires. Variable W/m output heat tracing has a polymeric heating element sandwiched between the bus wires that decreases heat output as its temperature rises. Electric heat tapes may also be made to vary their output by changing the applied voltage. This is a desirable feature if the tapes are also to be used for thawing (Mace, 1985).

The main advantages of electric heat-tracing systems is their ease of installation and low maintenance and space requirements. Disadvantages include low system efficiencies where electric costs are high (compared with boiler-fired glycol systems), and for buried installations, keeping the wire and connections of the circuits in a dry environment. Because of concern with fire and safety, ground-fault interrupters should also be used in all installations. Furthermore, because of the voltage drop along resistance circuits, power must generally be applied to points about 100 to 200 m apart on electric heat-traced pipelines. Longer lengths are possible; more detailed information on electric heat tapes and their controls can be obtained from manufacturers.

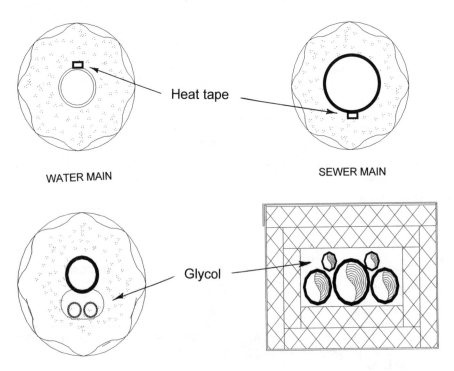

FIGURE D-4   TYPICAL HEAT-TAPE AND GLYCOL TRACING

Combination Glycol and Electric Heat-Tape Systems. In above-ground systems where glycol is used on long lengths, an electric heat tape may also be included. The heat tape can be sized for warming the glycol after an extended freeze-up, to allow the glycol to be pumped. In some situations, a combination system can also be used to minimize the costs, the operating expense, and the demands placed on the existing electrical generating capacity. Figure D-4 shows typical schematic heat tape and glycol configurations for arctic pipe and combination utilidors.

Other Methods of Heat Tracing. Electric resistance heat tapes and glycol heating loops make up the majority of water and sewer utility heat-tracing systems in the North. Special circumstances though, may warrant investigation into other types of pipeline heat addition methods.

Skin effect electrical current tracing and electrical impedance heating are commonly used in industrial and commercial applications. They work only on metal pipes. This technique is used for some copper service connections. Information on these methods is available from most major pipeline suppliers.

Friction heating (hydraulically induced), while very inefficient, may be applicable where small amounts of heat make a design concept feasible. Information on this subject can be found in Section 4.

D.3 Thawing

Once a system has frozen up, thawing may occur during the following summer (i.e., in above-ground pipes and in the sections of those buried in the active layer). However, to thaw frozen pipes expeditiously at any time, appropriate measures must be incorporated in the original design.

Key concerns in designing a thawing system are the kind of carrier pipe material, insulation and outer jacket configuration (as applicable), accessibility, and temperature of the surrounding soil or heat sink. Occasionally, heat tracing, as described above, is chosen as the primary means of pipe thawing, especially where the heat tracing is also the primary means of heat maintenance in the pipeline. Just as often though, to avoid extra initial expenditures, no thawing hardware (heat tape or thaw loop) has been laid with the pipe. Arrangements must then be made to access the pipe for thawing by special machines brought to the site.

An important analysis in designing thaw systems is to determine how long thawing will take. The energy supplied to thaw will initially go to heating up the pipe material, insulation, and perhaps the surrounding soil before the pipe contents completely melt.

The period of time required is a function of the heat-trace rating of the system and its thermal inertia. This can be thought of as the increasing thermal momentum a system attains from the heat flux. The functional variables of pipe core temperature are influenced by the following, (assuming a boundary at the insulated pipes outer jacket):

- specific or volumetric heat capacity of the system (pipe contents, pipe material, insulation etc.);

- mass or volume of the system;

- thermal conductivity of the system;

- heat sink temperature;

- latent heat capacity of the fluid contents during phase change; and

- system geometry.

A time-to-thaw analysis for a constant output heat-traced, insulated pipeline is shown schematically in Figure D-5. Such methods are very adaptable to computer solutions. The length of time other thawing methods take is addressed in the following

Metal Pipes. All of the heat-tracing methods described in Section D.2 as well as steam and hot-water thawing methods have been used successfully to thaw metal pipes of many sizes and lengths. However, the most common method of thawing small-diameter metal pipes is by passing an electrical current through it (Nelson, 1976). Either portable gasoline or diesel-driven generators, welders, or heavy service electrical transformers (110 or 220 V, i.e., "buzz-boxes") have been used. Figure D-6 shows several possible alternative circuits for thawing metal pipes. Either AC or DC current at high amperage and very low voltage (seldom more than 15 V) can be used.

When reasonable flow of water has been obtained, the welding machine or electrical transformer can be turned off, and the water flow can finish thawing the line.

With electrical thawing methods the warmer water from the mains should be used to finish the thawing. To prevent refreezing of the line, the water must

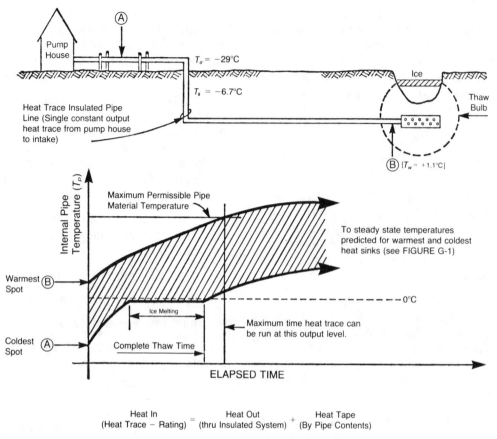

FIGURE D-5   SCHEMATIC COMPUTATION OF THAW TIME FOR A CONSTANT OUTPUT HEAT-TRACED PIPELINE

be left running until the danger of freezing has passed. A frozen underground pipe indicates that a frost line extends below the pipe. The pipe remains susceptible to refreezing unless the water is permitted to run slowly to replace heat loss to the soil.

The amount of heat generated when a current is passed through any electrical conductor, such as a metal pipe, can be expressed by the following power equation:

$$P = I^2 R$$

where,

P = heat or power, watts

I = current, amps

R = resistance, ohms

Hence, the rate of thawing a frozen pipe is a function of the square of the current applied and the resistance offered by the pipe itself; i.e., pipe resistance, R, equals the effective electrical resistance of the pipe material times the volume of material of the pipe being thawed where the volume is equal to the cross-sectional area times the length of pipe. For example, doubling the current, I, increases the heat generated by a factor of four. Of course, longer and larger diameter metal pipes require higher currents and more time to thaw.

The approximate time required to thaw various sizes of steel pipe using different currents is given in Figure D-7. Table D-4 lists the thaw times for typical service line configurations as a function of number of welding machines used (a typical 800 A welder

TABLE D-4   THAW TIMES FOR TYPICAL SERVICE LINE CONFIGURATIONS AS A FUNCTION OF NUMBER OF WELDING MACHINES USED*

Water Line Length (m)	Steel Pipe Diameter (mm)	Number of Thaw Machines	Time to Thaw (hours)
60	100	3	1 to 2
60	100	4	0.5 to 1.5
60	150	4	1 to 2
75	40	2	0.5 to 1
75	40	1	5

*800 amp welders running continuously at 50 amps

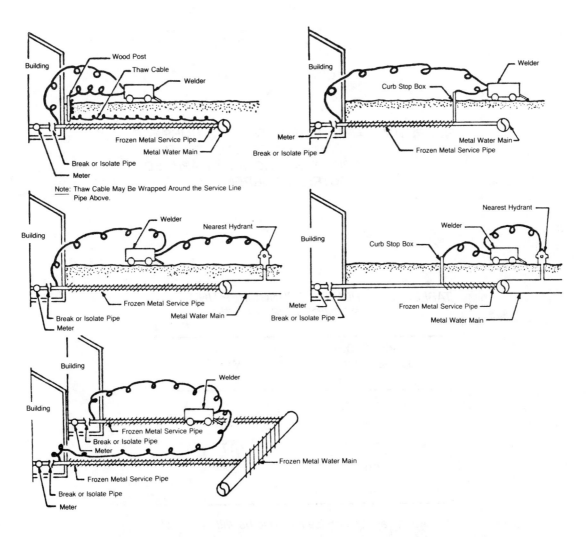

FIGURE D-6   ALTERNATIVE CIRCUITS FOR ELECTRICAL THAWING OF METAL PIPES

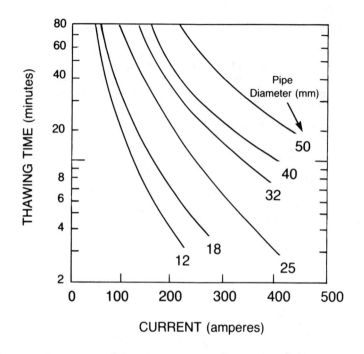

FIGURE D-7   APPROXIMATE THAWING TIMES FOR
VARIOUS DIAMETERS OF STEEL
PIPES AS A FUNCTION OF APPLIED
ELECTRICAL CURRENT

TABLE D-5   POWER RATINGS PER UNIT
SERVICE LINE LENGTH*

Length of 20 mm Copper Service Pipe (m)	W/m
12	24
24	18
36	17
50	15
60	13

*Thawing various service line lengths with a typical
300-A, 6.3-V electrical service transformer

TABLE D-6   RECOMMENDED CABLE SIZES FOR ELECTRIC THAWING CURRENT DISTANCE FROM WELDING MACHINE OR TRANSFORMER TO PIPE CONNECTION

Current	Distance (m) from Welding Machine or Transformer to Pipe Connection											
(A)	15	23	30	38	46	53	61	69	76	91	107	122
100	2	2	2	2	1	1/0	1/0	2/0	2/0	3/0	4/0	4/0
150	2	2	1	1/0	2/0	3/0	3/0	4/0	4/0	4/0	2.2/0	2.3/0
200	2	1	1/0	2/0	3/0	4/0	4/0	4/0	2.2/0	2.3/0	2.3/0	2.4/0
250	2	1/0	2/0	3/0	4/0	4/0	2.2/0	2.2/0	2.3/0	2.3/0		
300	1	2/0	3/0	4/0	4/0	2.2	2.3/0	2.3/0	2.4/0			
350	1/0	2/0	4/0	4/0	2.2/0	2.3	2.3/0					
400	1/0	3/0	4/0	2.2/0	2.3/0	2.3/0						

can run continuously at 450 A, burning fuel at a rate of 7.5 L/h).

Table D-5 shows the power rating per unit service line length when thawing with a typical "buzz-box".

Another factor to be considered in electrical thawing of metal pipes is the resistance and length of the cable used to connect the current-producing device to the pipe that is being thawed. Generally, the cable should be large enough (cable resistance less than pipe resistance, i.e., cross-sectional area of a copper cable greater than that of the copper pipe) so that it does not get hot and create a safety hazard. Table D-6 gives recommendations for cable sizes as a function of current applied and cable length.

Alternatively, the electrical cable may be deliberately undersized and wrapped around the buried service line during installation. The cable heats up, warming the surrounding soil and frozen pipe. An electrical cable laid with the service line during construction allows a smaller capital investment because the ideal cable size is smaller. The cable resistance is greater than the pipe resistance, i.e., the cross-sectional area of a copper cable is less than that of a copper pipe.

Finally, the following precautions should be taken when thawing metal pipes electrically:

1. Use on underground or protected pipe only (not indoor plumbing).

2. Do not use high voltage. Twenty volts with 50 to 60 A for service lines is sufficient. (Do not use a constant voltage power source because there is usually no control for limiting the current.)

3. Make good, tight connections to the pipeline.

4. When conventional arc welders are used for thawing, do not operate them at their maximum rated amperage for more than five minutes. Only use them at about 75 percent of the rated amperage if greater times are needed.

5. Disconnect electrical wires grounded to the water pipes in the buildings, or disconnect the service pipe from the house plumbing. Failure to do this could cause a fire.

6. Remove meters that may be in the service line.

7. Turn on the thaw unit only after the connections have been made secure. Check them again once the current has been applied.

8. A problem may be encountered with the thawing current jumping from the water service line into nearby gas or other lines. These should be separated by a 25-mm wood wedge.

Plastic Pipes. The common methods of thawing plastic pipe are by heat tapes or with hot water. Heat tapes have been discussed previously.

Hot-water thawing of both large- and small-diameter plastic pipes has developed through experience in the field. It is a manual operation that can also be used on metal pipes of all diameters. A smaller-sized

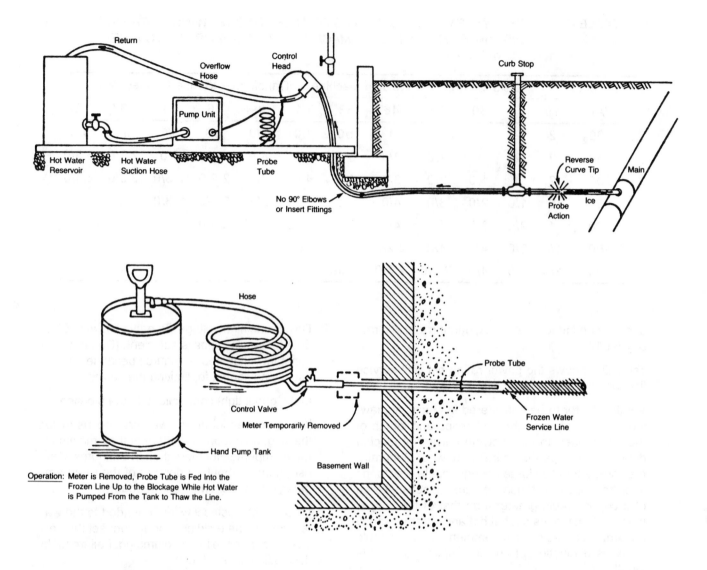

FIGURE D-8   THAW TUBE METHODS FOR SMALL-DIAMETER PIPES (Currey, 1980)

hose or flexible tube (called a probe), generally with a leading nozzle, is pushed through the frozen pipe while hot water is pumped through the probe.

For thawing service lines, water pressure can be obtained from a nearby building, or a conventional hand pump filled with warm water may be used. Thaw rates of under a minute to several minutes per metre of service line are typical depending on size and location of the freeze plugs. Figure D-8 shows a schematic of both a commercial unit and a field-improvised unit.

This method of thawing service lines is reported to be about 50 percent successful. Most failures have occurred because the thaw tube could not be inserted due to mineral buildups, sharp bends, and kinks in the service pipe. The success rate would be much higher if the pipes were installed with this thawing technique in mind (Currey, 1980).

Because of the practical limitations on probe insertion distances (about 150 m), access points (i.e., access holes on gravity sewers or thaw ports) must be provided on long lines.

Low-temperature steam thawing has been used, and does not appear to diminish the long-term strength of these pipes (Whyman, 1973). However, with uninsulated plastic pipe (especially PE) and in some instances with flexible insulated thin-walled plastic pipe, permanent damage (i.e., flattening, puncture by sharp rocks) may occur as a result, in areas of poor bedding or high external loading. Special precautions should be taken if this approach is considered.

D.4   References

Corwin, B. 1984. Waterproofing and heat trace provisions for pre-insulated polyethylene piping systems. In: *Proc. Cold Regions Engineering Specialty Conference,* Canadian Society for Civil Engineering, April 4-6, 1: 523-536.

Currey, J.J. 1980. Thawing of frozen service lines. In: *Utilities Delivery in Northern Regions.* Environmental Protection Service Rep. No. EPS 3-WP-80-5, Ottawa, Ontario, 310-313.

Dow Chemical Company. 1990. *Engineering and Operating Guide for Dowtherm SR-1 and Dowtherm 4000: Inhibited Ethylene Glycol-based Heat Transfer Fluids.* Midland, Michigan.

Mace, W. 1985. *Temperature Prediction Model for A Thaw Wire Inside an Insulated Polyethylene Pipe.* Thesis, MS. Arctic Engineering, Univ. Alaska, Anchorage.

Nelson, L.M. 1976. Frozen Water Services. *Journal of the American Water Works Association* 68(1): 12-14.

OEH&E. 1987. *Internal Tracing of PE Pipe with Thaw Wire.* Alaska Area Native Health Service, Office of Environment, Health and Engineering, Anchorage, Alaska.

OEH&E. 1984. *Thawing Device for the Chalkytsik Circulating Water Main.* Alaska Area Native Health Service, Office of Environment, Health and Engineering, Anchorage, Alaska.

Peskator, H.W. 1980. Frozen pipes: causes and cures. *Water and Sewage Works*, 127(1): 36-37, 60.

Whyman, A.D. 1973. *Effect of Steam Thawing on Sclair PE Pipe.* DuPont Canada Customer Technical Center, Kingston Pipe Testing Laboratory, Ontario.

APPENDIX E

ENERGY MANAGEMENT DATA

Appendix E   Table of Contents

List Of Figures

List of Tables

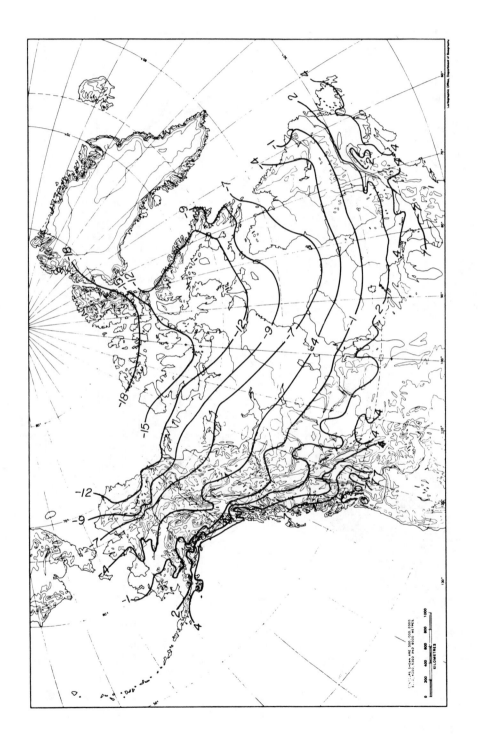

FIGURE E-1   MEAN ANNUAL AIR TEMPERATURE, °C (after Thomas, 1953; Hartman and Johnson, 1978)

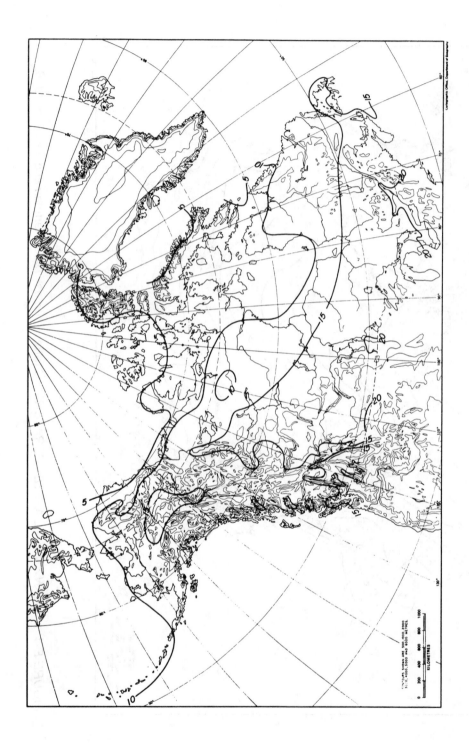

FIGURE E-2   MEAN JULY AIR TEMPERATURE AT 1.5 m ABOVE SURFACE, °C (from Bryson and Hare, 1974)

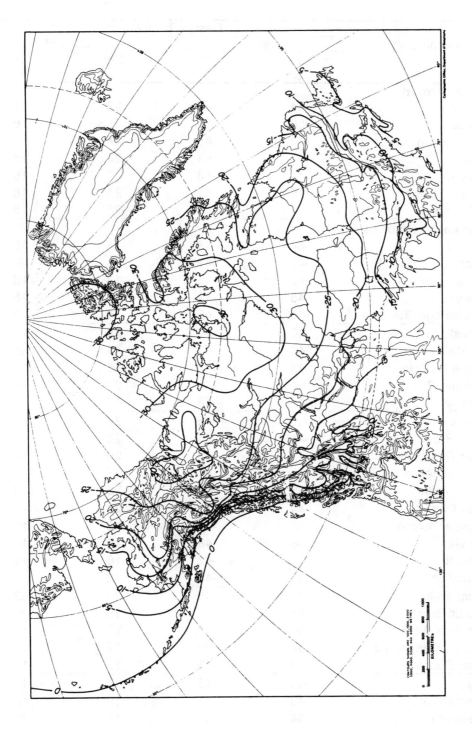

FIGURE E-3   MEAN JANUARY AIR TEMPERATURE AT 1.5 m ABOVE SURFACE, °C (from Bryson and Hare, 1974)

TABLE E-1   BUILDING DESIGN TEMPERATURES FOR SELECTED COMMUNITIES
IN CANADA (after NRC, 1977; NRCC, 1990)

Community	Design Temperature			
	January		July 2.5%	
	2.5% °C	1% °C	Dry °C	Wet °C
Yukon Territory				
Aishihik	-44	-46	23	16
Dawson	-50	-51	26	16
Destruction Bay	-43	-45	24	15
Snag	-51	-53	23	16
Teslin	-41	-43	25	16
Watson Lake	-46	-48	26	16
Whitehorse	-41	-43	25	15
Northwest Territories				
Aklavik	-44	-46	24	16
Alert	-43	-45	13	9
Arctic Bay	-43	-45	14	10
Baker Lake	-45	-46	21	15
Cambridge Bay	-45	-46	16	13
Chesterfield Inlet	-40	-41	20	14
Clyde	-41	-43	15	9
Coral Harbour	-41	-43	18	13
Eskimo Point	-40	-41	21	16
Eureka	-47	-48	12	9
Fort Good Hope	-46	-48	27	17
Fort Providence	-44	-46	24	18
Fort Resolution	-42	-44	26	18
Fort Simpson	-45	-47	27	18
Fort Smith	-43	-45	28	19
Hay River	-41	-43	26	18
Holman Island	-43	-45	18	12
Inuvik	-46	-48	25	16
Iqaluit	-40	-42	16	11
Isachsen	-46	-48	12	9
Kugluktuk	-44	-45	20	13
Mould Bay	-45	-47	10	8
Norman Wells	-46	-47	27	17
Nottingham Island	-38	-40	14	13
Port Radium	-44	-46	22	16
Rae	-44	-46	24	17
Rankin Inlet	-40	-41	20	15
Resolute	-44	-45	11	9
Resolution Island	-35	-37	8	7
Tungsten	-49	-51	26	16
Yellowknife	-43	-45	25	17

TABLE E-2   BUILDING DESIGN TEMPERATURES FOR SELECTED COMMUNITIES IN ALASKA
(after Strock and Koral, 1965)

Community	Design Temperature				Community	Design Temperature			
	January		July 2.5%			January		July 2.5%	
	2.5% °C	1% °C	Dry °C	Wet °C		2.5% °C	1% °C	Dry °C	Wet °C
Adak (Joint Unit)	-5	-7	14	13	Kogru River AFS	-42	-44	12	10
Anchorage	-29	-32	22	16	Kotzebue	-38	-39	18	14
Aniak	-43	-47	22	17	McGrath	-42	-44	24	18
Annette	-11	-12	21	17	Middleton Island AFS	-6	-8	16	14
Anvile Mountain AFS	-34	-36	15	12	Murphy Dome AFS	-34	-37	22	14
Attu	-5	-7	12	11	Naknek	-31	-33	21	16
Barrow	-41	-43	12	11	Naptowne AFS	-28	-32	19	16
Barter Island	-42	-44	11	9	Neklason Lake AFS	-28	-31	21	16
Bear Creek AFS	-38	-42	22	16	Nikolski	-6	-7	13	12
Bethel	-33	-36	21	17	Nome	-33	-36	17	13
Bettles	-42	-46	24	17	Northeast Cape AFS	-29	-31	18	15
Big Delta	-41	-43	24	16	North River AFS	-34	-39	18	15
Big Mountain AFS	-33	-36	18	13	Northway	-46	-49	24	17
Boswell Bay AFS	-23	-26	17	15	Ohlson Mountain	-21	-24	17	14
Cape Lisbourne AFS	-36	-38	13	11	Pedro Dome AFS	-37	-39	23	16
Cape Newenham AFS	-24	-26	14	12	Petersburg	-17	-19	19	15
Cape Romanzof AFS	-26	-27	16	13	Pillar Mountain AFS	-12	-14	17	14
Cape Sarichef AFS	-11	-12	16	15	Port Heiden AFS	-17	-19	17	14
Cold Bay	-13	-16	14	13	Port Moller	-13	-14	15	14
Cordova	-22	-25	19	16	Rabbit Creek AFS	-32	-34	17	13
Diamond Ridge AFS	-20	-23	18	15	Richardson, Fort	-28	-31	21	16
Driftwood Bay	-9	-11	18	16	St. Paul Island	-17	-19	11	11
Dutch Harbor	-8	-9	17	16	Shemya Island	-5	-7	12	11
Eislson AFB	-44	-46	26	17	Sitkinak AFS	-12	14	18	14
Fairbanks	-46	-47	26	17	Soldotna AFS	-28	-32	19	16
Fairbanks AFS	-42	-44	24	16	Sparrevohn AFS	-33	-35	21	15
Fort Yukon	-48	-53	26	17	Tanana	-42	-46	26	17
Galena	-43	-45	24	17	Tatalina AFB	-34	-36	23	15
Granite Mountain AFS	-40	-42	22	10	Tin City AFS	-34	-36	13	11
Gulkana	-41	-44	24	16	Umiat	-48	-49	21	18
Homer	-18	-22	19	16	Unalakleet	-34	-37	19	16
Indian Mountain AFS	-33	-34	18	12	Unalakleet AFS	-34	-39	18	14
Juneau	-20	-22	22	18	Utopia Creek AFS	-0	-42	23	16
Kalskaket Creek AFS	-43	-45	24	17	Wainwright, Fort	-43	-45	26	17
Kenai	-28	-32	19	16	Wildwood Station	-28	-32	19	16
Kodiak	-11	-22	19	16	Yakutak	-18	-21	17	14

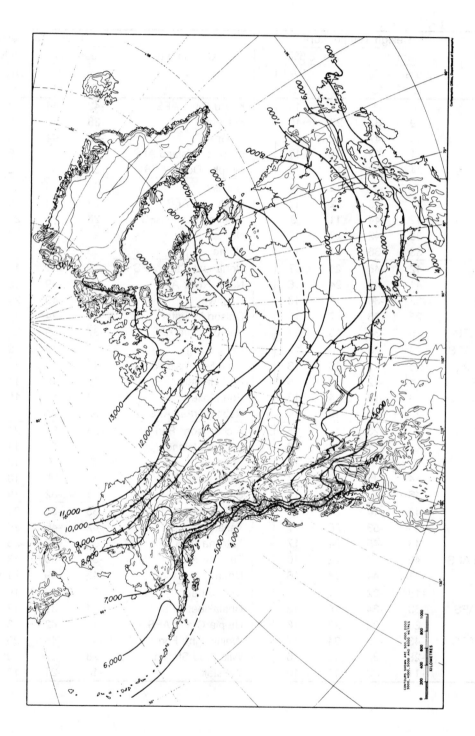

FIGURE E-4   MEAN ANNUAL HEATING INDEX (°C·d) (from Bryson and Hare, 1974)

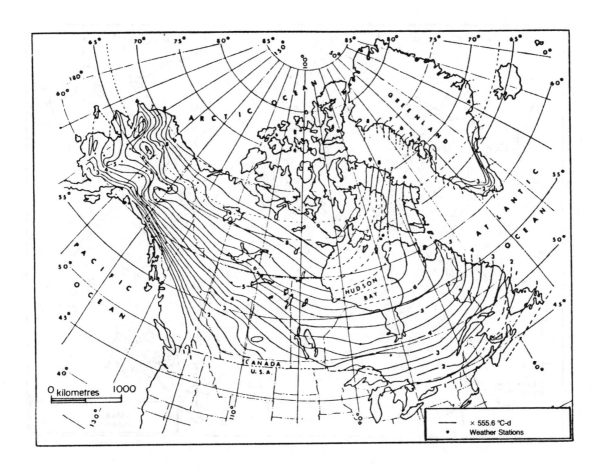

FIGURE E-5   MEAN ANNUAL AIR FREEZING INDEX (°C•d)

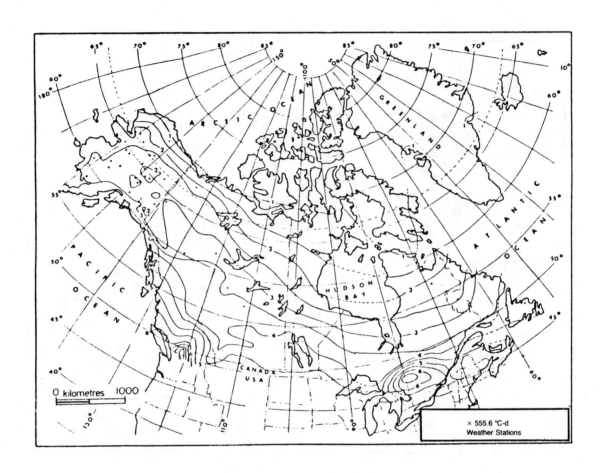

FIGURE E-6   MEAN ANNUAL AIR THAWING INDEX (°C•d)

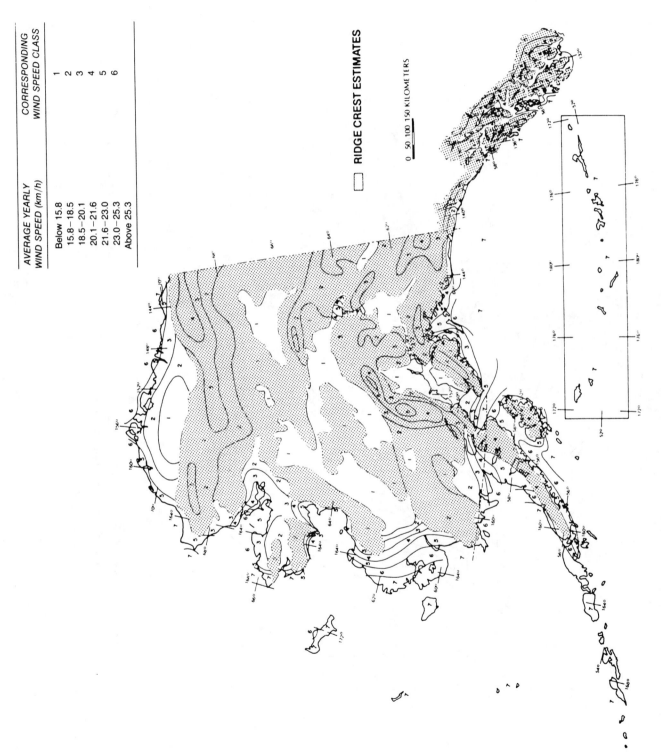

AVERAGE YEARLY WIND SPEED (km/h)	CORRESPONDING WIND SPEED CLASS
Below 15.8	1
15.8 – 18.5	2
18.5 – 20.1	3
20.1 – 21.6	4
21.6 – 23.0	5
23.0 – 25.3	6
Above 25.3	

RIDGE CREST ESTIMATES

0 50 100 150 KILOMETERS

FIGURE E-7(a) YEARLY AVERAGE WIND SPEEDS IN ALASKA (from De Harpporte, 1983)

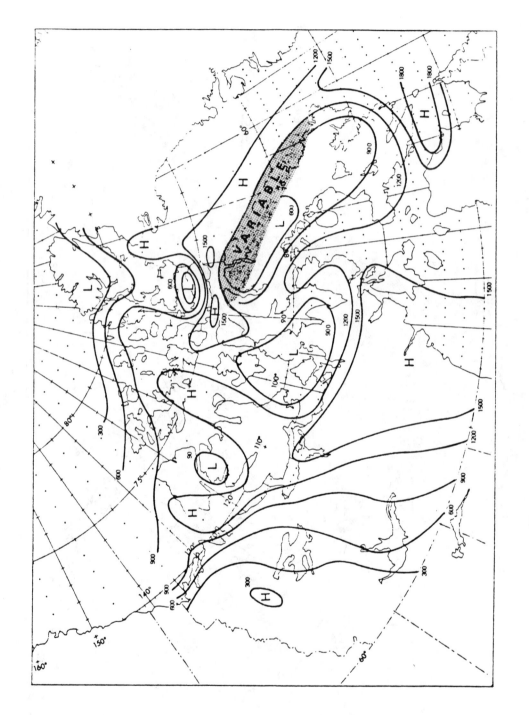

FIGURE E-7(b) MEAN ANNUAL WIND ENERGY POTENTIAL FOR THE CANADIAN NORTH (kWh/m² at 10 m for winds 1 to 58 km/h) (from Janz et al., 1982)

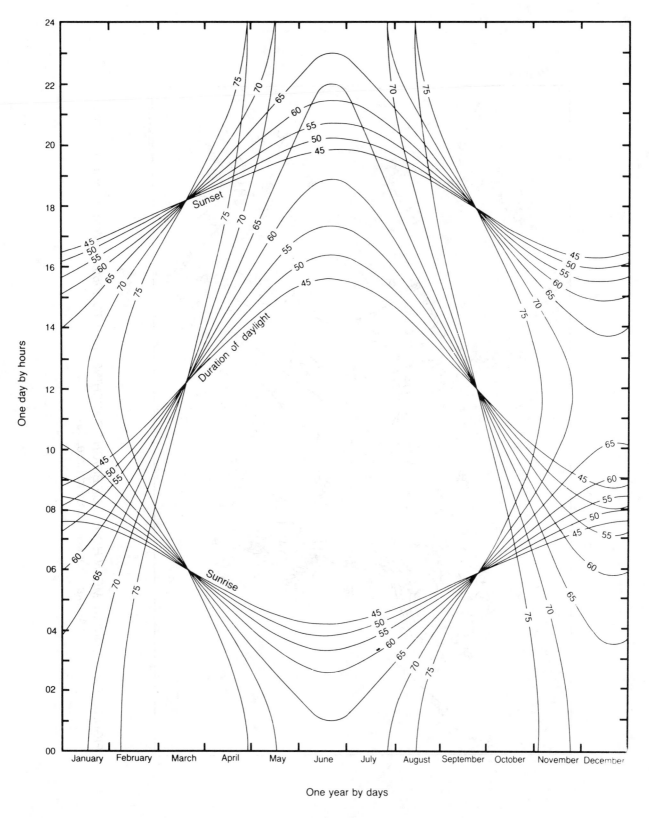

FIGURE E-8   HOURS OF DAYLIGHT AT VARIOUS LATITUDES

E-12

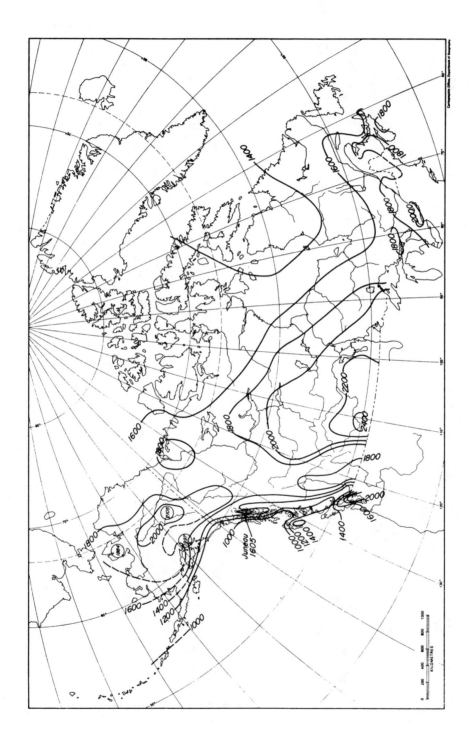

FIGURE E-9   ANNUAL MEAN HOURS OF BRIGHT SUNSHINE, 1931 TO 1960. Isopleths very approximate over Alaska. (from Bryson and Hare, 1974)

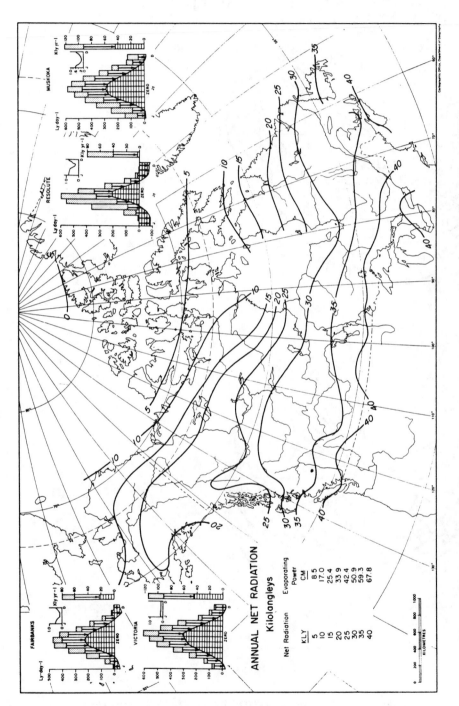

FIGURE E-10   MEAN ANNUAL NET RADIATION, 1957 TO 1964 (from Bryson and Hare, 1974)
Isopleths are in kly year⁻¹, with equivalent evaporating power in cm in table.
Inset diagrams for selected stations give:
• global solar radiation (entire column above zero);
• reflected solar (stippled);
• net long-wave cooling, downward arrow; and
• net radiation, ruled, month by month.
Smaller inset shows equivalent albedo variation. Note: 1 kly/year = 1.326 W/m²

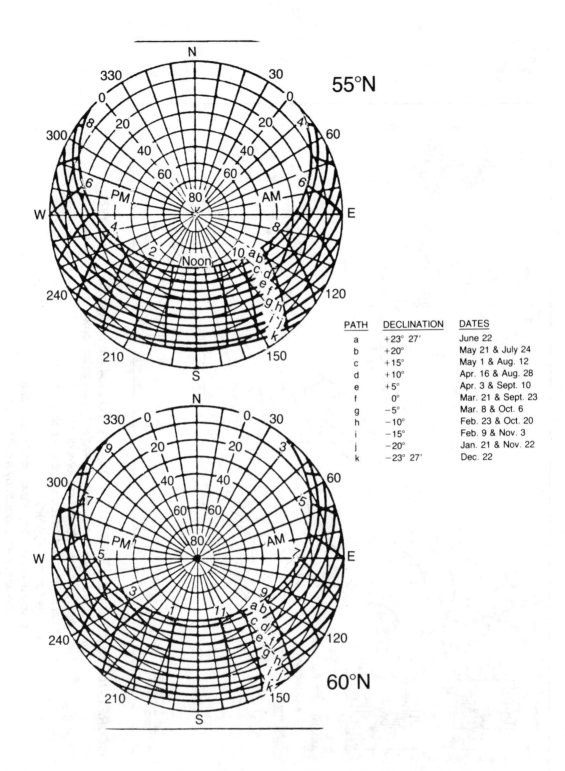

PATH	DECLINATION	DATES
a	+23° 27′	June 22
b	+20°	May 21 & July 24
c	+15°	May 1 & Aug. 12
d	+10°	Apr. 16 & Aug. 28
e	+5°	Apr. 3 & Sept. 10
f	0°	Mar. 21 & Sept. 23
g	−5°	Mar. 8 & Oct. 6
h	−10°	Feb. 23 & Oct. 20
i	−15°	Feb. 9 & Nov. 3
j	−20°	Jan. 21 & Nov. 22
k	−23° 27′	Dec. 22

FIGURE E-11   SUN PATHS AT VARIOUS LATITUDES AND DATES (from Hartman and Johnson, 1978)

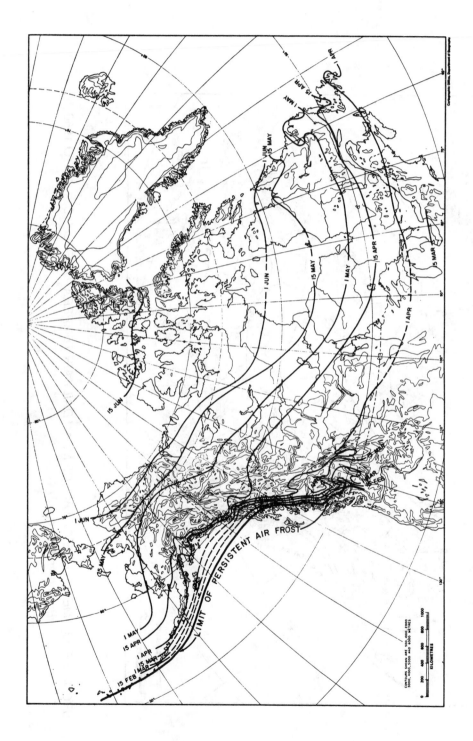

FIGURE E-12 MEAN DATE OF RISE OF MEAN DAILY AIR TEMPERATURE TO 0°C, 1931 TO 1960. Note the early warming of western districts, especially inland Alaska (from Bryson and Hare, 1974).

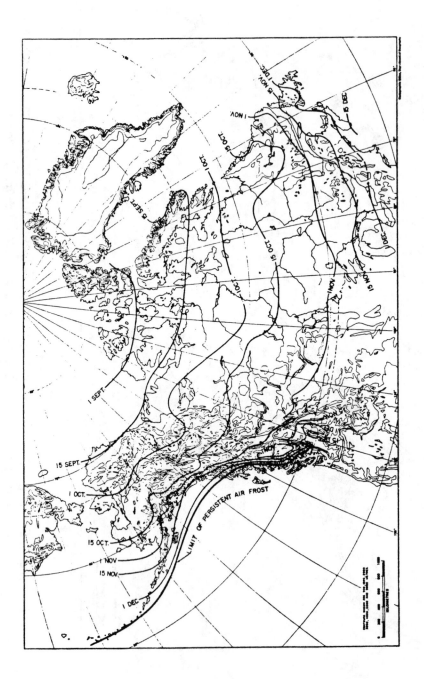

FIGURE E-13   MEAN DATE OF FALL OF MEAN DAILY AIR TEMPERATURE TO 0°C, 1931 TO 1960 (from Bryson and Hare, 1974)

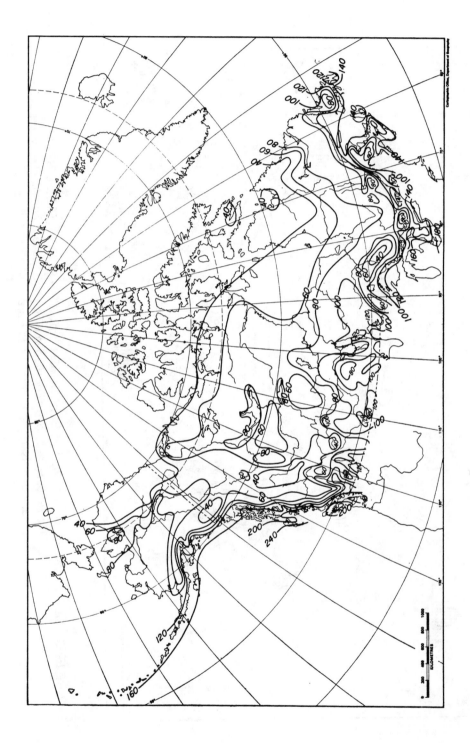

FIGURE E-14   MEAN LENGTH IN DAYS OF FROST-FREE PERIOD. Not rigorously standardized as to period, and heavily simplified in Alaska and much of the western Cordillera (from Bryson and Hare, 1974).

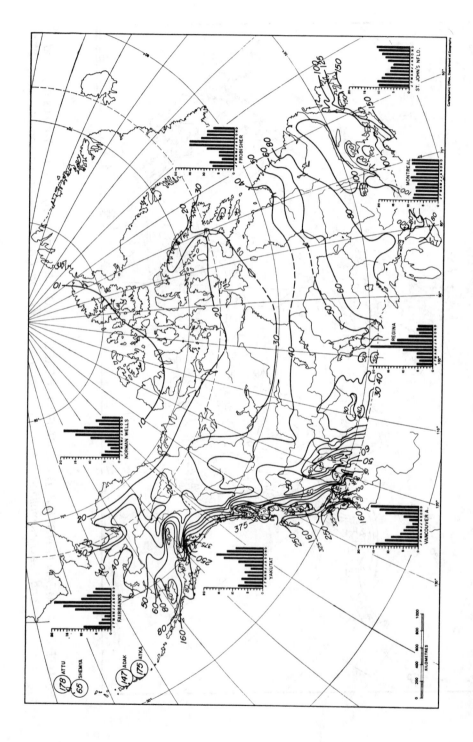

FIGURE E-15   MEAN ANNUAL MEASURED PRECIPITATION (cm) WITH PERCENTAGE DISTRIBUTION BY MONTH, 1931 TO 1960. Heavily oversimplified in western Cordillera and Alaska, where irregular isohyetal interval has been used (from Bryson and Hare, 1974).

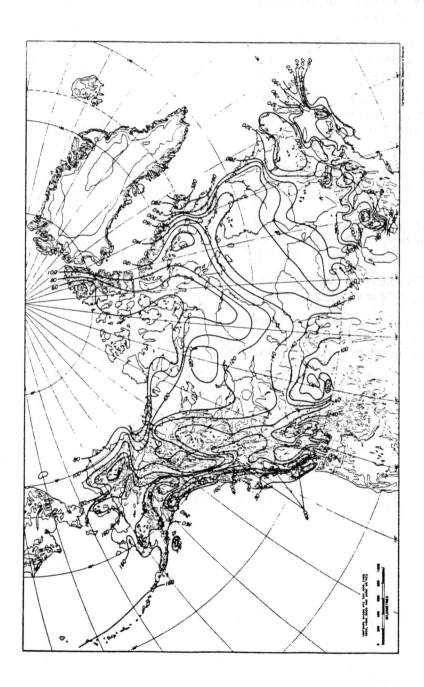

FIGURE E-16   MEAN ANNUAL MEASURED SNOWFALL (cm), 1931 TO 1960. Very approximate (and with irregular isopleth interval) in Alaska and the western Cordillera (from Bryson and Hare, 1974).

E.1 References

Bryson, R.A. and F.K. Hare. 1974. *World Survey of Climatology, Volume 11, Climates of North America*. Elsevier, New York. 420 pp.

De Harpporte. D. 1983. *Northwest, North Central and Alaska Wind Atlas*. Van Nostrand Reinhold, New York. 122 pp.

Hartman, C.W. and P.R. Johnson. 1978. *Environmental Atlas of Alaska*. University of Alaska, Fairbanks, Alaska. 95 pp.

Janz, B., D.G. Howell, and A. Serna. 1982. *Wind Energy in the Northwest Territories*. Government of Northwest Territories, Yellowknife, NWT. 108 pp.

NRC. 1977. *Climatic Information for Building Design in Canada*. National Research Council, Ottawa, NRCC No. 1556. 24 pp.

NRCC. 1990. *National Building Code of Canada* and *Supplement to NBCC 1990*. National Research Council of Canada, Ottawa, Ontario.

Thomas, M.K. 1953. *Climatological Atlas of Canada*. National Research Council, Ottawa. 256 pp.

Strock, C. and R.L. Koral (Editors) 1965. *Handbook of Air Conditioning Heating and Ventilating*. Second Edition. Industrial Press Inc., New York, NY.

APPENDIX F

CONVERSION FACTORS

Appendix F   Table of Contents

APPENDIX F   CONVERSION FACTORS

F.1   English to SI Metric

Multiply	By	To Obtain
Length		
mile, mi	1.609	kilometre, km
yard, yd	0.9144	metre, m
foot, ft	0.3048	metre, m
inch, in	25.40	millimetre, mm
Area		
square mile, mi^2	2.590	square kilometre, km^2
acre, ac	0.4047	hectare, ha
	4047.	square metre, m^2
square yard, yd^2	0.8361	square metre, m^2
square inch, in^2	645.2	square millimetre, mm^2
Volume		
acre foot, ac ft	1234.	cubic metre, m^3
cubic yard, yd^3	0.7649	cubic metre, m^3
cubic foot, ft^3	0.02832	cubic metre, m^3
	28.32	litre, L
US gallon, gal	3.785	litre, L
Velocity		
foot/second, ft/sec	0.3048	metre/second, m/s
foot/minute, ft/min	0.00508	metre/second, m/s
Flow		
million US gallons/day, mgd	3785.	cubic metre/day, m^3/d
	43.81	litre/second, L/s
US gallons/minute, gpm	5.450	cubic metre/day, m^3/d
	0.06309	litre/second, L/s
cubic foot/second, cfs	0.02832	cubic metre/second, m^3/s
Mass		
ton (2,000 lb)	0.9072	tonne, t
	907.2	kilogram, kg
pound, lb	0.4536	kilogram, kg
	453.6	gram, g

continued

Multiply	By	To Obtain
Density		
pound/cubic foot, lb/ft^3	16.02	kilogram/cubic metre, kg/m^3
BOD loading rate		
pound/1,000 cubic foot/day, lb/(1,000 ft^3/d)	16.02	gram/cubic metre•day, g/(m^3•d)
pound/acre/day, lb/(ac•d)	0.1121	gram/square metre•day, g/(m^2•d)
	1.121	kilogram/hectare•day, kg/(ha•d)
Solids loading rate		
pound/cubic foot/day, lb/(ft^3•d)	16.02	kilogram/cubic metre•day, kg/(m^3•d)
pound/square foot/day, lb/(ft^2•d)	4.883	kilogram/square metre•day, kg/(m^2•d)
Hydraulic loading rate		
million US gallons/(acre•day), mgad	0.9353	cubic metre/square metre•day, m^3/(m^2•d)
	9353.	cubic metre/hectare•day, m^3/(ha•d)
US gallon/square foot/day, gal/(ft^3•d)	0.04075	cubic metre/square metre•day, m^3/(m^2•d)
US gallon/square foot/minute, gpm/ft^2	0.6790	litre/square metre•second, L/(m^2•s)
US gallon/cubic foot/day, gal/(ft^3•d)	0.1337	cubic metre/cubic metre•day, m^3/(m^3•d)
Concentration		
pound/million US gallons, lb/mil gal	0.1198	milligram/litre, mg/L
Force		
pound force, lb	4.448	newton, N
Pressure		
pound/square foot, psf	0.04788	kilopascal, kPa
pound/square inch, psi	6.895	kilopascal, kPa
Torr, mm Hg at 0˚C	6.8948	kilopascal, kPa
Aeration		
cubic foot (air)/pound (BOD), cu ft/lb	0.06243	cubic metre/kilogram, m^3/kg
Vacuum		
Torr, mm Hg at 0˚C	0.13332	kilopascal, kPa

F.2   SI Metric to English

Multiply	By	To Obtain
Length		
kilometre, km	0.6214	mile, mi
metre, m	3.281	foot, ft
	1.094	yard, yd
millimetre, mm	0.03937	inch, in
Area		
square kilometre, km^2	0.3861	square mile, mi^2
hectare, ha	2.471	acre, ac
square metre, m^2	10.76	square foot, ft^2
	1.196	square yard, sq yd^2
square millimetre, mm^2	0.00155	square inch, sq in^2
Volume		
cubic metre, m^3	264.2	US gallon, gal
	35.31	cubic foot, ft^3
	1.308	cubic yard, yd^3
litre, L	0.2642	US gallon, gal
	0.03531	cubic foot, ft^3
Velocity		
metre/second, m/s	3.281	foot/second, ft/sec
	196.8	foot/minute, ft/min
Flow		
cubic metre/day, m^3/d	0.1835	US gallon/minute, gpm
cubic metre/second, m^3/s	35.31	cubic foot/second, cfs
litre/second, L/s	15.85	US gallon/minute, gpm
	0.02283	million US gallons/day, mgd
Mass		
tonne, t	1.102	ton (2,000 lb)
kilogram, kg	2.205	pound, lb

continued

Multiply	By	To Obtain
Density		
kilogram/cubic metre, kg/m^3	0.6242	pound/cubic foot, lb/ft^3
BOD loading rate		
gram/cubic metre•day, g/(m^3•d)	0.06243	pound/1,000 cubic foot/day, lb/(1,000 ft^3/d)
gram/square metre•day, g/(m^2•d)	8.921	pound/acre/day, lb/(ac•d)
kilogram/hectare•day, kg/(ha•d)	0.8921	pound/acre/day, lb/(ac•d)
Solids loading rate		
kilogram/cubic metre•day, kg/(m^3•d)	0.06243	pound/cubic foot•day, lb/(ft^3•d)
kilogram/square metre•day, kg/(m^2•d)	0.2048	pound/square foot•day, lb/(ft^2•d)
Hydraulic loading rate		
cubic metre/square metre•day, m^3/(m^2•d)	24.54	US gallon/square foot/day, gal/(ft^2•d)
	1.069	million US gallons/acre/day, mgad
cubic metre/cubic metre•day, m^3/(m^3•d)	7.481	US gallon/cubic foot/day, gal/(ft^3•day)
litre/square metre•second, L/(m^2•s)	1.473	US gallon/square foot/minute, gpm/ft^2
Concentration		
milligram/litre, mg/L	8.345	pound/million US gallons, lb/mil gal
Force		
newton, N	0.2248	pound force, lb
Pressure		
kilopascal, kPa	0.1450	pound/square inch, psi
kilopascal, kPa	0.14504	Torr, mm Hg at 0˚C
Aeration		
cubic metre (air)/kilogram (BOD), m^3/kg	16.02	cubic foot/pound, ft^3/lb
Vacuum		
kilopascal, kPa	7.5008	Torr, mm Hg at 0˚C

APPENDIX G

GLOSSARY OF ABBREVIATIONS AND TERMS

Appendix G   Table of Contents

G.1 Abbreviations

A $\quad$ = $\quad$ area, m^2

$A_{a,s,o}$ $\quad$ = $\quad$ amplitude terms, $^\circ C$

a $\quad$ = $\quad$ thaw factor = T arccosh (Zp/rp)

ABS $\quad$ = $\quad$ acrylonitrile-butadiene-styrene

AC $\quad$ = $\quad$ alternating current

ADEC $\quad$ = $\quad$ Alaska Department of Environmental Conservation

AEP $\quad$ = $\quad$ Alberta Environmental Protection

AK $\quad$ = $\quad$ Alaska

ASTM $\quad$ = $\quad$ American Society for Testing Materials

AWWA $\quad$ = $\quad$ American Water Works Association

B $\quad$ = $\quad$ geothermal gradient, $^\circ C/m$

b $\quad$ = $\quad$ breath, m

BOD $\quad$ = $\quad$ biochemical oxygen demand

BOD5 $\quad$ = $\quad$ five-day biochemical oxygen demand

Btu $\quad$ = $\quad$ British thermal unit

C $\quad$ = $\quad$ specific volumetric heat capacity, $\dfrac{kJ}{m^3 \cdot ^\circ C}$

$^\circ C\cdot d$ $\quad$ = $\quad$ degree-days Celsius

$^\circ C\cdot h$ $\quad$ = $\quad$ degree-hours Celsius

COD $\quad$ = $\quad$ chemical oxygen demand

CPE $\quad$ = $\quad$ chlorinated polyethylene

CSA $\quad$ = $\quad$ Canadian Standards Association

D $\quad$ = $\quad$ diameter, m

CD $\quad$ = $\quad$ direct current

DEW $\quad$ = $\quad$ defense early warning

DFC $\quad$ = $\quad$ Dominion Fire Commission

DO $\quad$ = $\quad$ dissolved oxygen

DOT $\quad$ = $\quad$ Department of Transportation

DWV $\quad$ = $\quad$ nonpressure drainage, waste and vent work

e $\quad$ = $\quad$ $\left(z^2 - r^2\right)^{1/2}$, m

EPA $\quad$ = $\quad$ U.S. Environmental Protection Agency

EPS $\quad$ = $\quad$ Environmental Protection Service (Canada)

FI $\quad$ = $\quad$ air freezing index, $^\circ C\cdot d$

f $\quad$ = $\quad$ correction modulus for emission of radiated heat

F/M = food to microorganism ratio

g = gravity

HI = heating index, °C•d

h = height, thickness, m

h/a = hours per annum

hp = horse power

k = thermal conductivity, $\dfrac{W}{(m \cdot °C)}$

kg/ha = kilogram per hectare

kg/d = kilogram per day

kg/m³ = kilogram per cubic metre

kPa = kilopascal, kN/m²

kWh/d = kilowatt hour per day

L = volumetric latent heat of fusion, $\dfrac{kJ}{m^3}$

Lpm = litres/minutes

L/(p•d) = litres/(person•day)

m³ = cubic metre

mg/L = milligram per litre

mh/day = manhours per day

MJ/h = megajoule per hour

MJ/kg = megajoule per kilogram

MLSS = mixed liquor suspended solids

m³/h = cubic metre per hour

MW/km² = megawatt per square kilometre

n = $\dfrac{SFI}{FI}$ or $\dfrac{STI}{TI}$

nfs = non-frost-susceptible

NWT = Northwest Territories

NBC = National Building Code of Canada

O&M = operation and maintenance

OSHA = Occupational Health and Safety Administration (U.S.)

O.D. = outer diameter

P = perimeter, m

p = period, d, h, s

PHS	=	Public Health Service (U.S.)
PVC	=	polyvinyl chloride
Q	=	quantity of heat, J
	=	volumetric flow rate, m^3/s
q	=	heat transfer rate, W
R	=	thermal resistance, k/k, $m \cdot °C/W$
r	=	radius, m
S	=	Stefan-Boltzman constant, $W/(m^2 \cdot °D^4)$
SFI	=	surface freezing index, $°C \cdot d$
SSU	=	saybolt universal seconds
Ste	=	Stefan number
STI	=	surface thawing index, $°C \cdot d$
T	=	temperature, $°C$
SDR	=	sidewall diameter ratio
TDS	=	total dissolved solids
TI	=	air thawing index, $°C \cdot d$
t	=	time, d, h, s
TOC	=	total organic carbon
TKN	=	total Kjeldahl nitrogen
U	=	coefficient of thermal expansion, $m/(m \cdot °C)$
USA CRREL	=	U.S. Army Cold Regions Research and Engineering Laboratory
v	=	velocity, m/s
w	=	water content, %
WHO	=	World Health Organization
UL	=	Underwriters Laboratory
x	=	direction vector
z	=	depth from surface, m
α	=	thermal diffusivity, m^2/s
β	=	thermal ratio in modified Berggren equation
π	=	3.14159
λ	=	correction coefficient in modified Berggren equation
ρ	=	dry density, kg/m^3
φ	=	constant in Neumann equation
μ	=	mean conduction heat transfer coefficient at solid/liquid interface, $W/(m^2 \cdot °C)$

G.2 Subscript Symbols

a	=	air
af	=	air film
C	=	conduit
c	=	conduction
d	=	convection
E	=	exterior casing
f	=	frozen
g	=	ground
I	=	insulation
i	=	interior conduit
L	=	thermal lining of utilidor
ma	=	mean annual air
ms	=	mean annual surface
o	=	freezing, 0°C
p	=	pipe
r	=	radiation
s	=	surface
T	=	total
t	=	thawed
U	=	utilidor
u	=	unfrozen
w	=	water
wf	=	water film
∞	=	infinity

G.3 Glossary of Terms

active construction	Method of construction in which perennially frozen soil (permafrost) is thawed and kept thawed. Thawed, unstable soil or permafrost with excess ice is often excavated or replaced by sand and gravel.
active layer	The layer of soil that freezes and thaws annually as seasons change (also called seasonal frost).
alluvia	Clay, silt, sand, gravel, or similar material deposited by running water.
anadromous fish	Fish that journey up rivers from the sea at certain seasons for breeding (salmon, shad, etc.).
anchor ice	See Ice, anchor.
arctic	Regions where no mean monthly temperature is greater than 10°C and where at least one month has a mean monthly temperature of 0°C or colder.

Arctic Circle	Where on at least one day the sun does not set in the summer or rise in the winter (latitude 66° 31'N).
ATV	All-terrain vehicle.
aufeis	Ice that is formed as water flows over a frozen surface.
beaded streams	Streams that contain enlargements or "beads" that are caused by the melting of blocks of ground ice along its course.
bentonite	A clay-type substance used as a drilling mud; it has the ability to expand when water is added to it.
blackwater	Wastewater which contains only human toilet waste and a small amount of flushing water.
bleeding	The continuous running of water through taps to maintain a flow in the mains, service lines, and sewers to prevent freezing of the pipes.
BOD	Biochemical oxygen demand; a measure of the amount of oxygen required by bacteria to oxidize organic matter aerobically to carbon dioxide, water, and other relatively stable end products.
BOD_5	The amount of oxygen required by bacteria during the first five days of the BOD test (at 20°C).
bog	A wet peatland which is extremely nutrient-poor, acidic, and has a tree cover of less than 25 percent of its area.
bog soil	A wet spongy soil composed of decayed mass and other vegetable matter. (Soil in its thawed state has almost no bearing strength.)
brackish water	Saline or mineralized water with a total dissolved solids concentration of about 1,000 mg/L to about 10,000 mg/L.
break-up	The melting time at which a) ice on rivers breaks and starts moving with the current, b) lakes can no longer be crossed on foot, and c) previously frozen mud is soft, and most of the snow is gone.
cat train	Trailer trucks or large sleds hitched together in a train-like manner which are then pulled by a snow plow or tractor.
central facility	A community facility where one or more sanitary services (washrooms, showers, laundry, etc.) are available.
central water points	A potable water supply centrally located within a community where hand-carried containers and/or water trucks are filled.
COD	Chemical oxygen demand; a measure of the amount of oxygen required to chemically oxidize the organics in wastewater.
cold region	Those regions where the design of water and sanitation facilities must consider the thermal design.
coliforms	Rod-shaped bacteria, which are gram-negative and will ferment lactose at 35°C, producing gas within 24 hours. Two classifications of importance here: total coliform and fecal coliform. The latter are from fecal matter of warm-blooded animals.
continuous permafrost	An area underlaid by permafrost with no thawed areas except under large lakes and rivers that never freeze solid.
degree-days	A quantity expressed as the product of "degrees variation from a base" and "time in days." Example: If the temperature averages 5°C for 10 days, there is

an accumulation of 50 degree-days of "thaw." (Base for freezing and thawing degree-days is 0˚C and base for heating degree-days usually is 18˚C.)

demurrage charges	The payment rates for detaining a freighter beyond a reasonable time for loading and unloading.
discontinuous permafrost	An area underlain mostly by permafrost but containing small areas of unfrozen ground, such as on south-facing slopes.
DO	Dissolved oxygen; the amount of oxygen dissolved in water.
evapotranspiration	The passage of water as a vapor into the atmosphere through the processes of evaporation and transpiration.
facultative lagoon	A lagoon that treats wastes through a combination of aerobic or anaerobic processes.
fill point	In a truck-haul system, this is the location where a water truck fills its water tanks; it also refers to the point on individual houses where ice or water is delivered.
flotation tire	A large tire that is filled with air; it allows wheeled vehicles to stay on top of the snow because of a very low unit area force.
F/M ratio	The ratio of food (organic matter) to microorganisms by weight; applies to a wastewater process.
frazil ice	See Ice, frazil.
freeze-rejection concentrate	The natural process whereby water slowly freezes, excluding impurities and forming crystals of pure water; impurities rejected from the ice are concentrated in the remaining liquid.
freeze-up	The transition time when moisture at the ground surface freezes forming a hardened surface.
freezing index	The integrated number of degree-days colder than the freezing point in a winter session.
frost creep	A gradual movement usually downhill of soil, clay, or loose rock due to alternate freezing and thawing.
frost heaving	The expansion of soil due to the growth within it of an extensive ice lens, which causes the displacement of the soil surface.
frost jacking	When soil bonded to an object moves upward through frost heaving and carries the object with it; upon thawing, the object does not return to its original elevation.
frost mounds	A microrelief of about 1 m formed by intense frost heaving in the active layer.
frostshield	An insulated shield that is used to deter the advancement or penetration of frost into the area in question.
frost-susceptible soil	A soil that retains and permits migration of large amounts of water, encouraging the growth of ice lenses during freezing, from which frost heaving develops; also defined as a soil passing more than three percent through a No. 200 sieve.
glacial rock flour	Finely powdered rock material produced by the grinding action of glacier on its bed.
glacial silt	Particles of crushed rock deposited by glacial stream. The particle size of this silt lies in the range of 0.002 to 0.06 mm.

glaciering	See icing.
glacier	A field or body of ice formed in a region where snowfall exceeds melting.
glaciolacustrine sands	The sands of a lake that come from the deposits of a melted glacier.
greywater	Wastewater from kitchen sinks, showers, and laundry, excluding human toilet wastes.
heat capacity	The quantity of heat required to raise the temperature of a mass by one degree. Therefore, the heat capacity of a body is its mass multiplied by its specific heat.
heating index	The integrated number of degree-days colder than some base figure (usually 18°C) during a heating season.
heat-trace system	An electrical system having thermostats and sensing bulbs along the pipe in question, thereby keeping it from freezing on a continuous or on an as-needed basis.
honey bag	A plastic or heavy paper bag that fits into a bucket toilet.
honey bucket	A plastic or steel bucket that fits into a bucket toilet.
Ice	
anchor	Ice formed below the surface of a body of water; the ice attaches itself either to a submerged object or to the bottom.
coastal	Formations that, regardless of origin, exist between land and sea on the coast.
frazil	Ice crystals that form in flowing, supercooled water and collect on any channel obstruction; usually frazil ice formation occurs at night because of the high rate of heat radiation away from the water.
iceberg	Huge mass of ice broken or calved from a glacier.
ice crypt	A sub-ice chamber or vault found in icebergs and glaciers.
ice field	An extensive sheet of sea ice that can be several square kilometres in area.
ice fog	Fog composed of particles of ice, usually caused by moisture or steam released into the cold environment or by a large open body of water exposed to air.
icing	Mass of surface ice formed by successive freezing of sheets of water which seeps from the ground, a river, or a spring. When the ice is thick and localized, it may be called an icing mound. Icing (also known a glaciering or aufeis ice) can produce ice 3 m thick and 0.8 km long.
ice-rich ground	Soil containing ice in excess of its thawed voids-volume.
ice scour	the marring effect of ice as it moves over the river or stram bed.
ice wedges	Vertically oriented V-shaped masses of relatively pure ice occurring in permafrost. The head of the wedge is on top and can be up to 4 m wide, whereas the wedge itself can be 10 m in height.
indigenous people	People originating in and characterizing a particular region or country.
infrastructure	Permanent installations and facilities belonging to a community.
insulating layer	A layer of sand, gravel, wood, or other low-heat-conductive material for the purpose of reducing heat transfer. Frequently related to the protection of permafrost.

intrapermafrost water	Groundwater within the permafrost; usually has high concentration of minerals which keeps it from freezing.
lateral thermal stresses	The thermal expansion of ice, upon a temperature increase towards melting, causes lateral thrusting (stresses).
leaching field	Plot of land used for disposing of sewage and other liquid wastes; by allowing it to percolate through the soil, treatment of the wastes is accomplished.
lignin	A long-chain organic molecule which causes a brown color in water; normally from mosses and other organic materials.
muskeg	An Indian (Algonquin) word for bog or peatland.
non-frost-susceptible soil	A soil that does not retain water, thereby not encouraging the growth of ice wedges.
northern communities	Those communities that lie in the Arctic and subarctic regions.
northern temperate zone	The northern part of the zone lying between the Tropic of Cancer and the Arctic Circle in the Northern Hemisphere.
open burning	Uncontrolled burning of wastes in an open dump.
organic matter	More or less decomposed material in soil derived from organic sources, usually from plant remains or animal and human waste.
package treatment plant	A treatment system available as prefabricated "packaged" units.
passive construction	Method of constructon that preserves the permafrost for its structural value.
pathogenic bacteria	An organism capable of producing disease.
peat	Highly organic soil, 50 percent of which is combustible, composed of partially decayed vegetable matter found in bogs (muskegs). It is very frost susceptible.
permafrost	Soil, bedrock, or other material that has remained below 0˚C for two or more years.
permafrost table	The dividing surface between the permafrost and the active layer.
pitorifice circulation	Continuous circulation of water through the water pipes of a house because of a pitorifice; the pitorifice, which is located in the water main and connected to the supply and return service lines, causes a higher water pressure on one side (for the supply line) and a lower pressure on the other side (for the return line).
polygonal ground	Patterned ground with recognizable trenches or cracks along the polygonal circumference (a surface relief); produced by alternative freezing and thawing of the surface soil above the permafrost.
potable water	Water suitable for drinking; physically, biologically, chemically, and radiologically safe water.
pressure ridge	Ridge produced on floating ice by buckling or crushing under lateral pressure of wind or tide.
reach of a river	A specified portion of a river.
sand spit	A narrow, sandy point of land projecting into a water body.
seasonal frost areas	Areas where ground is frozen by low seasonal temperatures and remains frozen only through the winter; in permafrost this refers to the active layer.

self-haul system	A system where water is carried in containers from a central water point to the home for use or storage.
service bundle	A set a service lines supplied to the house; these lines are bundled together and usually enclosed by a cover (see utilidette).
shore-fast ice	A wall or belt of ice frozen to the shore; it has a base at or below the low-water mark; it is formed as a result of the rise and fall of the tides, freezing spray, or stranded ice.
sink hole	a hole formed in soluble rock or melted permafrost by the action of water going from the surface to an undergound passage.
slump	A depression or landslide on the land due to the removal of the natural vegation which causes the underlying massive ground ice in the permafrost to melt
snowdrift	a mound or bank of snow formed by the wind.
sludge	A semi-liquid substance consisting of settled sewage solids combined with varying amounts of water and dissolved materials.
snow or ice road	A road made of snow and/or ice that exists only in winter. These roads, which melt each spring and are reconstructed each winter using the icy surface on the lakes as highways, make travel by land vehicles possible.
snow fence	A barrier erected on the windward side of a road, house etc., to encourage the dropping of snow and therefore serving as protection against drifting snow.
soil bonding	The bonding of the individual soil grains by the freezing of water between them; if a pipe is interwoven with soil of this nature, it can become locked in place.
sporadic permafrost	Isolated masses of permafrost located within an area generally thawed during the summer.
stick built	An expression referring to on-site construction; all raw materials are brought to the site and the structure is constructed there.
subarctic	Regions adjacent to the Arctic in which one to three calendar months have a mean monthly temperature above 10°C and at least one month that has a mean monthly temperature of 0°C or colder.
subpermafrost layer	The layer below the permafrost; it may contain some permafrost islands.
suprapermafrost layer	The layer betweeen the ground surface and the permafrost table; this layer contains the active layer, year-round thawed areas (taliks) and temporarily frozen areas (pereletoks).
talik	A layer of unfrozen ground between the active layer and the permafrost; the term applies also to an unfrozen layer within the permafrost, as well as to the unfrozen ground below the permafrost.
tannin	A bluish black or greenish black dye from plant leaves.
thaw bulb	A thawed section in the permafrost due to the warming effect of a house, river, lake, etc.
thawing index	The yearly sum of the diffrences between 0°C and the daily mean temperature of the days with means above 0°C.
thermal erosion	The undercutting of a frozen bank or shore by melting of the soil from exposure to running water and/or wave action.
thermal inertia	The degree of slowness with which the temperature of a body approaches that of its surroundings.

thermal insulation	Insulation to resist the transmission of heat.
thermal resistance	The resistance of a body to the flow heat.
thermal stratification	The layering effect of temperature in an enclosed body of water or air due to lack of mixing.
tideland	Land alternately exposed and covered by the ordinary flow of the tide; the only vegetation present is salt-tolerant bushes and grasses.
tundra	Term applied to the treeless areas in the Arctic and subarctic; consists of mosses, lichens, and small brush.
unstable permafrost	Permafrost that is not physically stable in its thermal environment; it does not melt.
utilidette	The enclosure for a bundle of service lines supplied to the house from the utilidor; may or may not be insulated.
utilidor	An above- or below-ground conduit (not necessarily insulated) that acts as an enclosed corridor for a network of pipes and cables that supply community services to individual homes and businesses.
vehicle-haul system	A vehicular (truck or tractor) system which transports water or ice from a source to individual buildings, and/or collects wastewater and transports it to a treatment or disposal point.
volumetric heat of fusion	The amount of heat required to melt a unit volume of a substance at standard pressure.
"warm" permafrost	Arbitrarily defined as permafrost that has a temperature of -4°C or greater.
water wasting	See bleeding.
watering point	A central point for users to obtain potable water for domestic purposes.
wetland	General term, broader than muskeg, to name any poorly drained tract, whateer its vegetation cover or soil.
whiteout	Blowing snow so thick that visibility is limited to a metre or two.
windchill factor	The cooling effect of both temperature and wind on a body, expressed as a temperature which is equivalent to the heat lost per unit time in quiescent air.

APPENDIX H

INDEX

INDEX